The F-35 Lightning II: From Concept to Cockpit

The F-35 Lightning II: From Concept to Cockpit

Edited by

Jeffrey W. Hamstra, Lockheed Martin Corporation

Volume 257
Progress in Astronautics and Aeronautics

Timothy C. Lieuwen, Editor-in-Chief
Georgia Institute of Technology
Atlanta, Georgia

Published by
American Institute of Aeronautics and Astronautics, Inc.
12700 Sunrise Valley Drive, Suite 200, Reston, VA 20191-5807

American Institute of Aeronautics and Astronautics, Inc., Reston, VA.

ISBN 978-1-62410-566-1

TABLE OF CONTENTS

FOREWORD

My love for aviation—specifically high-performing aircraft—went a mirror opposite route than the title of this book: I went from cockpit to concept. As a Naval aviator I had the privilege to fly 38 different types of military aircraft and logged more than 3,000 flight hours.

After my active duty flying days, I moved to the industry side of aviation, working on several military aircraft programs for Lockheed Martin, including leading the team building the F-22 Raptor, followed by leading the F-35 Joint Strike Fighter program through its pre-competition concept development phase and competitive contract selection.

The F-35 program, at the highest level, encompasses a complex problem statement addressing much more than developing the next fighter.

For the last several decades, the operational model for long term combat and peace keeping operations revolved around tri-service, coalition participation for both operational and financial burden sharing reasons. No single country can go it alone any longer.

The reality of this concept, however, has historically been complex and unrealistic due to underlying technology, lack of platform interoperability and political differences among and between participants. This was the basis for the complex undertaking called the Joint Strike Fighter.

As you read the history of the F-35 Joint Strike Fighter team, think about the challenges they faced from the start:

1. Meet a very complex operational requirement: Design, develop, test, field and sustain a family of highly common first line fighters designed to fully incorporate the unique basing requirements of conventional air force operations, expeditionary operations off small ships and unprepared fields and projecting power from the sea from large deck aircraft carriers.
2. Make that platform stealthy, supersonic and embed the most sophisticated suite of multi-spectral sensors ever employed in a fighter.
3. Incorporate high volume production capabilities that can capture economies of commonality and scale.
4. Ensure our closest allies can participate with the goal of true interoperability and burden sharing in future combat and peace keeping operations.
5. Revitalize a global industrial base that has atrophied over the years by implementing revolutionary new manufacturing capability based on very high precision, automation and motion-based production systems, and transferring technology that enables the industries of our allied partner nations to compete on a global world stage with the best of the best.
6. Assemble a world class prime contractor partnership and global supply chain that can perform to the unprecedented production and sustainment requirements of the F-35 Joint Strike Fighter program objectives.

The F-35 Lightning II: From Concept to Cockpit is the seminal archive for the journey that the F-35 team experienced. It is a remarkable collection of lessons learned on a very unique venture to change the acquisition paradigm for tri-service and coalition allied procurement of front-line combat capability. I was privileged to lead this transformational team through the challenges of the first decade and I salute the men and women that delivered this incredible capability.

Capt. Charles Thomas (Tom) Burbage, U.S. Navy, Ret.
F-35 Joint Strike Fighter Program
Executive Vice President and General Manager, Ret.

PREFACE

In the years since the F-35 Joint Strike Fighter development program was awarded to Lockheed Martin, countless articles have appeared in the popular media discussing the merits and challenges of the program. F-35 information has also occasionally appeared in the scientific literature through AIAA or other technical venues. However, an integrated story of the F-35's technical development has heretofore never been presented in a single work.

The purpose of *The F-35 Lightning II: From Concept to Cockpit* is to tell that story—a full-spectrum history of the design, development and verification of the F-35 Lightning II as described by the engineers, scientists and managers intimately involved throughout the development program. The journey to achieve the F-35's extraordinary capability was long with untold challenges encountered on the path. The reader will find many of the key technical challenges, and the innovative solutions that resulted, discussed herein.

The reader should realize this work is not intended as an overall assessment of the F-35 or the Joint Strike Fighter program. Rather, it is an engineering development story from the perspective of insiders, many of whom dedicated the best part of their career to F-35 and are justifiably very proud of their work.

The book is based on 18 technical papers presented in a two-day F-35 track at the 2018 AIAA AVIATION Forum. The "From Concept to Cockpit" subtitle appropriately summarizes the contents, with the chapters grouped into three major sections:

- F-35 Program Overview, which begins with an overall history of the program and further presents discussion of aircraft configuration design, weapons integration, key technologies brought into the program, and overviews of aircraft manufacturing and the flight test program.
- F-35 Air Vehicle Design, which presents discussion of key aircraft systems, including airframe structure, flight controls, propulsion, subsystems and mission systems.
- F-35 Test and Verification, which presents discussion of the flight and ground test programs undertaken to verify the F-35's airworthiness, flight performance and mission capability.

The chapters are in a logical progression but are not interdependent, and so one may read the contents in any order without confusion.

Although much information is included in this work, there are several important topics that have been left out. Given the sensitive nature of military products, not every technology or system can be discussed.

I wish to express appreciation to the staff at AIAA for their invaluable help and guidance regarding this publication as well as for their encouragement and leadership in bringing the F-35 program to an AIAA event.

Special thanks is also due to several Lockheed Martin individuals for their outstanding contributions in championing, leading, and supporting the Concept to Cockpit project, from initial discussions through the presentations at the AVIATION Forum and preparation of this work:

- Ron Bessire and Lee Sampson, Engineering & Technology
- Russel Baldonado, Daniel Buck, Alyssa Campbell, Karen Hagar, Kathleen Hearons, Ken Ross and Dennis Soultaire, Communications
- Rick Mange, J.D. McFarlan and Joey Sikorski, F-35 Program

Our greatest thanks must go to the 60 plus individuals from Lockheed Martin, our F-35 partner companies and suppliers, and the F-35 Joint Program Office who contributed the 18 chapters of this work. The authors brought to bear their knowledge, experience and significant personal time to develop the content. Most of all, they brought a passion to tell the integrated F-35 engineering story and help advance the state of the aerospace profession by doing so. Without their expertise and dedication, this work would not have been possible.

Jeffrey W. Hamstra
Sr. Fellow, Lockheed Martin Corporation

F-35 Program History: From JAST to IOC

Arthur E. Sheridan*
Lockheed Martin Aeronautics Company, Fort Worth, TX

Robert Burnes†
F-35 Lightning II Joint Program Office, Arlington, VA

The Joint Strike Fighter program leading to the Lockheed Martin family of F-35 aircraft has been unprecedented in terms of scope and challenge. This paper reviews the background and need for the air system. It summarizes the environment, objectives, approach, and results of each of three distinct development phases, and highlights some of the most significant challenges encountered and solutions achieved. It also covers initial production and sustainment achievements in parallel. Despite the ambitious goals and numerous challenges, the development program is drawing to a close, and a system is now being produced and sustained that meets its customers' warfighting requirements.

I. BACKGROUND

The origins of the Joint Strike Fighter (JSF) program can be traced to the long-standing commitment of the U.S. Marine Corps (USMC) and United Kingdom (UK) Royal Air Force (RAF) and Royal Navy (RN) to develop a Short Takeoff and Vertical Landing (STOVL) strike fighter, and to the end of the Cold War. Drastic defense budget reductions after the Cold War, together with aging fleets of fighter aircraft in the United States and across the west, demanded a new level of cooperation in development and production. The U.S. Department of Defense (DoD) Bottom-Up Review in 1993 cancelled previously separate fighter/attack development plans of the U.S. Air Force (USAF), U.S. Navy (USN), and USMC that aimed at replenishing U.S. fleets but became viewed as unaffordable. The need for new aircraft procurement was compelling, however, due to the end of production of legacy fighters (Fig. 1). Furthermore, the large number of aircraft types in use by the United States and its allies could not be

*Program Management Principal, F-35 Customer Programs, AIAA Associate Fellow.
†Director, Program Operations and Director, Program Management.

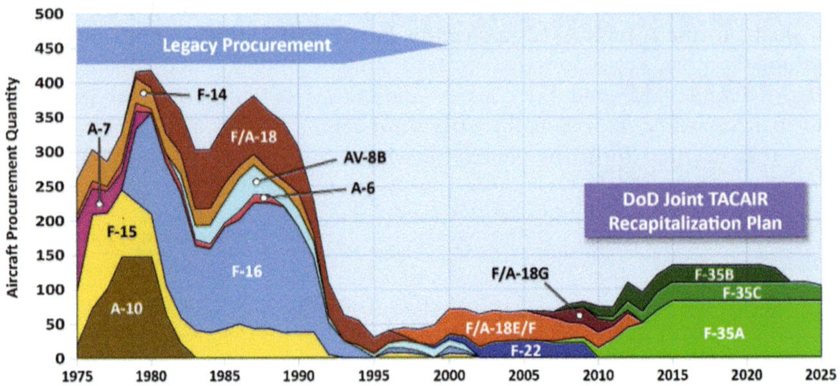

Fig. 1 Historical and projected U.S. fighter procurement profile (circa 2001).

affordably maintained (Fig. 2). Trends toward joint operations and coalition warfare required significant improvements in interoperability. In this environment, service leaders in the United States and United Kingdom agreed to develop a single program to address the next generation of affordable strike platforms. Additional affordability strategies contributing to the environment were acquisition reform initiatives advocating performance-based specifications and concurrent development, as well as the desire to exploit the digital revolution with simulation-based acquisition, digital design, and paperless commerce.

Existing U.S. service strike fighter requirements were widely disparate, ranging from the U.S./U.K. Advanced STOVL (ASTOVL) (a small supersonic STOVL

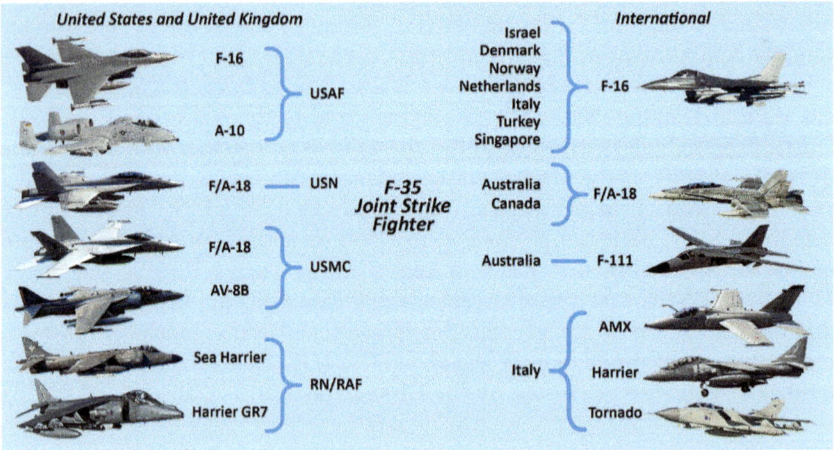

Fig. 2 Legacy fighter types expected to be replaced by JSFs.

airplane for the USMC and United Kingdom with a maximum empty weight of 24,000 lb) to the Navy's A-12 (a stealthy carrier-based, twin-engine, long-range medium bomber), to a low-cost fleet-structure fighter to succeed the USAF's F-16. References [1] and [2] provide summaries of U.S. precursor programs and their sequence, as well as the early development of STOVL propulsion concepts that together form the genesis of what is now the F-35 program. At the time, the industry had great doubt that a single aircraft could be designed to satisfy the needs of all services. For this reason, and lacking common air-vehicle requirements, DoD did not approve the creation of an aircraft acquisition program; rather, the initial program mandate was to invest jointly in technologies that could be applied irrespective of a specific aircraft configuration, and to perform configuration studies to determine whether a common family of aircraft could meet service needs.

II. JOINT ADVANCED STRIKE TECHNOLOGY

The services formed a JSF Program Office (JSFPO) drawing from the Naval Air Systems Command (NAVAIR), USAF Aeronautical Systems Command (ASC), and U.K. Ministry of Defence (MoD), and created the Joint Advanced Strike Technology (JAST) program in 1994. The JSFPO was (and remains) located in Arlington, Virginia, where top personnel from each systems command are co-located. The leadership structure was established with roles of the Senior Acquisition Executive (SAE), Program Executive Officer (PEO), and deputy PEO positions alternating among the departments of the USN and USAF, and rotating at nominally two-year intervals; that is, a USAF SAE would be served by a Department of the Navy PEO (USN or USMC officer) and a USAF Deputy PEO. The pattern would then reverse upon the change of command on a nominally two-year basis.

A. TECHNOLOGY DEMONSTRATIONS

The new program issued initial contracts to industry under the JAST banner, with separate contracts for individual technology maturation and demonstration efforts. Contracts were issued to all eventual JSF competitors, but they were to collaborate in planning the efforts and were required to share the results across the industry. JSF Integrated Subsystems Technology (J/IST) is a prime example of such an effort, and within that, the More-Electric Actuation program is an example that was led by Lockheed Martin and eventually incorporated into the F-35 configurations. Other proprietary technologies were pursued separately by competitors, a Lockheed Martin example being the diverter-less supersonic inlet that performs as a mixed-compression inlet that avoids boundary layer ingestion through the use of innovative shaping with no moving parts, providing a lightweight and smooth configuration with favorable signature integration.

Both the J/IST electric power and actuation concept and the diverter-less inlet were demonstrated on separate F-16 platforms [3, 4].

B. CONCEPT DEFINITION AND DESIGN RESEARCH

In parallel with these technology programs, the JAST program also issued Concept Definition and Design Research (CDDR) contracts. These efforts were begun to conceive of specific aircraft configurations and specific performance requirements. In general terms, the requirements were to provide configuration variants to serve the services' differing basing needs: Conventional Takeoff and Landing (CTOL) for USAF, STOVL for the USMC and RAF/RN, and a Carrier Variant (CV) for USN operations with catapults and arresting gear. The up-and-away performance requirements for range/payload and maneuverability were to be common among the variants and approximately equivalent to combat-equipped current F-16s and F/A-18s. Signature levels, mission systems, and weapons capabilities were to be studied over a wide trade space. An affordability target of 28 million dollars in 1994 base-year economics was established for the CTOL variant—roughly equivalent to the price of an F-16 Block 50, including targeting and electronic warfare pods and external fuel tanks. The different variants were to be as common as possible to take advantage of economies of scale and interoperability.

In light of budget constraints, the U.S. services recognized the value of commonality and saw that overly specific requirements could negate those benefits. So, they re-examined basic assumptions embedded in their previous development projects and adopted some important changes. Most notably, for the first time since the A-7, the USN accepted the single-seat and single-engine requirement, which was essential to achieve a practical STOVL configuration for the USMC.

The result of the CDDR phase was that the configuration concepts and corresponding requirements trade studies instilled sufficient confidence in the JSFPO and the participating services to proceed to the next phase to develop a family of air systems to be known as the JSF, designed to satisfy a Joint Operational Requirements Document (JORD) that would also be developed in the next phase: the Concept Demonstration Phase (CDP).

C. INDUSTRY COMPETITORS

Three industry competitors participated in the program, each with configuration families based on different STOVL propulsion concepts. Reference [2] gives an overview of the evolution of STOVL concepts preceding JAST. The Lockheed Martin CDDR designs were all based on the Shaft-Driven Lift Fan (SDLF) described below.

The Boeing concept was based on direct lift, which relied on diverting the majority of engine exhaust to Harrier-like swiveling nozzles at the center of gravity for hover. Exhaust flow was abruptly switched to and from an aft-mounted vectoring nozzle during wingborne/jetborne transitions. A series of remote nozzles provided attitude control, and a jet screen was located near the main inlet to reduce hot-gas ingestion. At this stage in the program, the Boeing configuration was a delta-wing arrangement that was essentially common for all three variants.

The McDonnell Douglas (later acquired by Boeing)/Northrop Grumman/ British Aerospace (now BAE Systems) concept was a conventional wing-tail arrangement with conventional propulsion for the CTOL and CV variants. The STOVL variant employed the lift-plus-lift/cruise propulsion system similar in arrangement to the Russian YAK-38 and YAK-141 aircraft. It was to have a single lift engine mounted forward and a combination of swivel nozzles and aft conventional nozzle for main-engine exhaust, similar to but much shorter than the Boeing exhaust system.

D. THE LOCKHEED MARTIN AIR VEHICLE CONCEPT

The Lockheed Martin air vehicle concept centered on the SDLF propulsion system [1, 4] illustrated in Fig. 3, which was key to both STOVL capability and family commonality among variants. Together with a vectoring lift-fan nozzle and a continuously vectoring three-bearing swivel nozzle (3BSD) for the engine exhaust, the system inherently addressed the major challenges and typical causes of failure for previous supersonic STOVL concepts.

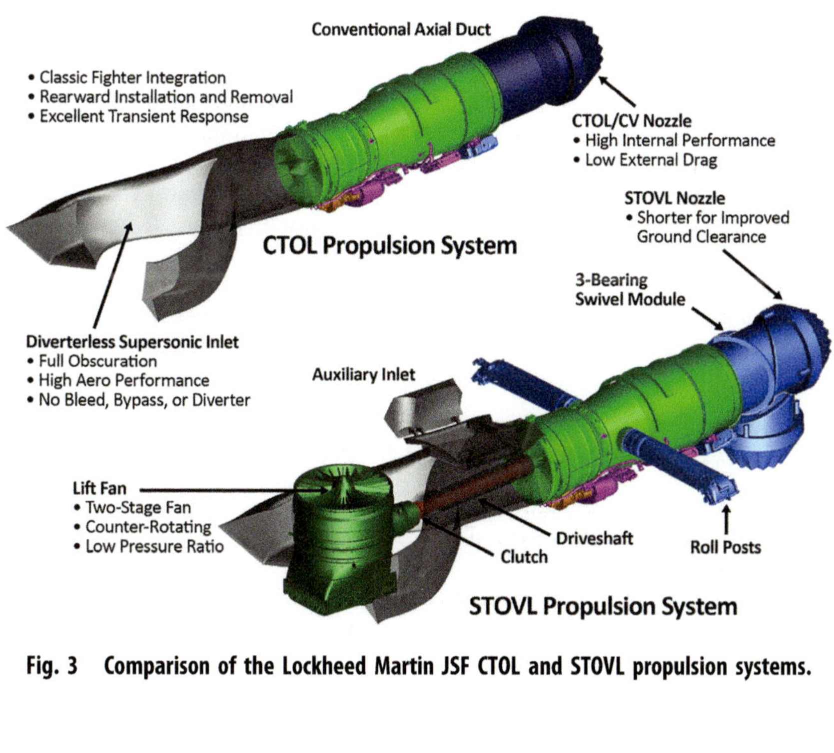

Fig. 3 Comparison of the Lockheed Martin JSF CTOL and STOVL propulsion systems.

1. VERTICAL THRUST-WEIGHT MARGIN

The lift fan augments thrust relative to the basic nonafterburning engine by approximately 40%, achieving thrust more efficiently with higher mass flow and lower exhaust velocity.

2. HOVER BALANCE/TRIM

The SDLF arrangement creates natural vertical-thrust posts around the aircraft center of gravity, so pitch and roll control is achieved simply through shifting of upward vertical thrust among the four inherent nozzles. Importantly, the high thrust capability of the forward-placed lift fan allows the aft nozzle and engine to be placed at the aft end of the fuselage, permitting conventional arrangements of aerodynamic configuration, structure, and systems. This is key to the efficient conventional configuration that is well-suited to all three variants.

3. CONTINUOUS TRANSITION

Continuously vectoring lift-fan and engine exhaust nozzles permit a smooth transition (wingborne to/from jetborne flight) without requiring a propulsion-system mode change during transition. This simplifies transition and reduces risk. The propulsion system converts to STOVL mode prior to downward transition and converts out of STOVL mode after upward transition.

4. INDUCED EXTERNAL ENVIRONMENT

The forward lift fan exhaust is generated solely by the compression of air by the fan. Although the exhaust is still energetic, it is much cooler and slower than engine exhaust. In the aft, engine exhaust is more benign due to the extraction of energy to drive the lift fan by the low-pressure turbine. Exhaust from the roll nozzles is similar to that from the lift fan and is produced by the main engine fan. The result is an acceptable induced thermal, acoustic, and flow environment imposed on the aircraft, deck, or ground surface, as well as surrounding personnel, aircraft, or equipment. Furthermore, the much cooler lift-fan exhaust blocks the warmer engine exhaust from reaching the forward parts of the aircraft in hover, minimizing hot-gas ingestion in the inlets.

At Lockheed Martin, two series of configurations were developed [5]. The 100 series was derived from the Defense Advanced Research Projects Agency's ASTOVL and Common Affordable Lightweight Fighter (CALF) programs. These were canard-configured designs with an SDLF for the STOVL variant, like that tested under the ASTOVL program in the National Aeronautics and Space Administration Ames Research Center's full-scale wind tunnel [6]. The commonality strategy for the CTOL and CV variants was simply to remove the SDLF system and replace the aft vectoring nozzle with a two-dimensional vectoring nozzle, as implemented on the F-22. The 200 series of configurations used the same propulsion system with a conventional wing-tail arrangement. The commonality approach was to use identical aerodynamic configurations for the CTOL and STOVL variants. The CV variant was to be common as well, but at this stage had greater wing area due only to enlarged wing leading- and trailing-edge flaps and an extended wing tip, so the airfoil shape was thinner and the profile somewhat compromised to facilitate a common wing box with the other variants.

A conventional wing-tail arrangement was selected for two primary reasons. First, extensive research by Lockheed Martin leading to the F-22 design concluded that the conventional arrangement produced the best transonic turning maneuver performance and maximum lift for desired longitudinal stability levels, with competitive supersonic drag characteristics [7]. Second, although in this phase there was no explicit requirement, the need for low Carrier Powered-Approach Speed (V_{PA}) indicated much lower risk for a wing-tail design compared to a delta-wing or canard configuration, both from a lift perspective (maximum lift and at approach Angle of Attack [AOA]), and from a low-speed control perspective (control-power and adverse control coupling). The canard configuration carried through the ASTOVL and CALF programs was designed for CTOL/STOVL capability without regard for carrier compatibility with a CV variant.

At this stage, commonality among the variants was very high within the configuration family, including airframe structure and Vehicle Systems (VSs). It was expected that the benefits of extensive commonality would outweigh the costs of improvements in performance that might be gained by optimizing the structure and systems to specific service requirements. In later stages of the program, the airframe structure and some VSs did evolve away from commonality at the detail level, as described later in the paper. Mission Systems (MSs) remained nearly 100% common throughout.

III. CONCEPT DEMONSTRATION PHASE: JOINT STRIKE FIGHTER

The JAST program to date and CDDR results were deemed promising enough to proceed to the next phase of an air system acquisition program. The priorities, or pillars, of the program were established as lethality, survivability, supportability, and affordability.

The CDP program had ambitious objectives. First, the Preferred Weapons-System Concept (PWSC), representing the future production air system, would be developed and matured. Second, the requirements set would be refined iteratively with the PWSC. Third, key technologies would be matured and demonstrated, including flight test of a full-scale Concept Demonstrator Aircraft (CDA).

A. THE COMPETITION

In 1996 the three competitors (Lockheed Martin, the McDonnell Douglas/British Aerospace/Northrop Grumman team, and Boeing) submitted proposals for the

CDP phase. By this time the high stakes of the program were clear, so the government maintained a fair competition by placing limits on spending for the program to keep contractors from attempting to buy the competition. In early 1997 Lockheed Martin and Boeing were selected for the CDP program.

The CDP contract values were each just more than $1 billion, including the CDAs and engine development. After McDonnell Douglas's loss, both Northrop Grumman and British Aerospace were still interested in participating in the program, and both clearly had valuable technical capability, so there was a courtship period during which both companies considered and were being considered for joining Boeing or Lockheed Martin as teammates. Ultimately both companies teamed with Lockheed Martin. Soon after the CDP down-select, McDonnell Douglas merged into the Boeing Company, making the McDonnell Douglas JSF resources available to the Boeing CDP effort.

On the Lockheed Martin team, both new teammates were full-fledged aircraft prime contractors, and each brought unique strengths to the team. Northrop Grumman had extensive experience with low observables and a long legacy in carrier suitability, and British Aerospace had a legacy and unique capabilities relative to STOVL aircraft and extensive capabilities in precision fabrication. Teaming agreements were established outlining the teammates' responsibilities and work share, as well as provisions for sharing intellectual property within the program. During this phase, the development team functioned as a single unified team, largely co-located in Fort Worth, Texas, and numerous key leadership positions were filled by personnel from Northrop Grumman or British Aerospace.

B. REQUIREMENTS DEVELOPMENT

Establishing requirements was a key objective of the CDP program, resulting in the JORD, a JSF Model Specification (JMS), and Key Performance Parameters (KPPs). The JORD was preceded by a series of JSF Interim Requirements Documents (JIRDs) that were released on a roughly annual basis. Requirements maturation was closely overseen by service representatives, with frequent reviews by the Operational Advisory Group (OAG) and Senior Warfighters Group (SWG).

As the trivariant configurations matured, requirement trade studies were conducted in parallel (separately by both competitors) to determine which combinations of capabilities were achievable and affordable. Indeed, requirements management was the principal affordability lever applied during this phase. The basic aircraft sizing was determined through several iterations of Cost and Operational Performance Trades (COPTs), addressing aircraft performance (e.g., mission, maneuver, basing). These results fed JIRD-I and JIRD-II between 1995 and 1997, establishing initial aerodynamic performance, low-observability requirements, and overall supportability and avionics targets. The COPTs were followed by formal Cost-as-Independent-Variable (CAIV) studies using

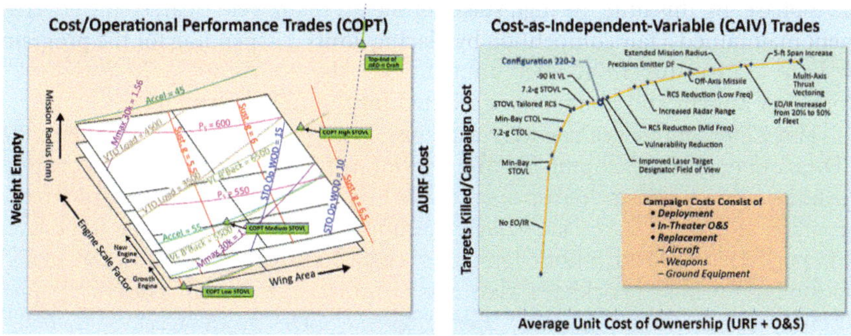

Fig. 4 Typical JSF COPT and CAIV trade study data.

campaign-level operational analysis measures of merit to determine the most cost-effective combinations of sensors, weapons, signature, and maneuver/flight envelope capabilities. These fed JIRD-III in 1998 and the draft JORD in April 1999 with more detailed avionics and supportability requirements (Fig. 4).

JSF was the first program to apply CAIV in a quantifiable way. Dozens of trade studies of individual capabilities were conducted to quantify operational benefit and impact on remaining life-cycle cost. These trade results were prioritized and plotted as the CAIV curve (Fig. 4). The joint requirements authors understood that affordability would restrain imagination when it came to capabilities, and the CAIV curve facilitated reconciliation of personal biases and warfighting value. Some capabilities are difficult to quantify in terms of campaign analysis, such as the gun. These cases were settled using consensus techniques within the OAG, SWG, and JSFPO, and among contractor operational analysts.

Although, on the surface, performance requirements did not vary significantly through the phase, there were subtle ground rule changes and a few added parameters that drove commonality among variants apart, particularly in the airframe. For example, revisions to the USAF design-mission penetration altitude and Mach number drove required fuel volume in the CTOL variant, and increasing the vertical load factor requirement to 9*g* for the CTOL caused most structural members and actuators to diverge from the corresponding STOVL parts. The addition of the V_{PA} requirement as a KPP for the CV variant directly resulted in increased CV wing area and movement to a noncommon wing-box planform and cascading changes into other systems, such as actuators.

Within the industry team, a rigorous systems engineering process was maintained to decompose the top-level performance-based specification requirements to lower tiers. Decomposition was documented in requirements work packages and tracked via IBM Rational DOORS software. Affordability requirements were imposed as design-to-life-cycle-cost targets for each design area.

C. PREFERRED WEAPONS SYSTEM CONCEPT

Lockheed Martin PWSC configurations were designated as the 230 series. The first released configuration family, 230-1, was designed to satisfy JIRD-I requirements and retained a common wing-box planform for all three variants. This became the point of departure for the CDAs, later designated X35A, B, and C, so the external lines and other design data were transferred to the CDA design team in Palmdale, California. Subsequent design iterations through 230-5 were driven by wing, propulsion, and internal arrangement changes [5]. In general, the design team strove to make the configurations as compact as possible. Overall fuselage dimensions were heavily constrained by the propulsion system and weapons bays. Wing area and span for CTOL and STOVL were being driven up by subsonic maneuver requirements, but the design team was constrained by hover weight. One major breakthrough was the adoption of a Variable-Area Vane-Box Nozzle (VAVBN) for the lift-fan exhaust. Previously, the lift fan exhausted through a Telescoping Vectoring Exhaust Nozzle (TVEN), referred to as the pram hood by British Aerospace personnel. This nozzle, when sized for Short-Takeoff (STO) thrust at aft vector angles, had too great an effective area at the vertical hover condition, limiting available hover thrust [4]. The variable area feature of the VAVBN, however, allowed thrust optimization at all vector angles, improving Vertical Landing (VL) capability, which in turn permitted a larger wing area for combat turn performance.

The external configuration matured to balance subsonic vs supersonic vs basing performance, stability and control, signature characteristics, and sensor locations and fields of regard. This was supported by extensive wind-tunnel and radar-range testing in addition to extensive Computational Fluid Dynamics (CFD) and computational electromagnetics analyses. As the configurations evolved, sometimes subtle changes created significant effects. In one example just before the 230-1/X-35 lines were to be frozen, a slight increase in wing incidence was incorporated to improve STO performance that also required reshaping the upper fuselage over the inlet duct. This change showed the desired increase in lift in the low-speed wind tunnel, but in transonic tests showed an unexpected and unacceptable reduction in directional stability. This created an extra design iteration just before the X-35 configuration could be frozen. Perhaps the most significant evolution of the external configuration was in the CV wing size (and tails) driven by the addition of the V_{PA} requirement as a KPP. Iterations of the CV design included successively larger wing areas in order to maintain sufficient risk margin to the requirement. This caused the CV variant to abandon the common wing-box geometry with the other variants, although the substructure locations were retained to allow for common assembly tooling.

The internal arrangement, which directly affected the external lines as well, matured significantly as internal systems matured. The size of subsystem components often increased with each design iteration, but the biggest challenge was in integration, accounting for mounting provisions, connectors and couplings, routing of tubes and harnesses, bend radii, separation requirements, and maintainer access. Major geometric integration challenges included carriage and clearance of the long list of internal weapons, landing gear retraction and stowage, expendable countermeasures dispensers, Electro-Hydrostatic Actuation System (EHAS) actuators and electronic units, engine and accessories removal and installation, and inflight opening doors. Thermal management was also a continuing challenge. As would be expected of a 5th-Generation-class fighter, system heat loads are absorbed internally as much as possible, but the STOVL lift fan also rejects substantial heat, adding to the challenge. Also, each variant had unique center-of-gravity requirements, which again were most demanding for STOVL. The result is that, although many fuel-system components were common among the variants, the integrated fuel system arrangements became unique. Unique center-of-gravity requirements also significantly affected weapons suspension and release equipment design [8].

The need for hover (and takeoff) thrust-weight margin placed demanding requirements on nearly all aspects of the design family, but most strongly on STOVL. Refinement of the STOVL propulsion system to improve both thrust and aeromechanical operability continued throughout PWSC development, as well as weight reduction. Early in the phase, the lift-fan diameter was increased by 2 in. for added thrust. Significant changes were made to the primary, lift fan, and auxiliary air inlets for improved pressure recovery and distortion, the most visible being adoption of the aft-hinge lift-fan inlet door [4]. The main inlet was shortened by approximately 40 in. as a weight-saving measure, and the four-edge main-inlet aperture was replaced by the three-edge configuration to improve high-AOA distortion.

The PWSC manufacturing approaches for affordability also had a direct impact on the configurations. One approach was targeted at rapid assembly (five months) with minimal tooling. Major subassembly components were to be fully stuffed with subsystem components and joined during final assembly using discrete quick-mate joints with relatively few large fasteners accompanied by fluid and harness couplings at every joint. The final assembly mate planes were selected to minimize complex rigging tasks after mate. The goal was to avoid mate planes that would be crossed by the canopy, inlet duct, or weapons-bay doors, but because the inlet duct and canopy overlapped in fuselage station, a mate joint at Fuselage Station 270 was accepted that crossed the canopy.

Another goal was to maintain common tooling among all variants. Even if web or flange thicknesses varied between/among variants, pad-ups were envisioned to maintain the same interfaces with tooling. Very tight outer mold line tolerances at seams would be maintained by precision machining inner mold lines of the composite skins and the assembled substructure.

As the PWSC product definition and production system requirements firmed up during the CDP program, the teaming agreements between Lockheed Martin and teammates Northrop Grumman and BAE Systems were amended to define specific work content responsibilities for the teammates in coming detailed design and the future production and sustainment phases. The team organization, personnel assignments, and business practices migrated to more discrete subcontractor relationships. In development, both teammates were to have broad participation in numerous areas. In the production program, Northrop Grumman would have full responsibility for the center fuselage and inflight opening doors, notably the large weapons-bay doors, as well as the weapons-bay-door drive; arresting-gear, fire-protection, inertial-navigation, and global-positioning systems; landing-aid antennas; and MS common components. BAE Systems would be responsible for the aft fuselage, horizontal- and vertical-tail boxes, CV outboard wing boxes, and the crew-escape, fuel, life-support, and ice-detection systems.

The final PWSC configuration, 230-5, became the basis for the Engineering and Manufacturing Development (EMD) proposal in early 2001.

D. CONCEPT DEMONSTRATOR AIRCRAFT: THE X-35s

As mentioned previously, the CDA or X-35 aircraft were derived from the 230-1 PWSC configurations. These aircraft were crucial to demonstrate the SDLF system's capability and verify that variants with a high degree of commonality could satisfy the diverse requirements of each service. To that end, Lockheed Martin produced two aircraft with one basic design. Unique features for each variant could be installed on either aircraft. The Boeing X-32A and Lockheed Martin X-35A aircraft are shown in Fig. 5.

The Lockheed Martin demonstrators did not reflect the PWSC's airframe structure, nor most of the subsystems. Risk reduction for those elements was accomplished through other unique demonstrations. Rather, the X-35s represented the aerodynamic configuration, the full-scale propulsion system, and the flight control laws (not hardware). The airframes were built-up airframes as one-off prototypes. They were built as single assemblies, not from mated subassemblies. As much as possible, systems components were taken from existing

Fig. 5 Boeing X-32A and Lockheed Martin X-35A CDAs.

aircraft. The landing gear, for instance, were taken from the A-6 for the main gear and the F-15E for the nose gear.

Four key differences between the X-35B STOVL variant and the proposed PWSC were: 1) the X-35B retained the TVEN pram-hood lift-fan nozzle rather than the later VAVBN; 2) the side-hinged lift-fan inlet doors on the X-35B were replaced with a single aft-hinged door; 3) the center-hinge arrangement of the auxiliary air inlet doors was replaced by side-hinge doors; and 4) the X-35B retained the use of three-handle STOVL inceptors much like the Harrier arrangement, rather than the two-handle unified-control scheme adopted for the proposed PWSC.

The first airplane (Ship 1) first flew in October 2000 as a CTOL variant at Palmdale. It immediately transitioned to Edwards Air Force Base (AFB), California, to conduct up-and-away flight tests representative of both the CTOL and STOVL variants. After 28 flights in five weeks, the airplane returned to Palmdale for conversion to the STOVL variant by installing the SDLF, 3BSD, and roll-control ducts. During its brief test program, the X-35A was flown by Lockheed Martin, USAF, USMC, and U.K. pilots. It achieved 5*g*, 20-deg AOA, supersonic speeds, and aerial refueling. Quantitative results for specific range and maneuvering performance closely matched predictions. Solid Level 1 handling qualities were observed by all six pilots throughout the test envelope, and trouble-free propulsion performance demonstrated engine/inlet compatibility.

Ship 2 first flew in December 2000. It was configured as the X-35C with larger wing leading- and trailing-edge flaps and larger tail surfaces. It was flown by Lockheed Martin, USN, USAF, and U.K. pilots for approximately 40 flights at Edwards AFB, including Field Carrier Landing Practice (FCLP), aerial refueling, and supersonic flight, before ferrying across the country in February 2001 to Naval Air Station (NAS) Patuxent River, Maryland, for an additional 30 flights of aggressive field carrier landing tests over four weeks, accomplishing 258 FCLPs. Again, up-and-away tests supported prediction models, and excellent flying qualities were noted. Eight government and Lockheed Martin pilots flew the X-35C configuration (Fig. 6).

Ship 1 completed its conversion to the X-35B configuration in February 2001 and entered a series of ground tests prior to first flight. In parallel, the STOVL propulsion system completed accelerated mission testing, and the flight-control software underwent final regression tests. The first ground test involved mounting the airplane on struts just above a grated hover pit for restrained measurements of net forces produced by the installed propulsion system, and to demonstrate the functionality of the integrated flight and propulsion control system. Then the grate was replaced by a solid ground plane to determine the effects of close ground proximity on net aeropropulsive forces. Also, ground environment acoustic, thermal, and flow measurements were made on the aircraft's lower surface, on the ground plane, and at a 50-ft radius from the airplane.

Initial STOVL-mode flights were made in late June 2001 over the grated hover pit to establish hover performance limits and clear the bottom of the jetborne flight envelope. This established known conditions for the end point of subsequent decelerating transitions. After demonstrating conversions to/from STOVL mode at altitude, systematic build-downs to hover were performed, including semi-jetborne landings, leading to the first VL on the AM-2 landing pad at Edwards AFB. Successively slower STOs were made down to a rotation speed of 60 kt. Level-1 flying qualities were consistently seen during the tests. The engine was free of hot-gas ingestion, and infrared images confirmed that the forward lift-fan flow effectively blocked engine exhaust from the engine inlets and forward portions of the aircraft.

Aerodynamic, propulsion, and environment measurements from these ground and flight tests were better than predictions in nearly all cases, as conservative margins in prediction models were found to be unnecessary. During the first flight, a press-up (vertical takeoff from the grated hover pit), British Aerospace test pilot Simon Hargreaves pressed up to a stable hover 20 ft high when he only expected to barely break contact from the grated pit. In the lead-up to STOVL flight tests, the test team contemplated plans to follow Ship 2 to NAS Patuxent River in order to do hover tests at favorable conditions at sea-level elevation and cooler temperatures. In reality, the system produced robust vertical performance in the high desert at 2300-ft elevation and in the middle of summer, so the entire test program was completed there.

On 20 July 2001 the X-35B became the first aircraft in history to achieve both supersonic and hovering flight in a single sortie (Fig. 7). For more than 40 years prior to the X-35, U.S. and European aerospace enterprises had pursued Vertical and/or Short Takeoff and Landing (V/STOL) fighter technology, yielding only the subsonic Harrier as a truly operational system. Besides the X-35B, only three research vehicles had previously demonstrated both hovering and supersonic capability with the same configuration: the German VJ-101C, French Mirage

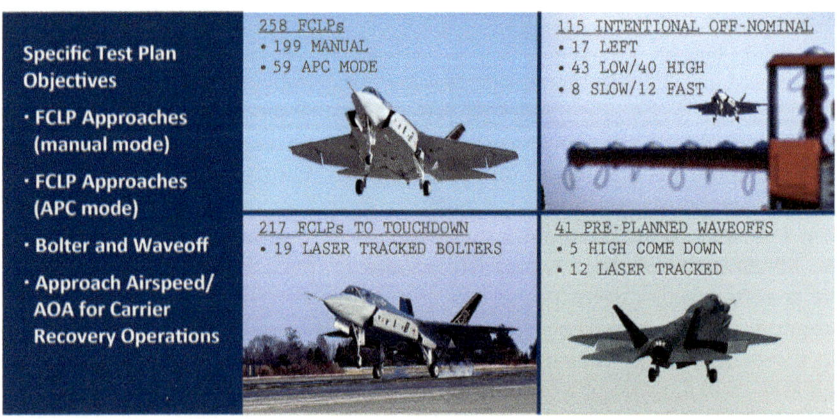

Fig. 6 X-35C carrier-suitability flight test accomplishments.

Fig. 7 X-35B performance of Mission X: STO, supersonic dash, and VL.

III-V, and Soviet Yak-141. However, none of these aircraft represented operational weapons platforms, and their limited performance prevented them from accomplishing hover and supersonic flight in a single mission. With that background, the X-35B team set out to execute Mission X, consisting of an STO, a supersonic dash, and a VL to conclusively demonstrate that the SDLF propulsion system would overcome the fundamental incompatibilities of supersonic and STOVL flight and enable the common configuration family.

IV. SYSTEM DEVELOPMENT AND DEMONSTRATION

The Lockheed Martin proposal for what was then called the EMD phase was submitted in February 2001, while the X-35 flight tests were still underway. At that time, only the X-35A CTOL had completed its testing and was still being converted to the X-35B STOVL. The X-35C CV was preparing to ferry from Edwards AFB to NAS Patuxent River to complete carrier suitability tests. Supplements to the proposal covering successful X-35 flight test results from all three variants were later submitted in April and August. Formal questions and answers were exchanged over the summer via evaluation notices and face-to-face discussions, and final proposal revisions were due in mid-September.

The proposal evaluation criteria included three equally weighted factors: affordability, EMD (each component of which received both proposal ratings and risk ratings), and past performance. The affordability factor addressed the proposed products, including subfactors for the air vehicle, autonomic logistics (sustainment system), and remaining life-cycle cost (procurement, operations, and support). The EMD factor addressed the proposed program, including subfactors for technical plans, management, and EMD cost.

The entire proposal evaluation process remained on schedule despite delays in planned STOVL flight tests experienced by both competitors and the 9/11 terrorist attacks on the Pentagon. The Defense Acquisition Board met as planned on 24 October to confirm that the JSF was ready to enter the next phase of development, now termed System Development and Demonstration (SDD). The dramatic public announcement of the winning contractor was also made on schedule on 26 October 2001 [9].

The announcement that Lockheed Martin had won the contract was made by:

1. The source selection authority, Secretary of the Air Force James Roche
2. Under Secretary of Defense for Acquisition, Technology and Logistics Edward C. "Pete" Aldridge
3. Secretary of the Navy Gordon England
4. U.K. Parliamentary Under-Secretary (Ministry of Defence) (Procurement) Lord William Bach
5. Maj. Gen. Michael Hough, USMC, JSF PEO

Secretary Roche stated that both competitors' proposals were "very strong," but that the Lockheed Martin team "emerged continuously as the clear winner" on a best-value basis, considering strengths, weaknesses, and relative risks [9]. Interestingly, the F-35 designation seemed to be determined and announced on the spot in response to a reporter's question, based on the X-35 designation of the Lockheed Martin demonstrator. In keeping with the demonstrator designations, the CTOL variant was designated F-35A, the STOVL variant was designated F-35B, and the CV variant was designated F-35C.

No time was wasted in executing the contract; the model contract submitted with the proposal was signed by the government the same day on which the downselect was announced.

A. THE PROGRAM

The objectives of the SDD program were to: 1) develop an affordable family of air systems (air vehicles plus autonomic logistic systems) that meet service requirements and significantly reduce life-cycle cost; 2) develop a life-cycle plan that supports production, fielding and operational support, and eventual disposal; and 3) demonstrate and implement affordability initiatives. In the interest of affordability and rapid development, the program was aligned with several acquisition-reform initiatives, including integrated product and process development, Performance-Based Specifications (PBSs) and contractor Total System Performance Responsibility (TSPR), Simulation-Based Acquisition (SBA), concurrent development, and Performance-Based Logistics (PBL). A JSFPO policy minimized the use of government-furnished property, except for full-scale ground-test and flight-test facilities and, notably, the propulsion system (development and hardware).

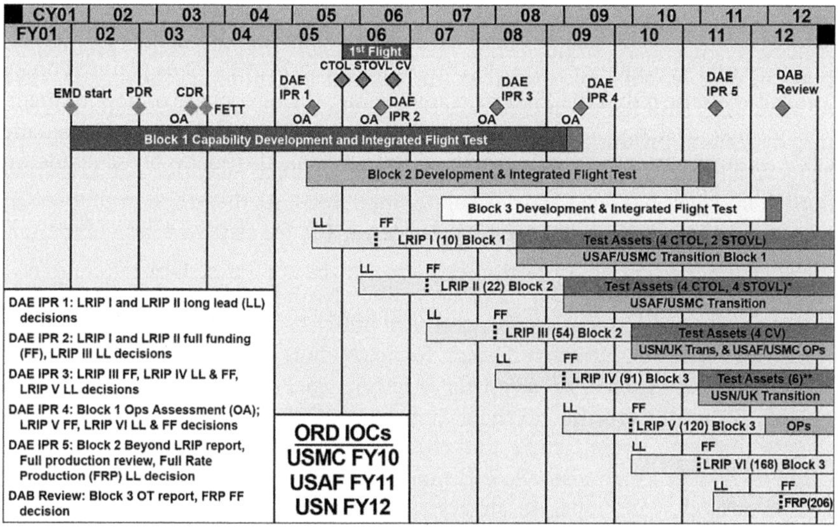

Fig. 8 Development and LRIP schedule originally specified in the EMD CFI.

The overall program plan was specified by the government in the EMD Call for Improvements (CFI), the title given to the request for proposals, as reproduced in Fig. 8. The overall program period of performance was to be an aggressive 126 months, with first flights of the CTOL, STOVL, and CV variants planned for 48, 53, and 62 months after go-ahead, respectively. The plan also specified certification of three blocks of progressive MS capabilities to execute mission vignettes with particular weapons loadouts.

Though not part of the SDD program, the Low-Rate Initial Production (LRIP) program was planned to be highly concurrent with development. Long-lead funding for the first LRIP lot was planned to be authorized before the first flight of the CTOL variant, with full funding to be authorized just after the STOVL first flight. Six annual lots were envisioned, with production quantities ramping from 10 Block-1 aircraft in the first lot to 168 Block-3 aircraft in the sixth.

Lockheed Martin was awarded a $19-billion cost-plus-award-fee contract. Pratt & Whitney was separately awarded a $4-billion contract. The two companies established an associate contractor agreement to govern the coordination and integration of the propulsion system in the aircraft. In the Lockheed Martin contract, the majority of the available award fee was tied to the customer's assessments of contract performance during each six-month period covering the affordability, developmental cost control, management, and technical categories, as well as an overall comprehensive rating. However, a significant portion of the contract fee, known as the Schedule B award fee, was to be determined by an objective comparison of actual LRIP production costs with specified

Affordability Improvement Curves (AICs) for each variant. The AICs corresponded to the cost trend, when averaged over the projected production program, that would correspond to the original \$28 million (fiscal year 1994 dollars) CTOL Unit Recurring Flyaway (URF) cost target, when adjusted for economic escalation, production quantity and rate, variant configuration, and scope changes. This device was intended to provide a direct incentive for achieving affordability goals but was very difficult to implement. As the program progressed and setbacks to performance were encountered, the fee structure was ultimately renegotiated.

The Lockheed Martin SDD program plan implementation through first flights of all variants aligned with the CFI, as shown in Fig. 9. In Lockheed Martin's plan, a single air system Preliminary Design Review (PDR) covering all three variants was set for 17 months after receiving Authority to Proceed (ATP), rather than only the 12 months specified in the government's CFI, compressing the timespan available to design, manufacture, and check out the CTOL variant after the PDR. Lockheed Martin's plan also allowed more time for each variant to produce the detailed designs, with separate air system Critical Design Reviews (CDRs) for each variant at 31, 37, and 44 months, respectively, rather than the specified 21 months for all variants in the CFI plan, greatly reducing the time available for manufacturing and/or requiring a greater degree of concurrency between design and manufacturing.

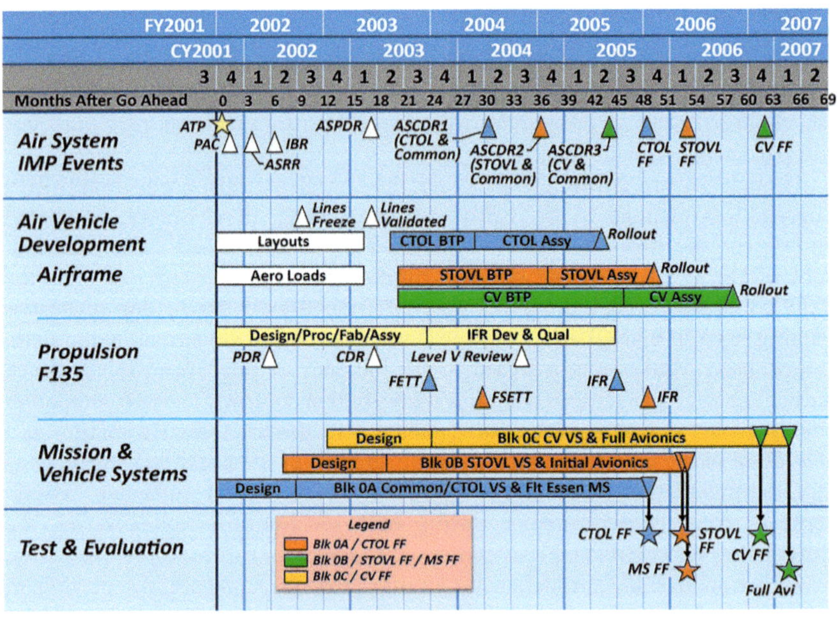

Fig. 9 Excerpt from Lockheed Martin's initial SDD schedule.

In order to accomplish the trivariant PDR event, lines-freeze was to be completed only nine months after ATP, barely enough time to conduct a single wind-tunnel test cycle (design, build, test, and analyze). A second milestone, lines validation, was established recognizing that additional testing may be required to finalize the geometry. However, structural layouts would have to be completed based on the initial lines freeze. The other pacing element to begin detailed structural design was aerodynamic loads. The time available was not enough to conduct a full loads wind-tunnel program, so the initial structural sizing was done using data from the X-35s, and adjustments were made based on varying levels of CFD analysis.

In summary, the early stages of the SDD plan depended on a high level of concurrency among aerodynamic lines and loads, structural design, and manufacturing. The final PWSC configuration from the CDP phase was well defined at the at the outset of SDD, but wind-tunnel testing of the specific configurations had not been completed in the CDP due to the need to focus resources on X-35 flight tests. The design team expected a few weeks to pass between the down-select and ATP, during which advance work would mitigate some of the concurrency risk, but as mentioned earlier, the two events actually occurred on the same day. In fact, the first critical challenge the team had to overcome was to ramp up personnel very rapidly, increasing the staff from 400 to 4000 people in the first eight months. Therefore, the early SDD schedule was very challenging.

B. SYSTEM REQUIREMENTS DEPLOYMENT

The SDD phase began with an operational requirements document taken from the draft JORD produced by the JSFPO and U.S. and U.K. services at the end of the CDP program. This established KPPs for the system, defining values for combat radius, CV recovery (V_{PA}), STOVL performance (flat-deck and ski-jump STO distance, and VL bring-back), interoperability, Radio-Frequency (RF) signature, mission reliability, sortie generation rate, and logistics footprint.

The JSF Contract Specification (JCS) aligned with the JORD and was included in the Lockheed Martin prime contract. It was an entirely capabilities-based PBS. Roles and missions for each of the U.S. services and the United Kingdom were defined along with mission usage. Performance, interface, and environment requirements were defined for lethality, survivability, basing, security, safety, reliability, supportability and training, situational awareness, and handling qualities, along with their corresponding definitions.

One of the initial tasks was to formalize requirements allocations to create performance and functional requirements at all tiers of the product hierarchy and organization. This was done through mission decomposition and requirements work packages. Requirements allocations were documented and managed through the use of a commercial requirements management tool tailored to the F-35's needs. PBSs for system suppliers could be generated directly by the tool.

C. KEY PROGRAM STRATEGIES

The development team recognized that the complexity and scope of the SDD program, combined with the aggressive schedule and affordability goals, would require extraordinary team performance. In addition to the broad acquisition reform initiatives, a number of engineering and management strategies were ingrained in the organization from the outset, some widely applied and some narrowly focused. The following are examples of the strategies used.

1. JSF FIRST AND JSF ENTERPRISE GUIDING PRINCIPLES

Program leadership continued to reinforce *JSF First!* as the program motto, reflecting the high stakes in the success of the program to U.S. and allied militaries, the prime teammates' and critical subcontractors' businesses, and individual careers. In other words, whatever selfish advantage one organization might gain within the program paled in comparison to the benefits of the highly functioning overall team. The Lockheed Martin program general manager also established 10 enterprise-guiding principles governing the way people worked together, categorized in support of five objectives:

1. Inspire excellence.
2. Expect the exceptional.
3. Seek to connect.
4. Foster trust and respect.
5. Value the individual.

2. AFFORDABILITY AND BEST VALUE

The affordability pillar permeated all aspects of the F-35 program. In SDD, the emphasis shifted from COPT and CAIV top-level requirements tradeoffs to the allocation of design-to-cost targets to the system, subsystem, and airframe component levels, and to the execution of affordability initiatives to progressively achieve the targets (Fig. 10). Integrated Product Teams (IPTs) were accountable to report progress against burndown plans on a monthly basis, and suppliers formally reported statuses quarterly on projected recurring cost and their own affordability initiatives.

3. TOTAL SYSTEM PERFORMANCE RESPONSIBILITY AND PERFORMANCE-BASED SPECIFICATIONS

TSPR was a DoD acquisition reform doctrine designed to save costs by eliminating redundant or unnecessary work between the government and prime contractor. In this approach, Lockheed Martin accepted TSPR and was free to make use of

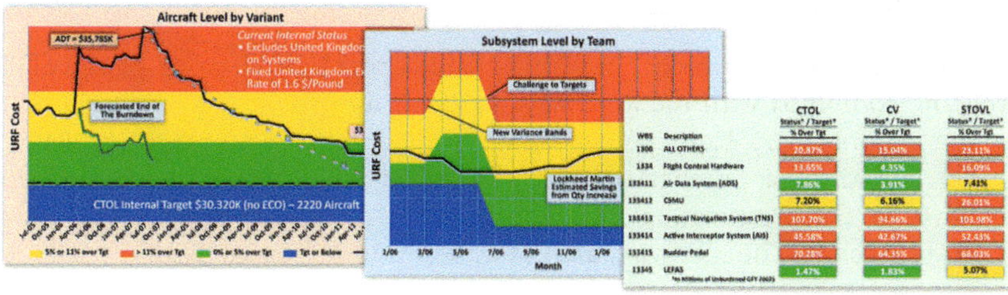

Fig. 10 Typical monthly design-to-cost data summary.

MIL-SPEC and MIL-STD standards but was not bound by them if there were industry or commercial equivalents that resulted in meeting overall requirements. In this approach, the government customer was afforded *insight* into the program without traditional government *oversight*. The JCS was a PBS dictating the required system performance capabilities (the *what*) but not specifying the methods by which the system was developed to achieve the required performance (the *how*). Lockheed Martin, in turn, employed PBSs in most of the major MS and VS subcontracts.

4. SIMULATION-BASED ACQUISITION

SBA was a DoD initiative to exploit advances in information technologies, specifically modeling and simulation, to enable better, faster, and cheaper weapons-systems acquisition. To oversimplify, the objective was to verify system behavior and performance vs requirements in a digital virtual environment. To these ends, Lockheed Martin and the JSFPO invested heavily in a robust system of laboratories dedicated to F-35 development and verification. The main laboratory facilities are depicted in Fig. 11. These produced high-fidelity simulations used to verify requirements. Extensive physical testing in various domains served not to verify requirements directly but rather to validate the models that comprised the virtual verification environment. The dedicated F-35 laboratory environment included the following:

- VS Integration Facility
- VS Processor/Flight Controls Integration Facility
- MS Integration Laboratories (including systems integration stations with simulated, RF-stimulated, and open-air [actual hardware in physical aircraft model] inputs)
- Air System Integration Facility
- Verification Simulator

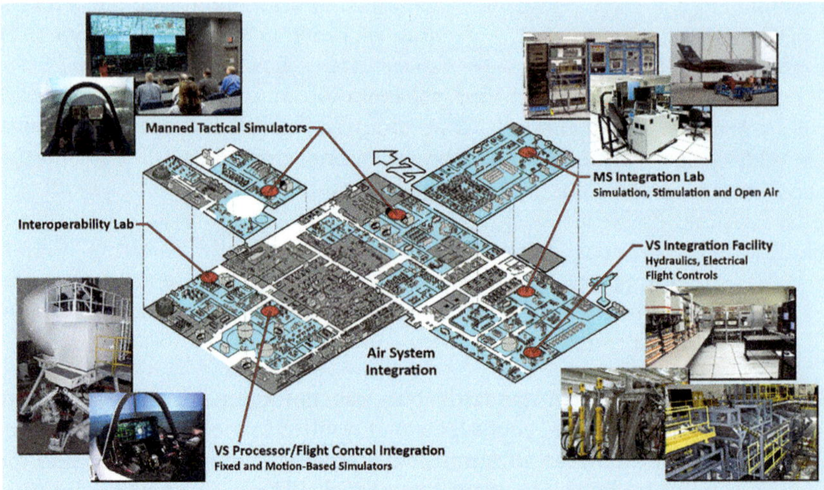

Fig. 11 Integrated, co-located high-fidelity laboratory complex.

Lessons learned from legacy programs drove the laboratory requirements and design. First, co-location in one geographic area of all major test and verification laboratory venues was determined to be essential, as opposed to the dispersed facilities of the F-22 program in Seattle, Washington; Fort Worth; and Marietta, Georgia. The more integrated and synergistic F-35 environment was available to the development team on a 24/7 basis. Second, in alignment with DoD SBA policy statements, reuse of software and data was found to save costs. Reuse was a strategic initiative from the very early stages of laboratory development. Reuse allowed the team to avoid duplicative efforts to develop the middleware/executive layers of software and aircraft/sensor models across all of the laboratory venues. The F-35 laboratories had two to three times greater capability than that of any preceding program by any measure, with unprecedented connectivity. These capabilities were key elements in the SDD plan, shaping the makeup of the flight test aircraft fleet and supporting a target of only 5000 flights. These flights were planned to provide a top-level validation of the integrated models, not to verify lower-level requirements.

5. COLLABORATIVE ENVIRONMENT AND THE DIGITAL THREAD

Another strategy to exploit advanced information technologies involved sharing needed data among a wide range of participants and among phases of the program life cycle. Considerable investment was made early in the program in tools to facilitate nearly instant access to data across the team, while at the same time reliably controlling access to restricted data for ITAR or other reasons. The JSF Data Library was established as a program-wide repository for collaborative file exchange and vaulting of program records of all types. For detailed designs, the product data manager, later known as the product lifecycle manager, facilitated near-real-time collaboration among nine primary design-release organizations spanning 17 time zones, as well as approximately 50 suppliers with design authority for their systems, and more than 100 build-to-print suppliers across the globe [10, 11]. Design information was comprehensive for each part and contained in Build-to Packages (BTPs), including CATIA solid models, drawings, analysis documents, specifications, manufacturing planning, and tooling information, all organized into the product data structure, including changes. The system ensures that consistent data are used in the as-designed, as-planned, as-built, and as-maintained configurations.

6. CONCURRENT ENGINEERING AND INTEGRATED PRODUCT TEAMS

The F-35 air vehicle, production system, and support system were developed simultaneously. Even the earliest flight test aircraft were built on production hard tooling at Lockheed Martin and, to varying degrees, at teammates' and suppliers' sites. Also, the manufacturing system makes use of the same support equipment developed or selected for use in the field. Design-for-manufacture and the supportability pillar were among the design team's primary goals from the outset. Multidiscipline IPTs were the norm, with representation from manufacturing and sustainment disciplines, and virtual simulations of assembly and maintenance tasks were done in parallel with the parts/assemblies definition [10].

7. SYSTEMS ENGINEERING AND REQUIREMENTS MANAGEMENT

Top-level requirements, either directly from the JCS or derived from mission decomposition, were allocated down to all tiers of the design team to be verified up to form the systems engineering *V*. Requirements management is addressed later in the paper.

8. RISK MANAGEMENT AND KEY SYSTEM DEVELOPMENT AND INTEGRATIONS

The program established and maintained a robust risk management process following the classical ISO 31000 approach, identifying and assessing risks and executing mitigation plans for moderate to high risks. This process was applied at each organizational level. Most often, risks were identified at lower tiers and evaluated using consequence thresholds (i.e., impacts on cost, schedule, or performance) tailored to that tier. Moderate or low risks were managed at that tier, but high risks were elevated for assessment at the next tier, and so on. In this way, for instance, a risk that could be managed within a Tier IV IPT's budget, that did not impact the critical-path schedule, or that only affected internally imposed requirements might be handled at Tier IV, whereas a risk with high cost impacts, affecting critical milestones or top-level performance requirements (e.g., a KPP) would be elevated to the program level. Risk reviews were held monthly or as needed, and new risks were processed. Existing risks, once mitigated, were retired. At the outset of SDD, the program-level risks included:

Risk 01: Software Executability

Risk 02: Lift System Hardware

Risk 03: Trained Manpower Availability

Risk 04: Supplier/Partner Management

Risk 05: MS Fusion Algorithms

Risk 06: Deck Crew Environment

Risk 07: Production URF

Risk 08: Process & Tool Performance

Risk 09: Air Vehicle Weight Control

Risk 10: Electro-Hydrostatic Actuator Development

Risk 11: Canopy Bird Strike Compatibility

Risk 12: Requirements Control under PB

For the most part, mature technologies were chosen for inclusion in the F-35 air system design, produced by the prior JAST and CDP phases or other prior programs. However, it was understood that successful development was highly dependent on complex cross-IPT integration of multiple technologies to an extraordinary extent, even among high-performance fighters. Integration challenges could be physical, functional, or both. An evaluation of integration challenges was done at the outset of SDD by mapping candidates' complexity vs potential program impacts to arrive at the Key System Developments and Integrations (KSDIs). Teams were created around each KSDI with a leader responsible to the program manager for planning and executing the cross-IPT effort, particularly in the early stages of SDD as the system requirements were finalized and the system became fully defined. The list of 22 KSDIs is presented here:

- Interoperability Integration
- Prognostics-and-Health-Management Development & Integration
- Outer-Mold-Line Definition
- Lean Manufacturing
- Low-Observables Aperture/Edge/Sensor Integration
- Virtual Weapons System/Simulation-Based Acquisition
- Subsystems—Airframe Integration
- Cockpit Integration
- Integrated-Core-Processor Development & Integration with Sensors
- Mission-Systems Software Development and Domain Integration
- Vehicle-Systems Software Development and Domain Integration
- Integrated Flight-Propulsion Control

- Integrated Subsystems Development and Integration
- STOVL Propulsion System
- Integrated First Flight Checkout
- Shipboard Suitability Integration
- Joint Distributed Information System Development & Integration
- Integrated Avionics
- Design for Reliability, Maintainability, & Supportability
- Factory-to-Field Support Equipment Commonality
- Air Vehicle/Training Software Commonality
- Air Vehicle Service Life

D. KEY CHALLENGES: WEIGHT GROWTH

One of the most significant challenges to the success of the F-35 program was that of air vehicle weight growth and control. Weight control is critical to any flying machine, but the STO and VL requirements for the F-35B are particularly unforgiving. During the first two years of SDD, many factors contributed to both real weight *growth* in all variants, attributed to meeting evolving requirements, as well as the *realization* of higher weights as estimating accuracy matured. The combined effects resulted in a projected failure to meet both STO and VL KPP thresholds that threatened the success of the entire program. Figure 12 presents current estimates of STOVL-variant empty weight through early SDD, compared to various target measures.

At the entry to the SDD phase, the three variants were at a preliminary design level of maturity. At that stage, weight estimates, particularly for the airframe structure, relied on Lockheed Martin's parametric prediction tools. With these, estimates for each component were based on regressions of statistical data from past aircraft for like components, based on the size, shape, material mix, and structural requirements (e.g., Mach number and load factor). Features of the F-35 configurations that were not present in the historical aircraft, such as low observables, variant commonality, integrated subsystems, and rapid-assembly features, were accounted for with incremental weight allowances added to the parametric predictions. In addition to the current estimate, a growth allowance to Initial Operational Capability (IOC) was also added for use in performance predictions and structural sizing, based on prior programs' history.

Throughout the CDP and early SDD phases (prior to PDR), JSFPO mass properties engineers produced independent estimates using similar tools, and the two organizations performed frequent weight reconciliations as the air vehicle configurations evolved. The two estimates remained separate until PDR, with the government's estimate always being higher, but the reconciliation

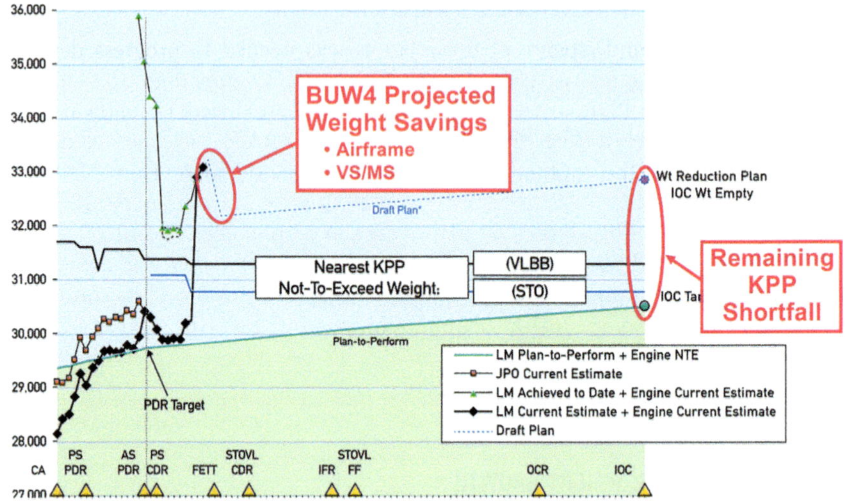

Fig. 12 Current weight-empty estimates as of year-end 2003 compared to plan and critical KPP requirements.

process facilitated a mutual understanding of each team's analyses. A unified JSFPO and Lockheed Martin weight-empty status was an entry criterion for PDR.

Completion of PDR in late March 2003, 17 months into the program, kicked off detailed design of the CTOL variant, with the first aircraft designated 2AA:0001 (abbreviated as AA-1). At that time, the adjusted parametric weight estimates (including estimated growth to IOC), together with aerodynamics and propulsion estimates, supported achievement of all KPP thresholds, and these estimates were used to allocate weight budgets to airframe components and subsystems for detailed design.

As the detailed design and analysis processes matured the AA-1's structure from preliminary design to sized layouts and eventually to released BTPs, periodic weight estimates based on a detailed accounting of parts became available, known as Bottom-Up Weights (BUWs). Unsurprisingly, the earliest BUW totals were significantly greater than both Lockheed Martin's and the JSFPO's parametric estimates, because the part details were not yet optimized. However, at PDR it was recognized that the roughly 4800-lb gap was too great to expect part optimization alone to close. More work was needed to improve the efficiency of the overall structural arrangement, and some concessions to commonality and rapid assembly had to be reconsidered due to significantly more weight impact than what was allowed. Furthermore, PDR identified a number of unresolved design integration issues primarily related to weapons clearance and loading in the internal weapons bay, integration of the internal gun on the CTOL variant, and clearance to parts of the STOVL propulsion system.

1. THE BLUE RIBBON ACTION TEAM

It was clear that rapid closure of these issues was needed to progress detailed design on the very aggressive release schedule. AA-1 CTOL first flight was targeted just 31 months after PDR, as well as the other variants that were to follow closely. To address the issues, a Blue Ribbon Action Team (BRAT) was assembled from technical leaders of the contractor team, the JSFPO, and experts from NAVAIR and ASC. The objective was to resolve open PDR weight and integration issues while preserving the baseline SDD plan as much as possible. The BRAT established a time-phased weight reduction curve with targets for June and September 2003, and closure to the CDR target weight by December 2003. They conducted detailed reviews of every major airframe component, every part weighing 5 lb or more, and a bottom-up estimate of smaller parts: tubes, harnesses, fasteners, brackets, and sealants.

Initial results from the BRAT reviews were good. Design decisions resolved the key integration issues, and the June 2003 weight estimates improved the June BRAT targets for all variants. The most significant configuration changes involved increasing the cross-sectional area to improve structural load paths and internal equipment volume (which penalized supersonic performance) and eliminating quick-mate joints between major assemblies (which nearly doubled the planned final assembly timespan). Work resumed on the AA-1 BTPs, but the combination of delayed BTP starts and added final assembly timespan resulted in a forecasted delay in AA-1 first flight by approximately seven months, into the second quarter of 2006.

By September 2003, however, although the optimization of large parts continued to show small improvements, a more complete accounting for small airframe parts and other parts revealed a large increase over the June BUW. Primary contributors to the other weight category changes from June were tubing and installation (clamps, brackets), systems installation, shims and spacers, and fasteners. The total CTOL airframe weight estimate increased about 800 lb from June to September. During the fall of 2003, calculated weights for released AA-1 BTPs also continued a slow but steady increase, as did the government-furnished propulsion system weight. At that time, the BUW estimates replaced parametric estimates as the current estimates of record.

2. THE STOVL WEIGHT ATTACK TEAM

The actions taken during the BRAT period saved significant weight and resulted in a CTOL configuration that, although not optimum, would have satisfied its driving mission-radius KPP. However, when the weights were projected onto the much more weight-sensitive STOVL variant, it failed to meet *all* of its KPP thresholds, and significantly would have essentially no usable VL capability. As this reality became quantified through the fall of 2003, program leadership on both contractor and customer sides decided that more fundamental changes in the configurations and the program plan were necessary. The first priority was to define a viable STOVL configuration with robust margins to the KPP requirements that still aligned with the program pillars of lethality, survivability, supportability, and affordability. It was decided to largely pause the program until confidence in the STOVL variant could be restored. Once that was accomplished, it was recognized that a major redesign, affecting all variants, would be required, as well as rebaselining the SDD program. It was also decided that the AA-1 detail design and production process was providing extremely valuable data and experience, so it would continue in a parallel effort producing a de facto prototype of the aircraft.

In December 2003, program leadership directed that a dedicated team, to be known as the STOVL Weight Attack Team (SWAT), be assembled to reestablish the configuration family. During the first quarter of 2004, leadership, membership, funding, and operating principles of the team were decided upon, and in early March the plan was approved by the program leadership and the DoD SAE. In mid-March the dedicated SWAT team crowded into unused office space at Lockheed Martin's Fort Worth headquarters; at teammate Northrop Grumman's El Segundo, California, site; and at teammate BAE Systems' Samlesbury, Lancashire (United Kingdom) facilities. The team was made up of roughly 550 full-time equivalent personnel. A kickoff meeting for the whole team in mid-March laid out the full extent of the problem, the stakes at risk, and the overall team approach. In addition to the dedicated SWAT, the entire program team was engaged, starting with a weight stand-down day and a program-wide weight incentive program for weight-saving ideas.

Operating principles for the team were designed to accelerate decision making but maintain systems engineering rigor and configuration control, so these disciplines were well represented on the team. Affordability and supportability remained prime objectives as well, with dollars-per-pound thresholds enforced and affordability, reliability, and maintainability impacts reported for every proposed change. A simple two-tier decision process was set up with a weekly tempo. The SWAT board had delegated authority and funding resources within certain constraints to enact changes, and a higher-level multiboard met the next day for decisions outside of those constraints. The multiboard consisted of the entire executive leadership team, including engineering, manufacturing, sustainment, business management and contracts, and others, whose presence was required each week.

Recognizing that weight reduction actions could compromise various other objectives, means were put in place to prevent hijacking of a weight-saving change by advocates for disciplines adversely affected by changes before they were aired at the SWAT board or multiboard. All inputs from the customer community were filtered through only two JSFPO senior leaders.

SWAT activities were organized into several thrusts.

1. *Optimize zero-fuel weight.* The highest volume of trades involved exhaustive optimization of existing airframe, MS, and VS components by examining

each component, assembly, or part for excess margins, nonessential redundancies or growth capability, inefficiencies, opportunities for low-risk technology insertion, or commonality penalties.

2. *Leave no stone unturned.* A review was made of past design or trade study decisions affecting weight in light of the now better-understood weight situation. Multiple previously disjointed weight savings were reviewed, updated, and consolidated. A team was set up to review new weight savings ideas from the employee weight incentive program. Potential weight increases were treated with the same rigor as weight savings. Design studies were pursued to avoid or mitigate the increases, if possible; if not, weight increases were accounted for exactly as weight reductions.
3. *Reduce fuel weight.* At the outset of the SWAT effort, STO performance was the driving KPP requirement, and of course mission fuel was the largest single element of takeoff gross weight. The most extensive CFD-based aerodynamic drag optimization to date on the program, accompanied by wind tunnel verification, was used to optimize mission performance and reduce the KPP mission fuel required.
4. *Improve STOVL performance.* STO maneuver trajectory and control effector usage were optimized to maximize gross weight capability, and VL control allowances were optimized to maximize available vertical thrust. Improvements were made in inlet and nozzle performance to increase available installed STOVL-mode thrust, but increases in core engine temperatures and uninstalled thrust were not considered.
5. *Challenge requirements.* Requirements at all levels were examined for areas that were overly specific. The primary focus was on internal self-imposed requirements by the contractor team, followed by conservative interpretations of JCS requirements. JCS requirements were prioritized in terms of value to the warfighter based on operational analyses. KPP threshold values were held to be sacrosanct, but interpretations of the definitions, ground rules, and assumptions underpinning them were challenged. Although requirements studies were conducted throughout the SWAT phase, relaxation of requirements was reserved until all other changes were implemented.
6. *Quash weight increases.* New processes for approving detailed design releases were instituted to ensure that each part was designed to the minimum acceptable weight.

Intensive activity in the SWAT phase lasted approximately seven months after the kickoff. More than 600 design changes were approved in the period, highlights of which are described in Ref. [5]. Most of the changes, amounting to about 2600 lb of weight reduction and 600 lb of installed thrust improvement, were made within the discipline teams, airframe, VS, MS, or propulsion. However, many key trade studies required the integration of the overall aircraft configurations, particularly requirements studies. These utilized the CAIV approach to identify capabilities that drove aircraft weight significantly but had relatively little operational benefit to the warfighter. Scout configurations were designed covering a broad range of weapons capabilities, combat flight performance, signature-control features, and MS functions and performance. Analysis of these configurations quantified weight impacts as functions of mission capabilities.

Meanwhile, mission-effectiveness evaluations across the design space were done in conjunction with the OAG. The most promising capability combinations were evaluated in a July 2004 manned simulation exercise known as Trial Buccaneer at the U.K. MoD's Cutlass simulation facility at Abbey Wood using USAF, USMC, USN, RN, and RAF pilots. The exercise concluded that the selected capability combination could effectively execute the stressing JORD missions. These combat requirements tradeoffs resulted in: 1) reducing the STOVL weapons bay capacity to a 1000-lb weapon (plus missile) required for the STOVL variant (previously, a common bay with the CTOL/CV was incorporated as a growth provision); 2) reducing outboard wing store stations to 1000-lb class; 3) limiting dual external air-to-ground loads and external tanks to subsonic; 4) removing unused wiring (growth provision) from air-to-air stations; 5) reducing structural limit speed to actual achievable airspeed; and 6) slightly reducing signature treatment. Together these changes contributed about 500 lb of weight reduction. Some of these changes affected ORD requirements, so they had to be approved by the SWG and Joint Requirements Oversight Council.

The other dimension of requirements relief used to finally close the thrust-weight gap was in definitions of the STO and VL requirements in conjunction with the USMC. STO mission fuel requirements were reduced by optimizing the ingress/egress altitude. Similarly, required fuel reserves for VL were reduced by revising the prescribed wave-off/go-around pattern, control allowances, and on-deck fuel reserves to better match the operations of the legacy Harrier fleet. These revisions increased allowable STO/VL gross-weight capability by approximately 700 lb, so the net gap closure attributable to requirements revisions totaled 1200 lb.

Figure 13 depicts weight-empty status through the SWAT phase as it was monitored weekly and reported to the team and to external stakeholders at all levels. The green curve shows the SWAT burndown plan derived from the probability-weighted list of planned design trades, their potential weight savings, and expected completion dates. This was the plan against which SWAT progress was measured. Black diamonds indicate the current estimate, including all approved changes to date. In some weeks, more weight increases were recognized than savings. Also, maturating AA-1 and engine designs continued to increase in weight, as indicated by red diamonds. These increases were also included in the current estimates as they were projected onto the STOVL variant. Each week a weight-empty projection was reported based on the in-work and remaining trade studies.

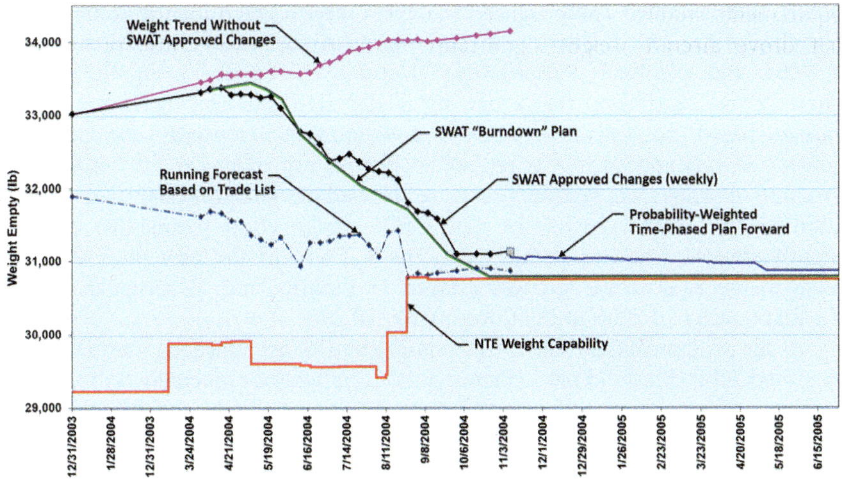

Fig. 13 Progress and forecast of weight-empty achieved and required during SWAT effort.

The red solid line in the figure represents the weight-empty Not-to-Exceed (NTE) target to satisfy the STO and VL KPP thresholds. Steps in the line reflect changes in propulsion, aerodynamic, or control characteristics, or changes in requirement ground rules. For example, the upward step in early August 2004 reflects the adoption of installed thrust improvements [5]. The difference between the forecast and the NTE target represents the expected gap once planned trade studies were complete. In late August 2004, the forecast and NTE target lines converge to eliminate the gap because the decision was made at the program level to adopt the proposed STOVL configuration as a baseline and to accept the requirements revisions described previously. Completing trade studies continued into October 2004, and ratification of the configuration, requirement adjustments, and approval to recommence the detailed design phase was obtained through a series of DoD decision boards, culminating in approval by the SWG, Configuration Steering Board, and finally the Defense Acquisition Board, also in October 2004.

3. POST-SWAT

Closure of the SWAT effort resulting in a viable STOVL configuration was a major success, overcoming the most significant technical challenge to date on the program, but it was also a major disruption of the program plan. Indeed, a complete replan of the SDD program was required, as described later in this paper. One of the most significant changes was in the sequence of variant-detailed design efforts. Fabrication and assembly operations were already underway on AA-1, but issues identified at PDR and addressed by the BRAT, together with slower than planned ramp up of resources, meant that first flight, originally planned for the fourth quarter of 2005, was then planned for the third quarter of 2006, and eventually slipped to 15 December 2006. Following the SWAT phase, it was recognized that a new detailed design was needed for the CTOL variant based on the new STOVL configuration. However, the previous crawl-walk-run approach was replaced with the principle, *Do the hardest one first, and the others will benefit.* Therefore, detailed design of the STOVL configuration became primary, with the new CTOL design effort planned to be only slightly staggered and in parallel with the STOVL. In the new plan, the first flight of the STOVL variant was delayed 18 months from its original date in the first quarter of 2006 to the third quarter of 2007.

As the program exited the SWAT phase, a new target approach was developed to protect KPP threshold performance at IOC, in light of expected weight growth and uncertainties remaining at the time. Figure 14 shows the STOVL NTE line established in conjunction with the JSFPO and Pratt & Whitney that included a 3% growth allowance for the Lockheed Martin – responsible aircraft as well as an engine weight-growth allowance. In addition, it was recognized that uncertainties and variability in weight estimation, propulsion performance, and aerodynamic effects combined to create substantial uncertainties in STOVL performance capability. Therefore, a Monte Carlo uncertainty analysis was conducted to determine additional margin to be imposed on weight-empty requirements. As the program matured through the detail design, manufacturing, and flight test phases, these

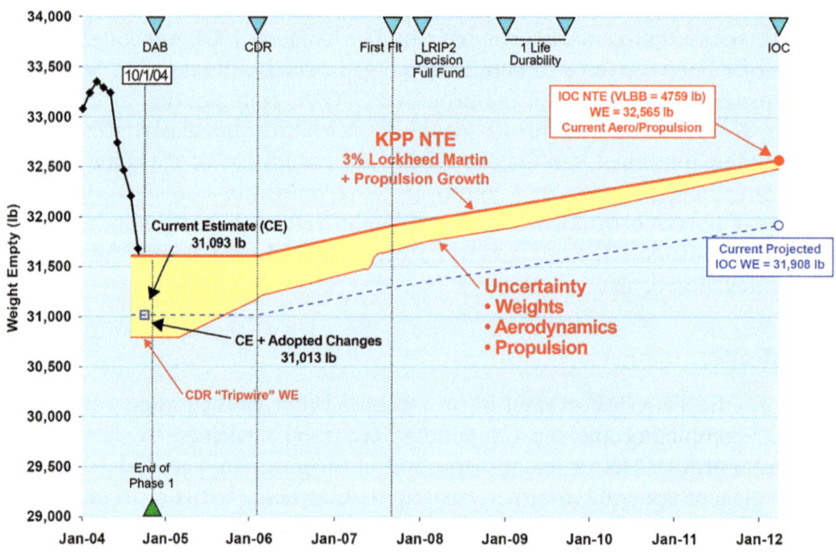

Fig. 14 Post-SWAT weight-empty management plan with design margins.

uncertainties were expected to lessen. Therefore, the uncertainty margin was computed as a function of time (Fig. 14). The combined growth and uncertainty margins established the Weight Tripwire line in the figure. As the program progressed, a weight status that exceeded the tripwire compelled additional weight reduction actions to be undertaken to offset the overage.

A predictable result of the weight challenge and program-wide effort to resolve it was a strong emphasis on weight management going forward, at both organizational and individual levels. A weight czar was appointed as part of the chief engineer's office, the mass-properties engineering staff was substantially increased, and more rigorous weight reviews were implemented in the BTP release and change management processes. In addition, the entire design staff became highly sensitized to the need for weight optimization. Designers were required to demonstrate that all parts were at or near the minimum acceptable weight as part of the BTP release reviews. Similarly, all change-board reviews began with an update of current weight-empty compared to the tripwire line (accounting for growth plus uncertainty) and a review of potential weight reductions that could be implemented if necessary to offset any weight increases predicted for changes under consideration that day. These measures were very successful in enforcing a zero-weight-growth policy; the STOVL weight-empty at CDR was only 90 lb greater than the final SWAT configuration (0.4% growth). However, this vigilance did increase pressure on BTP release schedules, affecting SDD's overall schedule performance.

The SWAT effort projected a reduction in URF cost, due mainly to the simplification and elimination of systems components, of about $700,000 per aircraft, relative to the SWAT starting point, and a substantial reduction in operating and support costs when accounting for fuel savings due to weight reduction. Supportability KPPs were virtually unaffected. STOVL weight-empty was reduced by more than 3000 lb (and CTOL and CV by approximately 2400 and 1900 lb, respectively), but the resultant current estimates were still roughly 2000 lb heavier than those at the outset of SDD. The post-SWAT configurations were measurably heavier and more complex than was recognized at the outset of the contract. Furthermore, commonality among the variants was reduced by an estimated 7% overall. These impacts on affordability created challenges in other aspects of the program.

E. KEY CHALLENGES: MAJOR SDD PROGRAM REPLANS—OVER TARGET BASELINE

Obviously, the extensive redesign effort to solve weight growth issues and reestablish KPP-compliant configurations had significant impacts on the overall program cost and schedule. The review and approval of the new 240-4 family of configurations was tantamount to a new PDR, roughly 18 months after the original event. A complete replan of the program was implemented immediately following the SWAT effort with a new variant firing order, new systems engineering, design engineering, and program management policies, as well as new program leadership and a new organizational structure. In order to distinguish pre- and post-SWAT designs, a new type-version designation system was established: 2AF for CTOL, 2BF for STOVL, and 2CF for CV. Therefore, the first STOVL aircraft would be 2BF:0001, abbreviated as BF-1. As previously stated, AA-1 was continued at its own pace; its CDR (now called a design integration maturity review) was held in April 2004, just as the SWAT was beginning, and first flight occurred in December 2006. The AA-1 effort provided valuable data and experience to the design team as a de facto prototype, but this was a major element of scope not included in the SDD program at contract award. Furthermore, most of the MS and VS suppliers had completed or begun CDRs of their own based on weight allocations that turned out to be too generous from the air system PDR. These were heavily influenced by the CTOL requirements because that was the most urgent schedule. Many of these systems were similarly set back to a PDR state of maturity. All these factors combined to require the first of what would eventually become three program replans (Fig. 15). This would come to be referred to as Over Target Baseline – 1 (OTB-1). After receiving the government's agreement and approval, the SDD contract was renegotiated, adding 18 months to the schedule and more than $5 billion to the cost.

1. POST-SWAT RESET

The post-SWAT replan still called for the production of the original 14 test aircraft and 7 full-scale ground-test articles built to the new design, in addition to AA-1, which was built to the pre-SWAT configuration. The flight test program was to include five STOVL, five CTOL, and four CV aircraft. Three of the

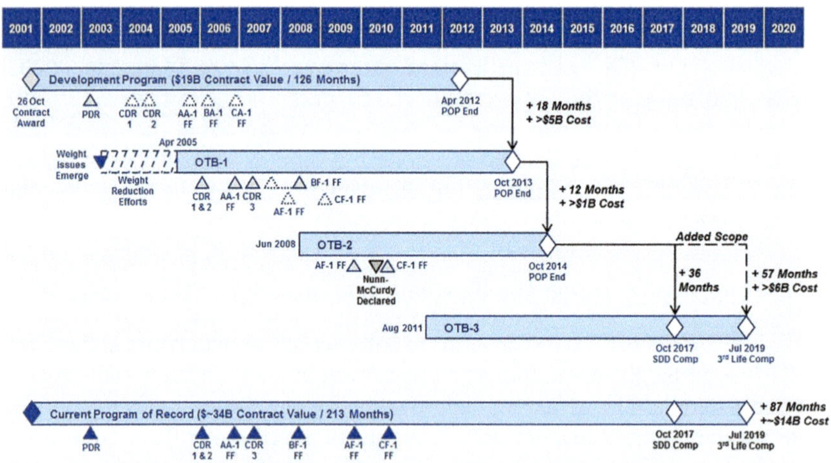

Fig. 15 Progression of SDD replans and milestone schedules.

CTOL and one each of the STOVL and CV aircraft were to be fully equipped with avionic systems and low observable treatments and dedicated to MS and signature flight tests. The remaining flight test aircraft were referred to as Flight Sciences aircraft and were equipped with instrumentation for performance, flying qualities, and structural tests. Static and durability test airframes were planned for each variant, as well as a dedicated test airframe for CV drop and barrier tests. Original SDD production plans included a factory-built CTOL airframe for Radar Cross-Section (RCS) pole-model tests. However, at this time it was determined that this was impractical because the integration of the rotator would negate any fidelity benefit of using the factory-built airframe. Instead, a more conventional model structure would be utilized for the pole tests, but the model would include actual production components for key features, such as the canopy, flight-operable doors, and flight control surfaces.

2. MID-COURSE UPDATE

By the fall of 2006, a great deal of progress was made in the SDD program with KPPs intact, but adverse cost and schedule trends were building that would result in a second program replan. As cost and schedule pressures increased over the next two years, program leadership instituted a series of intensive efforts to contain costs and achieve key milestones that supported the on-schedule completion of SDD and the beginning of production.

Technical progress was steady during the period. CDRs were completed for all three variants. Beginning with a first flight in December 2006, AA-1 flight tests were validating modeling, simulation, and analysis tools. Laboratories and flying testbeds were verifying the performance of system components, as well as their integration. Many of the ground-test and flight-test aircraft were in assembly with fit and quality that confirmed the effectiveness of the digital thread from design to manufacturing. Flight and laboratory software were demonstrating better-than-legacy stability. KSDIs were successfully achieving milestones in cross-discipline integration. All but two of the initial program-level risks were retired, and those that were being elevated to the program level were tied more to cost/schedule than technical performance.

This technical progress, however, was proving to take more effort and more time than allotted by the SDD plan. Cost and schedule threats and pressures emerged in many elements in the program.

In the airframe design release process, weight optimization required more design analysis iterations and resulted in less commonality among the variants than had originally been assumed. The structural arrangements and external lines remained highly common among the three optimized variants, but they ended up sharing only very few specific airframe part numbers. In order to obtain enough resources to develop and release airframe BTPs, engineering personnel from across the team were employed, in addition to subcontracted resources. Airframe design locations spanned 17 time zones, from Australia to

Europe to California. This effort was successful due to the common collaborative digital thread design environment and toolsets, but the rigors of the digital thread and the complexity of overseeing the worldwide operation contributed to cost and schedule pressures. The results of these factors were that more BTPs were required than originally planned, and each BTP required more design and analysis effort than originally planned.

Manufacturing of the SDD test aircraft, both fabrication and assembly, became the largest cost and schedule driver during this period. Significant changes were made to the initial tooling system used for AA-1, and they were still in progress as assembly of the weight-optimized aircraft was underway. A number of design and tooling features aimed at easing producibility were eliminated in the weight-optimization process. Prolonged BTP release schedules, combined with longer-than-planned turnaround at fabrication suppliers, resulted in inefficient assembly due to out-of-sequence and out-of-station work, and sometimes work stoppage while awaiting parts. Suppliers' initial fabrication efforts were hampered by complexities associated with weight-optimized parts, both in machine programming and in actual fabrication times. The parts outsourced included some of the most challenging parts ever produced by industry; for example, the wing carry-through/landing gear attachment bulkhead on the CTOL and CV variants are machined from the largest titanium forgings yet produced.

The volume of engineering change traffic per BTP was well predicted in the program plan, but the number of BTPs was increased, and the cost of processing and implementing changes affected both engineering design and manufacturing.

Systems suppliers also contributed significantly to the threats and pressures. Overall, a large number of system suppliers required management reserve funding to cover overruns to their development contracts, creating the second largest cost driver. For example, the EHAS, originally highly common, became largely unique for each variant. The EHAS was pioneered and demonstrated in the preceding J/IST program using an F-16, but requirements for the F-35 had diverged substantially, and major technical challenges had to be overcome related to motor design/regenerative power, thermal management, seals, pumps, weight, and high-voltage separation. Similarly, the integrated CNI system scope was initially underestimated, ultimately requiring development of 22 hardware items and approximately 1.4 million software lines of code. Some communications, navigation, and identification functions required new technology invention, for example, the Multifunction Advanced Data Link.

Throughout this timeframe, managers at all tiers, together with the Affordability team, worked to identify, quantify, and mitigate these pressures within their own spans of control, but the trends at the program level indicated that the current program plan could not be executed within the OTB-1 budget or schedule. However, clear direction was received from the JSFPO that no additional funding would be available for the program.

In a manner reminiscent of the SWAT effort, Lockheed Martin and JSFPO program managers jointly chartered a special team to reexamine original premises underlying the SDD program plan, in light of progress to date and remaining risks. The objective was to realign remaining funds with essential tasks to enable the successful completion of SDD *within the OTB-1 budget and schedule.* Timing for action was urgent because the budget was being consumed at a high spend rate, but the budget profile was declining steeply. Therefore, decisions had to be made quickly to protect the remaining budget.

This effort was known initially as the Mid-Course Risk Reduction, and later as simply the SDD Completion Plan. The team was composed of major IPT leads or senior members of their staffs from both Lockheed Martin and the JSFPO. The charter was to ensure that essential program objectives were accomplished, while still achieving affordability and weight targets, and establishing the training center, international production, and test sites. The essentials included:

1. Complete the design definition (drawings and software).
2. Build required ground-test and flight-test aircraft.
3. Validate and certify the design (developmental/operational tests and systems qualification).
4. Verify JSF contract specification requirements.

Like the SWAT effort, the team took a multipronged approach. First, scope was defined, and realistic cost and schedule estimates were made for all known cost threats. These, together with all remaining baseline tasks, were comprehensively reviewed by the joint leadership team to determine which tasks were essential and to what extent. Recommendations to add, reduce, or eliminate tasks were made by the entire multi-IPT team in order to preserve balance. Although this continuing effort identified on the order of $1 billion in task reductions, the cumulative increases from cost threats more than negated the savings in the end.

Second, the team developed a set of candidate cross-cutting initiatives that challenged the underlying premises, ground rules, and assumptions built into the original SDD plan. Candidate initiatives ranged widely, from a management organizational structure to certification practices to manufacturing quantities. The majority of remaining resources at the time were devoted to produce and test the last of the flight-test aircraft and structural test airframes, so premises and roles for those efforts were scrutinized. This effort resulted in the most prominent changes that were recommended and approved, namely the elimination of two flight-test aircraft: AF-5 and CF-4.

Finally, the third prong involved a return to the CAIV process, this time focused on the remaining development costs. The air system designs were nearly complete or very well defined in this period, so design changes would only increase costs. Therefore, the CAIV tradeoffs involved primarily the extent to which the designed-in capabilities would be certified in the SDD program;

that is, the military utility of specific capabilities (weapons loadouts, employment flight envelopes, and corresponding MS software functions) were ranked against the cost, timespan, and number of test flights required to certify the capability. Although some options defined along that spectrum did yield substantial cost savings and schedule improvements, no significant capability deferrals were accepted in the recommended mid-course update to the SDD completion plan.

Despite the intensity and priority of these efforts, they did not identify enough cost savings to fully offset realistic estimates of known cost pressures. Furthermore, because the SDD test aircraft were manufactured on the same assembly lines as the LRIP aircraft, schedule delays for the test aircraft would result in similar delays for production. This, in turn, would delay operational testing, so that a schedule extension for SDD was going to be required. Finally, a new 2008 program baseline was approved, OTB-2, which recognized a 12-month extension to the SDD schedule and more than $1 billion in additional cost (Fig. 15).

3. NUNN-McCURDY BREACH AND TECHNICAL BASELINE REVIEW

The need to re-baseline and add significant funds to the SDD program for a second time, combined with the funding required to begin production, resulted in increased scrutiny of the program by the government beyond the JSFPO. In the aftermath of OTB-2, continuing schedule challenges with both manufacturing and flight tests, and evolving policies and priorities within the government, eventually led to a critical breach of Nunn-McCurdy cost-growth thresholds and a third major OTB.

The F-35 PEO initiated an Independent Manufacturing Review Team (IMRT) in 2009 led by former F-35 PEO RADM (Ret.) Craig Steidle. This team consisted of government and industry experts and was chartered to evaluate current program plans, funding, manning, and facilities and to assess whether they were adequate to achieve planned production ramp-up and sustain the predicted maximum production rates. The IMRT review was to encompass program management, product definition, parts fabrication, assembly and test, supply chain management, and global sustainment. The IMRT conducted reviews in 2009 and 2010, and their recommendations included a more formal production integrated master plan, increased emphasis on affordability, more timely funding, and improved management metrics. Overall, the IMRT affirmed the production system and plan but concluded that production quantities should be limited to no more than a 50% increase year to year.

As the OTB-2 program plan was being formally approved and implemented as the new baseline, several DoD agencies were developing independent cost and schedule estimates. In order to improve clarity of the results and minimize resources drawn from the JSFPO and industry teams, the PEO insisted that these agencies join forces to form a single independent review team with a single outcome. An acquisition decision memorandum was issued naming the Office of the Secretary of Defense (OSD) Cost Analysis Improvement Group (CAIG) as the lead of a joint team that included other OSD offices, NAVAIR, the Air Force Cost Accounting Agency, and ASC. The joint team, which became known as the Joint Estimate Team (JET), was initially chartered to estimate program costs and requirements for fiscal years 2010 through 2015 in the Future Years Defense Plan and the president's FY2010 budget, but later the charter was extended to cover the costs of the entire program, and the team eventually had an important role in the 2010 Nunn-McCurdy recertification. The team consisted of approximately 25 personnel with expertise in cost estimating, scheduling, and various technical disciplines. A similar joint assessment team was separately chartered in 2009 to review the F135 engine program at Pratt & Whitney and Rolls-Royce.

Throughout 2008 and 2009, the JET conducted a series of contractor visits to collect data for its independent cost and schedule estimates. Detailed reviews at Lockheed Martin (with teammates in attendance) were typically two to three days in duration and occurred at three- to six-month intervals, with extensive data exchange and numerous side meetings between reviews. The JET also made site visits to the seven highest-value system suppliers, both engine manufacturers and the Edwards AFB test site. As the JET estimates evolved, feedback was given to Lockheed Martin with the objective of confirming the JET's understanding of the facts presented. The data reviews were focused on development at this time, and were quite comprehensive, covering technical risk, engineering staffing, drawing productivity and change volume, software productivity and growth, laboratory capacity, supplier staffing, flight-test productivity, schedule risk, test-aircraft manufacturing productivity and timespan, labor rates, and material and systems procurement costs.

Although JET SDD estimates were not public, they were significantly and unsurprisingly greater than those built into the joint Lockheed Martin/JSFPO OTB2 baseline due to the differing bases for the estimates. The JSF program was initially founded on a number of acquisition reform practices designed to separate it from historical programs—this out of necessity for a rapid, affordable recapitalization of the western fighter fleet. In contrast, the JET estimates were firmly based on those same historical programs.

However, by October 2009 cost and schedule pressures against the OTB-2 baseline were again building, and it was evident within the program that more software and flight test resources were needed. Again, detailed joint reviews were held, but this time with a different mandate. In the efforts preceding OTB-2, the mandate had been to not provide more money, but the new program priority was the low-risk completion of the remaining program. Joint teams worked through the summer of 2010 to recommend a new baseline that included an additional software test line, an additional dedicated flight test aircraft, and the temporary use of several production aircraft for flight tests.

Although the focus of the JET reviews and data gathering had been on the SDD program, the CAIG also maintained independent estimates of production URF costs on a routine basis using its own methodology. These estimates were used within the OSD and were typically conservative (i.e., high) compared to

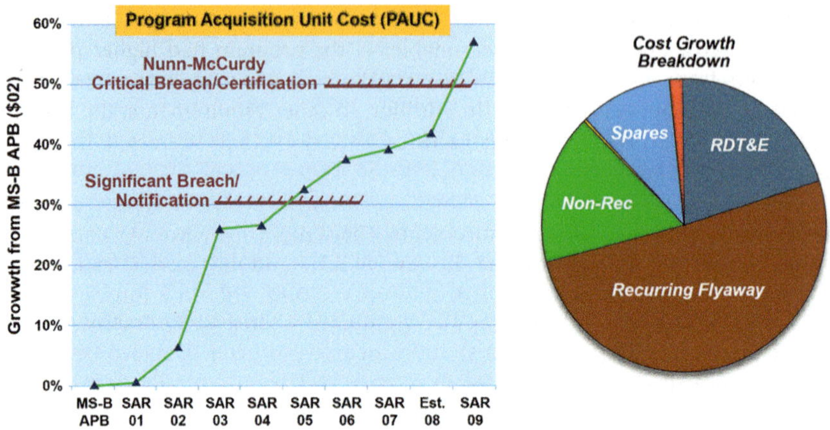

Fig. 16 Cost growth history and breakdown.

program estimates reported in the annual F-35 Selected Acquisition Report (SAR). However, when Congress enacted the Weapons System Acquisition Reform Act of 2009, the Cost Analysis and Program Evaluation (CAPE) organization was established. This organization absorbed the former CAIG and assumed the cost-reporting responsibility for major defense acquisition programs like the F-35. As a result, the 2009 SAR cost estimates, based on the JET SDD estimate and the CAIG production estimate, increased significantly compared to the prior year's estimate, despite the fact that F-35 aircraft configurations had actually remained stable since 2005.

Figure 16 depicts the trend of reported F-35 Program Acquisition Unit Cost (PAUC) over the period of the SDD program to date, compared to the original acquisition program baseline. PAUC includes development, production (URF), and support system costs on a per-unit basis. The trend of increasing costs consists of three distinct periods. In the early program, air vehicle weight and complexity had increased until the SWAT effort arrested and partially reversed the trend, resulting in a net increase in URF that drove a steep increase in PAUC. In the middle period, SDD costs rose significantly (OTB-1 and OTB-2), causing a steady but shallower increase in PAUC. Finally, more conservative estimating methodology in the CAPE estimates for both SDD and production resulted in a discrete upward step in PAUC reflected in the 2009 SAR, driven primarily by URF estimates rather than SDD. In April 2010, this step, along with other similar criteria, triggered the formal declaration of a critical breach of cost-growth limits defined by the Nunn-McCurdy Act and resulted in the third program re-baseline.

A Nunn-McCurdy critical breach has major implications for a program [12], subjecting it to a detailed review for potential termination. If not terminated, the DoD was required to certify to Congress that: 1) the program was essential to national security; 2) there were no lower-cost alternatives; 3) the CAPE had determined new cost estimates to be reasonable; 4) the program had higher priority than others from which funding would be taken; and 5) management was able to control additional cost growth. In addition to other requirements, the statute required that the program be restructured in a way that addressed the root cause of the cost growth.

Immediately following the critical breach, the DoD established IPTs to address each of the five recertification requirements, chartered a sixth team to conduct a comprehensive Technical Baseline Review (TBR), and initiated a root-cause analysis. The recertification requirements were satisfied, and on 2 June 2010 the DoD issued an acquisition decision memorandum certifying the F-35 program. Immediately afterward the TBR began to engage, with more than 100 government personnel organized into five teams: Air Platform, Mission Systems, Test and Evaluation, Service Integration, and System Acceptance.

The TBR's stated objectives were:

1. Assess the planning baseline to ensure that cost and schedule planning reflected the technical scope of SDD and was adequate to execute the program.
2. Assess the technical planning for gaps (i.e., risks, issues, or other concern areas) to ensure that resolution or mitigation was covered on a technical basis.
3. Provide a final assessment three to four weeks prior to the November 2010 Defense Acquisition Board (DAB) meeting.

The team addressed the Lockheed Martin air system, the Pratt & Whitney propulsion system, and other government costs for SDD. TBR products for each team's domain were three-point cost estimates (best case, most likely, and worst case), a recommended schedule and risk assessment, and technical findings regarding gaps and risks. Although the TBR objectives were aligned in general with the ongoing replanning activity already well underway within the program, the combined result constituted the largest (and last) replan of the program (OTB-3). It stretched the completion by an additional 36 months for tasks within the SDD contract scope. The TBR also identified gaps in the SDD scope and added tasks, most notably a third lifetime of structural durability testing. This resulted in the addition of a further 21 months to the SDD period. The overall increase for the Lockheed Martin contract cost was greater than $6 billion (Fig. 15). Following this third replan, implemented in 2011 as the program baseline, performance tracked well to the plan.

F. OTHER SELECTED KEY CHALLENGES

A development program of the magnitude and complexity of SDD encounters and must overcome more challenges than can be covered in a paper such as this.

However, the following are brief summaries of some issues encountered that are representative of the types of problems overcome and that had significant effects on the course of SDD or high-profile impacts on other program elements.

1. STOVL PROBATION

As program personnel were working through the implementation of the OTB-3 baseline in early 2011, the CTOL and CV test aircraft were exceeding planned flight rates. However, the STOVL fleet experienced low rates due to a combination of unrelated development issues associated with the STOVL-unique propulsion system. Around the same time, the STOVL structural durability test airframe experienced cracking in the main wing carry-through bulkhead within the first lifetime of testing. Figure 17 locates components affected by these issues in the aircraft. Concerns about these issues themselves, as well as potentially unacceptable consequences to weight/performance and maintenance time/cost, prompted the government to impose a two-year probation on the STOVL variant. In January 2011, then-Secretary of Defense Robert Gates announced the probation, or "period of increased scrutiny," and stated that, if the issues could not be resolved in that time, the STOVL variant should be cancelled [13]. As a result, the CTOL and CV test programs would no longer depend on the STOVL as the lead variant

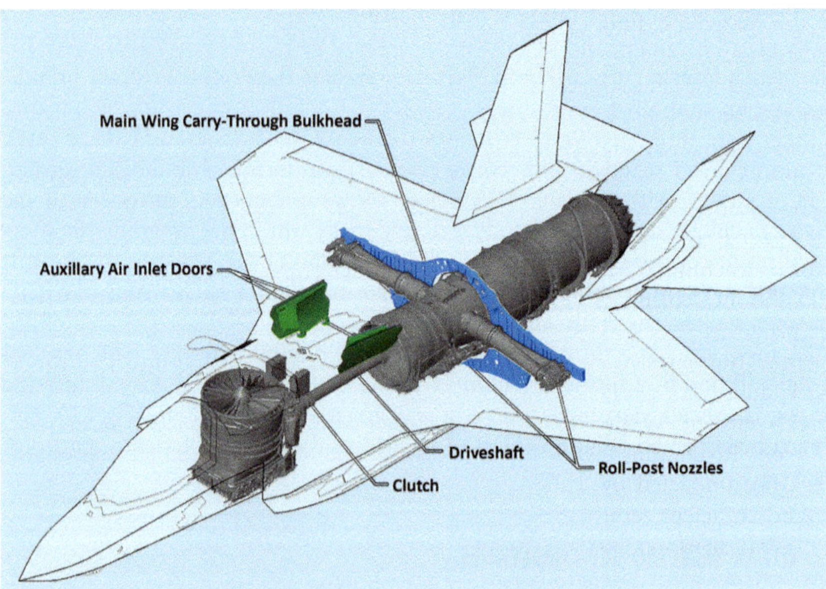

Fig. 17 F-35B components with issues triggering STOVL probation.

in development. In addition, the production quantity for STOVL in the current LRIP-5 contract was cut from 13 aircraft to only 3, a reduction of 16 aircraft from the previous contract.

Rapid progress in resolving the identified issues and improving the STOVL flight rate, together with a highly successful first deployment of two STOVLs to an L-class ship, resulted in lifting of the probation after just one year. On 20 January 2011, Gates's successor Leon Panetta, at the NAS Patuxent River test site, stated, "The STOVL variant is demonstrating the kind of performance and maturity that is in line with the other two variants of the JSF." Panetta said, "The STOVL variant has made sufficient progress so that as of today, I am lifting the STOVL probation" [14]. Summaries of the key development issues that led to the probation and their resolutions follow.

The STOVL durability test article developed a significant crack on the main wing carry-through bulkhead early in the first life of testing. Prior to the start of durability testing, Lockheed Martin predicted the number of findings expected during the test program based on legacy aircraft of each variant type. The actions called out in the government's probation letter were already part of the durability test plan. All findings in the test were evaluated and correlated to determine the life of the part against the design spectra. Parts with life deficiencies were redesigned to full life and incorporated into the first available production lot, following the standard configuration and weight management procedures. Any required modification designs for aircraft already fielded were created and programmed for implementation prior to the effective flight hours being reached in the field. In this case, the finding was correlated to the test spectra to calculate an updated service life for the component. A design change involving sculpting the area where the crack initiated to reduce stress concentration was incorporated into production at LRIP-4. The change had a negligible effect on weight. For the 13 production aircraft already built, a modification was retrofitted. This involved similar blending of material at the stress concentration point, as well as the addition of external straps to reduce gross stresses. These added approximately 70 lb to weight-empty. The retrofit was scheduled for each affected aircraft prior to reaching 577 flight hours. Modification on the SDD test aircraft required only the local blending to provide adequate life extension to finish the flight test program.

In 2010 the test team identified a vibration issue with the Auxiliary Air Inlet (AAI) door during high-speed semi-jetborne testing. A redesign of the AAI door quickly ensued. While the AAI door was being redesigned in 2011, the NAS Patuxent River team continued to expand the STOVL mode flight envelope in areas not impacted by the door vibration issue. In fact, the team was able to expand sufficient semi-jetborne and jetborne flight envelope that year to allow the initial developmental sea trial aboard the U.S.S. *Wasp* in October 2011 with two test aircraft [15]. After completing the successful sea trial, a redesigned AAI door system was installed on BF-1 later that same year. The redesign was successfully verified through regression flight testing.

During flight testing, also in 2010, an analysis of thermal data projected that the roll-post-nozzle actuators would exceed their maximum temperature capability when the aircraft was subjected to the required 1% hot day (120°F) conditions. The additional heat was attributed to higher-than-expected leakage around the roll-post nozzle. A series of parallel actions was taken to protect the aircraft during flight test and alleviate operational limitations:

1. Establish lower temperature thresholds on flight test monitored aircraft.
2. Establish operating time and ambient temperature limits on unmonitored flight test and fleet aircraft.
3. Add insulation to the actuator housing.
4. Redesign the roll-post actuator for higher temperature capability.

The insulation was introduced to production aircraft in 2011 and subsequently removed in 2015 when Rolls-Royce introduced the increased-capability actuator to production. Despite extensive operations conducted in one of the hottest areas of the United States (Yuma, Arizona), there have been no temperature-related failures associated with the roll-post-nozzle actuators.

Throughout the development of the SDLF concept, driveshaft linear and angular deflection was a driving design consideration. The flex couplings at both ends of the driveshaft are designed to flex as the aircraft maneuvers and thermally expand and contract during a mission. The combination of higher-than-expected thermal growth and manufacturing variation resulted in conditions under which the driveshaft flex couplings were predicted to stretch or compress beyond the designed axial deflection limits. For the SDD flight test aircraft, until a design fix could be put in place, a gap measurement between the engine flange and driveshaft flange was performed to establish a manufacturing variation datum for axial deflection. Using various models, prediction tools, and real-time aircraft and engine flight test instrumentation, an axial tension and compression margin estimate was made available for real-time monitoring in the flight test control room. This monitoring typically constrained operations when the aircraft was hot (high fuel temperatures) and the engine was at a low power setting. An interim design improvement was introduced in 2011 in which a classed spacer was placed between the engine and driveshaft flanges to optimize the gap. Although this solution was relatively simple and quick to implement, it introduced undesirable logistics and maintenance complexity. In 2016 a new, increased axial capability flex coupling was introduced that accommodated deflections throughout the full F-35B operating range and alleviated the additional maintenance burden of the classed spacers.

During F-35B conventional mode flight testing as early as 2008, elevated clutch housing temperatures were observed that correlated with uncommanded lift fan rotation. A Lockheed Martin, Pratt & Whitney, and Rolls-Royce team conducted a root-cause/corrective-action activity and concluded that the uncommanded lift fan rotation had been caused by tight tolerances between clutch plates on newly built clutches. The resultant friction between the plates caused the clutch-case temperature to increase throughout the flight and potentially exceed design limits. With no immediate solution available to keep the clutch plates separated, Lockheed Martin developed a passive cooling modification to the existing active cooling system. In 2011, a clutch thermal monitoring system was also added to provide pilot awareness of clutch thermal state during both the uncommanded lift fan rotation and pilot-commanded STOVL conversions. In parallel with the cooling modifications, Rolls-Royce initiated an effort to thin the clutch plates to reduce clutch drag. This change was incorporated into production clutch deliveries beginning in 2014. Because this change reduced clutch life, Rolls-Royce implemented a more durable clutch plate material in 2016 that restored the maximum number of clutch engagements to exceed the specification level. Since these design changes were implemented, there have been no reported up-and-away clutch heating events.

2. CV ARRESTING HOOK REDESIGN

One of the fundamental features of the CV variant allowing arrested landings on a carrier is the Arresting Hook System (AHS). The F-35C AHS was among the most challenging systems to design and one of the last to mature sufficiently to complete CDR. The requirement to completely enclose the system when retracted, combined with the relatively forward location of the engine nozzle, resulted in a configuration with the hook point longitudinal position considerably closer to the main landing gear than in any legacy aircraft. When roll-in testing of the system began at the USN's Lakehurst, New Jersey, facility during 2011, and following initial fly-in tests, results demonstrated an unacceptably poor engagement rate. The low rate was due to the hook point bouncing over the arresting wire, caused by the unique physical dynamics of the system. Nose and main landing gear tires excited a wave motion in the wire that resulted in the center of the wire being on or close to the deck just as the hook point passed. Together with a relatively low hold-down force in the system that allowed the hook to bounce upward, and a blunt hook point, the hook often failed to catch the wire.

This problem led to a major redesign of the AHS [16], but fortunately did not have a significant effect on the surrounding airframe, systems installations, or door arrangement. The changes aimed chiefly to flatten and sharpen the leading edge of the hook-point shoe to improve the pickup of the wire and significantly increase the hold-down force on the hook to reduce bounce-up. A new CDR was conducted for the system, and testing resumed in early 2014, leading to ship trials later that year with excellent engagement rate results. By the end of 2014, 16 CV airplanes had already been delivered. Those, and an additional two aircraft, would ultimately be delivered before the change was implemented on the production line, requiring retrofits of a total of 18 aircraft.

3. FUEL TANK INERTING REDESIGN

Challenges related to the Onboard Inert Gas Generation System (OBIGGS) also began to emerge in 2012 that eventually affected SDD flight tests, LRIP production, and retrofit modifications that paced USMC and USAF IOCs.

F-35 air vehicles rely on the OBIGGS for lightning protection in order to avoid weight penalties associated with passive systems. Unlike aircraft with aluminum skins that create a Faraday cage, keeping currents induced by lightning strikes on the exterior of the aircraft, a composite-skinned aircraft like the F-35 experiences those currents on the external surface as well as within the aircraft structure. This is primarily due to the fact that lightning-level currents will attach to the nearest metal fastener and penetrate the skin. This was well-understood early in the aircraft design, and features were included to protect against direct effects due to lightning strikes. Indirect effects were also mitigated to ensure that components withstood currents induced in the aircraft wiring system. The remaining lightning-strike risk is the ignition of fuel system ullage (vapor above the fuel level) if arcing occurs as a result of a lightning strike. Available passive measures to prevent arcing would have resulted in unacceptable weight, so requirements were placed on the OBIGGS to maintain low enough oxygen concentrations in the fuel tank ullage to prevent ignition. These requirements were more demanding than other vulnerability requirements, so flow rates were adjusted to ensure that the ullage remained below the oxygen concentration requirement. However, at a special CDR for the system in 2012, it was determined that the system did not meet the requirement at some locations for short-duration transients. Therefore, several modifications to the design, already in production, would be needed to satisfy the program's safety hazard risk-index requirements for all variants.

The required changes involved redesigns of valves and orifices and the addition of wash lines for nitrogen-enhanced air to certain locations in fuel tanks. Although changes to the system architecture were modest, some of the new components were installed in locations that were difficult to access once the aircraft was assembled. Furthermore, some of the components were entirely new, requiring a complete design-qualification cycle in addition to procurement and manufacturing spans before the new hardware would be available. Furthermore, fuel-system software changes were needed. Therefore, the change implementation was divided into two phases. The first phase covered the components that were the most difficult to access and was incorporated at the soonest possible effectivity in LRIP-6 production, while the remaining hardware was targeted at LRIP-7 effectivity. A retrofit plan was developed for all prior aircraft.

However, in 2014, a new issue related to the OBIGGS emerged. It was determined that, at certain fuel states and during high-*g* maneuvers, the arrangement of the OBIGGS and fuel siphon tanks could result in fuel tank pressures that exceeded design limits. The immediate effect of this discovery was the imposition of maneuver and weather limitations on the fleet. Fuel-system software could be modified to reduce tank pressures, but at the expense of inerting performance. For the F-35B STOVL aircraft, such a software change resulted in small pockets with oxygen concentration above the limit. However, ignition laboratory tests were performed at Wright-Patterson AFB, Ohio, that produced pressure data confirming that the structure could withstand such ignition events. This solution was verified using the fuel-system simulator and flight tests, but production implementation was delayed until LRIP-8. The verification of the solution, acquisition of parts, and modifications of 10 F-35Bs were just in time for USMC IOC declaration in July 2015.

For the F-35A CTOL configuration, the tank overpressure condition was actually more severe than on the STOVL due to differences in the fuel system arrangement and maneuver requirements between the variants, and its schedule was only slightly less pressing. The CTOL benefitted from the work done on STOVL, but the CTOL solution required the incorporation of an all-new software-controlled pressure relief valve, and a new wash line routed from the outboard wing tank all the way to near the center of the fuselage. This change would not be incorporated in production until midway through LRIP-9 production. Like the STOVL, depot modifications to the required number of F-35A aircraft were only completed just before the USAF IOC declaration in August 2016.

The F-35C CV fuel system configuration is similar to the F-35B, but the ultimate OBIGGS modification was somewhat different than either the STOVL or CTOL. Production effectivity for the change is LRIP-10, and as with the other variants, retrofits will be installed in previously delivered F-35Cs to relieve flight restrictions.

As a result, the F-35 is now fielded with no inflight restrictions related to lightning.

4. HELMET AND SEAT REDESIGNS

The F-35 Helmet-Mounted Display (HMD) system [17] and US16E ejection seat [18] each has unprecedented capabilities (Fig. 18). Each has overcome significant development challenges in its own right, and the progress of each development effort has been intertwined with the other.

The HMD serves as the pilot's primary display, providing integrated flight reference information, tactical and navigational display integration, and digital imagery, including night vision. The system development during SDD has involved several iterations. Early versions faced functional performance issues with night vision acuity, tracking/boresight alignment and latency, display jitter, green glow, and obscuration by the canopy bow frame. In addition, in tests of the ejection seat, issues arose with the security of the visor attachment and yawing moments exerted on the pilot's head due to the oxygen hose. Finally, qualification tests identified the need to modify the helmet transmitter unit and helmet/vehicle interface cable. Based on the Gen II version of the system, the system was deemed deficient to requirements and the 2010 TBR

Fig. 18 HMD and US16E ejection seat.

found the system not suitable to complete SDD or fleet operation. Work continued on the system to develop a Gen III version to address these issues, but as a risk mitigation fallback, Lockheed Martin pursued an alternate system using separate night vision goggles.

Numerous changes incorporated into the Gen III design successfully addressed most of the issues, and the fallback option was eventually dropped. Night vision capability was improved with a new camera and software changes, helmet display unit hardware improvements, and the addition of a fixed-camera assembly. Tracking improvements resulted from the addition of a boresight reticle unit to allow the pilot to see alignment status, optical trackers to the HMD and fixed camera, and an inertial measurement unit to the helmet. In addition, several other hardware and software improvements were made. The visor attachment was strengthened, and an oxygen hose attachment was added to minimize head turning. Finally, the helmet transmitter unit and helmet/vehicle interface cable were strengthened. Although these changes together addressed the functional performance shortfalls of the Gen II design, they also added head-borne mass to the helmet and shifted the center of mass forward.

The US16E ejection seat was also designed to meet an unprecedented level of requirements. The system is designed to function safely over a wide range of flight conditions, from static hover very near the ground to high altitudes and very high equivalent airspeeds, and over virtually any attitude. Moreover, the system was designed to accommodate a very wide range of pilot shapes and sizes, from a 103-lb female to a 245-lb male. Simply stated, the key design challenge for the seat is that high forces are required to eject and then decelerate heavy pilots at extreme conditions, but these forces impose the risk of neck injury on light pilots. These risks are increased with increased mass of the pilot's helmet and center of mass that is not aligned with the seat forces. High neck loads can occur during the ejection and parachute opening phases.

Prior to mid-2015, the ejection seat had been qualified with a Gen II helmet, but a repeat test at low speed with a light pilot, performed in support of redesigning the seat sequencer, showed that neck injury criteria were exceeded, in contrast to earlier tests. A review of the earlier test data revealed that the test mannequin's head was being supported by the parachute riser at the critical load condition, giving misleadingly low measured neck loads. In late August 2015, U.S. services imposed a minimum weight limit of 136 lb for F-35 pilots. In response to this, two improvements were developed to the ejection seat. First, a lightweight aircrew switch was added to the seat. Selecting the lightweight pilot position adjusts the timing of the drogue chute sequence, lowering the speed of main chute deployment. Second, a fabric head support panel was added between the parachute risers to prevent hyperextension of the neck (whiplash).

Both of these changes mitigate neck loads in the parachute opening phase of seat operation but do not affect the catapult stage. To address the catapult stage, a weight limit was established for the HMD that would satisfy neck-injury criteria for light pilots. In May 2015 the JSFPO directed that a Gen III Light version of the HMD/helmet be developed. The primary weight reduction is achieved by the introduction of a missionized visor. This removes the external tinted visor and introduces a two-visor system with one clear display visor and one tinted display visor. The pilot will be able to swap the visor on his or her helmet while in flight to adjust to changing environments.

The production effectivity for the revised ejection seat is LRIP Lot 10, which commenced aircraft deliveries in January 2018, and the Gen III Light HMD began production deliveries.

G. KEY RESULTS: NOTABLE FULL-SCALE TEST ACCOMPLISHMENTS

Although the ambitious SDD program endured numerous significant technical, schedule, and cost challenges, the F-35's equally ambitious 5th-Generation system capabilities have been successfully achieved, based on extensive full-scale test results.

1. FLIGHT TESTS

Reference [15] provides a complete summary of F-35 flight tests. The test program was conducted by the F-35 Integrated Test Force (ITF) composed of engineering, flight operations, maintenance, and management personnel from Lockheed

Martin, Pratt & Whitney, USAF, USN/USMC, international partners, and suppliers, as needed, in a single integrated organization. Two primary test sites provided extensive base and test-range infrastructure: Edwards AFB and NAS Patuxent River. Numerous other test locations provided specialized test capabilities, including L-class amphibious assault ships and CVN-class carriers. Figure 19 highlights a few significant flight test milestones.

In April 2018 the final SDD test flight was completed. Overall, more than 9000 test flights accomplished more than 65,000 test points in more than 17,000 hours of testing over nearly 10 years. In recognizing the joint government/industry team, Lockheed Martin's program manager stated that the F-35 flight test program represented the most comprehensive, rigorous, and safest developmental flight test program in aviation history. SDD flight testing highlights included full flight envelope performance and flying qualities, high AOA, STOVL development testing, ship trials, 183 weapon separation tests, 42 weapons delivery accuracy tests, and 33 mission effectiveness tests, which included numerous multiship missions of up to eight F-35s against advanced threats.

2. FULL-SCALE GROUND TESTS

In addition to flight testing, numerous ground-based tests using full-scale hardware contributed to validating models and verifying requirements. A sample of ground tests performed is illustrated in Fig. 20.

For each variant, two full-scale airframes were produced for structural tests on the same assembly line as that used for flight test and production aircraft. One of

Fig. 19 Notable flight test milestones: AA-1 first takeoff, BF-1 first vertical landing, X-35B bomb drop, and successful F-35C arresting wire engagement.

Fig. 20 Notable full-scale ground tests: F-35A gun-fire, antenna model, F-35C drop, F-35A live fire, F-35B climatic, and RCS model.

each variant's airframes was used for static testing to confirm the strength and stability of the structures. The other was dedicated to durability testing, which was extended to three lifetimes as a scope addition included in OTB-3. The CTOL tests were performed at BAE Systems' Brough, United Kingdom, facility, as was the CV durability test. The STOVL tests were done at Lockheed Martin's facility in Fort Worth, as was the CV static test. The CV static test article was also used for drop testing to verify structural integrity for extreme-sink-rate carrier landings at Vought Aircraft Industries in Grand Prairie, Texas [11].

Full-scale live-fire vulnerability tests were performed at the Naval Air Warfare Center Weapons Division at NAS China Lake, California, using the fully equipped AA-1 CTOL aircraft, the CV structural static-test airframe with an engine installed, and the STOVL structural static-test airframe. Ballistic testing of the STOVL propulsion system was also conducted.

Extensive climatic tests were performed in 2014 and 2015 on a fully equipped STOVL aircraft at the McKinley Climatic Laboratory located at the Eglin AFB, Florida [19]. These tests covered a wide range of climatic conditions, but also a wide variety of F-35 flight conditions, including simulated hover for the STOVL variant.

Although they did not include a representative airframe structure, other full-scale tests involving salient flight hardware were accomplished using pole models for signature and antenna aperture testing. The full-scale signature model included flight-representative versions of all salient features and was tested at the Lockheed Martin Helendale, California, facility. Full-scale integrated antenna tests were conducted at the USAF Research Laboratory Rome Research Site in Rome, New York.

H. ALTERNATE ENGINE PROGRAM

In addition to the lead engine, the Pratt & Whitney F135, in SDD the JSFPO contracted Lockheed Martin to integrate into the F-35 air system an alternate General Electric/Rolls-Royce F136 engine. The rationale was that competitive engine offerings would spur continuous development and reduce price as the respective engine manufacturers competed for work share, as has been exhibited with F-16 engine competitions. To those ends, the F-35 was designed not only to be compatible with the F136, but also to allow the physical interchange of the F135 and F136 engines in the field, operating in such a way that propulsion operations were transparent to the pilot regardless of engine type.

Therefore, the F-35 inlet was sized to support either engine, with growth margin. Engine interfaces were closely managed to ensure compatibility with either engine. Lift-system hardware (lift fan, driveshaft, and roll posts) and even the CTOL and STOVL engine exhaust ducts and nozzles were intentionally established as common hardware, provided by Pratt & Whitney/Rolls-Royce for both the F135 and F136. Specifically, the F135 and F136 shared a number of identical interfaces, such as the front engine mounts, hydraulic mounts, and Engine Starter/Generator (ES/G) pumps. Some flexibility in F-35 interfaces was designed in to accommodate minor location differences in the mounting locations of the rear engine side mount and power feeders to the ES/G. Portions of the traditionally airframe-owned main fuel line and electrical panels were provided to the propulsion system contractors to be engine-mounted to allow freedom in the terminal routing of those lines.

To provide Level 1 flying qualities, both propulsion system contractors were responsible for meeting nontraditional control requirements, especially for STOVL operations. For example, when operating in STOVL mode the pilot commands forward and vertical acceleration with stick and throttle inputs. F-35B control laws command the engine system to provide a required thrust level, thrust split between the main-engine nozzle and lift fan and the roll-post command. In many respects the propulsion system is now treated like any other aerodynamic effector. Accuracy, bandwidth, rate capability, and system response to failure are managed by the propulsion system. For example, although different engines may operate at different rpm at a common control point, the effector output is a closed loop managed by the F-35 propulsion system to provide predictable performance to the pilot.

Success in these endeavors was established at the JSF program's August 2004 alternate engine readiness review.

Under separate JSFPO contracting, the General Electric/Rolls-Royce Fighter Engine Team took a pre-SDD F136 engine into conventional testing in July 2004, followed by initial STOVL engine testing in 2005. After transitioning to SDD, the F136 completed its PDR in 2006 and CDR in February 2008. The first SDD F136 began ground testing in February 2009, followed by STOVL ground testing in November 2010. By the end of 2010, the F136 had accrued more than 1000 test hours, with flight test engines planned for delivery in 2011 and first flights planned for each of the three variants soon thereafter.

Though cancelled in 2011 after a long and highly publicized government funding debate, technologies and lessons learned from the F136 program live on in the joint government/industry Versatile Affordable Advanced Turbine Engines program.

V. INTERNATIONAL PARTICIPATION

The scope and complexity of international participation in the F-35 program has been both an asset and a challenge. From the earliest roots of the joint U.S./U.K. ASTOVL program in the 1980s, then supported by the Nunn-Quayle Research and Development Initiative, the program has involved international partners. Although DoD regulations require acquisition managers to pursue international cooperation for most programs, the F-35 SDD program became a cooperative development arguably like no other.

For several decades, the operational model for long-term combat and peacekeeping operations has revolved around tri-service, coalition participation for both operational and financial burden sharing. However, the historical reality of this concept had been difficult and was limited by technology, a lack of legacy platform interoperability, and political differences among/between the services and allied air forces. In the early 1990s, the U.S. services faced severe budget pressures, and there was an emerging view that significant potential gains could come from having a common platform. The elimination of interoperability barriers and reduction of duplicative training and maintenance infrastructure would reduce both procurement and operational budget requirements. At a lesser level, the same dynamics existed in most of the participating allied nations.

A. BACKGROUND

In the early 1990s, each of the U.S. services was entering development programs for the replacement of their respective frontline tactical fighter aircraft. The USMC was furthest along and was developing prototypes for its ASTOVL replacement for the venerable AV-8B Harrier. The Harrier was a British design that had been improved and manufactured in the United States for the USMC. The United Kingdom and United States were co-developing the ASTOVL concept and were joint signers of the formal ORD that defined the next-generation requirements. The Italian Navy, a strong U.S. ally in coalition operations, also operated the Harrier and was expected to join the replacement program.

The USAF was in the early stages of developing operational requirements for a multirole fighter to replace the multirole fighter inventory, which consisted of the F-16, A-10, and potentially the F-117. The F-16 had been widely deployed as an international fighter in 21 countries as the primary Foreign Military Sales (FMS) offering by the U.S. government to allied air forces. Key allies requiring modernization of their air forces were prime candidates for joining the JSF program.

The USN had been through several attempts to replace the F-14/A-6/A-7 inventory, including a derivative of the USAF Advanced Tactical Fighter (which became the F-22), A-12, and AX/AFX competitions. All had been aborted in favor of a less risky approach to upgrade and modernize the F/A-18 platform. The F/A-18 had achieved limited foreign air force acceptance with six international air forces, including principal allies Canada and Australia. Those allies were also expected to have an interest in joining the JSF program.

B. THE INTERNATIONAL COALITION

The United Kingdom joined the JAST program as a founding member in 1995 under the premise that British industry would suitably share in the industrial benefits accruing from participating early in a development program that would lead to substantial production and sustainment industrial opportunities. Italy, the Netherlands, Denmark, Norway, Canada, and Turkey joined the program at various times as observers during the CDP program.

At the time of the SDD contract award, the United States and United Kingdom were committed participants, and the JORD was cosigned by both U.S. and U.K. senior government officials. The United Kingdom became the only Tier 1 partner, with an investment in the SDD phase of about $2 billion. Over the following year, seven additional allies joined the program (Fig. 21). Italy and the Netherlands joined the SDD phase as Tier 2 partners, each with an investment of more than $1 billion, and Norway, Denmark, Canada, Australia, and Turkey joined as Tier 3 partners at about $150 million each. Based on their financial level of participation, those allies were allowed to provide representatives in leadership positions in the JSFPO. Tier 1 and Tier 2 partners would also have the opportunity to participate in Operational Test and Evaluation (OT&E) of the

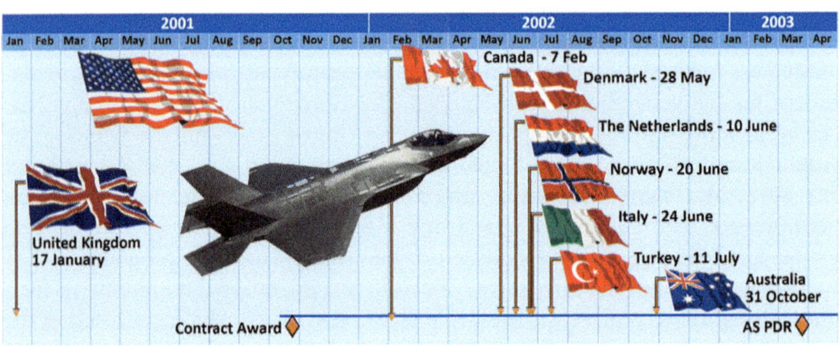

Fig. 21 Timeline for international cooperative partners joining the F-35 program.

F-35. Each of these government-to-government bilateral agreements also had unique elements, referred to as side agreements, that identified unique national requirements of the participating nation. All partner countries would be integrated into the baseline production program seamlessly and benefit from the economies of scale from larger procurement quantities.

All participating nations would be allowed to participate industrially in the SDD phase of the program and later phases on a best-value, competitive basis. This was a significant departure from the traditional offset-based program in which procuring nations would receive economic benefits to offset their procurement costs. Lockheed Martin and its teammates conducted a comprehensive survey of industries in partner nations for unique development capabilities that could enhance the supplier base already selected during the CDP program.

In early 2006, the U.S. government secured the commitment of the nine partner nations to continue in the JSF program via the Production, Sustainment, and Follow-on Development (PSFD) Memorandum of Understanding (MoU) signed by the parties between November 2006 and February 2007. This agreement eliminated the level-based designations in the SDD agreements so that all nine participating nations held equal status. The JSF Executive Steering Board (JESB), with representation from all participants, assumed overall governance of the program.

The program management complexity of this program structure differed from any previous program in DoD's history in terms of the degree of challenges involved to design, develop, test, and produce the most advanced-technology fighter in history, integrate the operational requirements of three separate U.S. services with unique operational environments, include eight international partners in the entire development and production process on a best-value basis, ensure that all U.S. technology and export control requirements are adhered to, and maintain cost and schedule commitments made in a competitive proposal that did not include these complexities when the contract was initiated.

From the perspective of the JSFPO, the challenge was to convince the participating nations that the governance process was fair and that their unique requirements would be met within the boundaries of the funded program. For unique requirements (e.g., a drag chute for Norway for landing on icy runways), separate funding would be required. Participating nations would have representation in the JSFPO in line with their level of investment. Participating nation acquisition leaders would be full participants in the governance bodies directing program decisions. Participating nations would procure their F-35s as an integrated element of the U.S. annual buys, eliminating traditional international procurement processes.

From the perspective of the Lockheed Martin program management team, the challenge was to identify international industrial partnerships that could provide best value to the F-35 supply chain while meeting individual national economic and political requirements for the financial justification to continue with the program. All economic benefit to the procuring nation was required to be direct work on the F-35, as opposed to historic offset-based programs that were allowed to use indirect or nonassociated trade. All participants wanted high-value work on a program that was a relatively small platform. To meet this challenge, significant partnering requirements were flowed down to all teammates and major suppliers. Figure 22 illustrates the breakdown of airframe component coproduction from U.S. teammates and International Industrial Participation (IIP) suppliers.

From the perspective of the participating nations, the requirement for IIP was imbedded in the PSFD MoU as a fundamental principle of the F-35 program, with IIP country targets tied to the quantities procured. The agreement addressed the best-value principles to be used and mandated that the contractors provide opportunities to partner nation industries. All partner nations insisted that Lockheed Martin sign industrial participation Letters of Intent (LoIs) with each nation's ministries of defense and economic affairs before they would sign the government-to-government agreements.

The LoIs identified four categories of IIP opportunities in the production program:

1. Continuation as a result of being selected in the SDD phase
2. Competitive across all nations
3. Competitive strategic source-directed procurement to a nation's industry but subject to best-value, competitive pricing
4. Country-unique for specific capabilities that only one country required

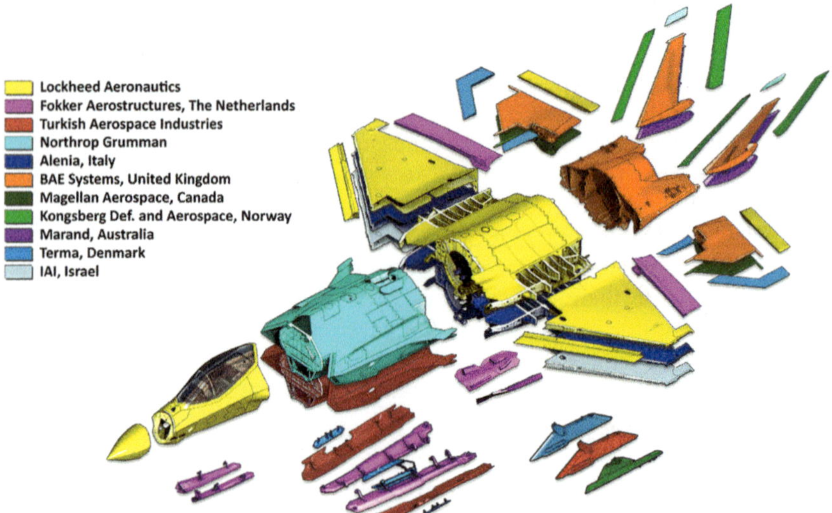

Fig. 22 International production/coproduction of major airframe components.

Fig. 23 International suppliers.

The LoIs also required Lockheed Martin to report semiannually on the IIP performance of the team, which consisted of Lockheed Martin as the prime contractor, the prime teammates (Northrop Grumman and BAE Systems), and major system and subsystem manufacturers. Figure 23 shows the international supply chain created by the F-35 IIP program. Most of these suppliers have greatly increased their capabilities, facilities, and equipment, as the F-35 is recapitalizing not only the partner nations' fighter fleets, but also their defense industries.

C. BUILDING THE F-35 INTERNATIONAL SUPPLY CHAIN: A FEW EXAMPLES

The Tier 1 partner (United Kingdom) and Tier 2 partners (Italy and the Netherlands) were given the opportunity to procure test jets and participate in OT&E. The United Kingdom procured three STOVL aircraft and the Netherlands procured two CTOL aircraft. Italy decided to forgo the investment in OT&E jets and instead invested in a Final Assembly and Checkout (FACO) facility in Cameri, Italy, to produce Italian and Dutch operational jets. The facility was designed to be converted into a maintenance, repair, and overhaul facility for European F-35s following the production program (Fig. 24).

1. FOLLOW-THE-SUN ENGINEERING

The United Kingdom, Italy, the Netherlands, and Australia were significant participants in the F-35 digital thread design toolset. Other partner countries could participate on a more restricted basis. The establishment of this virtual design toolset was a complex and highly controlled infrastructure, but it allowed

Fig. 24 Italy's FACO (left) and wing assembly building (right).

continuous design across multiple time zones, leveraging the time dimension of global participation. Of note was GKN, an Australian enclave of stress engineering experts who were significant contributors to the design activities of the program.

2. WORLD-CLASS COMPOSITE MANUFACTURING FACILITIES

F-35 manufacturing tolerance control and significant expansion in the use of advanced composite structures demanded a next-generation manufacturing capability. New composite manufacturing facilities were established across several of the participating nations. Turkey, the Netherlands, Denmark, Norway, Canada, and Australia all invested in building new world-class composite production facilities.

D. FOREIGN MILITARY SALES

Although neither part of the cooperative development program nor members of the JESB, other nations have placed orders under the FMS program. These began to play a significant role with the first production commitments by Israel, Japan, and the Republic of Korea. Israeli and Japanese industries are already participating. Israel is providing wing panels and electronic components. Japan has already rolled out the first aircraft from its own FACO. In the future, other countries may also be allowed to participate through the traditional FMS processes.

E. IMPACT AND CHALLENGE

Keeping the *Why F-35?* value proposition in alignment across changing environmental and political factors, and sustaining that value proposition across key stakeholders, continue to be major strategic challenges that depend on sustaining a clear economic benefit to participants. In many ways it is a government-to-government commitment but is generally regarded as an industry responsibility to sustain it, from an industrial or economic perspective.

Combining this requirement with the intense day-to-day execution challenges of a major, next-generation-technology engineering development, testing, and production global program is a unique and unprecedented effort in its complexity and scale.

VI. TRANSITION TO PRODUCTION

A. BACKGROUND

The JSF production system is absolutely central to achieving the program goals of affordability and a high production rate. Much like the F-35 air system is recapitalizing and transforming a large portion of the west's air-power capability, the F-35 production system is recapitalizing and transforming the aerospace industry. From early in the CDP program, as Northrop Grumman and BAE systems joined the Lockheed Martin team and numerous countries, in addition to the United States and United Kingdom, became engaged in the program, it was clear that the production system would be global. Within the program, Global Production System was the name given to the manufacturing organization. The envisioned one-a-day rate would require substantial investment in plants and equipment. Lean manufacturing was embraced as a core strategy, along with automation and the digital thread. Producibility was a major emphasis in the air vehicle design, although the specifics of both the air vehicle and the production system evolved significantly over the CDP and SDD programs.

In the early concept stages of the program, there was a strong emphasis in the air vehicle design on rapid final assembly, minimizing part count and tooling, commonality, and the use of a precision-manufacturing capability to achieve the geometry controls needed for 5th-Generation capabilities. As described earlier, many air vehicle design changes had a significant impact on the production system, but the overall objectives stayed constant, prompting the continuing evolution and refinement of the production system as the aircraft configuration stabilized.

In the CDP program, preliminary development and risk reduction of the production system paralleled the air vehicle design. In SDD the system was actually built, because one of the objectives of the program was to build even the first test aircraft on production hard tooling. In fact, there was a strong financial incentive built into the original SDD award-fee structure based on the actual cost of building SDD and early LRIP aircraft relative to the desired cost-improvement curve extended over the entire production program. Reference [10] describes the F-35 production system and ongoing advances.

B. EVOLVING QUANTITY PROFILE

A high degree of development-production concurrency and a steep production ramp to high rate were built into the program from the outset of SDD, as specified

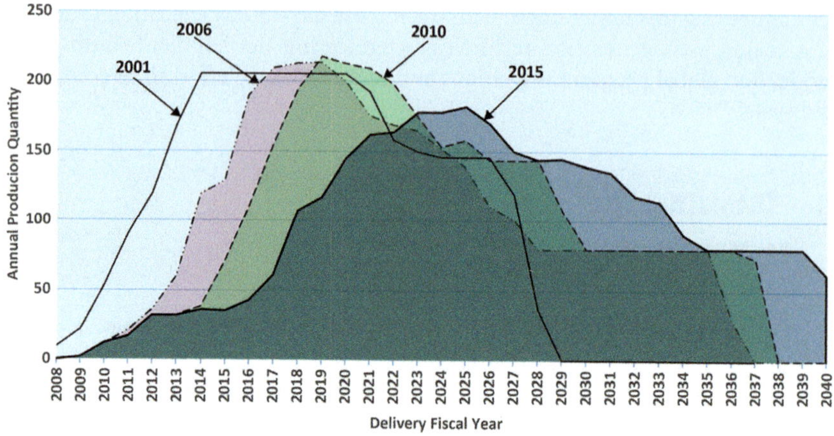

Fig. 25 Evolution of production quantity profile.

in the EMD CFI (Fig. 25). Within two years of the last variant's (CV's) first flight, the production rate was originally to be 54 per year, including all three variants. The production rate was planned to be 168 per year in the sixth and final year of the LRIP phase. Based on these steep production ramp rates, the prime teammates and the suppliers invested aggressively in plant capability, and the JSFPO funded a significant amount of special tooling and test equipment.

As described earlier, a number of challenges with the engineering design, production, and testing of the early test aircraft significantly prolonged the SDD program, and each program replan attracted additional critical scrutiny. Several factors acted to progressively delay and stretch out planned production-rate increases in each planning year. First, the governments became wary of buying significant numbers of aircraft before they had been more completely tested, for fear that test discoveries would require engineering changes, which would in turn require high-cost modifications to already fielded aircraft. The U.S. Government Accountability Office was strongly critical of these potential concurrency costs in its annual reports to Congress [20], as was the DoD Director of OT&E. Second, funds committed to SDD replans were then not available in the acquisition budget. Further limitation to procurement rates were related to the worldwide financial crisis that strained many participating governments' budgets.

C. PRODUCTION FACILITIES

The F-35 program has required very large facility investments across the globe in new facilities, equipment, and tooling. New facilities have been constructed in Australia, Japan, European countries, Canada, and the United States.

Fig. 26 Lockheed Martin F-35 production facility in Fort Worth, Texas.

Lockheed Martin production is centered at the Lockheed Martin Aeronautics Company headquarters in Fort Worth (Fig. 26). Throughout SDD and the first 10 LRIP lots, F-35 wing and forward fuselage assembly, mate, and FACO operations have progressively displaced continuing F-16 production. In 2017 Lockheed Martin produced the final Fort Worth–produced F-16 and announced that future F-16 production would be done at its Greenville, South Carolina, facility, essentially devoting the entire Fort Worth production facility to the F-35 and completely revamping the arrangement and equipment of the nearly mile-long assembly building. Major new construction at the site included component and aircraft final finishes facilities, an acceptance test facility, and a hover pit, as well as a complete refurbishment of the flight line run stations. Major wing carry-through assemblies are produced at the Lockheed Martin Marietta plant formerly used for F-22 production. Wing- and tail-surface edges with special treatments and embedded sensors, as well as radomes, are produced at the Palmdale plant. Subassembly work is also done at the Johnstown, Pennsylvania, and Pinellas, Florida, locations.

Northrop Grumman assembles center fuselages and weapons bay doors at its Palmdale facility, formerly used for B-2 production. This facility features an integrated assembly line (Fig. 27) that uses automated guided vehicles to progress

Fig. 27 Northrop Grumman F-35 integrated assembly line in Palmdale, California.

Fig. 28 BAE Systems F-35 production facility in Samlesbury, United Kingdom.

assemblies through the flow line. Production of the complex main-inlet duct is accomplished at the advanced fiber placement facility in El Segundo.

BAE Systems has developed its Samlesbury site with numerous all-new facilities dedicated to the F-35 program (Fig. 28). The site produces aft fuselage and horizontal- and vertical-tail box assemblies for all variants. The CV outboard wing box is produced for BAE Systems by an IIP supplier in Canada. At Samlesbury, an all-new assembly hall was constructed in three phases and contains three overhead-rail flow lines for aft fuselage, horizontal tails, and vertical tails, with a shared complex of precision milling machines. Each line accommodates all variants. Assembly of the specialized STOVL aft-nozzle-bay doors is done in a new facility adjacent to the hot-forming facility where a superplastic-forming/diffusion-bonding process produces the door detail. Composites are produced in a preexisting building that was completely gutted and reequipped. An all-new highly automated hard-metal machining facility was constructed on the site, as well as a new office building to house management, engineering, and business operations.

D. AFFORDABILITY: URF IMPROVEMENT

Once the air system design was frozen in the SDD phase and production began, the affordability program again evolved. Rather than air system requirements tradeoffs of conceptual design or design-to-cost targets of detailed design, affordability became driven by manufacturing performance, business performance, continuous improvement, and business cases for change. These factors are in play at the prime teammates' sites, but—importantly—within the supply base as well. Approximately 70% of F-35 URF cost is accounted for by teammates and suppliers. As successive LRIP lots progressed, strong cost-performance incentives were created in the form of price-focused contract negotiations and fixed-price-incentive-fee contract types, both at the prime and subcontract levels. In fact, the prime contract for LRIP-4 was structured as a fixed-price-incentive-fee type, beginning with the 32nd production aircraft. Negotiations for this lot would have been conducted before any production aircraft were delivered.

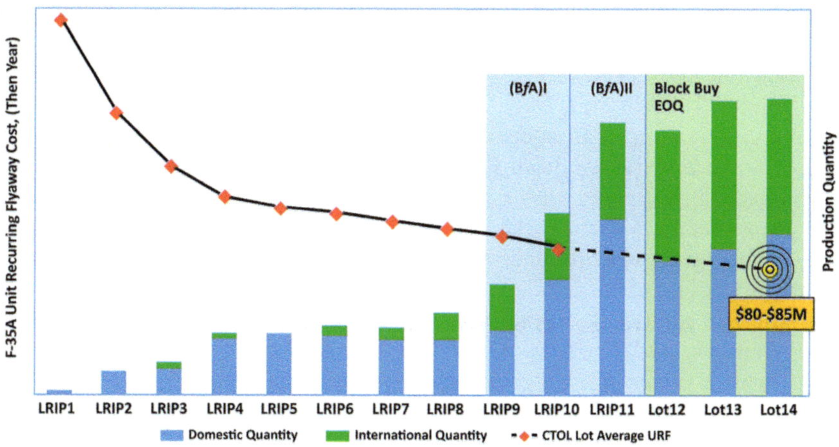

Fig. 29 F-35A URF trend and BFA projection overlaid with overall production quantity profile.

Figure 29 illustrates the steady URF cost reduction achieved by the program, superimposed on the production quantity profile.

Manufacturing performance is improving through the use of producibility improvements and proactive management of traditional metrics, such as defect rates, labor hours per unit, material availability, rework, and traveled work. Business performance involves managing internal cost structure (rates, overheads, etc.), as well as supplier engagement. Continuous improvement using lean principals has been applied to manufacturing, engineering, and business operations.

Another element of the affordability strategy has been investment in cost-saving design improvements. These are termed *investments* because significant costs are incurred up front to develop and implement an engineering change, but the cost savings accrue little by little in recurring production over the long term, eventually reaching a break-even point, where they begin to produce returns. The large planned procurement quantities for the F-35 often make business cases for such changes compelling, with calculated overall return multipliers ranging from 20:1 to 50:1 over the length of the planned production program. During the SDD program, such affordability initiatives were implemented as an explicit part of the contract scope. Once that scope was completed, however, it became less certain how the costs of such investments and benefits of the returns would be balanced between the government and industry parties, and what would be the source of investment funds. The savings for a given change may stretch over a long period of time and be blended with the effects of other changes, so it can be impractical to validate actual cost savings due to that specific change. With annual procurement lots, often the timespan and quantities in a single production contract are not sufficient to implement a change and reach a break-even point. This dampened interest in continuing to invest in affordability initiatives by industry and the government, despite the obvious long-term benefits.

In July 2014, however, DoD, Lockheed Martin, BAE Systems, and Northrop Grumman announced an agreement, known as the Blueprint for Affordability (BFA), intended to drive down F-35 production costs. The agreement provided for industry investments, with a path to recover those investments as savings to the government are accrued. The stated objective of the program was to reduce the purchase price of a 5th-Generation F-35 to the equivalent of a 4th-Generation fighter by the end of the decade, with a specific target of an F-35A URF of between \$80 and \$85 million (then-year dollars) (Fig. 29). Together, Lockheed Martin, BAE Systems, and Northrop Grumman funded \$170 million from 2014 to 2017, with a projected savings of \$4 billion over the life of the program. The program was renewed in 2017, BFA-II, with initial government investment funds added to those from industry.

E. THE GROWING FLEET

The first production F-35A aircraft was accepted by the USAF in May 2011. In July that same year, the first aircraft was delivered to Eglin AFB to begin pilot and maintainer training. A year later, in July 2012, the first F-35B STOVL production aircraft were delivered to both the USMC and the United Kingdom, and that November the first F-35B was delivered to the first USMC operational base at Marine Corps Air Station (MCAS) Yuma. By December 2013, 100 F-35s had been produced.

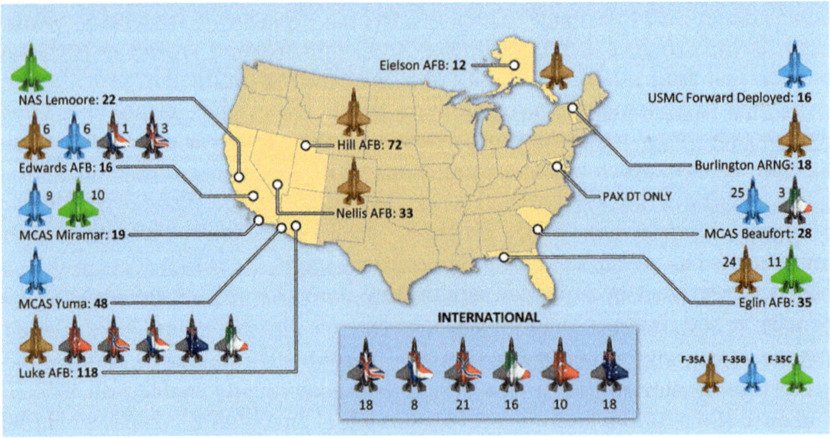

Fig. 30 F-35 global basing projection for 2020.

As of January 2018, nearly 300 F-35s had been delivered or were in flight status preparing for delivery in Fort Worth. Aircraft were operating at eight operational or training bases in the United States and at five international bases. The largest contingent was at Luke AFB, Arizona, which housed 118 aircraft for joint international pilot and maintainer training, nearly four years after the first F-35A arrived there.

Aircraft deliveries and base standups are climbing rapidly at the time of this writing. Figure 30 illustrates firm plans through 2020 to increase U.S. bases to 10 and international bases to 6, with double the aircraft fleet, totaling more than 500 aircraft.

VII. OPERATIONS: ROAD TO INITIAL OPERATIONAL CAPABILITIES

In defense acquisition, the IOC is a point in time during the production and deployment phase when a system is determined to meet the minimum operational (threshold and objective) capabilities for the service's stated need. The operational capability includes support, training, logistics, and system interoperability within the DoD operational environment. IOC is a good gauging point to see whether any refinements are needed before proceeding to full operational capability. The F-35 is unique in that multiple DoD services and international partners and FMS participants had a need to declare IOC. One set of criteria could not satisfy the needs of each customer, so each service defined its unique requirements for IOC [21].

A. USMC F-35B IOC

In January 2012 the USMC took delivery of its first fleet aircraft at Eglin AFB for the Marine Fighter Attack Training Squadron 501 (VMFAT-501), the Warlords. Flight operations began later that year with the initial cadre of USMC pilots. VMFAT-501 was designated as an F-35 Fleet Replacement Squadron (FRS) in 2010. In July 2014 the Warlords relocated to MCAS Beaufort.

Marine Fighter Attack Squadron 121 (VMFA-121), the Green Knights, were redesignated as an F-35B squadron in November 2012 in Yuma and became the first operational F-35 squadron. The Green Knights accepted their first three F-35Bs and began working toward satisfying the requirements for IOC.

USMC headquarters issued a letter in 2013 identifying criteria to declare F-35B IOC. The requirements stated the USMC required 10 to 16 aircraft and that U.S. Marines be trained, manned, and equipped to conduct Close Air Support (CAS), offensive and defensive counter-air, air interdiction, assault support escort, and armed reconnaissance in concert with Marine Air-Ground Task Force resources and capabilities. These requirements enabled the USMC to declare IOC with an interim capability standard known as Block 2B. Similarly, requirements were defined for the interim-capability deployable Autonomic Logistics Information System (ALIS). The initial aircraft selected for the IOC squadron were produced in LRIP-4 and had a number of retrofit modifications required due to engineering changes implemented prior to LRIP-4, including the invasive OBIGGS change described earlier. Estimates for completing the modifications fell well into 2016, a year later than needed by the USMC. Therefore, a series of reassignments of aircraft was made in April 2014 in order to use later LRIP-4 and LRIP-5 aircraft, which reduced the aircraft modification workload.

In order to accomplish the extensive aircraft modifications, the program expanded modification bays at the Fleet Readiness Center – East in Cherry Point, North Carolina, from two to six and added a contractor field team depot capability for three aircraft at MCAS Yuma. Two intermediate airworthiness releases were added prior to the Block 2B fleet release to enable aircraft modifications and support training with an expanded flight envelope without weapons. The squadron went through operational testing aboard the L-class amphibious assault ship U.S.S. *Wasp* (LHD 1) in May 2015 (Fig. 31). Additionally, they completed an operational readiness inspection to ensure that they were capable of performing the five IOC operational mission scenarios.

The USMC declared IOC on 31 July 2015. A comment from General Joseph Dunford, the Commandant of the Marine Corps, provides a summary of the events leading up to IOC from the USMC's perspective.

> I am pleased to announce that VMFA-121 has achieved initial operational capability in the F-35B, as defined by requirements outlined in the June 2014 Joint Report to Congressional Defense Committees. VMFA-121 has

Fig. 31 First F-35B OT&E session aboard U.S.S. *Wasp*.

> ten aircraft in the Block 2B configuration with the requisite performance envelope and weapons clearances, to include the training, sustainment capabilities, and infrastructure to deploy to an austere site or a ship. It is capable of conducting close air support, offensive and defensive counter air, air interdiction, assault support escort and armed reconnaissance as part of a Marine Air Ground Task Force, or in support of the Joint Force. Prior to declaring IOC, we have conducted flight operations for seven weeks at sea aboard an L-Class carrier, participated in multiple large force exercises, and executed a recent operational evaluation which included multiple live ordnance sorties. The F-35B's ability to conduct operations from expeditionary airstrips or sea-based carriers provides our Nation with its first 5th generation strike fighter, which will transform the way we fight and win. The success of VMFA-121 is a reflection of the hard work and effort by the Marines in the squadron, those involved in the program over many years, and the support we have received from across the Department of the Navy, the Joint Program Office, our industry partners, and the Under Secretary of Defense. Achieving IOC has truly been a team effort. [22]

Since declaring IOC, USMC F-35B squadrons have been on an increasing tempo of deployments over the past few years, including large-force exercises like Red Flag and Cope Thunder. They have conducted shipboard and expeditionary operations and, in 2018, they executed their first operational ship deployments aboard the U.S.S. *Wasp* and U.S.S. *Essex*, both in the Pacific. VMFA-121 relocated to Iwakuni, Japan, in January 2017, where it is now permanently based with 16 aircraft. The USMC currently has three operational squadrons: VMFA-121, VMFA-122, and VMFA-211, in addition to the training squadron VMFAT-501. Formal F-35B OT&E will be conducted by VMX1 with a detachment of six F-35Bs stationed at Edwards AFB.

B. USAF F-35A IOC

Air Combat Command (ACC) produced the USAF IOC criteria in June 2013 and updated them in January 2015. USAF defined a required schedule for IOC declaration by 1 August 2016 (objective date) and no later than 31 December 2016 (threshold date). USAF requirements defined a set of mission capabilities that required the interim capability standard, Block 3i, a later standard than that required by the USMC for IOC. The mission set included basic CAS, interdiction, and limited Suppression/Destruction of Enemy Air Defense (SEAD/DEAD) in a contested environment. The requirements called for the SDD program-of-record weapons. Inherent in these mission definitions was the capability to fly throughout the Block 2B flight envelope, day and night, in or out of adverse weather, and to carry and employ AIM-120 missiles, 2000-lb joint direct-attack munitions, and 500-lb laser-guided bombs. A total of 12 to 24 deployable aircraft were required with associated support equipment, spares, verified technical manuals and training programs, and trained pilots and maintainers.

The 34th Fighter Squadron at Hill AFB, Utah, began receiving F-35s in the third quarter of 2015. The squadron's aircraft, to meet the minimum quantity of 12, were delivered between the third quarter of 2015 and second quarter of 2016. Seven of these aircraft were LRIP-7 jets and the remainder were LRIP-8 jets. In the months leading up to IOC declaration, several challenges were addressed by the F-35 enterprise, including on-schedule completion of the aircraft modifications required to support the Block 3i configuration and development, and the testing of a Block 3i software release with software stability characteristics deemed acceptable for warfighting needs, along with the release of the next version of ALIS, 2.0.2.

A key challenge to achieving IOC was overcoming instabilities experienced with the Block 3i MS software that required resets of the software. The MS software team worked through Block 3i issues and ended up releasing 11 software versions to integrate Block 3i and address the stability concerns near the end of development. The final software release for IOC, however, had excellent stability characteristics.

The 12 aircraft identified for IOC declaration required a series of updates to bring the configuration into alignment with the IOC declaration configuration. The engineering design release, procurement of modification kits, and execution of the on-aircraft modifications spanned 18 months, including the OBIGGS upgrade. In late June 2016, the final modifications on 12 aircraft were completed at Ogden Air Logistics Complex, Utah [23] (Fig. 32), and the aircraft ferried back to the 34th Fighter Squadron at Hill AFB. Outstanding workmanship and focus from this team provided all 12 jets with just four weeks of flight time available prior to the USAF IOC objective date.

USAF conducted several operational demonstrations prior to IOC declaration. The 34th Fighter Squadron at Hill AFB deployed to Mountain Home AFB, Idaho,

Fig. 32 Final F-35A for IOC fleet delivered at Ogden Air Logistics Complex, Utah.

with the newest hardware and software and demonstrated the initial combat capabilities of the F-35A with 88 sorties scheduled and flown. The 422nd Test & Evaluation Squadron also conducted an operation test IOC readiness assessment. This assessment included the execution of CAS, interdiction, and SEAD/DEAD missions. The 422nd delivered a report to the Commander of ACC to guide his decision in declaring IOC.

After notifying Congress, Commander of ACC Gen. Herbert "Hawk" Carlisle signed off on the IOC declaration on 2 August 2016. Carlisle said, "The F-35A will be the most dominant aircraft in our inventory, because it can go where our legacy aircraft cannot and provide the capabilities our commanders need on the modern battlefield" [24]. Air Force Chief of Staff Gen. David Goldfein stated, "The F-35A brings an unprecedented combination of lethality, survivability and adaptability to joint and combined operations, and is ready to deploy and strike well-defended targets anywhere on Earth" [24].

C. USN F-35C IOC

The USN took delivery of its first fleet F-35C in the summer of 2013 at Eglin AFB. CF-6 was delivered to the Grim Reapers of Strike Fighter Squadron 101 (VFA-101), which was reestablished as the first F-35C FRS in 2012, seven years after being disestablished as an F-14 FRS. The new VFA-101 trained the initial cadre of Navy and USMC instructors and operational test pilots flying the F-35C. In January 2017 the USN reestablished VFA-125, the Rough Raiders, a former F/A-18 FRS at NAS Lemoore, California, to become the USN's west coast FRS and take on the role of expanding its training capacity to support the transition of the first deployable USN F-35C squadron, VFA-147. VFA-147 was in the early stages of transition in early 2018.

USN IOC was first planned for 2015 but was delayed due a combination of issues, most notably delays in software development and the AHS redesign that delayed shipboard testing. The revised USN IOC window defines an objective of August 2018 and a threshold of February 2019. Requirements for F-35C IOC were delineated in a 2011 Commander, Naval Air Forces letter that stipulated IOC declaration when the following conditions were met:

1. Ten fully Block 3F capable aircraft are available to the first fully manned, trained, and equipped operational squadron. (Block 3F is the final capability standard developed by SDD, to which all variants will be upgraded.)
2. The required ship infrastructure, including ship alterations, is in place to support F-35C carrier-based operations.
3. The required shore infrastructure, including tools, spares, technical repair and flight series data, and support equipment, is available to support sustained training operations.
4. Initial OT&E is complete and declared operationally effective and suitable.

Although the USN initially required the completion of OT&E as one of the key elements of IOC declaration in early 2018, the demonstration of adequate warfighting capability ahead of official OT&E completion was determined to be an acceptable alternate criterion. The objective of the criterion is to ensure that the aircraft delivers expected capabilities prior to first deployment in a carrier strike group.

Assessment of the Gen III HMD and associated green glow in the demanding night carrier landing environment also caused the USN to tie IOC to green glow resolution to the extent it would allow relatively inexperienced pilots to safely conduct night carrier landings in all conditions experienced at sea. The solution to green glow has been demonstrated in ground testing as described previously, and flight trials are planned for later in 2018.

VIII. UPCOMING PLANS: FUTURE DEVELOPMENTS

The F-35 Follow-On Modernization program provides capabilities that ensure that the F-35 maintains a dominant combat edge in future years, adheres to future civil aviation mandates, and continuously implements sustainment and supportability improvements. The program has implemented a requirements governance process that identifies the upgrades required to ensure relevance throughout the air system. This approach provides all F-35 operators with an opportunity to be involved in the F-35 roadmap and supports maintaining commonality and interoperability across all F-35 users. A continuous capability development and delivery framework is used to provide timely, affordable incremental warfighting capability improvements to maintain joint air dominance against evolving threats to the United States and its allies.

IX. SUMMARY AND CONCLUSIONS

The JSF program, beginning with JAST, emerged from a unique set of circumstances in the early 1990s that combined widespread needs to recapitalize western governments' fighter/attack fleets with similarly widespread acute budget constraints. This created an opportunity for a common platform that could satisfy many or most of the requirements, if feasible. The early phases of the program demonstrated that the SDLF-based propulsion system uniquely enabled a common family of configurations to provide the very demanding STOVL capabilities needed by the USMC and U.K. forces without compromising the configurations or capabilities of CTOL and CV variants. The Lockheed Martin F-35 family built upon the unique propulsion system by applying 5th-Generation low observables and MS capabilities to provide a weapons system with transformational lethality and survivability. Commonality among the variants, including their support systems, provides unprecedented interoperability among U.S. and allied services, together with economies of scale that make the system affordable.

Besides its daunting and disparate technical requirements, the program faced highly ambitious development and production schedule requirements. Initial plans for the SDD program relied on significant benefits from acquisition reform strategies, as well as technical tools and processes that were unproven at this scale, and a great deal of concurrency to achieve the schedule. As a result, the development and early production efforts encountered repeated challenges that slowed—but did not stop—progress. Schedule and cost estimates proved to be inadequate, forcing three replans of the SDD program, each adding cost, stretching the schedule, and delaying production. Throughout the development phase, however, the promise of the system capabilities and the need for the system remained strong. Although many specific design, manufacturing, and management features, practices, and approaches were changed over the course of development, the fundamentals of the propulsion concept and air vehicle configurations remained constant. The missions and requirements also remained almost wholly intact, with only minor adjustments in areas with little operational impact.

Massive investments in state-of-the-art production facilities across the globe enabled nearly 300 aircraft to be produced by early 2018 for all three U.S. services and eight international customers. More than 170 aircraft per year are expected to be produced in the mid-2020s. The USMC and USAF declared F-35 IOCs in 2015 and 2016, respectively, and have active operating squadrons in the United States and overseas. The aircraft have been very successful in live training exercises, according to service officials.

Although it has been a longer process than planned, with many problems encountered and solved, the F-35 family is on track to fulfill the original vision for a lethal, survivable, supportable, and affordable 5th-Generation weapon system that is interoperable among services and allies.

APPENDIX: ACRONYM LIST

AAI	Auxiliary Air Inlet
AFB	Air Force Base
AHS	Arresting Hook System
AIC	Affordability Improvement Curve
ALIS	Autonomic Logistic Information System
AOA	Angle of Attack
APC	Approach-Power Compensation
ASC	Aeronautical Systems Command
ASCDR	Air System Critical Design Review
ASPDR	Air System Preliminary Design Review
ASRR	Air System Requirements Review

ASTOVL	Advanced Short Takeoff/Vertical Landing
ATP	Authority to Proceed
BfA	Blueprint for Affordability
BRAT	Blue Ribbon Action Team
BTP	Build-to Package
BUW	Bottom-Up Weight
CAIG	Cost Analysis Improvement Group
CAIV	Cost as the Independent Variable
CALF	Common Affordable Lightweight Fighter
CAPE	Cost Analysis & Program Evaluation
CAS	Close Air Support
CDA	Concept Demonstrator Aircraft
CDDR	Concept Definition and Design Research
CDP	Concept Demonstration Phase
CDR	Critical Design Review
CFD	Computational Fluid Dynamics
CFI	Call for Improvements
COPT	Cost and Operational Performance Trades
CTOL	Conventional Takeoff and Landing
CV	Carrier Variant
DAB	Defense Acquisition Board
DAE	Defense Acquisition Executive
DEAD	Destruction of Enemy Air Defense
DoD	U.S. Department of Defense
EHAS	Electro-Hydrostatic Actuation System
EMD	Engineering and Manufacturing Development
ES/G	Engine Starter/Generator
FACO	Final Assembly and Checkout
FCLP	Field Carrier Landing Practice
FETT	First Engine to Test
FF	Full Funding or First Flight
FMS	Foreign Military Sales
FRP	Full-Rate Production
FRS	Fleet Replacement Squadron
FSETT	First STOVL Engine to Test
HMD	Helmet-Mounted Display
IBR	Integrated Baseline Review
IFR	Initial Flight Release
IIP	International Industrial Participation
IMRT	Independent Manufacturing Review Team
IOC	Initial Operational Capability
IPR	Interim Progress Review
IPT	Integrated Product Team
J/IST	JSF Integrated Subsystems Technology

JAST	Joint Advanced Strike Technology
JCS	JSF Contract Specification
JESB	JSF Executive Steering Board
JET	Joint Estimate Team
JIRD	Joint Interim Requirements Document
JORD	Joint Operational Requirements Document
JSF	Joint Strike Fighter
JSFPO	Joint Strike Fighter Program Office
KPP	Key Performance Parameter
KSDI	Key System Developments and Integration
LL	Long Lead
LoI	Letter of Intent
LRIP	Low-Rate Initial Production
MoD	Ministry of Defence
MoU	Memorandum of Understanding
MS	Mission Systems
NAS	Naval Air Station
NAVAIR	Naval Air Systems Command
NTE	Not-to-Exceed
OA	Operational Assessment
OAG	Operational Advisory Group
OBIGGS	Onboard Inert-Gas Generation System
OSD	Office of the Secretary of Defense
OTB	Over-Target Baseline
PAUC	Program Acquisition Unit Cost
PBS	Performance-Based Specifications
PDR	Preliminary Design Review
PEO	Program Executive Officer
PSFD	Production, Sustainment, and Follow-on Development
PWSC	Preferred Weapons System Concept
RAF	Royal Air Force
RCS	Radar Cross-Section
RF	Radio Frequency
RN	Royal Navy
SAE	Senior Acquisition Executive
SAR	Selected Acquisition Report
SBA	Simulation-Based Acquisition
SDD	System Development and Demonstration
SDLF	Shaft-Driven Lift Fan
SEAD	Suppression of Enemy Air Defense
STO	Short Takeoff
STOVL	Short Takeoff/Vertical Landing
SWAT	STOVL Weight Attack Team
SWG	Senior Warfighters Group

TBR	Technical Baseline Review
TSPR	Total System Performance Responsibility
TVEN	Telescoping Vectoring Exhaust Nozzle
URF	Unit Recurring Flyaway
USAF	U.S. Air Force
USMC	U.S. Marine Corps
USN	U.S. Navy
VAVBN	Variable-Area Vane-Box Nozzle
VFA	USN Strike Fighter Squadron
VL	Vertical Landing
VMFA	Marine Fighter Attack Squadron
VMFAT	Marine Fighter/Attack Training Squadron
V_{PA}	Carrier Powered-Approach Speed
VS	Vehicle Systems
3BSD	Three-Bearing Swivel Duct/Nozzle

ACKNOWLEDGMENTS

Many people contributed to the completion of this paper. Particularly notable were Charles T. (Tom) Burbage for the International Participation section, J.D. McFarlan for the Transition to IOC section, Jeff Catt for the Alternate Engine Program section, Mark Middlebrook for the Program Replans section, and Greg Walker for the STOVL Probation section. Others providing significant data included Bruce Bullick, Wade Cross, Bob Ellis, Carl Fink, David Ford, Jim Gigliotti, Dean Hayes, Don Kinard, Steve Kopp, Jeff McConnell, Philip Mosley, David Rapp, Kevin Renshaw, Drew Robbins, Ken Seeling, Kevin Smith, and Art Tomassetti. Special thanks go to Suzie Pate for cheerfully uncovering many years' worth of data in the form of briefing slides.

REFERENCES

[1] Bevilaqua, P., "Genesis of the F-35 Joint Strike Fighter," *Journal of Aircraft*, Vol. 46, No. 6, Nov./Dec. 2009, pp. 1825–1836. doi: 10.2514/1.42903

[2] Maddock, I. A., and Hirschberg, M. J., "The Quest for Stable Jet Borne Vertical Lift: ASTOVL to F-35 STOVL," AIAA Paper 2011-6999, Sept. 2011. doi: 10.2514/6.2011-6999

[3] Wiegand, C., Bullick, B. A., Catt, J. A., Hamstra, J. W., Walker, G. P., and Wurth, S. P., "F-35 Air Vehicle Technology Overview," AIAA Aviation Forum, June 2018 (to be published).

[4] Wurth, S. P., Smith, M. S., and Celiberti, L. A., "F-35 Propulsion System Integration, Development & Verification," AIAA Aviation Forum, June 2018 (to be published).

[5] Counts, M. A., Kiger, B. A., Hoffschwelle, J. E., Houtman, A. M., and Henderson, G., "F-35 Air Vehicle Configuration Development," AIAA Aviation Forum, June 2018 (to be published).

[6] Eshleman, J., "Large Scale Testing of the Lockheed Martin JSF Configuration," *Proceedings of the International Powered Lift Conference*, SAE P-306, Society of Automotive Engineers, Warrendale, PA, 1996, pp. 319–340.

[7] Nicholas, W. U., Naville, G. L., Hoffschwelle, J. E., Huffman, J. K., and Covell, P. F., "An Evaluation of the Relative Merits of Wing-Canard, Wing-Tail, and Tailless Arrangements for Advanced Fighter Applications," *14th Congress of the International Council of the Aeronautical Sciences*, ICAS-84-2.7.13, Toulouse, France, 10–14 Sept. 1984, pp. 771A–771L.

[8] Hayward, D., Duff, A., and Wagner, C., "F-35 Weapons Design Integration," AIAA Aviation Forum, June 2018 (to be published).

[9] Aldridge, E. C., "Transcript of Briefing on the Joint Strike Fighter Contract Announcement—Oct. 26, 2001— 4:30 p.m. EDT," U.S. Department of Defense [online database], http://archive.defense.gov/Transcripts/Transcript.aspx?TranscriptID=2186 [retrieved 4 March 2018].

[10] Kinard, D. A., "F-35 Digital Thread and Advanced Manufacturing," AIAA Aviation Forum, June 2018 (to be published).

[11] Ellis, R., Gross, P., Yates, J., Casement, J., Chichester, R., and Nesmith, K., "F-35 Structural Design, Development, and Verification," AIAA Aviation Forum, June 2018 (to be published).

[12] Schwartz, M., and O'Conner, C. V., "The Nunn-McCurdy Act: Background, Analysis, and Issues for Congress," Congressional Research Service, CRS 7-5700, 12 May 2016.

[13] Gates, R. M., "Transcript of DOD News Briefing with Secretary Gates and Adm. Mullen from the Pentagon—January 06, 2011," U.S. Department of Defense [online database], http://archive.defense.gov/transcripts/transcript.aspx?transcriptid=4747 [retrieved 25 April 2018].

[14] Marshall, T. C., Jr., "Panetta Lifts F-35 Fighter Variant Probation," *DoD News/American Forces Press Service* [online database], http://archive.defense.gov/news/newsarticle.aspx?id=66879 [retrieved 25 April 2018].

[15] Hudson, M., Glass, M., Hamilton, T., Somers, C., and Caldwell, R., "F-35 SDD Flight Testing at Edwards Air Force Base and Naval Air Station Patuxent River," AIAA Aviation Forum, June 2018 (to be published).

[16] Wilson, T., "F-35 Carrier Suitability Testing," AIAA Aviation Forum, June 2018 (to be published).

[17] Lemons, G., Carrington, K., Frey, T., and Ledyard, J., "F-35 Mission Systems Design, Development, & Verification," AIAA Aviation Forum, June 2018 (to be published).

[18] Robbins, D., Bobalik, J., De Stena, D., Plag, K., Rail, K., and Wall, K., "F-35 Subsystems Design, Development, and Verification," AIAA Aviation Forum, June 2018 (to be published).

[19] Rodriguez, V. J., Flynn, B., Thompson, M. G., and Brelage, S., "F-35 Climatic Chamber Testing," AIAA Aviation Forum, June 2018 (to be published).

[20] "Joint Strike Fighter: Additional Costs and Delays Risk Not Meeting Warfighter Requirements on Time," U.S. Government Accountability Office, GAO-10-382, March 2010.

[21] "F-35 Initial Operational Capability," *Report to Congressional Defense Committees*, U.S. Navy, U.S. Marine Corps, U.S. Air Force, June 2013.

[22] Story, C., "U.S. Marines Corps Declares the F-35B Operational," *Headquarters Marine Corps* [online database], http://www.marines.mil/News/News-Display/Article/611657/us-marines%20-corps-declares-the-f-35b-operational/ [retrieved 15 May 2018].

[23] Lloyd, A. R., "Ogden Complex Delivers F-35s for Units to Reach Combat Readiness," *U.S. Air Force News/Ogden Air Logistics Complex* [online database], http://www.af.mil/News/Article-Display/Article/837054/ogden-complex-delivers-f-35s-for-units-to-reach-combat-readiness/ [retrieved 5 May 2018].

[24] "Air Force Declares the F-35A 'Combat Ready'," *U.S. Air Force News/Air Combat Command Public Affairs* [online database], http://www.hill.af.mil/News/Article-Display/Article/883234/air-force-declares-the-f-35a-combat-ready/ [retrieved 15 May 2018].

CHAPTER 2

F-35 Air Vehicle Configuration Development

Mark A. Counts[*], **Brian A. Kiger**[†], **John E. Hoffschwelle**[‡] **and Adam M. Houtman**[§]
Lockheed Martin Aeronautics Company, Fort Worth, TX

Greg Henderson[¶]
Lockheed Martin Aeronautics Company (retired), Beulah, WY

Configuration development of the multinational three-variant F-35 Lightning II fighter program is traced from its earliest beginnings in the Common Affordability Lightweight Fighter (CALF) effort through to the F-35 System Development and Demonstration program. Technologies that were identified and matured during the Joint Advanced Strike Technology (JAST) program are explored as well as the parallel development of the X-35 demonstrator aircraft and the Preferred Weapon System Concept aircraft. Development and maturation activities along the path to a production configuration, including many of the key design and integration challenges faced by the design team, are examined, and discussions are included on the STOVL Weight Attack Team as well as the unprecedented weight management effort.

I. INTRODUCTION

The U.S. multiservice strike fighter development program that evolved into the multinational F-35 program held many unique challenges for the air vehicle configuration design team. The program was developed around four pillars: affordability, lethality, survivability, and supportability. Effectively balancing the unique requirements of each service and operator while optimizing performance in each of the pillar areas proved to be a challenge unprecedented in modern fighter development history. Examining the evolution of the F-35 air vehicle configuration

---

*Senior Manager, F-35 Engineering Project Office.
†Program Management Senior Staff, Advanced Development Programs.
‡Principal Systems Engineer, Lockheed Martin Aeronautics.
§Aeronautical Engineer, Advanced Development Programs.
¶Director (retired), F-35 Mass Properties.

Fig. 1 The three F-35 variants: F-35C (left), F-35B (center), and F-35A (right).

provides an understanding of the integration challenges faced while developing a multiservice platform composed of complex, highly integrated systems.

In order to best understand the F-35 development, we will follow a chronological timeline of the configuration development and expand key events along the timeline. We will also explore the initial development of the weapon system, the Concept Demonstration Phase (CDP) that included demonstration aircraft, and System Development and Demonstration (SDD) that led to production configurations of the three variants: F-35A Conventional Takeoff and Landing (CTOL), F-35B Short Takeoff and Vertical Landing (STOVL), and F-35C Carrier Variant (CV). The three F-35 variants are shown in Fig. 1.

II. INITIAL DEVELOPMENT: TECHNOLOGIES FOR TACTICAL AIRCRAFT

The Joint Advanced Strike Technology (JAST) program was created in 1993 as a result of a U.S. Department of Defense (DoD) Bottom-Up Review. The major tactical aviation results of the review were to continue the ongoing F-22 and F/A-18E/F programs, cancel the Multi-Role Fighter (MRF) and the A/F-X programs, curtail F-16 and F/A-18C/D procurement, and initiate the JAST program. The JAST program's purpose was to define and develop aircraft, weapons, and sensor technology that would support the future development of tactical aircraft. The final goal was to develop a common family of aircraft to replace aging U.S. and U.K. inventory aircraft.

The JAST program planned to develop shared engines, avionics, munitions, and other features that could be incorporated into variants of a stealthy airframe suitable for the individual needs of each service. The goal was to achieve high commonality across all variants, regardless of structural configuration. This approach was predicted to greatly reduce the overall cost of the program.

When the Defense Advanced Research Projects Agency (DARPA) STOVL programs were integrated into the JAST program, they brought with them unique STOVL technology that would ultimately lead to the world's first operational 5th-Generation supersonic STOVL aircraft (Fig. 2).

Boeing, Lockheed Martin, McDonnell Douglas, and Northrop Grumman were each awarded 15-month Concept Definition and Design Research (CDDR) contracts in December 1994. Shortly after the CDDR contracts were awarded, Northrop Grumman teamed with McDonnell Douglas/British Aerospace (BAE). Each of the contractors refined their designs and performed a number of risk reduction activities such as wind tunnel tests, powered-model STOVL tests, and engineering analyses.

The JAST technologies were available to the Weapon System Contractors (WSCs). These included programs such as the Air Force Research Laboratory's Advanced Compact Inlet Systems (ACIS); Subsystem Integrated Technology (SUIT); Joint Strike Fighter (JSF)/Integrated Subsystems Technology (J/IST), which looked to save both weight and cost by taking a more integrated approach to aircraft subsystems; Advanced Lightweight Aircraft Fuselage Structure (ALAFS), which looked at concepts to reduce structural weight and cost; and Contracted Research and Development (CRAD) studies.

The "family of aircraft approach" (Fig. 3) was an important element of the Lockheed Martin commonality story. Commonality is a key enabler to affordability when implemented in an effective way. Previous development efforts on multiservice aircraft have had only limited success as a result of design inefficiencies and

Fig. 2 JAST technologies led to the F-35B, the world's first 5th-Generation supersonic STOVL aircraft.

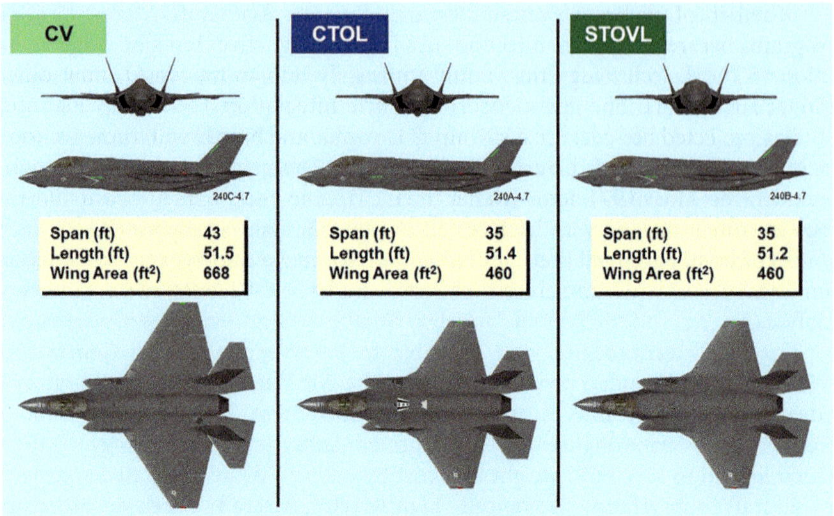

Fig. 3 The F-35 family of aircraft.

scar weight. The challenge was to develop a commonality approach that reduced cost in key areas without adding complexity and weight to each variant. This approach separated every part of the aircraft into three categories: common, cousin, and unique parts.

Common parts were essentially identical parts used for each variant whereas unique parts were unique to a single variant. Cousin parts were identified as parts that were not exactly the same but could be designed to take advantage of similarities in order to maximize cost savings. The cousin part concept can be explained with a simple example. Different variants with common Outer Mold Line (OML) might have different loads that, if optimized, would result in unique composite lay-ups and unique skins. If these skins were made common, the variant with the lower loads would be saddled with unnecessary scar weight. In the cousin approach, the OML tooling could be common whereas the composite skin lay-up would be unique for each of the cousin parts. A significant savings to tooling costs could be realized without adding scar weight to either aircraft. A similar approach was used for internal bulkheads and structure. Component commonality across all three variants reduced requirements for unique spares as well as the logistics footprint. The cousin approach also reduced manufacturing and assembly line footprint and tooling.

Aircraft subsystems have traditionally been designed with a federated approach that consists of a number of independently designed subsystems. Effectively integrating key subsystems can significantly improve aircraft affordability and provide warfighting benefits through increased performance and a reduction in the number of dedicated standalone components. The J/IST demonstration program matured integration technologies for aircraft subsystems to enable transition to the Engineering and Manufacturing Development (EMD, now called SDD) program [1]. The JSF-sponsored Vehicle Integration Technology Planning Studies projected life-cycle cost savings that could be achieved with these technologies vs 1995 state-of-the-art federated subsystems. Integration technologies included the Thermal/Energy Management Module and its integration with the engine (Integrated Power Package [IPP] and fan duct heat exchangers), 270 VDC power management and distribution, electric-powered hydraulic flight actuation, and associated controls. These technologies were identified by Lockheed Martin, Boeing, and McDonnell Douglas/Northrop Grumman/British Aerospace as providing substantial cost savings and warfighting benefits to JSF weapon system concepts. The technologies encompassed in the J/IST program allowed 13 major subsystems to be replaced with 5. Hardware and software components were integrated into major subsystems ground and flight demonstrations in the 1997–2000 timeframe.

The goals of the ALAFS program were to reduce the cost of the aircraft build by 30% and reduce the weight by 20%. These goals were predicted to translate into 6–8% life-cycle cost savings for the JSF program. Technologies and focus areas encompassed materials, structural design concepts, and manufacturing processes for improved fabrication and assembly. This program identified and developed concepts and methodologies to allow much greater integration of advanced composite structures. Unitized composite design concepts were explored to enhance structural integrity at reduced weight while employing lower cost production concepts and creating volumetric efficiencies. Full-scale test articles were fabricated to allow direct comparison with traditional structural designs.

The propulsion system used many unique design elements that can be combined into two categories: those associated with the STOVL lift system and those associated with conventional propulsion systems. For the conventional system, Lockheed Martin developed the diverterless supersonic inlet to maintain low signature and reduce weight and cost. With no moving parts and no diverter, this results in a much simpler and better structural integration of the inlet system. Lockheed Martin also incorporated the Low Observable Axisymmetric Nozzle (LOAN) that provided a lightweight, low observable nozzle for the engine [1].

For the STOVL lift system, Lockheed Martin developed a concept that uses a vertically oriented Shaft-Driven Lift Fan (SDLF). A two-stage low-pressure turbine on the engine provides the horsepower necessary to power the Rolls-Royce–designed LiftFan®. The LiftFan generates a column of cool air that provides approximately 20,000 lb of lift using variable inlet guide vanes to modulate the airflow. Along with the LiftFan, an equivalent amount of thrust from the downward vectored rear exhaust provides balanced lift to the aircraft. The LiftFan utilizes a clutch that engages the shaft drive system for STOVL operations. Because the LiftFan extracts power from the engine, exhaust temperatures

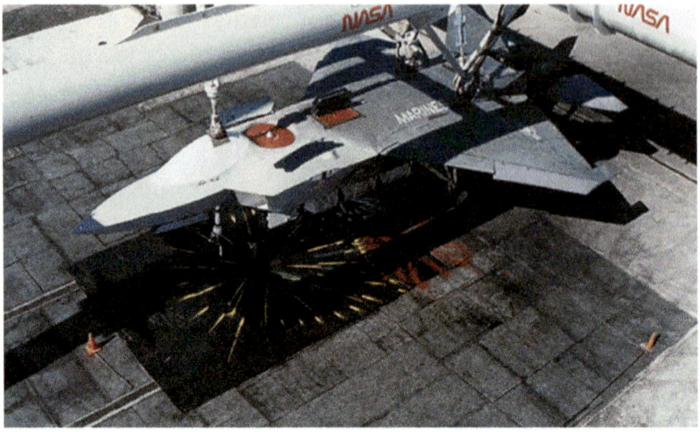

Fig. 4 LSPM testing confirmed the capability of the STOVL propulsion system.

are typically lower than traditional direct lift STOVL systems. The Pratt & Whitney (P&W) F135 engine leverages the existing F119 core technology with a larger fan and integrated fan-duct heat exchanger to reduce the number and size of secondary inlets, exhausts, and heat exchangers. This lift system was proven using a Large Scale Powered Model (LSPM), shown in Fig. 4, to validate both the design and supporting analysis [2].

The F-35 aircraft incorporated cutting-edge technologies in the sensor suite to provide unprecedented pilot awareness. The Distributed Aperture System (DAS) provides the pilot with a unique spherical view around the aircraft for enhanced situational awareness, missile warning, and day/night pilot vision. The Multifunction Advanced Data Link (MADL) provides low probability of intercept ship-to-ship communications that includes both voice and data sharing. The internally mounted Electro-Optical Targeting System (EOTS) provides extended range detection and precision targeting against ground targets, plus long-range detection of air-to-air threats. Additional information is provided to the pilot through an integrated Communications, Navigation, and Identification (CNI) suite, an Active Electronically Scanned Array (AESA) radar, and an integrated Electronic Warfare (EW) suite [3]. Sensor data are fused together by the core processor to provide a seamless integrated picture of the battlespace, giving the pilot unsurpassed situational awareness, positive target identification, and precision strike capability under all weather conditions [4]. The weapon system also incorporates a Helmet-Mounted Display System (HMDS), which replaces the traditional Head-Up Display (HUD), to deliver information directly to the pilot's helmet.

The JAST program provided the risk reduction needed to mature these new technologies to the level necessary to be considered for incorporation into the JSF design [5]. The combination of these new technologies coupled with outstanding aerodynamic performance gives the F-35 unique capabilities and a commanding edge over any other fighter in the battlespace.

III. CONCEPT DEMONSTRATION PHASE

In March 1996, a Request for Proposal (RFP) was released for the Concept Demonstration Phase (CDP) of the JSF program. Design teams from Lockheed Martin, Boeing, and McDonnell Douglas (with Northrop Grumman and BAE) entered the competition and put forward uniquely different designs in their proposal offerings.

Lockheed Martin's conventional four-tail design was said to resemble a single-engine version of the F-22 Raptor (Fig. 5). The STOVL version featured a shaft-driven LiftFan behind the cockpit plus a three-bearing vectored engine exhaust nozzle aft and two roll control ducts extending from each side of the engine. The nozzles for the roll control ducts were located on the lower surface of the wings.

The Lockheed Martin F-35 configuration development lineage, as shown in Fig. 6, can be traced back to Configuration 100. This configuration was developed in response to the government's Common Affordable Lightweight Fighter (CALF) initiative in the early 1990s that was intended to develop a highly common aircraft to replace the U.S. Marine Corps (USMC) AV-8B and F/A-18 aircraft, as well as provide a new fighter platform for the U.S. Air Force (USAF). This was a single-engine configuration with a delta wing, canards, and twin vertical tails that was developed with the intent of being a starting point on the path to the LSPM demonstrator. A series of trade studies matured and refined the design even though the basic planform remained largely unchanged.

This effort resulted in Configuration 140, which became the baseline for the LSPM. Configuration 150 was established as the baseline to support wind tunnel testing. The 150 series of configurations evolved into Configuration 160, which relocated the vertical tails and changed the control surfaces, but still retained the basic delta wing/canard planform.

When the CALF program merged with the JAST program, the configuration took on the added requirement of providing a carrier-based platform for the U.S.

Fig. 5 Lockheed Martin's X-35 Concept Demonstrator Aircraft.

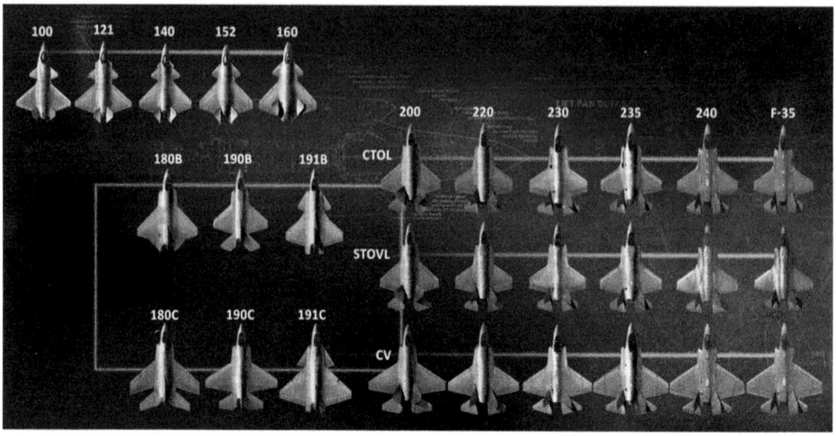

Fig. 6 F-35 configuration evolution.

Navy (USN). This programmatic merge was really a turning point in the configuration design evolution. The CALF configurations had struggled to stay below the specified weight limit of 24,000 lb, so Configuration 180 adopted several significant changes in an effort to reduce weight and improve performance. The caret inlet was dropped in favor of the diverterless supersonic inlet, and the three-bearing swivel nozzle replaced the 2D nozzle in the STOVL variant. For weight savings, the canards were dropped, the twin vertical tails were replaced with a single tail, and the delta wing planform was modified. These changes were made for both CTOL and STOVL variants, which received the designations A (CTOL) and B (STOVL). For the USN variant (C), it was quickly determined that the baseline delta wing planform would not be able to provide the low-speed handling qualities needed for a carrier-based aircraft. It was therefore decided that Configuration 180C would adopt a conventional wing/tail planform.

With weight-related challenges on the A and B variants from the CALF effort, the addition of the C variant, and the desire for commonality, it was not at all clear which would be the best planform to carry forward. The 190 series of configurations was developed to answer that question. Configuration 190 was a conventional wing/tail planform with a single vertical tail for all three variants. Configuration 191 was a trivariant delta wing/canard planform also with a single vertical tail. The configurations were sized to the same requirements, and analysis was performed to compare critical design characteristics including mass properties, aero performance, systems integration, basing, and ship suitability. The conventional wing/tail planform emerged from this competition as the winner and was carried forward into the 200 series of configurations. Configuration 200 kept the traditional boom-mounted horizontal tails, but went back to the twin trapezoidal-style vertical tails that had been featured up through Configuration 160. With the basic planform set, attention was next focused on trade studies to mature the integration of subsystems and weapons. The 210 series of configurations evolved the internal weapon arrangement as well as changed the trapezoid vertical tail to a swept tail design.

The 220 series of configurations made commonality improvements, which included the adoption of common wing structure as shown in Fig. 7. The wing and fuselage were joined as a single structural entity from tip to tip on CTOL/STOVL and wing fold to wing fold on CV. The design team continued to mature the configuration with updates to systems integration details, and Configuration 220-2 became the baseline for Lockheed Martin's CDP proposal. The focus on commonality is apparent in the Configuration 220-2 internal arrangement cutaways for all three variants, which are shown in Fig. 8.

The McDonnell Douglas team offered a relatively conventional-looking aircraft design (Fig. 9), except that it replaced conventional horizontal and vertical tail surfaces with a canted control surface similar to the Northrop/McDonnell Douglas YF-23 offering for the USAF Advanced Tactical Fighter (ATF) competition. The STOVL version employed a separate lift engine that was installed aft of the cockpit.

Boeing offered a delta wing design with twin vertical tails and a single scoop inlet under the nose of the aircraft (Fig. 10). The STOVL version delivered thrust

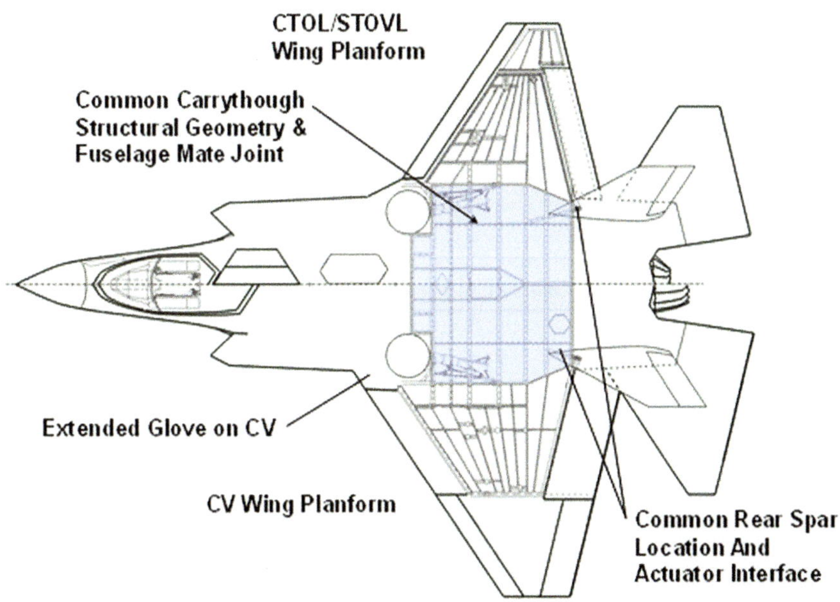

Fig. 7 Wing commonality approach.

Fig. 8 Configuration 220-2 internal arrangement: CTOL (top), STOVL (center), and CV (bottom).

from the engine to vectored lift nozzles located on the lower surface of the aircraft, and the nose intake scoop hinged forward to allow increased airflow.

The government's CDP acquisition strategy was predominately threefold: 1) maintain the competitive environment prior to engineering and manufacturing development down-select while providing for two different STOVL approaches and two different aerodynamic configurations; 2) demonstrate the viability of a multiservice family of variants with high commonality among CTOL, STOVL, and CV variants; and 3) provide affordable and low-risk technology transition to JSF EMD in FY 2001 [6].

Fig. 9 Model of the McDonnell Douglas team's CDP design.

In fall 1996, it was announced that Lockheed Martin and Boeing were each awarded Concept Demonstration Phase contracts, with the McDonnell Douglas team being eliminated from the competition.

IV. CONCEPT DEMONSTRATION AIRCRAFT

Each CDP contract called for building two Concept Demonstration Aircraft to demonstrate the three different configurations of the JSF: CTOL, STOVL, and CV. In addition, each contractor developed in parallel the Preferred Weapon System Concept (PWSC), which would be the basis for the contractor's System Development and Demonstration (SDD) proposal offering.

The contractors were responsible for defining the ground and flight demonstrations they believed were critical for their Concept Demonstration Aircraft. Each contractor was expected to demonstrate STOVL hover, STOVL transition, and low-speed handling qualities of their CV concept. The results of the demonstration program along with the PWSC offering would be the basis of the downselect to one contractor for the SDD phase.

Fig. 10 Boeing's X-32 Concept Demonstrator Aircraft.

Along with Lockheed Martin and Boeing, Pratt & Whitney received a contract to provide hardware and engineering support for both Weapon System Concept Demonstration efforts. The primary propulsion systems being designed for the JSF program were derivatives of the F119-PW-100 engine that powers the F-22 Raptor. The propulsion system concepts for Lockheed Martin and Boeing used a P&W F119 core (compressor, combustor, and high-pressure turbine). Both propulsion system concepts utilized new fan and Low-Pressure Turbine (LPT) designs that were based on F119 designs, materials, and processes. P&W's JSF119-PW-611 was developed to power the Lockheed Martin demonstrator aircraft, and the P&W JSF119-PW-614 was developed to power the Boeing demonstrator aircraft. Fabrication of the P&W engine designs began in March 1997, and engine testing started in June 1998.

Lockheed Martin made the decision to design and build the CDA vehicles at the company's Skunk Works facility in Palmdale, California. Starting with Lockheed Martin's current PWSC design, internally called Configuration 230-1, the Skunk Works team focused on the key demonstrations needed to prove the concept. Using a commonality approach, the team developed a methodology to demonstrate all three aircraft with only two demonstration vehicles. The fuselage for both vehicles was designed to accept the STOVL lift system with the CTOL and CV versions incorporating a large fuel tank in the location where the lift fan was installed on the STOVL aircraft. This concept allowed either demonstrator aircraft the flexibility to be converted to demonstrate STOVL capability. Key demonstrations included short takeoffs, vertical landings, conversions for STOVL, carrier approaches for CV, and vehicle flight performance for CTOL. With the basic design approach set, Lockheed Martin launched forward on developing the demonstrator aircraft (Fig. 11).

Because the X planes were intended as demonstrator aircraft only, there are many differences between the X-35 and the F-35 production configuration [7]. These differences predominately fall into three major categories: changes to the STOVL lift system, operational needs of the F-35 that did not need to be demonstrated, and vehicle improvements that were incorporated after the demonstrator design and flight test. Some of the most notable differences were in the STOVL vehicle, which the team continued to mature and refine as the demonstrator vehicles were being built and tested (Fig. 12 and Fig. 13) [7]. The lift fan inlet doors of the X-35 were side-hinged bi-fold doors, selected for their light weight and low-profile design. This design had higher distortion than expected, which led to changes that were incorporated into the production design. The F-35 lift fan uses a single aft-hinged inlet door. Although this design is heavier than the X-35 design, it reduced distortion and improved performance, resulting in a better overall design. The auxiliary inlet can provide as much as 60% of the air flow to the engine during low-speed operation. The X-35 auxiliary inlet used a small two-door design that was hinged on the aircraft's center line. Although simple in design, this concept limited air flow at low speeds. The production design moved the hinge line to the outboard side of the inlet, which yielded

Fig. 11 Lockheed Martin's X-35 aircraft during assembly.

Fig. 12 Plan views of X-35 STOVL variant (left) and PWSC STOVL variant (right).

Fig. 13 Bottom views of X-35 STOVL variant (left) and PWSC STOVL variant (right).

improved flow characteristics at low-speed conditions. The X-35 lift fan nozzle used a segmented extending hood configuration to vector the lift fan exhaust flow. This design was changed to a Variable Area Vane Box Nozzle (VAVBN) for the F-35 PWSC design. The VAVBN provided not only weight savings because of more efficient integration into the airframe, but also improved thrust characteristics by independently controlling the vanes to regulate and vector the air flow. The STOVL airplane has a roll nozzle outboard of the main landing gear under each wing to provide roll control during STOVL flight modes. The roll nozzles on the X-35 used butterfly valves at the engine attachment to control the roll nozzle flow, and the roll nozzles themselves were completely uncovered to reduce weight and complexity on the demonstrator. The F-35 production design removed the butterfly valves and used the roll nozzles themselves to control the flow. Aerodynamic improvements in up-and-away flight were made by adding doors on the lower surface of the wing to cover the roll nozzles.

Many operational capabilities of the F-35 were not required to be demonstrated in the CDA vehicles. The X-35 had no radar, but instead used this space to install the vehicle's flight test boom. Other key Mission Systems components that were not included in the demonstrator aircraft were the EOTS and the DAS. The EOTS is a targeting system that can be seen in the chin area of the lower forebody on the F-35. The DAS includes a set of sensors that, when combined with the pilot's HMDS, provides the pilot with unprecedented all aspect day/night vision. Lockheed Martin took advantage of the fact that weapon carriage was not a demonstration requirement, and used the weapon bay volume for systems installation and to stow the modified A-6 main landing gear. The X-35 also incorporated in-flight refuel capability. The design team could have installed a nonfunctional refuel system to demonstrate flight characteristics during aerial refueling, but the team decided to incorporate a functioning in-flight refuel system as an enabler to increase flight test tempo. This proved to be a great decision because the in-flight refueling capability was used on numerous occasions to extend flight test time.

To meet stringent timelines and minimize cost, the X-35 demonstrator aircraft used off-the-shelf components and conventional systems whenever possible. The X-35 cockpit used off-the-shelf parts from other aircraft for displays and incorporated a conventional Head-Up Display (HUD). Conventional hydraulic flight actuators, a conventional Environmental Control System (ECS), and a conventional Auxiliary Power Unit (APU) were also used. Leveraging the technologies developed under JAST, the F-35 incorporated a more electric aircraft approach that utilized an Engine Starter/Generator (ESG) that provides generator and engine start functionality, as well as Electro-Hydrostatic Actuators (EHAs) that greatly reduced hydraulic system requirements [8, 9]. A unique Integrated Power Package (IPP) provides ECS, APU, and emergency power functions all in one package. An external heat exchanger was incorporated on the lower center fuselage of the STOVL demonstrator to ensure adequate heat rejection capability for flight testing in the summer timeframe at Edwards Air Force Base. The X-35 utilized modified main landing gear from the A-6 Intruder and nose landing gear from the F-15 Eagle, while the F-35 developed unique landing gear tailored to save weight and meet packaging requirements. The X-35 used a modified F119 engine with a new low-pressure fan and turbine combined with a conventional nozzle. The F-35 engine (F135) shares a common core with the F119 but has a higher bypass ratio and uses a LOAN nozzle design that was developed during JAST. Fan duct heat exchangers added to the engine bypass flow path provide a heat rejection path for the IPP [8]. The X-35's "four-sided" supersonic diverterless inlet was revised to a "three-sided" design to reduce weight and drag as well as to improve high angle of attack performance of the F-35 production configuration. The X-35 CTOL and STOVL versions had a wing area of 450 ft^2 and wing span of 33 ft, whereas the F-35 production versions of CTOL and STOVL increased the size to 460 ft^2 of area and a 35-ft span. The X-35C wing area was 540 ft^2 with a span of 40 ft, whereas the production F-35C increased the wing area to 668 ft^2 and span to 43 ft in order to provide improved approach speed characteristics for ship-based aircraft.

After meeting the unique design and build challenges of the CDA aircraft, Lockheed Martin was ready to start flight testing (Fig. 14). The X-35A CTOL demonstrator took to the skies for the first time on 24 October 2000. This first flight was quickly followed by subsequent successful flight envelope expansion

Fig. 14 The two X-35 aircraft awaiting a fast-paced flight test program.

tests, and the X-35A CTOL flight test effort was completed with all objectives achieved or exceeded in November 2000. Per the Lockheed Martin plan, the CTOL vehicle (Aircraft #301) was then modified to the STOVL configuration. In December 2000, the X-35C vehicle (Aircraft #300) completed first flight. In parallel to the CV flight test effort, the team continued to work on modifications to Aircraft #301 to turn it into a STOVL vehicle.

The Lockheed Martin JSF team completed installation of the X-35B's flight-ready propulsion system, including the shaft-driven lift fan (Fig. 15) and

Fig. 15 X-35B lift fan installation.

engine, in May 2001. Following these modifications, the aircraft was towed to the hover pit, and BAE test pilot Simon Hargreaves began the STOVL test effort. The X-35B conducted its first press-up in June 2001, marking the first time that a shaft-driven lift fan propulsion system had lifted an aircraft into the sky. One of the most exciting events of the X-35B flight test effort was the successful completion of what was referred to as Mission X. This particular flight was piloted by USMC Maj. Art Tomassetti and included a short takeoff, supersonic dash, and vertical landing. This was the first time in aviation history that a single aircraft had demonstrated all three capabilities on the same flight. Mission X showcased the revolutionary capabilities of the X-35 aircraft and the incredible benefits of the shaft-driven lift fan design concept. On 6 August 2001, Aircraft #301 completed its 66th and final test flight. The aircraft left the runway at Edwards Air Force Base and completed a 3.7-hour flight that included six aerial refuelings and six touch-and-go landings at Palmdale. The X-35B totaled 48.9 hours of flight time. Upon completion of flight testing, the X-35B eventually found its way to a permanent display at the National Air and Space Museum (Fig. 16). The X-35 CDA flight test program successfully demonstrated all three variants of the Lockheed Martin design and was one of the most successful and efficient flight test programs ever.

V. PREFERRED WEAPON SYSTEM CONCEPT

The configuration development effort leading up to the CDP proposal resulted in a solid configuration to carry forward into the Concept Demonstration Phase.

Fig. 16 X-35B and the STOVL propulsion system on display at the National Air and Space Museum.

The basic configuration was in the mold of a classic fighter design. It had a conventional wing/tail arrangement with four tails for high maneuverability, aft engine placement with a long inlet and short exhaust, and a large internal weapons bay for lower drag and improved signature. In addition to these key characteristics, another critical factor in the design of a modern 5th-Generation fighter aircraft is stealth. Stealth capabilities are difficult to add to an aircraft after it is designed, providing only limited benefits, and a 4th-Generation aircraft cannot be turned into a 5th-Generation aircraft. In order to maximize the aircraft's effectiveness, stealth must be incorporated into the configuration from the very earliest conceptual designs, and that is exactly what the Lockheed Martin design team did. In addition, affordability, supportability, survivability, and lethality were the four overarching program pillars that influenced design development activities throughout the PWSC as well as the SDD phase of the program. Every design decision was carefully assessed to ensure that impacts to the pillars, either favorable or unfavorable, were understood before incorporation. The four pillars became ingrained as a mindset that guided the team throughout the design effort.

The Configuration 220-2 design, which was the basis for the CDP proposal, was matured with fresh wind tunnel data and systems integration refinement leading to Configuration 230-1. This new configuration represented the fork in the road where development of the Concept Demonstrator Aircraft separated from the development of the Preferred Weapon Systems Concept. The Skunk Works team took the 230-1 configuration as a baseline and developed the two X-35 aircraft while a separate team at the Fort Worth facility focused on maturing the operational design concept. The PWSC team was also bolstered in this timeframe by the addition of teammates BAE and Northrop Grumman following the dissolution of the McDonnell Douglas team, which occurred as a result of the CDP down-select.

The government's plan was to issue an interim requirements document with three successive iterations, followed by a draft of the final requirements, and then ultimately the final requirements document on which the contractor's SDD proposal design would be based. The government had a formidable task in getting to the final set of requirements because it not only had to address the needs of the three U.S. services, but also had to include the needs of the international partners that had joined the program. Another factor that even further complicated the government's task was that in order to maintain the competition, it had to be mindful of issuing requirement values, particularly Key Performance Parameters (KPPs), that couldn't be met by one of the contractors. The government had an ongoing struggle during this time period to combine all of its customers' needs into a single set of requirements. To assist in this process, several series of Cost and Operational Performance Trades (COPT) were performed. COPT was an iterative circular process that included the warfighters, simulation-based acquisition, and weapon system–level trade studies and resulted in a carpet plot of cost as a function of weight, performance, and effectiveness. The resulting data were iterated back through the process again with a new set of trades and associated results. The iterations continued until requirement parameters were identified. The COPTs led to requirement updates, and each successive release of requirements yielded changes to range, payload, aerodynamic/maneuver performance, flight envelope, and other critical parameters.

The Lockheed Martin team continued to mature the design while evolving it to align with each requirement update. Wing area is an excellent example to help understand how evolving requirements influenced the design. Configuration 220-2 was the baseline when the First Interim Requirements Document (JIRD I) was released, and the wing area for 220-2 was 450/540 ft^2 (the first number is the area for CTOL and STOVL and the second number is the area for CV). The Second Interim Requirements Document (JIRD II) was released about a year later. This led to Configuration 230-2 in which the wing size grew to 500/600 ft^2 in order to improve maneuver performance, as well as improve Short Takeoff (STO) and CV approach speed (V_{pa}), both of which were KPPs. The Third Interim Requirements Document (JIRD III) was released several months after JIRD II and led to Configuration 230-4. The wing area was reduced to 412 ft^2 for CTOL and STOVL while staying at 600 ft^2 for CV to maintain V_{pa} performance. The smaller wing area reflected an emphasis on lighter aircraft weight, which included smaller internal weapons. The release of the draft final requirements document (Draft JMS) brought the focus back to a more capable aircraft, and Configuration 230-5 was developed to address the Draft JMS with a wing area of 460/620 ft^2. The final requirements document (JMS) was released several months later and was the basis for the SDD proposal. Configuration 235 was developed as the SDD proposal configuration and maintained the 230-5 wing area of 460/620 ft^2.

Wind tunnel testing was used throughout the PWSC design period in conjunction with extensive Computational Fluid Dynamics (CFD) to optimize the Outer Mold Line (OML) as well as identify performance improvements that provided additional thrust and control to the STOVL propulsion system. In addition to refining the external surfaces, this time was also critical for the development of the aircraft's internal structure. Major load paths were defined and structural components including bulkheads, shear webs, and longerons were modeled. Analysis was performed, and the component models were iterated to revise load paths where needed and to support material selection. The Configuration 230-2 structural arrangements are shown in Fig. 17.

Packaging of the major systems was coordinated with the wind tunnel test plan to ensure that the configuration was sized large enough to accommodate all of the necessary systems, payload, and fuel volume, yet small and agile enough to meet all of the maneuver and mission radius requirements. The internal weapons arrangement was a critical packaging challenge during this timeframe. From a fuselage station perspective, the maximum cross-sectional area is influenced by the internal weapons, the engine, and the main landing gear. Reducing cross-sectional area typically results in better aerodynamics, including lower drag,

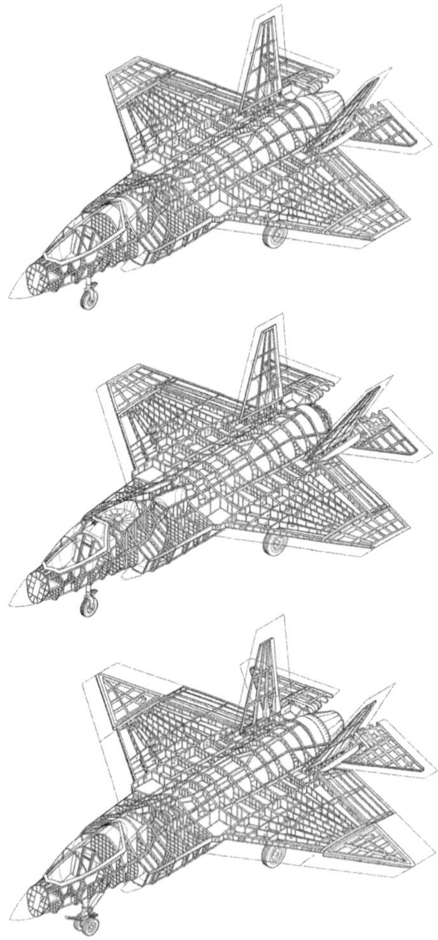

Fig. 17 Configuration 230-2 structural arrangements: CTOL (top), STOVL (center), and CV (bottom).

longer range, better maneuver performance, and better acceleration. Integrating the large set of required internal weapons [10] while trying to minimize cross-sectional area proved to be particularly challenging. The weapons were arranged in a staggered manner to avoid the alignment of maximum wing span or tail span on adjacent weapons. The packaging breakthrough came by toeing the nose of the weapons inboard while in the staggered arrangement. The resulting arrangement allowed hinge lines and weapon bay doors that met fall line requirements, freed up critical volume for door drive mechanisms, and yielded the smallest possible cross-sectional area.

The CTOL internal gun influenced several PWSC design iterations. The initial internal gun was an advanced single barrel gun. After a series of trade studies, the decision was made to delete the gun in favor of weight and cost. Primarily as a result of warfighter feedback, the gun was readopted as part of the baseline configuration, and a trade study was initiated to look at the single barrel gun vs a Gatling-type gun. The Gatling gun emerged as the study winner, primarily on cost, and the GAU-12 five-barrel 25-mm gun was selected [10]. Significant air vehicle level impacts resulted from selection of the GAU-12. The cross-section was larger than the single barrel gun, so a larger external OML bump was needed. This was an air-to-ground gun, so a negative installation bias was preferred that pulled the breech up higher resulting in an even larger bump with steep slopes. The team finally compromised on a zero-bias installation angle as a balance between the drag of a larger bump vs projectile trajectory. It should be noted that a subsequent trade study performed during SDD, and as a part of the STOVL Weight Attack Team (SWAT), deleted one barrel for weight savings. The resulting four-barrel gun was designated GAU-22.

Integrating key technologies demonstrated in JAST, as well as new and unique mission systems, added to the PWSC design challenges. The more electric concept of a 270-V DC aircraft has many advantages; however, each DC motor requires a controller to be installed in close proximity to avoid excessive wire weight, and that can present packaging challenges when numerous motors are spread throughout the aircraft. EHAs also offer many advantages including a reduction in overall hydraulic system weight and a reduction in vulnerable area, but they are typically larger than similar load-sized hydraulic actuators, and therefore require more installation space that often results in larger OML bumps and associated drag impacts. Another new technology that was incorporated was nonpyrotechnic weapon ejection racks [10]. These racks were developed for both internal and external weapon carriage and offered improved supportability and combat turnaround times.

The SDD proposal was prepared and submitted based on the Configuration 235-1 family of aircraft and supported by test data from the X-35 demonstrator aircraft. In October 2001, it was announced that Lockheed Martin's proposal had been selected as the SDD winner, and the company was awarded the contract to develop the aircraft that would be designated F-35.

VI. SYSTEM DEVELOPMENT AND DEMONSTRATION

The design that was matured during the PWSC effort retained its classic fighter characteristics and proved to be an excellent configuration to carry forward into SDD. The design, as shown in Fig. 18, had a conventional wing/tail arrangement, bifurcated long inlet duct, aft engine placement with a fighter optimized cycle, internal weapons bay, and mid-wing location that took advantage of legacy-derived shaping.

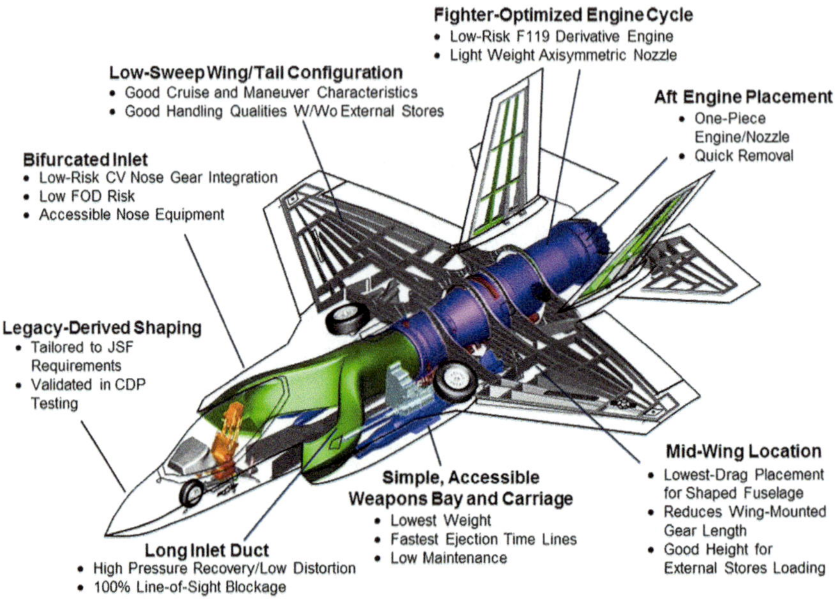

Fig. 18 F-35 design approach.

Immediately following the SDD contract award, Configuration 240-1 was established as the baseline configuration. This configuration primarily incorporated maturation activities that had occurred in the timeframe between proposal submittal and contract award. Changes included standing up the vertical tails by four degrees for improved signature performance, as well as hydraulic system sizing changes and revisions to the engine gearbox.

Freezing the OML surfaces (lines freeze) is a critical point in any aircraft development program. For the F-35, this milestone occurred with Configuration 240-1.1. The lines freeze configuration was the baseline for fabrication of wind tunnel models as well as the basis for design-to loads. There were also two key configuration changes that occurred with 240-1.1. The first was the adoption of the common weapon bay for STOVL. Optimization of the STOVL variant during PWSC resulted in a unique weapon bay sized for 1000-lb (1K) class weapons as opposed to the CTOL/CV weapon bay, which was sized for 2000-lb (2K) class weapons. Adoption of the 2K bay for the 240-1.1 STOVL configuration provided increased capability for the warfighter and improved commonality across the variants, but the cost to STOVL was a weight increase of approximately 240 lb. The second change was an increase of 7 in. to the fuselage length for all variants: 5 in. in the forward fuselage and 2 in. in the tail boom. The increased length was needed to provide space for equipment packaging in the forward fuselage and actuator installation in the tail boom.

The next step in the F-35 design evolution was Configuration 240-2, which was the Preliminary Design Review (PDR) configuration. Several key changes were incorporated into 240-2 including the deletion of dedicated speed-brakes underneath the engine on the lower fuselage. A three-month intensive co-located air vehicle integration effort culminated in the deletion of dedicated speed-brakes and the addition of access panels on the inboard wall of the weapons bay, both of which improved maintenance access to the engine. The decision reduced the weight and complexity associated with dedicated speed-brakes while continuing to provide speed-brake functionality through the use of control surfaces. A critical feature of the design, which was added to the upper right side of the center fuselage, was a fuel/air heat exchanger to provide the aircraft with greater heat rejection capability. The design team was also deeply engaged with finding adequate routing paths for hydraulics, fuel, air, and wiring harnesses, with the STOVL variant being particularly challenging in the areas adjacent to the lift fan. The final key change for Configuration 240-2 was the adoption of a single piece lift fan inlet door as opposed to the previous pair of bi-fold doors. The single door provided improved airflow characteristics to the lift fan, and two open settings were adopted to provide 1) max airflow when needed and 2) lower drag when the max airflow was not required. Following PDR, the baseline was updated to Configuration 240-2.1, which was defined to support the start of Build-To-Packages (BTPs) for CTOL, and subsequently Configuration 240-2.2 which was defined to support the start of STOVL BTPs.

In late 2003, it started to become apparent that the gap between parametric weight estimates and bottom-up weight estimates was not closing. As a result, the SWAT activity, which will be discussed in the next section, was initiated. While SWAT was being planned, Configuration 240-3 was defined and a major program review was held in early 2004. A key decision coming out of that review was to continue with the design and build of the first CTOL aircraft (designated AA-1), but to not continue with detailed design activities on the STOVL or CV variants until a more weight-efficient configuration was defined. Although not weight optimized like the post-SWAT configurations, it was determined that AA-1 could still be a valuable flight test asset to the program while the post-SWAT configurations were being designed and fabricated. However, the fate of the program as a whole, and particularly the STOVL variant, was in the hands of the SWAT team. Configuration 240-4 was defined at the end of SWAT and incorporated all of the design changes, refinements, and improvements that were identified by the SWAT team. The exceptional work of the SWAT team put the program on a solid trajectory toward the Critical Design Review (CDR).

The SWAT effort highlighted several persistent driving requirements that influenced a large number of design decisions throughout SDD. Chief among these were mission radius, signature, Vertical Landing Bring Back (VLBB), STO, V_{pa}, internal and external stores, mission systems functionality (including sensor integration), transonic acceleration, 1% hot day environment, unlimited angle of attack, internal gun for CTOL, 9 g for CTOL, and powered approach handling qualities for CV. Two other key areas also drove many design decisions. The first area was providing the maintainer with the best possible ground access to the aircraft. The design team took advantage of in-flight opening doors wherever possible to avoid the time associated with removing fasteners on fixed panels. Examples of areas where this philosophy was used are the weapon bays, the nose wheel well, and both main wheel wells. A unique feature of the forebody is the EOTS hinge that rotates the system down to provide maintainer access to EOTS and other forebody components. Another unique feature is the forward hinged canopy that enables seat removal without having to remove the canopy. The second area that drove many design decisions was basing and ship suitability. Each variant had unique basing requirements. All three had to be land based, with CTOL having unique shelter requirements. The STOVL aircraft had basing requirements on several types of smaller deck ships, and CV aircraft had basing requirements on large deck Nimitz-class aircraft carriers. Ship-based aircraft also had many unique requirements that impacted the design including compatibility with the hangar decks and elevators on each type of ship, providing an adequate field of view to the pilot for landing approaches to the ship as well as being able to see personnel and ship features while on the flight deck, self-contained ladder access, and compatibility with catapult and arresting gear systems, as well as many other requirements.

BTP activity for the STOVL variant was defined by Configuration 240-4.1 and started shortly after SWAT in fall 2004. A few months later, CTOL BTP activity kicked off and was defined by Configuration 240-4.2. CDRs for both the STOVL and CTOL variants were held concurrently in early 2006. Configuration 240-4.3 was a STOVL variant–only configuration that was defined to support STOVL CDR. Similarly, Configuration 240-4.4 was a CTOL variant–only configuration that was defined to support CTOL CDR. The internal arrangement of Configuration 240-4.3 is shown in Fig. 19. Both CDR events were successful with the decision to proceed toward completion of detailed design and fabrication of the SWAT-optimized configurations. Following quickly on the heels of the STOVL and CTOL CDR configurations was Configuration 240-4.5 which defined BTP activity for the CV variant.

While the BTP design effort on the SWAT optimized configurations was in full swing, fabrication, assembly, and checkout of AA-1 progressed leading up to a successful first flight in December 2006 with Lockheed Martin Chief Test Pilot Jon Beesley at the controls. Flight test activity with AA-1 continued in Fort Worth, gaining critical insight into numerous systems and features including handling qualities, fuel dump, electrical system, and flight controls.

Configuration 240-4.7 was defined to support the Critical Design Review for CV. Although primarily intended as a CV-focused configuration, 240-4.7 included updates to the baseline for the other two variants as well. The Configuration 240-4.7 internal arrangement cutaways for all three variants are shown in Fig. 20. CV CDR was held in June 2007. The event was successful and the decision was made to continue detailed design and fabrication of the SWAT-optimized CV variant.

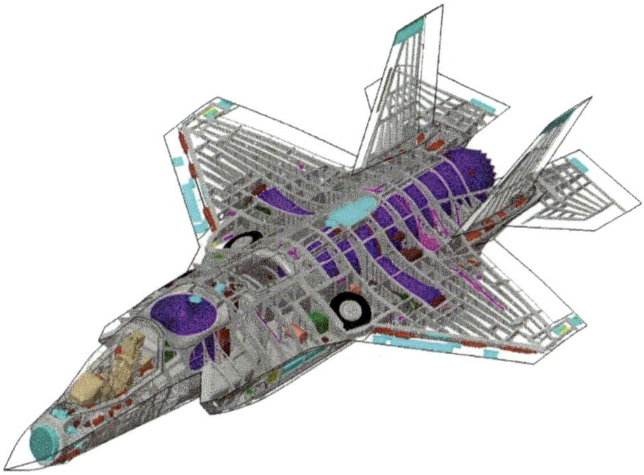

Fig. 19 F-35B internal arrangement (Configuration 240-4.3).

Fig. 20 F-35 variants: CTOL (top), STOVL (center), and CV (bottom).

Throughout SDD, as with any aircraft development program, design and integration challenges were encountered ranging from relatively simple issues that were solved quickly to highly complex issues that required extensive efforts and coordination across multiple teams, organizations, sites, and customers over an extended period of time. The F-35 fuel system is highly complex, with each variant having many unique features including unique feed tanks. Managing Center of Gravity (CG) through the fuel burn sequence across the variants with the full range of possible weapon loadings proved to be quite challenging, particularly for the STOVL variant, which is highly sensitive to CG location while in jetborne flight. Flight test data identified concerns with the in-flight stiffness of the open auxiliary inlet doors, necessitating a redesign to improve stiffness of this key component of the STOVL propulsion system. Early testing of the CV tailhook identified concerns regarding its ability to capture the arresting wire under all required conditions. A significant redesign effort was undertaken on the tailhook, and the results were shown to be highly effective during sea trials [11]. Nacelle ventilation was a function that saw several iterations occur from the start of SDD through Configuration 240-4 as data from multiple wind tunnel tests were combined with analysis and the maturation of the surrounding areas of the aircraft. A nacelle vent fan was added for low speed and ground idle conditions, and improved flow conditions drove relocation of the nacelle vent inlets from the upper fuselage to the lower side of the aircraft near the wing/fuselage intersection. Abrupt Wing Stall (AWS) was identified as a potential concern early in the SDD program based on experience with legacy aircraft. The team conducted numerous high-speed and low-speed wind tunnel tests to characterize critical areas of the flight envelope, test flight control settings, and test potential configuration changes. After an intensive effort, it was determined that control law changes could be used to adequately address potential AWS conditions [9]. Another challenge that had concerns rooted in legacy experience was the impact of IPP exhaust on the ground environment. Numerous tests and analyses were performed to show that the ground temperatures that resulted from running the IPP according to the prescribed procedures were compatible with ground materials used for runways, taxiways, hold short areas, hangars, and maintenance areas. One of the most challenging issues turned out to be fuel dump. For low observable aircraft that have fuel dump capability, dumping tends to be used in case of emergency as opposed to routine daily use. However, the operation of STOVL and CV aircraft from ships requires the pilot to dump fuel often in order to control landing weight (usually on every flight) without requiring postflight maintenance of the fuel dump system. The combination of day in and day out routine usage coupled with low observable aircraft requirements drives the need for a robust fuel dump design. The team conducted several wind tunnel tests followed by extensive flight testing across a wide flight envelope on all three variants to identify a design that effectively balanced all of the requirements to meet the needs of the warfighter.

VII. STOVL WEIGHT ATTACK TEAM

Most aircraft programs experience weight growth in the early stages of design and development, and the F-35 program was no exception. In anticipation of this reality, a weight management plan was instituted following contract award that conducted periodic bottom-up weight estimates. This period is always a challenging time in any program's lifecycle as weight prediction methodologies transition from a parametric weight estimate basis to a calculated weight estimate based on preliminary design drawings as airframe layout data become more detailed.

Parametric estimates typically make certain assumptions regarding weight changes, both growth and savings, as the design database matures, so each bottom-up weight estimate was compared to the parametric estimate in order to provide a clearer picture of the estimated weight of each aircraft variant. Following the third bottom-up estimate, it was starting to become apparent that some of the predicted improvements were not being fully realized, particularly in the areas of fasteners, shims, harnesses, tubing, and system installations.

There were many contributing factors that led to weight issues in the original F-35 design. As it turns out, the parametric weight prediction analysis tools were overly optimistic when compared to the bottom-up estimates. The parametric tools predicted weight assuming a highly optimized configuration with conventional design, fabrication, and assembly techniques; however, the configuration at this point in time was not optimized to the level that the parametric data assumed. This resulted in a gap between the estimated values and the achieved to date values that couldn't close. Primary load paths are a good example of this gap. The prediction methodology assumed relatively simple and straight optimized load paths, when in fact the load paths were much more complex and inefficient. Much of the complexity and inefficiency were driven by previous decisions that had been made on fabrication and assembly techniques, as well as an overall philosophy to maintain commonality between variants whenever possible. For example, the wing carry through was not highly optimized from a loads/weight perspective because some of the bulkheads lacked enough depth to function efficiently. The wing box also had some inefficient components, particularly where the trailing edge spar attached to the fuselage.

The major mate joints added to the weight and structural inefficiencies. Too much emphasis had been given early in the program to manufacturing efficiency and takt time at the expense of weight and load path simplicity. The forward and aft mate joints were termed "quick mate" joints in the original assembly concept. These joints were held together by a small number of large diameter fasteners. The joints were fast to assemble, but this concept drove additional weight into the surrounding webs and flanges, which was not adequately predicted by the parametric weight analysis tools. The wing-to-fuselage mate joint was on a waterline and incorporated the pi preform technique that essentially aligned a tab in a slot to be bonded with only a minimal number of fasteners. The wing mate turned out to be not only complex and time consuming to assemble, but also heavier than originally predicted. The original upper wing skin design was a one-piece tip-to-tip concept that turned out to be thick and complex to manufacture, as well as providing only limited access to systems installed underneath.

Weight was also adversely impacted in this time period by configuration features and details associated with maturing systems. Issues associated with trying to maintain a common weapon bay across all three variants drove weight into the design, particularly for the STOVL variant. Propulsion system contractor Pratt & Whitney was maturing the propulsion design in parallel to the Lockheed Martin air system design maturation. A parallel development effort like this is a highly collaborative and iterative environment, and there were a few unexpected weight/CG issues related to the propulsion system that also had to be accounted for. Other integration issues that occurred during this period that impacted weight included thermal management, nacelle ventilation, battery technology, flight control actuators, canopy, and electronic rack components, as well as several other smaller and miscellaneous issues.

Finally, the parametric methodology didn't adequately account for weight in the area of routing. The actual routing details (tubes, lines, and harnesses) were discretely added instead of being parametrically predicted; however, certain routing details and accommodations were included in the parametric data, which assumed relatively simple and optimized routing paths. In fact, the actual routing was a significant challenge due to more complex configuration features, particularly around the highly integrated lift fan bay on the STOVL variant and the tightly wrapped engine nacelle on all three variants. In order to get routing around the nacelle, a cross-ship routing tunnel was incorporated, which added weight as well as displaced fuel volume.

The design team recognized that the parametric tools assumed an optimized design, and as a result attempted to include factors where F-35 unique features varied from traditional design techniques. However, by the fourth bottom-up estimate it was determined that a fundamentally different approach to weight needed to be implemented.

SWAT was formed in early 2004 to address aircraft weight issues that were identified in the bottom-up estimate. Weight was a concern for all three variants, but it was particularly critical for the STOVL variant. The simple and overriding fact of STOVL aircraft design is that thrust has to be greater than weight or the aircraft does not work. The primary SWAT effort was focused on the STOVL configuration, but the CTOL and CV configurations also benefitted from weight improvement as a result of commonality in both systems and structure. The SWAT activity was a focused six-month intensive effort intended to reduce weight, improve performance, and close on an optimized KPP-compliant STOVL design. A dedicated, co-located team was identified, and program leadership gave the team the time necessary to effectively plan and execute the effort. The initial focus was on the STOVL and common design, and a broad range of trade space was explored including airframe and systems weight optimization, air vehicle performance, propulsion installed performance, and capability tradeoffs. Trade study results flowed to the other variants when appropriate, and every effort was made to preserve commonality where possible. The SWAT team was tasked with the fundamental responsibility of instilling confidence in a viable STOVL design while developing a solid plan to recover performance on all three variants.

The SWAT attack plan had six key areas of focus:

1. *Reduce zero fuel weight.* Based on the most recent installed propulsion performance, the bottom-up weight estimate suggested weight reduction and/or thrust performance improvements on the order of 2800 lb were needed at the air vehicle level to recover VLBB performance. Weight optimization was needed for airframe components (such as bulkheads, shear webs, longerons, brackets, skins, and access panels), vehicle systems components (such as landing gear, flight controls, the thermal management system, and fuel), and mission systems components (such as avionics racks, cards, and fire control and stores).
2. *No stone unturned.* Consolidate the Weight Savings/Threat database, renew and strengthen the Weight Incentive Program (WIP), and review all previous trade decisions. No weight savings idea was too small to be considered.
3. *Reduce required fuel weight.* Improve drag and aero performance so that less fuel is required to perform the mission. Less required fuel onboard at the start of a mission results in a lighter air vehicle at takeoff, which leads to shorter flat deck Short Takeoff (STO) and ski jump STO distances, both of which are KPPs.
4. *Increase performance.* Optimize STO techniques, optimize Vertical Landing (VL) control allowances, and improve installed thrust.
5. *Challenge requirements.* Scrub internal requirements, ground rules, and assumptions. Prioritize and quantify Joint Contract Specification (JCS) impacts. Maintain KPPs and quantify impacts associated with ground rules and assumptions.
6. *Quash weight increases.* Mitigate weight threats and manage growth to zero. Do not allow weight increases.

Support, participation, and buy-in from the customer were crucial for the SWAT effort to succeed. The customer community understood the critical nature of the SWAT effort, and they were great partners throughout the process. They gave key insight into the relative importance of air system capabilities and provided a prioritized list of operational attributes that was used to guide the trade study process. Cost as an Independent Variable (CAIV) is a commonly used design tool that helps determine the relative value of individual items vs their cost. Plotting the CAIV data typically shows which items offer the largest benefit for the least amount of impact. The SWAT team adopted the CAIV concept for weight by creating a Weight as an Independent Variable (WAIV) curve, which clearly identified which trade study items offered the largest weight savings

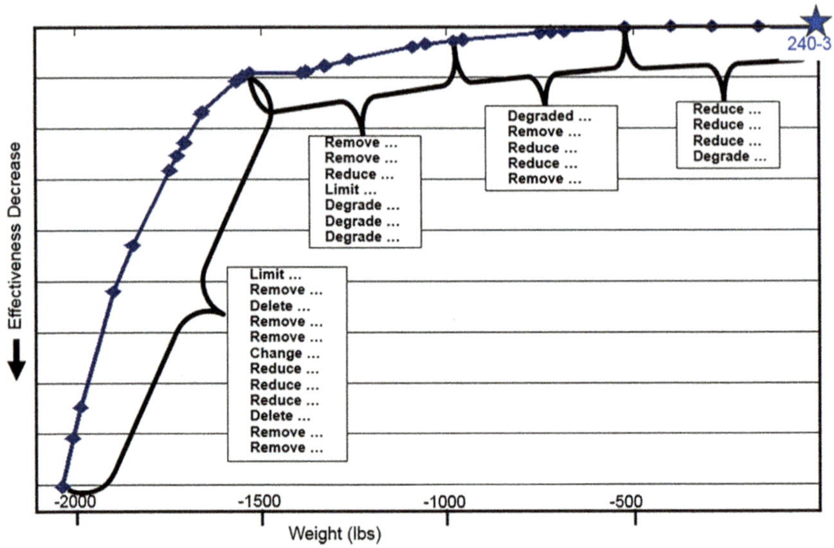

Fig. 21 WAIV curve.

potential with the least effectiveness impact. A conceptual example of a WAIV curve is shown in Fig. 21. The best WAIV items were identified, and teams were assigned to quickly address these high payoff trade studies.

The SWAT team had to move fast in order to build confidence, meet critical path schedules while still maintaining the program's four pillars (affordability, supportability, survivability, and lethality), and ensure that a rigorous change management process was followed. Potential trade studies were triaged and prioritized with full cross-team coordination that included the customer, and comprehensive study tracking was coordinated weekly. Another area that was key to the SWAT team's success was the implementation of a streamlined decision process. A two-tier decision board was stood up that consisted of the SWAT Technical Coordination Meeting (TCM) where technical details of trade studies were coordinated, and a multiboard that combined several decision-making boards into a single board that had overall decision responsibility for engineering changes, contractual matters, and business matters. The multiboard concept allowed the SWAT team to incorporate changes much faster than would have been possible by using the existing established program decision board structure.

A. AIRFRAME

Airframe-related trades were the most numerous of all SWAT trade studies. More than 500 studies were completed accounting for approximately 1900 lb of weight savings. These studies ranged in scope from fastener changes to major load path changes with values ranging from more than 100 lb all the way down to a fraction of a pound. Forward fuselage improvements included skin, fastener, landing gear attachment, and STOVL F-1 fuel tank changes. Center fuselage changes included optimization of upper skins and fuel tank covers, fuel floors, lift fan door drive mechanism, keel beams, and weapon bay door hinge fittings. Improvements to the wing included changes to the upper wing skins, substructure, bulkheads, nacelle materials, fasteners, and engine rail. The aft fuselage improvements included longerons, vertical tail attachments, and the F-5 fuel tank. The empennage changes optimized vertical tails and horizontal tails, and eliminated fuel vent space in the vertical tail.

B. VEHICLE SYSTEMS

Vehicle systems-related trade studies numbered in excess of 80 and accounted for approximately 490 lb of weight savings. These studies included technology updates, material revisions, design optimizations, and margin analysis. Flight control actuation improvements included changes to the leading edge flap drive, actuator reservoirs and valves, as well as Electromagnetic Interference (EMI) filters. Optimization of the electrical power system included changes to the starter/generator, 28-V battery, power panels, inverter, and electrical distribution unit. Power and Thermal Management System (PTMS) changes were dominated by closed loop system architecture and component optimization. Landing gear savings included optimization of the STOVL nose gear and STOVL main gear as well as material changes. Hydraulic and utility actuation improvements included changes to the lift fan inlet door actuation, ground maintenance pump, hydraulic accumulator, weapon bay door drive, and materials. Other vehicle systems changes included modifications to fuel system tubing, probes, valves, and pumps, as well as optimization of wiring.

C. MISSION SYSTEMS

There were more than 40 mission systems-related trade studies, and they accounted for approximately 100 lb of weight savings. These studies included architecture changes, installation changes, technology updates, material changes, and design maturation. Improvements to sensors, processors, and common components included changes to racks and antennas, and repackaging of countermeasure installations. Pilot systems changes included optimization of the canopy, canopy actuator, and ejection seat. Fire control and store improvements included changes to the ejector racks and adding STOVL unique adapters for the ejector racks.

D. AIR VEHICLE

Improvements made at the air vehicle level often spanned multiple teams and stakeholders, and many were among the most beneficial of all SWAT weight

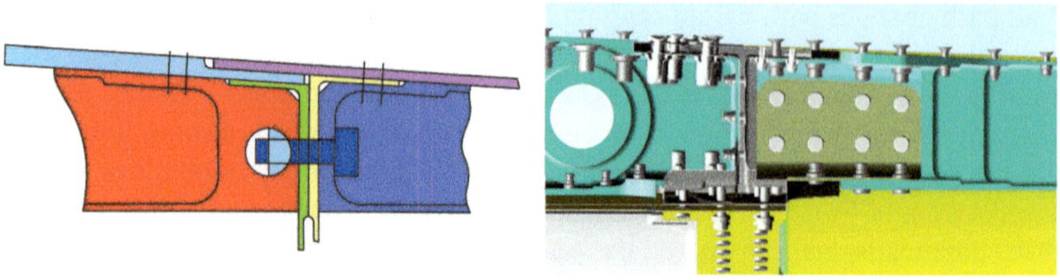

Fig. 22 Quick mate (left) and integrated mate (right).

changes. Mate joints are perhaps the most discussed SWAT change and are fundamental to how the aircraft is assembled. As previously discussed, the pre-SWAT design used quick mate joints at the forward and aft joints. These joints were changed to a more conventional integrated design, as shown in Fig. 22. The integrated design improved the load path across the joint and offered the opportunity to save weight in the adjacent webs, flanges, and stiffeners.

The pre-SWAT waterline mate of the center fuselage and wing is shown in Fig. 23. This joint turned out to be more difficult and time consuming to complete than originally thought. As a result, the center fuselage was split with the portion aft of the engine face being combined with the wing. The new mate joint utilized the conventional integrated joint philosophy similar to the forward and aft joints. The resultant post-SWAT air vehicle mate concept is shown in Fig. 24.

STOVL control surface sizing was another area that saw significant change. The horizontal tail planform was changed while maintaining area. The flaperon area was slightly reduced, and the vertical tail area was reduced while the rudder area slightly increased. These changes resulted in weight savings of approximately 200 lb split about equally between airframe and vehicle systems flight control actuation.

Fig. 23 Waterline mate (left) vs conventional mate (right) joint.

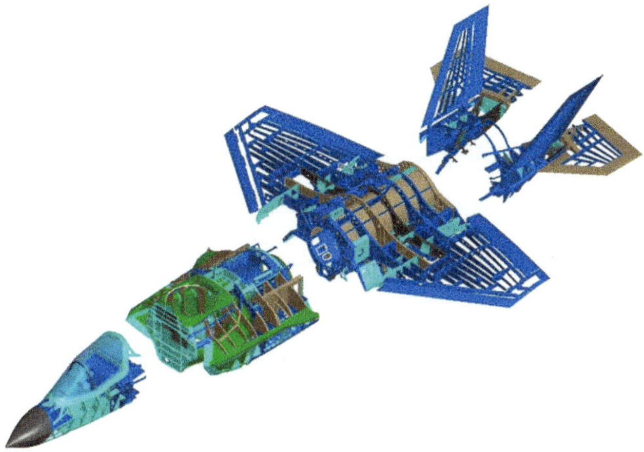

Fig. 24 Revised mate joints.

Numerous configuration changes were made to reduce drag, increase internal fuel volume, reduce routing, and improve equipment installations. Examples of these items include relocation of the nacelle vent inlet, OML changes to the fuselage and to the wing-to-body fairing, nose landing gear door changes that enabled a reduction in vertical tail size, reorienting the IPP exhaust, relocation and integration of countermeasures, engine feed tank, upper wing access panels, and many other smaller changes that are too numerous to mention here.

E. PROPULSION

Increased thrust provides performance improvements in jetborne and semi-jetborne flight that are just as effective as weight reduction. Propulsion-related improvements were made in several places including modifications to the STOVL exhaust that lowered weight while reducing aft body drag. The auxiliary inlet door actuation system was repackaged and the air surface geometry was slightly modified to smooth the airflow, which improved pressure recovery. The main engine inlet throat area was increased by 5%, which also smoothed the airflow while providing increased mass flow and improved inlet/engine compatibility. The final major propulsion improvement incorporated roll post modulation. The pre-SWAT short takeoff design kept the roll posts flowing throughout the takeoff roll. The modification to add modulation closed the roll posts during the early portion of the takeoff, increasing axial thrust through the main nozzle and therefore allowing the aircraft to build takeoff speed earlier in the roll. Later in the takeoff roll, the roll posts opened to provide roll thrust prior to the aircraft leaving the deck. The combination of these propulsion improvements resulted in an increase to hover thrust of approximately 700 lb, a reduction in flat deck STO distance of approximately 100 ft, and an improvement to mission radius of approximately 26 miles as a result of reduced drag and improved fuel burn at cruise.

F. GROUND RULES AND CAPABILITIES

Three ground rules were revised in order to provide KPP performance improvements in VLBB. The Hover Weight Ratio (HWR) at the wave-off point in the pattern was increased by 1% from the more conservative baseline value to better align with legacy experience and NASA studies. The vertical landing recovery pattern distance was slightly reduced along with a reduction in the on-deck fuel reserve. The combination of these ground rule changes resulted in an effective weight improvement of 745 lb. One additional ground rule was revised to improve KPP performance in STO. This changed a portion of the ingress and egress profiles to take better advantage of optimum Mach and altitude, which resulted in less fuel required to perform the mission and therefore a lighter weight aircraft at the start of the mission for short takeoff.

Prior to SWAT, the weapon bay size was common across all three variants. The CTOL and CV variants had requirements that specified 2000-lb-class weapon carriage whereas the STOVL variant had a 1000-lb-class carriage requirement. If the STOVL variant was to remain viable, it was clear that the beneficial exceedance of 2000-lb-class carriage could no longer be maintained for the STOVL aircraft. A smaller weapon bay was incorporated that retained the capability to carry two 1000-lb-class stores with two missiles. The resized STOVL weapon bay still met the weapon carriage requirements but allowed for more weight-efficient load paths and more efficient subsystem installations. External store capability on Stations 2 and 10 was reduced to 1500 lb, which resulted in improved flutter characteristics and enabled the horizontal tail to be resized. In addition, external fuel tanks and dual external store air-to-ground loadings were limited to subsonic carriage, which enabled reductions in flight control actuators. Unused MIL-STD-1760 wiring at the air-to-air stations was eliminated, and certain external coating features were modified. The final capability change was a reduction in limit speed for STOVL. The combination of these capability trades resulted in weight savings of 540 lb, a portion of which was included in the airframe and vehicle systems totals previously discussed.

G. SWAT RESULTS

In just six months, in excess of 600 design changes were incorporated capturing approximately 2600 lb of weight savings, 700 lb of thrust improvements, and 745 lb of improvements associated with ground rule revisions. With these documented results, it's clear that the SWAT team returned the technical design back to a solid foundation and was highly successful in restoring confidence in the viability of the STOVL variant as well as providing significant improvements to the

CTOL and CV variants. In addition, an important point to recognize is the interdependency and commonality of the variants. At program award, CTOL was the lead variant followed by STOVL and then CV. Upon exiting SWAT, the lead variant was switched to STOVL followed by CTOL and then CV. This decision was fundamental to the success of the entire program. By putting the STOVL variant first, some of the most difficult program challenges had to be addressed immediately, which in some cases led to the discovery of other unknown issues that had simply been lying in wait. Making this firing order change resulted in a better STOVL aircraft design earlier in the design sequence than previously planned, and it also resulted in better CTOL and CV aircraft designs due to commonality and similarity benefits.

VIII. WEIGHT MANAGEMENT

The F-35 program incorporated a highly successful weight management program, and the results of that program are unprecedented in modern fighter design history. Government and industry experts predicted a specified weight growth curve based on previous experience. Following SWAT, the Lockheed Martin team adopted a growth curve that was significantly below what was recommended by the government and industry experts. The team consistently performed at or below the adopted curve until the final weight was verified, which covered a period of approximately 12 years.

Concurrent with and following SWAT, Lockheed Martin, in coordination with its DoD teammates, implemented numerous changes to effect weight reduction on the F-35 aircraft. Figure 25 shows the weight-empty status of the F-35 STOVL variant from the period prior to SWAT to the Initial Operational Capability (IOC) timeframe. Early on, bottom-up weight assessments were conducted to obtain the best possible weight-empty estimate of the aircraft, but it was determined to be too heavy by approximately 2800 lb in order to meet the vertical landing requirement. One team of independent reviewers anticipated that at best, only two thirds of the excess weight could be redesigned out of the aircraft. The weight goal was achieved, however, primarily through the efforts of SWAT. An often-overlooked accomplishment is also inherent in the data. Weight increases that were encountered as a result of maturing the design were offset with SWAT items, so the gross weight savings associated with SWAT was actually much greater than the 2800 lb net savings that was needed to meet the vertical landing requirement.

Perhaps even more challenging than weight reduction was keeping the weight off as the program progressed. Some experts thought it would be impossible to take off so much weight and prevent the typical future weight growth. History has shown that all aircraft gain weight in the layout and detail design phases of a program, and it is not unusual to see weight growth curves on the order of 8% or more. Further, it is typical for the weight to grow throughout the lifecycle

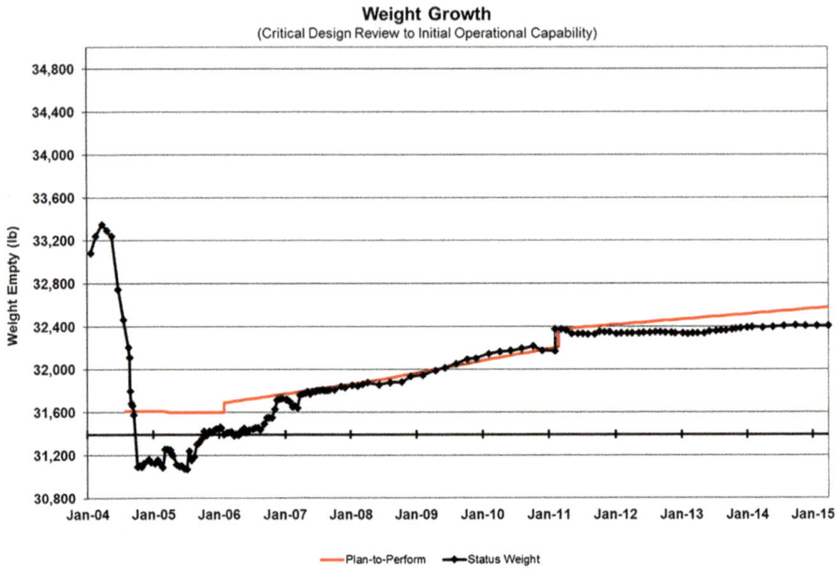

Fig. 25 STOVL plan to perform weight.

of the design. Given that the F-35 program had a STOVL variant incorporating a vertical lift design, the weight-empty growth curve had to be much more aggressive. Lockheed Martin decided to adopt a weight growth curve of 3%, which was unprecedented in military aviation for a major development program like the F-35. Using historical data from numerous military aircraft, the U.S. Navy Mass Properties Division's firm opinion was that a weight growth curve of less than 6% was unlikely to be achieved. No other "clean sheet" military fighter aircraft design (discounting aircraft mods) had ever come close to this challenge. As shown, the F-35 program not only achieved the 3% weight growth curve, but also did so even after incorporating a 200-lb increase later in SDD that was the result of recategorizing a piece of mission systems equipment from missionized weight to empty weight.

The unprecedented STOVL results shown in Fig. 25 are superimposed onto Fig. 26 to illustrate comparative results for several military jet fighter aircraft. In addition, the CTOL variant has held the weight-empty curve flat for over 70 months. The CV variant has also demonstrated similar results.

Another area of concern common to many programs is the credibility of the mass properties database. In the early stages of the program, there was skepticism regarding the accuracy of the database. Compounding this perception was the fact that tailored software for large program weight databases is generally not commercially available and must be developed in-house. Quality data are fundamental to an effective weight management system, and Lockheed Martin has expended a significant effort in development of the database for the F-35 program. The robustness of the mass properties database allowed the F-35 program to use the database as a repository for trade study results. Armed with these data, management decisions on design tradeoffs were made routinely and efficiently. This was further enhanced by a weighted analysis that balanced unit flyaway cost, implementation time, and other factors.

The analysis methods for estimating the weight of individual parts were initially validated by actually weighing the parts. After weighing approximately 35,000 lb of parts, the validation practice was terminated. By using high fidelity digital solid models, the results of the estimation methods were so accurate that it became clear further weighing of parts didn't warrant the expense. Furthermore, at completion of the first production aircraft, the discrepancy between actual weight of the aircraft and the database derived estimate was only 0.16% across the approximate 40,000 records per aircraft. The Lockheed Martin database is now considered among the best robust mass properties databases in existence.

Moving on to a few of the methods used to achieve these results, one of the first steps beyond SWAT was the total system approach strategy defined for weight management and control. The strategy called for active engagement by senior management, and in order to emphasize that engagement, a review of the weight management and control strategy was kicked off by the company president outlining the importance of weight reduction to the program. Senior management engagement continues to this day.

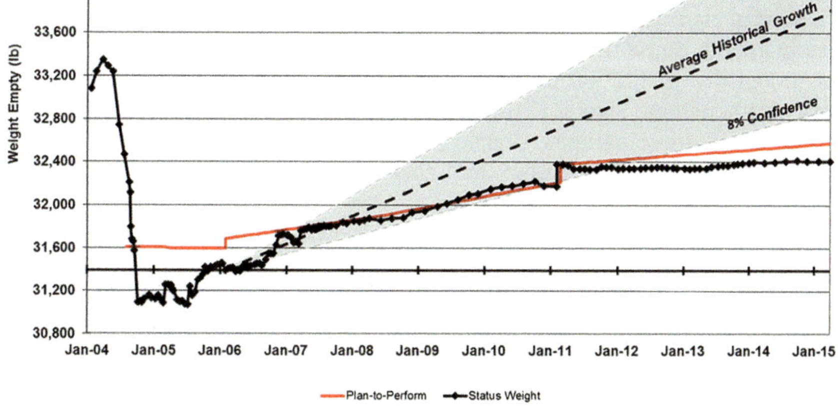

Fig. 26 STOVL plan to perform weight vs the historical average.

A critical activity that led to weight reduction was the Weight Incentive Program (WIP). Over its four-year existence, 12,140 ideas were submitted by employees of Lockheed Martin, Northrop Grumman, BAE Systems, and other companies under contract. A total of 1148 ideas were reviewed for incorporation, and 855 ideas approved. Weight reductions for individual ideas ranged from just a few ounces to more than 70 lb. Employees were monetarily rewarded for their ideas that were incorporated. Suppliers were similarly incentivized through contractual actions.

Partnering with the customer was critical. Joint evaluation of design changes often involved compromises in other areas. Impacts were evaluated and approved or disapproved by joint government/contractor weekly meetings. Paul Kachurak, NAVAIR Mass Properties Engineering Branch Head, later said of the joint effort:

> Reaction to the weight growth problem was swift and decisive—nearly the entire government and industry team focused on weight reduction resulting in a more than ten percent net decrease in weight empty in about ten months, and a program that was back on track to success. Government and industry leadership emphasis on fixing the problem set the team on the right path. They instilled confidence and focused efforts on success in spite of the negativity that assailed the program.

In considering the above methods for weight reduction, it is important to recognize the interdependency of each activity. Lockheed Martin Aeronautics Senior Manager of Mass Properties Vearl Durrington noted that "It was the exceptionally strong willingness to execute, combined with the unprecedented data available that allowed the program to excel in managing weight." Indeed, the cultural change that occurred as a result of the emphasis on weight continues to this day.

IX. TEST AND EVALUATION

An essential element of successfully delivering 5th-Generation capabilities to each customer, both domestic and international, was the maturation of the air vehicle design and the verification of capabilities during wind tunnel, ground test, and flight test activities. Figure 27 highlights the maturation of the configuration design for all three variants aligned with the associated wind tunnel testing from contract award in October 2001 through the last Critical Design Review (CDR) in June 2007.

The Lockheed Martin team successfully executed major wind tunnel test programs for both the CDP and SDD phases of the program. CDP wind tunnel testing was used to develop the demonstrator configurations of all three variants as well as support X-35 flight testing. CDP wind tunnel testing totaled 19,472 hours. Maturation of all three variants during the SDD program required an additional 49,984 hours spanning 17 wind tunnels in the United States, Italy, Germany, Netherlands, and the United Kingdom (Fig. 28). Details of the wind tunnel User Occupancy Hours (UOH) for SDD are provided in Table 1 [12].

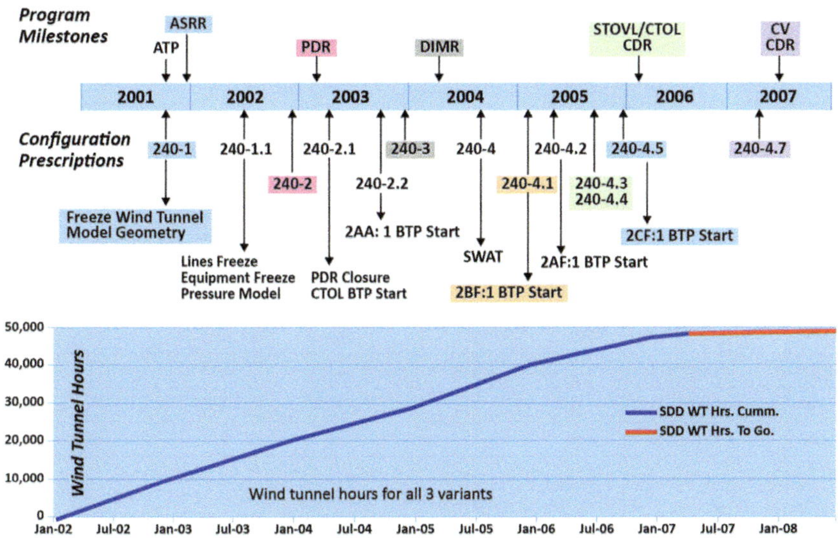

Fig. 27 Testing from ATP to CDR.

The team also used water tunnel testing, as shown in Fig. 29, to supplement wind tunnel activities in both CDP and SDD. The combined wind tunnel test total of 69,456 hours coupled with the extensive use of CFD and flight test data has resulted in a highly accurate and correlated aerodynamic database that is unsurpassed by any other fighter program.

Examples of wind tunnel test findings that led to configuration refinements include the fuselage chine, aft-hinged lift fan inlet door, lift fan inlet bellmouth geometry, revised wing camber and twist for CV, and fuselage contour changes to increase fuel volume while optimizing drag for all variants. Additional

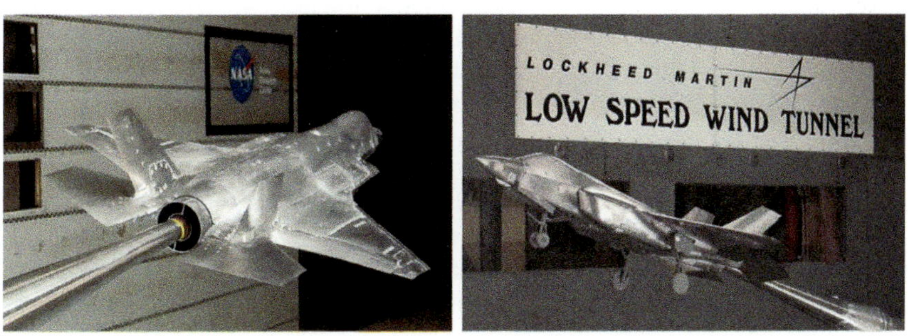

Fig. 28 F-35 wind tunnel testing.

TABLE 1 SDD WIND TUNNEL TEST SUMMARY

User Occupancy Hour Summary (Actuals thru Mar. 30, 2010)

Year	Actual Hours per Test Discipline						
	Aero Analysis	Stability & Control	External Environment	Store Separation	Propulsion Aero	Flutter	Total
2002	3975	1435	868	1175	2269	192	**9914**
2003	3464	3168	677	578	2060	–	**9947**
2004	1991	2582	810	373	2892	–	**8648**
2005	4054	3319	556	732	2935	–	**11,596**
2006	2677	2315	468	342	1253	–	**7055**
2007	780	–	308	445	250	–	**1783**
2008	–	–	–	264	–	–	**264**
2009	–	–	–	–	–	–	**–**
2010	–	–	777	–	–	–	**777**
Actual thru Mar. 30, 2010	**16,941**	**12,819**	**4464**	**3909**	**11,659**	**192**	**49,984**
Plan	**16,941**	**12,819**	**4464**	**3909**	**11,659**	**192**	**49,984**
Complete	**100.0%**	**100.0%**	**100.0%**	**100.0%**	**100.0%**	**100.0%**	**100.0%**

changes identified through wind tunnel testing include wing area changes to the CV variant for improved approach speed, contour refinements for drag improvements, horizontal and vertical tail planform changes, and the split nose landing gear door, which was incorporated during SWAT for all three variants, enabling a reduction in vertical tail size.

CDP full scale structural testing subjected the X-35 aircraft to 100% Design Limit Load (DLL) prior to flight through a total of 12 tests. Maturation of the three variants to support SDD requirements totaled 547 strength tests and included 115% DLL, 150% DLL, and store loads greater than 150% DLL [13]. Durability testing of the CTOL and STOVL variants has completed and CV is scheduled to complete in 2018. The summary of structural testing is provided in Table 2.

CDP flight testing began with the X-35A in October 2000 followed by the X-35C starting in December 2000 and concluded with the X-35B between June and August 2001. The demonstrator flight test program totaled 139 flights with variant totals shown in Table 3. SDD flight testing encompassed a significantly larger scope in order to provide the data necessary to verify program performance requirements and to support flight certification for the three U.S. services, international partners, and Foreign Military Sales (FMS) customers. The SDD flight test program has

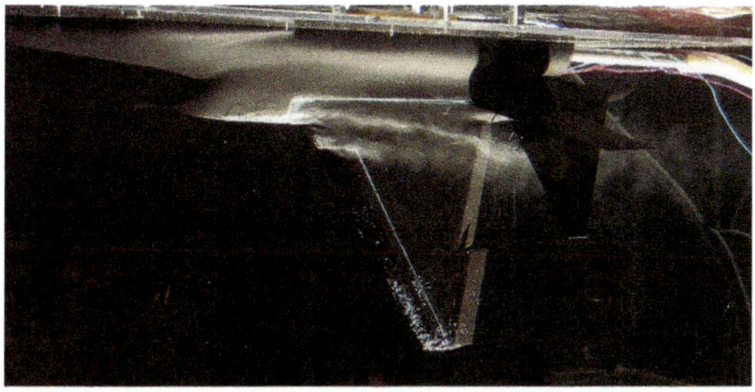

Fig. 29 F-35 water tunnel testing.

totaled 9213 flights through 9 February 2018 utilizing 18 flight sciences and mission systems flight test aircraft. Details of the SDD flight test program are provided in Table 3. Testing primarily occurred at the Edwards Air Force Base, California, and Naval Air Station Patuxent River, Maryland, flight test centers and also included ship deployments for both the STOVL and CV variants [14].

X. SUMMARY

The Lockheed Martin F-35 design team successfully met the challenge to develop an excellent performing multinational three-variant configuration composed of complex, highly integrated systems and solidly anchored by the four pillars of affordability, lethality, survivability, and supportability. The team effectively balanced the unique requirements of each service and customer while optimizing

TABLE 2 SDD STRUCTURAL TESTING SUMMARY

Number of strength tests performed on full-scale static test articles:	
F-35A, AG-1	174
F-35B, BG-1	214
F-35C, CG-1	159
Total Tests	**547**
Simulated flight hours performed on full-scale durability test articles:	
F-35A, AJ-1	24,000
F-35B, BH-1	16,000
F-35C, CJ-1	18,761 (as of 2 March 2018)

TABLE 3 FLIGHT TEST SUMMARIES OF CDP AND SDD

CDP Flight Test Summary		SDD Flight Test thru 9 Feb 2018	
Variant	**Flights**	**Variant**	**Flights**
X-35A	27	F-35A	3538
X-35B	39	F-35B	3579
X-35C	73	F-35C	2096
Total	**139**	Total	**9213**

performance to produce a balanced design that took advantage of commonality where possible and uniqueness as needed to meet the needs of the individual customer in order to deliver a game-changing 5th-Generation platform. The configuration methodically evolved from JAST to CDP and then on to SDD while taking advantage of an impressive set of test activities and an unprecedented weight control program to deliver an outstanding family of three variants that includes the first 5th-Generation supersonic STOVL aircraft and the first 5th-Generation carrier-based aircraft.

ACKNOWLEDGMENTS

The authors would like to thank Kevin Renshaw for his insight into the early stages of the program, Joshua Harris for the use of his X-35 photograph taken at the National Air and Space Museum, and Doug Moore for creating the pictorial lineage of the configuration development. The authors would also like to thank Dan Sturdevant and Jonathon Curtis for their assistance in the development of this paper and Thomas Mellies for coordinating the Subject Matter Expert content review by the F-35 Joint Program Office.

REFERENCES

[1] Wiegand, C., et al., "F-35 Air Vehicle Technology Overview," AIAA Aviation Forum, June 2018 (to be published).

[2] Wurth, S., Smith, M., and Celiberti, L., "F-35 Propulsion System Integration, Development & Verification," AIAA Aviation Forum, June 2018 (to be published).

[3] Lemons, G., Carrington, K., Frey, T., and Ledyard, J., "F-35 Mission Systems Design, Development & Verification," AIAA Aviation Forum, June 2018 (to be published).

[4] Frey, T., Aguilar, J., Engebretson, K., Faulk, D., and Lenning, L., "F-35 Information Fusion," AIAA Aviation Forum, June 2018 (to be published).

[5] Sheridan, A., and Burnes, R., "F-35 Program History—From JAST to IOC," AIAA Aviation Forum, June 2018 (to be published).

[6] Steidle, C. E., "The Joint Strike Fighter Program," *Johns Hopkins APL Technical Digest*, Vol. 18, No. 1, 1997.

[7] Hehs, E., "X to F: F-35 Lightning II and Its Predecessors," *Code One*, Second Quarter, 2008.

[8] Robbins, D., et al., "F-35 Subsystems Design, Development, and Verification," AIAA Aviation Forum, June 2018 (to be published).

[9] Harris, J., and Stanford, J., "F-35 Flight Control Law Design, Development and Verification," AIAA Aviation Forum, June 2018 (to be published).

[10] Hayward, D., Duff, A., and Wagner, C., "F-35 Weapons Design Integration," AIAA Aviation Forum, June 2018 (to be published).

[11] Wilson, T., "F-35 Carrier Suitability Testing," AIAA Aviation Forum, June 2018 (to be published).

[12] Parsons, D., Eckstein, A., and Azevedo, J., "F-35 Aerodynamic Performance Verification," AIAA Aviation Forum, June 2018 (to be published).

[13] Ellis, R., Gross, P., Yates, J., Casement, J., Chichester, R., and Nesmith, K., "F-35 Structural Design, Development, and Verification," AIAA Aviation Forum, June 2018 (to be published).

[14] Hudson, M., Glass, M., Hamilton, T., Somers, C., and Caldwell, R., "F-35 SDD Flight Testing at Edwards Air Force Base and Naval Air Station Patuxent River," AIAA Aviation Forum, June 2018 (to be published).

CHAPTER 3

F-35 Air Vehicle Technology Overview

Chris Wiegand*, Bruce A. Bullick†, Jeffrey A. Catt‡, Jeffrey W. Hamstra§, Greg P. Walker¶ and Steve Wurth**

Lockheed Martin Aeronautics Company, Fort Worth, TX

The Lockheed Martin F-35 Lightning II incorporates many significant technological enhancements derived from predecessor development programs. The X-35 concept demonstrator program incorporated some that were deemed critical to establish the technical credibility and readiness to enter the System Development and Demonstration (SDD) program. Key among them were the elements of the F-35B short takeoff and vertical landing propulsion system using the revolutionary shaft-driven LiftFan® system. However, due to X-35 schedule constraints and technical risks, the incorporation of some technologies was deferred to the SDD program. This paper provides insight into several of the key air vehicle and propulsion systems technologies selected for incorporation into the F-35. It describes the transition from several highly successful technology development projects to their incorporation into the production aircraft.

I. INTRODUCTION

The F-35 Lightning II is a true 5th-Generation, trivariant, multiservice air system. It provides outstanding fighter class aerodynamic performance, supersonic speed, all-aspect stealth with weapons, and highly integrated and networked avionics. The F-35 aircraft features many technological enhancements in air vehicle and propulsion subsystems derived from predecessor programs. These include the Subsystems Integration Technology (SUIT) studies [1–7], Joint Advanced Strike Technology (JAST) program, Air Force Research Laboratory's (AFRL's)

---

*Systems Engineering Senior Staff Specialist, F-35 Power and Thermal Management System.

†Director, F-35 Air Vehicle.

‡Systems Engineering Senior Manager, F-35 Vehicle Sciences and Systems.

§Lockheed Martin Senior Fellow, F-35 Air Vehicle.

¶Systems Engineering Senior Manager, F-35 Propulsion and Flight Controls.

**Lockheed Martin Fellow, F-35 Propulsion and Flight Controls.

Advanced Compact Inlet Systems (ACIS) program, Vehicle Integration Technology Planning Studies (VITPS) studies [8, 9], More-Electric Aircraft (MEA) studies [10], and Joint Strike Fighter (JSF)/Integrated Subsystems Technology (J/IST) demonstration program [11 – 20]. Additionally, numerous Independent Research and Development (IRAD) and Contract Research and Development (CRAD) projects were completed that formed a part of the F-35 design basis [21 – 36]. Many of these technological enhancements were not incorporated into the X-35 demonstrator due to schedule constraints to complete the flyable demonstrator aircraft. They were also postponed due to the results of a technical risk assessment that made them undesirable candidates for the X-35. Instead, the full development and integration of these technologies were deferred for incorporation during the Engineering and Manufacturing Development (EMD) phase (later known as the System Development and Demonstration [SDD] program). This resulted in the final F-35 design configuration.

Enormous efforts from these less well-known predecessor projects produced many of the significant technical achievements that provided necessary technical maturity and risk reduction. This allowed Lockheed Martin to proceed with them confidently in the EMD proposal. The production F-35 incorporates a highly integrated air vehicle subsystems architecture that reduces overall aircraft size and takeoff gross weight. It does so by replacing the federated, individual subsystems used in other legacy aircraft. Low Observable (LO) technologies are incorporated into the engine inlet and exhaust nozzle, and the F-35B Short Takeoff and Vertical Landing (STOVL) propulsion system, with its revolutionary integrated flight propulsion controls, provides unprecedented capabilities. The system represents a revolutionary step-increase in vertical lift, compared to predecessor aircraft. Its fault-tolerant controls are seamlessly integrated with the aircraft control laws, minimizing pilot workload across the entire flight envelope from hover to supersonic flight [36].

The key air vehicle and propulsion systems technologies selected for incorporation are depicted in Fig. 1. This paper focuses on the evolutionary path followed to develop these technologies. The final F-35 aircraft subsystems [35], propulsion [36], and mission systems [37], as well as the SDD program, are discussed in greater detail in supporting publications. Each of the items featured in Fig. 1 represents a significant aircraft capability enhancement that added to the overall success of the F-35 configuration. Successes in the associated development programs for these led to their incorporation into the F-35 design baseline entering the SDD program.

The various development projects that evolved into the systems configurations used in the F-35 spanned the 1990s during the period preceding the winner of the JSF competition. Figure 2 provides key development milestones leading to the incorporation of selected technologies into the F-35 program.

The J/IST integrated subsystems development occurred in parallel with the Concept Development Program (CDP). Interestingly, in it the various JSF competitors cooperated in a collaborative environment, sharing all results and data. This

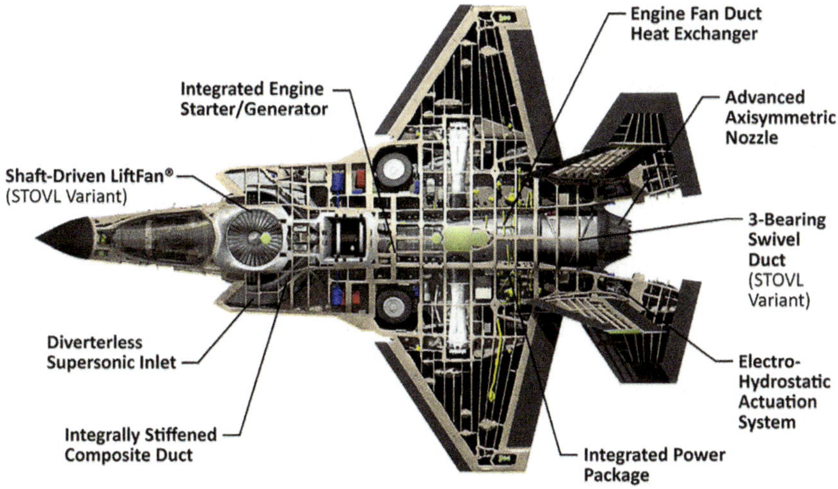

Fig. 1 Advanced technologies selected for F-35 air vehicle and propulsion systems incorporation.

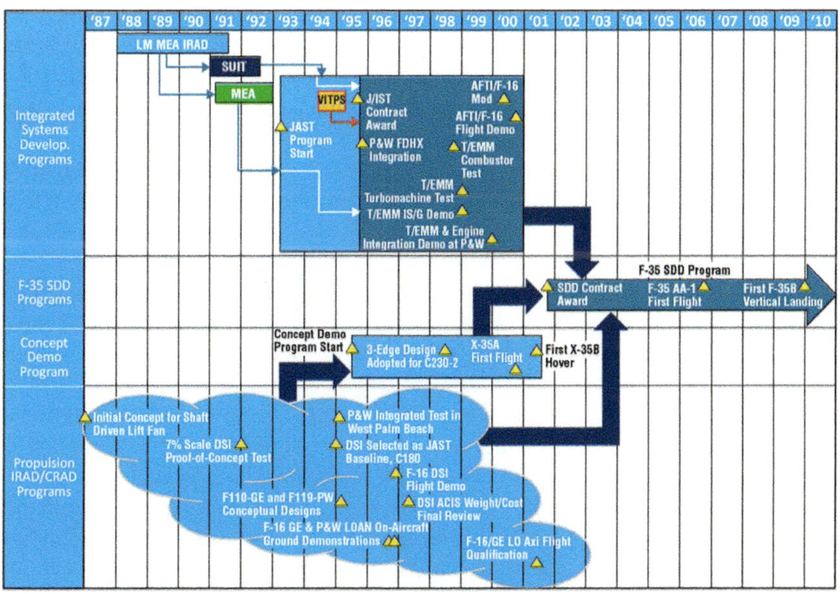

Fig. 2 JSF program air vehicle and propulsion systems technology development roadmap.

allowed the risk reduction activities associated with the integrated vehicle systems to be pursued without the need to encumber the Concept Demonstrator Aircraft (CDA) schedule, and enabled the final results and lessons learned to be incorporated into the F-35 at the outset of the SDD program.

During the same period, numerous IRAD and CRAD studies evaluating potential propulsion innovations continued. As with the J/IST results, several of these were incorporated into the F-35 after the SDD contract award. Significant technical risks associated with the Diverterless Supersonic Inlet (DSI) and LO axisymmetric nozzle, and STOVL propulsion system configurations were retired in parallel with the CDA work, culminating in flight demonstrations showing the maturity and efficacy of the concepts. As an example, dual-redundancy features of the STOVL exhaust nozzle were developed in parallel with the CDA program and incorporated during SDD.

II. INTEGRATED AIR VEHICLE SUBSYSTEMS

A. EARLY DESIGN STUDIES

Aircraft subsystems have traditionally been designed using a federated approach consisting of a number of independently designed subsystems. JSF-sponsored studies showed the potential for significant benefits from integrating these functions. This applied to three subsystems: flight controls and actuation systems, electrical and auxiliary power systems, and the Environmental Control System (ECS). Results from the studies showed that the effective integration of these three key subsystems could significantly improve aircraft performance and dramatically reduce the amount of equipment required. In doing so, it would provide improved affordability and warfighting benefits essential to F-35 program goals.

In conventional systems, electrical, hydraulic, pneumatic, and mechanical power are generated and distributed from the engines and auxiliary power system. The government-funded SUIT [1–7], MEA, and VITPS [8–9] studies showed the potential benefits of eliminating or shrinking the conventional centralized hydraulic system. They also showed the potential benefits of reducing engine bleed air extraction [1–7]. AFRL sponsored SUIT in the late 1980s and early 1990s to look into what could be gained from more efficient integrations of selected air vehicle subsystems [39]. The original SUIT concept was to replace single-function subsystem equipment with multifunctional hardware, potentially reducing volume and weight while increasing overall efficiency. Later, the objective of using the engine's fan air stream as a sink for waste heat from the subsystems was added. Concurrently, AFRL's propulsion laboratory was independently pursuing MEA technology, including electrically powered flight control actuation and robust electric power generation and distribution system concepts. As a result, MEA system and component technologies were undergoing development and testing through several separate and independent initiatives.

Between 1994 and 1995 the JAST program identified key technology building blocks to support the development of an advanced strike capability. The idea was to screen candidate technologies for their applicability based on their respective payoffs with regard to the four JSF program pillars: affordability, lethality, survivability, and supportability. At that time, three Weapons Systems Contractors (WSCs) were actively competing to win JSF: Lockheed Martin, Boeing, and the McDonnell Douglas (later Boeing St. Louis)/Northrop Grumman/BAE Systems team. The candidate JSF configuration was expected to be a single-seat, single-engine strike aircraft, largely due to affordability considerations. Originally, the JSF platform focused on Air Force and Navy customers; however, during JAST the government concluded that the Advanced STOVL (ASTOVL) concept development should be rolled into JAST/JSF. Thus, the STOVL jet was added to the JSF design space [39]. JAST initiated the VITPS studies [8, 9]. All three WSCs concluded that SUIT and MEA technologies could be combined synergistically in a strike aircraft. Accordingly, they recommended that the SUIT/MEA combination be pursued under JAST. Each WSC advocated pursuing integrated subsystem technology, and from that advocacy the J/IST demonstration program was conceived.

B. JSF INTEGRATED SYSTEMS TECHNOLOGY DEMONSTRATION PROGRAM

The goals of the J/IST demonstration program were to set the stage for the JSF SDD program and sharpen the focus for the target JSF platform configuration. Ultimately, that configuration was shaped by the winning proposal for what would become the largest acquisition program in Department of Defense history. The J/IST program was to be executed by all three WSCs competing for the JSF. Each WSC was supporting its own JSF proposal team while simultaneously being responsible to the other JSF WSCs for the technical results of the J/IST work.

The J/IST government team ensured that each participating WSC team was contractually accountable to and executing on behalf of the other WSCs. Each participant was treated by the government as customers for the technologies being demonstrated. This arrangement provided a level playing field using WSC input to make all key decisions within the scope and resources of the program. This fostered trust, a sense of fair play, and government responsiveness within the otherwise highly competitive JSF program environment. In turn, that collaborative environment enabled the results of each element of the studies to be capitalized by each competing contractor. Direct involvement of the three WSCs was fortuitous, because it compelled expanded involvement and cooperation beyond what otherwise might have been the case [39]. Each WSC benefitted from the combined J/IST demonstration results, because these provided information to be used in each WSC's preferred weapons system configuration aircraft proposals.

The focus of the J/IST demonstration program was to reduce the risks associated with subsystem integration technologies. This applied specifically to MEA technologies, including Switched-Reluctance (SR) Starter/Generators (S/Gs), Electro-Hydrostatic Actuation System (EHAS) integration, and Thermal/Energy Management Module (T/EMM) integration through a series of maturation efforts. The work included developing and flight testing prototype versions of the SR S/G and Electro-Hydrostatic Actuation (EHA) flight control actuators, successfully reducing the risks in these technologies from high to low for the SDD program. The work was divided into three main focus areas: one focusing on the electrical power and flight control actuation system (led by Lockheed Martin), one focusing on the development of T/EMM-related technologies (led by Boeing St. Louis), and one dedicated to independent benefit assessments (led by Boeing at its Seattle facility).

Figure 3 compares a conventional federated subsystems architecture to the F-35 integrated subsystems architectures. The integrated architectures reduced system parts count, which led to smaller, lighter, and lower-cost aircraft. Based on those conclusions and the configuration design developed for the SDD program, these systems were incorporated into the F-35 baseline.

The J/IST program was instrumental in reducing the risk of integrating these technologies prior to entering the JSF SDD program in 2001. The J/IST conclusions indicated an overall reduction in Life-Cycle Costs (LCCs) projected to be 2–3%, compared to the LCCs of a similar legacy configuration [19]. The benefits of these integrated technologies are highlighted in Fig. 4.

1. J/IST ELECTRICAL POWER SYSTEM AND ELECTRO-HYDROSTATIC ACTUATION SYSTEM

The Lockheed Martin–led team achieved the following key accomplishments in the J/IST program:

a. Development and flight test of an SR, dual-channel S/G system to provide a fault-tolerant, redundant 270-VDC electrical power source for a single-engine fighter aircraft

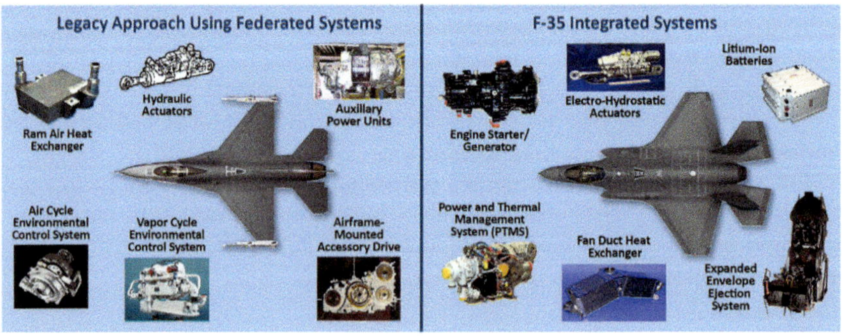

Fig. 3 F-35 integrated vehicle systems.

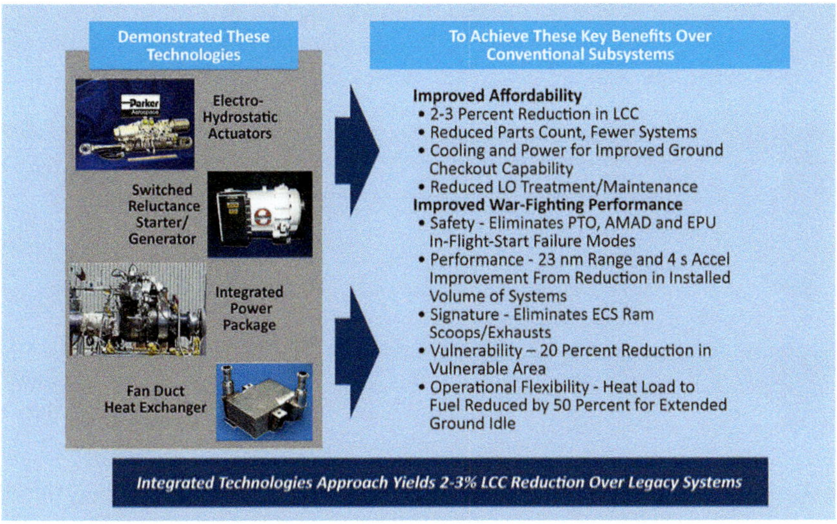

Fig. 4 Benefits of integrated subsystems technologies.

b. Development of a dual-redundant, flight-qualified electric actuation system
c. Development of a flight-qualified 270-VDC battery system to provide uninterruptible electrical power to the flight-critical 270-VDC Electrical Power System (EPS)
d. Development of a flight-qualified emergency power system to provide a secondary power source
e. Completion of specific technology demonstration tests on the S/G to verify operation under certain fault conditions
f. Modification, integration, and flight test of the above technologies in a single-engine Advanced Flight Technology Integration (AFTI)/F-16 aircraft (Fig. 5)
g. Validation of MEA technologies with an AFTI/F-16 demonstration for the F-35 subsystems suite

J/IST Electrical Power System

Immediately following J/IST contract award, Hamilton Sundstrand (now a UTC Aerospace Systems [UTAS] company) was selected to develop and qualify the SR generator system. UTAS was involved in the initial Air Force SR technology development and proposed using the same 250-kW, 270-VDC generator design. However, UTAS rated it for 80 kW and packaged it for the F-16 application. The primary focus was on using an SR generator system, but the fault tolerance and redundancy of the 270-VDC power were emphasized as well. This included

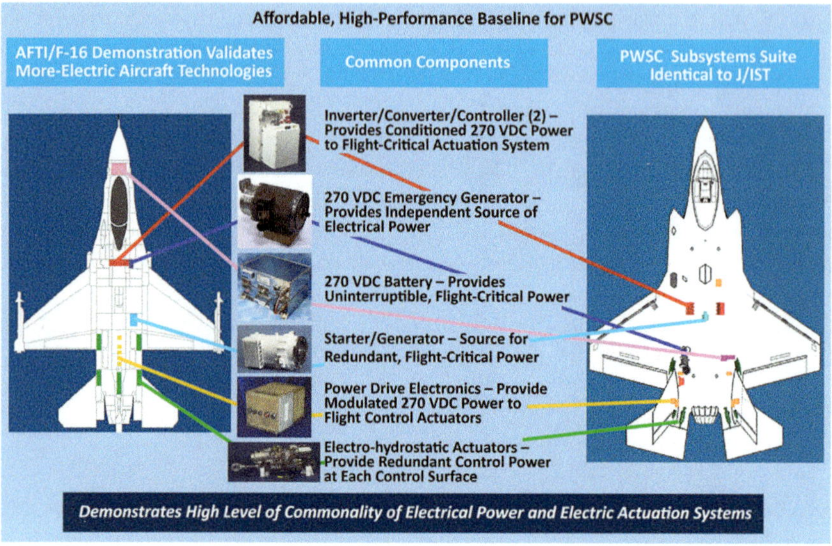

Fig. 5 AFTI/F-16 J/IST aircraft MEA technologies and transition to F-35.

an emergency generator and a 270-VDC battery. The AFTI/F-16's EPS was modified significantly to support the J/IST program's MEA technologies. These consisted of one 270-VDC EHAS for the five primary flight control surfaces, two 270-VDC fuel pumps, one 270-to-28-VDC converter, and one 270-VDC-to-115-VAC inverter.

AFTI/F-16's pre-J/IST electrical system was a combination of an F-16 Block 15 production system and a digital flight control power system (production Block 40), receiving only 115-VAC and 28-VDC power. Therefore, to support the legacy F-16 equipment and MEA systems, the electrical system was modified to provide 270-VDC, 115-VAC, and 28-VDC power. During the initial program's design phase, it was determined that the MEA systems would consume the most power. Accordingly, the primary power type would be 270-VDC power. Also, because the EHAS is a flight-critical system, the 270-VDC system would be designed to be fault tolerant and provide limited uninterruptible power. For EPS integration, the primary challenges were:

- SR S/G flight certification
- Fault-tolerant power generation and distribution
- 270-VDC power system stability with multiple variable power loads
- 270-VDC fill-in battery operation
- Electromagnetic Interference (EMI) and Electromagnetic Compatibility (EMC)

The following combination of components provided the baseline design for the J/IST Engine S/G (ES/G) system. The Inverter/Converter/Controller (ICC) for the ES/G system was taken from the UTAS LV100 SR S/G system. The power electronics converter and the SR generator were taken from the UTAS/General Electric Integrated High-Performance Turbine Engine Technology (IHPTET) research program. The ES/G system was required in two tests: the AFTI/F-16 flight test and a ground test. For the flight test it was used to demonstrate generation capability in an MEA application. For the ground test, it demonstrated motor, start, and generation capabilities. The ground test application included a demonstration of starting a Pratt & Whitney F119 engine and transitioning to generate mode. The ES/G system was designed to accommodate all applications; however, the ICC package design was driven by the AFTI/F-16 installation. Both the flight and ground demonstrations used nearly identical hardware, with minor changes made to adapt to their respective operating environments. The resultant J/IST MEA architecture is shown in Fig. 6.

The ground demonstrations exercised all the capabilities of the ES/G system, specifically its engine motoring and starting, power generation, and fault tolerance. The T/EMM motor/generator system provided 270-VDC power to the ICCs for engine motoring and starting. The ES/G system was transitioned from start to generate to provide 270-VDC power to two primary motor buses. Modifications were made to the ES/G and ICCs from the AFTI/F-16 configuration to enable bidirectional power flow and increased speed range operation.

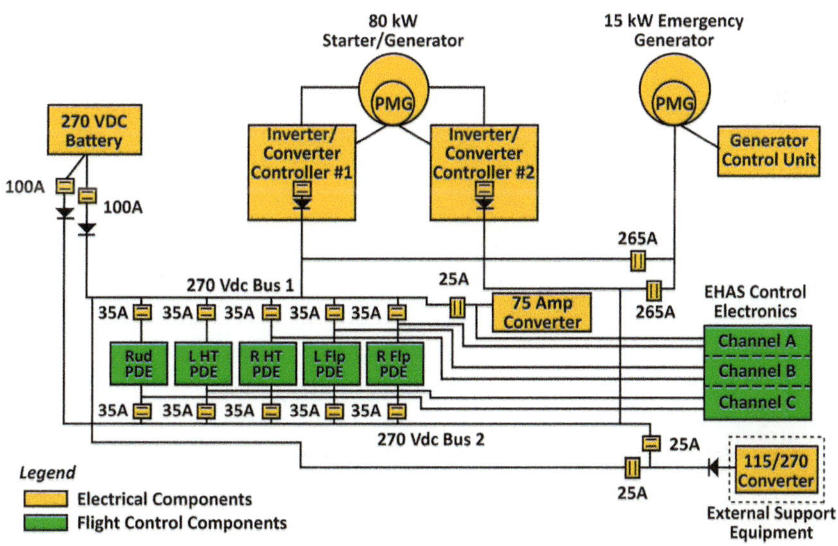

Fig. 6 AFTI F-16 J/IST electrical power and actuation flight demonstration architecture.

J/IST Electrically Powered Flight Control System

Parker Aerospace was selected to provide the EHAS for the F-16. The company had previously developed prototype EHAs for various ground and flight test programs; among them were the flight tests of an aileron actuator on the Lockheed Martin C-130 high-technology testbed. Previous electric actuation integration programs focused on single-surface operation or non-flight-critical control surfaces. Two examples are the F/A-18 Electrically Powered Actuation Development (EPAD) and F-15 Fly by Light Advanced Systems Hardware (FLASH) programs. By contrast, J/IST focused on integrating all primary flight control surfaces, with no hydromechanical backup system. This bold approach was essential to convincingly prove that the MEA concept supported the JSF technology transition criteria.

Lockheed Martin worked closely with Parker Aerospace to define the EHAS system-level requirements, integration test requirements, and software configuration management. The development and testing of the EHAS was the most challenging of the contracted tasks. The selection of the F-16 as the flight vehicle provided known air vehicle and system-level requirements, which made the development task easier. However, multiple hardware and software development challenges were encountered that provided valuable lessons learned for the SDD program.

The Lockheed Martin team conducted a detailed design of two EHASs: one to replace the F-16 flaperon and horizontal tail Integrated Servo-Actuators (ISAs), and one to replace the F-16 rudder ISA (Fig. 7). The designs incorporated a common power electronics package for all five actuation systems. The Lockheed

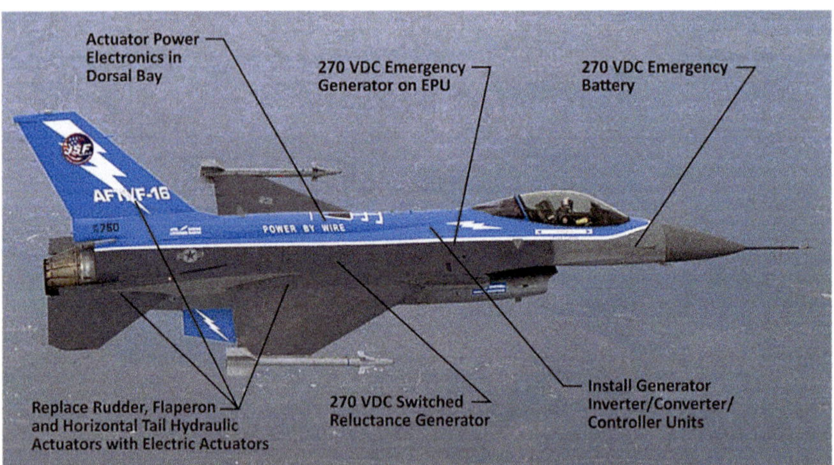

Fig. 7 Key AFTI F-16 modifications.

Martin team also provided an analog interface with the existing F-16 Digital Flight Control Computer (DFLCC). To do so, it developed and integrated a separate interface box that required no major flight control software or hardware changes to the DFLCC. This approach resulted in four major component designs: an F-16 flaperon/horizontal tail dual-tandem EHA, an F-16 rudder dual-tandem EHA, a common power drive electronics package, and an EHA interface controller electronics unit.

EHAS integration testing of all five actuators with electronics was successfully completed at Parker Aerospace's facility. The testing verified that the hardware and software met all design requirements. Further, it provided detailed visibility into the system's operation and added confidence in the EHAS software. It also reduced risk in the integration of the system into the AFTI/F-16 aircraft. The few technical issues discovered were resolved in the laboratory environment, avoiding the risks of discovery on aircraft. Throughout the on-aircraft flight control integration and test, minimal EHAS changes were required.

The power and actuation flight demonstration provided key technical proof to mature MEA technologies for the JSF EMD phase. This demonstration tested the external S/G, the 270-VDC power distribution system, and the EHAs in a realistic aircraft environment. It provided valuable integration and installation data for thermal environment, EMI/EMC, and supportability, among other areas. The AFTI/F-16 modifications were successfully demonstrated in the aircraft, with the key highlights including:

- Nine flights accomplished totaling 13 flight hours, performing realistic JSF-like mission profiles
- Flight envelope including Mach 1.3 (600-kt engine limitation), up to 30,000 ft altitude (ferry)
- Test pilots reporting no observable difference in handling qualities, compared to a baseline F-16

For its achievements, the AFTI/F-16 flight demonstration won *Flight International* magazine's 2000 Aerospace Industry Award for Engineering, Maintenance and Modification, presented at the Paris Air Show in 2001 [40].

2. J/IST THERMAL/ENERGY MANAGEMENT MODULE

Boeing St. Louis (formerly McDonnell Douglas) led an industry team responsible for the subsystem architectures and ground demonstrations of power and cooling integration, and electric actuation and power system integration [11]. Major participants in the team were Honeywell Aerospace (formerly AlliedSignal Aerospace Inc.), BAE Systems (formerly Astronics Corporation), Moog Inc., Northrop Grumman, and Pratt & Whitney. The team achieved the following key accomplishments in the J/IST program:

1. Development of an integrated T/EMM system:
    a. T/EMM turbomachine
    b. High-temperature air/fuel heat exchanger
    c. Air/liquid heat exchanger
    d. Engine fan duct air-to-air heat exchanger
    e. Dual-mode recuperator heat exchanger
    f. T/EMM integral S/G
    g. T/EMM controller and vehicle management computer interface
2. Development of the T/EMM system controls and modes of operation
3. Demonstration of integrating the power and cooling subsystems into a stand-alone integration environment
4. Demonstration of the integration of the T/EMM integrated subsystems with the engine

The team selected a generic, JSF-like aircraft for assessing and defining the J/IST requirements. These included requirements for power, cooling, and actuation systems of the aircraft for ground, flight, and emergency operating conditions. The subsystem architecture that was developed resulted in the consensus architecture configuration depicted in Fig. 8. This architecture used the engine fan duct as a heat sink and integrated the functions traditionally performed by the ECS, Auxiliary Power Unit (APU), and Emergency Power Unit (EPU). The system was designed to support requirements for electrically powered flight control actuation and electric start of the main engine using T/EMM power to drive the ES/G.

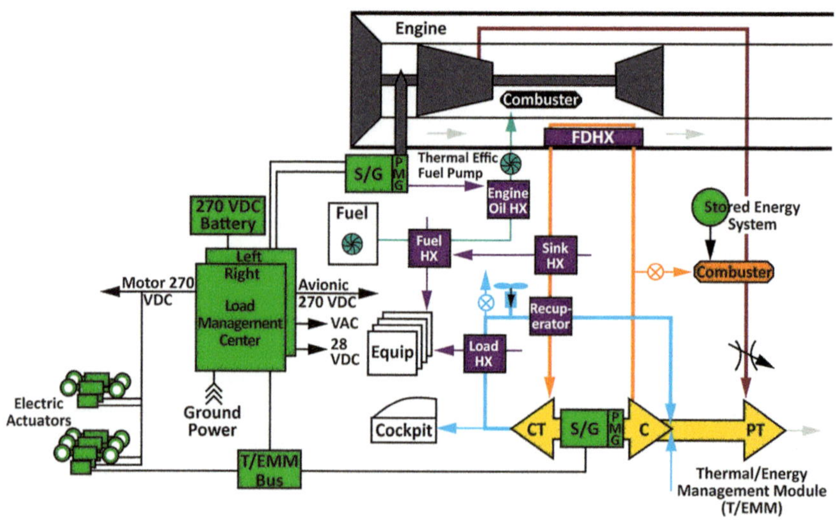

Fig. 8 J/IST T/EMM architecture.

Thermal/Energy Management Module Requirements

The T/EMM system architecture was intended to perform all functions normally accomplished by a traditional aircraft ECS, auxiliary power system, and emergency power subsystems. The T/EMM was required to provide additional capabilities expected to be required in the F-35, including cooling high-powered electronics. This requirement drove the requirement in J/IST for a backup cooling capability to accommodate inflight failures and provide active cooling for Navy hangar deck maintenance. The operating requirements created a need to develop multiple modes of operation and reconfigure the system between these modes as required to perform the differing functions. Consequently, the following operating modes were developed and demonstrated in the J/IST effort:

- Mode 1.0: Electrically powered hangar deck cooling
- Mode 2.0: Stand-alone T/EMM start
- Mode 3.0: Cooling and electrical power for ground maintenance
- Mode 5.0: Electrical power for main engine start
- Mode 6.0: Engine bleed air-driven cooling and triplex electrical power (normal flight)
- Mode 7.0: Engine bleed air-driven cooling and emergency electrical power (failure of the ES/G)
- Mode 8.0: Emergency electrical power (failure of the ES/G)
- Mode 9.0: Stand-alone emergency electrical power—stored air (engine failure at high altitude)
- Mode 10.0: Stand-alone emergency cooling and electrical power—ambient air (engine failure at low altitude)
- Mode 11.0: Shutdown
- Mode 12.0: Emergency cooling—fuel heat sink (not demonstrated in J/IST)

Thermal/Energy Management Module Turbomachine and Component Development

The design and development of a full-scale T/EMM turbomachine was accomplished by Honeywell Aerospace in Phoenix, Arizona. The turbomachine configuration incorporated a power turbine, compressor, SR Integral S/G (IS/G) rotor, and cooling turbine on a single spool that operated at speeds up to 60,950 rpm (Fig. 9). The compressor and power turbine were mounted on a single shaft, and the IS/G and cooling turbine were mounted on a separate single shaft. The two shafts were connected by a floating quill shaft, resulting in a design that provided sufficient margin from high-speed shaft bending modes. This split allowed each section to be developed and tested independently. The power turbine used a variable geometry stator with movable vanes that allowed for performance optimization at the various operating conditions. The turbomachine included a canister-type combustor developed specifically for the T/EMM that featured three distinct modes (tri-mode combustor) and was subsequently patented by Honeywell.

Additional components making up the T/EMM configuration requiring significant development were the SR I/SG configuration, T/EMM controller, and various other heat exchangers, valves, and related components.

In parallel with the Honeywell T/EMM development projects, Pratt & Whitney integrated the Honeywell-designed fan duct heat exchangers into a modified F119 engine. This was another key success supporting the claim that the T/EMM concept was viable and supported the JSF technology transition criteria.

T/EMM System Demonstrations

Several significant demonstrations were accomplished in support of the J/IST team effort. Figure 10 summarizes the major demonstrations, building up from the component level to successively higher levels of integration. One of the three most significant demonstrations was the integration of the power and cooling subsystems into a stand-alone integration environment. Another of the three was the electrical integration of T/EMM and ES/G, high-power EHA integration testing, and EMI testing. The third was the demonstration of the integration of the T/EMM integrated subsystems with the engine.

The stand-alone integration demonstration began in the summer of 1999 at Honeywell in Torrance, California. The demonstration, which simulated a mission profile, included ground maintenance operation with 25 kW of electrical

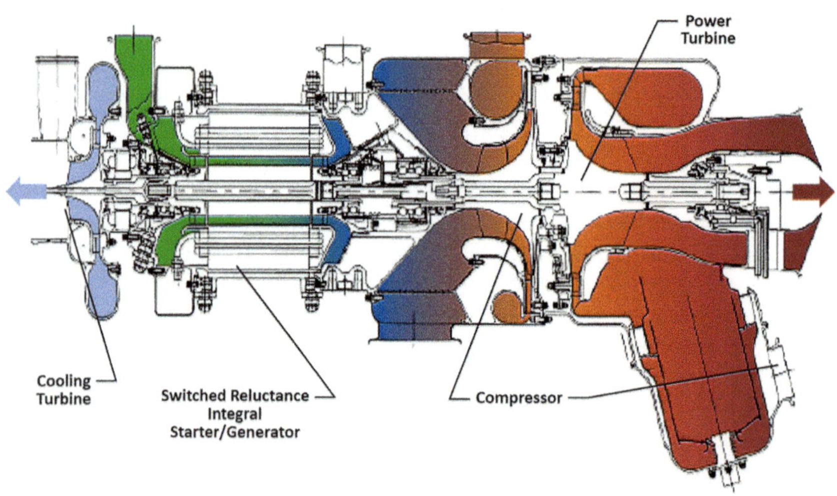

Fig. 9 T/EMM turbomachine cross-section.

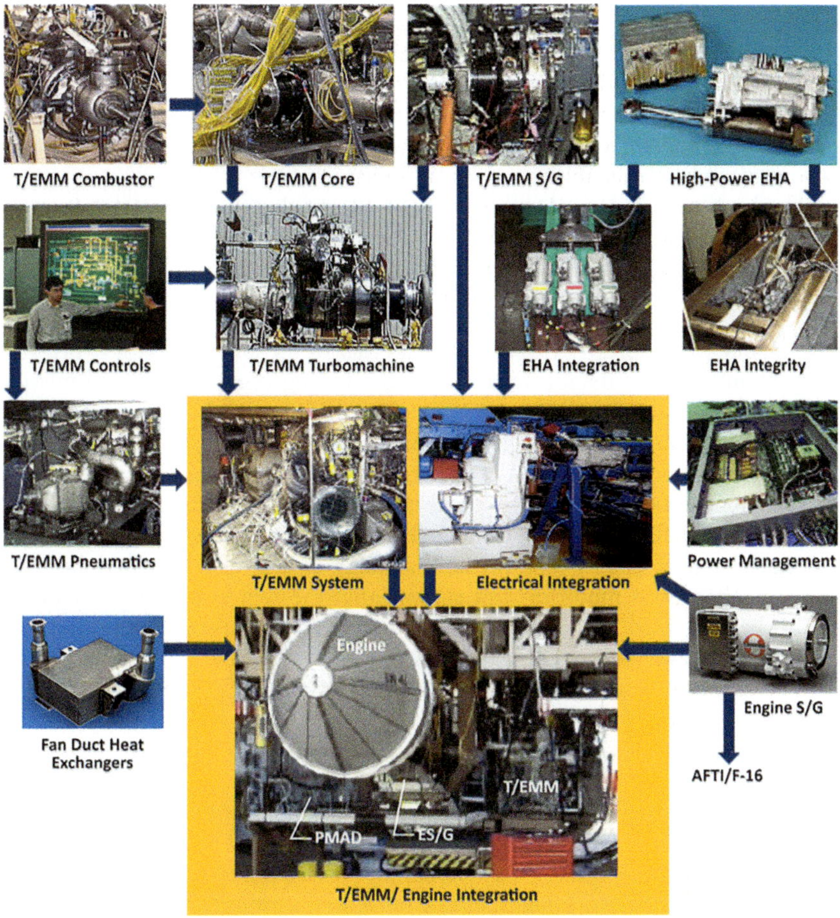

Fig. 10 Key T/EMM-related J/IST demonstrations.

power and 12 kW of cooling. It also included normal-engine-bleed air-driven operation during taxi, takeoff, acceleration, climb, cruise, loiter, descent, and landing. In addition, it showed emergency operations that simulated the loss of the engine and ES/G failure. The results demonstrated the capability to perform all the required modes and successfully reconfigure between modes as required. Post-test comparisons of test data with the dynamic model confirmed the model's ability to predict results. The comparisons also reinforced the value of the virtual modeling effort prior to commencing hardware operations.

The electrical power management integration of T/EMM and ES/G was conducted at Northrop Grumman's facility in El Segundo, California. The testing demonstrated that the IS/G-integrated EPS could provide the needed capability to power the ES/G using IS/G power to start the F119 engine. The testing also showed that the system could provide the necessary emergency power requirements, and that its generators could handle the dynamic loads imposed by the high-power EHAs.

The final engine integration demonstration was performed in 2000 at Pratt & Whitney's facility in West Palm Beach, Florida. It brought Honeywell's T/EMM system, Northrop Grumman's EPS, and Hamilton Sundstrand's (UTAS's) ES/G together with a modified Pratt & Whitney F119 engine. With this combination, it validated the integrated subsystems concept. The Hamilton Sundstrand (UTAS) ES/G was connected to the engine using a speed-increaser gearbox. The Honeywell T/EMM system was connected to the engine with low-pressure drop plumbing. The demonstration included testing using a simulated mission profile incorporating operations in all ground, flight, and emergency modes. Mission segments included taxi, climb, cruise, loiter, descent, dash, and combat. The demonstration validated the integrated subsystems concept by operating the system successfully in all required conditions. It showed that the T/EMM could be successfully driven by engine bleed air, and it demonstrated the successful integration of engine and airframe systems. It also demonstrated electrical engine starting and motoring and showed that the required emergency power could be provided within 50 milliseconds without using a battery. The maximum starting torque of 131 lb/ft using electric power furnished from the T/EMM was also shown.

C. TRANSITION TO THE F-35 PROGRAM

J/IST was one of the largest and most successful subsystems technology demonstration programs ever accomplished. Every major element of the J/IST architecture was transitioned to the F-35's design [39]. The program was composed of a series of cascading development and demonstration projects of increasing complexity. Each demonstration built upon a prior one to gradually reduce component, subsystem, and air vehicle–level risks. These efforts showed the various technology elements in the most appropriate environment, considering the desired risk reduction, available resources, and perceived potential. The two final J/IST demonstrations were the most significant. One was a comprehensive ground demonstration of the engine and relevant (associated) subsystems. The other was a flight test of an F-16 with all primary flight control surfaces powered electrically.

The F-35 employs the key elements derived from J/IST, including the EHAS and EPS architectures. It also incorporated numerous elements from the T/EMM architecture and equipment. The Pratt & Whitney F135 engine uses integral fan duct heat exchangers and additional bleed manifolds to support the Power and Thermal Management System (PTMS) configuration. The F-35 EPS features the Hamilton Sundstrand (UTAS) ES/G and General Electric 270-VDC and

28-VDC lithium-ion batteries. The Flight Control System (FCS) features Moog/Parker Aerospace EHAS hardware.

The resultant F-35 PTMS uses many of the control modes and requirements developed in J/IST. The components used in the system derive directly from the J/IST configuration. Many lessons learned in J/IST were applied to the design, development, and test of the turbomachine and its subcomponent systems, the heat exchanger arrangements, the valve designs, and other system aspects.

The F-35 turbomachine was redesigned to eliminate the variable area nozzle and replace the SR generator with a permanent magnet generator. The F-35 PTMS has a robust, highly reliable electrical power and cooling system, providing an electrically driven engine start system and supporting inflight emergencies. The F-35 aircraft subsystems enable stand-alone ground maintenance power and cooling for all systems maintenance and checkout operations. The single-stage power turbine was replaced with a two-stage radial/axial configuration, and the variable area power turbine nozzle was eliminated. The SR generator was replaced by a permanent magnet generator configuration. Numerous design lessons learned were incorporated into the lubrication system, sump sizing, rotor element design and clearances, system controls, and operation. As a result of J/IST, Honeywell developed significant improvements for the modeling and simulation techniques applied when developing the SDD program. The F-35 PTMS heat exchanger configuration used a thermal cycle similar to that in the J/IST system. However, significant optimization in the packaging was made by Honeywell. This resulted in a highly integrated compact heat exchanger configuration with multiple cores in a single assembly that proved essential to meeting the installation design requirements. F-35 design requirements also permitted the elimination of the stored-energy system and avoided the need to integrate the T/EMM turbomachine exhaust into the main engine exhaust. The requirements also mitigated other complexities discovered during J/IST.

1. F-35 FLIGHT CONTROL ACTUATION SYSTEM

The F-35 FCS uses the EHAS to power its primary and secondary flight control surfaces. This is a departure from predecessor legacy combat aircraft powered by hydraulics. The conventional system design was reliable and had a mature design concept, but it added substantial weight and volume and drove the hydraulics system's sizing and redundancy. The SUIT and MEA studies showed that the more-electric architecture promised significant improvements [4]. The F-35 flight control actuation system provides control to position both of the primary and secondary flight control surfaces. The FCS architectures are largely common for all three F-35 variants for the primary control of horizontal tails, flaperons, rudders, and leading edge flaps. However, the F-35A and F-35C variants also incorporate horizontal tail centering actuators, and the F-35C variant also incorporates conventional hydraulically powered ailerons. The architecture of the F-35 flight control actuation system derived from the MEA studies and J/IST demonstration program. Integrating the flight control actuation system with the power and cooling systems was key to the overall success of the F-35 flight control development. Additional discussion of this can be found in Ref. [35].

2. F-35 ELECTRICAL POWER SYSTEM

The F-35 EPS provides the generation, distribution, control, and protection of electrical power for various equipment. Its key power management functions for the air vehicle include primary flight control power, power to numerous aircraft systems, main engine starting, emergency power control and distribution, and power for ground maintenance functions. Key system elements include a single ES/G, two ICCs, 28-VDC and 270-VDC batteries, and the additional elements required to provide the uninterruptable power sources required for safety of flight. The EPS provides 270-VDC, 28-VDC, and 115-VAC power. The architecture of the F-35 EPS was derived from the MEA studies and J/IST demonstration program. Integrating the main engine starting function and flight control power sources matured risk reduction activities during the F-35 development program. One major change made during the SDD program was the replacement of the original SR machine ES/G with a synchronous generator to accommodate a broader power generation capability. The F-35 EPS design and development is further explored in Ref. [35].

3. F-35 POWER AND THERMAL MANAGEMENT SYSTEM

The F-35 PTMS, sometimes referred to as the Integrated Power Package (IPP), integrates the conventional functions of the ECS, emergency power system, engine starting system, and auxiliary power systems into a single, highly integrated system. The system features two primary modes of operation: stand-alone combusted-mode operation and bleed-driven operation when the F135 engine is running. The primary power and cooling enables stand-alone ground maintenance with no required external power and cooling carts. The PTMS also provides primary power for on-ground engine starting, followed by seamless reconfiguration into bleed-driven operation to support flight operations. The PTMS likewise supports inflight emergencies and automatically reconfigures into combusted-mode operation to support flight control, emergency electrical power, and inflight engine-start assist power.

The PTMS provides 270-VDC and 28-VDC electrical power, as well as forced-air cooling for flight-essential systems. It also provides liquid cooling for aircraft avionics systems and pressurization for the cockpit, fuel system, and other aircraft systems. During flight, waste heat generated by onboard systems is rejected overboard via engine-mounted heat exchangers embedded within the F135 engine fan air duct. This eliminates the weight and volume penalties associated with conventional ram air heat-sink systems.

The system incorporates many elements of the architectures developed in the J/IST demonstration program and MEA studies; however, substantial additional system design maturation was accomplished during F-35 development. Several of the concepts studied in J/IST were not adopted because they were not needed to meet the program requirements. Among these were a stored-energy backup system and an integrated T/EMM exhaust system integrated with the main engine exhaust [35]. The resultant configuration provides improvements in aircraft range due to reduced bleed air consumption and improved thermal management. The integrated PTMS eliminates the need for the separate aircraft-mounted accessory drive gearbox, air turbine starter, EPU, APU, and ECSs employed in legacy aircraft. The fan duct heat exchanger configuration avoids the increased weight and drag penalties and LO impacts associated with a dedicated ram air circuit.

III. PROPULSION TECHNOLOGIES

A. OVERVIEW

The conventional F-35 propulsion system features critical propulsion/airframe integration technologies that began with technology development programs. These technologies were used in large-scale demonstrations, incorporated onto the X-35 demonstrator, and transitioned to the production F-35 program. Key propulsion features (Fig. 11) include a DSI and LO axisymmetric nozzle configuration (LO Axi) common among all variants. The F-35B STOVL variant includes a

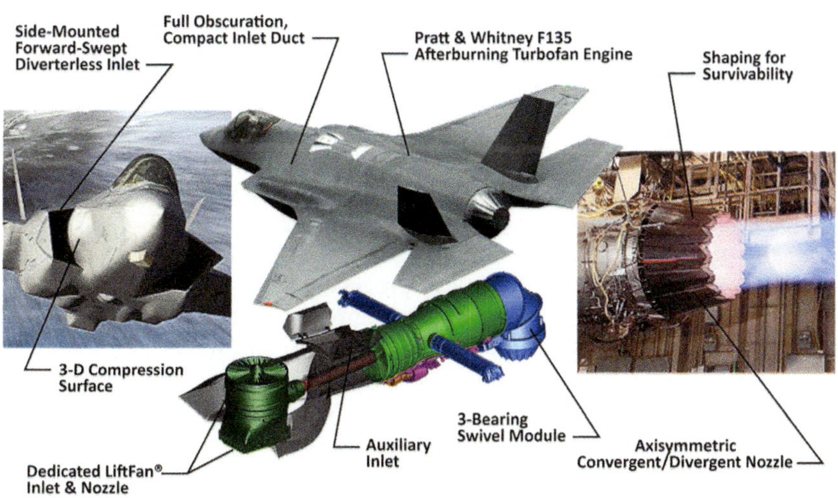

Fig. 11 F-35 advanced propulsion system technologies.

three-bearing swivel module (3BSM) that integrates the LO Axi configuration with thrust vectoring, providing vertical lift. The lift is augmented by a shaft-driven lift fan system providing additional vertical lift. The DSI design is characterized by a detailed, shaped compression surface and forward swept cowl, with twin inlet apertures feeding a bifurcated, serpentine duct. The design eliminates the need for boundary-layer diverter or bleed-system inlet subsystems, reducing cost and weight. The LO Axi configuration minimizes radar reflections by using serrated trailing edges, a serrated interface with the airframe, and tight gap and seam control. It also has specialized high-temperature coatings on internal and external surfaces to provide an excellent balance of signature, weight, and performance requirements.

The F-35B STOVL propulsion system utilizes the same main engine turbomachinery as the Conventional Takeoff and Landing (CTOL)/Carrier Variant (CV) configurations with the addition of a shaft-driven lift fan, a roll control system, and an auxiliary inlet. The 3BSM adds thrust vectoring to the LO Axi nozzle. This permits engine exhaust to either pass straight through for forward propulsion in conventional flight or be deflected downward to provide aft vertical lift. The 3BSM can move through 95 deg of motion seamlessly with no change in engine operation. The nozzle also provides yaw control during hover and transitions to hover. Additionally, the F-35B STOVL variant incorporates Rolls-Royce's shaft-driven LiftFan® system. In lift mode, horsepower is extracted from the F135 engine's low-pressure turbine via a drive shaft, clutch, and gearbox, and is used to drive the LiftFan. The exhaust is discharged through a thrust vectoring nozzle on the underside of the aircraft to provide balanced lift with the 3BSM. Engine fan air ducted to outboard roll nozzles provides roll control.

B. F-35 DIVERTERLESS SUPERSONIC INLET

The F-35's primary engine air induction system features a DSI design conceived and matured through several technology programs and then transitioned to the JSF program. The DSI is characterized by a 3-D compression surface and forward-swept cowl (Fig. 12). These features enable high aerodynamic perforce, boundary layer diversion, and inlet stability without using a boundary layer diverter or bleed system. Eliminating these inlet subsystems reduces cost and weight compared to prior state-of-the-art designs. Although many different configurations were explored in the technology programs, the F-35's DSI is embodied in twin, side-mounted inlets that feed a compact bifurcated duct. A detailed history of the DSI concept can be found in Refs. [21] and [22].

1. BACKGROUND

Tactical aircraft have always posed a formidable challenge for engine inlet designers, and the incorporation of modern affordability and survivability requirements have made the challenge more difficult [23, 24]. The inlet must

Fig. 12 F-35 engine air induction system features.

provide the engine with high-quality (high pressure, low distortion) airflow over a wide range of speeds, altitudes, and maneuver conditions. At the same time, it has to accommodate the full range of engine airflow from idle to maximum afterburning power. The inlet designer must also consider the constraints imposed by configuration features, such as nose landing gear, weapon bays, equipment bays and access panels, and forebody shaping.

In addition to these general considerations, two key aerodynamic requirements are at the forefront in the design of any supersonic inlet system. The first requirement is for flow compression. The inlet system must reduce the airstream's speed while increasing its static pressure as airflow approaches the engine. For combat aircraft, this is usually done with a series of external shockwaves and internal flow area expansion. As freestream speeds approach Mach 2, elaborate compression schemes, including movable compression ramps, were historically used to reduce losses and enable high inlet efficiency.

The second key issue is Boundary Layer Control (BLC). This is the means by which the inlet system will account for a layer of low-energy air that forms on the surface of the fuselage and compression surfaces. This must be managed at both subsonic and supersonic speeds. The boundary layer can create chaos when disturbed by shockwaves created during flow compression. Shockwave/boundary layer interaction can lead to severe airflow distortion at the engine face, which may subsequently lead to engine stall. Several methods can be used for BLC. The inlet can be physically isolated from the fuselage by a boundary layer diverter, a feature found on most of today's combat aircraft. Another primary technique is boundary layer bleed. Bleed systems may be fully fixed or involve mechanical variation, such as movable exit louvers, to optimize performance. Many of today's tactical aircraft use a combination of bleed systems, diverters, and compression ramps.

Variable compression and bleed systems can provide the aerodynamic functionality required for a high-performance inlet. However, such features also introduce mechanical and structural complexity, weight, and cost into the system [24].

2. DIVERTERLESS SUPERSONIC INLET CONCEPTUAL DEVELOPMENT

In the early 1990s Lockheed Martin began an IRAD project to develop a Mach 2 class combat aircraft inlet concept. The concept would embody traditional aero-performance levels and advanced survivability features. Further, it would improve affordability (reduced cost and weight) compared to state-of-the-art design concepts. To meet these goals, the concept would need to incorporate flow compression and BLC functionality, advanced shaping, high structural efficiency, and minimal or no moving parts. These studies were conducted primarily with Computational Fluid Dynamics (CFD) tools augmented with small-scale wind tunnel testing.

The DSI emerged as the preferred concept early in the IRAD program. It is distinguished by two main physical features: a fixed, 3D compression surface (bump) and an edge-aligned, forward-swept cowl. The bump compression surface derives from the flow field produced by a reference axisymmetric body in supersonic flow. The reference body (virtual cone) may be a simple cone, a double or isentropic cone, or any of these bodies at incidence angle to the oncoming stream. In the latter case, the flow field is 3D, not axisymmetric. A set of CFD particle traces are released along a locus of points representing the intersection of the shock field and aircraft surface. As the particles travel into the shock field, they are deflected away from the virtual cone by internal flow field pressure gradients. A 3D contour is then defined by a surface faired through the particle traces. When introduced into an identical supersonic flow field, this contour produces a shock structure identical to that of the virtual cone. The bump surface not only achieves flow compression, but also creates a span-wise static pressure gradient that assists with boundary layer diversion (Fig. 13).

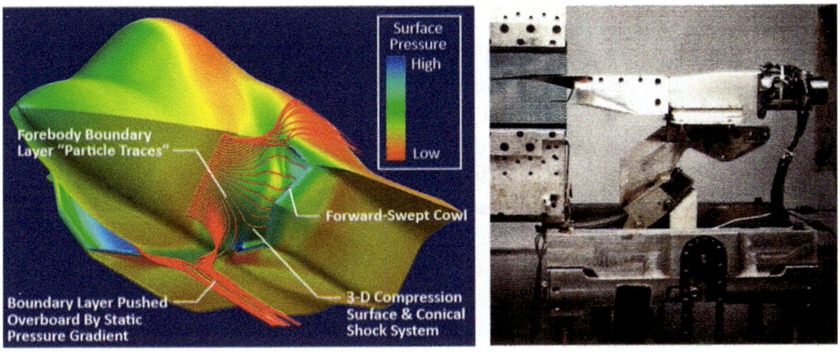

Fig. 13 CFD simulation of DSI supersonic boundary layer diversion.

The second physical feature distinguishing the DSI is a forward-swept cowl. *Forward swept* denotes that the cowl structure is cantilevered from the aircraft and closes against the forebody at its aft-most points. This geometry enables low-pressure boundary layer air to be spilled out the side of the inlet as mass flow ratio is decreased. Many CFD studies showed the effects of combining a bump compression surface and forward-swept cowl. The studies demonstrated that the combination could provide the aerodynamic functionality of traditional inlets without a boundary layer diverter or bleed system.

Small-scale testing of isolated inlet aperture models (Fig. 13) was conducted throughout the conceptual development program to augment CFD studies. Whereas CFD was best at providing detailed flow physics at a few specific operating conditions, testing produced key operating data at a broad range of conditions. These tests were used to evaluate inlet pressure recovery, distortion, and shock-wave instability (buzz) boundaries for the key bump design parameters.

3. INTEGRATED AIRCRAFT DESIGN AND VALIDATION

Having verified the basic DSI principles, the focus shifted to integrating the inlet into a realistic aircraft configuration and assessing air vehicle system-level impacts. These activities were accomplished with synergistic studies on the JAST and AFRL ACIS programs. JAST studies investigated the best way to integrate the unique DSI features. As with the eventual F-35 configuration, the JAST aircraft featured twin, side-mounted inlet apertures with a bifurcated duct and single engine. On ACIS, Lockheed Martin performed three tasks that helped to build the overall knowledge base. We analyzed the benefits, cost, weight, and maintainability of the DSI vs a reference system. We also assessed the performances of different subsonic diffuser concepts integrated with the DSI. In addition, we performed a large-scale wind tunnel validation of a forebody/inlet/duct configuration, shown in Fig. 14.

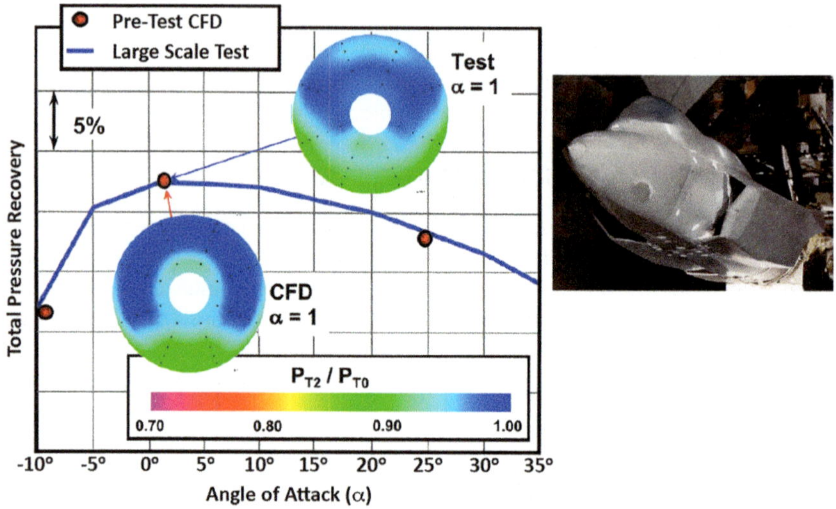

Fig. 14 Large-scale wind tunnel testing validation of CFD design methodology and CFD test model (inset).

System-level trade studies were performed to quantify the weight, cost, and benefits of the DSI, compared to more conventional inlets (e.g., F-22 and F/A-18E/F caret inlet systems). In these studies, a 30% inlet weight reduction was estimated for the DSI, relative to the reference caret inlet. The largest contributing factor was the elimination of the bleed and bypass systems. Studies performed by other ACIS contractors [25] indicated similar savings for diverterless/bleedless systems.

The ACIS subsonic diffuser study evaluated the DSI with both straight and serpentine diffusers and different flow area curves. Results from this task provided guidelines on how to best shape the diffuser, given the rather unique incoming flow field from the DSI.

A 0.125-scale inlet/forebody model of the ACIS configuration was tested at subsonic, transonic, and supersonic conditions in two test entries. A wind tunnel test was conducted at the NASA Langley Research Center 16-Foot Transonic Tunnel. Another was conducted at the Lockheed Martin High Speed Wind Tunnel, which is a 4-by-4-ft tri-sonic blow-down wind tunnel. Test data were obtained over a wide range of angle-of-attack and angle-of-sideslip conditions at speeds up to Mach 2. Test results indicated that the integrated inlet configuration satisfied typical goals of high pressure recovery, manageable distortion, and good supersonic stability. Pressure recovery for the DSI with a serpentine duct was shown to be comparable to that of the F-16 modular common inlet duct recovery. This was at static conditions and speeds ranging from Mach 0.6 to 1.2. At supersonic speeds, the DSI's recovery was shown to exceed F-16 recovery.

The development of the DSI was based on CFD. As such, it was necessary to verify the accuracy of CFD to predict the complex inlet flow fields associated with integrated forebody, aperture, and duct geometries. In general, excellent agreement between the CFD and test data was noted (Fig. 14).

4. DIVERTERLESS SUPERSONIC INLET FLIGHT DEMONSTRATION PROGRAM

The next step in the development of the DSI was to conduct a flight demonstration on a relevant combat aircraft platform. A Block 30 F-16 powered by an F110-GE-129 engine was selected for this demonstration due to its modular inlet construction and consequently low cost of modification. One objective of this effort was to demonstrate engine/inlet compatibility throughout the combat aircraft envelope. Another was to demonstrate stable inlet operation up to Mach 1.8. Although integration in the F-16 chin location dictated an inlet design different from the twin, side-mounted ACIS/JAST design, the aerodynamic design principles were identical.

Prior to the flight test, an inlet wind tunnel model was fabricated and tested at one of the Arnold Engineering Development Complex propulsion wind tunnels in November 1995. Wind tunnel test results verified CFD-predicted inlet performance and demonstrated engine/inlet compatibility. Inlet lines were frozen based solely on CFD results, and flight hardware fabrication had begun prior to any wind tunnel test verification of the CFD-based design. The flight test program consisted of 12 flights flown in nine days in December 1996. The flight tests covered the entire F-16 flight envelope and achieved a maximum speed of Mach 2 (Fig. 15). With this flight demonstration, the viability of the DSI as a Mach 2 class, diverterless, bleedless, highly survivable inlet concept was proven.

Fig. 15 DSI technology was matured through rigorous F-16 flight demonstration.

5. TRANSITION TO THE JSF PROGRAM

Risk had been significantly reduced through IRAD efforts, ongoing work and the flight demonstrations on ACIS, and extensive program trade studies. As such, the DSI system was selected as the baseline inlet for Lockheed Martin's JAST program in early 1995, replacing the previous F-22-like inlet. Configuration refinement continued through the lines freeze on the X-35 concept demonstrator aircraft. Because of the sophisticated yet simple DSI design, it was possible to support this milestone with only a single additional up-and-away inlet wind tunnel entry.

C. LOW OBSERVABLE AXISYMMETRIC NOZZLE

Before the development of the F-35 Low Observable Axisymmetric Nozzle (LOAN), signature demands typically drove nozzles to fixed, structurally integrated affairs (e.g., F-117). They had to have very high aspect ratio designs (e.g., F-117) or highly capable but heavy two-dimensional systems (e.g., F-22), as illustrated in Fig. 16. Departing from what was then the state of the art, industry and CRAD efforts developed multiple nozzle configurations to create a LOAN for the F-35. The F135 engine with a LOAN balanced the requirements of LO and efficient aeromechanical performance. This resulted in a lightweight configuration with reduced radar cross-section.

1. BACKGROUND

The F-16's and F-15's exhaust systems are examples of what was used in classic 4th-Generation tactical supersonic aircraft. These systems are axisymmetric for low weight and structural efficiency, with variable geometry to maintain stable and efficient engine operation. Specifically, afterburning requires a large increase in the minimum nozzle flow area (throat area) to retain engine stability. This is particularly necessary when fuel is being dumped into the exhaust and the resultant flow density decreases. These convergent/divergent nozzles are composed of overlapping flaps and seals. The exit area and throat area are mechanically linked and scheduled with a power setting for efficient flow expansion. Externally, overlapping flaps provide a fairing between the air vehicle aft body and nozzle exit for

Fig. 16 A spectrum of fighter aircraft nozzle designs.

reduced drag. However, the nozzles were not generally considered capable of meeting LO requirements.

The subsonic F-117 stealth fighter and B-2 bomber exemplify how exhaust system designs can be dominated by LO features. This effect results in nontraditional exhaust systems that are driven by air vehicle shaping. The F-117 incorporated an airframe-mounted, structurally integrated, fixed exhaust system with planform-aligned edges. The exhaust system transitions from axisymmetric to a high-aspect two-dimensional design and forgoes the variability needed to accommodate afterburning. Although these highly integrated designs can reduce drag, they are less aerodynamically and structurally efficient than axisymmetric designs.

IRAD and CRAD investments were critical to positioning the industry to develop world-class solutions targeted at major program innovations. Well before the JSF X-planes were contracted, multiple vertical lift and exhaust system configurations were developed and matured. These were then available on call to meet the evolving needs of both new and existing aircraft designs. Some of the nozzle technologies explored were conformal fully fixed aperture nozzles, fixed aperture nozzles, and LOANs.

Complexly shaped, fixed-geometry nozzles were matured for reduced drag and signature. These designs allowed flexibility to implement full line-of-sight obscuration. With few moving parts, they allowed for a great diversity of LO and thermally tolerant materials.

Innovations in convergent section mechanical manipulation and shaping were explored as well. These were intended to reduce leakage, weight, and structural integration penalties while enabling the nozzle exit to remain motionless. Throat skewing and inducing shocks in the divergent section were assessed for thrust vectoring potential. Variable cycle engines were evaluated to reduce the need for large-scale nozzle throat area control, even during augmentation.

During this period the notion that axisymmetric nozzles were not amenable to LO (or even thrust vectoring) was being challenged. A new generation of LO nozzles emerged that was characterized by shaping features to minimize radar reflections. Namely, they used a serrated trailing edge, serrated interface with the airframe, and interleaved external seals to complement the external flaps. They retained tight control of gaps and seals and had specialized high-temperature coatings on internal and external surfaces. In the years leading up to the X-35, various LOAN configurations were developed, ground tested, and flight tested on F-16s. Testing was done in concert with both Pratt & Whitney F100 and General Electric F110 engines. These versions incorporated a nod to an ejector feature that introduced nacelle bay airflow near the nozzle throat to cool divergent flaps. This improved divergent seal durability and reduced the infrared signature.

2. LOW OBSERVABLE AXISYMMETRIC NOZZLE RAPID PROTOTYPE EVALUATION

In the early 1990s, engineers at Lockheed Martin used CFD and an astatic thrust measurement facility to perform rapid prototype evaluation and 3D printing. This

Fig. 17 Lockheed Martin rapid nozzle prototyping capability in Fort Worth, Texas.

allowed for a rapid assessment of nozzle aerodynamic performance (Fig. 17). The facility had a flow-through six-component balance housed within an altitude (pressure reduction) chamber. Varying ambient backpressure allowed continuous, low-load testing with very high nozzle pressure ratios with minimal variation in mass flow and minimal model loads. This also allowed the balance to operate within an optimum band of its calibrated mass flow and force measurement range. It had excellent overall accuracy and repeatability and reduced the variation in the Reynolds number. Multiple airflows could be independently controlled and metered via a bank of critical-flow venturis.

3. AIRCRAFT AERODYNAMIC INTEGRATION

Intentional aft body and nozzle integration is critical to balance drag and weight, which are driven by aft body and nozzle boattail angles. Finding the optimum length and external shape is important to net propulsion performance. Additionally, the aft body is typically not axisymmetric, and integrating an axisymmetric exhaust may result in large aft body boattail angles and base regions. This would introduce drag, effectively stealing from the net thrust of the propulsion system. This is evident in most twin-engine configurations in which a base or dead region is normally found between two closely spaced nozzles. In certain circumstances, the low local pressure endemic to these base regions can serve as an exit for secondary flow systems. For example, base regions at the root of the F-16 horizontal and vertical tail surfaces are used to promote nacelle ventilation.

4. GROUND AND FLIGHT TESTING

The LOAN configuration developed by Pratt & Whitney was developed under a precursor to the JSF contract to evaluate advanced, affordable technologies applicable to the F-35. Ground testing was completed in 1996 (Fig. 18). A Lockheed Martin/Pratt & Whitney team modified an Air Force F-16 and F100-PW-200 engine with a bailed LOAN from the F-35 Joint Program Office. This was used for the ground test in a two-day rapid prototype operation. During tests from idle to maximum afterburner, infrared images, nozzle temperatures, and nacelle inlet pressures and airflow velocities were measured. The Pratt & Whitney LOAN configuration significantly reduced radar cross-section and infrared signature emissions from the engine, as well as maintenance costs. The result was a low-cost nozzle system that reduced the chance of radar and infrared detection and applied to both retrofit and new-production aircraft.

The solution developed by General Electric was designated the LO Axi Nozzle, demonstrated by Lockheed Martin on an Air National Guard F-16C (Fig. 18). Dramatic temperature reductions provided by the LO Axi Nozzle were expected to greatly improve F-16 exhaust system durability. F-16 flight certification was conducted at Edwards Air Force Base in the summer of 2001. That system was offered as an F-16 upgrade option, reinforcing the vibrancy of the F-16 platform.

These F-16 LO nozzle configurations included an ejector feature to address thermal degradation of nozzle divergent section seals and flaps. That remains a

Fig. 18 General Electric LO Axi Nozzle (left) and Pratt & Whitney LOAN (right) ground and flight testing on the F-16.

Fig. 19 Radio frequency test fixture with the F-35 LO nozzle at the Pratt & Whitney facility in West Palm Beach.

top driver of maintenance on modern jet engines. The ejector provided effective film cooling to reduce nozzle temperatures and improve component life using engine nacelle bay bypass air. These techniques were expected to double or quadruple nozzle divergent flap life, resulting in significant maintenance cost savings.

5. RADIO FREQUENCY TESTING

For the F-35, the ability to see before being seen provided a fulcrum upon which to balance sensor/radar capability and LO technologies; strengths in one area provided flexibility in the other. The strength of emerging technologies in the radar and sensor suites opened up options for using the LO nozzle. This was comparable to the manner in which the development of high off-boresight missiles reduced the appetite for thrust vectoring and nose pointing.

Radar testing was employed for the JSF program to evaluate the unique shape and special coatings evident in the Pratt & Whitney LOAN. The configuration achieved stealth through a combination of technologies, including geometric shaping, advanced cooling, and special coatings on internal and external surfaces. One major radio frequency model (Fig. 19) developed under IRAD was suitable for integrating inlets, nozzles, apertures, edges, and other subsystems. It was designed for installation on the 30,000-lb rotator at the Meridian Antenna Test Range in Meridian, Texas.

6. TRANSITION TO THE JSF PROGRAM

For the F-35, the Pratt & Whitney F135 engine and LOAN balanced the requirements of LO and efficient aeromechanical performance. It offered a lightweight (especially for the F-35B STOVL variant), low-cost configuration. The F-35A and F-35C variants use the same nozzle configuration. A shorter version was readily configured for compatibility with the F-35B STOVL 3BSM to meet ground clearance needs while landing vertically. Because the engine exhaust system is the primary contributor to aft sector infrared signature, engine and nozzle design needed to incorporate effective methods to reduce infrared emissions. This was accomplished using reduced radar cross-section-compatible techniques, including hiding, shaping, and temperature control. The F-35 exhaust system employs a cooled turbine face blocker, effectively eliminating the temptation to employ more impacting techniques like a serpentine exhaust duct. The F135 exhaust system does use a cooled nozzle to significantly reduce the aft sector infrared signature. With these techniques, the cooled blocker and nozzle tail-on infrared signature is significantly less than the signature of an uncooled exhaust system.

D. F-35B STOVL LIFT SYSTEM

1. BACKGROUND

For more than 50 years, fighter aircraft designers have vigorously pursued the speed and range of a conventional jet while achieving the basing flexibility of Vertical Takeoff and Landing (VTOL). Numerous STOVL concepts have been developed over the decades, all with compromises that limited the effectiveness of the aircraft. The F-35B STOVL lift system successfully achieved a breakthrough that redefines the relationship between conventional thrust and vertical propulsive lift. Moreover, it achieves that with major increases in performance, efficiency, and safety. This elegant integration results in a relatively simple engine-driven LiftFan. It has an enabling engine powerful enough to achieve a lift-to-thrust ratio of approximately 1.5-to-1 (Fig. 20)—a significant increase over direct lift designs. The shaft-driven LiftFan provides high levels of thrust augmentation

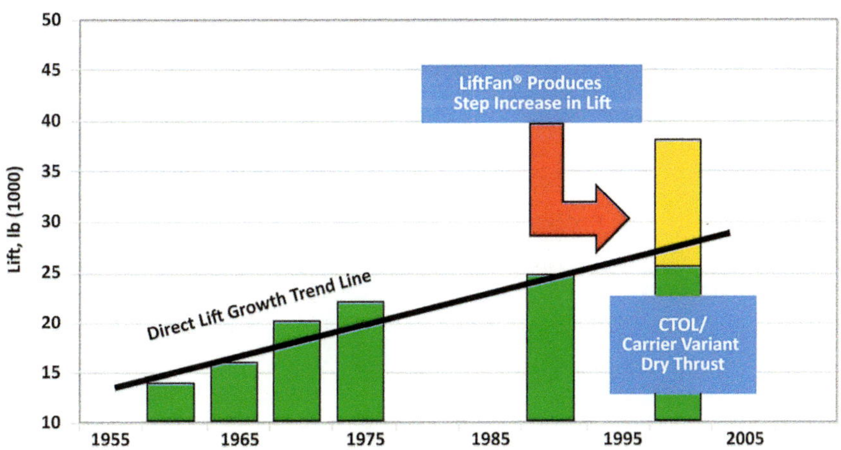

Fig. 20 Revolutionary step increase in vertical lift.

with a cool, low-pressure footprint, sufficient control power, and efficient packaging in the airframe design. The main engine is primarily optimized for conventional flight, so the propulsion system performance is not compromised for its vertical lift capability. The LiftFan augments vertical flight similarly to the way an afterburner augments high-speed performance [31]. The LiftFan provides an additional ingenious benefit: the (relatively cool) thrust exhaust protects the main engine inlet and forward portions of the aircraft from hot gas re-ingestion or damage.

In addition to achieving powerful lift thrust, a STOVL aircraft must achieve sufficient control power in each axis to successfully transition through the wingborne, semi-jetborne, and jetborne flight phases. The F-35B STOVL lift system accomplishes this through several key components (Fig. 21):

1. *LiftFan clutch and driveshaft:* To selectively transfer power from the main engine to the LiftFan
2. *Variable Area Vane Box Nozzle (VAVBN):* To control the LiftFan exit area and fore-aft thrust vectoring
3. *Roll post nozzles:* To redirect main engine fan air through under-wing nozzles for roll control
4. *3BSM:* To vector the main engine nozzle fore–aft and laterally for yaw control

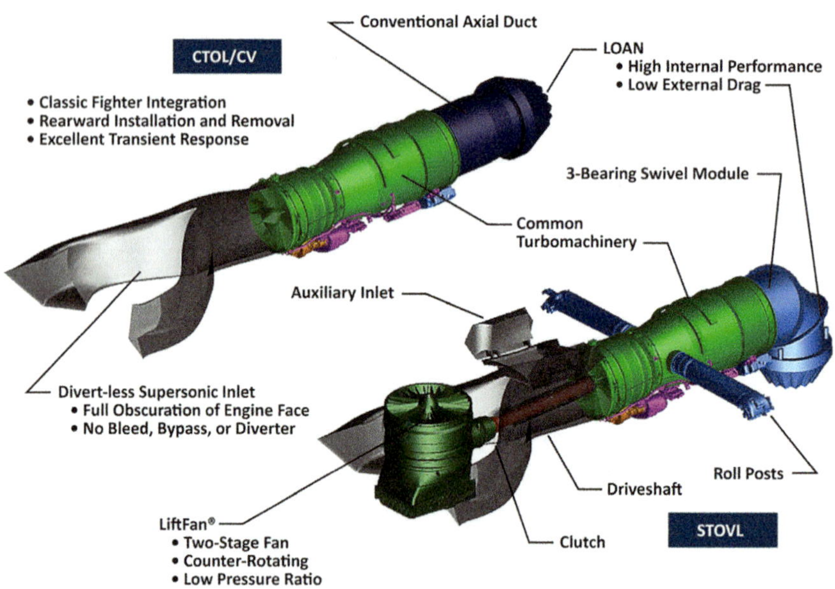

Fig. 21 Comparison of F-35 conventional and F-35B STOVL lift systems.

2. LIFTFAN DEVELOPMENT

The F-35 LiftFan system is the overarching characteristic of the F-35B STOVL variant. It underwent years of technology development and maturation by Lockheed Martin and F-35 propulsion contractors Pratt & Whitney and Rolls-Royce. Initial work began in the late 1980s with STOVL JSF studies sponsored by the Defense Advanced Research Projects Agency. Lockheed Martin, General Dynamics, Boeing St. Louis (then McDonnell Douglas), and Boeing all developed concepts with different technologies for generating vertical lift [34]. These studies led to the ASTOVL competition that Lockheed Martin won with the shaft-driven LiftFan propulsive concept. This effort eventually evolved into the JSF concept demonstration phase resulting in the X-35B flight demonstration.

Rolls-Royce's LiftFan is a novel, counterrotating concept with a bladed disk (blisk), two sets of stationary vanes, and a set of Variable Inlet Guide Vanes (VIGVs). The VIGVs provide the thrust variation from maximum to idle necessary for the VTOL application. The gearbox distributes 29,000 hp to the LiftFan rotor stages. The load capacity and envelope characteristics were key to providing an industry-first 30-to-1 horsepower-to-weight ratio. The previous norm (in earlier aircraft) was a ratio of no more than 15-to-1, which was then doubled.

The gearbox is integral to the LiftFan unit and employs counterrotating output shafts to simplify geometry and reduce gear and bearing loads. VIGVs on the first fan stage provide thrust modulation. Lubrication for the LiftFan bearings and gearbox is provided by the LiftFan lubrication system, which is independent from the main engine lubrication system. The Rolls-Royce LiftFan is designed to operate throughout the entire speed range of the main engine.

One of the key challenges in transitioning the concept development to production was in the LiftFan's aeromechanical rotor modes. These caused operating restrictions (time at certain LiftFan speeds) on the X-35B. The spatial pressure distortions in the inlet flow field excited resonance modes in the LiftFan turbomachinery, becoming a high-cycle fatigue or aeromechanics concern. This was addressed in the F-35B by redesigning the upper LiftFan door configuration to reduce flow angularity and distortion. It was also addressed by redesigning the LiftFan rotor (hollow blades, blisk) that attenuated the modal responses.

3. LIFTFAN CLUTCH DEVELOPMENT

Shaft/clutching functionality was achieved with both hardware and software functionality. Pioneering shaft, clutch, and gearbox designs permitted the development of a lightweight, high-speed (8000 rpm) drive train. A unique closed-loop clutching system provides precise control resulting in smooth, reliable power transmission to the LiftFan. This innovative clutch design, leveraging aircraft brake technology, produced a dry clutch plate arrangement. This achieved the required rapid engagement performance time, while providing durability that exceeded program requirements. The clutch is mounted to the LiftFan, with the input directly coupled through the main drive shaft and couplings to the main engine low-pressure rotor shaft. The clutch consists of a pack of dry disk plates. When driven together by aircraft-powered hydraulic actuation, the pack couples the main engine low-pressure rotor via the drive shaft to the LiftFan. Lubrication for the clutch bearings is provided by the LiftFan's lubrication system. The driveshaft couplings can flex to take up misalignment between the main engine and the clutch.

The LiftFan clutch allows the engagement and disengagement of the LiftFan from the main engine. It achieves this through two devices, each providing a torque path from input to output. During engagement, speed synchronization and acceleration of the fan rotors at low power is achieved by applying pressure to a pack of five carbon-carbon plates, operating dry. Subsequent engagement of a locking spline is required for high power transmission. Engaging the splined lock requires synchronizing the clutch input and output shaft speeds within a few rpm. An indexing mechanism insures against a failure to engage due to mating splines contacting end to end. During disengagement, the clutch plate pack unloads the splines to enable them to be retracted.

One of the key challenges experienced during the X-35 development was obtaining smooth clutch engagement with minimal transition time. Early clutch control design encountered a chatter phenomenon as the clutch plates came in contact. Through innovative closed-loop control modes, a combination of clutch clamping force and longitudinal position feedback solved the chatter problem, permitting smooth and precise engagements. Continued maturation during the F-35 program intended to complete the conversion in the minimum time (operational flexibility) and obtain a full-life clutch (minimized maintenance interval). The F-35 clutch can complete an engagement cycle within nine seconds from command to engage. With improved clutch plate material, the system will accommodate more than 1500 engagements.

4. VARIABLE AREA VANE BOX NOZZLE DEVELOPMENT

Prior to the F-35 development phase, the X-35 LiftFan nozzle vectoring was accomplished via a three-hooded telescoping nozzle. Although very precise in directing the LiftFan thrust vector, it was heavy, required a lot of volume, and was difficult to integrate into the aircraft signature. This prompted the pursuit of a more compact design through a series of parallel vanes that could be hidden behind lower fuselage doors. The development of the F-35 VAVBN capitalized on five years of prior efforts by Rolls-Royce on vane box nozzles. These had been developed for earlier lift engine concepts that were considered in the JAST program [35]. It was tested in a 27%-scale demonstration (Fig. 22).

Many nozzle design variables were studied including duct geometry, and the number, spacing, and profiles of the movable vanes. Additional design parameters that influenced nozzle integration with the LiftFan included the gearbox profile, the location of six support struts, and the size and location of

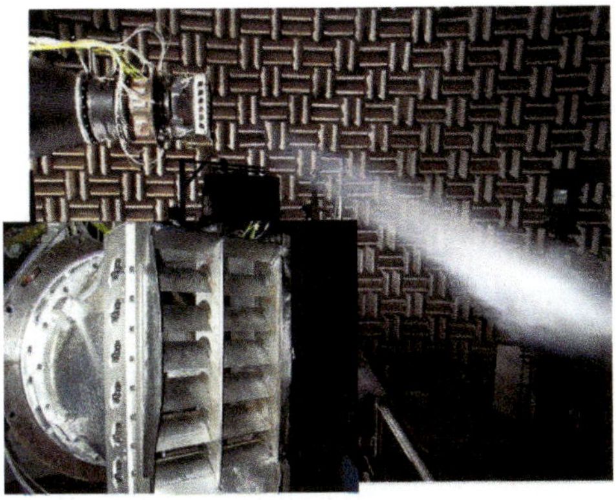

Fig. 22 Twenty-seven-percent-scale F-35B STOVL VAVBN test, with (inset) VAVBN close-up.

the vane actuator mechanism. Studies were also conducted to evaluate the tradeoffs of performance, vane actuation, and airframe integration. From these it became apparent that the vane box configuration with six highly cambered vanes with a low thickness-to-chord ratio was most promising. In addition to supporting flow path pressure, vane aerodynamics, and actuation loads, the nozzle box is designed to contribute to the airframe structural stiffness. The nozzle vane box is airframe-mounted, with the vane box sidewalls serving as aircraft structure keel members.

The VAVBN (Fig. 23) provides directional control of the LiftFan thrust vector and an additional effector to the VIGVs for turndown. *Turndown* refers to the commanded position of the lift fan inlet guide vanes used to control lift fan thrust. Three VAVBN vanes are driven by dual-tandem, linear, hydraulic actuators. Drive is transferred to the other three vanes through bar linkages. With this system, nozzle thrust may be directed in an arc of 41.75–104 deg (fore/aft aircraft coordinate system), at a rate of 40 deg/s. Independent control of the three VAVBN actuators provides the capability to vary the nozzle throat area independent of the vector angle. VIGV and VAVBN area variation are both used to control LiftFan performance, manage the LiftFan stall margin, and minimize thrust-thrust split coupling effects.

Fig. 23 F-35 VAVBN as seen from below.

Thrust split is defined as the ratio of main engine thrust over lift fan thrust, typically used to represent the propulsion system pitching moment applied to the aircraft.

5. ROLL POST NOZZLE DEVELOPMENT

The F-35B STOVL lift system uses roll nozzles in each wing to provide roll control in powered-lift operation. The roll nozzle controls thrust by varying the nozzle area using two hinged flaps. Unlike the reaction control systems on legacy Vertical and/or Short Takeoff and Landing (V/STOL) aircraft, the F-35 roll posts produce about 10% of the vertical thrust through redirected engine fan air. The port and starboard side roll post systems are part-number common and interchangeable, providing improved maintenance flexibility. Actuation for the nozzle flaps is provided by a twin-motor, hydraulic, rotary actuator. External, hydraulically actuated aircraft doors on the underside of the wing are opened in powered-lift operation to provide an exit aperture for roll post thrust.

A key challenge in transitioning to the F-35B was providing adequate roll control authority and rate for store asymmetries and fuel imbalance. The roll post nozzles were positioned as far outboard as the internal wing structure would allow to maximize the moment arm. Main engine fan air to the roll posts was increased to the extent possible while maintaining adequate flow to cool the exhaust liner. Architectural changes in full-authority digital engine control were made to minimize the time delay from roll moment command to roll nozzle actuator response.

6. 3BSM DEVELOPMENT

The original design for the primary nozzle on the ASTOVL was a two-dimensional single expansion ramp nozzle. In this design, one nozzle flap is longer than the other. The nozzle vectors the primary thrust by deflecting the upper flap through at least 90 deg. To control the nozzle exit area in hover, the lower flap was designed as a sliding panel that would retract as needed to adjust the backpressure on the engine. This was a critical control needed to make the shaft-driven LiftFan turbine work.

As Lockheed Martin began construction and tests of the nozzle, the shortcomings of the design became more apparent. The abilities to turn the flow through 90 deg under high loads and control the nozzle exit area would have resulted in a very heavy design. This resulted in the pursuit of a lighter design that traced its roots to an early 1970s nozzle design from the proposed Convair Model 200 V/STOL fighter aircraft concept. A three-bearing swivel nozzle was developed by Pratt & Whitney and became part of the Convair Model 200 design that never continued into development. Following joint studies by Pratt & Whitney and Lockheed Martin, the 3BSM concept was integrated into the X-35B design and shown to be lighter. It also provided a very efficient means for turning the aft thrust post with minimal losses [36].

The F-35B 3BSM consists of a STOVL LOAN and a three-bearing swivel mechanism. The mechanism can deflect the exhaust flow through 95 deg in the pitch axis and ± 12.25 deg in the yaw axis as a function of pitch angle. The 3BSM can vector up to 23,900 lb of thrust at the maximum rearward thrust split. The 3BSM forward (No. 1) bearing is powered by twin fueldraulic actuator motors through a gearbox and drive train. The middle (No. 2) bearing is likewise powered by a twin fueldraulic actuator motor and gearbox/drive train system. A transfer gearbox links the middle and aft (No. 3) bearings with an efficient, compact, epicycle gear train. The twin actuator motors on the No. 1 and Nos. 2 and 3 bearings, respectively, are designed with a fail-degraded capability (full torque, half rate). This is one of the key differences between this design and that of the X-35B. In the X-35B, the Nos. 2 and 3 bearings were braked following a first failure, with no ability to continue vectoring the aft thrust post. This did not satisfy operational requirements requiring an ability to perform a shipboard vertical landing following a first failure. The dual redundancy on the fueldraulic motors enabled that fault tolerance on the F-35B.

7. TECHNOLOGY DEMONSTRATION PROGRAM AND TRANSITION TO THE JSF PROGRAM

The X-35B STOVL lift system completed more than 1200 hours of ground testing, culminating in the successful concept flight demonstration in August 2001. The aircraft accomplished this impressive performance under demanding hot, high-desert conditions and substantiated the robust performance capabilities of the shaft-driven LiftFan concept. Particularly impressive were the precise aircraft dynamics enabled by the responsive and accurate control of the STOVL lift system. Thirty-nine flights were conducted on the X-35B, including 22 hovers, 17 vertical takeoffs, 18 short takeoffs, 57 STOVL mode transitions, 27 vertical landings, 116 conversions (95 ground, 21 inflight), 63 clutch engagements, and 21.5 flight hours. This performance far surpassed the vertical operation goals and demonstrated sufficient maturity to proceed to production aircraft development.

The transition to a production F-35B principally centered on evaluating the STOVL lift system design changes and demonstrating a full-life propulsion system. More insights into the transition and full system development for the production F-35 system can be found in Ref. [38].

IV. CONCLUSION

The F-35 combines numerous technologies that have significantly advanced the state of the art in combat aircraft. This is particularly pronounced in the areas of integrated air vehicle subsystems and propulsion systems. The resultant aircraft provides exceptional performance with unparalleled capabilities, enabled by the air vehicle and propulsion systems.

The integrated air vehicle subsystems architecture selected for incorporation was based on a continuum of progressively refined development projects. Each of these further refined the concepts and validated the approach. The SUIT and MEA studies from the early 1990s gave the JSF contractor teams confidence in the concepts. The J/IST studies then provided the final proof of the viability of the designs. They also validated the conclusion that the overall air vehicle takeoff gross weight and cost could be reduced by 2–3%. The T/EMM system development project in J/IST contributed to the development of the turbomachine, fan duct heat exchangers, and other key elements used in the current F-35 PTMS. Without these elements, the chosen configuration might have been deemed too risky to pursue in the SDD program. Likewise, without the great successes of these development programs, many elements of the F-35's integrated systems, EHAS, and EPS would likely have been substituted with more conventional federated configurations. In such an instance, the benefits of the integrated systems might never have been realized. Instead, the resultant systems incorporated into the F-35 have been proven to provide excellent technical performance and reliability. They also provide a backbone for future systems growth through the expected long life of the F-35 program.

The F-35 propulsion systems incorporating the numerous technology upgrades have driven an unprecedented capability in performance. This has enabled the aircraft's unique performance capabilities, particularly in the F-35B STOVL variant. The final F-35 configuration incorporated a DSI, LO axisymmetric engine thrust nozzle, and unique STOVL propulsion system integrating the LiftFan and three-bearing swivel nozzle. These systems enabled the development of the F-35 variants, each providing exceptional performance and serving as the basis for long-term growth and capability improvements.

ACKNOWLEDGMENTS

The authors thank Jeff Hamstra, Rick Mange, and J.D. McFarlan for their leadership in pursuing the entire family of related papers. The authors give special thanks also to graphic artists Daniel Buck and Mark Lawhon, and to Joey Sikorski for help in pulling together a lot of details that went into this report.

REFERENCES

[1] Barhydt, J., Carter, H., and Weber, D., "Subsystem Integration Technology (SUIT) Phase I," WL-TR-93-3031, July 1993.

[2] Blanding, D., Schlundt, D., and Aldana, J., "Subsystem Integration Technology (SUIT)," WL-TR-92-3105, December 1992.

[3] Hill, B., Stem, P., Ho, Y., and Martin, D., "Subsystem Integration Technology (SUIT)," Executive Summary, Vol. 1, WL-TR-92-3115, October 1992.

[4] Hill, B., Stem, P., Ho, Y., and Martin, D., "Subsystem Integration Technology (SUIT)," Phase I Report, Vol. 2, WL-TR-92-3116, October 1992.

[5] Hill, B., Stem, P., Ho, Y., and Martin, D., "Subsystem Integration Technology (SUIT)," Phase II Report, Vol. 3, WL-TR-92-3117, October 1992.

[6] Hill, B., Stem, P., Ho, Y., and Martin, D., "Subsystem Integration Technology (SUIT)," Phase III Report, Vol. 4, WL-TR-92-3118, October 1992.

[7] Carter, H., Rupe, K., Mattes, R., and Wiese, D., "Subsystem Integration Technology (SUIT)," Phase II Report, WL-TR-96-3065, May 1997.

[8] Terrier, D., Hodge, E., Eicke, D., and Roberts, J., "Vehicle Integration Technology Planning Study (VTIPS)," WL-TR-96-2077, April 1996.

[9] Weber, D., Hill, B., and Van Hom, S., "Vehicle Integration Technology Planning Study (VTIPS)," WL-TR-97-3036, February 1997.

[10] Eicke, D., "Joint Strike Fighter Integrated Subsystems Technology (J/IST) Power and Actuation Flight Demonstration Program," AFRL-PR-WP-2001-2036, January 2001.

[11] Carter, H., "Joint Strike Fighter (JSF)/Integrated Subsystems Technology (J/IST) Demonstration Program," Executive Summary, Vol. 1, AFRL-VA-WP-TR-2001-3028, June 2001.

[12] Rupe, K., and Carter, H., "Joint Strike Fighter (JSF)/Integrated Subsystems Technology (J/IST) Demonstration Program," Vol. 2, Program Integration, AFRL-VA-WP-TR-2001-3029, June 2001.

[13] Heydrich, H., and Roddiger, H., "Joint Strike Fighter (JSF)/Integrated Subsystems Technology (J/IST) Demonstration Program," Vol. 3, Thermal/Energy Management Module (T/EMM) Turbomachine, AFRL-VA-WP-TR-2001-3030, June 2001.

[14] Dastur, N., and Roddiger, H., "Joint Strike Fighter (JSF)/Integrated Subsystems Technology (J/IST) Demonstration Program," Vol. 4, Thermal/Energy Management Module (T/EMM) System Integration, AFRL-VA-WP-TR-2001-3031, June 2001.

[15] Anderson, G., and Waffner, W., "Joint Strike Fighter (JSF)/Integrated Subsystems Technology (J/IST) Demonstration Program," Vol. 5, High Power Electric Actuation, AFRL-VA-WP-TR-2001-3032, June 2001.

[16] Torchia, D., and Rupe, K., "Joint Strike Fighter (JSF)/Integrated Subsystems Technology (J/IST) Demonstration Program," Vol. 6, Power Management Integration, AFRL-VA-WP-TR-2001-3033, June 2001.

[17] Anderson, J., and Roddiger, H., "Joint Strike Fighter (JSF)/Integrated Subsystems Technology (J/IST) Demonstration Program," Vol. 7, Engine Integration, AFRL-VA-WP-TR-2001-3034, June 2001.

[18] Eicke, D., "Joint Strike Fighter Integrated Subsystems Technology (J/IST) Power and Actuation Flight Demonstration Program," Executive Summary, AFRL-PR-WP-2001-2077, January 2001.

[19] Iya, S., Harlan, R., and Hawkins, L., "LCC and War Fighting Assessment of J/IST Suite," AFRL-VA-WP-TR-2000-3016, March 1999.

[20] Hamstra, J. W., McCallum, B. N., McFarlan, J. D., and Morehouse, J. A., "Development, Verification and Transition of an Advanced Engine Inlet Concept for Combat Aircraft Application," NATO RTO Paper MP-121-P-43, October 2003.

[21] Hehs, E., "Diverterless Supersonic Inlet," *Code One*, July 2000, pp. 8–13.

[22] Scharnhorst, R. S., "An Overview of Military Aircraft Supersonic Inlet Aerodynamics," AIAA Paper 2012-0013, January 2012.

[23] Sobester, A., "Tradeoffs in Jet Inlet Design: A Historical Perspective," *Journal of Aircraft*, Vol. 44, No. 3, May 2007, pp. 705–717.

[24] Gridley, M. C., and Walker, S. H., "Inlet and Nozzle Technology for 21st Century Aircraft," ASME Paper 96-GT-244, June 1996.

[25] Gridley, M. C., and Cahil, M. J., "ACIS Air Induction System Trade Study," AIAA Paper 1996-2646, July 1996.

[26] Catt, J. A., Welterlen, T., and Reno, J., "Decreasing F-16 Nozzle Drag Using Computational Fluid Dynamics," 29th Joint Propulsion Conference and Exhibit, 1993. doi: 10.2514/6.1993-2572.

[27] Catt, J. A., Welterlen, T., and Reno, J., "Evaluating F-16 Nozzle Drag Using Computational Fluid Dynamics," 32nd Aerospace Sciences Meeting and Exhibit, 1994. doi: 10.2514/6.1994-22.

[28] McWaters, M., "F-35 Conventional Mode Jet-Effects Testing Methodology (AIAA Paper 2015-2404)," 31st AIAA Aerodynamic Measurement Technology and Ground Testing Conference, 2015. doi: 10.2514/6.2015-2404.

[29] Miller, D., and Catt, J. A., "Conceptual Development of Fixed-Geometry Nozzles Using Fluidic Injection for Throat Area Control," 31st Joint Propulsion Conference and Exhibit, 1995. doi: 10.2514/6.1995-2603.

[30] McFarlan, J. D., McMurry, C. B., and Scaggs, W., "Computational Investigation of Two-Dimensional Ejector Nozzle Flow Fields," 26th Joint Propulsion Conference, 1990. doi: 10.2514/6.1990-2148.

[31] Bevilaqua, P. M., "Inventing the F-35 Joint Strike Fighter," AIAA Paper 2009-1650, Jan. 2009.

[32] Maddock, I., and Hirschberg, M., "From ASTOVL to JSF: Development of the STOVL Propulsion Systems," 11th AIAA ATIO Conference, AIAA Centennial of Naval Aviation Forum, Aug. 2011.

[33] Simons, T., and Sokhey, J., "Performance Optimization of VAVBN the Variable Area Vane-Box Nozzle for JSF F-35B LiftFan Propulsion System," International Powered Lift Conference, Philadelphia, PA, Oct. 2010.

[34] Renshaw, K., "F-35B Lightning II Three-Bearing Swivel Nozzle," *Code One*, Aug. 2014. https://codeonemagazine.com/article.html?item_id=137

[35] Robbins, A., Bobalik, J., De Stena, D., Plag, K., Rail, K., and Wall, K., "F-35 Subsystems Design, Development, and Verification," AIAA Aviation Forum, Atlanta, GA, June 2018 (unpublished).

[36] Wurth, S., Smith, M., and Celberti, L., "F-35 Propulsion System Integration, Development & Verification," AIAA Aviation Forum, Atlanta, GA, June 2018 (unpublished).

[37] Lemons, G., Carrington, K., Frey, T., and Ledyard, J., "F-35 Mission Systems Design, Development & Verification," AIAA Aviation Forum, Atlanta, GA, June 2018 (unpublished).

[38] Counts, M., Kiger, B., Hoffschwelle, J., Houtman, A., and Henderson, G., "F-35 Air Vehicle Configuration Development," AIAA Aviation Forum, Atlanta, GA, June 2018 (unpublished).

[39] Dietrich, R., "Brief History and Perspective on J/IST," JSF Joint Program Office (Ref: DM #444383) (approved for public release, distribution unlimited) (JSF14-304), Distribution A.

[40] "Lockheed Martin Receives Prestigious Aviation Award for Joint Strike Fighter Technology," Lockheed Martin Aeronautics Company [online database], https://news.lockheedmartin.com/2001-06-20-Lockheed-Martin-Receives-Prestigious-Aviation-Award-for-Joint-Strike-Fighter-Technology [retrieved 11 May 2018].

F-35 Digital Thread and Advanced Manufacturing

Don A. Kinard*

Lockheed Martin Aeronautics Company, Fort Worth, TX

The F-35 program has delivered more than 270 aircraft to the U.S. Air Force, Marine Corps, Navy, and partner countries. This was accomplished during the Low-Rate Initial Production (LRIP) phase of the program. LRIP facilitated the completion of structural and flight testing, and enabled pilot and maintainer training. Arriving at LRIP for this quantity of F-35 jets surpasses the total production quantities achieved at the same stage for the F-117 and F-22. Also, initial operating capability has already been reached for the F-35 by the Air Force, Marine Corps, and Israeli Air Force. The F-35 program is now advancing toward full-rate production and reinforcing key strategy tenets. Included among them are lean manufacturing deployment, the use of low-risk materials and supportable low observables, and the implementation of digital thread technologies.

I. INTRODUCTION

In 2001, when the Joint Strike Fighter System Development and Demonstration (SDD) program commenced, the challenges ahead for manufacturing were daunting. Not since the days of the McDonnell Douglas F-4 Phantom II in the 1960s and 1970s had a program attempted to produce three variants of an advanced fighter aircraft to satisfy requirements for the Air Force, Navy, and Marine Corps on a single assembly line (Fig. 1). Not since the early F-16 days had advanced fighter aircraft been scheduled to be produced at the rates envisioned by the F-35 Joint Program Office (JPO). Affordability was a fundamental challenge, as was the need to support international participant aircraft configurations, manage a global supply chain, and initiate Final Assembly and Checkout (FACO) facilities in three countries. As the newest 5th-Generation fighter, the F-35 introduced and incorporated many evolutionary and revolutionary technologies. These included sophisticated avionics, supportable low observables, an unprecedented

---

*Senior Fellow, F-35 Production Operations, AIAA Sr. Member, don.a.kinard@lmco.com—Approved for public release 5/17/18, JSF18-402.

Fig. 1 F-35 multivariant final assembly line in Fort Worth, Texas.

amount of aircraft and support software, and complex vehicle systems developments, such as the LiftFan® and Electro-Hydrostatic Actuation Systems (EHASs).

To date, Lockheed Martin, our partners, and our suppliers have delivered more than 270 aircraft, and there are currently approximately 175 additional new aircraft in work around the world. These aircraft were built during the Low-Rate Initial Production (LRIP) phase of the program to allow the completion of testing and pilot and maintainer training. The F-35 LRIP deliveries are already greater in number than the total production aircraft for either the F-117 or F-22 programs. Further, the Marine Corps, Air Force, and Israeli Air Force have all declared initial operating capability with the F-35.

The Fort Worth facility's layout, nearly one mile in length, is shown in Fig. 2. Wing components and the fuselage are produced in Fort Worth. These Lockheed Martin components are then mated to the center fuselages from Northrop Grumman and the aft fuselages from BAE Systems along with the center wing assemblies built at Lockheed Martin in Marietta, Georgia, before going to final assembly. This is followed by coatings, radar cross-section testing, and fuel testing before finally sending them to the flight line for delivery operations. The complex F-35 supply chain is illustrated in Fig. 3. There are more than 1400

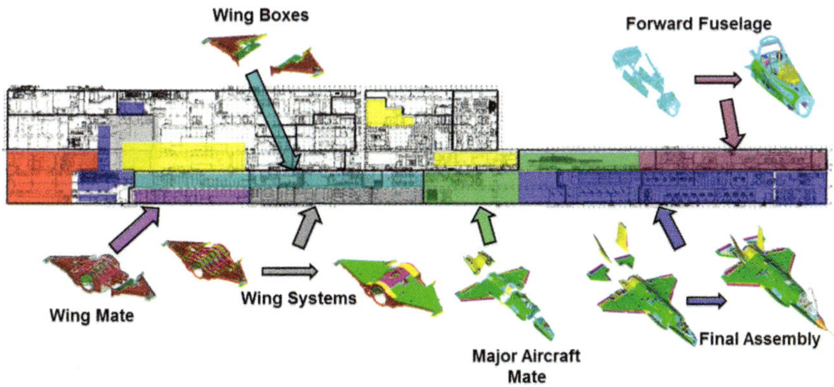

Fig. 2 F-35 facility layout in Fort Worth.

suppliers, more than 80 of which are in non-U.S. countries, as well as three FACO facilities. Orchestrating this supply chain requires 24/7 operations and other efforts to satisfy the requirement to support and supply parts to the more than 270 aircraft already delivered to the field.

The F-35 aircraft is completing development testing, and changes have been incorporated into the design. The aircraft is meeting its key performance points, and trained pilots and maintainers complement the aircraft's performance and capabilities well. The F-35 is now starting to transition operations from LRIP (Fig. 4) to full-rate production. The engineering, manufacturing plan, and supply

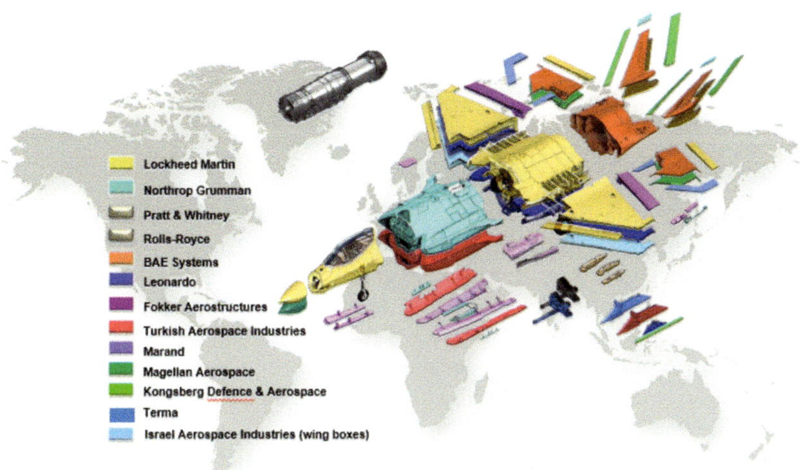

Fig. 3 F-35 global supply chain.

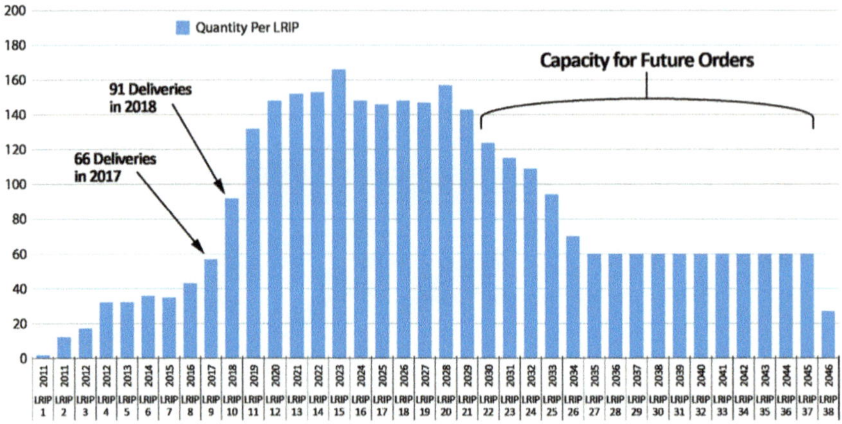

Fig. 4 F-35 commencing the climb to full-rate production with approximately 1400 suppliers.

network have been tested in low-rate production in preparation for the transition to full-rate production, but there are many remaining production challenges. Thousands of employees around the world need to be hired, and the last of the required tooling and capital needs to be installed. In addition, efforts must continue to improve quality and reduce costs, and to climb the rate production curve.

Several key strategies have been implemented as part of the overall F-35 fighter production system. Lean manufacturing principles have been adopted, including flow-to-takt production lines, point-of-use materials delivery, and emphasizing standard work. A low-risk approach has been taken to select materials, structures, and supportable low observables technologies. The development and implementation of engineering and manufacturing technologies has been enabled by the digital thread.

II. F-35 LEAN MANUFACTURING

The Lockheed Martin F-35 assembly lines in Fort Worth; Palmdale, California; and Marietta, Georgia, implemented a flow-to-takt lean manufacturing philosophy [1–3]. Lockheed Martin wanted to create rhythm and reinforce the sense of urgency in the factory. To do this, it moved the individual components at the same pace as the average delivery rate. Takt is the pace of the factory expressed in manufacturing days (i.e., calendar days not counting weekends or holidays). For example, five aircraft per month equates to one aircraft every four manufacturing days, which is a takt time of four.

On the F-35 program, this sometimes requires large material handling systems to efficiently transfer components from one station to another at this takt pace. An example is shown in Fig. 5 for the Fort Worth wing box assembly line. This wing box line is effectively duplicated in Italy at the Leonardo plant in Cameri, and in Israel at the Israel Aerospace Industries plant. Lean assembly lines are also the standard at Northrop Grumman and BAE Systems sites, at sites in participating nations, and at supplier locations. Lockheed Martin is responsible for ensuring that all components produced for the F-35 fighter production system conform to U.S. government engineering requirements. Other examples of F-35 lean manufacturing are:

- Point-of-use materials delivery
- Tasks broken into four-to-eight-hour segments
- A factory-wide radio frequency identification system for tracking part kits and tooling
- Automated dispensing units for providing perishable and hand tooling

Lockheed Martin anticipated that single-piece flow manufacturing was initially going to be difficult, and this turned out to be the case. Flow requires a stable manufacturing system, and the initial F-35 flow was disrupted often by multiple factors. Among these were:

- Normal developmental corrections to the Build-to Packages (BTPs)
- Implementation of changes to correct deficiencies found during structural and flight testing

Fig. 5 Flow-to-takt assembly line with highly common tooling and material handling.

- Natural progression of the learning curve reducing the number of hours per unit by 75% since production start
- Startup and expansion of the global supply chain
- Variable production rates

Instability in the manufacturing system and the disruptions often result in out-of-station work being sent downstream in the flow-to-takt areas, increasing cost and span [4]. Lockheed Martin expects to realize the benefits of flow after incorporating SDD-related changes, stabilizing the production rate, and fully maturing the supply chain.

The trivariant production line was less disruptive than expected, and the trivariant assembly line exhibited a shared learning curve as a result of high parts commonality. There was nearly 100% commonality among the mission systems, 70% among the vehicle systems, and 20% among the airframes. In this way, each variant contributed to an overall reduction in the learning curve. The F-35C Carrier Variant (CV) and F-35B Standard Takeoff and Vertical Landing (STOVL) variants in particular experienced a reduction in hours per unit from this commonality. They benefitted from the F-35A Conventional Takeoff and Vertical Landing (CTOL) variant's higher production rate. This was likely due to the common materials, processes, and assembly, which led to similar work being done in each manufacturing position regardless of variant type.

Lockheed Martin determined another arrangement that increased production efficiency: spreading the F-35B/C models among the F-35A models. The F-35B and F-35C models each constitutes 15% of the total build and requires more work to produce, and building them in blocks would disrupt staffing and learning.

Also contributing to production efficiency, Lockheed Martin was able to use many common tools, such as the large assembly tool holding the wing in Fig. 5. The tool's gold and blue stripes indicate that it can be used for both F-35As and F-35Bs. Building all three variants on a single assembly line with high commonality saved an estimated 30% in overall production costs, including those of capital and tooling. This was an improvement over the traditional approach, with which each aircraft is built in a different factory, by a different company.

III. F-35 MATERIALS AND STRUCTURES: SUPPORTABLE LOW OBSERVABLES

As far as material and process selections are concerned, the F-35 program adopted a low-risk approach using well-characterized materials and processes. In some respects, this decision was a foregone conclusion because of the lack of material (metallic and composite) development in the aerospace industry today. The industry is looking at novel processing technologies, such as additive manufacturing, but the basic composites and metals available in the marketplace have been there for more than 20 years. The cost of qualification and inherent risk aversion on the part of customers and companies have reduced the emphasis on development. Exceptions to the low-risk philosophy for the F-35 were the aluminum forgings in Fig. 6 and the formed titanium engine doors from BAE Systems on the F-35B. Another exception was the incorporation of the latest in low observable materials and structures technologies. In general, the F-35 program used aluminum wherever possible and titanium wherever necessary, such as in hot areas and areas of concentrated load introduction. Composites were used wherever weight savings were sufficient to justify the cost.

Another fundamental aspect of the F-35 involves the technologies controlling the Outer Mold Line (OML) of the aircraft or simply the panel-to-panel and panel-to-skin mismatches. Mismatches on the surface of a 5th-Generation fighter (F-22 and F-35) are detrimental to the aerodynamics and low observable performance. Therefore, the F-35 applied advanced manufacturing technologies to control part dimensional tolerance tightly and thus control the mismatches.

As an example, controlling surface mismatches requires tightly controlling the skin thickness on both sides of the panel/skin joints. Composite parts typically have a cured ply thickness variation insufficient to control F-35 mismatch tolerances. Therefore, we must correct for this thickness variation in the cured parts. One approach was to add sacrificial plies to the Inner Mold Line (IML) of the composite skins and machine the IML using very tight tolerance Numerically Controlled (NC) machines. These NC machines feature laser-compensated

Fig. 6 Aluminum forgings from Alcoa for unitization and weight savings.

Fig. 7 Laser scanning the thickness measurement of composite parts.

calibration and temperature control to produce the final skin thickness to the desired tolerance. Another approach, patented by Lockheed Martin, is to use a laser radar system (Fig. 7) to measure the composite thickness after cure. This is done where the skins are designed to be at or below the final required tolerance. These thickness measurements are automatically sent to a ply-cutting machine to produce compensation plies. Compensation plies are located on the skin IML using laser projection systems and then the part is recured to provide the final thickness-controlled part. Both of these approaches may seem expensive, but the goal was to produce an aircraft that was supportable and affordable long term, for Lockheed Martin customers.

Trade studies indicated that it was beneficial to the overall F-35 air system to reduce maintenance requirements over the life of the aircraft, specifically the maintenance hours per flight hour. In addition, because OML is controlled so tightly, Lockheed Martin was able to develop simplified coating applications. These not only reduced the cost of initial application, but also significantly reduced the repair costs when a skin incurred cosmetic or maintenance damage in the field.

IV. THE DIGITAL THREAD

The F-35's development and early production benefitted significantly from the phased adoption of a digital thread philosophy. Designers produced 3D solid models, constructed them to support factory automation, and facilitated their consumption by downstream manufacturing and sustainment functions. Recently, technology has allowed the rapid validation of as-designed to as-built configuration verification through the use of laser scanning and structured light technologies.

The term *digital thread* was reportedly coined by the Air Force Research Laboratory (AFRL) and Lockheed Martin during the early days of F-35 development [5]. The overall F-35 digital thread philosophy is depicted in Fig. 8. Lockheed Martin defines it as the creation, use, and reuse of the 3D models by engineering and downstream functions, including manufacturing and sustainment. In Phase 1 of the digital thread implementation, engineering produced exact 3D engineering models and 2D drawings. Partner and supplier models, 3D tool designs, drawings, specifications, and related analysis data were released into a common product life-cycle management system for accessibility and configuration integration. Manufacturing produced 3D models for tools and factory layouts that improved facilities development and installation. For many of the airframe parts, engineering was able to produce reduced-dimension drawings that decreased engineering costs and facilitated supplier NC machining. Fiber placement was used for composites based on the digital thread. Coordinate-measurement machines' inspection points were programmed directly into the solid models because the solids contained the master engineering data. These models also supported the supportable low observable structures processes discussed earlier involving the machining of the IML/OML and cured laminate compensation.

Three-dimensional models were used for virtual mockups, manufacturing, and sustainment simulations. Significant successes of the 3D solids led to large reductions in the quantities of engineering and tooling changes. Solid models reduced engineering changes, compared to historical numbers, because of the ability to provide accurate surfaces and improved integration between parts. Because of solid model engineering and solid model tooling, Lockheed Martin was able to reduce tool design changes due to tool interferences with released

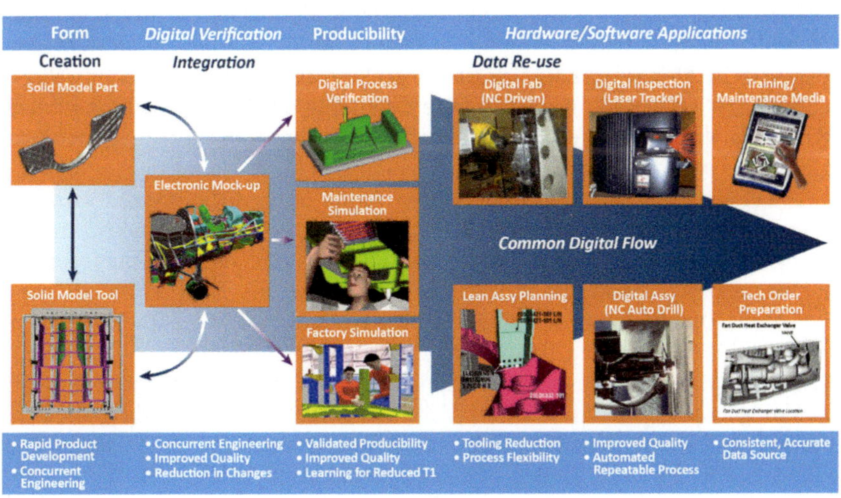

Fig. 8 The F-35 digital thread.

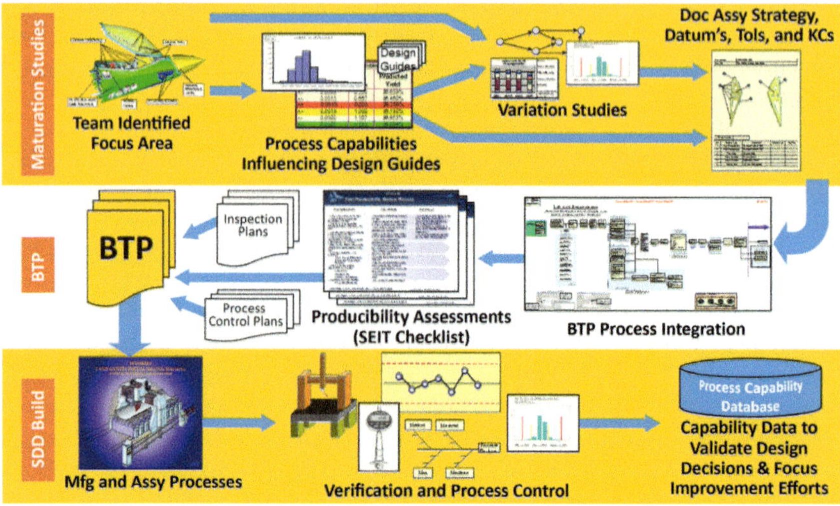

Fig. 9 Producibility analysis and key characteristics.

parts. Interferences between parts were also reduced, compared to nonsolid model programs, and suppliers were able to produce and validate machined parts to the released solid model masters. These improvements were especially important to the F-35 because there were three variants to design. Assembly interface control drawings, typical of legacy programs, were not required because the solid models facilitated coordination among partners and suppliers.

Virtual manufacturing simulations, encouraged by the digital thread technology, turned out to be time consuming and expensive, so only a few areas were even attempted. This is one area requiring tool development for future programs. Physical mockups were used for specific bays depending on complexity. Sustainment engineering was able to use virtual reality helmets to assist in simulating maintenance actions with the completed 3D engineering models.

The digital thread allowed significant increases in producibility and variation management analysis during the F-35 SDD program (Fig. 9). Geometric dimensioning and tolerancing was introduced on the F-35, and variation management analysis was performed by manufacturing engineering. This included using specialized 3D software to perform complex assembly variation studies. Critical installations were identified, followed by the collection of process capabilities facilitating the variation studies and the creation of variation management documents containing assembly datum schemes. Variation analysis led ultimately to the definition of engineering tolerances that were flowed into the models, drawings, and tooling, and to the identification of Key Characteristics (KCs).

A KC [6] is a feature of a material, process, or part (including assemblies) whose variation within the specified tolerance significantly influences product fit, performance, service life, or manufacturability. The F-35 identified many KCs early on, but would likely revise selections for future programs. KC selections should be made with the understanding that their selection will drive costs into the production system. They will have this effect by requiring the development and imposition of KC management plans, the formal collection of data by fabricators and assemblers, and reporting and analysis of deliverables. KCs should only be selected under certain conditions. Plans must first be in place to alter the engineering or manufacturing BTP designs. Similarly, requirements must first be identified and implemented to better control or take advantage of the variation reduction. KCs that simply accentuate but do not drive changes to engineering accept/reject criteria or BTPs are not KCs. For example, KCs were put on the diameters of holes on previous programs, but there was never a plan to change the tolerances. As such, these should not have been regarded as KCs.

Phase 2 of the digital thread transformation is about constructing the engineering data to support factory automation. Examples are automated drilling (Fig. 10) and robotic coatings applications (Fig. 11). Automated drilling is used by all the F-35 partners, and we drill 20% of the total holes using automation. This includes 80% of the accessible OML holes. Automated drilling is about four times faster than manual drilling and its quality is nearly perfect, with remarkable repeatability.

Lockheed Martin uses automated drilling for the wing boxes, forward fuselage skins, and upper skin to center wing skins. We also use it for the center wing at Marietta, Georgia. Northrop Grumman uses a metrology-assisted robot to drill the narrow inlets on the F-35. BAE Systems drills its empennage skins and structure separately using its high-precision machining centers. This is a remarkable feat of high-precision machining, considering the bolt-to-hole tolerances it requires. There is a plan in place to implement even more automated drilling in the future as part of the continual effort to drive costs down and improve quality.

Other automation used for the F-35 includes fiber placement technology, which Lockheed Martin, Northrop Grumman, and other suppliers use to lay up complex

Fig. 10 Automated drilling is enabled by the digital thread.

Fig. 11 Robotic coatings applications.

inlets, nacelles, and large wing parts. Recently BAE Systems also introduced a robot to countersink its composite skins (Fig. 12).

Phase 3 provides the digital thread directly to the mechanics to create such products as work instruction graphics. These graphics were facilitated by the 3D solid models, which can be used to create graphics through visualization software tools. Ideally, they would visually instruct mechanics on the floor or maintainers in the field and reduce the time it takes them to understand their tasks; however, for production this intended benefit on the floor ended up being very difficult to maintain. This was because graphics are static images that are incapable of

Fig. 12 BAE Systems robotic countersinking.

affordably being updated as engineering or manufacturing changes. For the F-35 program specifically, with concurrent engineering and manufacturing development, graphics must frequently be updated to accommodate a steep initial learning curve and significant engineering, tooling, and planning changes.

An unexpected factor in this process was the impact of flow-to to-takt manufacturing on graphics creation. As the takt time (production rate) changed, new tool positions were added and hours per unit decreased. As a result, the manufacturing sequencing needed to be adjusted constantly by breaking up planning cards and redoing graphics. In an effort to circumvent static graphics costs, graphics can be made available on the factory floor. This is done by granting the mechanics access to the visualization tools from their work terminals. Harness installations are a good example of how this can work because harness routings are especially difficult to understand from a 2D drawing. Early in SDD, large television monitors were placed in some of the work areas, and the mechanics used these until experience was gained with the installations.

Another downside of graphics is that the mechanics typically only need them for a short time. Despite this, however, graphics do help mechanics who are new to the F-35 program during ramp-up to full-rate production. Recently, movies of the critical installations were produced that could be accessed as needed through the electronic work instruction terminals that mechanics use on the floor.

One of the unique ways for mechanics to use the digital thread is through optical projection technology. As Fig. 13 demonstrates, mechanics can use this to visually project work instructions directly onto the aircraft. In the figure, fastener locations and part numbers are being projected onto the inlets being built

Fig. 13 Optical projection of work instructions directly onto the work surface.

at Lockheed Martin's Marietta facility. The conventional procedure requires mechanics to look at drawings and write down part numbers and inlet locations to do their work. With the digital thread's procedure, mechanics can instead view the projected instructions while performing their work.

Continuing work is being done to capture the actual fastener grip during automated drilling operations. This information will be used to eliminate grip validation time and support fastener projection. In addition, it will be used to kit, clean, and promote fasteners and deliver these to the point of use.

An additional example of Phase 3 is the technology for laser ply projection in the composites shop. This was one of the first Phase 3 digital thread technologies used in the aerospace industry. Bulkhead marking was initiated on the F-35 where ink jet markings for bracket locations were printed directly onto large bulkheads. This saved span and costs and eliminated thousands of tools that had to be designed and maintained.

The production of tools for the mechanics and sustainment modifications using additive manufacturing (Fig. 14) is another digital thread success. Lockheed Martin has produced more than 5000 tools for the floor and field using Fused Deposition Modeling (FDM) of polymers. FDM provides a fast, low-cost approach to producing temporary tools (temporary because of the FDM material's durability issues) to assist the mechanics. The development of more durable FDM materials continues in the industry. The objective is to enable producing permanent tooling and replacing more expensive metal, fiberglass, and other typical hole-drilling tooling.

Phase 4 of the digital thread is the validation of engineering as-designed to the as-built structure using advanced noncontact metrology techniques, including laser scanning and structured light (Fig. 15). This technology can identify

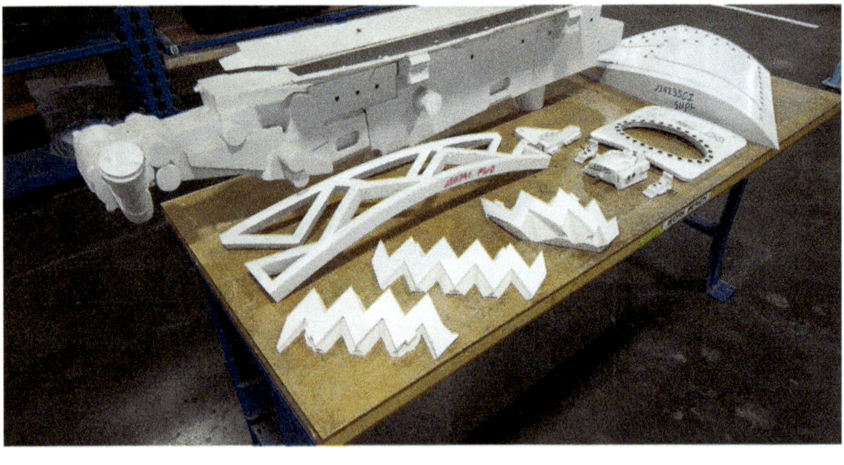

Fig. 14 Additive manufacturing of tools.

Fig. 15 Noncontact metrology.

deviations from engineering early in the build or fabrication process and rapidly correct them, reducing cost by stopping defects from traveling downstream. A truly revolutionary technology, it may eventually replace coordinate measuring machine inspections and become a requirement for suppliers prior to shipping parts, tools, and equipment.

Additively manufactured tools configured to mimic various F-35 weapons normally take about a full shift to be installed and pass clearance checks. Now, a laser can scan the bays and compare the as-built aircraft to the engineering models, which takes only a few hours. As another example, when there is a fit problem with a tube on the floor, the cause is not readily apparent. The tube could be bad, the bracket location could be incorrect, or there might be a problem with the structure. To overcome the problem, the tube can be brought to a scanner and rapidly validated, or equipment can be brought to the aircraft to have the brackets and structure checked.

Current scanning technologies typically depend on targets being placed on the aircraft or parts. However, this will eventually be replaced by feature-based recognition as the digital thread connections to the 3D scanner technology matures. This scanning technology can also replace the thousands of manual mismatch, gap, and flushness measurements required today (Fig. 16). Further, it can inspect detail parts, tooling, and assemblies on a first-article basis today. Identifying defects during first-article inspections will significantly reduce the cost and flatten the learning curve for concurrently developed products like military aircraft. Early identification will also reduce the recurring cost for measurements. In the future, the use of 3D noncontact metrology for recurring real-time validations in an assembly, as well as for supplier acceptance, may become routine.

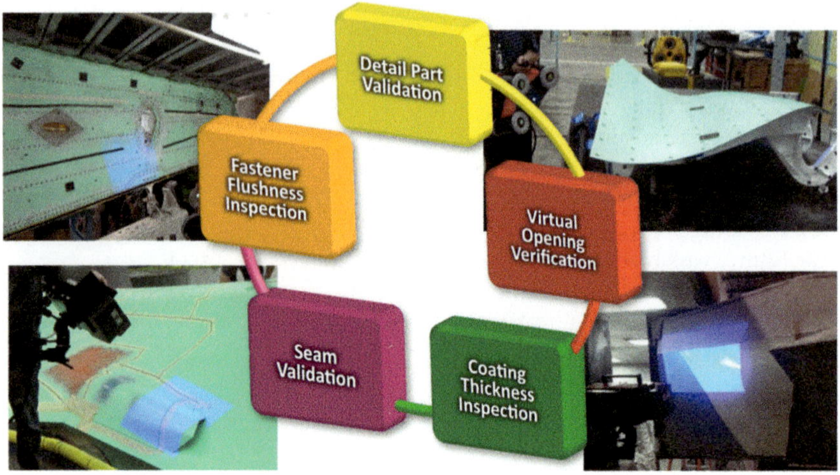

Fig. 16 Applications of noncontact metrology.

Applying the digital thread on the F-35 program has brought significant benefits, including:

- Development of the BTPs
- Use of automation to reduce touch labor and improve quality
- Integration of the digital thread on the factory floor
- Opportunity to validate the configuration using digital thread technologies (e.g., laser scanning and structured light)

There has been an explosion of digital technologies in the past five years and a tremendous amount of continued development in the industry. A future additional phase of the digital thread is discussed in the following section. It refers to the ascension of Industry 4.0 [7], the fourth industrial revolution: the revolution of data.

V. THE FUTURE OF ADVANCED MANUFACTURING

Affordability is and will continue to be a focus for the F-35 program in the years to come. Lockheed Martin, our suppliers, and the customer community have invested significantly in affordability through the Blueprint for Affordability Program (BFA). As BFA Phase 1 concludes, it is showing an impressive return on this collective investment, as measured by dividing dollars saved ($2 million per aircraft) by dollars invested. BFA Phase 2 began in 2018 and will provide funding for additional projects (Fig. 17).

The F-35 program aims to achieve an $80-million aircraft cost at full-rate production. As with most programs like F-35, the supply chain is responsible for more than 70% of the aircraft's cost. This is why Lockheed Martin is developing and transferring technology to simplify the supply chain. Doing so enables its suppliers to take advantage of manufacturing technology advancements.

One of the features of the F-35 program is its long-term production forecast (Fig. 4). This gives the F-35 the ability to develop and implement new technologies for cost savings now and for years to come. Multiyear buys are anticipated in the near future, which will also help to reduce the cost of the aircraft.

One particular focus is on how the mechanics consume engineering data on the factory floor. Three-dimensional digital models and access to the 2D paper drawings created from them are provided to the mechanics on the floor. Work instructions have graphics in some instances but can be expensive to maintain and tend to be limited to complex installations, as previously discussed. One remedy is to provide engineering data in augmented-reality-style glasses (Fig. 18). The ability to provide work instructions hands-free with voice-activated commands has shown promise in improving data accessibility; however, it is complicated by the same problems mentioned earlier for static graphics: obsolescence and maintenance. In the example in Fig. 18, the mechanic reads the wire number aloud and the pin location lights up in the pin diagram view in his glasses. (In the future, the glasses could be reading the wire number directly.) In cases like harness routing, mechanics need 3D views of the engineering models in order to route the harnesses. Movies for complex installations are now available to the mechanics at their work instruction terminals and eventually perhaps in glasses. The efficient creation, consumption, change, and utilization of the engineering data may provide affordability savings for current and future programs. Such efforts

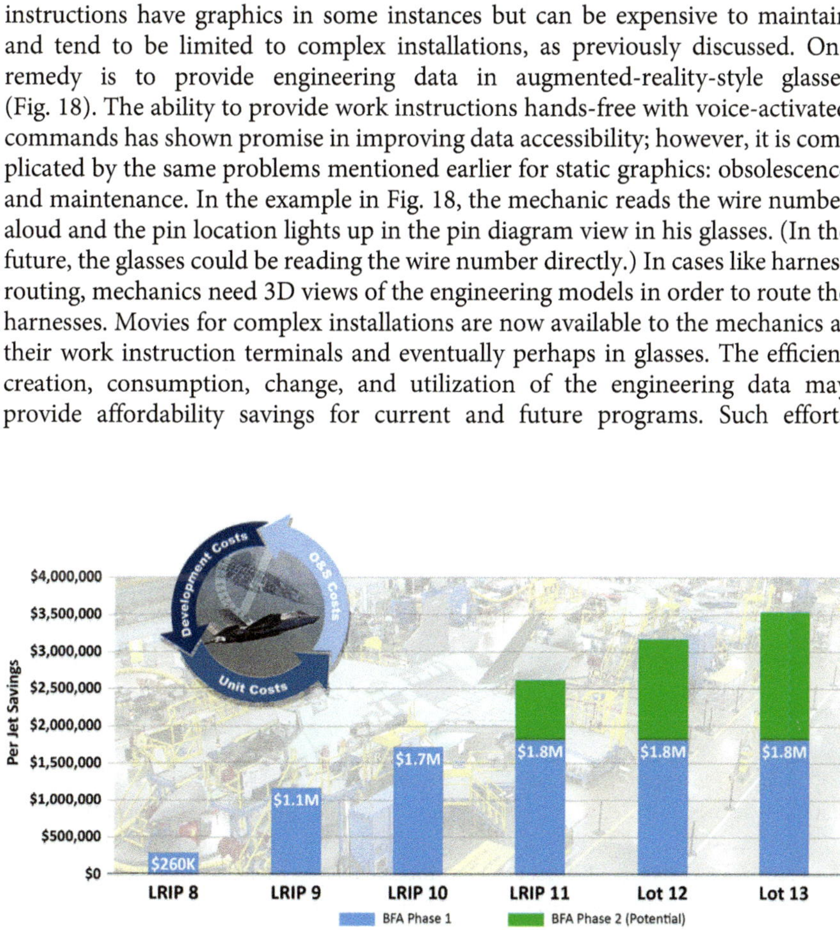

Fig. 17 Blueprint for Affordability investments reducing aircraft cost.

Fig. 18 Consumption of the engineering data using hands-free, voice-activated glasses.

include the development of digital twin technologies by manufacturing and sustainment.

The advent of Skype®, Facetime®, and similar applications has opened up the potential for real-time communication among suppliers, customers, and engineering from remote locations. Support personnel are typically located onsite with the customers and are often sent to supplier and customer locations to investigate and resolve problems. Communication through the use of services like Skype, using real-time photos and two-way video links, provide opportunities for cost savings and rapid response times. Figure 19 shows experts located at remote locations who can interact with maintenance, production, or supplier personnel onsite to resolve issues faster and at a lower cost. The key issue with this technology is data security, which is high on the customer's concern index,

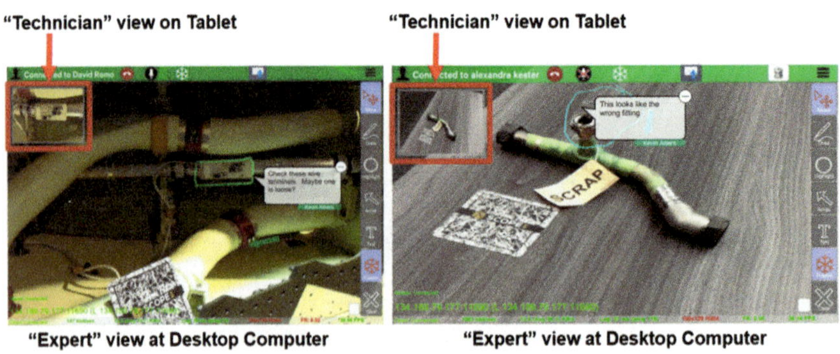

Fig. 19 Remote augmented reality.

especially for active bases. This will likely be resolved in the future because of customers' pronounced interest in faster, lower-cost support and the private sector's focus on data security solutions.

The digital thread, noncontact metrology, vision systems, artificial intelligence, and machine learning technologies will bring new opportunities for automation and robotic applications. This is especially the case for routine operations, such as sealant and coatings application, fastener installation, hole drilling, and inspections. Also, drones are being developed to perform external aircraft inspections. Additive manufacturing has been extensively used to produce tools as well. Its uses in fracture-critical structures are still years away due to certification requirements; however, there are opportunities for additive manufacturing in support equipment. It can also be used to produce nonstructurally critical parts that cannot be produced using conventional technologies.

Lockheed Martin has been working with 5ME to implement cryogenic machining technologies (Fig. 20) for titanium part cost savings. Liquid nitrogen is applied at the cutting surface and has shown potential for increasing tool life and machining speeds. This technology is a result of Creare LLC's work on the U.S. government Small Business Innovation Research (SBIR) program. It is an example of Lockheed Martin's interest in developing technologies to support cost savings in the supply base. SBIR projects have also developed and implemented the following:

- BTG Labs surface energy measurement system for nutplate bonding
- Creare fastener fill measurement systems
- Twin Coast Metrology and Delta Sigma Company fastener projection systems

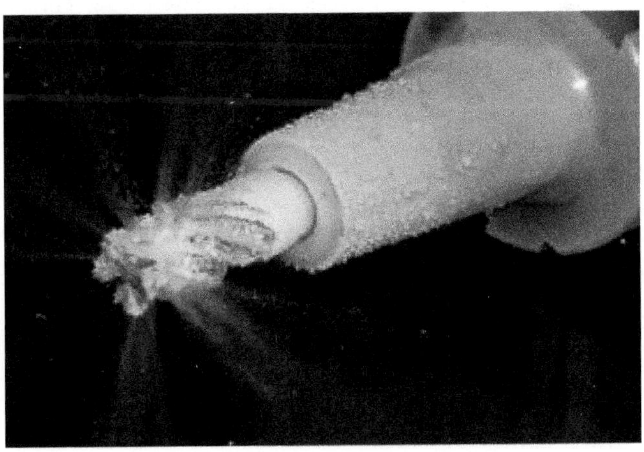

Fig. 20 Cryogenic machining of titanium.

The F-35 JPO, AFRL, Naval Air Systems Command, and Office of Naval Research have collaborated extensively for such efforts. Their cooperation has been instrumental in securing support for SBIR and other contracted research and development efforts for technology development.

Another innovation that could impact future F-35 affordability is the advent of Industry 4.0, which is the fourth industrial revolution: the revolution of data. Steam and water power drove the first industrial revolution, electrical power drove the second, and computers drove the third. The fourth is about the strategic and tactical use of data resident in our systems.

Industry is recognizing that the data in IT systems can provide tactical insight to drive efficiency in operations. It can also lower support costs for data collection, analysis, and performance visibility and transparency. The connected enterprise depicted in Fig. 21 is the key to enterprise efficiency. It improves the integration of systems data, facilitates automated data collection and dashboard metrics, and supports descriptive, predictive, and prescriptive analytics. It also connects factory equipment with the IT systems, driving efficient usage and secure data transfer.

As previously discussed, data security is one of the enabling technologies for the connected enterprise. This revolution is in its early stages but will rapidly drive industry to become more efficient in above-the-factory-floor functions. Further, it

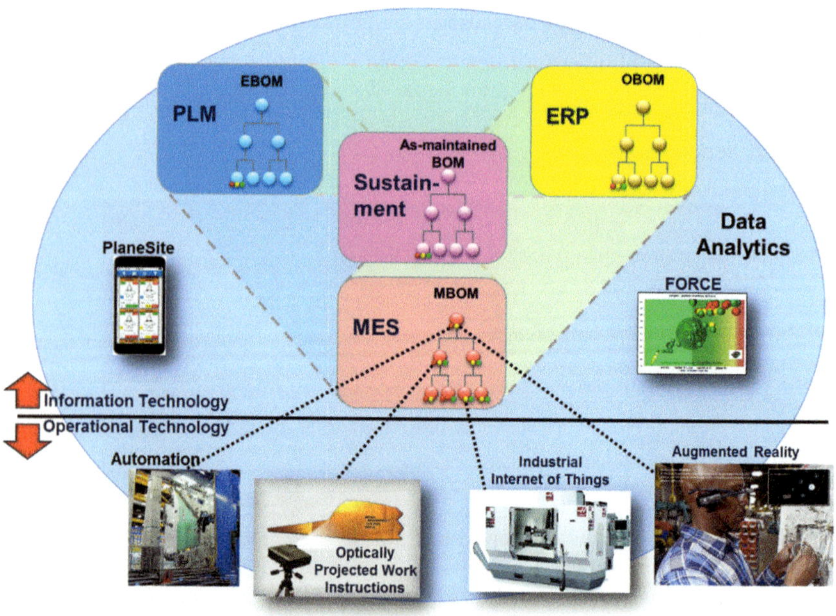

Fig. 21 The connected enterprise.

will provide needed insight for continued productivity gains on the factory floor and in the supply base.

Lockheed Martin has already deployed a phone/computer application for aircraft on the production floor. It provides information about locations, schedule performance, part shortage, nonconformances, and other factors for every aircraft and component in work. This pertains to not only those in the Fort Worth factory, but also the aircraft and components in Italy and Japan. The application is being combined with a factory-wide part kit and tool Radio Frequency Identification System (RFID). Ultimately, it will be able to automate updates of each aircraft on monitors located at each tool position with status and performance data.

VI. CONCLUSION

The F-35 has benefitted greatly from new technologies and is well-positioned to invite further technology development. It will do so through affordability investments and nonaerospace and defense industry commercial developments in artificial intelligence, augmented reality, machine learning, and other areas, as well as the rise of Industry 4.0.

With a focus on future technologies, it will also benefit the broader defense industry and other Lockheed Martin programs by helping them to maintain a technological edge. Manufacturing technologies have applications in the sustainment of aircraft in the field, including automated measurement technologies, drone inspections, and data integration. The F-35 program will continue to innovate with the addition of warfighter capabilities and the continued reduction in cost through advanced manufacturing implementations. It will also provide significant benefits in sustaining delivered aircraft.

REFERENCES

[1] Ohno, T., *Toyota Production System: Beyond Large-Scale Production*, Productivity Press, Portland, OR, 1988.

[2] Womack, J. P., Jones, D. T., and Roos, D., *The Machine that Changed the World: The Story of Lean Production*, Harper Perennial, New York, 1991.

[3] Spear, S., and Bowen, H. K., "Decoding the DNA of the Toyota Production System," *Harvard Business Review*, Sept. 1999, pp. 97–106.

[4] Cochran, D., Kinard, D., and Bi, Z., "Manufacturing System Design Meets Big Data Analytics for Continuous Improvement," *Proceedings of the 26th CRIP Design Conference*, Stockholm, Sweden, 2016.

[5] Kraft, E., "Expanding the Digital Threat to Impact Total Ownership Cost," U.S. Air Force Presentation, 2013 NIST MBE Summit, National Institute of Standards and Technology [online database], https://www.nist.gov/sites/default/files/documents/el/msid/1Kraft_DigitalThread.pdf [retrieved 8 May 2018].

[6] Aerospace Standard AS9103, "Variation Management of Key Characteristics," SAE International, Warrendale, PA, 2001.

[7] Smit, J., Kreutzer, S., Moeller, Carolin, and Carlberg, M., "Industry 4.0 Analytical Study," Policy Department A: Economic and Scientific Policy, Directorate General for Internal Policies, European Parliament [online database], http://www.europarl.europa.eu/RegData/etudes/STUD/2016/570007/IPOL_STU(2016)570007_EN.pdf [retrieved 8 May 2018].

F-35 Weapons Design Integration

Douglas M. Hayward* **and Andrew K. Duff†**
Lockheed Martin Aeronautics Company, Fort Worth, TX

Charles Wagner‡
F-35 Joint Program Office, Arlington, VA

Tactical aircraft operations in a high-threat, hostile environment require an internal weapon carriage capability that supports stealth operations while ensuring aircrew survivability and mission success. Once the threats have been reduced, a multirole, increased external carriage capability is then required to prosecute an expanded target set. The F-35 meets both roles with internal carriage of air-to-air and air-to-surface stores. It has a robust external carriage capability of multirole stores, an internal 25-mm Gatling gun for the F-35A, and a missionized 25-mm gun pod for the F-35B and F-35C. It also has a stores management system designed to control current and future store interfaces. The F-35 weapon carriage design is based on experience gained from developing our successful 4th-Generation F-16 and 5th-Generation F-22 fighter aircraft. The design is highly common across the three F-35 variants and provides interoperability and operational flexibility to the theater commanders to engage a wide array of targets. This paper describes the weapons requirements, design development, evolution, and trades performed to define and mature the weapon carriage layout. It also details the associated systems required to load, carry, release, and interface with F-35 weapons.

I. INTRODUCTION

The design and integration of weapon carriage provisions on a stealth aircraft, such as the F-35, are a fundamental part of the overall aircraft design. As such, they must be considered during the earliest stages of configuration development and layout. The weapons suite and carriage requirements are key drivers for both the fuselage and wing designs. Internal carriage affects fuselage sizing and

---

*Director, F-35 Systems Engineering.
†Senior Manager, F-35 Armament and Stores Management.
‡GS-14, F-35 Weapons Integrated Product Team Lead, F-35 Joint Program Office.

structural arrangement, and external carriage affects wing load and flutter requirements. The weapons also define the interfaces required to employ the weapons, which drives system capabilities from a targeting, wiring, power, and computing perspective. The mass properties and resultant weapon inertias are drivers in the overall flight control system and impact control surface definitions. Even the landing gear design and placement are affected by ground clearances and center-of-gravity relationships. In spite of these design constraints, the system does not contradict the F-35 program's pillars of affordability, lethality, survivability, and sustainability.

II. F-35 WEAPONS SUITE AND GENERAL ARRANGEMENT

F-35 weapon requirements were defined by the F-35 Joint Program Office (JPO) in the Joint Contract Specification (JCS). They were later refined during the Concept Definition and Design Research (CDDR) and Concept Development Program (CDP) phases, prior to the System Development and Demonstration (SDD) production aircraft selection. The initial weapons suite from the SDD contract contained effectively all stores in the U.S. and U.K. inventories, as well as smart and older, unguided weapons. Early in the SDD program, we refined the weapons suite to eliminate some of the older weapons that were becoming obsolete, such as the AIM-120/B Advanced Medium-Range Air-to-Air Missile (AMRAAM), naval mines, current generations of the AGM-84 Harpoon anti-ship missile, and the AGM-88 High-Speed Anti-Radiation Missile (HARM).

Newer weapons, such as the Small Diameter Bomb (SDB) and Guided Bomb Unit (GBU) GBU-49 Enhanced Paveway II munitions, were added to the requirements list. This keeps the F-35 weapons suite as modern, relevant, and flexible as possible. The suite enables the F-35 to perform multirole missions, providing flexibility to the operational commander to prosecute a wide variety of targets. Figure 1 shows the resultant internal and external weapons suite.

In addition to the weapons the F-35 is required to carry, the JCS also included other specific requirements that drove the weapons carriage arrangement and weapon station capacities, such as being able to carry external fuel tanks. It also issued mixed parent and dual-carriage requirements for each service. Other influential factors were the variant-dependent mission requirements, such as the F-35A's need for an internal gun and the F-35B/C's needs for a missionized gun. Additional requirements focused on safety and flexibility, such as the requirement for MIL-STD-1760 interfaces at all stations and the ability to safe the weapons systems from outside the cockpit. These design requirements resulted in four internal and seven external weapon stations. Figure 2 shows the general arrangement of the weapon carriage, and Fig. 3 shows the station type and weight capacities.

The commonality of the weapon carriage design across the three aircraft variants was maximized to reduce the differences in the certification process and to minimize cost. The internal weapon bays are common on the F-35A and F-35C

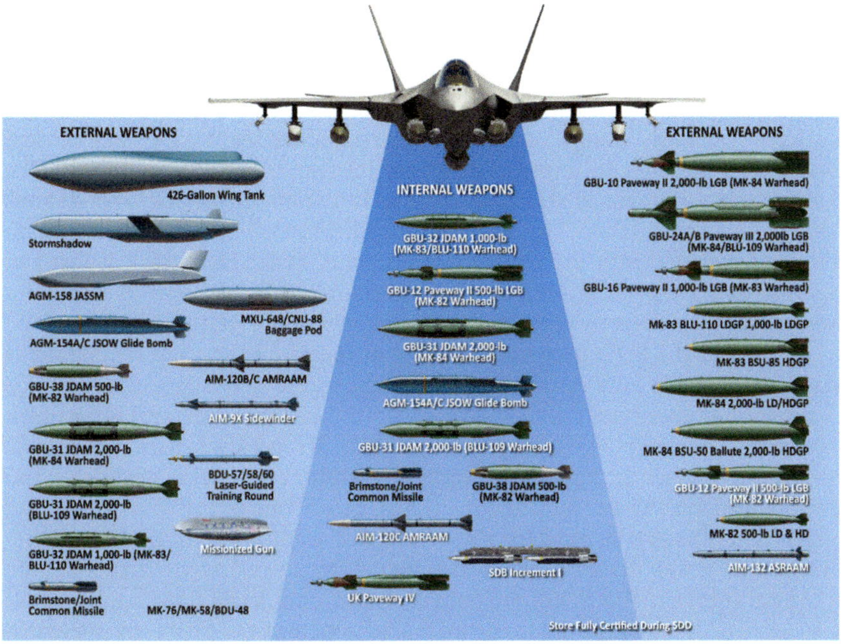

Fig. 1 F-35 internal and external weapon carriage requirements.

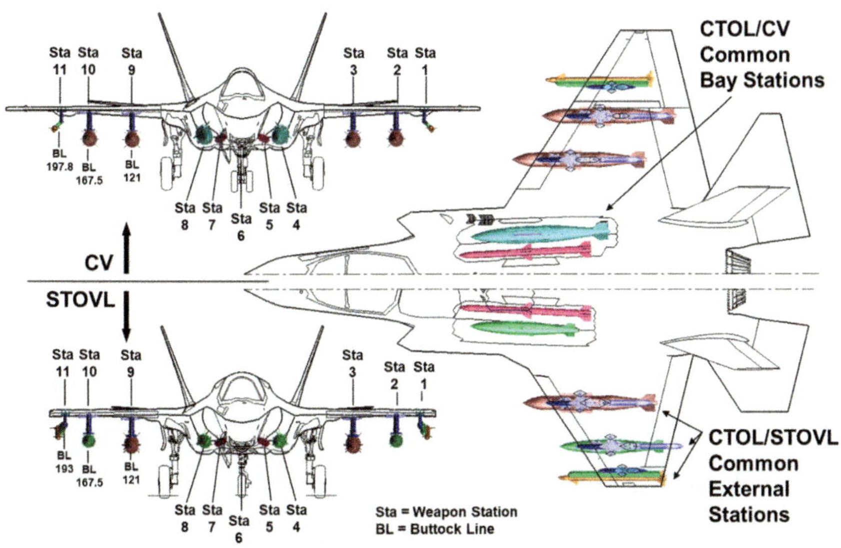

Fig. 2 General weapon carriage arrangement.

variants. These bays are capable of carrying one internal 2000-lb class store with one AIM-120C AMRAAM missile in each bay. The F-35B bay is unique and capable of carrying one 1000-lb class store with one AIM-120C missile in each bay. Air-to-ground or air-to-air stores are carried on adapters on the roof of the bays. Each bay design includes a door-mounted AIM-120C station that rotates with the door to improve separation clearance and loading access. The external wing stations are located on nearly common buttock line stations across all three variants. The suspension and release equipment is common across all variants, with the addition of a unique 14-in. bomb rack tailored to the 1000-lb class internal bays for the F-35B aircraft. The Stores Management System (SMS), interfaces, safing features, and system operation are common.

As part of the SDD program, a subset of these weapons and requirements was required to be fully certified. The weapons in the subset needed to provide a warfighting capacity in time for each service's declaration of initial operational capability. The weapons and specific loadouts we certified during the SDD program are shown in Fig. 4. The certification effort for these capabilities and specific loadouts to be fully certified was based on extensive analysis and testing. The design was developed and validated using a combination of three-dimensional (3D) design tools, wind tunnel tests, and computational fluid dynamics analyses. The F-35's preliminary layout and detail designs were developed using a 3D digital thread to ensure that all systems were effectively integrated and complete. Wind tunnel testing formed the backbone of the validation tests and included 11 test entries with more than 3900 user-occupancy hours and more than 1700 air-on hours. In these tests, we assessed the weapon bay design from both separation and acoustic perspectives. We also defined performance

Station	11	10	9	8	7	6	5	4	3	2	1
Store	Air-to-Air	Air-to-Air Air-to-Surface	Air-to-Air Air-to-Surface	Air-to-Air Air-to-Surface	Air-to-Air	Gun	Air-to-Air	Air-to-Air Air-to-Surface	Air-to-Air Air-to-Surface	Air-to-Air Air-to-Surface	Air-to-Air
Capacity CTOL/CV (lb)	300	2,500	5,000	2,500	350	1,000	350	2,500	5,000	2,500	300
Capacity STOVL (lb)	300	1,500	5,000	1,500	350	1,000	350	1,500	5,000	1,500	300

Over 22,000 lbs of Payload Capacity

Fig. 3 Weapon station capacity.

Block	SDD Certifications Loadings	11	10	9	Int Bays 8	Int Bays 7	6	Int Bays 5	Int Bays 4	3	2	1	Effectivity CTOL	Effectivity STOVL	Effectivity CV
Block 2B	Int: (2) GBU-31 BLU-109 + (2) AIM-120 – Ext: none				□	✕		✕	□				✓		✓
Block 2B	Int: (2) GBU-32 + (2) AIM-120 – Ext: none				□	✕		✕	□					✓	
Block 2B	Int: (2) GBU-12 + (2) AIM-120 – Ext: none				□	✕		✕	□				✓	✓	✓
Block 3F	Int: (4) AIM-120 – Ext: none				✕	✕		✕	✕				✓	✓	✓
Block 3F	Int: (2) GBU-32 + (2) AIM-120 – Ext: none				□	✕		✕	□						✓
Block 3F	Int: (4) AIM-120 – Ext: (2) AIM-9X				✕	✕		✕	✕				✓	✓	✓
Block 3F	Int: (2) GBU-12 + (2) AIM-120 – Ext: (4) GBU-12 + (2) AIM-9X		□	□	□	✕		✕	□	□	□		✓	✓	✓
Block 3F	Int: (2) GBU-31 Mk-84 + (2) AIM-120 – Ext: none				□	✕		✕	□				✓		✓
Block 3F	Int: GBU-39 + (2) AIM-120 – Ext: none	Ӿ			⊞	✕		✕	⊞			Ӿ	✓		
Block 3F	Int: (2) AGM-154 + (2) AIM-120 – Ext: none	Ӿ			□	✕		✕	□			Ӿ			✓
Block 3F	Int/Ext: Block 2B/3F CTOL loadouts + Internal Gun		□	□	□	✕	▽ INT	✕	□	□	□		✓		
Block 3F	Int/Ext: Block 2B/3F STOVL/CV loadouts + Missionized Gun Pod (Excludes UK Stores.)		□	□	□	✕	▽ POD	✕	□	□	□			✓	✓
Block 3F	Int: (2) Pvwy IV + (2) AIM-120 – Ext: none				□	✕		✕	□					✓	
Block 3F	Int: (2) Pvwy IV + (2) AIM-120 – Ext: (2) AIM-132	Ӿ			□	✕		✕	□			Ӿ		✓	
Block 3F	Int: (4) AIM-120 – Ext: (2) AIM-132	Ӿ			✕	✕		✕	✕			Ӿ		✓	
Block 3F	Int: (2) Pvwy IV + (2) AIM-120 – Ext: (4) Pvwy IV	E	□	□	□	✕		✕	□	□	□	E		✓	
Block 3F	Int: (2) Pvwy IV + (2) AIM-120 – Ext: (2) AIM-132 + (4) Pvwy IV	Ӿ	□	□	□	✕		✕	□	□	□	Ӿ		✓	

□ Air-to-Ground Store Ӿ AIM-9X/AIM-132 ✕ AIM-120 ⊞ SDB ▽ Gun

Fig. 4 SDD certified weapon loadouts.

requirements for the Suspension and Release Equipment (S&RE) and validated our separation models to support flight test continuation criteria. The weapon certification effort culminated in the successful separation of 110 air-to-ground and 73 air-to-air stores during the flight test program. The effort also led to 46 Weapon Delivery Assessment (WDA) engagements in which the total end-to-end systems were exercised to validate the overall air system employment's accuracy. The details of the weapon separation and certification efforts are described in Ref. [1].

III. INTERNAL WEAPON BAY DESIGN

The final F-35 configuration was heavily influenced by choices made with respect to the weapon bay design. From a weapon arrangement standpoint, designers often discuss packing efficiencies associated with rotary, stacked, flat, and tandem weapon bays; however, it is not until these arrangements are integrated into an aircraft concept that the resultant system can be evaluated. As with all designs, a balance of competing requirements must be struck. From a weapons standpoint, the final configuration must result in:

- Tight packaging solutions while also optimizing weapon loading
- Open bay acoustic environment

- Employment timelines
- Station failure immunity
- Separation risk
- Loadout flexibility

Embedded in the arrangement decision are lower-level impacts that may significantly influence air vehicle weight, handling qualities, and performance. These include weapon bay door configuration, the need for acoustic suppression devices, provisions for maintainability, and mission reconfiguration. Finally, the F-35 must provide these 5th-Generation stealth features while also providing a solution across three separate variants and for a variety of users. This implies additional constraints on loading solutions and structural load considerations.

Key to making F-35 weapon arrangement decisions was the list of weapons required for internal carriage. Many of the tightly packed arrangements often advertised by weapon suppliers and advanced design advocates depict a limited number of weapon configurations. This allows more efficient options to be employed when it comes to rotary or stacked bays. However, once the full list of weapons is explored, many of these options eliminate themselves. Additionally, rotary and stacked configurations often imply a deep weapons bay. Historically, deep weapons bays drive higher acoustic levels. This leads to the need for acoustic suppression devices to comply with weapon qualification levels, and adversely impacts the structural integration in a fighter [2]. Early F-35 wind tunnel testing evaluated the acoustic environment associated with the relatively shallow bay and intended operational envelope. The testing indicated that the environment would allow for the incorporation of all required weapons without needing acoustic suppression devices. Because the use of ramps and spoilers adds weight, complexity, and bay length, the decision was made to move forward without dedicated acoustic suppression devices. This option could have presented some risk if the actual environment proved to be incompatible with the weapon qualification levels; however, the benefit to the aircraft was sufficiently significant and long lasting to justify taking the risk. Further, the qualification levels of most weapons are based on the existing aircraft environment. Accordingly, increasing the weapon qualification (if required) is typically a matter of retesting to higher levels, and does not necessarily indicate an inherent limitation of the weapon.

Aside from rotary and stacked weapon bay configurations, tandem weapons bays were also an option. However, from a weapons standpoint, these bays are less desirable due to their impacts on loading. They are especially problematic when loading the Navy's F-35C Carrier Variant (CV) aircraft with the tail parked over water. Additionally, tandem bays cause undesirable aircraft center-of-gravity excursion, given the weight of the weapons to be carried. The final F-35 configuration is a roof-mounted air-to-ground weapon and a door-mounted air-to-air missile that offers a good mix of each weapon bay type. It has the efficiency of stacked bays without the acoustic and loading issues, and the flexibility of a purely flat bay. After the overall configuration was selected, detail decisions needed to be made with respect to design constraints to establish the final aircraft sizing. One unique feature of the F-35 weapons bay is the air-to-ground stores are angled nose inboard. This was done to improve the integration of the bays with the fuselage and was validated through wind tunnel testing.

One of the most critical and controversial decisions to be made involved weapon envelope size. This refers to the volume to be allocated and protected around the weapons to allow for adequate installation, carriage, and separation. Again, the large mix of weapons to be carried by the F-35 played a major role in these decisions. Historically, aircraft weapons bays were pristine boxes with fixed S&RE locations. The bays were not encumbered with hydraulic, pneumatic, and electrical routing or an abundance of vehicle system components. However, overall volume in a fighter aircraft is at a premium. As such, insisting that all routing and systems be located outside the weapons bay significantly increases the size of the aircraft. Ultimately, the decision was made to allow systems and weapons to coexist in the weapons bay. This required the qualification of all systems to weapon bay environmental levels. The resultant weapon bay design is highly integrated with systems, routing, and weapons to ensure the smallest package and highest aircraft performance possible, as Fig. 5 illustrates.

Fig. 5 Bottom view looking into the weapons bay.

Sharing the weapon bay volume necessitates tight controls to ensure that the evolving systems and routing do not intrude on space needed for weapons. To minimize the weapons package volume, we performed a tradeoff between location commonality and bay length. If a singular location was allocated in the bay, this would provide an easy and efficient installation from a structural standpoint. This is because all loads would be carried by the same structure in the same way, no matter which weapon was installed. However, weapons come in all shapes and sizes, and hook locations are not always located in the middle of the weapon. Consequently, having a common hook location would result in a relatively large weapon volume.

By contrast, we could locate all weapons in the minimum volume required to allow the resultant hook and bomb rack location to be a fallout. In this way, the bay volume and overall size of the aircraft could be minimized. Unfortunately, this would create a need for a variety of Alternate Mission Equipment (AME) adapters and hook locations within the bay. However, even with the weight and logistic penalties for this arrangement, the benefit to the overall aircraft performance was undeniable. Therefore, the final weapon configuration embodied multiple bomb rack locations. To accommodate this solution, we identified several adapters and developed a system to quickly latch them into the bay. We designed the Ordnance Quick Latch System (OQLS) to permit quick bay reconfiguration by latching and snubbing adapters in a manner similar to that used by bomb racks to latch and snub stores. Figure 6 shows the OQLS layout and a typical bomb rack adapter when clipped into the OQLS.

With the location of the stores established, the final volume required for carriage and employment could be established. First, however, we needed to decide how much volume was needed around each store as a stay-out zone. Key factors to consider were the volumes required for maintenance, weapon motion (deflection) during carriage, and uncertainty associated with weapon separation. Historically, military specifications, such as MIL-STD-1289 and MIL-I-8671, would be used to establish clearance requirements. However, these specifications are often

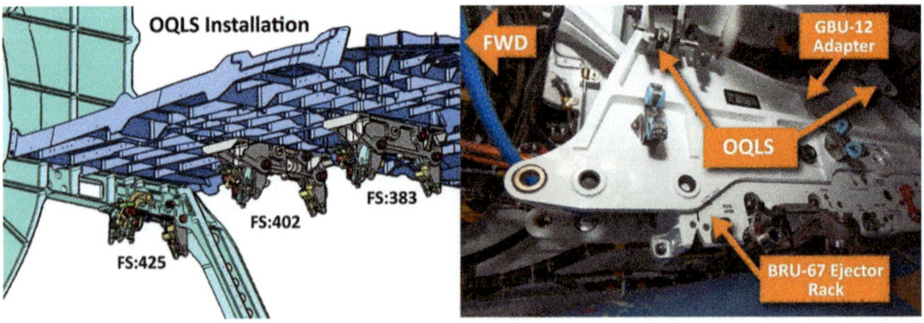

Fig. 6 OQLS arrangement and typical bomb rack adapter installation.

conservative in some areas and vague in others, resulting in larger bay volumes. In general, the aircraft performance reductions that would be driven by strict adherence to military standards were deemed unacceptable. The F-35 JCS did not require the use of historical specifications for weapon spacing. However, early versions of the JCS did require the maintenance of a 2-in. stay-out volume around all weapons. Some exceptions were allowed between the hook points where relative motion would be at a minimum. As the program progressed, there were multiple identified and proven instances in which this blanket requirement would significantly penalize the air vehicle if strictly followed. Therefore, the specifications were ultimately changed to preclude store-to-store or store-to-aircraft contact, with no specific clearance requirement identified. This approach required the development of complete, specific design guidance to establish and protect the weapon stay-out zones.

Once again, a balance of risk and benefit was required in establishing the weapon stay-out zones. Too much volume allocated to the weapons would result in an overly large and heavy aircraft, decreasing performance as a result. Too little volume would result in high-risk weapon carriage and employment, with the potential for aircraft operational envelope limitations or configuration restrictions. The general approach chosen for the F-35 was to create a volume that reflected tolerances, freeplay, and a deflection of the store throughout the operational envelope. In addition, we added a small margin to account for analysis fidelity. This margin was based on differences discovered between analysis accuracy and flight test verification, as proven on the F-22 program.

In the end, decisions made to optimize the overall air vehicle made the weapon volume allocation more difficult. This is because weapon attachment involves multiple interfaces (S&RE, adapters, OQLS) with associated flexibility and freeplay. Deflections associated with these interfaces are significant when fractions of an inch are important to the overall integration of the weapons.

Ultimately, the weapon bay volume will be challenged again and again until it has been reduced to its bare minimum. Therefore, we performed a detailed modeling of each weapon, to include loads associated with its specific carriage and employment envelope. The deflection of a store depends on where and how it is attached, so the envelopes that result from the deflection analysis are larger when farther away from the attach points. When viewed from above, the final weapon envelopes appear narrow in the middle, near the bomb racks, and larger toward the nose and tail of the stores. As a result, these envelopes are referred to as bow ties. We rolled together and consolidated the bow ties from each weapon to create the overall weapon volume. This clearance approach, shown in Fig. 7, is a unique design feature of the F-35.

When looking aft at the volumes, the ejection sweep was largely influenced by the design of the S&RE, which constrains the store during the ejection stroke. The stay-out volumes also reflected a narrow band of deflection at the top that flared to a 10-deg sweep at the bottom. However, the width assumptions made for the weapons bay required that the lateral motion of

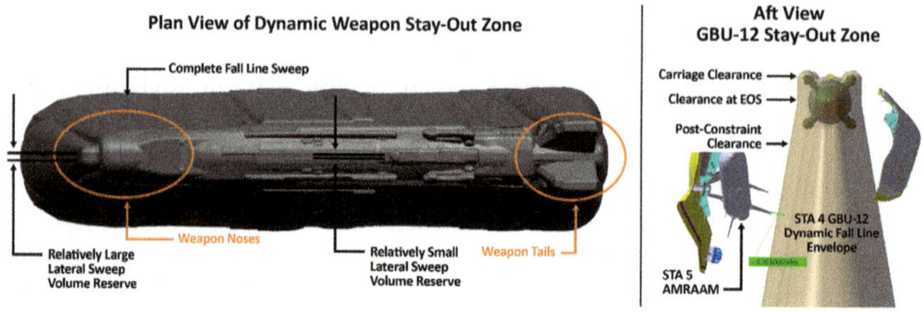

Fig. 7 Weapon stay-out volume bow tie arrangement.

the stores never exceed 0.75 in. total at End of Stroke (EOS). This was particularly critical between the suspension hooks in order to accommodate tolerance, freeplay, and deflection. This resulted in a significant design stiffness driver for the S&RE.

After establishing the volume, we needed to address other issues due to the highly integrated relationship between the required weapon stay-out volume and the volume used by systems and structure. Namely, we needed to determine how to ensure adherence to the volumes during the design, manufacture, and field service phases. Each phase of the volume life cycle required individual treatment. The design phase required establishing configuration control processes, to include design release requirements and volume conflict dispute resolution boards. For the manufacture phase, the central question was whether standard manufacturing tools and processes could ensure adherence to the volumes. This was especially important in areas where there is some allowed variation in harness routing and support sizing. Stay-out volume maintenance during in-service use also needed to be accommodated to prevent clearance conflicts due to normal aircraft operations, modifications, and maintenance. To ensure that adequate clearance was retained, we created a physical tool that included all required weapons, AME, deflections, and fall clearances.

The tool we created for the field service phase is now used on every F-35 that leaves the factory to ensure that the volume is conflict free. Figure 8 shows one of the fit-check tools used to inspect the bay and verify the bay volume. One key feature of the tool is its inclusion of removable sections, which enables access to visualize and measure specific interfaces. Currently, we are investigating the efficiency of using a laser mapping of each bay as an alternative to the physical tool.

Another concern is that maintenance activities in the field may impact the final configuration of systems near the stay-out zones. This is especially difficult for maintainers in the field to recognize because not all weapons are installed during maintenance, making it difficult to understand the potential impact. To address this, aerospace equipment instructions have been created to identify critical areas for inspection as the aircraft is maintained.

The ability to load weapons was a key consideration when establishing the configuration for 5th-Generation internal carriage aircraft, especially for the limited volumes of internal weapon bays. Loading considerations for the F-35 were complicated by the fact that many different services and countries would be using the aircraft with a variety of support equipment. One of the toughest loading requirements was associated with shipboard operations. In this situation, push-up loading equipment (e.g., the legacy Air Force MJ-1 munitions jammer) is not suitable due to space requirements. It is also ill-suited due to the need to better control the store being loaded during rough weather conditions. Historically, the solution for shipboard loading has been the Single Hoist Ordnance Loading System (SHOLS). However, the use of SHOLS within the limited confines of the F-35 weapons bay is difficult due to volume and access requirements. Access to both sides of the weapon is required to permit the installation and disconnection of the cable system. Originally, this was to be accommodated on the F-35 through access panels located in the main wheel wells; however, as the design progressed and more weapon locations were defined, this arrangement was no longer feasible.

As a result, the Ordnance Hoist System (OHS) was developed, originally conceived as a permanently installed onboard hoist system. The OHS would raise the adapter, bomb rack, and weapon into the bay using straps located at the front and aft ends of the adapter. The hoist spools were to be driven by small, handheld battery-powered motors by two maintainers. Coupled with the bomb racks in adapters, the OHS would allow the weapon to be fully mounted

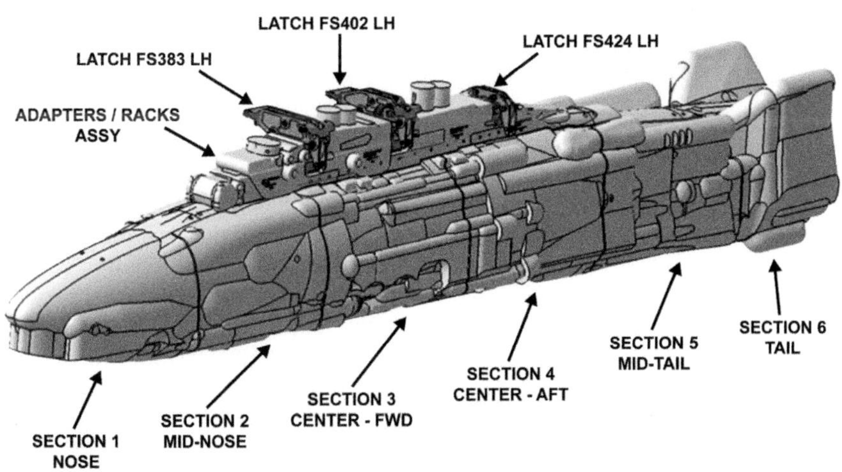

Fig. 8 Weapon bay clearance fit-check tool.

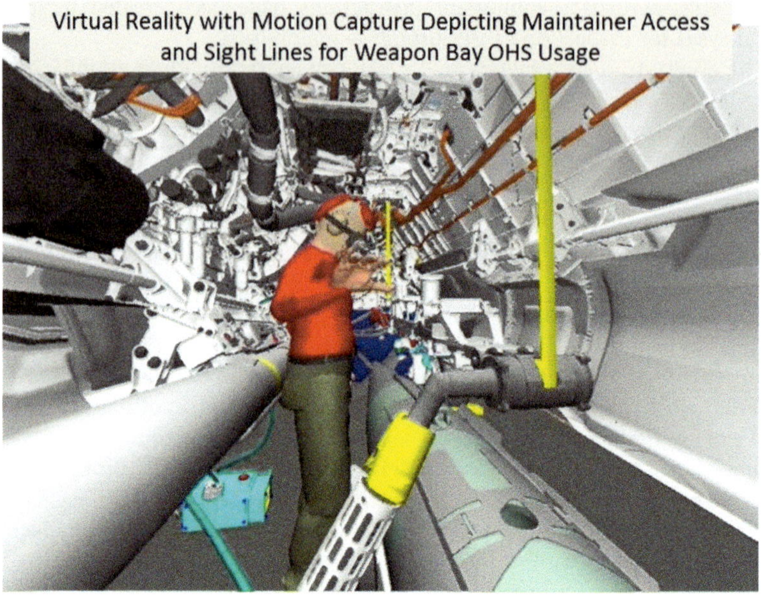

Fig. 9 Weapon bay maintainer access study.

on the rack and lanyards rigged outside the weapons bay. Once completed, the whole assembly would be raised into the bay and quickly latched into the OQLS. The system worked so well in demonstrations for the F-35C and F-35B aircraft that the OHS was incorporated onto the F-35A as well. This was because the MJ-1 loading of the stores left little room for maneuvering error. While exploring various AME, S&RE, and weapon loading equipment, the F-35 team used computer simulations and mockups, along with virtual reality, extensively. Figure 9 depicts one of the early studies showing that maintainers had room to conduct required operations with an onboard hoist system.

As the design of this system progressed, additional requirements for shipboard operation evolved. The concept of using handheld drill motors for power was lost, and a requirement for a stand-alone 270-V battery was added. The battery was sized to load multiple aircraft and included a sophisticated controller that automated many aspects of the loading process. This increase in complexity, along with changes made to receive a Conformité Européene (CE) marking, ultimately led to the elimination of the OHS from F-35A aircraft. Following this, a manual drive replacement was developed for the powered components of the OHS. Although the OHS has evolved, it remains the only approved method for loading weapons shipboard. With the OHS removed from the F-35A, a new Length-Extender Table Adapter (LETA) was developed to install on the MJ-1 load table and improve clearances.

IV. EXTERNAL WEAPON CARRIAGE DESIGN

The emphasis of the F-35 is on internal weapon carriage, but there will always be ferry and lower threat missions that benefit from external carriage. As a result, all F-35 variants include seven external hardpoints: three on each wing and one on the fuselage centerline. As noted previously, stores specified for external carriage include a variety of weight classes and types. Station type and weight capacities (shown in Fig. 3) reflect the differences in station capabilities of the F-35B aircraft for structural weight considerations. Fuel tank carriage is accommodated at the inboard stations, and each of the four inboard stations can accommodate both parent and multiple-store carriage. The outboard stations are for air-to-air weapons and were designed to improve stealth performance over traditional external pylons and launchers, while still balancing the cost and serviceability of traditional AME. Figure 10 depicts an F-35A heavy external mission loadout with external air-to-air missiles on the dedicated missile stations. It also shows 2000-lb air-to-ground parent carriage stores at the multirole air-to-ground or air-to-air stations.

From a configuration standpoint, the locations of the pylons were established in a manner similar to that for legacy fighter aircraft; fowever, the fact that the design needed to accommodate three variants meant that additional factors

Fig. 10 Heavy weapon mission loadout.

needed to be considered. For example, all external designs need to consider locations to accommodate wing flaps, landing gear doors, and ground lines for all the landing gear and pitch angle variations. Owing to this, the addition of the seaborne F-35B/C variants added some challenges. Specifically for F-35C, the deck of the carrier has multiple obstructions that can become an issue. These obstructions must be considered when evaluating high roll angles and large stores. Furthermore, the deck obstructions, cables, and launch shuttle can create additional clearance challenges for some of the designs. Examples are the gun pod on a centerline station and outboard pylon clearances. Additionally, the plume impingement from the F-35B roll post is another design driver that has to be considered. These are columns of hot gas ejected from the engine at the wing-to-fuselage intersection used to control roll attitude when in vertical flight. It is important to minimize store heating in the basic positioning of the stores and pylon design. Additionally, the carriage pitch angle of the stores is usually established to eliminate drag in cruise flight conditions. Therefore, the stores are nose down to accommodate Angle of Attack (AoA) and can be nose inboard to accommodate wing wash. The mix of variants, missions, and required stores made all these compromises a challenge.

One of the tools frequently used in F-35 development from a human factors standpoint was virtual reality. Many of our questions concerning maintainer access, vision, and safety with conducting operations were addressed using traditional mockups. For the remainder, we supplemented those assessments with virtual assessments. Figure 11 depicts a maintainer in a mixed (i.e., having some of the stores and physical constraints actually connected with virtual hardware) virtual environment. The F-35 weapons team used this tool extensively early in the concept evaluation phase to validate the acceptability of initial design.

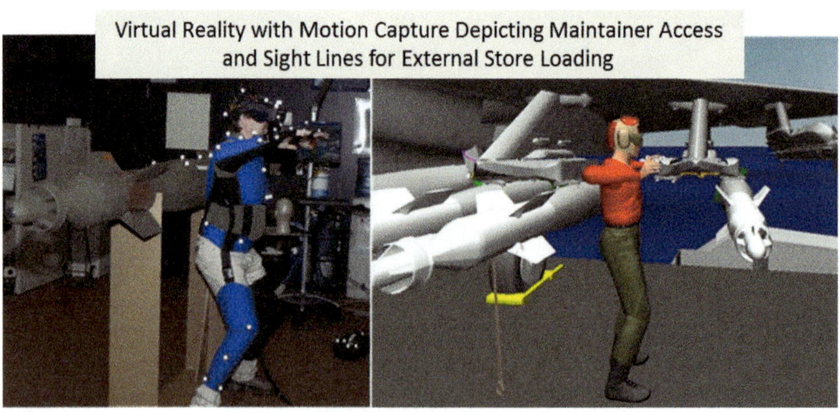

Fig. 11 Maintainer virtual environment clearance study.

There is a high degree of commonality among the external AME. However, there were differences required due to the fact that the F-35C wing and flutter requirements were significantly different from those of the F-35A/B wing. This meant that pylons used on the F-35C aircraft needed to carry the stores farther forward, with respect to the wing hardpoints, than the F-35A/B pylons. Despite this difference, however, there remained a high degree of commonality; for example, the same pylons are used at inboard and outboard stations. Rather than changing the pylon structural core to accommodate the significant differences in wing twist, we provide an adapter. Other commonality-driven features include the fact that the sway braces can be reversed to allow the installation of the pylon on either the left or right wing. In addition, such components as the pivot fittings, fairings, and the entire Pneumatic Power Source (PPS) are also common. Commonality in the design has significant advantages for provisioning and spares, as well as cost reductions that come from manufacturing economies of scale. Affordability has always been one of the foundations of the F-35 program.

The air-to-air pylons at Stations 1 and 11 presented a unique challenge, and commonality was once again a key consideration. Ideally, from a structural standpoint, the pylon would accommodate a launcher directly below the hardpoints. However, adjacent store station clearance and structural provisions on the three variant wings meant that the missile could not be located directly below the hardpoints and needed to be offset. Rather than creating a unique canted pylon, the pylon was designed with two launcher attach points at its base. This allowed for launcher offset while also allowing the same pylon to be used on the left or right wing by simply switching the location of the launcher. Additionally, the same pylon could be used on all three variants' wings—despite significant angular differences—by simply changing a set of bushings used during the attachment.

The rail launcher used at the base of the pylon was originally designated as a non-low observable LAU-148 and LAU-149. The only difference between the two was that only the LAU-149 contained a gas generation system. This was needed to provide continuous cryogenic cooling to the AIM-132 Advanced Short Range Air-to-Air Missile (ASRAAM) seeker head. By contrast, the LAU-148 contains ballast. Either launcher can be used with an AIM-9X. The fact that the LAU-148 used ballast was done to create similarity between the two launchers such that extensive flutter and loads testing would not need to be repeated. As the F-35 program matured, the desire for improved Low Observability (LO) of the air-to-air mission evolved. This new requirement resulted in the design of the LAU-151 and LAU-152 rail launchers and an associated upgrade to the air-to-air pylon (now designated as a SUU-96). Figure 12 depicts these designs. To limit the financial impact of potential regression flight testing, a significant effort was expended to design the new AME to be very similar (structurally) to the initial design. The cost savings of this approach over a completely new qualification and certification program were significant.

Fig. 12 SUU-96 pylon and launcher installation.

V. SUSPENSION AND RELEASE EQUIPMENT DESIGN

F-35 S&RE is a family of components that carries and employs air-to-ground and air-to-air weapons. Safe release and separation is achieved through two eject pistons powered by 5000-psi nonpyrotechnic pneumatics. The primary air-to-ground bomb rack is the BRU-68/A, which can carry stores from 500- to 2000-lb class using either 14- or 30-in. conventional suspension hooks. It is used internally on the F-35A/C variants, and on all three variants in the air-to-ground pylon. The BRU-67/A is similar to the BRU-68/A but is only used in the F-35B internal weapons bay, and it can carry 500- to 1000-lb class stores using conventional 14-in. hooks. The BRU-67/A is a product of the Short Takeoff and Vertical Landing (STOVL) Weight Action Team (SWAT) program, and it reduced air vehicle weight by 20 lb. The LAU-147/A is specifically designed to eject-launch the AIM-120 missile from any of the internal weapons stations on all three F-35 variants. Since the LAU-147/A is uniquely designed to eject only the AIM-120 missile, the sway brace is customized to the diameter of the missile. It incorporates an anti-rotation feature in the sway braces to decrease missile roll tendencies discovered during early wind tunnel testing.

S&RE also has unique store control requirements to accommodate the tight internal carriage volume allocations. Historically, two methods have been employed in S&RE to ensure that store clearances are maintained during the ejection event. The first is to push the store hard enough to achieve high-EOS velocity, and the second is to restrain the store's lateral motion during the ejection event.

The first concept is limited by the structural loads that are imparted to the aircraft and store, as well as the physical constraints of the stored energy package. The second concept relies on the stiffness of the S&RE and backup structure to provide lateral constraints. Reducing deflections by increasing system stiffness increases weight.

A hybrid approach combining both the restraint and velocity concepts was chosen for the F-35 for the following reasons. The variety of weapon locations in the internal weapon bays requires movement of S&RE. This requirement to translate the S&RE necessitates locating the pneumatic power storage bottle in the S&RE housing. This constrained the maximum volume of gas and resultant energy available for ejection. Whereas actual EOS performance is a function of many variables, two of the primary drivers are ejection stroke length and stored energy content. Packaging issues tend to drive the stroke length. As a result, original versions of the bomb racks contained multistage pistons to achieve the desired stroke length while fitting in the rack housing. To constrain the stores, the sway braces were designed to move with the store and provide lateral restraint through inertia only. The alternative was to constrain the store throughout the stroke and release the hooks at EOS. Historically, racks with true lateral constraint have been complex, heavy linkages that open the store attach hooks at EOS. The F-35 hybrid approach eliminated much of the weight concern from these linkages by using simple pistons with moving sway braces; however, it still required significant tailoring of the design to accommodate the required stiffness. We built an analytical model of the weapons bay structure and S&RE, and we performed complex dynamic simulations. With these simulations, we calculated the lateral deflection of stores during the ejection event throughout the operational envelope. Unfortunately, it was shown that the BRU-67/68 could not achieve the required clearance with multistage pistons. This is the reason for which the ejector towers of the BRU-67 and BRU-68 were extended beyond the rack housing to eliminate the multistage pistons. The LAU-147 retained the original two-stage piston because the door-mounted location required minimal launcher height. Moreover, the available volume could accommodate the AMRAAM ejection path. This hybrid design provides the mix of performance and constraint needed to live within the very strict lateral confines of the F-35 bay.

S&RE was designed to accommodate nonpyrotechnic ejection while conforming to the conventional bomb rack load and form requirements of MIL-A-8591 and MIL-STD-2088. Because the air-to-ground stores vary in diameters (ranging from 8 to 23 in.), the bomb rack units incorporate a semi-automatic pivoting sway brace. This sway brace allows the maintainer to quickly adjust the rack to the many F-35 stores. This feature simplifies store installation and removal procedures in the limited space of a tight weapons bay. All three S&RE units incorporate the same reversible in-flight lock that provides mechanical blocking and electrical interrupts to prevent inadvertent store ejection. These enhanced safing features provide at least two levels of protection whenever the maintainer is working around the rack.

TABLE 1 F-35 SUSPENSION AND RELEASE EQUIPMENT

	BRU-68	BRU-67	LAU-147
		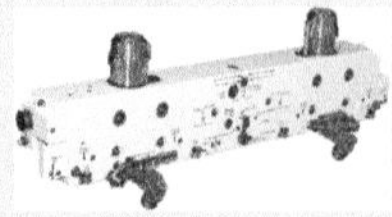	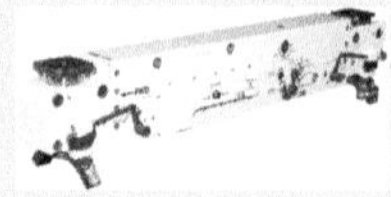
Envelope (L × H × W) (in.)	36.0 × 10.9 × 4.0	32.0 × 10.9 × 4.0	36.9 × 6.9 × 4.0
Specification Weight (lb)	87.5	67.5	63.9
Store Class (lb)	500/1000/2000	500/1000	350
EOS Velocity (fps)	20/15/11	20/15	25
Piston Type/ Length (in.)	Single/6.9	Single/6.9	Dual/7.5

The level of Prognostic Health Management (PHM) incorporated into the S&RE is consistent with maintainability goals for the F-35. Ejector technology continues to advance, and other options to resolve 5th-Generation fighter integration issues may appear within the next few years. However, F-35 S&RE (Table 1) represents a significant departure and advancement from 4th-Generation pyrotechnic systems.

Nonpyrotechnic ejection was a key requirement for F-35 S&RE. Legacy ejector systems incorporated in 4th-Generation (and earlier) aircraft relied on pyrotechnic cartridges to provide ejection power. By changing to a nonpyrotechnic power source, many cost and maintenance issues can be eliminated; for example, pyrotechnic cartridges require frequent rack disassembly to remove debris and replace damaged parts from the burning combustibles. Furthermore, inconsistent burn profiles caused by slow- or fast-burning cartridges generate performance anomalies that can be eliminated through the use of stored gas systems. In addition, shipping and storing pyrotechnics creates safety issues and an expensive logistics footprint. These issues were evaluated in a variety of feasibility studies performed in the early 1990s. The Life-Cycle Cost (LCC) advantages of pneumatic racks for the F-35 program were estimated at between $2.3 and $2.6 billion in savings for internal carriage [3]. If external pylons were also considered, an additional $1.67 billion in savings was estimated. As a result, the F-35 air vehicle was required to incorporate a nonpyrotechnic power source into the S&RE.

There are many ways to generate and maintain the high gas pressures needed for store ejection. The solutions generally comprise three types of systems:

1. Stored gas systems that must be charged on the ground

2. High-pressure gas generation systems that provide onboard pumping
3. Hybrid systems that use hydraulic intensified gas systems, both closed and open loop

Aside from the requirement to provide a high-pressure volume of gas, the thermal variations present in a high-performance aircraft will drive pressure variations that need to be accommodated as well. Given the temperature differences between sea level and 40,000 ft, the system loses as much as one-third of the pressure when flying at altitude. Therefore, the system must be either regulated or supplemented to accommodate the reduction in pressure.

The time required for the system to recharge or be serviced in between missions is also a design driver. For the F-35, the time allotment for an integrated combat turn did not accommodate servicing each weapon station's pneumatic power module. Therefore, we decided to eliminate the need for ground charging of all racks and launchers, both internally and on the wing stations. As a result, the F-35 has an onboard PPS that generates the required 5000 psi source pressure for the racks and launchers. The components of this system are depicted in Table 2.

Although the pumps and controllers needed to accommodate this requirement add weight and complexity to the system, the task and support equipment benefits to the maintainer outweigh the costs. To minimize the system sizing, the PPS was designed as a maintenance system. In it, the pump sizing and power requirements assume that S&RE reliability only requires the PPS to pump four empty racks at once, four times in the life of the aircraft. Most pressurizing operations aim to maintain pressure due to low temperatures, or to repressurize the partially discharged S&RE following a store release. The PPS concept of operations is to recharge all racks following a store release to full operating pressure during the return flight. This must be done before the aircraft is chocked to begin the integrated combat turn. This operational concept requires that the

TABLE 2 F-35 PNEUMATIC POWER SOURCE EQUIPMENT

	Motor Compressor	Filter/ Manifold	Pylon Filter	Electronic Control Unit

Envelope (L × H × W) (in.)	12.4 × 4.3 × 4.4	13.1 × 4.1 × 5.3	5.6 × 3.1 × 4.7	8.5 × 3.9 × 4.4
Weight (lb)	11.0	7.0	3.0	2.7

racks be normally charged even when in storage. This concept was a major driver of S&RE requirements because the racks had to be qualified for handling, storage, and shipping while fully charged to 5000 psi.

The PPS was sized primarily by the needs of the internal weapon bay. A single PPS consisting of a motor/pump assembly, electronic control unit, and filter/manifold assembly provides high-pressure pure air to all four internal weapon stations. To maintain performance at any altitude, the PPS receives inlet air from the aircraft's Power and Thermal Management System (PTMS). The PTMS typically provides air at around 18 psi, no matter what the ambient air pressure. The pump charges a filter/manifold assembly that subsequently distributes the air to each weapon station through a series of valves, as directed by aircraft commands. Each station is isolated through specific valving and sequential distribution rather than plumbing to a common bottle. This isolation eliminates the potential to lose charging capability for all stations due to a failure (leak) at one. For added reliability, each rack contains a check valve to ensure that it is not possible to back-flow out to the PPS once engaged.

Compression of ambient air results in significant amounts of water expulsion because the relative humidity will exceed 100% at high pressure. As a result, the system contains devices to clean, dry, and expel excess moisture from the compressed air. Removing this moisture prevents contamination of the racks that may prevent operation from icing. For commonality, the motor/pump assembly and electronic controller used for the weapons bay is the same motor/pump and electronic control unit used in the air-to-ground pylon. Because the PPS supports a single rack in the air-to-ground pylon, the filter assembly does not incorporate a manifold or isolation valves, allowing air to flow directly into the rack reservoir. Unlike the weapons bay with PPS inlet air from the PTMS, the pylon PPS inlet air is from ambient. Accordingly, pressure varies as a function of altitude. Although the pylon PPS performance is reduced due to altitude effects, recharge times are equivalent to charge times for all the main bay racks because it only services one rack. Therefore, the pylon and all internal racks are fully recharged by the time the aircraft returns to the chocks to begin an integrated combat turn.

Mission reconfiguration is a key requirement for weapons in the bay, and there is a wide variety of S&RE locations within the bay. To accommodate these, multiple flexible hoses are required to connect the S&RE to the PPS after the adapter and bomb rack have been loaded. The hoses are equipped with Quick Disconnect (QD) fittings and incorporate a safety device that prevents de-mating the QD until the hose has been depressurized. As testing of the F-35 progressed, it was apparent that even the smallest of air leaks can have a profound impact on overall system performance. As a result, the system design has evolved over time. Improvements incorporated range from design changes in the hose QD fittings to process changes in PPS manufacturing methods. To improve the maintainer experience, enhanced courseware is being deployed explaining system operation, troubleshooting, and normal operational expectations.

VI. GUN SYSTEM DESIGN

The gun system installation was defined by the requirements of each aircraft variant. The F-35A Conventional Takeoff and Landing (CTOL) aircraft required an internal gun system. The F-35B STOVL and F-35C CV aircraft required a missionized gun system that could be installed when necessary to meet specific mission requirements. The primary requirement of the gun system is that of a Close Air Support (CAS) mission, but it must also be able to accomplish an air-to-air combat role. The program specification includes an open fire range of 9000 ft, a specified strafe dive angle and speed, and a requirement for three effective passes per ammunition loadout. Due to the dive angle and speed requirement, a limited time is available to complete each effective pass.

During the competition phase of the F-35 program, a trade study was completed to determine the best gun system to meet the aircraft specification requirements. The trade study reviewed five possible gun solutions that included three Gatling guns: the M61A2 20-mm, the GAU-12 25-mm, and the advanced cased-telescoped round CT-525 25-mm. The possible solutions also included two single-barrel guns: the BK-27 27-mm and the Aden 25-mm. Each system was compared with respect to its ability to be integrated onto the aircraft. This was based on weight, volume, interface loads, round capacity, and power requirements. It also factored in the ability to be integrated internally into the Air Force aircraft and into a missionized pod for the Marine Corps and Navy.

The systems were evaluated with respect to performance capabilities that included system firing rates, round accuracy, and round effectiveness against the target set stipulated in the aircraft system specification. Development and Unit Recurring Flyaway (URF) costs were also compared for each of the systems. After reviewing the data, we determined that the Mauser BK-27 27-mm system was the best system for integration for both the internal gun system and missionized gun system. This determination was based on the following factors:

- It is the lightest-weight solution.
- It requires no hydraulic power from the aircraft.
- It has the lowest predicted development and URF costs.
- It requires only 152 rounds.
- It has high round accuracy and effectiveness.

The cased, telescoped CT-525 had the potential advantages of increased lethality and reduced weight; however, the design was not mature enough for the F-35 and would have introduced unacceptable risk in terms of development and cost if selected.

During the early detailed design phase of the production aircraft, a concern surfaced with respect to the LCC and lethality of the BK-27 and its ammunition. As a result, we conducted another trade study to ensure the selection of the best gun system solution. The new trade study included the BK-27, GAU-12, and GD-425 (a four-barrel variant of the GAU-12). The trade study evaluated the three gun systems based on the same criteria used in the initial trade study that had led to the BK-27 being chosen; however, this study focused on lethality and LCC as well, along with the availability of 25-mm and 27-mm rounds.

The results of the trade study showed that the GAU-12, shown in Fig. 13, was the best solution. It had lower development costs, lower URF costs, significantly lower LCC, and improved lethality. Further, the 25-mm rounds were available to the services. The GAU-12 selected was to be hydraulically driven, fire 4000 shots per minute, and use the 25-mm rounds already in the United States' inventory. The GAU-12 is a reverse-clearing gun system. It reverses direction after each burst and brings unfired rounds from the last burst back through the gun and into position to be fired on the next burst. This reverse-clearing procedure ensures that the highest number of rounds in the ammunition handling system will be fired during normal-burst firing.

GAU-12 development design and testing was in progress when the F-35 program entered into the SWAT weight-reduction program. To support SWAT program goals, the GD-425 four-barrel Gatling gun was reviewed to determine the weight savings that could be achieved over the five-barrel GAU-12 system. It was determined that a weight savings of 35 lb could be achieved by utilizing the GD-425 system for the F-35A. Although this change did not affect the empty weight of the F-35B variant, it was determined to be a good trade for the

Fig. 13 GAU-12 gun.

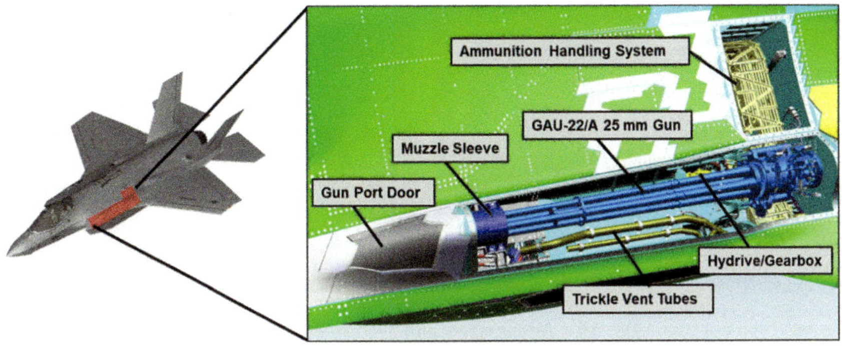

Fig. 14 F-35A internal GAU-22/A and ammunition handling system installation.

overall F-35A configuration. The weight reduction was accomplished by removing one barrel and one bolt assembly and its supporting guides. We also reduced the diameter required for the new gun housing and rotor. The GD-425 (officially referred to as GAU-22/A) is a hydraulically driven gun with a firing rate of 3000 shots per minute. The ammunition feed system is a hydraulically driven single-bay, linear-linkless system capable of supplying a 181-round total system capacity. It enables three operational employment opportunities of 60 rounds each.

Figure 14 shows the F-35A internal GAU-22/A and linear-linkless Ammunition Handling System (AHS) installation. The AHS was originally designed to accommodate three types of rounds: PGU-20/U armor-piercing incendiary, PGU-23/U target practice, or PGU-25/U high-explosive incendiary. Later, more rounds were certified, including PGU-32 semi-armor-piercing, high-explosive incendiary; PGU-47 armor-piercing explosive; and PGU-48 fragmented armor piercing. The gun has a rotary action mechanism into which the percussion-primed 25-mm ammunition is fed from the AHS by the linkless carrier into the breech bolt assemblies. The rounds are chambered and fired, and spent cases are extracted and ejected back into the AHS as the gun rotates. The gun is reverse rotated to clear unspent rounds after burst firing. It includes provisions for automatically arming and safing the gun on each firing burst.

Gun system boresighting is accomplished with software rather than hardware adjustments. Tooling is used for boresighting to inspect and record the exact aiming and alignment of the airframe installation interface points during aircraft manufacture. Any airframe aim point corrections are recorded and stored in the aircraft software. The gun assembly aim point and dynamic impact point of the rounds are measured and recorded during acceptance test procedures by the gun system manufacturer. This aim point and dynamic impact point are supplied with each gun for installation into the aircraft software. The gun and attach point correction factors are then coupled with canopy and other correction maps through an algorithm to establish the correct aiming queues for the pilot.

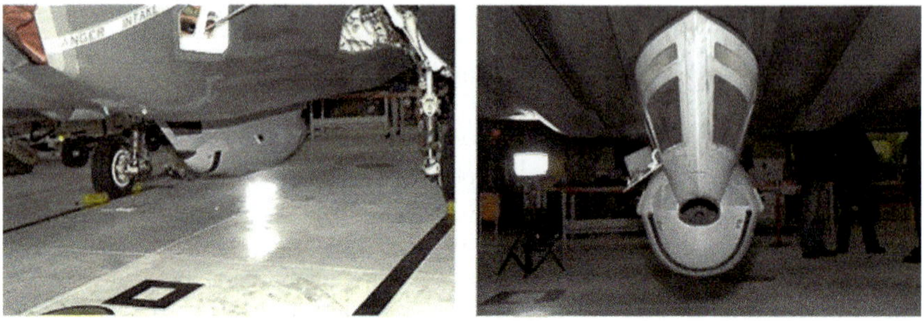

Fig. 15 Missionized gun system installation.

The Marine Corps and Navy require the gun system to serve as AME. Under this requirement, the gun system is only carried on an aircraft during missions requiring its use. For use on the F-35B/C it has been designed into a Missionized Gun System (MGS) pod. As Fig. 15 shows, the gun pod is installed on the aircraft centerline, between the weapon bay doors. This location is limited in space because both the inboard weapons bay doors must be allowed to fully open when the MGS is installed on the aircraft. The lower surface of the Navy's MGS is also limited by the need to clear the launch shuttle used during carrier launch of aircraft. The limited volume of the pod required the supplier to develop a helical feed system to store the ammunition. The helical system is double helical, surrounds the gun assembly barrels, and allows for the carriage of 220 rounds of ammunition. The MGS is installed on the centerline store station on the bottom of the aircraft, with two guide pins and a four-bolt interface.

A requirement was levied early in the program to make as many components as possible common between the internal gun system and the MGS. This requirement resulted in common gun assemblies and gun system control units, and a common hydraulic drive motor. The MGS has the same reverse-clearing operation of the internal gun system and uses the same technique for boresighting. It uses a gun assembly boresight number supplied by the manufacturer and a correction number based on the MGS installation hardpoint alignment. There remained differences in the aircraft surfaces, door contours, and airframe structures, however. As such, minor changes to the aft fairing, door bumpers, and access panel were required for the F-35B MGS and F-35C MGS.

VII. STORES MANAGEMENT SYSTEM DESIGN

The F-35 SMS benefits from the evolution of a standardized weapon electrical interface defined by MIL-STD-1760. This standard enables a shift in the operating premise of the weapon electrical interface from an energy control mechanism to a data message control mechanism. Although many of the weapon detail functions remain energy-driven mechanisms, the actual control circuits have become small and inexpensive enough to implement within the weapon itself. This enables the electrical interface between the weapon and the aircraft to become functionally generic and flexible.

Safety interlock features are preserved via dedicated function circuits within MIL-STD-1760 such that hazard probabilities are maintained at acceptable levels. The aircraft/weapon interface is primarily a data message mechanism, so most weapon delivery functions are controlled by software programs in both the aircraft and the weapon. This has established a basis for further standardizing that data message interface such that a weapon plug-and-play mechanism can be defined. This is now known as the Universal Armament Interface (UAI). The F-35 will be hosting the UAI in incremental aircraft software releases as the development program concludes and follow-on block upgrades are fielded.

Architecturally, the F-35 SMS is composed of interface units installed near the weapon point of use and designed as slaved equipment items. The items' serial data interfaces to the aircraft fire control software are located in the avionic computers, which constitute the Integrated Core Processor (ICP). The ICP supplies all SMS processing resources and connectivity with other mission systems and vehicle management systems functions, as well as appropriate redundancy for these functions. Figure 16 shows the architectural arrangement of the SMS components.

The SMS Fuselage Remote Interface Unit (FRIU) and Missile Remote Interface Unit (MRIU) are hardware components that form an input/output structure.

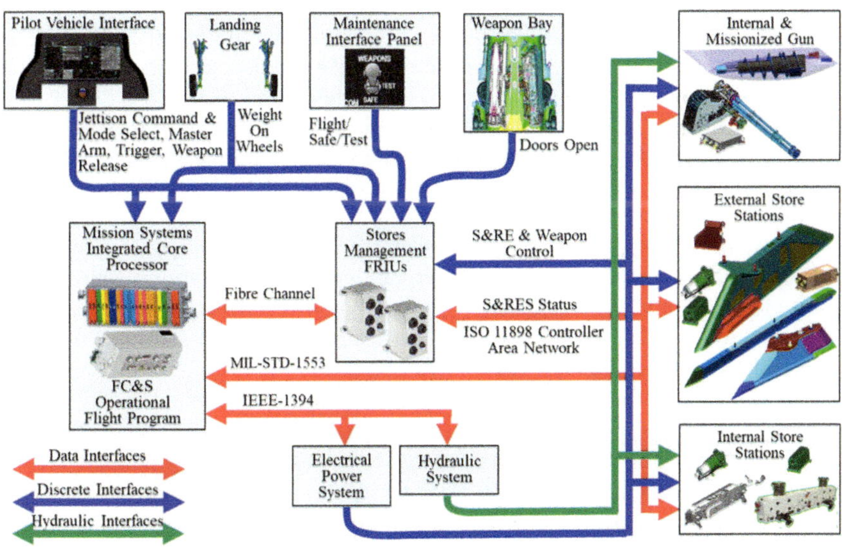

Fig. 16 Stores management system architecture.

Through this structure, the Fire Control and Stores (FC&S) software controls the weapons complement. The FRIUs provide hardware interfaces to control the S&RE ejector racks and pneumatic equipment, as well as the MIL-STD-1760 circuits (except for power). The FRIUs are located in the main landing gear wheel wells as a central location relative to the weapon stations, minimizing the lengths of wire harnesses. The MRIUs provide hardware interfaces for peculiar air-to-air missile signals and are installed within the LAU-151/A rail launchers.

Weapon MIL-STD-1553 data message interface hardware is located on the general-purpose input/output modules within the ICP. This location affords minimum communication latency between the FC&S software threads and smart weapons. The traditional approach is to employ a single MIL-STD-1553 bus controller for the entire aircraft weapons suite, but the F-35 SMS implements eight controllers instead. It employs a separate controller for each air-to-surface weapon station (six channels), and another controller for the internal and external dedicated air-to-air weapon stations (one controller channel for four stations). Adding to this, it employs a final controller for the internal gun and centerline station, totaling eight controllers. This enables significantly faster bandwidth for MIL-STD-1760 weapons.

Operational benefits that arise from the multiple-controller approach include the rapid initialization and targeting of weapons, flexibility to support multiple carriage subracks (e.g., the BRU-57), and an enhanced ability to simultaneously employ multiple weapons. The approach also provides a functional isolation benefit. Through it, the FC&S domain software that manages a store at any station can execute independently of FC&S domain software that manages stores at other aircraft weapon stations. This feature enables superior consistency of weapon event timelines, as well as ease of integration.

For the electrical circuit, the F-35A/C implement a full Class 1 interface, as defined in MIL-STD-1760, at all weapon stations. The SWAT weight reduction exercise implemented early in the development program resulted in the high-bandwidth signals HB1, HB2, HB3, and HB4 wiring being omitted from the F-35B at the air-to-air missile weapon stations. It also led to omitting the 115 AC power and low-bandwidth wiring from the centerline weapon station. This achieved a 10-lb weight savings at the aircraft level. All corresponding electronics were preserved, and the aircraft design provided for restoring the omitted wiring. The retained F-35B MIL-STD-1760 Class 1 circuits at these stations enable all required weapons and a UAI capability.

Single-point safing, nonpyrotechnic S&RE, and the elimination of ground support equipment were other F-35 requirements that significantly influenced the SMS architecture. Single-point safing requires the entire armament system to be capable of being safed from a single location on the aircraft. This drove the SMS design to be implemented such that an absence of electrical energy would safe the weapon articles (bomb racks, missile launchers, and gun). This design aspect affected equipment articles as well, compared to those of legacy equipment. Specifically, the new design prohibited the use of REMOVE BEFORE FLIGHT pins, levers, or mechanisms at the weapon stations. Therefore, the F-35 equipment was designed to require specific activation energy to solenoid mechanisms in the bomb racks, missile launchers, and gun from the SMS. This was needed to advance from a safe state to an employment state.

To satisfy the nonpyrotechnic requirement for S&RE, the bomb racks in the F-35 are operated via pneumatic solenoid mechanisms. The SMS electrical interfaces for those solenoids are tailored specifically for solenoid operation. They incorporate tickle-test features that enable periodic confirmations of end-to-end circuit connectivity.

Another primary requirement for the F-35 air system was the elimination of most ground support equipment. This presented a challenge to the SMS design because most weapon interface circuits are open-ended wires at the weapon stations, with aircraft electronics attached to the internal end. Further, there is a companion F-35 SMS requirement that the aircraft be capable of validating the integrity of the armament system without support equipment. During discussions with the F-35 JPO, we observed that the open-ended aircraft wires required by MIL-STD-1760 presented a validation impossibility without some form of aircraft support equipment. Our solution for the F-35 was to incorporate a voltmeter capability into selected circuits of the MIL-STD-1760 signal set and wrap adaptor support equipment. These wrap adaptors are small, low-complexity items with internal passive components that wrap the voltages from circuits in the MIL-STD-1760 signal set back to the circuit with the voltmeter capability. This design approach confirms the integrity in the aircraft wiring and, to a lesser degree, the functionality in the MIL-STD-1760 signal electronics. Over the life of an aircraft, a primary root cause of downtime is wiring issues, such as open circuits and shorts to structure. The F-35 weapons wrap test is designed to detect and isolate these problems.

VIII. FUTURE WEAPONS GROWTH

The F-35 weapons systems will be operational for many years and have been designed to evolve to counter ever-changing threats. The internal bays have been designed to maintain the maximum volume available for future weapons growth, and can accommodate the integration of new stores. Maximum bay volume is important because the F-35 will be the primary internal weapon carriage attack aircraft in any conflict. As such, future weapons will be largely designed to be compatible with the F-35 weapon bays. Only 500-lb-class air-to-ground weapons were certified externally during SDD. However, the F-35 and its requisite AME are structurally designed to accommodate up to 2000-lb-class munitions and external fuel tanks. This includes dual carriage of 1000-lb munitions and single and dual carriage of external air-to-air weapons to augment this flexibility. Figure 17 illustrates a potential maximum air-to-air loadout.

MIL-STD-1760 compatibility at all stations provides the electrical and logical interface for the current and future stores that implement it. For legacy gravity weapons and their carriage systems, the design accommodates a pylon remote interface unit that converts applicable MIL-STD-1760 signals to analog discretes recognized by legacy carriage systems. External air-to-air carriage is accomplished by integrating external missile adapters onto the existing F-35-unique air-to-ground pylons and rail launchers. The external missile adapters provide for single or dual missile carriage of existing or future missiles.

To further simplify future weapon incorporation, the F-35 is incorporating a UAI: a functional and logical interface definition between platforms and stores. It is intended for decoupling the integration of new stores (weapons, carriage systems, and sensors) on platforms (aircraft, ships, ground vehicles) from platform Operational Flight Program (OFP) update cycles. The UAI standard is being certified in SDD through laboratory integration with the UAI Certification Tool. The SDB II will be the first weapon to be integrated using a UAI in the post-SDD Follow-on Modernization (FoM) program. Figure 18 illustrates other FoM weapons being considered for future integration.

The UAI is platform-agnostic and independent of program specifics; however, F-35 personnel hold key positions in several UAI teams: Systems Engineering, Platform Store, Mission Planning, and Launch Acceptability Region. The teams' F-35 insight into the interfaces and processes they developed will enable efficient and effective integration as new weapons are brought into the program. To further enhance future growth capabilities, we assessed current and future developmental weapons that could enhance weapon system effectiveness. Additional volume reserves were identified, and we are modifying the weapons bays to accommodate the new weapon volume, where required.

Fig. 17 Potential maximum air-to-air weapon loadout.

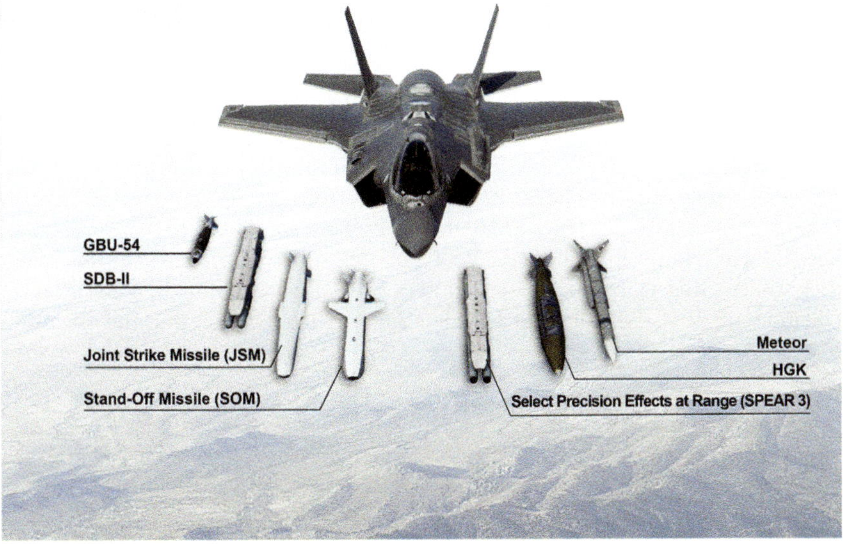

Fig. 18 Candidate follow-on weapons.

IX. CONCLUSION

F-35 weapon carriage systems have been designed and validated through a successful SDD certification program. They provide tactical commanders with the flexibility to prosecute a wide array of targets in either a stealth mode with internal weapons or using an enhanced carriage configuration with external weapons when the threat level allows it. We designed the internal weapon bays for maximum flexibility for weapons and aircraft systems integration to limit aircraft performance impacts across the variants. In addition, we incorporated loading access into the design with the inclusion of an onboard hoist system. Supporting this, we mounted air-to-air missiles on the weapon bay door mechanism, thereby opening up maintainer access to load the weapons. We also included robust external weapons provisions to enable the carriage of a wide variety of weapons on seven hardpoints. Internal and missionized gun systems, tailored to service needs, provide lethal target engagements for close combat with inherent air-to-air capability. Nonpyrotechnic bomb racks and missile ejectors provide consistent separation performance over a wide range of temperatures, providing significant LCC savings over legacy pyrotechnic systems. The adaptable SMS we designed provides the conditioning and control of all required weapons. Further, its flexibility accommodates the development of a UAI to easily integrate future weapons. As F-35 weapons development and integration continue, growth provisions will enable future smart weapon certification to respond to evolving threats and necessary weapon/target pairings.

ACKNOWLEDGMENTS

Many members of the Weapons Integration, Graphics, and Multi-Media teams provided input for this paper to capture the history, design evolution, and lessons learned. Douglas Hayward and Andrew Duff would like to thank the Lockheed Martin team of Jubal Arey, Pete Paciorek, Bill Whitener, Keith Spalding, Rick Wild, Greg Cochran, Keith Schulz, Daniel Buck, Dennis Soultaire, David Gilland, and Russel Baldonado for their support in documenting the evolution of F-35 weapons.

REFERENCES

[1] Hetreed, C., Carroll, M., Collard, J., and Snyder, R., "F-35 Weapons Separation Test and Verification," 2018 AIAA Aviation Forum, Lockheed Martin, Fort Worth, Texas, 2018 (submitted for publication).

[2] Dix, R. E., and Bauer, R. C., "Theoretical Study of Cavity Acoustics," Arnold Engineering Development Complex, AEDC-TR-99-4, 2000.

[3] Rhodes, L. R., "Weapon Carriage Technology (WCT), Life Cycle Cost Study Phase II for Wright Laboratory/MNAV," F-35 Joint Program Office, Oct. 1996.

CHAPTER 6

F-35 System Development and Demonstration Flight Testing at Edwards Air Force Base and Naval Air Station Patuxent River

Mary L. Hudson* **and Michael L. Glass**[†]

Lockheed Martin Aeronautics Company, Edwards Air Force Base, CA

Lt Col Tucker "Cinco" Hamilton[‡]

United States Air Force, Edwards Air Force Base, CA

C. Eric Somers[§] **and Robert C. Caldwell**[¶]

Lockheed Martin Aeronautics Company, Naval Air Station Patuxent River, MD

Since the deliveries of the first F-35s to Naval Air Station Patuxent River (PAX) in late 2009 and Edwards Air Force Base (EDW) in mid-2010, more than 9000 flights, 16,000 flight hours, and 65,000 test points have been completed. This huge feat utilized 18 flight test aircraft to demonstrate the flight sciences and mission systems capability of all three F-35 variants. An Integrated Test Force team composed of government, military, and contractors orchestrated test ranges, tankers, chase aircraft, flight clearances, run cards, parts, support equipment, control rooms, and aircraft configuration. Extensive flight test instrumentation collected the required data to verify modeling and simulation and support the certification of F-35 capabilities. This paper describes the many aspects of flight test successes, challenges, and lessons learned at EDW and PAX to position future F-35 flight testing for continued success and improved efficiency.

I. INTRODUCTION

The F-35 is the first-ever stealth dual-role fighter designed and built to conduct all fighter mission sets, both air-to-air and air-to-ground: Offensive Counter-Air

*Lockheed Martin Sr. Manager, F-35 Test Operations.
[†]Lockheed Martin Director, F-35 Test Operations.
[‡]Commander, 461 FLTS, AIAA Sr. Member.
[§]Lockheed Martin Sr. Manager, F-35 Test Operations.
[¶]Lockheed Martin Director, F-35 Test Operations, AIAA Sr. Member.

(OCA), Defensive Counter-Air (DCA), Close Air Support (CAS), air interdiction, maritime interdiction, and Suppression of Enemy Air Defenses (SEAD). To conduct these missions, the F-35 needed to be able to penetrate enemy air defenses virtually undetected. Additionally, the F-35 actually comprises three distinct aircraft: the F-35A Conventional Takeoff and Landing (CTOL) variant; the F-35B Short Takeoff and Vertical Landing (STOVL) variant, which can land on aircraft carriers previously occupied by AV-8B Harriers; and the F-35C Carrier Variant (CV) for use on large aircraft carrier ships currently occupied by F/A-18 Hornet aircraft. In order for the F-35 to conduct these mission sets, the airframe and mission systems imbedded in the airframe needed to be developed, integrated, and proven. Each variant required rigorous testing to ensure that it could safely fly, maneuver, and land with all types of stores. This scope of flight test is defined as flight sciences. In addition, the aircraft needed to prove it could conduct each of these missions against the most advanced threats and in concert with other friendly battlespace platforms. This scope of flight test is defined as mission systems testing. The first of each flight test variant is shown in Fig. 1.

To accomplish delivering this 5th-Generation capability, a flight test fleet was established coast to coast, as shown in Fig. 2. At Edwards Air Force Base (EDW) and Naval Air Station Patuxent River (PAX), the Integrated Test Force (ITF) teams tested all three F-35 variants with one goal: to deliver unrivaled capabilities to the warfighter. Flight testing at EDW and PAX provided data to multiple disciplines for both model verification and demonstration [1–11]. From flight sciences to mission systems testing, the critical work of the F-35 test teams cleared the way for capabilities to be delivered to the training and operational F-35 fleets. Flight testing obtained the data required to verify the F-35s' key performance capabilities and validated extensive modeling and simulation work. This supporting evidence was required in the packages delivered with the F-35 aircraft to define and authorize warfighter capability. EDW and PAX ITF teams successfully executed the ground and flight tests required to deliver U.S. Marine Corps

Fig. 1 Firsts of each flight test variant: AF-1, BF-1, and CF-1.

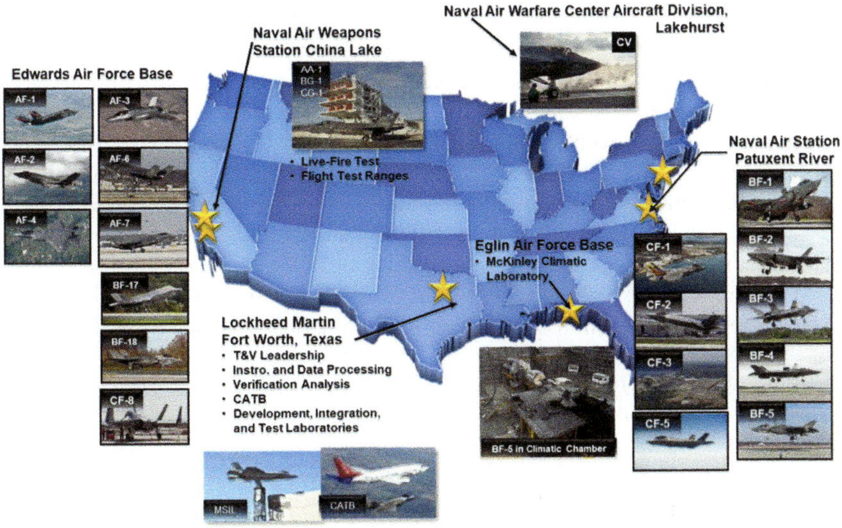

Fig. 2 F-35 Integrated Test Force coast to coast.

Initial Operational Capability (IOC) with Block 2B software in 2015 and U.S. Air Force IOC with Block 3i software in 2016. They also completed flights supporting the end of the System Development and Demonstration (SDD) phase in 2018 and delivered U.S. Navy IOC in 2019 with Block 3F software. Three IOCs in this period of time are unprecedented for any weapons system.

Eighteen F-35 aircraft at EDW and PAX have tested flight sciences and missions systems capabilities since 2009. Over a period of extensive replan, with redefined engineering requirements and revised flight test metrics, nine aircraft at each test site were determined as the right-sized flight test fleet to complete flight test for the SDD phase of the F-35 program. In addition to the number of flight test aircraft, the team to maintain the aircraft and execute the flights had to be the right size and have the right composition. The F-35 ITF team included military, government, and contractor personnel working as one team to plan and execute flights. Every employee on the ITF team had a unique function that utilized his or her specific skillsets. The ITF conducted flight test missions every day, and the criticality of the work, coupled with pushing the limits, required performance at the highest level from the entire team. This realization, along with a seemingly overly constrained task, created a very dynamic and challenging work environment. Every day, the team had to be flexible and adaptable to deal with changes in range availability, cancellation of tankers or test assets, weather, engineering discoveries, and a host of other constraints that often made it seem a miracle to execute the mission.

The F-35 Joint Program Office (JPO) was also an integral part of delivering F-35 capability and interacted daily with each test site. The F-35 JPO had an impact on the day-to-day operations and overall processes that dictated flight test. Through this unique test structure, the team discovered some key lessons:

- The F-35 JPO needed to identify one conduit to prioritize tasks to the test sites.
- The F-35 JPO needed to understand and respect the autonomy of the test sites and not dictate changes too close to execution.
- Clear lines of responsibility between the F-35 JPO and test sites needed to be defined and a process put into place that checked that balance.

Given 18 flight test aircraft stationed at two test sites, nearly 1000 employees at each test site, and all the possible flight test constraints, the F-35 flight test team accomplished more than 9000 flights, 16,000 flight test hours, and 65,000 test points. Although flight test metrics are not the end goal, they lead to the ultimate result of demonstrated and delivered warfighter capability. The following assessment from Exercise Northern Edge/Distant Frontier 2017 best exemplifies the flight test effort and resultant capability that each ITF team member contributed to F-35.

> During NE [Northern Edge Alaska], the F-35 repeatedly demonstrated the capability to use its high fidelity sensors and Link-16 capabilities to significantly increase the situational awareness of 4th Generation assets. This support was a major force multiplier and enabled 4th Generation assets to be more lethal and survivable. In many cases, the F-35B was the last remaining aircraft in the fight (often even after it's simulated A/A or A/S [air-to-air or air-to-surface] load out was expended). F-35s were able to remain on station to support other assets non-kinetically and still perform in the aforementioned "quarterback" role. . . . Regardless of ordnance load out, the F-35 is a key node in an A2/AD [any access/area denial or contested] environment and its mere presence on station makes the entire package more lethal and survivable. [12]

Over the eight years of flight tests at EDW and PAX, the following key flight test capabilities were demonstrated and summed toward the delivered capability to the warfighter:

- High Angle-of-Attack (AoA) missions
- Aerial refueling qualification with KC-10; KC-130; KC-135; F/A-18; and U.K., Australian, and Italian tankers
- High sink rate and arrested landings
- Flutter/loads/buffet flights
- Numerous vehicle systems and mission systems software release regression flights
- Ground and airborne gun fire on all three variants
- Weapons environment, store separation, and Weapons Delivery Accuracy (WDA)

Fig. 3 F-35 testing at EDW.

- Shore-based ship integration with jet blast deflector, catapult, arrestment gear, and ski jump
- F-35B deployments onboard USS *Wasp* (LHD 1) and USS *America* (LHA 6)
- F-35C deployments onboard USS *Nimitz* (CVN 68), USS *Dwight D. Eisenhower* (CVN 69), and USS *George Washington* (CVN 73)
- Climatic laboratory testing in the McKinley Climatic Laboratory at Eglin Air Force Base (AFB), Florida

These and other major test events are shown in a timeline depiction at EDW and PAX in Figs. 3 and 4, respectively. The EDW and PAX test sites have unique capabilities allowing unique testing and accomplishments. The data obtained at each site were required together to certify warfighter capability on each F-35 variant. This was truly an integrated test force with not only cross-teams, but also cross-country data gathering and analyses from California to Maryland and Texas and in between. The expanse of aircraft, flights, data, and people led to the F-35 ITF's selection as a finalist for the 2017 Collier Trophy.

II. EDWARDS AIR FORCE BASE

EDW has an incredible history of flight test firsts, and the F-35 is proud to join this history of engineering innovation demonstration. EDW is a unique flight test location in the Mojave Desert with excellent year-round weather and dry lake beds providing a natural extension to the base's runways (Fig. 5).

Nine F-35 flight test aircraft resided at EDW to support SDD flight testing. Three CTOL aircraft were dedicated to flight sciences testing: AF-1 to obtain

Fig. 4 F-35 testing at PAX.

store separation data, AF-2 to obtain loads data by expanding the envelope, and AF-4 to understand the complex fuel system. Six aircraft, using all three variants, comprised the tails supporting mission systems flights: AF-3, AF-6, AF-7, BF-17, BF-18, and CF-8. Each tail had some claim to a flight test first; for example, AF-1

Fig. 5 Aerial view of F-35 AF-1 over EDW.

Fig. 6 EDW ITF team with the nine flight test aircraft.

was the first F-35 aircraft to arrive at EDW to begin SDD flight testing on 19 May 2010. AF-2, aka Work Horse, holds the record for flights and flight hours at 689 and 1509, respectively (as of December 2017), accomplishing loads, flying qualities, buffet, dry and wet runway directional control and antiskid braking, air data tower fly-bys, propulsion, software regressions, and gunfire. AF-3 was the first to conduct countermeasures testing. The mission systems aircraft were the first of their variants to employ weapons using their complete avionics suites. In addition to firsts, these nine F-35 aircraft accomplished more than 4500 flights, 10,000 flight test hours, and 30,500 flight test points.

The people were the key to orchestrating and executing the incredible number of flights, supporting the nine F-35 flight test aircraft, and delivering 5th-Generation warfighting capability. The nine flight test aircraft and ITF team at EDW are shown in Fig. 6. At the peak of F-35 SDD flight testing at EDW, there were nearly 1000 men and women in the ITF made up of government civilians, military service members, and contractors (Lockheed Martin, Northrop Grumman, BAE Systems, Raytheon, Boeing, and Pratt & Whitney). The F-35 test pilots came from the military, U.S. government contractors, and Lockheed Martin. The ITF organization evolved over time to improve team communication and test efficiency. The ITF was organized into Flight Sciences Engineering, Mission Systems Engineering, Maintenance, Sustainment/Logistics, Test Operations, Instrumentation/Data Processing, and Test Pilots. The successes, challenges, and lessons learned of each team are summarized in the following sections.

A. FLIGHT SCIENCES ENGINEERING

Flight sciences engineers test the aircraft's critical systems, analyze data, and man the control rooms advising test pilots when they reach the edge of the envelope or providing situational awareness of system functionality. Engineering works in conjunction with the Test Operations team to develop the Joint Test Plans (JTPs) that define the test objectives, test plan, and required flight test points. There are JTPs and engineers specializing in the aircraft fuel system, loads and

Fig. 7 Loads testing with different weapon loadout configurations.

dynamics, flying qualities, landing gear, hydraulics, electrical power system, propulsion, mass properties, store separation, and low observables.

Flight sciences envelope expansion testing commenced in summer 2010. During this phase, the aircraft performed maneuvers under a variety of configurations to experience aerodynamic forces to test the limits of the design envelope. Testing factored in environmental conditions, such as temperature, pressure, altitude, and airflow. The aircraft were subjected to high-speed/high-*g*-load conditions and flutter and buffet to characterize structural responses, as shown in Fig. 7.

To validate the aircraft's flight control laws and handling qualities, high-AoA testing was performed, as shown in Fig. 8. The tests characterized the aircraft's performance beyond the operational envelope and thus demonstrated the ability to recover from uncontrolled flight.

The aircraft's vehicle subsystems were also tested to their limits in both ground and flight scenarios. Some of the critical tests included inflight aerial refueling (Fig. 9), inflight engine restart, high sink rate landing, crosswind landing, high brake energy, wet runway performance, arresting hook, and ground gunfire (Fig. 10).

A key lesson learned from these teams was test planning; for example, several of the most difficult-to-acquire test points remaining at the closure of SDD were

Fig. 8 High angle of attack: aircraft flies at various uncontrolled flight conditions.

Fig. 9 Aerial refueling tests from different tankers, clockwise from upper-left: nighttime refueling using a KC-10 tanker, F-35 refueling from an Italian KC-767, and refueling from an Australian KC-30.

Fig. 10 Ground testing on the F-35, clockwise from upper-left: arresting hook, wet runway performance, and ground gunfire test.

those for flying qualities crosswind testing. This testing can be especially difficult because of the limited number of days on which the atmospheric conditions meet the test objectives. A handful of test points could take months to complete because of uncooperative weather, namely wind speed and direction. While waiting for the weather, it became apparent that overly constrained test plans and tight airfield restrictions had driven this predicament. The test cards could have been prepared and carried to every sortie to catch the rare opportunity for these tests. The test plan also specified tight tolerances on weights and Center of Gravity (CG) that significantly decreased test efficiency. In the first attempts to capture test points, the weight band was burned through so quickly that aerial refueling was often needed. Upon further data review, the weight and CG conditions were deemed irrelevant for this testing. These changes to the JTP were doable but incurred schedule impact due to the lengthy change process and approvals. Updating the test plans was a common occurrence because the testing requirements changed as the system matured and became more understood.

There were discoveries on a near-daily basis: exceedances, hardware issues and software issues that led to new constraints limiting testing, analysis that reprioritized the schedule, or new understandings of how the systems operate that were reason to pause and reevaluate planned points. A lesson learned was to review the entire test point requirements to most efficiently missionize test points and to write a test plan that allows as much flexibility and agility as possible early in the program.

B. MISSION SYSTEMS ENGINEERING

Mission systems discipline engineers verify and validate the mission systems capabilities, such as communication, navigation, and interrogation; datalinks and interoperability; displays and core processing; weapons integration and accuracy; electro-optics and infrared; offboard mission planning; radar; fusion; electronic warfare and countermeasures; and signature and survivability. They document anomalies and work with the software engineers and teams on root-cause analysis and resolution. The mission systems engineer also works directly with the maintenance and control engineers to troubleshoot and resolve aircraft anomalies. Here are a few lessons learned from the Mission Systems engineers.

1. SOFTWARE RELEASES

Inflight discoveries, related to verifying millions of lines of code enabling F-35 mission systems capability, triggered numerous software releases. A flexible, agile flight clearance process was required to support a fly-fix-fly concept for revised software. The Quick Reaction Capability (QRC) flight clearance process was developed to address this issue and allow flight clearance turnaround in a couple days vs several weeks for a full release. If the software revision did not touch any airworthiness software, a board reviewed and approved a QRC flight clearance that enabled the fly-fix-fly concept and greatly improved flight test efficiency while maintaining safety.

2. WEAPONS SURGE

A new test concept was used for complex weapons testing called *surge* due to the numerous assets and personnel required for a short period of time. The F-35 flight testing showed that for complex testing, such as WDA flights, maximum test efficiency was achieved with a focused surge approach including exclusive use of ranges, dedication of tankers, secure communication and datalinks for control rooms, chase and photography, targets, off-site staffing, and a seven-day team workweek. During Block 3F, there were two such surge events that directly supported the Air Force declaring IOC and the closeout of SDD. During the first surge event in August 2016, 25 missions, composed of 12 WDAs and 13 weapon separation tests, were executed during a month-long period. This required laser focus by the entire test team to ensure that the aircraft were available every day and to maintain priority for use of the range and support assets. Historically, the execution of such events would happen once a month given the extensive coordination required. Prior to the surge, the highest number of weapons events accomplished in a single month was three in November 2014 during the Block 2B software testing. During the unprecedented period in

August 2016, a total of 30 weapons were dropped or fired in 31 days, including the Joint Direct Attack Munition (JDAM), Advanced Medium-Range Air-to-Air Missile (AMRAAM), Small Diameter Bomb (SDB), AIM-9X Sidewinder air-to-air heat-seeking missile, and GPS/laser-guided munition.

The second surge event occurred in August 2017. Unlike the prior surge event in 2016, there were only a handful of missions required to complete the F-35 developmental test for Block 3F. These final missions demonstrated the complex air-to-air capabilities of the F-35 with the AMRAAM, showing the full capability of the F-35. The AMRAAM missiles used during these test missions were equipped with live motors and guidance systems, but the warheads were exchanged for telemetry units. The open-air testing was conducted over the water at Naval Air Station Point Mugu's range in Ventura County, California (Fig. 11). The WDA testing was the graduation exercise before the aircraft was delivered to the operational test organizations to prove combat readiness.

3. MISSION EFFECTIVENESS

The final and most complicated demonstration of the F-35 5th-Generation capability was the mission effectiveness, multiship missions (Fig. 12). These missions required maximum coordination for successful execution. As many as 15 airborne assets participated in a single test mission including F-35s, F-16s, F/A-18s, F-15s, KC-10 tankers, KC-135 tankers, and airborne command and control platforms. The various mission types executed during Block 3F included CAS, SEAD, Destruction of Enemy Air Defenses (DEAD), OCA, and DCA. The mission effectiveness missions highlighted the interoperability of the F-35 and its ability to command the technical situation. This ability was further demonstrated in several Large-Force Exercises (LFEs) by providing command and control situational awareness of the battlefield using the various arrays of onboard sensors.

Fig. 11 Weapons delivery accuracy flights.

Fig. 12 Four-ship mission effectiveness flight.

C. AIRCRAFT CONFIGURATION AND MAINTENANCE

Maintainers and Flight Test Control Engineers (FTCEs) repair, inspect, configure, and modify the aircraft to ensure safe and correct functioning during every flight. Aircraft maintenance was performed using the tail-team concept composed of an aircraft supervisor, FTCE, quality assurance inspector, field and service mechanics, and avionic technicians. The FTCE ensured that the aircraft was properly configured to perform a flight test mission by providing a liaison between engineering and maintenance. They used many tools, including the Autonomic Logistics Information System (ALIS) and Product Data Management (PDM) and worked with maintainers to ensure that parts that had been replaced on the aircraft were within the proper aircraft effectivity. The team reviewed the air vehicle's status and developed the near-term maintenance requirements during a daily maintenance scheduling/coordination meeting at the beginning of each shift. This meeting set the tone for the day and provided critical aircraft status information to ITF leadership. Below are some details of lessons learned regarding aircraft configuration, maintenance procedures, inspections, and ALIS.

1. AIRCRAFT CONFIGURATION

The required aircraft configuration for each flight was defined by the Mission Prep Sheet provided by the Flight Test Engineer (FTE) and specified the software versions, weapons loadout, instrumentation configurations/settings, and mass properties Form F defining the alternate mission equipment. The FTCE was the focal point to coordinate with all involved teams to ensure that the aircraft was configured per the Mission Prep Sheet requirements prior to releasing the aircraft for flight.

Perhaps the most valuable and accurate tool to ensure proper aircraft configuration and tail effectivity of parts was PDM. For example, if a -0007 part was issued to an aircraft, it was the FTCE's responsibility to ensure that the part was applicable to that aircraft. After checking in PDM, if the -0007 part number was not listed, the FTCE reverted to the previous dash number (-0006) and checked whether a service data note was present stating that a -0007 could be used for all replacements of a -0006. Then, the -0007 was authorized to be installed on the aircraft. The Configuration Management team ran a Release Authorization Notice report before every flight to verify that the aircraft was in an authorized configuration.

2. MAINTENANCE PROCEDURES AND INSPECTIONS

The ability to properly maintain a new aircraft from Day One is important and requires correct technical data to exist early on. Joint Technical Data (JTD) define maintenance procedures and must be verified and validated before use. Dedicated aircraft were intended to provide early verification of JTD but were not available early in the program. This required continued use of Aeronautical Engineering Instructions (AEIs), which are the unverified source data for JTD, to accomplish tasks on the aircraft in a timely manner. This delayed the use of some JTD to operational test and fielded units, requiring them to use field service representatives with instructions in the form of an action request that could delay an answer.

Another challenge early on was well-documented and communicated inspection requirements that are defined in Engineering Inspection Requirements (EIRs) for SDD aircraft and in Production Aircraft Inspection Requirements (PAIRs) for Low-Rate Initial Production (LRIP) aircraft. Both EIRs and PAIRs are designed to provide inspection criteria and timelines for parts that may have been identified with unknown life limits, manufacturing, or flight failure/damage issues seen in testing or fleet operations. In addition, Time Compliance Technical Directives (TCTDs) specify inspections or parts replacements/modifications. TCTDs are released in JTD dataset releases and managed and installed by ALIS administrators. The FTCE then performs the defined action based on aircraft scheduling/availability and time compliance definition. It is critical that an accurate and reliable tool exist to track all the inspection requirements.

3. AUTONOMIC LOGISTICS INFORMATION SYSTEM

The F-35 is the first tactical aviation system to have sustainment tools engineered in concert with the aircraft for efficiency and cost effectiveness. Compared to previous aircraft, a higher fidelity of information about the F-35 fleet is tracked

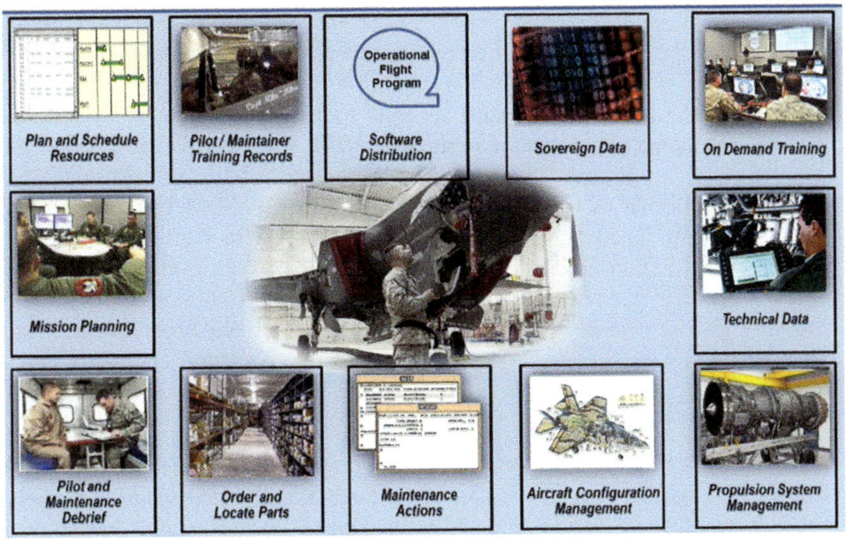

Fig. 13 ALIS: integrated application suite supporting multiple air system functions.

within ALIS to reduce operations and maintenance costs and increase aircraft availability. ALIS consists of the system, application, and network infrastructures required to provide global integrated and autonomic support, as illustrated in Fig. 13. ALIS integrates a broad range of capabilities, including operations, maintenance, prognostics, supply chain, customer support services, training, and technical data. It gives F-35 operators the ability to plan ahead and to maintain and sustain its systems over the life of the air vehicle. A single, secure information environment provides users with up-to-date information on any of these areas using web-enabled applications on a distributed network.

Just like the airframe, ALIS needed time to mature in order to reach full capability during SDD. The ITF used ALIS as a tool to manage aircraft maintenance while concurrently evaluating it and identifying issues and improvements. This dual task made it challenging to use a development tool for reliable aircraft maintenance. The FTCE or support FTCE used ALIS to check whether a part had been replaced per a PAIR, and whether a TCTD was valid for the assigned aircraft. Because of downtime during ALIS development, backup techniques were developed, including a paper system with which all the data were entered into ALIS when it was back up and running. A lesson learned with developing ALIS was to have users involved with developers from the beginning. Technicians could tell the developers what they needed to effectively accomplish their job, leading to an improved initial concept.

The Sustainment team at the flight test site used a specific module of ALIS to manage the location's inventory in support of aircraft maintenance. This team received, stored, issued, and moved retrograded F-35 assets in direct support of flight operations, back shop, and flight line maintenance. ALIS periodically received software upgrades to fix anomalies and add capabilities, so additional tools and processes were developed to keep parts flowing and aircraft flying. To adjust spares levels based on requirements, assets from the EDW ITF needed to be reallocated. To manage these limited assets, EDW developed a tracking tool for top degraders and a series of metrics to provide a snapshot of wellness in the following areas: top degraders (batteries, wheels, cartridge/propellant-activated devices), target stock levels, inventory effectiveness, historical usage comparison, support equipment analysis, tool room analysis, and a warehouse effectiveness summary. The use and feedback of ALIS performance in a flight test setting was critical to its development, resulting in a superior fielded product.

D. SUSTAINMENT/LOGISTICS

Parts and spares were an ongoing critical requirement and issue to support the pace of SDD flight testing. The Sustainment/Logistics team managed the Material Master Catalog, spares/repairable parts, sustainment data, transportation inventory, property, and shelf life. They were responsible for initial provisioning as well as demand forecasting. This team assisted with issues related to purchase orders, consignments, asset transfers, inventory audit and management, electronic equipment logbooks, and action requests; ensured that material was processed promptly to avoid potential work stoppages; clarified original equipment manufacturer/supplier data to clear frustrated cargo; and served as the liaison between the F-35 Maintenance team and the Sustainment Modification Kit Management team in Fort Worth, Texas, for supply-related issues. ALIS was also an integral part of the sustainment supply and logistics concept for supporting F-35 operations as described previously.

1. WHEEL AND TIRE MITIGATION PLAN

For flight test, tire tread limits are stringent to support test execution, causing Wheel and Tire (W&T) assemblies to be changed more often than in a standard squadron. The F-35 initial concept was to accomplish the buildup at the supplier facilities, which took weeks to complete a refurbish turnaround. To meet the demand, the Sustainment team developed a plan to train and certify the local tire shop to have the ability to refurbish F-35 W&T assemblies. With this plan in place, the refurbish turnaround was reduced from weeks to three to five days.

2. TOOL CONTROL

The Sustainment/Logistics team was also responsible for tool control. There are strict processes for Foreign Object Damage (FOD) and tool control, but a process and comprehensive training were required for visiting mod teams and vendors. Five documents were combined into one all-encompassing standard operating instruction for visiting mod teams and vendors to review and acknowledge prior to being released to work on the aircraft. Additionally, all tooling was processed through the tool room and went through a process of accountability that included quality team buy-in prior to assets being released to the flight line or hangar floor. The new single-document direction and comprehensive training effectively emphasized the importance of FOD and tool control and highlighted unique aspects of working at the test site. FOD and tool control are critical concerns of flight test safety and can cause not only safety issues, but also program delays if not given due attention.

E. TEST OPERATIONS

Test Operations consisted of a group of FTEs and schedulers who converted data requirements into test missions and executed these missions to obtain the required data. Test procedures and run cards were developed for the pilots to execute based on requirements in the JTPs. FTEs were key players in developing the JTPs and were skilled in recognizing what-ifs and requirements. They were a vital link between the pilots and engineering and were the gatekeepers from test planning through actual execution to ensure safe, smart, and efficient testing.

To execute a flight test mission, FTEs must be aware of limitations defined in multiple sources: test safety package, flight clearance, system release memo, and aircraft operating limitations. They must coordinate physical assets, including: test aircraft configuration, mission control room, instrumentation, telemetry, radio frequencies, airspace, test ranges, tanker, chase/target aircraft, weapons, ground vehicle targets, radar reflectors, terrain types, air temperature, wind speeds, day/night, and moon condition. To orchestrate these many moving parts, some key factors were identified to safely, successfully, and efficiently execute daily F-35 flight tests.

1. TEAMWORK AND TRAINING

Development work and daily flight test execution are challenging and difficult by nature and are best accomplished by a dedicated group of people who work together, respect each other, have a solid work ethic, and care about the work and each other. Teamwork has been shown in the form of knowledge, skill, and trust. The Test Operations team gained knowledge through detailed training; successful training on a program of this size required a lead instruction coordinator to oversee the effort. Per the EDW Lockheed Martin Chief Test Pilot David "Doc" Nelson: "The person training FTEs, test conductors, and test directors should be a leading world class expert in the role they are instructing. There is no amount of study or theory that can replace the knowledge and lessons learned by successful and especially unsuccessful experience" [13].

FTEs were trained to be active leaders in premission preparation including card writing, maneuver development, mission planning, and execution. Training was accomplished on not only the aircraft systems, but also the basics of flight test operations as well as the unique nature of testing the F-35. The control room training program worked well in establishing a capable group of Test Conductors (TCs), Test Directors (TDs), and discipline engineers. The program was successful in taking individuals with no control room experience and preparing them for control room duty (Fig. 14) through emergency procedure simulations and mentorship and guidance from instructors. Their skills were proven through real test missions to the point that trust was gained. The trust was then knowing that, no matter the challenge, the team would work seamlessly together to accomplish goals that could not be achieved alone. Some lessons learned were to thoroughly train the team, empower them, listen to them, and trust them.

Despite being part of an excellent team, personnel attrition rates occurred with one cause being the stress and workload. There were averages of six sorties a day of testing of all types and disciplines. Maintenance personnel often worked weekends, and overtime was common. The departure of trained, independent workers resulted in a loss in the organization's knowledge base, which required veteran personnel to focus on training new personnel rather than test planning and execution. Future programs should determine what workload is sustainable for the number of personnel over the course of the entire program and may consider offering incentives (e.g., financial, promotions) early in the program to help create and sustain a large group of veteran personnel.

Fig. 14 EDW control room.

2. ROLES AND RESPONSIBILITIES

With the large number of F-35 personnel with multiple affiliations (military, government, and contractors), it was especially important to define roles and responsibilities. Lacking this definition can lead to duplicated work, missed tasks, and miscommunication among groups, which leads to increased workload and personnel conflict. The Test Operations and the discipline engineering groups experienced some conflicts based on their differently perceived roles and responsibilities. With high pressure to succeed, it was tempting for personnel to develop an attitude of doing whatever it takes to get results, which disrupts the organization's leadership and schedule stability. Future programs of similar size and of a joint organizational nature need to define and establish the roles and responsibilities of their groups early on to avoid conflict and the impact on testing.

3. COMMUNICATION

Good communication was an essential key to success, both internally within the ITF and externally at other test sites, Fort Worth, JPO, vendors, and foreign partners. This is especially critical during large developmental test programs for which any printed schedule is soon inaccurate due to the dynamic nature of the work. Two lessons learned were to make the effort to communicate more and to communicate effectively. Communication success resulted from the core communication elements of speaking and listening to each other and using clear, concise, and direct language. This proved invaluable, especially in communications with ranges and asset providers to understand how their organizations worked. Lines of communication must also be established among FTEs, relevant engineering groups, and leadership. As discoveries are made, solutions must be evaluated and decided upon. A designated integrator/decision maker who has the big picture and required authority will be key to moving forward with minimal test delay.

4. TEST REQUIREMENTS

Each F-35 variant had unique test requirements that presented flight test challenges. This was most evident in the flight sciences engineering disciplines due to the differing wing planforms between CTOL/STOVL and CV, and the different weapon bay sizes between CTOL/CV and STOVL. Also, each variant had unique fuel system and structural designs. All CTOL flight sciences testing was performed at EDW, and all STOVL and CV flight sciences testing was done at PAX.

The majority of mission systems testing was variant agnostic. F-35A, F-35B, and F-35C aircraft participated in multiship scenarios together, seamlessly executing tactics for which the aircraft were designed. At no point during SDD did the aircraft operating limits match among the CTOL, STOVL, and CV aircraft because flight sciences testing was concurrent with, not ahead of, mission systems testing. Extensive maneuver development at EDW and Fort Worth simulators was essential to execute edge-of-the-envelope testing. Even within a single variant, an individual tail's limits differed from one another due to hardware and software modifications. There also existed customer-specific requirements that constrained particular testing to certain variants, depending on which combination of variant and weapon the customer was purchasing. Although the F-35 concept of commonality was beneficial to manufacturing affordability, it challenged engineers with complex and intertwined test requirements.

5. PLANNING, SCHEDULING, AND EXECUTION

The optimal forecasting period for scheduling both mission systems and flight sciences testing was determined to be between four and six weeks out from the current date. This allowed the team to adequately plan the major test missions (i.e., those with multiple test aircraft, ranges, and assets), as well as logistically easier missions. This time also allowed outside resources, including ranges, air, ground, and sea assets, to be coordinated, scheduled, and configured as required. The EDW test wing locks in the test schedule approximately two weeks out, whereas many external ranges lock in their schedule at different times. By starting the initial coordination and scheduling efforts further out, changes and refinements can be made to the schedule as needed by the pace of the program, thereby having a solid plan at the time of scheduling.

New, creative ways to schedule, plan, and execute missions were developed to support the fast-paced F-35 flight test schedule. Single-ship range blocks at Point Mugu's sea range, the White Sands Missile Range, Naval Air Weapons Station China Lake and Echo ranges, and EDW Precision Impact Range Area evolved into concurrent operations sharing one piece of airspace as the program matured. F-16 chase and target aircraft, oftentimes in short supply, were shared among separate missions, shifting those missions' particulars to accommodate. Hot pits, the ability to refuel with the engine running and aircrew remaining in the aircraft, were stood up when tankers fell out. A summary of flight test scheduling and execution lessons learned is as follows:

- Brief the plan and fly the brief.
- Schedule and plan for both success and failure.
- Be ready to keep moving if things are going well, and be ready with backup plans if they are not.
- Take tactical pauses when the test team needs them, because pilot safety could depend on taking just one more day to study data.
- Spend time in the simulator with the engineers, aircraft, and the people who maintain it because the more you research and evaluate what-if's in advance, the better you can prepare for the unknowns.
- Understand everything that you can about the systems and the data needs of the customer.

- Ask questions, then ask more questions.
- Identify a drop-dead time to make changes and, once past that time, do not allow additional changes even if the mission has to cancel. (The team needs stability; if they are always on the hook for making changes, safety can be compromised.)

F. INSTRUMENTATION/DATA PROCESSING

Robust instrumentation and timely data processing are central to flight test. Flights are executed to obtain data to support engineering disciplines to understand and verify system design. The extensive number of flights and test points drove a huge number of test data requests. In 2016 more than 41,000 test data requests were processed, with a total end-SDD of more than 279,000. New data analysis tools were also required to evaluate 5th-Generation capability, such as the Fusion Analysis Tool—Measures of Performance (FATMOPS) processing capability, which was brought online at EDW in 2016. The following are lessons learned from instrumentation, with some specific examples.

1. INSTRUMENTATION CHALLENGES AND LESSONS LEARNED

Instrumentation on the F-35 provided a unique challenge for the flight test community. Not much has changed in the analog world of evaluating the strain on the landing gear and the robustness of the flaps. The analog Orange Wire solution is used time and time again for aircraft structural data with high success. The F-35 has an intricate digital computer network. For the first time in aviation history, the avionics system under test had multiple styles of high-speed data buses, and the cost of an instrumentation system to bulk-capture all the aircraft traffic, as was traditionally done on legacy aircraft, far exceeded the feasibility of an affordable solution.

The biggest challenge of the digital instrumentation on the F-35 was to meet the data demands of the flight test community. There were extensive essential data to be captured to validate the functions under test and additional data to evaluate any unexpected anomalies. The instrumentation system recorded about six times the amount of data that it was initially designed to record. This was partly due to technology-refresh upgrades made over time with better processing power, RAM, and throughput. The Data Acquisition, Recording and Telemetry (DART) instrumentation pod was in the left weapons bay of all mission systems aircraft, as shown in Fig. 15. Different DART recording modes were selectable, which allowed the aircraft to have multiple data maps in which some data were recorded for some testing modes and not in others, with the intention of making more bandwidth available. The data maps required intense mission planning across all aircraft disciplines. This proved a successful solution, but the analysts are always in need of more data.

Fig. 15 Data Acquisition, Recording and Telemetry (DART) instrumentation pod.

The aircraft systems themselves have also been modified over time to provide less bandwidth in specific cases. For example, high-rate diagnostic data sent to the recorders may have previously published all zeros when no diagnostic information was available. At design time, this made sense because the data requester would know that the system was still alive and working based on the fact that data was received, albeit zeros. These have since been modified to integrate the instrumentation system and to only publish data as required, freeing up immense bandwidth for other messages to be recorded. The designers embraced the mechanism of making the instrumentation system more efficient and stretched it to making the aircraft network more efficient.

For the future, these lessons learned will probably lead to an even more integrated solution of instrumentation digital buses. The pitfalls are now known and can be harnessed on future upgrades and for future avionics platforms. The flexibility and adaptability of the instrumentation system directly contributed to the fast-paced test plan of the F-35, and any future platform will need to keep expandability as a top priority.

2. MULTIPLE AIRCRAFT TELEMETRY SYNCHRONIZATION

Although the six original mission systems aircraft's telemetry systems were crafted as tail-specific constructs, the closure of Block 2B testing spawned the invention of the Multiple Aircraft Telemetry Synchronization (MATS) scheme. All six aircraft's instrumentation systems were rebaselined and made compatible with one another. This allowed for testing multiple aircraft simultaneously from one control room and swapping telemetry from one aircraft to another when needed. The turn time for pilots stepping from a ground-aborted aircraft to a crew-ready cold spare was reduced, as well as the turn time for the control room, which no longer needed to shut down and repower with the cold spare's unique telemetry parameters. The ITF's telemetry footprint was reduced by programming wingmen with the same telemetry frequencies. The benefits of the MATS effort continued to be realized in Block 3F. Resource limitations, such as control rooms, personnel, and frequencies, that would have otherwise prevented the successful completion of 18 or more multiship mission effectiveness narratives were now easily overcome.

G. TEST PILOTS' PERSPECTIVE

Test pilots push the envelope and fly the F-35 through every situation possible, doing it first to clear the way for the fleet pilots. The test pilots are convinced that as a result of their work, more fleet pilots will return successfully from combat missions. The first F-35 pilot to reach 1000 flight hours was David "Doc" Nelson at EDW on 6 January 2017 in AF-3. The following are some of his comments on F-35 flight testing at EDW [13].

A personalized objective of the F-35 test pilots has been to maintain solidarity with the operational pilot end-user. High risk testing has been conducted by test pilots at each edge of the F-35 envelope in a controlled and instrumented flight test environment so that the operational pilot will never have to experience something in tactical flying that has not been explored. Discipline, pilot, and engineering skill along with team-wide trust have been critical to the success of the testing.

Every test pilot who has flown the F-35 has tactically relevant experience allowing that pilot to suggest improvements or call for needed changes to make this stealthy attack aircraft successful in combat and able to return its pilot safely home. Success has come from using appropriately aggressive approaches to flight testing that preserve safety while making efficient use of available resources and time. High-fidelity simulation and a robust safety process have been instrumental in balancing and enabling an aggressive approach. Examples include:

- Reaching the maximum AoA limit on only the fourth mission of high-AoA testing. This has taken weeks on previous programs.
- Intentionally putting the aircraft out of control in different ways more than 150 times to confirm its ability to recover to controlled flight.
- Turning off, then restarting the single engine multiple times, including near Mach 1 at 40,000 ft to verify its restart capability. (An innovative Self-Contained

Underwater Breathing Apparatus [SCUBA] system was invented for air-start flight test to maintain cockpit pressure with the engine off.)

- Releasing dozens of weapons, including missiles, bombs, and bullets, and including partner country and multiservice weapons, to verify safe release and lethality.
- Repeatedly achieving maximum Mach and maximum *g* to expand the F-35 by more than double the first flight envelope.

Each of these achievements gives confidence that the fighter pilots who will fly the F-35 into combat have the greatest probability of success and survival throughout the entire flight envelope of the aircraft. Test pilots have stressed several innovative features that have their first significant application in the F-35. Examples include:

- The Helmet-Mounted Display (HMD): The helmet is fully integrated with aircraft sensors and allows well off-boresight use of target designator boxes for air-to-air designation and weapon aiming, as well as seamless visual designation for ground targets. It also includes an integrated night vision camera. The Gen III HMD is shown in Fig. 16. The main difference between the Gen III and Gen II is the larger aperture for the improved night vision and the sensors that talk to the fixed-camera assembly mounted on the glare shield.
- The electro-hydrostatic actuators used in the F-35 flight control system are the first to be used in a widely produced aircraft.
- Electronic flight series data are used exclusively on the F-35.
- The multifunctional distributed aperture system serves as a missile launch detector and provides a second means of night vision to the pilot in a separate spectrum.

Flight test of the F-35 has been a high-tempo, aggressive, long-term task for the test pilots (Fig. 17). As a result of the sustained effort over time and the resultant

Fig. 16 Gen III helmet-mounted display.

Fig. 17 F-35 flight test pilots at EDW.

increase in flight envelope and sensor performance, confidence in the airframe and systems increased; the services, partner, and participating countries were encouraged; the press became friendlier; and the unit cost of the aircraft continues to decrease.

III. NAVAL AIR STATION PATUXENT RIVER

Continuing a tradition begun in 1943 with the creation of the Naval Air Test Center (NATC) at PAX (Fig. 18), SDD flight test of the Navy's (F-35C) and the U.S. Marine Corps' (F-35B) first 5th-Generation aircraft commenced on 15 November 2009 with the arrival of BF-1 from Fort Worth.

STOVL envelope expansion testing began soon after, kicking off an ambitious flight test program. The PAX ITF conducted flight sciences and weapons envelope expansion testing of the STOVL and CV variants in a similar fashion as the team at EDW, with the additional challenge of demonstrating unique basing and ship suitability requirements. PAX was the natural location for this testing due to the unique facilities it offered: a shore-based TC-7 catapult, Mk-7 land-based arrestment system, and centerfield Vertical Takeoff and Landing (VTOL) complex. The VTOL complex includes a 1900-ft AM-2 expeditionary airfield, a hover pit, two hover pads, a concrete outline of an LHD with an adjoining 38-ft hover position indicator, and a shore-based ski jump. Additional shore-based catapult and arrestment facilities were used at nearby Naval Air Warfare Center Aircraft Division's (NAWCAD's) site in Lakehurst, New Jersey. Over the next eight years, nine

Fig. 18 Aerial view of the F-35 BF-1 over Naval Air Station Patuxent River.

permanently assigned aircraft supported up to six test missions per day to achieve multiple program milestones. Figure 19 shows two of the nine flight test aircraft with the PAX ITF team.

The STOVL and CV basing and ship suitability demonstrations (Fig. 20) were accomplished through a progression of shore-based testing and culminated with

Fig. 19 PAX ITF team with flight test aircraft.

Fig. 20 Ship integration testing of the F-35B and F-35C.

each variant conducting a series of three sea-based developmental testing sea trials of increasing complexity. This required the team to deploy to conduct testing at various off-site locations with specialized facilities. The PAX ITF supported major detachments, including three F-35C deployments to CVN carriers, three F-35B deployments to LHA/LHD carriers, and numerous detachments to Lakehurst and EDW. The team also conducted climatic chamber testing in the McKinley lab at Eglin AFB, Florida.

A. FLIGHT SCIENCES ENGINEERING

The PAX ITF was responsible for flight sciences envelope expansion testing for the F-35B and F-35C variants, with the addition of highly specialized STOVL and CV suitability testing. This led to four envelope expansion test programs that ran in parallel, each with its own lines of wind-dependent testing and temperature considerations for STOVL. The PAX ITF was also responsible for all weapons environment and safe separation testing for the F-35B and F-35C variants, with more than 50 total separations performed by the end of the SDD program. Test planning had to be adaptable to engineering discoveries, including various loads envelope expansion pauses, weapons integration issues, UHF antenna redesign, F-35B Auxiliary Air Inlet (AAI) door redesign, and F-35C tail hook and outer wing redesign. With an integrated team effort, all these technical challenges were overcome and successfully demonstrated.

In addition to conventional Stability and Control (S&C), flutter and loads expansion, and handling qualities testing, F-35 testing at PAX included STOVL mode. STOVL mode testing was a major effort that required significant planning and integration among propulsion, flying qualities, performance, and external environment disciplines. STOVL testing included both S&C and handling qualities testing to clear a STOVL mode envelope separate from the conventional flight envelope. A build-down approach was used to achieve the first hover and vertical landing. First, the aircraft achieved first conversion from conventional flight into STOVL mode and subsequently conducted Integrated Test Blocks

(ITBs) to evaluate the S&C at flight conditions decreasing in speed. Once sufficient ITBs were conducted at a certain speed, slow landing and short takeoff testing began at those speeds. Then, once jetborne ITBs were conducted clearing the hover, the aircraft could complete the first vertical landing. The envelope was then cleared for additional stores configurations. The shore-based envelope was then used as a basis for conducting STOVL expansion on LHA/LHD class carriers. Vertical takeoff and ski-jump short takeoffs were conducted later in the program.

A significant amount of S&C data were required for door-open testing. During conversion into STOVL mode, the STOVL doors open before the propulsion system fully converts to STOVL mode. The doors have a significant impact on aircraft stability during that transition. Collecting S&C data to accurately characterize the aerodynamics of the door-open state had to be done carefully. Frequent data reviews were required to expand the envelope in this transition condition and provide a complete aerodynamic database with door effects.

Takeoff and landing testing required more testing than a conventional program. Conventional aircraft typically take off and land at relatively fixed speeds. The STOVL propulsion system allows the aircraft to land at a wide range of speeds, but this affects the takeoff and landing dynamics. Additional testing was required to verify the flying qualities and aircraft loads at these conditions.

B. TEST OPERATIONS

The PAX ITF needed to collect data to characterize the operation and performance of the F-35 in the shipboard environment. This environment consisted of a combination of ship electromagnetic effects, ship motion (pitch, roll, yaw, surge, heave, and sway), coupled with ship burble and winds of varying magnitudes and directions. These test conditions were not easily manufactured shore-based due to the complex environmental factors of launching and recovering fixed-wing aircraft at sea. The data collected by the PAX team during these sea trials have been used to generate fleet launch and recovery bulletins, flight manual information, and ship/aircraft Naval Air Training and Operating Procedures Standardization (NATOPS) in use by operational F-35 squadrons deployed aboard U.S. Navy ships around the world today.

Conducting a developmental testing sea trial was no trivial matter due to the data requirements, safety risk, associated mitigations, required efficiency, and limited time available onboard a ship. The six sea trials conducted by the PAX ITF were supported by an average of 200 ITF personnel who were deployed onboard ship for sea trials from three- to five-week periods of time.

Because the nature of the developmental testing sea trials was so complex, integration with the ship's company was critical to ensure that the test operations at sea were safe and operated as efficiently as possible. These efforts kicked off with an Initial Planning Conference (IPC) held a year prior to the ship going underway with the F-35. Additional planning conferences followed the IPCs, as required. A portion of the ship's company would travel to PAX for an aircraft familiarization that consisted of training briefings conducted by the PAX test team. This trip would include an observation of live F-35 test events, such as short takeoffs and vertical landings for the F-35B and catapults and arrested landings for the F-35C. Hands-on operations and demonstrations were conducted on such items as chain-down procedures while noting consideration of such items as instrumentation on landing gear. Upon arrival at the ship, the PAX ITF would complete flight deck training, a ship familiarization, and become familiar with normal ship operations/drills in order to integrate with ship's company.

1. DETACHMENTS PLANNED AND ACCOMPLISHED

PAX testing included frequent and extensive detachments to other locations (Table 1) due to unique test requirements. These detachments required personnel to commit large amounts of time to plan and train. The ITF developed the skill to plan these detachments efficiently through repetition, with FTEs taking lead roles in this coordination. Ship detachments were very high paced, with two aircraft flying two flight periods a day, seven days a week, all done in a very complex and unforgiving environment. The detachments shown in Fig. 21 were all accomplished while continuing test operations at PAX.

2. STOVL SHIP SUITABILITY TESTING

U.S. Marine Corps requirements for the F-35B variant led to unique shore-based ship suitability and compatibility testing in the following areas: external environment (vibro-acoustics), short takeoff/short landing/vertical takeoff testing in multiple weapon configurations, and ski jump testing. Additional ship-based testing was performed, including the three major sea trials that proved initial capabilities and then expanded the wind and weapon envelopes for fleet delivery. Smaller ship testing efforts occurred for Joint Precision Approach and Landing System (JPALS) testing and certification and helmet night compatibility.

In January 2010, the BF-1 conducted the first airborne conversion from fully wingborne flight to semi-jetborne flight. This milestone was followed by the STOVL initial transitions test program, which featured the STOVL aircraft demonstrating its ability to fly at progressively slower speeds throughout the semi-jetborne flight regime. As the aircraft transitioned to slower airspeeds, it also demonstrated its ability to perform short landings and short takeoffs. On 17 March 2010, the BF-1 performed its first hover, which was followed by the program's first vertical landing the very next day. Since that day, more than 1400 vertical landings have been performed by PAX flight test aircraft.

One of the reasons to flight test is to identify issues early so that they can be fixed prior to entering full-rate production. In 2010 the PAX team identified a vibration issue with the AAI door during high-speed semi-jetborne testing. This flight test discovery, in part, led to the STOVL variant being placed on probation by the U.S. Secretary of Defense. A redesign of the AAI door quickly ensued. While the AAI door was being redesigned in 2011, the PAX team continued to expand the STOVL mode flight envelope in areas not impacted by the door vibration issue. In fact, the team was able to expand sufficient semi-jetborne and jetborne flight envelope that year to allow the Initial Developmental Test (DT-I) sea trial to commence onboard the USS *Wasp* (LHD 1) in October 2011 with test aircraft BF-2 and BF-4. During STOVL DT-I, the team demonstrated daytime STOVL operations in the clean configuration, as well as aircraft spotting, deck handling, MIL and MAX power short takeoff launches, and recovery vertical landings from stern to bow approaches. Logistics operations were demonstrated

TABLE 1 PAX DETACHMENTS

Year	Detachments
2011	1. Lakehurst, NJ: F-35C initial F-35C catapults and traps 2. STOVL DT-I (USS *Wasp*, East Coast)
2012	1. Lakehurst, NJ: F-35C traps 2. EDW: F-35B air-starts
2013	1. STOVL DT-II (USS *Wasp*, East Coast)
2014	1. Lakehurst, NJ: F-35C catapults and traps 2. EDW: STOVL wet runway and crosswinds 3. CV DT-I (USS *Nimitz*, West Coast)
2015	1. EDW: STOVL crosswinds and CV wet runway and crosswinds 2. Eglin AFB: McKinley Climatic Laboratory 3. CV DT-II (USS *Dwight D. Eisenhower*, East Coast)
2016	1. Lakehurst, NJ: F-35C catapults and traps 2. NASA Wallops: Ops temporarily moved due to PAX runway construction 3. CV DT-III (USS *George Washington*, East Coast) 4. STOVL DT-III (USS *America*, West Coast)
2017	1. Lakehurst, NJ: F-35C catapults and traps
2018	1. MCAS Cherry Point/MCALF Bogue Field, NC: Mode 4 sloped pad testing

Fig. 21 PAX detachments.

and external environment data were gathered to characterize the operation of the STOVL aircraft in the carrier-based environment.

After completing a successful initial sea trial, a redesigned AAI door system was installed on the BF-1 late that same year. The redesign was successfully verified through regression flight testing, which resulted in the STOVL probation being rescinded by the Secretary of Defense. Throughout 2012 and the first half of 2013, the team expanded the flight envelope into night operations and began carrying internal stores.

In August 2013, the team returned to the USS *Wasp* (LHD 1) with test aircraft BF-1 and BF-5 for the second Developmental Test (DT-II) sea trial. During this detachment, the team of more than 200 PAX personnel successfully demonstrated the ability to conduct 94 shipboard sorties in day and night operations while flying in a clean-wing configuration with internal stores.

The final SDD STOVL sea trial took place onboard the USS *America* (LHA 6) in the fall of 2016 with test aircraft BF-1 and BF-5 in the Pacific Ocean. The USS *America* was the first Navy vessel designed and built to accommodate the F-35. This final sea trial demonstrated day and night operations in the presence of higher sea states and deck motion. The test team demonstrated the ability to launch and recover the aircraft with both symmetric and asymmetric external store configurations. This sea trial was also significant in that we no longer enjoyed the luxury of an exclusive flight deck; our developmental test operations were conducted in conjunction with those of VMX-1, an operational test squadron performing carrier qualifications.

3. CV SHIP SUITABILITY TESTING

Navy requirements for the F-35C led to unique shore-based ship suitability and compatibility testing in the external environment (vibro-acoustics), Jet Blast Deflector (JBD) and flight deck compatibility, approach handling quality survey in multiple weapon configurations, roll-in and fly-in arrested landing structural

Fig. 22 F-35C shore-based catapult spotting.

survey in multiple weapon configurations, and a catapult structural survey in multiple weapon configurations (catapult steam ingestion testing).

In parallel to the STOVL ship suitability and envelope expansion testing program, the F-35C was conducting its own flight test program and facing its own unique challenges. The first CV F-35, CF-1, arrived at PAX in November 2010. Early the next year, it began catapult spotting evaluations using the TC-7 site at PAX, as shown in Fig. 22.

In June 2011, the CF-2 traveled to NAWCAD Lakehurst to complete JBD testing. This testing was important to verify that the installed engine function was not degraded by hot gas ingestion when the aircraft was placed in front of or behind the JBD (Fig. 23). This test included JBD modifications to change water cooling designed for the exhaust impingement of smaller twin-engine aircraft to accommodate the larger single-engine F-35. This was followed by a month-long period of testing at Lakehurst that encompassed a structural survey

Fig. 23 F-35C jet blast deflector integration testing.

using the C13-2, steam ingestion testing, and roll-in arrestments using the Mk-7 and E-28 arrestment systems.

The following year, the CF-3 conducted the first fly-in arrestments with the Mk-7 and E-28 arrestment systems at Lakehurst. The PAX team then discovered that the F-35 Arrestment Hook System (AHS) and shoe geometry were inadequate to reliably catch an arrestment wire. These test results triggered modifications and updated designs of both the AHS and the hook shoe. In January 2014, the CF-3 resumed roll-ins and arrested landings with a redesigned AHS at Lakehurst. These test results were sufficient to allow the team to progress to a Mk-7 structural survey in the clean configuration with internal stores.

In November 2014 the initial CV developmental test sea trials commenced onboard the USS *Nimitz* (CVN 68) with CF-3 and CF-5 test aircraft. This was a west coast sea trial, which further complicated the logistics of the east coast PAX detachment. During this sea trial, for which we enjoyed exclusive use of the flight deck, the ITF conducted 124 catapults/traps and 222 touch-and-go landings with no bolters (i.e., failure to catch an arresting cable when landing). In doing so, the ITF successfully demonstrated day operations, deck handling and logistics, launch and recovery handling, minimum end-speed testing, MIL power launches, and some initial night operations. The aircraft was launched from both the bow and waist catapults. The team did identify an adverse roll when the aircraft was launched from the outer waist catapult (Cat. 4) that would have to be resolved, but overall it proved to be a very successful initial sea trial.

The team resumed sea trials onboard the USS *Dwight D. Eisenhower* (CVN 69) in October 2015 (Fig. 24), but the team no longer enjoyed the luxury of having exclusive use of the flight deck. This at-sea detachment focused on operations with increased aircraft weight and forward CG. During DT-II, 66 catapult launches were performed from all four catapults, along with 66 arrestments and 40 touch-and-go landings. Operations demonstrated during DT-II included an ability to operate in day and night conditions, internal weapon carriage, MIL and MAX power launches, and powered aircraft operation in the hangar bay. Additionally, external environment data were collected and the team performed a risk reduction of the JPALS.

In 2016 the test team began the shore-based work-up with external stores. Between February and May, the Mk-7 structural survey and arrestments were evaluated with a variety of external store loadouts. In August, the third and final CV developmental test sea trial was performed onboard the USS *George Washington* (CVN 73), shown in Fig. 25. The team shared the deck with VFA-101, an operational squadron performing carrier qualifications. During this final sea trial, 121 catapults and traps were performed, along with 67 touch-and-go landings. Operations were conducted with full internal bays and external stores.

These highly successful sea trials were a result of the PAX ITF's thorough shore-based envelope expansion test planning and execution. During the last

Fig. 24 PAX ITF personnel and test aircraft onboard the USS *Dwight D. Eisenhower* (CVN 69).

Fig. 25 CF-3 catapult launch from the USS *George Washington* (CVN 73).

eight years of flight testing at PAX, the shore-based testing was very demanding and uncovered several issues prior to taking the F-35B and F-35C to operate on U.S. Navy ships. These issues, their fixes, and the additional learning that took place during these sea trials have made the F-35B and F-35C more capable platforms.

C. INSTRUMENTATION/DATA PROCESSING

Test instrumentation data are the blood flow of the flight test heart. The primary purpose of mission execution at PAX is to provide the truest and most accurate data required for air vehicle system verification and to demonstrate that the F-35B and F-35C variants operate efficiently, effectively, and safely. Keeping the data system operational and productive had its challenges, but over the past several years, this team overcame several obstacles.

Integrating commercial, off-the-shelf systems with unique data bus protocols that matured over the course of the SDD program resulted in several flight test data system upgrades to increase throughput and recording capabilities. PAX worked out a restrictive approach to data throughput by asking the integrated product teams to narrow their scope of measurements and sample rates to a mission-critical and safety data list.

With a limited amount of noninstrumented weapons available at PAX, instrumented weapons were flown continually for dummy stores. This led to multiple incidents in which strain gauges, transducers, and wiring were compromised. When instrumented weapons were required for safety of test/safety of flight, delays in mission readiness became an issue, often after weapons had been loaded onto the aircraft. Resolution of this issue required a cross-team effort and a written process for inspecting, repairing, and tagging instrumented weapons after every flight.

Unique airframe instrumentation, paired with a limited number of instrumented landing gear, LiftFan®, and engines, posed unique constraints. The required instrumentation configurations, particularly for carrier integration work, were not always available due to competing needs. This issue resulted in several gear swaps between/among aircraft during the SDD flight test program. Similarly, a limited number of instrumented LiftFans and engines posed test configuration constraints that had to be effectively managed, particularly when this instrumentation became damaged or needed repair off wing. It was not uncommon during SDD to have a variety of lines of test on different airframes competing for instrumented propulsion system hardware, sometimes necessitating the swap of hardware between/among aircraft.

Before sea trials commenced, some modifications were required to aid the team in gathering the required data. Mobile control rooms, such as the Mobile Instrumentation and Telemetry System (MITS-12) and a deployable debrief facility, were installed in the hangar bay to allow engineers to monitor aircraft flight test activities in real time. High-speed deck edge cameras were installed to gather launch data. Cameras were also installed about the stern of the ship to document where the aircraft touched down during recovery. A 30-ft anemometer was installed on the bow of the ship. This, along with a ship motion package, helped to characterize the dynamic ship environment that the F-35s would call home.

D. TEST PILOTS' PERSPECTIVE

PAX flight operations included a cadre of test pilots representing Lockheed Martin, BAE Systems, and the U.S. Navy, U.S. Marine Corps, and U.K. Ministry of Defence (Fig. 26). Beginning with the arrival of the first aircraft in 2009, 35 different pilots flew test missions at PAX, completing more than 4400 flights totaling 6600 hours and 35,000 test points. All pilots were qualified to fly all variants of the F-35, and most of them carried additional qualifications in the F/A-18. With no available off-site F-35 training and no experienced F-35 pilots coming into the ITF, all training was developed and conducted within the organization. The team's goal to fly 10 hours per pilot per month in the F-35 was never achieved; rather, the average was close to 7 hours per pilot per month. However, this rate was acceptable due to the mixture of pilot experience, extensive use of simulator-based preparation, and a significantly extensive suite of automated maneuvers built into the control laws.

Manning considerations provided a challenge throughout SDD. The historically useful pilot-to-test-aircraft ratio of 1.5-to-1 was used as the basis for manning decisions. With nine assigned aircraft, manning levels generally kept

Fig. 26 F-35 flight test pilots at PAX.

to 13 pilots flying at any time (4–5 contractors and 8–9 government pilots). Unique test requirements, specifically STOVL and CV envelope expansion, multiple long deployments, and the availability of flight time and requirements for chase support, led to occasional deviations.

There is a cultural and historical belief that safety chase is required for developmental flight test. As a result, early in the program, a safety chase aircraft was required for nearly every flight. It soon became clear that, in many cases, the presence of another aircraft in the airspace and in close proximity to the test aircraft added risk to the test event. One example involved chasing very slow-speed STOVL missions with F/A-18s. The missions were flown at low altitude in busy airspace at speeds that were incompatible with the F/A-18, which left the Hornet pilot in a compromised position. The ITF methodically went through the process of reevaluating each test plan for residual risk and applied appropriate mitigations for each one, which was rarely a chase asset. The most effective mitigators included mission preparation requirements, maneuver standardization, and control room monitoring of critical parameters and air vehicle health. The result was a reduction in the complexity of ITF testing, lower overall risk associated with each mission, and a significant cost savings.

PAX ITF flight operations were successful in meeting all requirements with the following major accomplishments:

- Successfully contributing in the proof of design for the first-ever use of the dynamic inversion model for the control laws.
- Extending the STOVL concept to deliver STOVL capability, including verifying the STOVL performance models and developing Level 1 handling qualities throughout the STOVL envelope in nominal conditions.
- Recognizing CV hook design and pitch pivot point problems early enough that they had timely resolution to meet SDD completion requirements.

The immense amount of work performed in a short time by the PAX ITF is extraordinary and unprecedented. This work has already enabled the Marine Corps warfighter to deploy aboard U.S. Navy L-class vessels today and will soon do likewise for the U.S. Navy onboard CVNs. These efforts also significantly contributed to the Marine Corps declaring F-35B IOC in 2015, and they will contribute to the Navy declaring F-35C IOC in 2019.

IV. CONCLUSION

The SDD phase of the F-35 program was in work for 16 years. During that time, eight years of flight sciences and mission systems flight testing with 18 aircraft at two test sites yielded and delivered an unprecedented 5th-Generation warfighter capability, summarized in Fig. 27. It also provided invaluable experience and many lessons learned: test plans must be flexible and agile early on; a surge concept with focused and concentrated resources accomplished complex testing

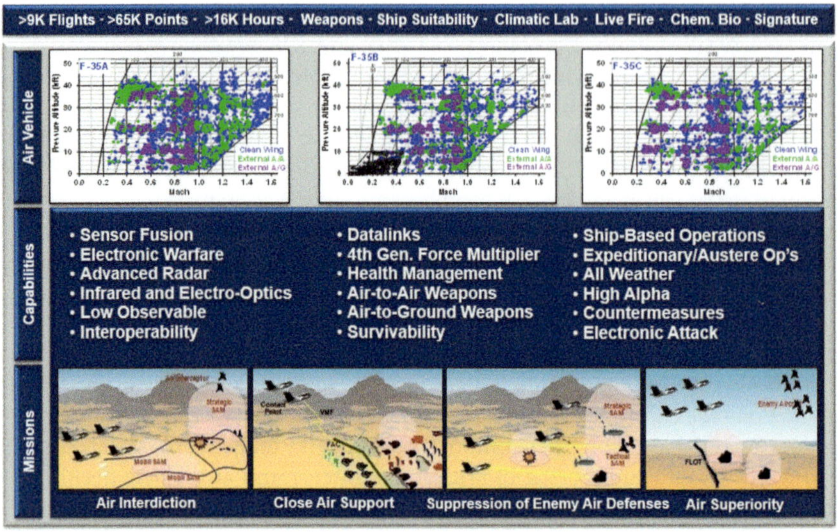

Fig. 27 The F-35 Block 3F delivered 5th-Generation warfighter capability.

in a short period of time; the optimized flight test scheduling forecast is four to six weeks out; instrumentation must be flexible, adaptable, and expandable; training by an expert trainer is key to successful test conductors; and teamwork, with clearly defined roles and responsibilities, and communication are essential. With these lessons learned, future F-35 flight testing can now expand on the historical accomplishments achieved during the SDD phase of the F-35 at EDW and PAX, and continue to deliver unrivaled F-35 capability.

ACKNOWLEDGMENTS

This paper is dedicated to the EDW and PAX Integrated Test Force (ITF) team who executed an unprecedented, historic flight test program. Special thanks to those who provided inputs to capture the history, flight test accomplishments, and lessons learned. Many thanks to the EDW team: David "Doc" Nelson, Dave Holcomb, Teo Avram, Brittany Galloway, Don Lewis, Ron Teague, Mike Shreeve, Tyler Sanders, Brian "Grumpy" O'Melia, Torrey Given, Bernie Taylor, Petar Simich, Jim Bishop, Ryan Mankin, Stan Bogumil, Rita Jones, Peter Liu, Brandon Dees, Jason McCoy, George Hicks, Lane Vaught, Caitlyn Fahey, Jen Schleifer, Ed Sabalburo, Wade Cross, Susan Bishop, and others. Many thanks to the PAX ITF team: Darrell Carney, Jerry Courville, Leigh DePiazza, Eric Faidley, Dan Levin, Trey Mangan, Larry Martin, Dane Wiedmann, Andy Wolfe, Mike Jackson, and others. Also, thanks to J.D. McFarlan for his leadership;

graphic artists Daniel Buck, Justin Simmons, and Laine Miller; and Chad Bellay for the awesome flight test video.

REFERENCES

[1] Ellis, R., et al., "F-35 Structural Design, Development, and Verification," AIAA Aviation Forum, June 2018 (to be published).

[2] Harris, J., and Stanford, J., "F-35 Flight Control Law Design, Development, and Verification," AIAA Aviation Forum, June 2018 (to be published).

[3] Wurth, S., Smith, M., and Celiberti, L., "F-35 Propulsion System Integration, Development, and Verification," AIAA Aviation Forum, June 2018 (to be published).

[4] Robbins, D., et al., "F-35 Subsystems Design, Development, and Verification," AIAA Aviation Forum, June 2018 (to be published).

[5] Lemons, G., Carrington, K., Frey, T., and Ledyard, J., "F-35 Mission Systems Design, Development, and Verification," AIAA Aviation Forum, June 2018 (to be published).

[6] Frey, T. L., Aguilar, J. C., Engebretson, K. R., Faulk, D. K., and Lenning, L. G., "F-35 Information Fusion," AIAA Aviation Forum, June 2018 (to be published).

[7] Wilson, T., "F-35 Carrier Suitability Testing," AIAA Aviation Forum, June 2018 (to be published).

[8] Parsons, D. G., Eckstein, A. G., and Azevedo, J. J., "F-35 Aerodynamic Performance Verification," AIAA Aviation Forum, June 2018 (to be published).

[9] Canin, D. G., McConnell, J. K., and James, P. W., "F-35 Weapons Separation Test and Verification," AIAA Aviation Forum, June 2018 (to be published).

[10] Hetreed, C. F., Carroll, M. D., Collard, J. E., and Snyder, R. C., "F-35 Weapons Separation Test and Verification," AIAA Aviation Forum, June 2018 (to be published).

[11] Parsons, D. G., et al., "F-35 STOVL Performance Requirements Verification," AIAA Aviation Forum, June 2018 (to be published).

[12] Commanding Officer, Marine Fighter Attack Squadron 121, "MAG-12 After Action Report for Exercises Northern Edge/Distant Frontier 2017," Marine Fighter Attack Squadron 121, U.S. Marine Corps Official Report, FPO AP 96310-7335, July 2017, p. 8.

[13] Nelson, D., "F35 SDD Lessons Learned," Lockheed Martin Aeronautics Company, F-35 System Development and Demonstration Presentation, Palmdale, CA, August 2017.

CHAPTER 7

F-35 Structural Design, Development, and Verification

Robert M. Ellis*, Philip C. Gross†, Joseph B. Yates‡ and John R. Casement§

Lockheed Martin Aeronautics Company, Fort Worth, TX

R.H.L. (Tad) Chichester¶

Lockheed Martin Aeronautics Company (retired), Fort Worth, TX

Kathryn Nesmith**

F-35 Joint Program Office, Arlington, VA

This paper will discuss the structural design, development, and verification of the F-35 Joint Strike Fighter, guided by the standard practices of MIL-STD-1530 Aircraft Structural Integrity Program (ASIP) to successfully produce a versatile air vehicle platform that meets the varied performance and airworthiness requirements of our worldwide customers. This approach resulted in the simultaneous development of three airframes that are exceptionally robust, and will meet the demanding requirements of war fighters worldwide that are choosing the F-35. The "Five Pillars of ASIP," as defined in MIL-STD-1530, provide the framework for development of an air vehicle platform from design through analysis, testing, and force management. ASIP applies throughout the entire aircraft life cycle from requirements definition through retirement. This paper will follow the progression through each task for the development of the F-35 used to enhance the development of this extraordinary product.

I. INTRODUCTION

This paper will discuss the structural design, development, and verification of the F-35 using the standard practices of MIL-STD-1530 [1] Aircraft Structural

*Senior Manager, F-35 Structures Development Integration.

†Deputy Director F-35 Structures Development and Chief Structures Engineer.

‡Aeronautical Engineering Principal, F-35 Service Life Analysis.

§Aeronautical Engineering Principal, F-35 Structures Development Integration.

¶Retired Senior Manager, F-35 Structures Development.

**Air Vehicle Lead, F-35 Joint Program Office.

Integrity Program (ASIP) to successfully produce a multiservice, versatile air vehicle platform that meets the varied performance and airworthiness requirements of our worldwide customers, as depicted in Fig. 1 showing unique requirements for each of the three variants. The "Five Pillars of ASIP," as defined in MIL-STD-1530, provide the framework for development of an air vehicle platform from design through analysis, testing, and force management. ASIP applies throughout the entire aircraft life cycle from requirements definition through retirement. This paper will follow the progression through each task as shown in Table 1, used to enhance the development of the F-35 product.

The Joint Strike Fighter (JSF) program, later to be designated the F-35, is the largest single defense acquisition program in U.S. government history. The program was awarded in a new era of acquisition reform. The primary tenant of this reform was the utilization of Performance-Based Specifications (PBSs), intended to give the awarded contractor unprecedented freedom in the development of the weapon system by merely specifying end-state capabilities, rather than utilizing legacy practices of employing layered military specifications, which dictated detailed requirements to the process of developing and verifying such capabilities. The approach gave the government's Joint Program Office (JPO) insight, but not oversight, of the efforts of the contractor, Lockheed Martin.

Although the approach had been successfully employed in the development of aircraft subsystems and other smaller acquisition programs, the use of performance-based specification to the development of an air vehicle structure was unprecedented. This precedent has been, and remains, the cornerstone in the certification of aircraft structures. To adhere to the intent of the PBS approach,

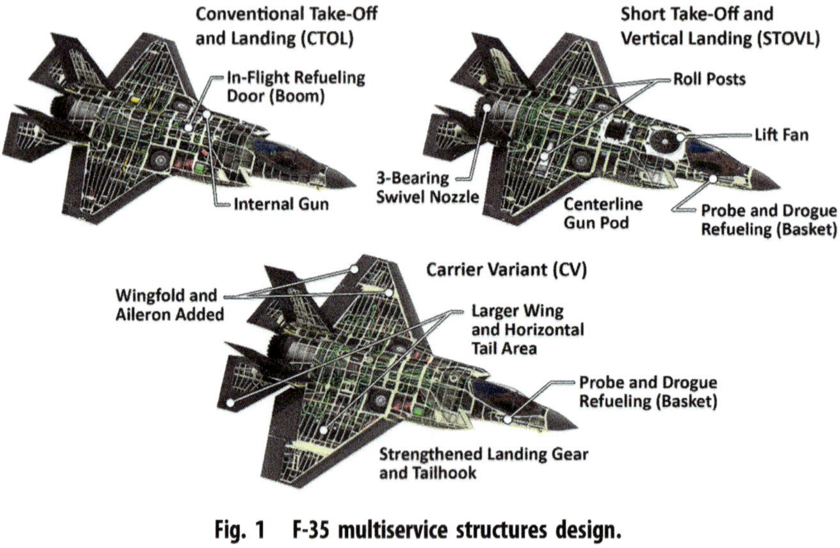

Fig. 1 F-35 multiservice structures design.

TABLE 1 FIVE PILLARS OF ASIP

Task I Design Information	Task II Design Analyses & Development Testing	Task III Full-Scale Testing	Task IV Certification & Force Management Development	Task V Force Management Execution
ASIP Master Plan	Material and Joint Allowables Testing	Static Tests	Certification Analysis	Individual Airplane Tracking (IAT) Program
Design Service Life and Design Usage	Loads Analysis	First Flight Verification Ground Tests	Strength Summary & Operating Restrictions (SSOR)	Loads/Environment Spectra Survey (L/E SS)
Structural Design Criteria	Design Service Loads Spectra	Flight Tests	Force Structural Maintenance Plan (FSMP)	ASIP Manual
Durability & Damage Tolerance Control Program	Design Chemical/Thermal Environment Spectra	Durability Test	Loads/Environment Spectra Survey (L/E SS) Development	Aircraft Structural Records
Corrosion Prevention and Control Program (CPCP)	Stress Analysis	Damage Tolerance Tests	Individual Airplane Tracking (IAT) Program Development	Force Management Updates
Nodestructive Inspection (NDI) Program	Damage Tolerance Analysis	Climatic Tests		Recertification
Selection of Materials, Processes, Joining Methods, and Structural Concepts	Durability Analysis	Interpretation & Evaluation of Test Results		
	Corrosion Assessment			
	Sonic Fatigue Analysis			
	Vibration Analysis			
	Aeroelastic & Aeroservoelastic Analysis			
	Mass Properties Analysis			
	Survivability Analysis			
	Design Development Tests			
	Production NDI Capability Assessment			
	Initial Risk Analysis			

JPO and Lockheed Martin worked together to develop the JSF Joint Contract Specification (JCS) to define expected structural performance parameters, while also clearly defining design criteria, verification analyses, and testing, which traditionally had been employed to achieve such performance.

In this joint government–contractor effort, development of a thoroughly integrated and executed ASIP became the foundation to achieve challenging performance requirements for airframe safety and service life. ASIP provides an umbrella framework that ensures coordination of critical engineering and management processes, specifications, and plans—such as quality assurance, manufacturing engineering, corrosion control, fracture control board, materials and processing, structures building block development test plan, and so forth—leading to air vehicle certification. The challenge facing the JSF program was to come to agreement on an ASIP that sufficiently addressed the concerns of the multiple service and partner country airworthiness authorities. Although each partner country had its own unique legacy to structural certification, all agreed that the most comprehensive foundational approach to the problem was the adaptation of MIL-STD-1530, with its prescription under the Five Pillars of ASIP, to define the certification path for the F-35.

The Five Pillars of ASIP are a planned series of tasks employed to ensure aircraft safety and structural integrity requirements are met and maintained throughout the fleet operational service life. Much of the affordability aspect of ASIP is accomplished by following a prescribed rigorous approach that attempts to avoid design shortfalls while building in robustness to avoid costly structural rework. The ASIP seeks to optimize cost of ownership and aircraft availability while maintaining aircraft safety. The foundation of ASIP is a set of test and analysis plans defined to meet the program requirements. The evidence that these requirements are met supports the structures integrity and certification position discussed herein.

II. TASK I: DESIGN INFORMATION

The Design Information task encompasses those efforts required to apply existing theoretical, experimental, applied research, and operational experience to specific criteria for materials selection and structural design. Elements of Task I include developing the ASIP Master Plan, Structural Design Criteria (SDC), and Durability and Damage Tolerance (DADT) control plan; design loads, selection of materials, processes, and joining methods; and defining the design service loads.

The primary Task I objective is to ensure the appropriate criteria and planned usage are applied to the design of the F-35 so the specific operational requirements will be met.

Design information and data used to develop the basic F-35 airframe structural design under the System Development and Demonstration (SDD) phase of the program is complete. Air system–level design requirements were specified in the F-35 JCS. The JCS is a performance-based specification that defines vehicle requirements in terms of required operational parameters like speed, range, or payload. The specific "how-to" actions to be performed to achieve the required performance are defined in lower level requirements documents developed by the contractor and JPO, and, in the case of structures, following the ASIP framework. Most of the structural requirements flow from the contractual requirements to provide a safe, durable, and damage tolerant air vehicle that will endure for 30 years or 8000 flight hours. Additional requirements are imposed by the need to meet performance-based requirements that encompass combat radius, payload requirements, austere basing requirements, flight performance characteristics, and so forth.

The F-35 Structural Design Criteria was one of the first and most fundamental of the structure requirements documents generated in Task I. The joint JPO–Lockheed Martin team employed the Joint Services Specification Guidance (JSSG-2006) to provide the basic framework for the SDC, tailoring the guidance of the vehicle-level performance requirements (speed, weapons, etc.) and anticipated peacetime training usage defined in the JSF JCS. The completed F-35 SDC document contains all the aircraft design weights, weapons carriage and employment, speeds, load factors, aeroelastic stability requirements, landing sink rates, ship compatibility parameters, and the like required to develop structural design loads that, when applied and verified, ensure each F-35 variant meets the specification performance. Of note is the early program policy decision to include projected empty weight growth-to-Initial Operational Capability values in the structural design weights, reducing the risk of "loads creep" and associated life and performance limits that have plagued other aircraft programs.

The F-35 SDC document also decomposes the required 8000 flight hour service life and design mission utilization specified in JSSG-2006 Appendix A-3 (anticipated mission profiles, mission mix, ship vs land basing, etc.) into a complete definition of the operational parameters required to determine the baseline load spectrum for each of the variants. The definition includes for each variant:

- Numbers and types of each mission per lifetime
- Mean and 90th percentile maneuver load factor exceedances per 1000 flight hours, as determined from analysis of measured usage data from applicable legacy aircraft
- Numbers and types of landings (vertical, shipboard arrested, etc.) as well as the distribution of landing sink rates
- Numbers and types of takeoffs including catapult launches, ski-jump launches, and short takeoffs
- Taxi and ground handling spectra for taxi, turning, braking, and runway roughness

The importance of the baseline usage and resulting spectrum cannot be overstated—every ASIP task, from detailed design to full-scale test to force life management, relies on or is impacted by this information as a foundational starting point.

Finally, the SDC defines the verification requirements for each of its requirements, including whether the verification method will be via inspection, analysis, demonstration, test, or a combination of these. These verification requirements are linked to the corresponding JCS requirements.

To support SDD contract closure, a Verification Cross Reference Matrix (VCRM) was established that details the verification objectives and jointly developed Success Criteria (SC) that verify each JCS requirement. Closure of the SC are tracked in the Verification Test and Evaluation (VT&E) database. SC closures support closure of the parent JCS paragraphs, which in turn support SDD contract closure, as shown in Fig. 2.

The establishment of the operating environment was critical for use in the design and analysis of the structure. Figure 3 highlights how the static and service external loads were developed using wind tunnel data and an established aerodynamic database. The PBS for the F-35 requires that 90% of the delivered aircraft meet the full service life requirement of 8000 hours and 30 years. This was accommodated by developing intentionally severe spectra based on legacy fleet usage severity factors such that a 90th percentile usage spectrum was developed for design and test. The analysis criteria on which the service life was to be determined was not specified. At the outset of the F-35 program, a common service life analysis criteria was proposed for the three variants; however, after discussions among JPO, U.S. Air Force, and U.S. Navy, no common criteria were found to be acceptable to all of the services. It was then decided to adopt service life analysis criteria that reflected the legacy experience base of the primary U.S. user for each variant. Thus, the F-35A was analyzed using crack growth-based criteria combined with a 90th percentile mission-based spectrum for durability analysis and a mean of 50th percentile spectrum for damage tolerance analysis—similar to previous Air Force aircraft. The F-35B and F-35C were designed using a crack initiation criteria coupled with a Critical Point in the Sky

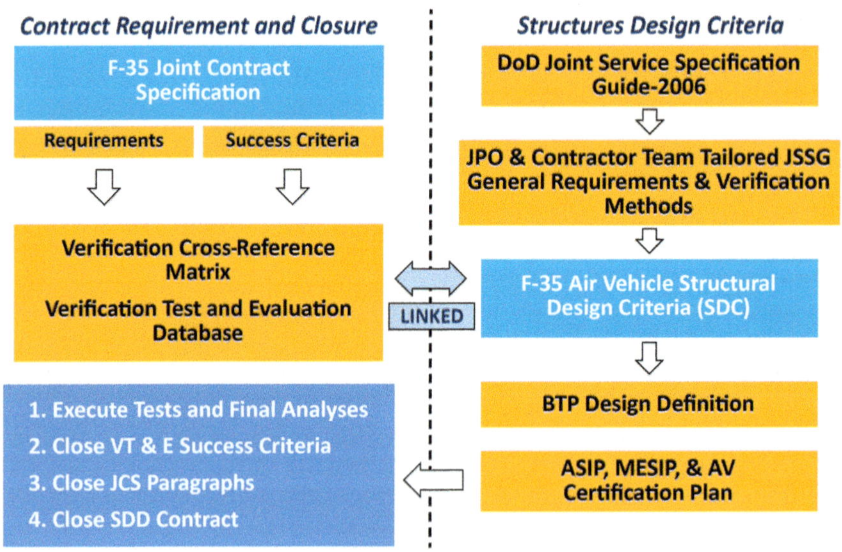

Fig. 2 F-35 structures certification process.

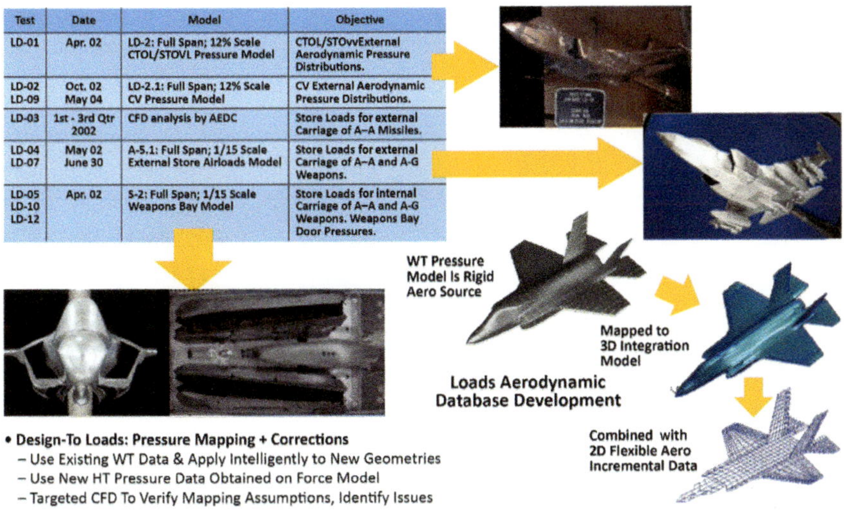

Test	Date	Model	Objective
LD-01	Apr. 02	LD-2: Full Span; 12% Scale CTOL/STOVL Pressure Model	CTOL/STOvvExternal Aerodynamic Pressure Distributions.
LD-02 LD-09	Oct. 02 May 04	LD-2.1: Full Span; 12% Scale CV Pressure Model	CV External Aerodynamic Pressure Distributions.
LD-03	1st - 3rd Qtr 2002	CFD analysis by AEDC	Store Loads for external Carriage of A–A Missiles.
LD-04 LD-07	May 02 June 30	A-5.1: Full Span; 1/15 Scale External Store Airloads Model	Store Loads for external Carriage of A–A and A-G Weapons.
LD-05 LD-10 LD-12	Apr. 02	S-2: Full Span; 1/15 Scale Weapons Bay Model	Store Loads for internal Carriage of A–A and A-G Weapons. Weapons Bay Door Pressures.

Fig. 3 Loads wind tunnel testing and aerodynamic database development.

(CPITS) based severe usage spectrum for durability analysis, and the damage tolerance analysis was performed using crack growth analysis coupled with the same severe CPITS spectrum.

The CPITS approach is based on a single critical point in the sky for each major airframe component, where a point in the sky is a unique Mach–altitude combination. A single reference aircraft weight, with multiple configurations and centers of gravity as shown in Fig. 4, was used for each such point in the sky. A damage reference level based on that single point in the sky was established for each major airframe component. Then, a single spectrum of multiple critical points in the sky was developed, with the goal that no single component's damage was less than 80% of its reference level. Catch-up cycles were added to bring each of the critical component damage levels as close to the 80% goal as feasible.

For the F-35B and F-35C, the CPITS approach resulted in a mission mix with 50% supersonic usage. This is far more supersonic usage than is expected in actual usage, which makes it especially severe for aft fuselage and empennage structure. The remaining usage includes 30% of the time spent at sea level, which is critical for the wing structure. In contrast to CPITS, the mission-based spectra used for the F-35A variant is based directly upon the Mach numbers, speeds, altitudes, and mass properties for the mission profiles provided in the JCS.

For the F-35 program to be successful under the new format of acquisition, it was imperative a robust ASIP Master Plan be developed. This plan outlined the

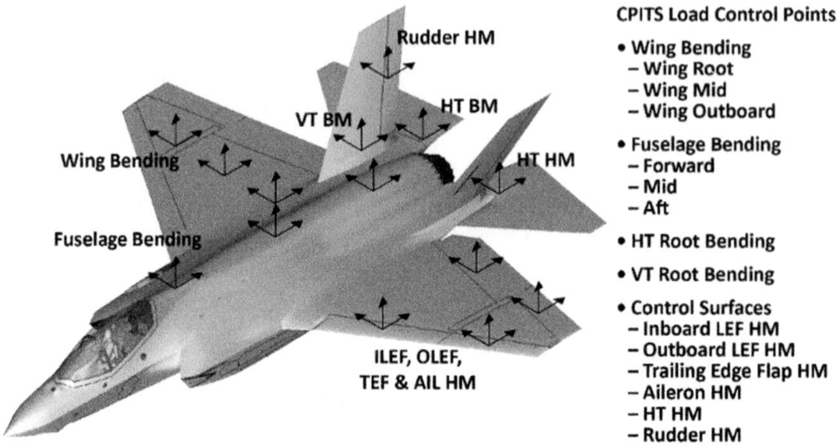

Fig. 4 CPITS load control points.

criteria for material selection, structural design and analysis, and certification testing requirements. The plan provided a roadmap to ensure all evidence for certification and verification was identified, planned, and controlled. The F-35 program was awarded to three companies with Lockheed Martin as the lead. Northrup Grumman and BAE Systems joined Lockheed Martin as the partners on the F-35. Lockheed Martin, as the lead, was responsible for overall program integration. The Fort Worth, Texas, facility held design authority for the forward and wing structure. The Palmdale, California, facility was responsible for all control surfaces and edges. The Palmdale teams worked closely with the BAE Systems teams who held design authority for the Horizontal Tails (HT), Vertical Tails (VT), and aft fuselage. Northrup Grumman was assigned design authority for the center fuselage structure.

The tri-company teaming arrangement for the F-35 program required significant emphasis on planning and coordination to ensure the establishment of consistent design and analysis criteria. F-35–specific design and analysis manuals and common toolsets were developed using the best practices of the three companies. Direction to the structures disciplines was provided through more detailed subordinate documents; for example, the Structural Analysis Methods and Design Criteria (SAMDC), DADT Guidelines and Control Plan, and Drawing Requirements Manual were published as unique F-35–specific direction. The SAMDC defines the structural analysis methods, policies, and criteria used by the development team to conduct structures analysis. A recurring coordination meeting was established to ensure all structural analysis disciplines from all three companies met and discussed resolution of issues with best practices selected. Frequent reviews were held at each partner site, as well as at major subcontractor sites, during the layout and design phases to review the design and analysis of the F-35 structure to ensure that the policies and criteria were being applied properly and that the resulting structure was the lightest weight structure that would meet the requirements. The multisite collaboration was enabled by the selection of a common design tool suite and carefully controlled interface control drawings. A centralized database, or Product Data Manager (PDM), was established to allow all partners to work on the latest designs. This "digital thread" allowed the three partners to develop and share digital models within the team and, upon release, to send them to suppliers for fabrication.

The design usage requirements in the SDC discussed previously were broken down further in the Environmental Description Document (EDD), which contained detailed usage and environmental data needed to flow down to any engineering group requiring it, particularly those responsible for procurement and qualification of mechanical and electrical systems for the aircraft. Lifetime and extreme usage information, such as total time in Mach-altitude and temperature bins, time at angle-of-attack, and so forth is defined in the EDD. That data, in turn, became the basis for usage and environments data for items such as subsystems design usage; vibratory and acoustic environments; temperature environments; and humidity, moisture, and precipitation environments.

The F-35 DADT Control Plan identifies and assigns responsibility for all of the fracture control tasks necessary to ensure compliance with the DADT design requirements and continued structural integrity. The Fracture Control Board ensures consistent implementation of the DADT Control Plan requirements at all F-35 team sites and provides guidance to the product teams with respect to the DADT requirements. Board membership includes the disciplines of stress, DADT, Materials and Processes (M&P), design, manufacturing, quality assurance, and an airframe Integrated Product Team (IPT) representative for a direct, authoritative tie to the subordinate IPTs. Similarly, the F-35 Corrosion Prevention Control Plan identifies and assigns responsibility for the corrosion control tasks. The Corrosion Prevention Advisory Board oversees the consistent implementation of the design requirements needed for corrosion prevention and control at all F-35 team sites. The NDI Requirements Review Board (NDIRRB) guides the development of NDI capabilities across the F-35 team and at all subcontractors to ensure that the quality requirements of the aircraft structure are met. It also ensures that the NDI capabilities needed to support the aircraft in service are well defined and available in a timely manner.

Materials, processes, and joining methods are defined, reviewed, and program approved for airframe structural applications. Building block tests of successively increasing complexity, as shown in Fig. 5, were conducted to provide engineering data used to demonstrate compliance with structures design requirements. A structures building block development plan was defined after contract award and is documented as part of the overall F-35 Systems

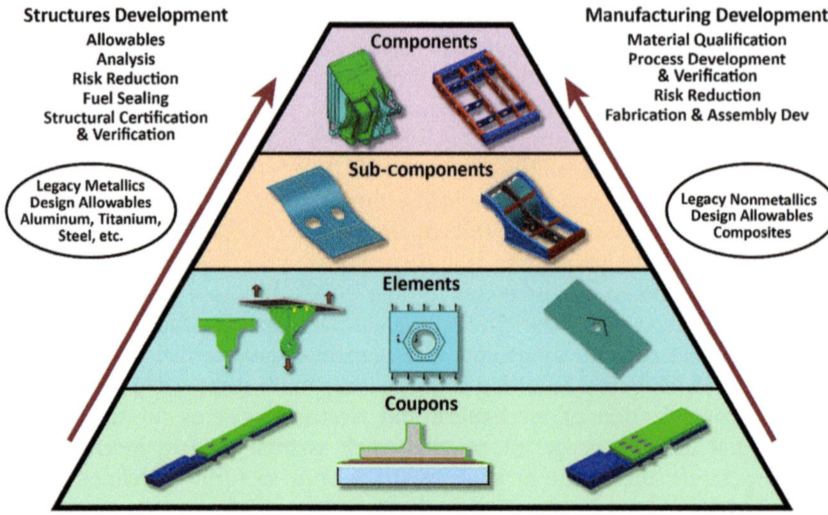

Fig. 5 Building block test approach.

Test Plan. Coupon- and element-level tests were conducted as part of the F-35 building block test program to generate material allowable properties, qualify materials and processes, characterize joining methods, and develop strength, durability, and damage tolerance properties for design-specific features of the F-35 aircraft.

Physical, mechanical, and chemical properties and attributes are documented and controlled by material specifications and verified through required incoming material receiving and quality assurance processes. All materials used in design applications are covered by released specifications. Processes used in airframe construction, from detail part fabrication through final assembly, are controlled by specification reference on the face of the drawing and by manufacturing work instructions or process bulletins.

III. TASK II: DESIGN ANALYSIS AND DEVELOPMENT TESTING

Once the framework for design and analysis procedures was established, the team was ready to move into structural layouts to mature the structural configuration.

The initial focus on affordability and structure commonality forced the aircraft design into a configuration that was too heavy and lacked efficient wing carry though structures load paths. The program went through a weight reduction effort as described by Counts et al. [2], which resulted in a greatly improved structures platform; however, this significant change in design created an extremely challenging schedule with all three variants concurrently in various phases of the design process, as represented in the schedule in Fig. 6. The tri-company teaming arrangement became a very important lever to the success of completing the three variant designs per schedule.

To enable this monumental task, the structures engineering team needed to find ways to engage engineering talent from a worldwide pool. Lockheed Martin and the partners hired engineers from multiple engineering firms around the world, as depicted in Fig. 7. A common saying on the program was, "The sun never sets on the F-35 program." This was enabled by the digital thread mentioned earlier. All data were continuously updated and available to allow work 24/7. The framework of ASIP as established in Task I allowed for the inclusion of engineers from multiple different companies, each with their own methods of design and analysis, to conform to the standard process used on the F-35 program. The enforcement of the defined criteria and plans ensured consistent design and analyses processes were used by all participating engineers.

A rigorous and disciplined design maturation process was followed as depicted in Fig. 8. Elements of those phases are discussed in this section.

The broad scope and complexity of the program presented an organizational challenge for the F-35 Structures Technology group, which was responsible for external loads, structural dynamics, and flutter, as well as all vehicle-level Finite Element Analysis (FEA) and internal loads. The first task was to determine how to efficiently and effectively transfer external and internal loads data for three variants to a large number of engineers on multiple teams in various locations around the world. The form and content of design loads databases had to be consistent across variants to avoid misinterpretations and ensure

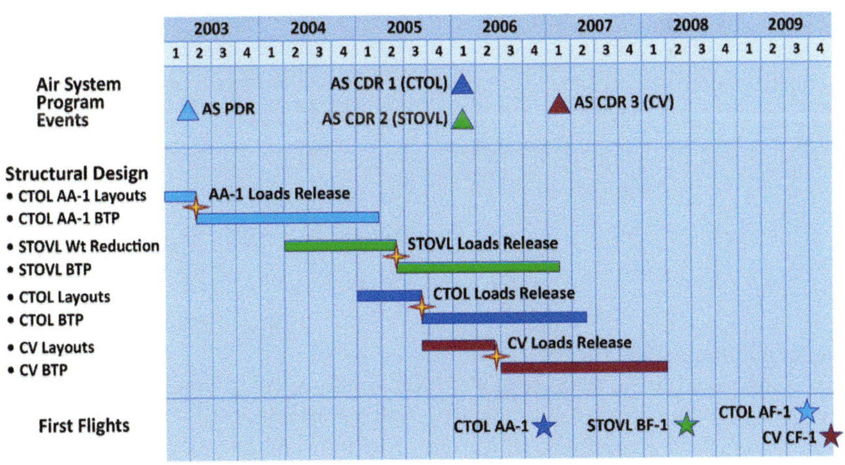

Fig. 6 F-35 structures design timeline.

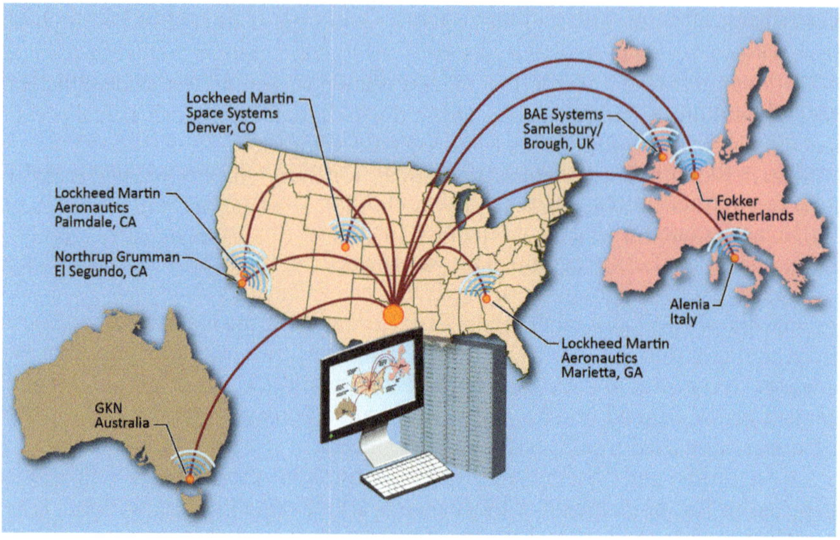

Fig. 7 Worldwide engineering effort.

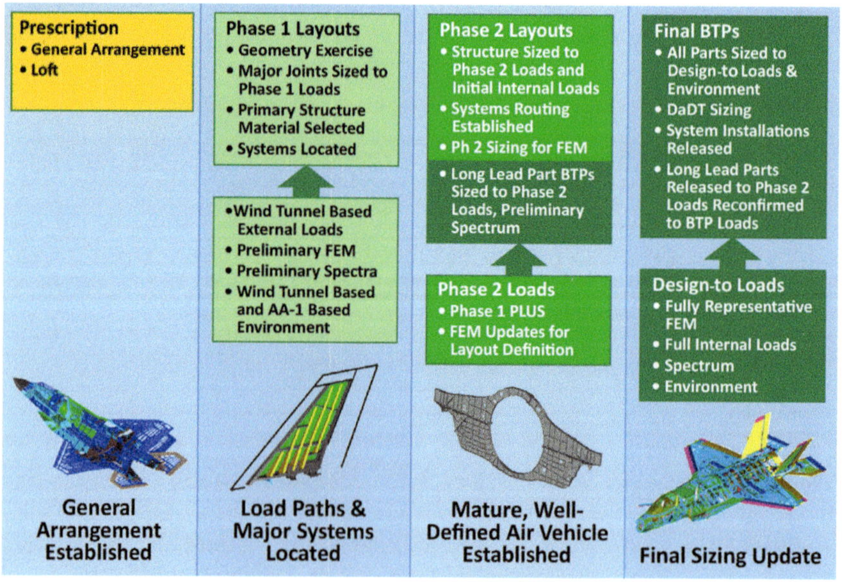

Fig. 8 Airframe structural maturation.

proper use. All types of design loads, regardless of the method used to produce them, had to be expressed quasi-statically so that differences were transparent to the design teams and to allow simple translation to ground test loads. Efficient ways of combining loading actions within the loads databases, such as fuel pressures in combination with air loads, were also developed.

Flutter prevention and aeroelastic stability were strongly emphasized early in the configuration development. Optimization of control surface and empennage planforms, hinge lines, surface thickness, hardpoint locations, and weapons pylon geometry was carried out based on flutter and divergence analysis using a combination of Commercial Off the Shelf (COTS) software and Lockheed Martin proprietary methods. A dedicated wind tunnel flutter model representing the F-35C aft fuselage, VT, HT, and rudder was tested in the NASA Langley Transonic Dynamics Tunnel (Fig. 9) to collect empirical data for correlation of the flutter analysis models. Data were obtained to help define HT and VT rotational stiffness requirements; characterize the unsteady aero interaction among the HT, VT, and rudder; develop transonic dip corrections for use in the flutter analysis; and allowable HT and rudder free play values. Once the required control surface rotational stiffnesses were established, flutter team members worked with design, stress, and flight control hardware engineers to allocate the required stiffness to each individual part in the most weight-efficient manner. This process also included the allocation of allowable weights including reserves for future repairs. Dedicated Finite Element Models (FEMs) for each variant optimized for dynamic analysis and derived from the master Air Vehicle (AV) FEMs were used throughout the program. By the Concept Development Review (CDR) for each variant, all of the designs were shown to be fully compliant with the aeroelastic stability requirements of the SDC.

Work also began early in the program ensuring that empennage, wing, and control surface buffet loads were fully accounted for in the initial design for both strength and life. Buffet caused by forebody vortex interaction with tails,

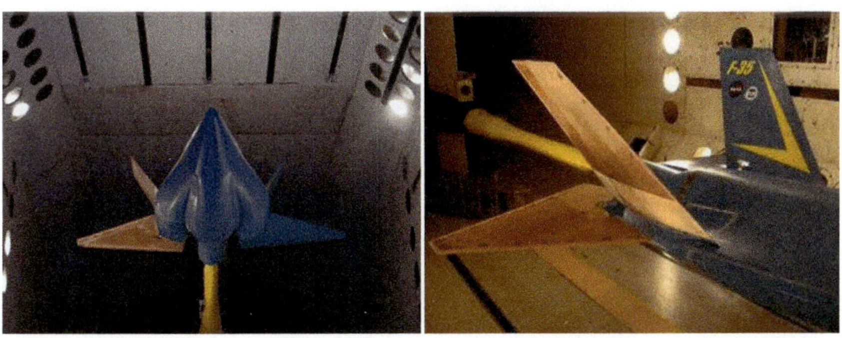

Fig. 9 F-35C aft fuselage and empennage flutter model in the NASA Langley Transonic Dynamics Tunnel.

as well as wing buffet from unsteady transonic flow, has challenged all recent legacy fighter designs. This has typically not been discovered until flight testing, making fixes expensive and disruptive. Early water tunnel testing was performed to characterize the aerodynamic drivers of tail buffet and point the way for more thorough wind tunnel tests. Wind tunnel tests were then conducted to measure unsteady pressures acting on the tail and wing surfaces. Summary of both water and wind tunnel testing used as the basis of buffet loads development is shown in Fig. 10. The unsteady pressures, after proper scaling, were used as the forcing function in a dynamic transient response analysis to calculate resulting buffet loads. These results were then combined with appropriate steady maneuver loads and used to generate internal loads for design. As the analysis matured, aircraft AA-1 was used to conduct a high force-level empennage Ground Vibration Test (GVT) to measure the extent that VT and HT buffet response propagated throughout the structure. Results of this test were used to define more representative structural damping at high force levels for use in design loads development.

A preliminary full airframe FEM was developed to provide internal load sets to the structures analysis teams to further mature the major structural load paths. At this stage in the process the primary focus is on defining a stable, airworthy platform; however, the service life of the aircraft is also a consideration as discussed in the Five Pillar process. During this phase, a preliminary service load spectra is developed to ensure features of the structural arrangement meet all Service Life requirements.

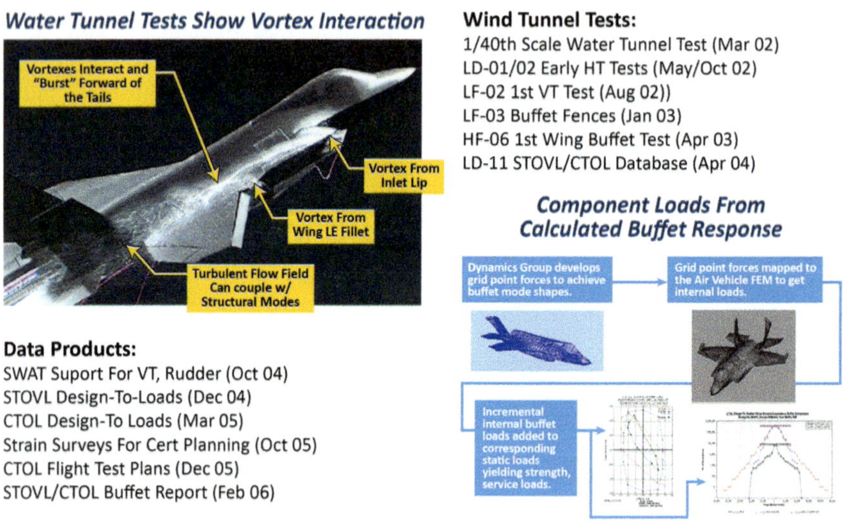

Fig. 10 Buffet loads development.

As the design teams matured the structural arrangement during Phase 1 layouts, part commonality was considered, where possible, to reduce the number of unique parts and save design span time for all three variants. Material selection for primary structure was also further developed. During this phase, the systems design teams received system component definitions from the Vehicle and Missions Systems teams and identified locations for major systems and reserved space for routing paths.

The exit criteria for Phase 1 was an airframe architecture with major load paths established and understood, as well as all major subsystems located and attachment configuration established.

The external static and service loads continued to be developed as part of Phase 2. A full aircraft preliminary FEM was constructed using the Phase 1 configuration and updated external loads to provide internal loads for use by the structural analysis teams to further develop the sizing of the structural parts. During this phase all subsystems Interface Control Packages (ICPs) were finalized to ensure adequate space and attachment features were identified. These ICPs also provide definition of electrical, hydraulic, fuel, and cooling requirements. The routing paths for these systems were established to ensure adequate clearances within the structural arrangement. This included wiring harnesses, fuel delivery tubes, hydraulic lines, and liquid cooling tubes. The effort in Phase 2 produced a mature, well-defined air vehicle architecture.

During the final phase, all loads information was finalized to incorporate environmental conditions, such as temperature and vibration. As the parts became more defined, a fully represented FEM was developed to incorporate all Phase 2 sizing of structures parts, as depicted in Fig. 11.

Combined with the updated external loads, a full set of internal loads were published to perform final sizing of all airframe structural parts and assemblies to facilitate release of the Build-to Packages (BTPs), or drawings. All systems installation drawings were also fully matured and released. The total number of drawings released for all three variants was over 55,000. The release of all drawings, as shown in Fig. 12, represented the closure of the structural maturation process and established the baseline SDD configuration produced and used for flight testing.

IV. TASK III: FULL-SCALE TESTING

The objective of Task III full-scale airframe ground and flight testing is to provide data necessary to validate the design strength, durability, vibration, flutter, aeroservoelasticity, and other structures analyses to provide a safe operating envelope for the warfighter and to provide the data needed to manage the service life of the fleet.

A rigorous structural test program was developed to verify contractual requirements and to produce certification evidence for the strength and durability of the airframe for each variant. The building block test approach mentioned

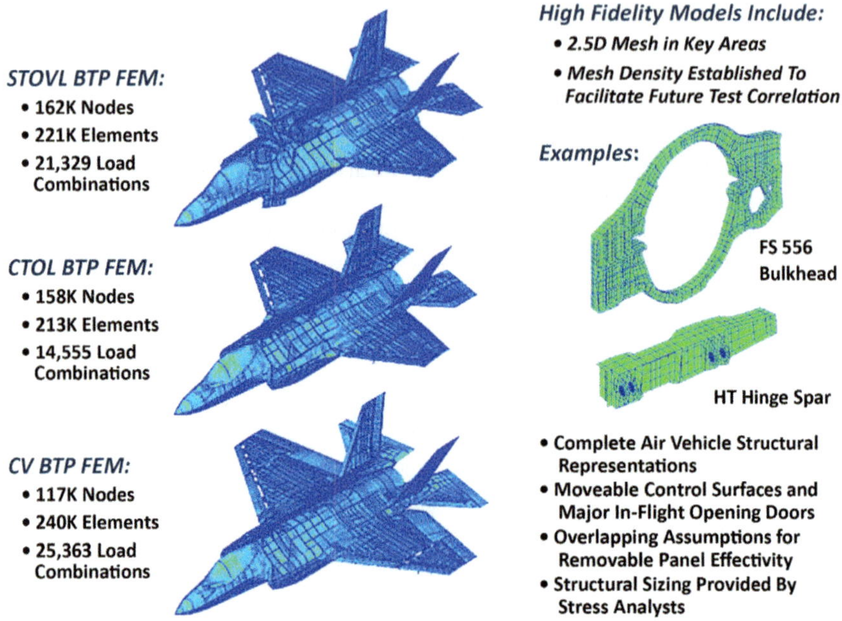

Fig. 11 Air vehicle finite element models for internal loads.

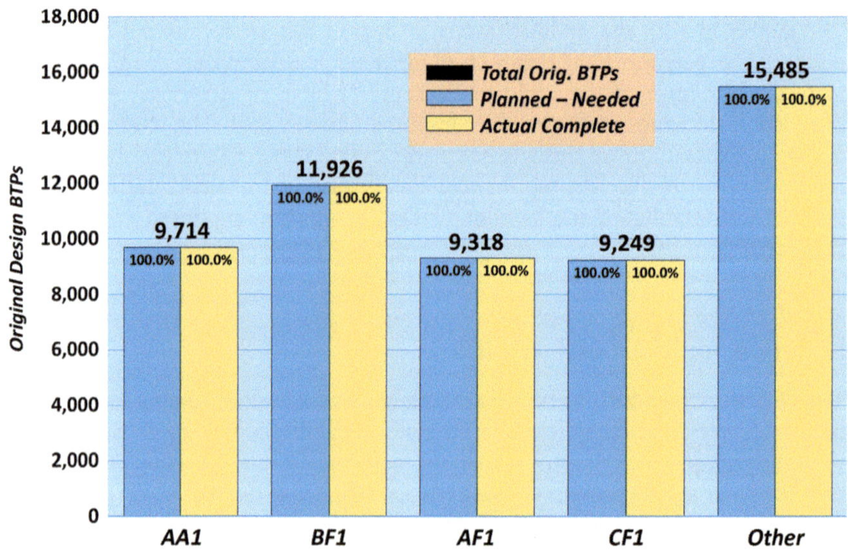

Fig. 12 Total original BTPs released.

previously sets the base leading to the full-scale ground and flight test programs. The test program was designed to include the following testing:

- Component-level certification tests
- Full-scale static and durability test articles for each variant
  - Three full-scale static test articles (fuselage, wing, and vertical tail)
  - Three horizontal tail static test articles
  - Three full-scale durability test articles (fuselage and wing)
  - Three horizontal tail durability test articles
  - Three vertical tail durability test articles
  - Full-scale nose and main landing gear static tests
  - Full-scale nose and main landing gear fatigue tests
- One combined drop test/barricade/live fire test
- Flight test aircraft
  - Proof testing and calibration of loads instrumentation: AF-2, BF-3, CF-2
  - Flutter, including GVT: AF-1, BF-2, CF-1
  - Flight loads, buffet, store ejection response: AF-2, BF-3, CF-2
  - STOVL doors and lift system, landing, and ski-jump loads: BF-1
  - Shipboard launch and recovery loads: CF-3
- Individual ground tests to aid in design development

These tests are designed to provide the necessary information to address material characterization development and risk reduction, manufacturing risk reduction, structural analysis correlation and calibration, material allowables, qualification, and certification.

The F-35 program tri-variant and tri-company organization and the program schedule provided unique challenges and opportunities for the full-scale static and durability testing. Seven dedicated airframes were initially planned as part of SDD. As the program matured, the dedicated drop test for the carrier variant, performed at Vought Aircraft Industries in Grand Prairie, Texas, was combined with the static test article, reducing the number of ground test articles to six. For the first time ever in an aircraft development program, three different variants were scheduled for testing concurrently. To accomplish this, full-scale testing was performed at two sites, as shown in Fig. 13. A further cost saving efficiency was to utilize the same test frame for both static and durability testing. A common test frame design was used for all static and durability testing with modifications as needed for unique local testing requirements.

A. STATIC TEST

The F-35 static test program consisted of three full-scale test articles as well as three independently tested horizontal tails. Each test article was subjected to a set of critical test conditions to evaluate the strength of the structure at limit

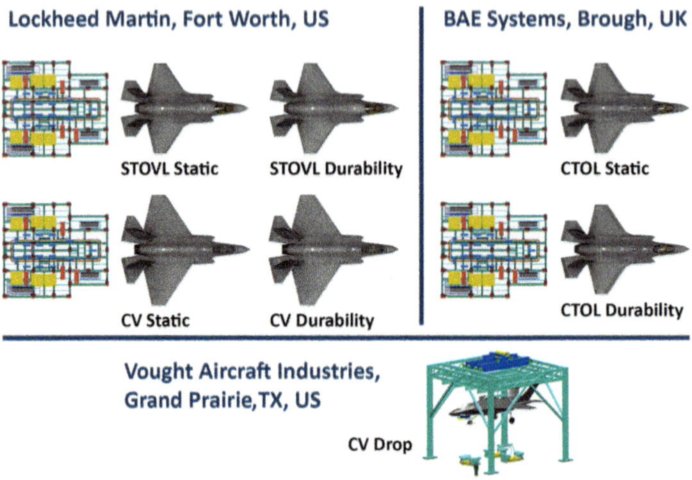

Fig. 13 Ground testing at multiple locations.

and ultimate load. In some cases, the measured deflection of the structure was also a major goal of the test. Each test article was instrumented with strain gages and load bridges during the build of the test article to collect data with which to correlate the FEM and to allow live monitoring of the test to ensure the safety of the article. The sequence of the test conditions was carefully optimized to support the needs of the flight tests and to reduce the cost of setup changes for the test where possible.

The requirements for flight clearance for the flight test program were also unique to the F-35. For multiple legacy fighter aircraft platforms, first flight clearance was based on an 80% envelope substantiated structural analysis. The F-35 program only allowed a 40% envelope for flight testing prior to proof or static test confirmation of the analysis. This resulted from a statistical analysis of the risk of a safety of flight failure that was based on prior ground test program results. To stay within the Program Executive authority for flight risks, the envelope was limited to 40% prior to confirmation of the analysis via test. To support the flight test schedule, proof tests were conducted on the loads-instrumented aircraft to provide the evidence needed to open the flight test envelope and to provide calibration data for the flight test instrumentation. Ground vibration and free play tests were also conducted on the loads and flutter flight test aircraft to assess the dynamic performance of the air vehicle.

To expand the flight testing envelope and support the flight clearance need dates, the static testing of all three variants needed to be completed as quickly as possible. Early decisions regarding the design of the common test frames and the deployment of real-time monitoring of the test at the partner sites supported the rapid pace of the tests. It was decided to include sufficient load rams in the test frame to be able to move from one test condition to the next with minimum reconfiguration. This investment in test equipment was offset by reductions in the duration of the test setup for each condition, thus enabling more efficient testing. It was possible to progress through many conditions with no physical configuration change to the test frame load introduction hardware. When possible, multiple critical conditions were combined into "hybrid" conditions and tested simultaneously. For example, a maximum hammer shock condition, which was critical for the inlet duct, could be combined with a maximum full airframe condition. Testing in the common test frames and testing concurrently provided the efficiency required to complete static test milestones well ahead of the flight clearance need dates. The F-35B test was completed first in Fort Worth, and the lessons learned from that test were used to accelerate the planned test rates for the F-35A and F-35C tests.

In the case of the F-35A test article, an early portion of the testing was accomplished at Fort Worth prior to shipment to BAE Systems in the United Kingdom to support the needs of the initial flight clearance. The shipment of the test article to the United Kingdom was a unique experience. During the transport from Fort Worth to the Port of Houston, there were many UFO sightings reports as the test article, wrapped to protect it during shipment, made its way along the highway. Further, the logistics of barging the article to a location where it could be removed from the barge (Fig. 14) required timing of the tides of the Humber River.

Fig. 14 Shipment of CTOL static test article to Brough, United Kingdom.

Figure 15 shows the pace of the F-35A testing as it accelerated and outpaced even the ambitious schedule needed to support flight test needs. This was due in part to the robust airframe, which allowed for a virtually trouble-free test program, as well as the lessons learned from the F-35B testing accomplished in Fort Worth.

Although there were some test findings on the article, all three full-scale static test programs were completed ahead of schedule with no major findings or any findings that would jeopardize safety of flight. The F-35B static test identified an interference issue between the inboard weapons bay door when it was under load that required a redesign for the production articles and a minor trim for the SDD flight test assets. The other noteworthy finding was the failure of the auxiliary air inlet door downlock mechanism, which carried less than 150% DLL prior to failure. This was redesigned for production and SDD flight test jets and retested independently. The F-35C static test had one noteworthy finding: The FS 503 frame segment cracked at slightly less than 150% DLL leading to a repair for the test article and a redesigned part for production and SDD flight test jets.

The successful early completion of the baseline static test programs allowed the F-35 program to perform additional expansion testing of the stores hardpoints to enable future expansion of the F-35's stores capability. Figure 16 shows the pace of all variants of the F-35 static test program relative to recent, relevant static test

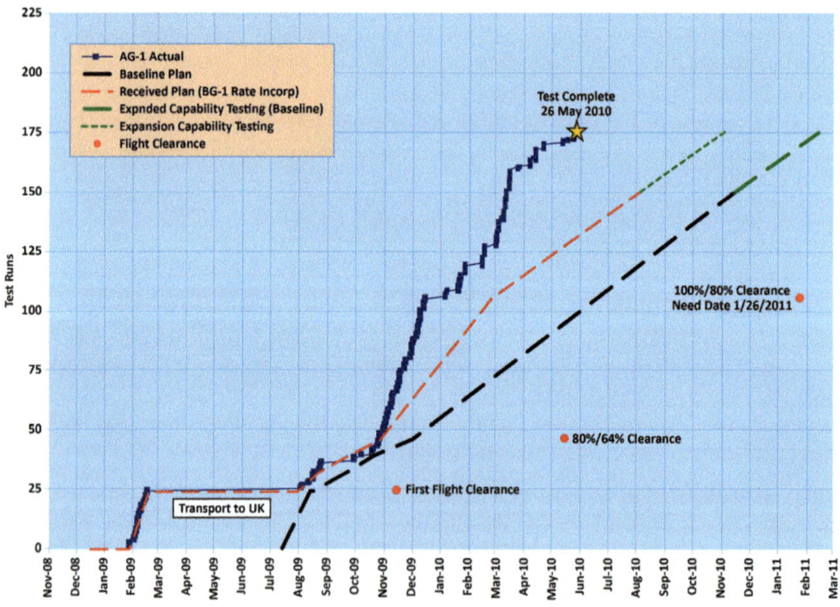

Fig. 15 AG-1 testing to support flight test need.

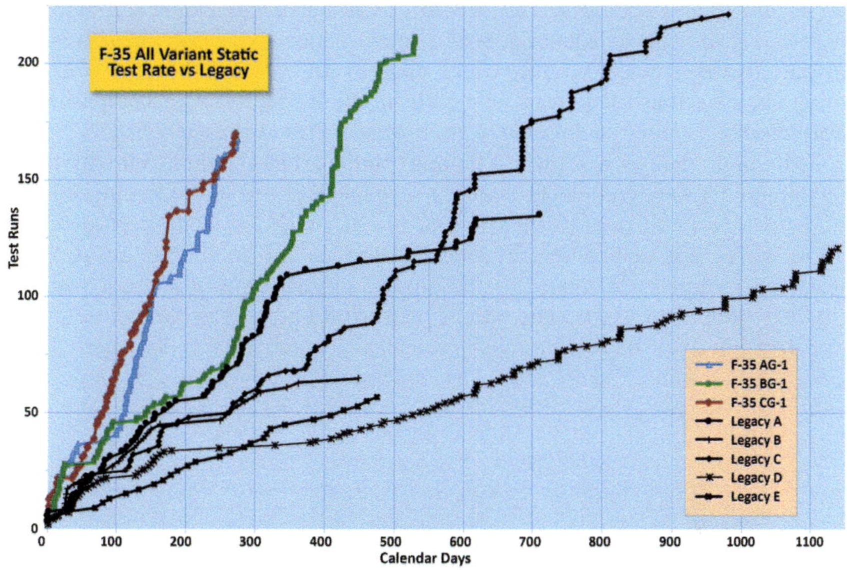

Fig. 16 F-35 all-variant static test rate compared to legacy aircraft.

programs. The pace of these tests could not have been achieved without a robust, well-designed airframe.

The three HT static tests were performed in two separate test rigs. This allowed these tests to be completed well in advance of the needs of the flight test. The success of the horizontal tail structure at withstanding the limit and ultimate loads during the test demonstrated the robustness of the structure and afforded the program the ability to stretch the test program and apply loads that were well in excess of the ultimate loads to support potential future flight test loads envelope expansion.

The F-35A and F-35C tails were taken beyond the 150% DLL ultimate loads up to 200% of DLL for multiple conditions. Figure 17 shows the F-35C HT being loaded to 200% of its design limit load. The HT for the F-35B variant was taken successfully to 200% of DLL as well and then the test was taken on to failure, which occurred at greater than 200% of DLL, significantly exceeding the strength requirements for the structure. This proved to be a very wise investment for the program because the data were utilized to expand the flight clearance envelope when later flight test discoveries demanded more capability out of the tails.

B. DURABILITY TEST

The F-35 durability structural test program included three full-scale airframe test articles, three full-scale horizontal tail component tests, and three full-scale

Fig. 17 F-35C horizontal tail maximum down bending test.

vertical tail component tests. It was realized early in the planning that the test program would need to address load cycling due to maneuver loads and buffet loads because both were significant contributors to the damage. Coupon-level testing, as described in the Building Block process, showed that the two components could be applied in separate, sequential blocks without significantly affecting the results. The program made the decision to test the full-scale articles with alternating maneuver and buffet blocks, which represented 1000 flight hours each so that when the two blocks were performed, the resulting damage represented the damage that would be accrued by 1000 hours of flight with both loading sources. In the case of the F-35C, the carrier variant, a third set of load cycling blocks, were applied, which represented three lifetimes of catapult loads and two lifetimes of arrestment loads. Each block represented 1000 hours of arrestments and 1500 hours of catapults. This was done to accelerate the testing program. It was faster to apply the buffet cycles in a block rather than interspersing them throughout the maneuver spectrum. Another test efficiency was replacing the VT on the full-scale airframe with a stiffer "dummy tail" that would not deflect as far under the applied load. This enabled the loads, in particular buffet loads, to be applied more efficiently to the vertical tail backup structure. As a further cost savings, a handful of specific features were tested on the full-scale test article as local tests.

Because of the timing of the durability tests relative to the delivery schedule for three F-35 variants and the inherent concurrency, the F-35 program had to address the need to develop both redesigns and retrofit solutions to findings on the test that were determined to be deficiencies. This concurrency was driven by program decisions regarding the production schedule for the F-35, which supported the affordability of the F-35. To address this concurrency, the program utilized data regarding the test findings from the recent, relevant durability tests to predict the quantities and types of findings that the tests would encounter. This was then used to create predictions of the number and magnitude of design changes that would need to be incorporated. This plan enabled the program to accurately estimate the budget needed for these changes and make plans to support the redesigns and to plan for modification periods to make the necessary changes to the delivered aircraft. These predictions of findings were then used as a basis for comparison for the F-35 durability test program. Another mechanism to address the concurrency was to invoke a rigorous inspection program and careful monitoring of the test instrumentation to ensure that test findings were discovered and acted upon at the earliest opportunity.

The F-35 HT durability tests were performed in the two static test fixtures following the successful completion of the HT static strength tests at the BAE Systems test lab in Brough, as shown in Fig. 18. These were also tested using intentionally severe test spectra that were applied in separate, repeating 1000-hour buffet and maneuver blocks. These tests were originally planned to include two lifetimes (16,000 hours) of testing; however, the Program Office issued a change to the contract specification to include additional cycling beyond two lifetimes. All three of the HT test articles successfully completed the three lifetimes of combined maneuver and buffet testing prior to being shipped back to the United States for teardown inspections.

The F-35 VT durability tests were performed in multiple fixtures at the BAE Systems Test Lab as well, as shown in Fig. 19. There was one maneuver test rig and two buffet/vibration test rigs. The maneuver loads were applied quasi-statically

Fig. 18 Horizontal tail test fixtures in Brough, United Kingdom.

Fig. 19 Vertical tail test fixtures in Brough, United Kingdom.

using hydraulic rams, and the buffet loads were applied in sound-dampening test chambers using electromagnetic shakers to excite the desired vibrational modes. These tests were also performed by alternating the buffet and maneuver loads, but due to the effort required to switch from one mode to the other, the F-35 program opted to use blocks that were 2000 hours long. Again, the three vertical tail tests successfully completed three lifetimes of combined maneuver and buffet testing and were shipped back to the United States for teardown inspections.

The F-35A, AJ-1 durability test encountered roughly half of the findings, as shown in Fig. 20, that were found on the most recent relevant durability test program. At the end of two lifetimes, the F-35A durability test had experienced a little over a third of the findings encountered on recent legacy aircraft. Following successful completion of the third lifetime of testing, this total had risen to just under half of the amount of findings experienced on the legacy aircraft durability test. This small number of findings is a testament to the diligence of the Structures team, the advancements made in the tools used to analyze the airframe structure, and the robustness of the crack growth–based service life analysis criteria used in the development of the CTOL variant. The discovery of these findings prior to teardown through the planned inspections has allowed the F-35 program to break in design changes in the early production aircraft to avoid more costly retrofits. Additionally, findings on each variant were used for inspections and design reviews on the other variants to determine if similar structural details were affected. This allowed for early break-in of changes on those aircraft as well.

The F-35B, BH-1, and F-35C, CJ-1, durability test program discovered findings consistent with the quantity and rate exhibited by recent legacy programs, also shown in Fig. 20. The most significant finding during the full-scale durability test program was on the primary carrythrough bulkhead on the F-35B variant, which is aluminum rather than titanium as on the other two variants Although this resulted in a crack through a significant portion of the bulkhead, the wing successfully carried limit load several times after the cracking occurred due to the robust load redistribution capability of the wing structure. Root cause analysis of

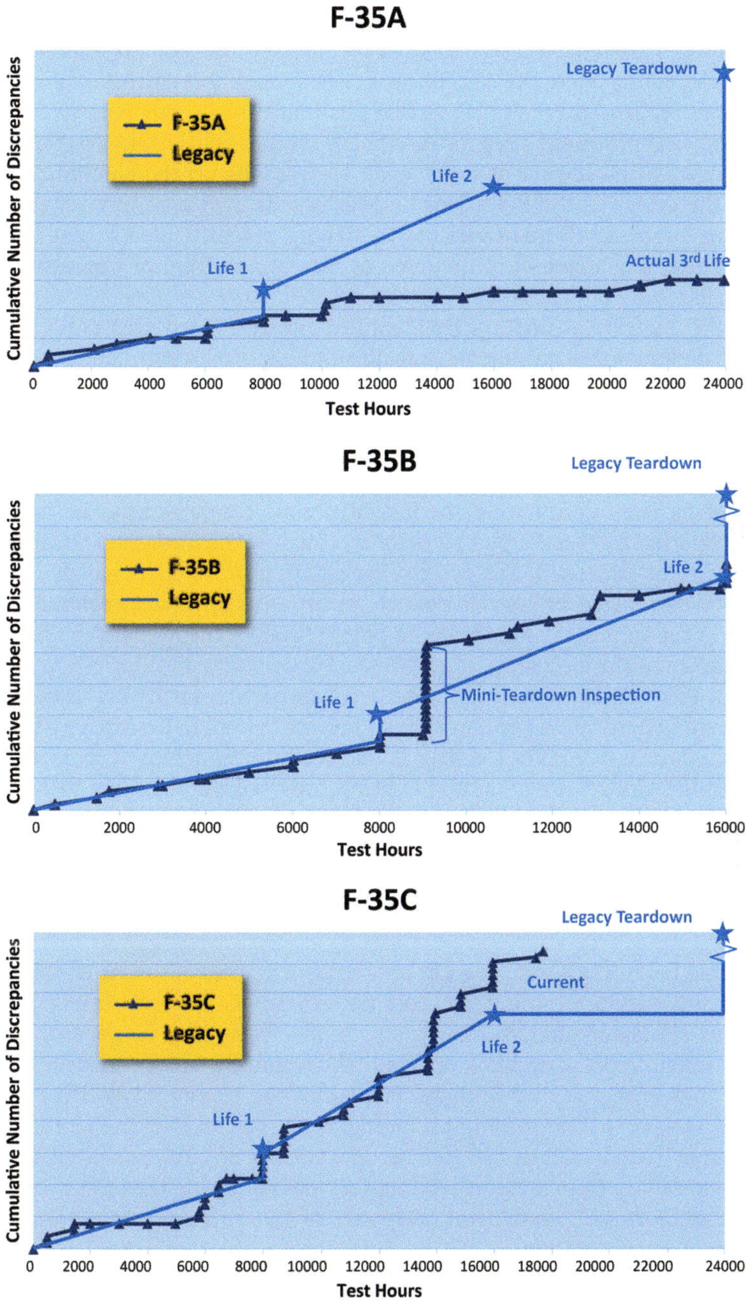

Fig. 20 F-35 durability test findings vs legacy predictions.

another crack finding on that part during the second life of testing led to the discovery that the etch process used to prepare the part for inspection during production, along with the anodizing process used for corrosion protection on aluminum, caused very small surface pits that were not adequately captured by the material properties used for service life analysis. This led to a test program to characterize the impact of these small surface pits on the aluminum alloys used in the design of the F-35. This knowledge was then used in a review of all aluminum parts on all three variants to identify potential life shortfalls and to break in design changes where required at the earliest point in production that was possible.

During the tests, as findings are identified and correlated back to test results, updated designs are incorporated into the earliest production configuration to reduce the need for modifications to fielded jets. Designs for modifications of existing aircraft structure are also released to be implemented on delivered jets. These modifications are then grouped together for the efficiency of installation and to maximize the availability of the aircraft for the warfighter.

Upon completion of testing, the test articles are carefully inspected and then completely torn down to inspect all parts and identify the final set of findings that may need to be addressed. The posttest inspection prior to teardown allows findings to be discovered as early as possible in order to introduce required design changes at the earliest possible break-in point to reduce the need for modification after delivery of the aircraft.

As of this writing, disassembly and detail inspection of the F-35B durability article commenced at the National Institute for Aviation Research (NIAR) in Wichita, Kansas. The F-35A article is currently being removed from the test fixture in the United Kingdom and is being prepared for shipment to NIAR in April 2018 to begin teardown of the article. The F-35C durability test article is still cycling in its third lifetime.

C. F-35C DROP TEST

A full-scale F-35C drop test program consisting of a series of simulated shipboard landings was conducted using the static test article CG-1, shown in Fig. 21. The drop tests simulated aircraft landing conditions—including sink speed, aircraft attitude (pitch/roll), wing lift force, and wheel spin-up speed—associated with extreme high energy carrier landings. Full-scale drop testing is typically required by the U.S. Navy to clear the aircraft for high sink rate landings in flight test. The test program, conducted in first and second quarter of 2010, consisted of more than 40 aircraft drops to measure landing gear loads and kinetic energy absorption characteristics as well as dynamic response of the airframe, stores, and other critical equipment locations.

High-level objectives of the full-scale drop test were to: 1) verify that the structural integrity and functional design requirements for shipboard landing impacts as specified by the F-35 air vehicle structural design criteria were met; and 2) to

Fig. 21 F-35C drop test.

collect loads and dynamics response data for correlation with analytical predictions. Specific test objectives were:

- Measurement of landing gear loads and energy absorption characteristics, including the effect of airframe flexibility, under simulated carrier landing conditions up to the maximum design sink speed.
- Measurement of dynamic response of the airframe, landing gear, stores, and other large mass items during simulated carrier landings.
- Confirmation of the functionality of the landing gear retract/extend, landing gear door open/close, and weapons bay door open/close mechanisms after application of critical carrier landing loads.
- Confirmation of fuel tank sealing after application of critical carrier landing loads.
- Measurement of internal weapons deflections during simulated carrier landings.

Results from the CG-1 drop test were excellent. The testing confirmed the validity of the landing gear and static airframe loads. Pretest landing gear load predictions showed excellent correlation with measured test values. No changes to the landing gear metering pin designs were required as a result of the test measurements, proving the value of the high-fidelity dynamic models developed in the design phase. A full account of the test and results are found in Ref. [3].

D. FLIGHT TESTING

As mentioned in a previous section, the requirements for flight clearance for the flight test program were also unique to the F-35. To stay within the Program Executive authority for flight risks, the envelope was limited to 40% prior to confirmation of the analysis via test. To support the flight test schedule, the required static testing and proof tests were conducted on the loads instrumented aircraft to provide the evidence needed to open the flight test envelope and to provide calibration data for the flight test instrumentation.

Structural flight testing for all three F-35 variants was conducted during the SDD program to collect the data required to correlate analytical models for final certification, as well as to demonstrate specification compliance with certain requirements that could not be verified analytically, such as control surface free play effects. Conduct of the flight testing program is described completely in Ref. [4]. The full scope of structural flight tests were:

- Flight flutter tests to verify requirements of SDC, including damping, and to collect data for dynamic model correlation
- External flight and ground loads measurements for correlation with models used for certification
- Aerodynamic buffet and other dynamic loads and response data, also for model correlation
- Measurement of vibroacoustic environments to verify levels used for design of structure and qualification of vehicle and mission systems components

Flutter flight tests were performed using AF-1, BF-2, and CF-1 after ground vibration tests verified aircraft vibration characteristics and provided data for dynamic model correlation. Dedicated flutter test points were conducted at increasingly critical airspeeds through an orderly buildup of Mach number at progressively decreasing altitudes. At each test point, critical structural modes were excited using a flutter excitation system that perturbed the airframe through forced oscillation of control surfaces at specific frequencies or at random. Decay of the structural response was measured to determine damping at that flight condition. Initial clean wing tests were done in parallel with flying qualities and propulsion flight tests as part of an integrated envelope expansion program. Flutter testing with various external stores was also conducted. Results of the flutter flight testing, as well the final flutter analyses based on correlated dynamic models show that all three F-35 variants are flutter free throughout the required envelope, including margins, and meet all of the damping requirements of the SDC. No Limit Cycle Oscillations (LCO) were observed with any of the external store configurations flown. Towards the end of the test program, flutter flight tests were conducted to significantly expand the horizontal tail free play limits, increasing the in-service interval between maintenance actions.

All maneuver loads flight tests were conducted using three dedicated instrumented and calibrated loads jets: AF-2, BF-3, and CF-2. Primary instrumentation comprised groups of strain bridges placed on structural components and airframe sections corresponding to the locations for which design loads were generated. The strain bridges were calibrated in a test fixture by applying distributed loads of known magnitude, measuring strain bridge output, and developing regression equations relating that output to the applied loads. These aircraft were also used for buffet and other in-flight dynamic response tests, particularly ejection of stores and weapons bay vibroacoustics. Flight loads tests were performed by executing a block of defined maneuvers (pullups, pushovers, sideslips, rudder kicks, 1-*g* and elevated-*g* rolls, etc.) at numerous Mach numbers and altitudes covering the entire flight envelope. Air loads on each structural component and section were then extracted and used to correlate the wind tunnel–derived air loads database. Results were continuously monitored as the flight envelope expanded, compared against the allowable strength envelopes, and action taken when adverse trends emerged. For example, early VT loads measurements indicated that limit loads would be exceeded as tests progressed into the transonic region. To counter this, the flight control laws were modified to introduce a rudder deflection bias in the critical Mach number range, relieving the VT air load moment and allowing testing to continue without having to modify the structure. The strategy of tailoring flight control laws to alleviate loads was used successfully on all three variants throughout the loads flight test program. Final loads analyses conducted using test-correlated air loads databases and the final optimized flight control laws show that loads on nearly all structural components were within limits, with a few notable exceptions described below.

Loads testing unique to the F-35B variant was conducted on BF-1, the test aircraft dedicated to Short Takeoff and Vertical Landing (STOVL) mode propulsion and flight controls. These tests were performed to measure loads and environment on the many in-flight operating doors that make up the STOVL mode propulsion system: Upper Lift Fan (ULF) door, Lower Lift Fan Doors (LLFDs), Auxiliary Air Inlet Doors (AAIDs), Roll Control Nozzle (RCN) doors, and three-bearing swivel module (3BSM) doors. Loads and dynamics testing were typically conducted concurrently because of the high noise and high vibration environment combining with steady air loads.

Some of the most challenging flight tests conducted during SDD were to measure buffet loads. Empennage, control surface, and wing buffet dynamic response were measured over the full range of flight conditions where buffet is present. Many of the required flight conditions were at high angles of attack that were difficult to reach and maintain long enough to gain statistical confidence in the high-frequency, random responses being measured. The tests were done with the loads jets using the loads instrumentation augmented with accelerometers in key locations on the wing, empennage, and control surfaces. Buffet data was normally collected during the maneuver loads tests, but many dedicated sustained windup turns and constant AoA descents were also conducted to gather data at constant flight conditions. Measured response was used for correlation of the dynamic models and results combined with the attendant steady loads to yield final results for certification and verification. The results were generally good with the measured frequencies and modal response well predicted; however, several cases of higher-than-predicted buffet loads emerged that required significant mitigation. For all three variants, rudder hinge moments at certain airspeeds at approximately 20 deg AoA were underpredicted and required extensive structural analysis and several flight control law changes to bring them within the limits of the existing structure. Measured buffet loads on the F-35B AAID were significantly higher than initially predicted and required structural redesign to mitigate. Redesign consisted of higher hinge line rotational stiffness to place the rotation mode frequency well above the excitation frequency as well as increased strength in the door and attach structure. Finally, significantly higher-than-predicted buffet loads on the F-35C wing tip AIM-9X hardpoint were measured during transonic windup turns. After limited progress with other mitigations, a modified wing tip hardpoint design of increased strength was developed and adopted for the final design. Weapons bay acoustics measured concurrently with the loads, buffet, and other structural flight tests were as predicted, alleviating early concerns that a spoiler at the front of the bay might have to be incorporated to control bay noise.

V. TASK IV: CERTIFICATION AND FORCE MANAGEMENT DEVELOPMENT

The final phase of the F-35 SDD program includes the verification of the contract requirements through final certification analysis products, development of fleet loads tracking (L/ESS) and Individual Aircraft Tracking (IAT) tools, and preparation of the Force Structural Maintenance Plan (FSMP) using the updated analysis to prepare for execution as defined in Task V.

Each of the F-35 contractual requirements has associated, predefined Success Criteria (SC) established at the beginning of Task I. Each of these SC is closed once all supporting reports are signed off by the contractor and F-35 JPO. In many cases, these reports are updates to Task II design analyses that have been validated using Task III test results.

The cornerstone of structural certification on the F-35 program is analysis that has been correlated to test results. There are many effects that are impractical to simulate throughout the structure during a full-scale ground test, such as system interface loads, distributed internal pressure loads including fuel head pressure, and thermally induced internal loads. Distributed aircraft inertia loads must be simulated by discrete loads applied to the surfaces of parts, which also leads to local differences between flight and ground test loads.

The design analyses performed during Task II were updated during Task IV by correlation to the results of the full-scale testing that was performed in Task III. First, the initial design external loads were updated by correlation to flight test measurements. Internal loads were verified by correlation of finite element model predictions to strain gauge measurements taken during the full-scale static tests. Next, the strength analysis was updated to reflect flight test correlated external loads and the results of the full-scale static tests. The flight test correlated loads and updated strength analysis were then used to update the Strength and Summary Operating Restrictions (SSOR) reports for each variant.

Similarly, the durability analysis from Task II will be updated to reflect flight test correlated external loads and findings during the full-scale durability tests. Root Cause Analysis (RCA) is performed on each full-scale durability test finding in order to understand what analysis updates are required. The updated durability analysis will be summarized in a final DADT Report at the conclusion of the SDD program.

When the updated Task IV analysis indicated that structural design requirements were no longer met, production design changes and retrofit modifications were undertaken as necessary to restore capability. These changes were typically minor perturbations of the baseline design and were analyzed using the latest test-correlated methods. Thus, they were able to be certified without additional testing of the revised configuration.

As part of this effort, the crack initiation life of aluminum structure in the STOVL and CV variants was reassessed using updated material properties that account for the impact of etching and anodizing, as discussed earlier. This correlated approach will be the basis for the certified service life analysis published in the final DADT report, and for the FSMP. Service life deficiencies resulting from this analysis update are being addressed as required using production design changes and structural retrofits.

Completion of the final SSOR and DADT analysis reports are key milestones that support closure of SC for structures requirements. The closure of all SC supports closure of the associated JCS paragraphs, which in turn enables SDD contract closure.

The L/ESS and IAT systems are developed and verified as part of Task IV. IAT equations are developed for critical control points that are verified or discovered during the full-scale durability tests. These equations will be used by the PHM system during Task V to support adjusting maintenance intervals on each aircraft based on their actual usage severity.

The initial FSMP for each F-35 variant was developed and put in place to provide the basis for the services to effectively manage the fleets. The FSMP defines when, where, and what type of maintenance is required, along with estimated costs. The report describes all durability and damage tolerance critical locations and provides a summary of their analysis inputs.

These L/ESS, IAT, and FSMP reports will be updated on a regular basis as part of the contractual ASIP process.

VI. TASK V: FORCE MANAGEMENT EXECUTION

The final pillar of ASIP is the execution of the Force Management Plan as depicted in Fig. 22. This is the culmination of all the efforts that have taken place leading up to this point. Real-life fleet usage data are collected and analyzed to provide more realistic life estimates for the services. As the pilots become more familiar with the capabilities of the aircraft, they find where in the flight envelope it performs best and how to best execute the missions that it was designed for. It may also take on new missions that affect the usage as new strategies and tactics are employed by the services. These changes may lead to usage that differs from the original usage spectrum used to project the life of the aircraft. When enough data are captured, the baseline operational spectrum can be updated for projecting the service life, inspection intervals, and maintenance timing. The FSMP, IAT, and L/ESS reports are updated using the analysis of this captured usage data and experience with the aircraft. The services will then use this information to effectively manage their fleets to ensure aircraft readiness with minimum downtime and reduced cost for preventative maintenance and repairs.

The rigorous, disciplined structural integrity approach followed by the F-35 program allowed the tri-company team led by Lockheed Martin to efficiently develop and provide certification evidence for three variants of the F-35 Lightning II. The continued application of the ASIP approach will allow the services to operate and maintain the structure of these aircraft to ensure their effectiveness and the safe operation throughout the 30-year/8000-hour service life of the F-35.

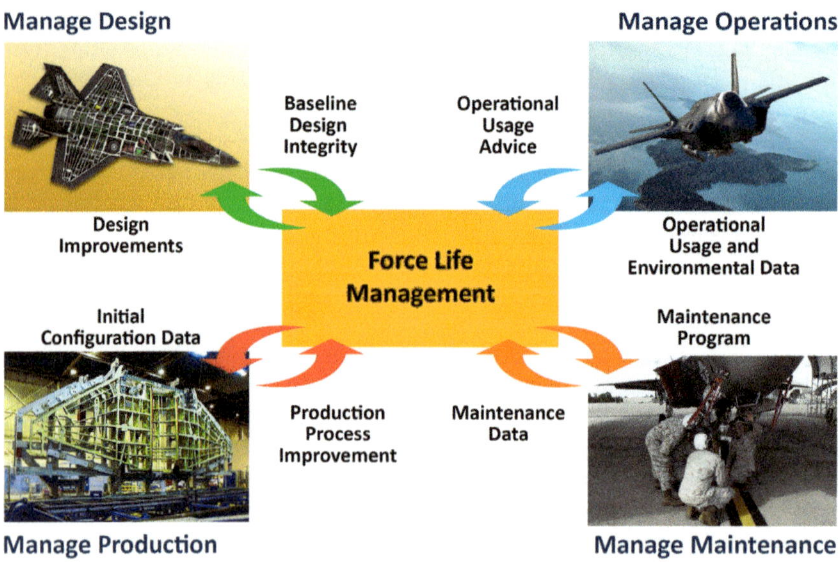

Fig. 22 Force life management execution.

VII. CONCLUSION

The development of the F-35 airframe structure and its lifecycle management plan has been unique in the annals of fighter airframe development history. The concurrent development of three variants with widely differing requirements, to be used by services with widely differing histories and certification philosophies, provided unprecedented opportunity for collaboration between contractor and government engineering organizations, as well as between worldwide industry engineering organizations. In this environment of such complex demands and global coordination, the prescription of airframe structural integrity given by MIL-STD-1530 provided the essential framework for executing the design, verification, and force management of the world's first 5th-generation multirole fighter weapon system.

Furthermore, with its multinational, multiservice demands driving a unique consideration of differing structural integrity philosophies, the F-35 program has gained a unique perspective. Comparison of the varied approaches through the verification phase of the program has allowed for an unparalleled understanding of the structures' capabilities, not only for its intended service usage, but also for whatever variations of service the aircraft may encounter.

With its expansive degree of ground and flight testing, along with development of state-of-the-art data-gathering and usage analytics, the F-35 program is positioned to provide decades of safe and effective use of a 5th-generation high-performance fighter platform for its many worldwide customers.

REFERENCES

[1] MIL-STD-1530C, "Department of Defense Standard Practice, Aircraft Structural Integrity Program (ASIP)," Wright-Patterson AFB, 1 Nov. 2005.

[2] Counts, M., Kiger, B., Hoffschwelle, J., Houtman, A., and Henderson, G., "F-35 Air Vehicle Configuration Development," AIAA Aviation Forum, June 2018 (to be published).

[3] Chichester, R. H. L., and Norwood, D., "Full Scale Drop Test Program for the F-35C Carrier Variant," AIAA Paper 2015-0459, 2015.

[4] Hudson, M., Glass, M., Hamilton, T., Somers, C., and Caldwell, R., "F-35 SDD Flight Testing at Edwards Air Force Base and Naval Air Station Patuxent River," AIAA Aviation Forum, June 2018 (to be published).

CHAPTER 8

F-35 Flight Control Law Design, Development, and Verification

Jeffrey J. Harris* **and James Richard Stanford**†

Lockheed Martin Aeronautics Company, Fort Worth, TX

The F-35 program will replace a number of aging fighter inventories and enhance others for the U.S. Air Force, Marine Corps, and Navy, as well as the inventories of partner/allied countries. Although the F-35A, F-35B, and F-35C have similar planforms and mission systems and share many common components, from a flight control system development perspective, they are quite unique. A novel approach to both control law design and software development was required to efficiently develop the control laws for three versions and ensure that the program met its demanding requirements. The use of nonlinear dynamic inversion as a viable control law methodology has been demonstrated on various flight control research platforms; however, the F-35 is the first production fighter to incorporate a dynamic inversion control law that operates from zero airspeed to supersonic speeds, controls the jet well beyond stall angle of attack, and provides exceptional flying qualities for operational tasks, ranging from short takeoff and vertical landing operations on U.S. Navy amphibious assault ships and large-deck carrier operations to basic fighter maneuvers. In addition, the F-35 control law software was developed based on a graphical model and auto-generated code. This was a key enabler to support software development for three F-35 versions, but it required a paradigm shift in the software development process at Lockheed Martin. This paper explores the in-depth effort of implementing a model-based control law methodology using dynamic inversion. It also provides an overview of the software development process used for F-35 flight control laws.

*Senior Manager, F-35 Control Law and Air Data Design.

†Principal Engineer, F-35 Control Law and Air Data Design.

I. NOMENCLATURE

A	aircraft dynamics matrix
B	control effectiveness matrix
B_{cond}	conditioned control effectiveness matrix
CV	control variable (vector)
CV_{dot}	control variable rate (vector)
d_{cmd}	desired acceleration command
$\Delta_{\delta\text{avail}}$	available change in effector command
$\Delta_{\delta\text{cmd}}$	change in effector command
$\Delta_{\delta\text{lim}}$	limit on change in effector command
$\Delta_{\delta\text{pref}}$	change in preferred effector command (scheduled)
$\Delta_{\delta\text{scale}}$	scaled change in effector command
u	control vector
W_d	weighting matrix, desired accelerations
W_{pref}	weighting matrix, preferred value
x	state vector
$\dot{x}$	state rate vector
$\dot{x}_{\text{des}}$	desired state rate vector

II. INTRODUCTION

The F-35 Lightning II Joint Strike Fighter (JSF) program was designed to provide the next-generation family of strike fighters. By combining stealth, advanced avionics, increased range with internal weapons carriage, and state-of-the-art prognostics and health management, the F-35 provides a lethal, survivable, connected, supportable, and affordable fighter platform. Although all three F-35 versions share the same general planform, core engine, avionics, and weapons, there are fundamental differences in the aircraft from a control law development perspective. As shown in Fig. 1, the aerodynamic differences in wing area and horizontal tail surface area among the versions, and the differences in basic stability and control power, are not trivial. Also, the uniqueness of the F-35B configuration for conducting Short Takeoff and Vertical Landing (STOVL) operations presented a challenge for the control law engineers to achieve an optimal blend between aerodynamic and propulsion system control effectors. All these aspects of the F-35 versions had an impact on the task of establishing a preferred control law structure.

Designing control laws for a new aircraft is a significant task. Designing control laws for three versions at the same time, without standing up independent development teams, required creative approaches in order to meet program schedules. A model-based, dynamic inversion approach using auto-generated code seemed like a viable option. One of the advantages of dynamic inversion is that the control law designer would not be required to linearize the system at set

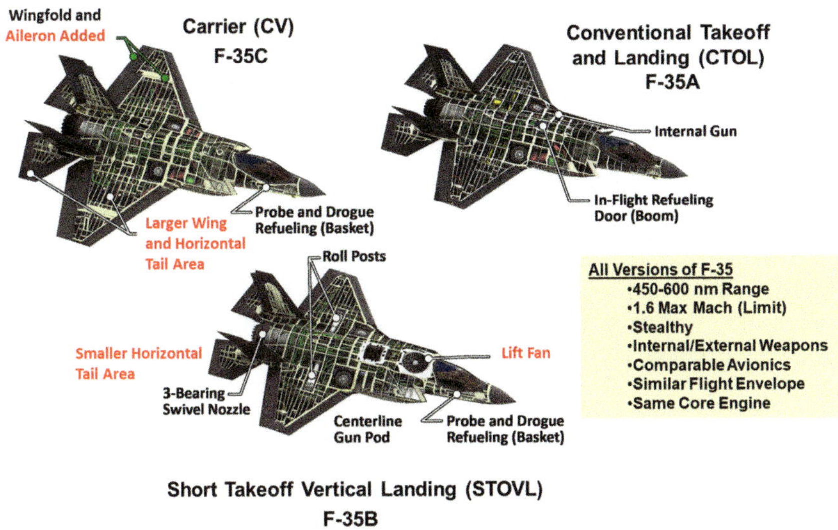

Fig. 1 F-35A/B/C general characteristics.

flight conditions and develop gain schedules to provide a desired closed-loop response, which can be very time consuming. Also, auto-generating code instead of relying on a separate software development team could result in a significant reduction in the software development/release cycle time. However, like any new methodology, it was imperative to demonstrate that the advantages would outweigh any risks, and that the new methodology could map directly to historical flight clearance/airworthiness products. The X-35 competition program provided an excellent opportunity to pursue this development approach. The X-35 Concept Demonstrator Aircraft program required an efficient control law design effort and a quick path to releasing flightworthy software. Program management was receptive to the approach because it allowed aggressive development schedules to be met without tripling the staff. Further, the customer was very open to exploring new control strategies that enabled all three versions to have similar flight characteristics.

Although early explorations of using dynamic inversion in flight control applications began appearing in research literature as far back as the early 1980s [1], there are practical limitations to implementing this method in flyable software, especially for highly nonlinear systems. If it were easy, we would not still see feedback control dominating the control laws in modern fighter aircraft. The most obvious challenge is the computing horsepower required to execute the algorithm. However, other challenges associated with invertibility of the system, numerical stability, and sensitivity to unmodeled dynamics also need to be addressed. The Lockheed Martin Control Law Design

Group has overcome all of those challenges to deliver a control system that provides exceptional flying qualities for all three production versions throughout their flight envelopes. This paper explores some of those implementation challenges and discusses the development process that was used for the F-35 flight control system software.

III. CONTROL LAW DESIGN METHODOLOGY

There have been dramatic changes in flight control system design over the past few decades. We have seen the expansion from analog systems with limited authority augmentation on stable platforms to full-authority digital flight control systems that are essential for allowing highly unstable jets to be safely flown. With the development of more powerful flight control computer processors, more advanced control methods that put a significant demand on computer memory and throughput are being developed and flown.

Traditionally, flight control laws were developed as linear controllers with gain scheduling used to provide desired closed-loop dynamics throughout the flight envelope. Gains would be developed based on a linear approximation of the aircraft and systems (aerodynamics, actuators, sensors, engine, and other subsystems) at various flight conditions under which the jet was expected to operate. This process can be complex because conventional fighter aircraft have expanded into highly nonlinear flight regimes, such as high-Angle-of-Attack (AOA) flight conditions. For the F-35, increased complexity was added with the STOVL capability on the F-35B, which involves operating down to zero airspeed, including a transition from aerodynamic controls to propulsion system controls, and a transition from pneumatic air data sensors to inertial sensors.

Faced with the challenge of designing control laws for three different versions operating conventionally, and a version with a STOVL capability, the F-35 Control Law team explored more direct control law methodologies with which a single control law structure could support all versions, directly accommodate system nonlinearities, and avoid the need for complex gain scheduling. Nonlinear Dynamic Inversion (NDI) was a very attractive option to pursue.

A. NONLINEAR DYNAMIC INVERSION

NDI, or feedback linearization, is quite simple in theory and is based on an understanding of the equations of motion of the aircraft. If one can predict the aircraft response to a control effector input (e.g., the aircraft pitch acceleration response to a horizontal tail deflection), then it should be possible to derive the required control effector input in order to achieve a desired aircraft response. This would be achieved by canceling (inverting) the original aircraft dynamics and defining the desired aircraft response. In linear, state-space form, the dynamic equations of motion can be expressed in Eq. (1) as

$$\dot{\boldsymbol{x}} = \boldsymbol{Ax} + \boldsymbol{Bu} \tag{1}$$

Where x is the system state vector (pitch rate, roll rate, etc.), A is the aircraft dynamics matrix (stability derivatives), B is the control effectiveness matrix (control power), and u is the control vector (horizontal tails, rudders, engine nozzles, etc.). Solving for the control vector u, using Eq. (2) yields

$$\boldsymbol{u} = \boldsymbol{B}^{-1}[\dot{\boldsymbol{x}} - \boldsymbol{Ax}] \tag{2}$$

Put in terms of deriving the required control vector to achieve the desired state rates, Eq. (2) becomes Eq. (3)

$$\boldsymbol{u} = \boldsymbol{B}^{-1}[\dot{\boldsymbol{x}}_{\text{des}} - \boldsymbol{Ax}] \tag{3}$$

Again, keeping with the simple theory, defining the control vector in this manner reduces the system dynamics to a simple integrator, as shown in Fig. 2, eliminating the need for gain scheduling.

The NDI aspect of the control laws and the plant model shown in Fig. 2 represent how the aircraft actually behaves, with all nonlinearities included. The desired state rates reflect how the control law designer wants the aircraft to fly in response to pilot inputs. The commands module translates the pilot's stick, throttle, and pedal inputs into desired aircraft movement. The regulator directly sets good low-order-equivalent system parameters to ensure classical handling qualities for the pilot while achieving this motion. This is generally defined in terms of traditional flying qualities parameters, such as short-period frequency and damping, or Tier II flying qualities metrics, such as pitch attitude bandwidth, Neal-Smith criteria, and so forth. Those flying qualities goals are embedded directly in the F-35 regulator because the remaining control architecture creates the inversion, as shown in Fig. 3.

An integral part of the NDI structure shown below is the Effector Blender (EB), which performs the control effectiveness matrix inversion and replaces the traditional control surface mixer in the control laws [2]. Whereas a traditional mixer uses a set of gain schedules to map an axis command to the control surfaces, the matrix inversion typically involves a real-time, pseudo-inverse iterative calculation and logic to account for surface rate and position limits and other performance constraints, as shown in Fig. 4.

B. IMPLEMENTATION CHALLENGES

As mentioned earlier, NDI is not a new concept. It has been successfully demonstrated working on a number of research aircraft, ranging from the F/A-18 Hornet fighter to helicopters [3, 4]. However, these research efforts were generally focused on very limited flight envelopes to demonstrate a concept implementation. The F-35 program adopted NDI as an approach for the full operating

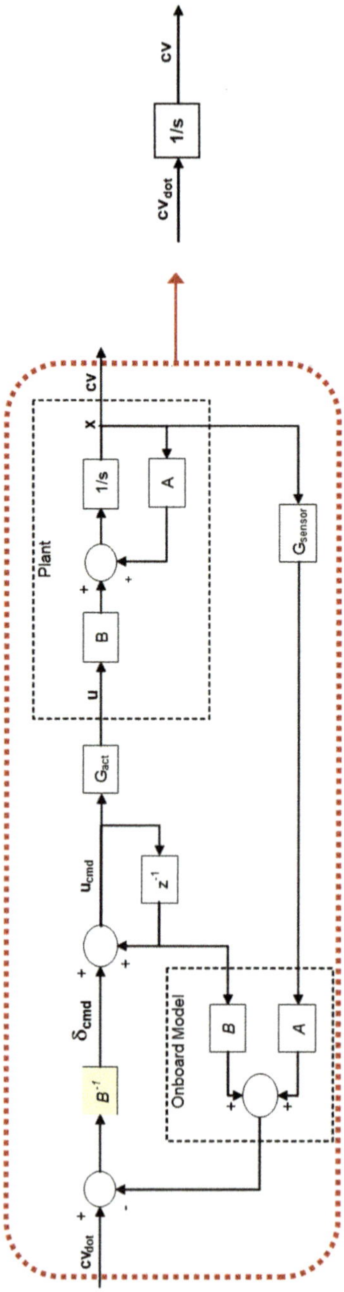

Fig. 2 Incremental nonlinear dynamic inversion decomposition.

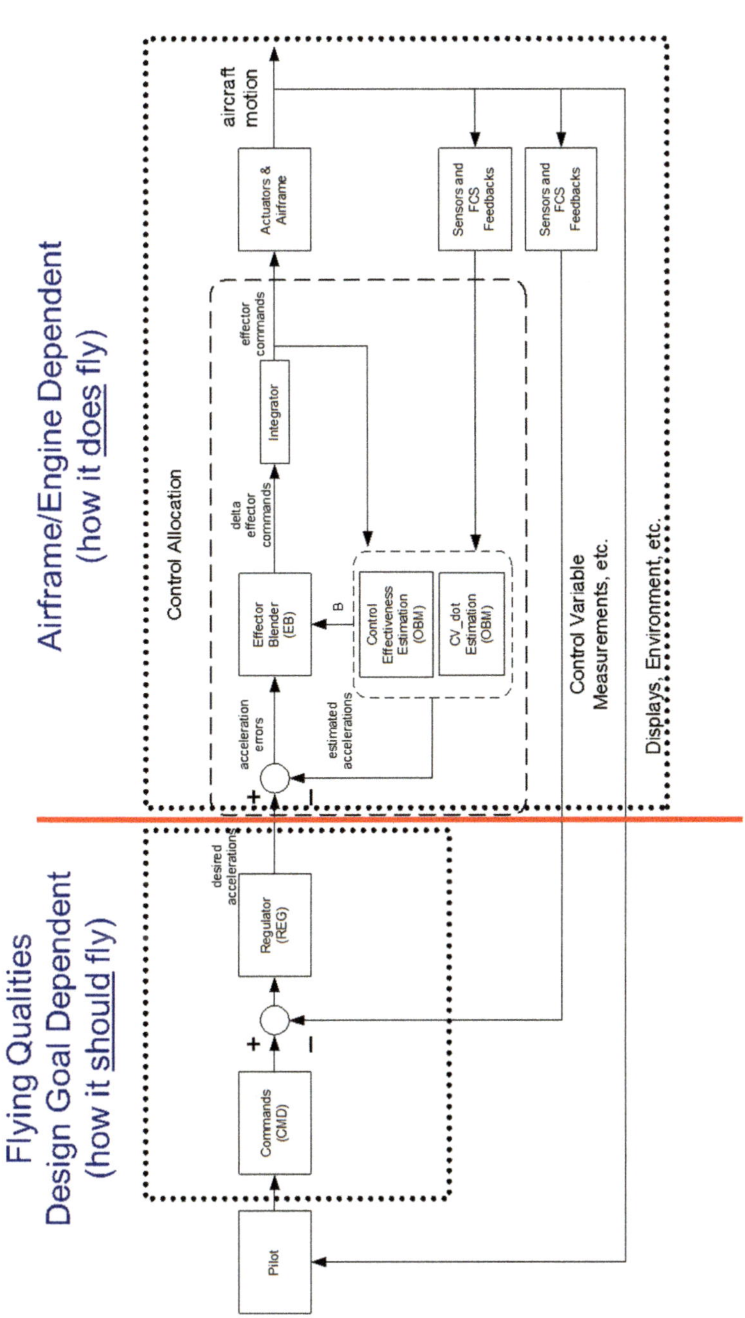

Fig. 3 F-35 control law structure.

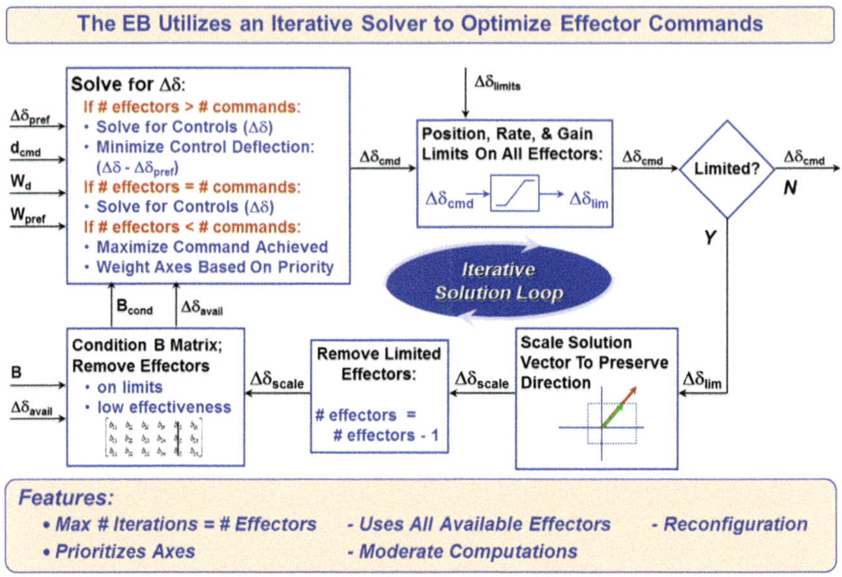

Fig. 4 Effector blender solution algorithm.

envelope, including high-AOA and STOVL operations, carrier landings, refueling, and similar operations. The design had to be robust to the aerodynamic effects of external store carriage and the significant range of a longitudinal and lateral Center of Gravity (CG) that was possible with the stores that the aircraft was designed to carry (Fig. 5). In addition, on-ground modes, reconfiguration for system failures, and out-of-control recovery modes would also need to be accommodated by the control law architecture, while addressing the practical aspects of real-time computer utilization for computationally intensive algorithms.

The decomposition of the NDI controller wrapped around the airframe (plant) model into a simple integrator relies on the precise modeling of the aerodynamics of the aircraft, perfect sensors, and the ability to invert the control effectiveness matrix B. The reality is that onboard models of the aerodynamics and propulsion system are never perfect, and higher-order effects can be difficult to accurately model. These effects will result in not achieving a perfect cancellation of plant dynamics, and this mismatch can have a significant impact on closed-loop dynamics, as discussed later in this paper. In addition, there are numerical challenges associated with an algorithm that requires the inversion of the control distribution matrix. These challenges would have to be overcome to successfully implement an NDI-based control law structure in a production aircraft. In addition, taking this design approach from the research environment to the production environment also has implications on system test requirements (both on-ground tests and flight tests), as well as airworthiness assessment impacts.

C. NUMERICAL CHALLENGES

Early in the development of the F-35 control system (prior to flight testing), chattering in the actuator commands was observed as toggling between solutions occurred. This was generally the result of ill-conditioned control power data (large changes in control effectiveness, local maxima/minima generated due to nonmonotonic changes in control effectiveness, singularities, etc.). Nonmonotonic control power data can result in zero control-effectiveness derivatives being generated by the onboard model, with the result being an effector not moving from a local maxima or minima, with potentially significant effects on aircraft flying qualities. Figure 6 shows a case in the simulator in which a local maxima in horizontal tail control power resulted in a pitch departure.

Control effectors with comparable effectiveness may result in singularities that lead to undesired control surface usage, such as two surfaces opposing each other and impacting maneuvering capability, as shown in the simulation data in Fig. 7.

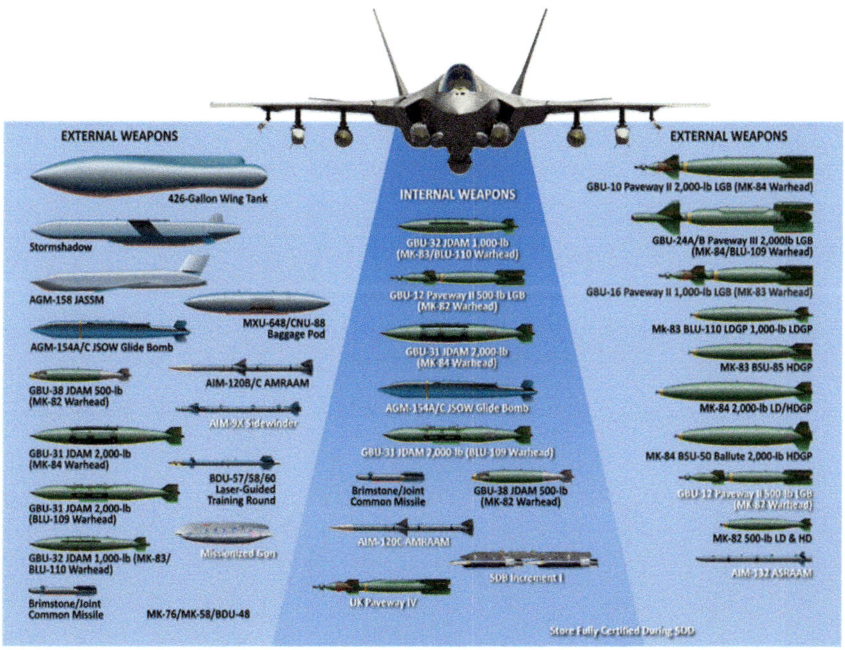

Fig. 5 The F-35 is designed to carry a large complement of internal and external weapons.

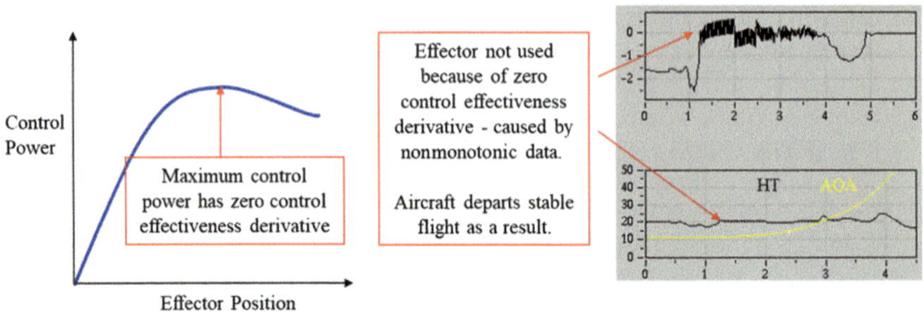

Fig. 6 Control effectiveness monotonicity.

All of these challenges required protections being put in place to preserve algorithm stability and achieve the desired flying qualities for the F-35 prior to starting the flight test program.

D. AIRCRAFT LOADS AND THE ELECTRO-HYDROSTATIC ACTUATION SYSTEM

The EB architecture allows the designer to utilize all of the available control power from the air vehicle. The designer no longer has to define a priori the optimal way to move the surfaces at each flight condition, but still has the freedom to define preferences of one surface over another based on other constraints, such as aircraft maneuver loads or aircraft performance. Integrating controller preferences into the EB architecture provides the control law designer with an easy means to reconfigure for system failures that result in the loss of surfaces by simply removing them from the solution. However, with that freedom comes a potential

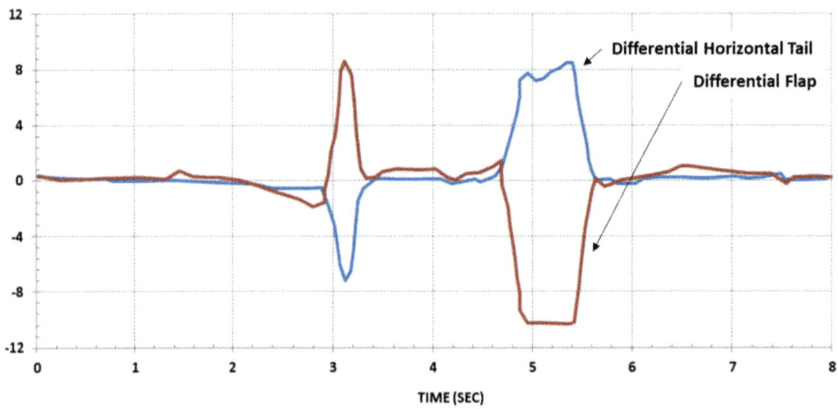

Fig. 7 Ill-conditioned control-effectiveness matrix.

challenge. For example, fighter aircraft have traditionally used their wing and tail surfaces in a fixed ratio to stay within airframe loads limits. The EB does not inherently obey that rule and had to be constrained to a fixed ratio in order to stay within airframe loads limitations, especially at high dynamic pressures where there is generally a delicate balance between maximizing maneuvering capability and remaining within structural envelopes. This constraint was achieved by simply forming virtual effectors, such as combining all the rolling surfaces (asymmetric flaperon, asymmetric aileron, asymmetric horizontal tail) into a single virtual rolling surface. This reduces the degrees of freedom in the solution and keeps the asymmetric tails, flaps, and ailerons in fixed ratios to each other. As loads increased on different surfaces in various portions of the envelope, the ratios could simply be changed to stay within allowable loads limits while preserving roll capability.

The F-35 is also the first production aircraft to incorporate an Electro-Hydrostatic Actuation System (EHAS) to replace conventional, hydraulically driven actuators, as shown in Fig. 8 for the F-35A version. This actuation system eliminated the need for a large, centralized hydraulic system by incorporating electrically driven, hydraulic power that is packaged as part of each individual actuator. This technology was a key enabler for the current aircraft configuration, and it yielded significant aircraft reliability and vulnerability improvements, as well as weight and cross-section reductions. However, from a control law perspective, the actuator load capability was significantly decreased compared to conventional

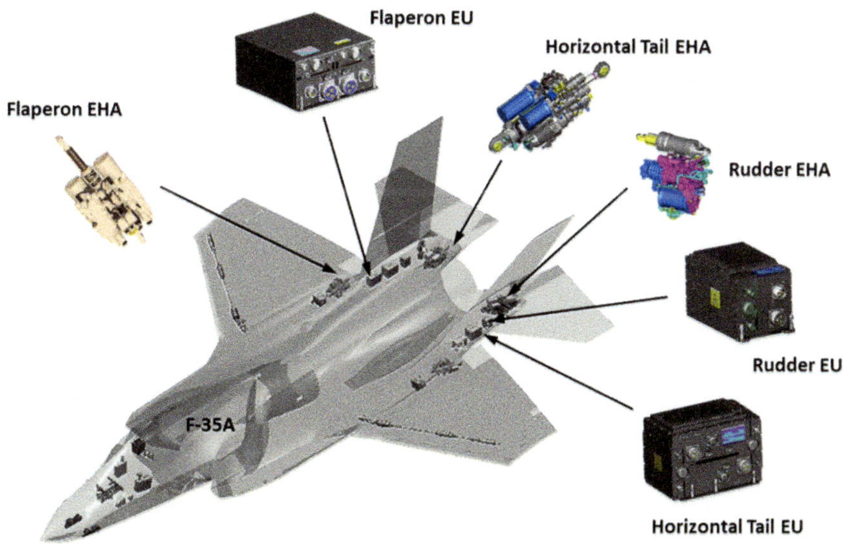

Fig. 8 Electro-hydrostatic actuation system (F-35A).

actuators, as shown in Fig. 9. In earlier-generation aircraft, actuators were typically sized to meet stiffness requirements, resulting in an excess load capability with less than 50% of the total stall capability used for general maneuvering. This provided large margins to actuator stall. The F-35 EHAS was designed to balance stiffness and load capability, the result being an actuator operating at closer to 90% of the capability for that same load, with a rapid loss of rate capability above 95% load. To accommodate these actuator characteristics, the F-35 control laws were tuned to use no more than 95% of the stall capability to achieve desired maneuvering performance. This was an additional constraint that required optimization of the EB to stay within actuator load restrictions.

The hinge moment loads were monitored during flight test, with updates to predicted hinge moments based on flight test results. Tailoring the response to only use up to 95% of the stall capability was only manageable due to the fact that the Electro-Hydrostatic Actuator (EHA) retained significantly more rate capability at 95% stall compared to a legacy actuator, as shown in Fig. 10. In some cases, surface command limits had to be tailored to stay within hinge moment limits as the flight test envelope was expanded during the course of the flight test program.

The EHAs also have significantly different characteristics when failed, compared to conventional actuators. In dual-redundant actuators, the imbalance in piston area generally causes the surface to drift up when failed until it balances with the external hinge moment loads. On the trailing edge flaps, the total upward deflection was manageable, but not on the horizontal tails. On the F-35A and F-35C versions, the failed tail would drift up more than what the unfailed tail could overcome, which could result in a departure. This situation was overcome by the addition of a small centering actuator that the control

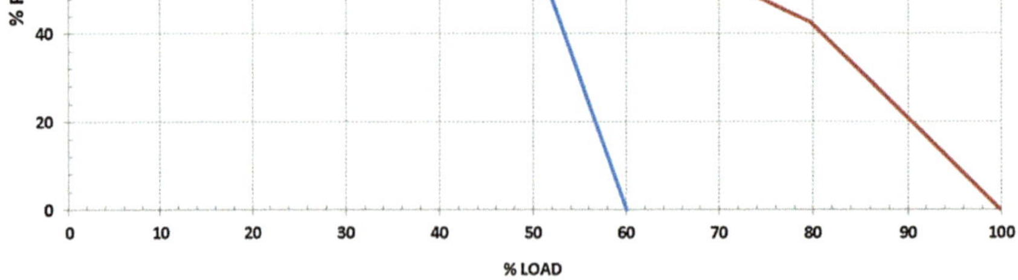

Fig. 9 Typical load-rate characteristics for an EHA vs a conventional actuator.

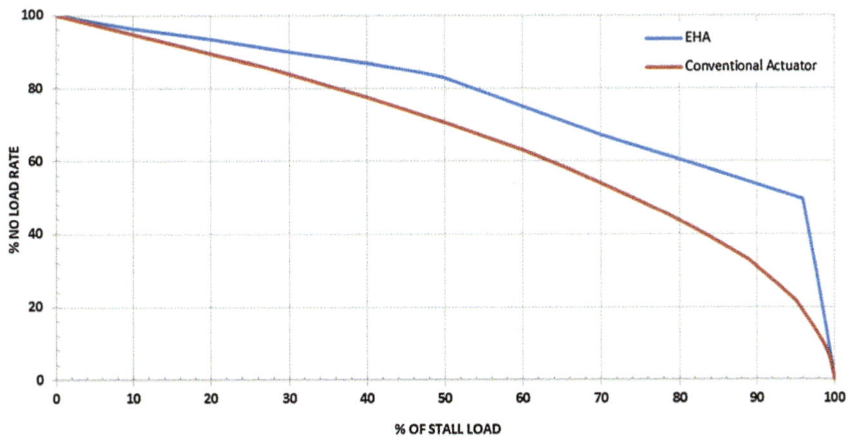

Fig. 10 Percentage of no-load rate vs percentage of stall for an EHA vs a conventional actuator.

laws could engage that would push the failed tail down to an optimal location at low speeds and then hold it in place. The F-35B version did not require a centering actuator because it can convert to STOVL mode and use the engine/lift fan thrust split to compensate for any pitching moment from the failed surface, eliminating the weight penalty of having to carry an additional actuator on the aircraft.

E. STRUCTURAL COUPLING CHALLENGES

One of the greatest challenges with the EB and onboard model architecture was preventing the control system from adversely responding to feedback sensor output resulting from structural deflection, called structural coupling. This is historically tested by setting the aircraft in the highest gain flight condition and perturbing the control surfaces/propulsive effectors to excite the structural modes, then measuring the loop characteristics of the path through the structure, sensors, and control system. In Fig. 11, random noise is injected into the actuators, and the effects on the aircraft sensors are measured as well as the total control law output. These measurements are used to assess stability when the loop is closed and verify the structural mode margin requirements in MIL-STD-9490.

For conventional control systems, the path gain from a sensor through the control laws is relatively easy to calculate. Control system gains are usually fixed at any given set of flight conditions, and determining the flight condition for maximum gain can be as simple as examining control law gain tables. For the F-35 control laws, the EB is iterating each frame to determine an optimized solution based on control surface effectiveness, as well as other constraints, such as effector rate/position limits and system failures, as shown previously in Fig. 4. At low airspeeds, where the control gains are usually the greatest, the EB

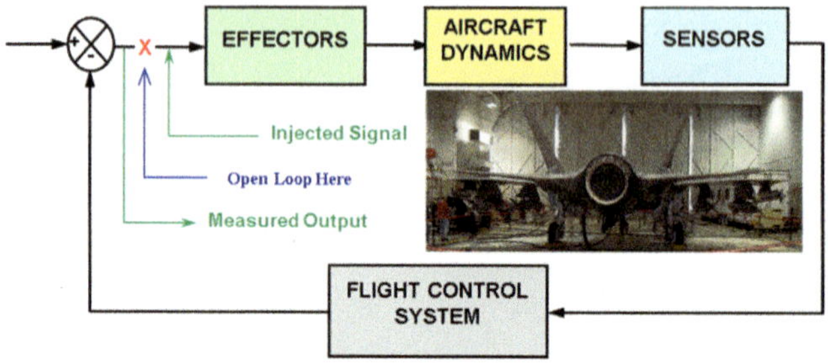

Fig. 11 Structural coupling testing.

has the freedom to change the surfaces it uses to control the aircraft and to increase the command on one surface when other surfaces hit a rate/position limit. This makes it nearly impossible to pick a single flight condition to evaluate during structural coupling tests. In order to ensure that the aircraft was free from structural coupling, actuator-to-sensor transfer functions were measured on the aircraft (like the one shown in Fig. 12 for pitch rate-to-horizontal tail) during testing and then linearly combined with possible control systems gains throughout the flight envelope posttest to verify structural margin. Very detailed linear models were generated and compared to on-aircraft test results to ensure accuracy of the airframe model. Figure 12 includes a comparison of the open-loop response measured during a structural coupling test to results achieved analytically by combining the derived airframe response with the linear model of the control system at the tested flight condition. With an excellent correlation between the analytical response and the measured response during testing, the Control Law team

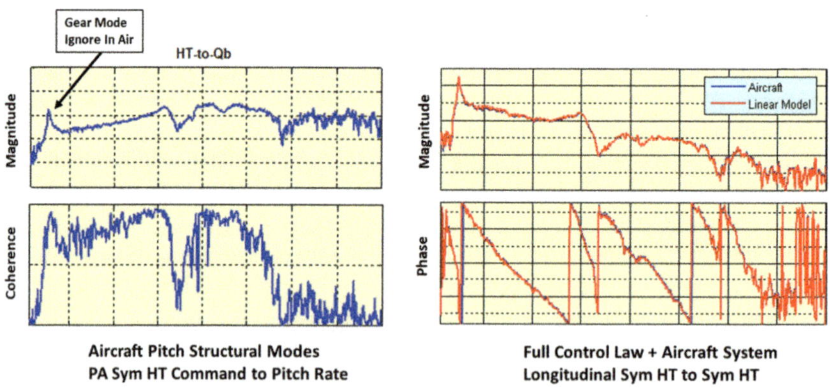

Fig. 12 Open-loop structural response.

successfully derived the required structural filters to achieve adequate stability margins. No instances of aero-servo elastic instabilities were observed during the extensive F-35 flight test program.

F. ACTIVE STICK AND THROTTLES

All three F-35 versions have unique mission requirements. Satisfying these missions with common hardware is an obvious benefit from a system architecture perspective, and also reduces maintenance and supply chain demands. In the cockpit, one key enabler to achieve this commonality was the use of active inceptors for the stick and throttle. The Active Inceptor System (AIS), shown in Fig. 13, allows the control law designer to program various features of the stick and throttle (e.g., setting force vs deflection, damping, soft stops, gates, and even ramps) as a function of flight phase in order to satisfy the unique missions of each version. The throttle motor is also used to back-drive the throttle to a commanded position when automatic command modes are engaged.

These features are used extensively on the F-35B for STOVL mode landing. Translational Rate Command (TRC) is an augmented STOVL command mode that uses the throttle inputs out of a center detent to command a change in forward velocity. With the throttle in the detent position, the control system is holding a reference velocity. This mode, along with vertical/lateral TRC modes on the stick, allows the pilot to easily fly STOVL approaches to a ship or shore-based landing pad. When the axial TRC mode is engaged, the throttle is programmed to back-drive to a center location, and ramps are programmed on each side. During approaches to a ship, the pilot enters the ship's speed as the

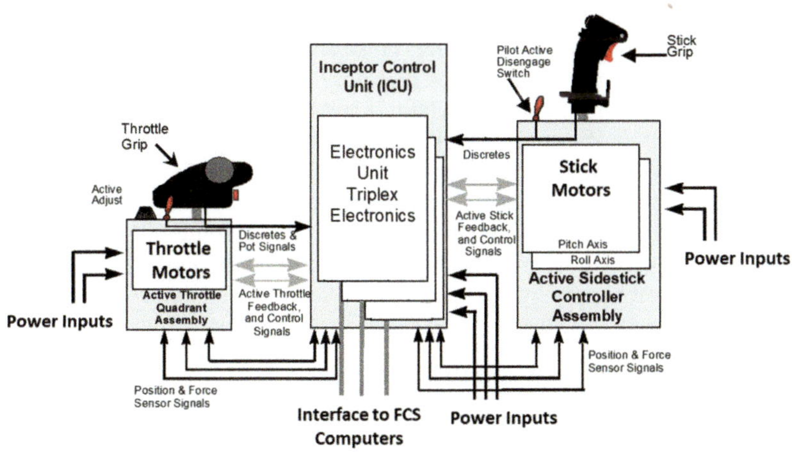

Fig. 13 Active inceptor system.

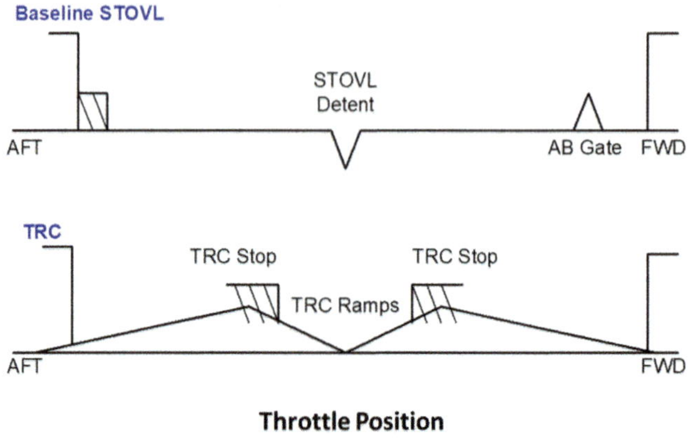

Fig. 14 Programmable throttle features for STOVL landing.

reference velocity for the center detent position. If the pilot pushes the throttle out of the detent and onto the forward ramp, an incremental increase in forward velocity is commanded. Pulling back on the throttle decreases the velocity. When the pilot gets to the desired position alongside the ship, simply letting go of the throttle will hold that position co-speed with the ship. The roll stick similarly holds lateral position when released and commands a change in lateral velocity when deflected. Fore–aft deflections of the pitch stick command changes in height rate. When the pitch stick is released, the control laws hold the current height. When touchdown occurs, the throttle automatically back-drives to the idle position where the control laws command the engine to ground-idle thrust. The programmable features of the throttle for STOVL landing operations are shown in Fig. 14.

Although the programmable features of the stick and throttle allow the control law designer a significant amount of flexibility to tailor the characteristics for various flight phases and modes, they also increase the complexity of the reconfiguration logic within the control laws to accommodate system failures. The F-35 control laws were designed to accommodate the stick or throttle going into passive mode, where all the programmable features are lost and the inceptors operate at a basic default spring rate. The control laws also have to accommodate instances in which the stick or throttle may become jammed and the pilot must be able to control the aircraft simply by applying force.

G. AERODYNAMIC MODELING CHALLENGES

As mentioned earlier, dynamic inversion relies on the cancellation of the plant dynamics, including system nonlinearities. At its core, the NDI controller contains a model of the aerodynamics of the aircraft (including entrained aerodynamics from the propulsion system) and a model of the aerodynamic and propulsion effectors. In order to minimize performance errors resulting directly from poor modeling, it was important to have a very detailed onboard model of the F-35 across the full operating envelope, and to account for discrete configuration changes associated with gear doors, STOVL doors, and weapon bay doors, as well as interactions between control surfaces. Parts of the flight envelope where aerodynamic characteristics can change rapidly (e.g., the transonic flight regime) require higher-fidelity modeling, with denser Mach and AoA breakpoints. Of course, even with improvements in the memory and throughput capacity of current flight control computers there are practical limitations on the size of models that can be embedded in the software. The current F-35 onboard models consist of approximately 3 million data points. The original models were derived from wind tunnel testing.

Although the tunnel testing for the F-35 was probably the most extensive testing conducted on a fighter program, there were still differences identified between the aerodynamic model derived from wind tunnel testing and the actual aircraft aerodynamics. One example of this was discovered during early envelope expansion testing in the transonic flight regime. While conducting an elevated g roll, a large sideslip excursion developed that led to a roll-reversal (Fig. 15), a maneuver response that was not predicted by simulation.

The primary cause of the excursion was traced to errors in the aerodynamic modeling of roll damping and control surface effectiveness in this part of the flight envelope, and updates to the truth aerodynamic models used in the simulation and the onboard model in the control laws were required to correct this response. Figure 16 shows the key aerodynamic model parameter changes that were made and the flight test response following the control law update compared

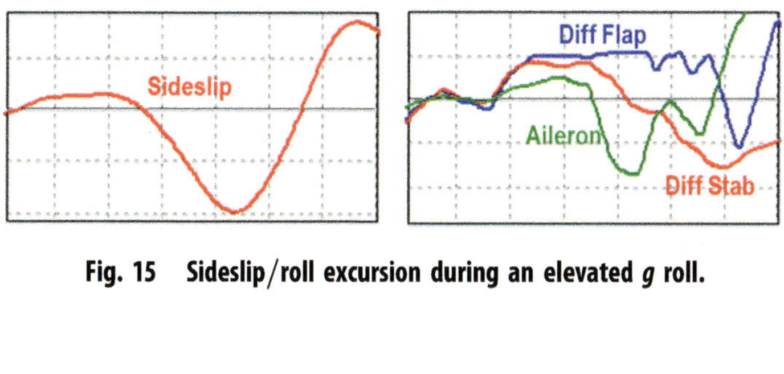

Fig. 15 Sideslip/roll excursion during an elevated *g* roll.

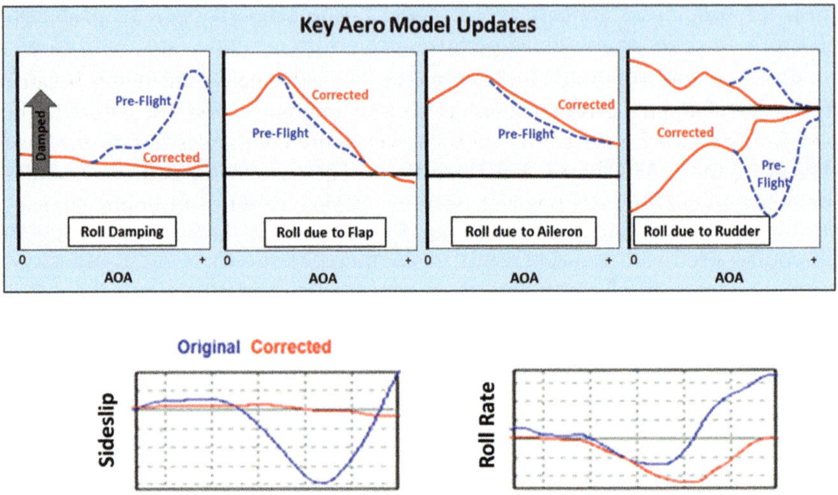

Fig. 16 Complex model updates were required to correct mispredictions.

to the original response, demonstrating that the roll-reversal was eliminated and the sideslip response was well-behaved following the software update.

Another potential challenge to an NDI implementation was associated with the practical limitations of modeling the complex aerodynamics that are typically encountered by fighter aircraft in the transonic, elevated G/AOA flight regime. This part of the envelope is associated with very complex flow fields that can lead to unsteady aerodynamics resulting in an asymmetric flow separation capable of inducing uncommanded yaw and roll motions. Abrupt Wing Stall (AWS) has been a challenge for fighter aircraft for decades, with documented examples from the 1950s-era F-84 Thunderjet to modern-day fighters [5]. This flow separation is typically characterized by discontinuous rolling/yawing moment coefficients at discrete AOA and/or sideslip, and hysteresis in those discontinuities across a sweep of AOA and/or sideslip. It would be impossible to capture these types of nonlinearities and discrete jumps in coefficients in the onboard aerodynamics model. If left uncorrected, however, the resultant transients could be quite noticeable to pilots, as shown in Fig. 17.

It is not possible to directly model these nonlinearities, so an alternate approach to improving the robustness of the algorithm was incorporated into the F-35 control laws. Approximations of the model error based on measured parameters were used to derive an additional acceleration error signal as an input to the EB, as shown in Fig. 18. This implementation was originally developed to accommodate modeling errors that were observed in the transonic flight regime during initial envelope expansion flight testing, but it has been shown to be effective at compensating for other areas of the flight envelope where modeling errors can be expected, such as high AOA, where airflow separation can challenge the effectiveness of an NDI-only control law.

Although the transonic flight regime can be extremely challenging because of rapidly changing aerodynamics as shocks migrate across the aircraft, the F-35 Control Law team had to deal with even more complex issues on the road to clearing the F-35B for STOVL operations. The conversion between conventional and STOVL modes involves opening/closing 11 different doors, some of which have an obvious impact on the aerodynamics of the aircraft, as Fig. 19 demonstrates.

Prior to engaging the lift fan system for the first time, it was important to demonstrate that the aircraft could be safely landed in conventional mode with the doors failed open. Initial flight testing with STOVL doors open revealed several modeling issues, including the basic directional stability of the jet, as shown in Fig. 20.

The level of model discrepency that was identified during this initial phase of testing had a significant impact on the planned flight test program and impacted the ability of the test team to proceed to engaging the STOVL propulsion system; however, it also highlighted just how complex the flow fields are around the aircraft in this configuration and how challenging the modeling task would be. An extensive effort was necessary to conduct the flight testing required to first capture the data that could be used to update the aerodynamic models and then resolve the issue. This required multiple fly-fix-fly cycles to eventually clear the full operating envelope for all required STOVL door configurations while allowing incremental envelope expansion to proceed.

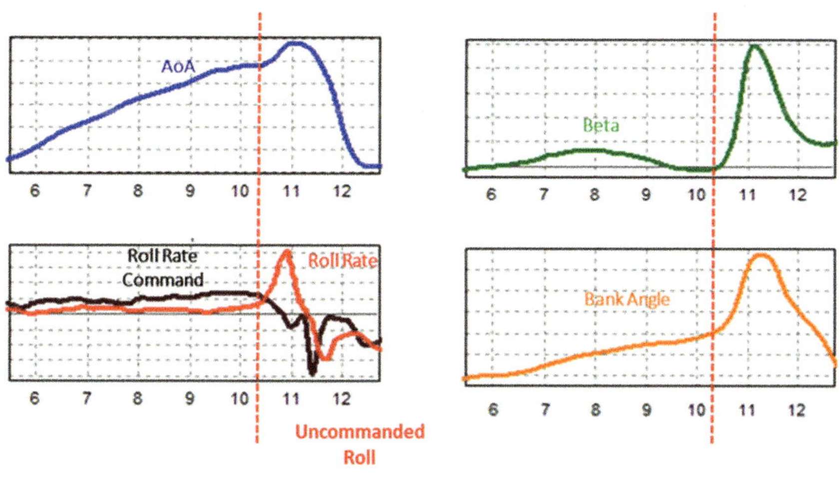

Fig. 17 Uncommanded roll/yaw as a result of unsteady aerodynamics.

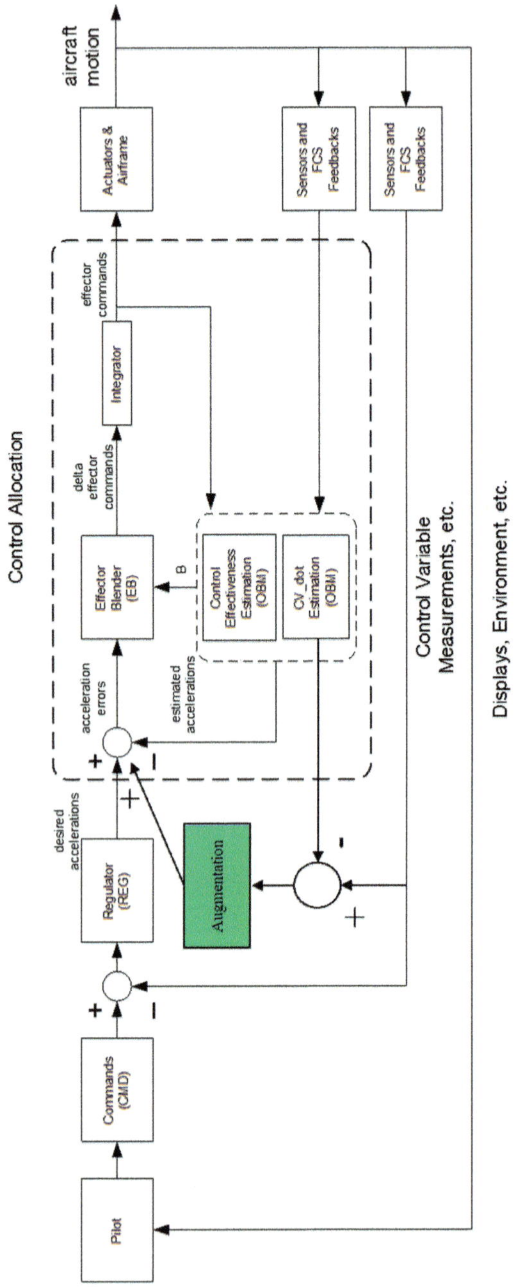

Fig. 18 Augmented NDI control law structure.

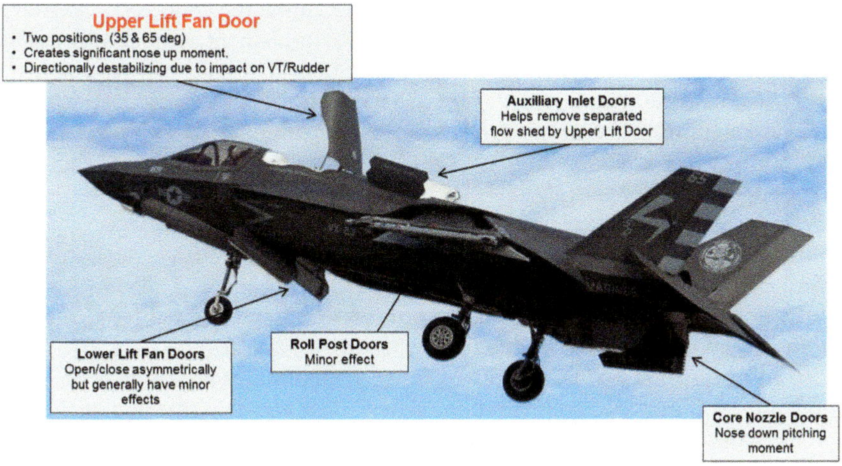

Fig. 19 F-35B STOVL doors.

H. SOFTWARE DEVELOPMENT PARADIGM

With the incorporation of the NDI controller structure into the F-35 control laws, the role of the control law engineer was fundamentally changed. The team was no longer focused on developing gain schedules based on linear models, which was a traditional control law engineer task. Now the team was much more deeply involved in understanding the aerodynamics of the aircraft, working closely with stability and control engineers to develop the onboard model, which is an

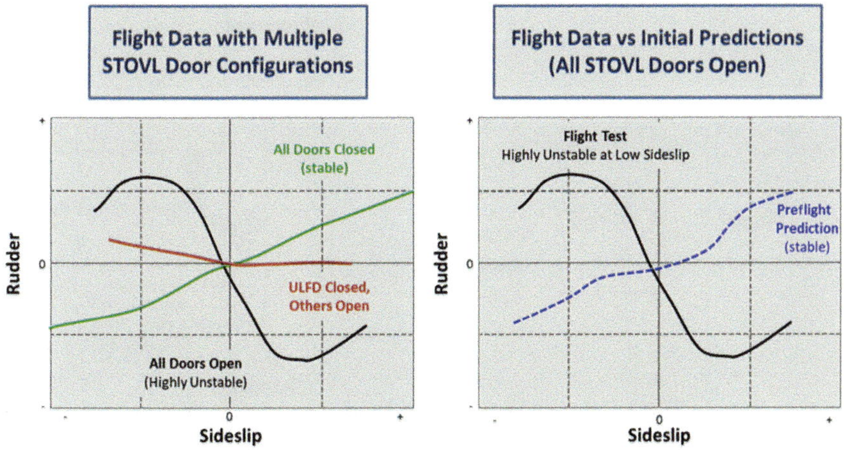

Fig. 20 Modeling challenge associated with STOVL doors.

integral part of the control laws. In order to streamline the control law software development process, the F-35 team also chose to implement a graphical, model-based approach to control law development and used MATLAB® Autocode to auto-generate code. The traditional approach of control law engineers developing S-plane diagrams and turning those over to a separate mechanization team to recode them into the Operational Flight Program (OFP) code loaded onto the flight control computers (Fig. 21) has been replaced with control law engineers who are also filling the role of the embedded software engineer, responsible for software development in accordance with established industry standards for safety-critical systems (Fig. 22).

In order to be successful in this approach, an unprecedented level of commitment and coordination by the Control Law, Stability and Control, and Software Development/Test teams was required. A rigorous approach to large-scale model development and configuration management was implemented, including the development of modeling standards, so that auto-generated code would comply with traditional software development guidelines (e.g., MISRA-C) and satisfy established software safety standards for airborne software (e.g., DO-178B, Defence Standard 00-56, MIL-STD-882) [6]. The overarching goal was to develop reliable, safe, and maintainable software.

The advantages of using a model-based software development process extended beyond just the cost savings associated with reducing the size of a separate team of software coders. Automatic code generation significantly reduced the frequency of coding defects that are typically introduced when manually translating requirements into design and code in a traditional software development

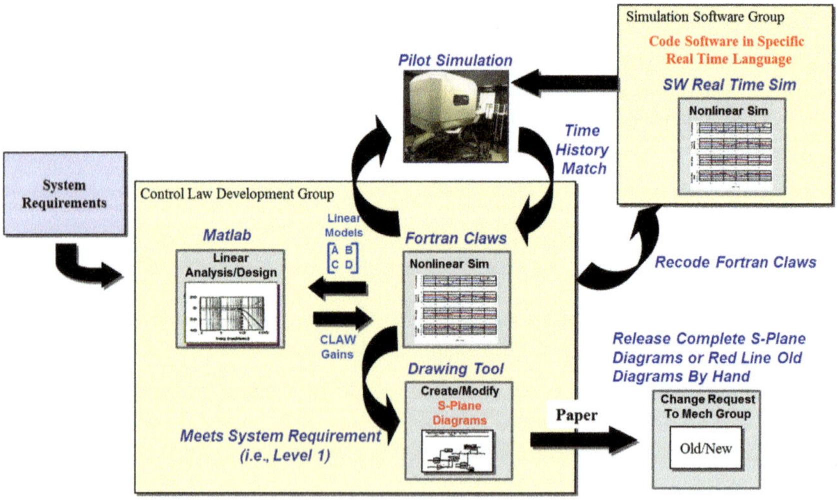

Fig. 21 Traditional software development process.

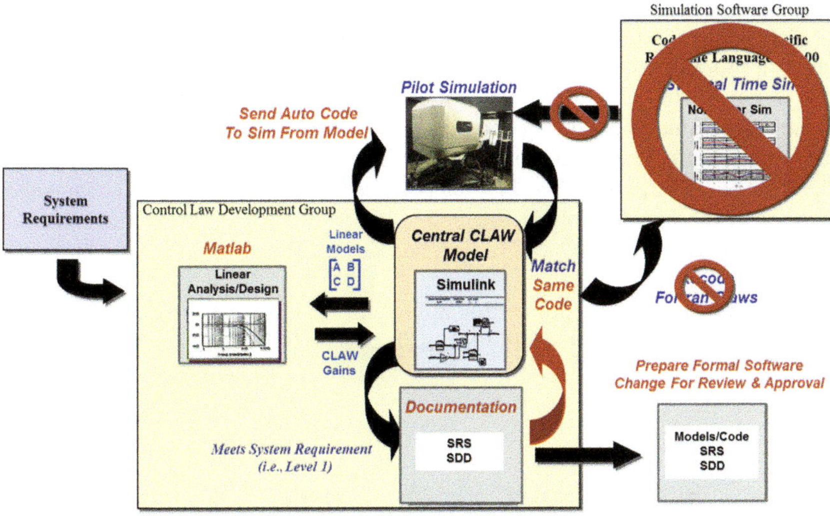

Fig. 22 F-35 control law software development process.

process. Also, the cycle time for developing code is reduced. Because the same code that is generated for the OFP is integrated into the offline and piloted simulations, verification of the control laws can be conducted much earlier in the development cycle and in multiple simulation environments.

These benefits were critically important during the JSF concept demonstration phase, when the Lockheed Martin team successfully executed a flight test demonstration program with the X-35 using a common control law structure for three aircraft versions using a single OFP. However, as critical as this development approach was to meeting the aggressive timelines associated with an X-plane program, it proved to be an even more significant enabler to meeting the demanding requirements associated with a production development flight test program for three versions.

IV. VERIFICATION

Verification of the F-35 control law development approach has been accomplished through a number of methods. Although flight testing of the F-35 is the most visible form of verification evidence, the original verification product is the desktop graphical model. As shown in Fig. 23, the validation of the design and testing at the desktop level forms the backbone of the OFP product that is then tested in simulators and hot-bench (hardware-in-the-loop) test facilities prior to being loaded onto a test aircraft and flown. All these venues have played an important role in validating the NDI structure of the F-35 control

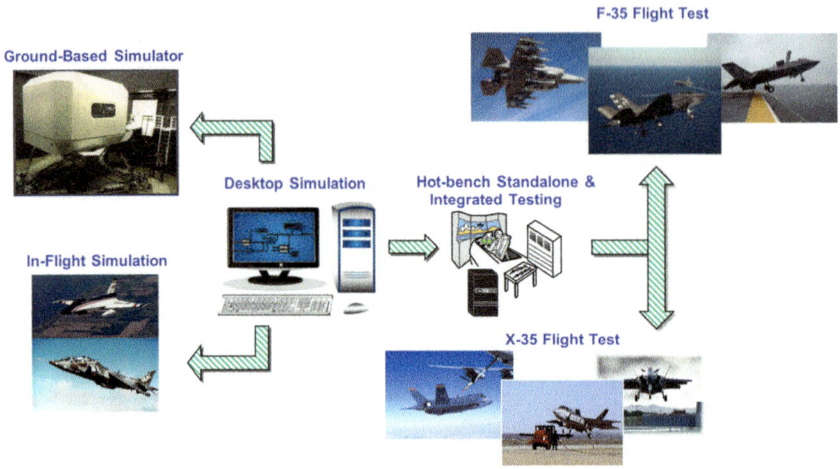

Fig. 23 F-35 software testing.

laws and the software development approach that includes auto-generation of the flight control OFP code from a graphical model of the control laws.

The elements shown in Fig. 23 span more than 20 years of development, either directly supporting the X-35/F-35 programs or supporting various technology demonstration programs and company research pursuits. The result of that effort is an air vehicle that meets all fighter requirements of all U.S. services, partner countries, and participating allied countries, while delivering significantly more capability than legacy aircraft.

V. CONCLUSION

The F-35 is a fighter program that consists of three versions, each with distinct mission requirements and aerodynamic characteristics. The STOVL version, in particular, has unique propulsion system characteristics as well. The Lockheed Martin Control Law design team implemented a novel and cost-effective approach to meet the demanding requirements of this program. The team was able to overcome a number of early design challenges to implement a robust, model-based dynamic inversion control law that provides exceptional flying qualities throughout the flight envelope, including high-AOA operations, supersonic flight, shipboard operations, and the demanding, low-speed envelope associated with STOVL operations.

ACKNOWLEDGMENTS

The authors would like to thank the entire Lockheed Martin Flying Qualities team for the work of its members over the years leading up to the delivery of the F-35 aircraft that is in operation around the world today. This team includes both stability and control engineers and control law design engineers. The stability and control engineers developed the aerodynamic models from wind tunnel tests and the onboard aerodynamic models that are such an integral part of the F-35 control laws. The control law design engineers were willing to embrace challenging new technologies for aircraft control and make them work! A special acknowledgment to members of the team whom we have lost over the years: John McCune and Donald Inman.

REFERENCES

[1] Enns, D. F., et al., "Control Design and Flight Hardware Implementation Experience with Nonlinear, Dynamic Inversion Control (NASA-2) for the F/A-18 HARV," *High-Angle-of-Attack Technology Accomplishments, Lessons Learned, and Future Directions*, NASA/CP-1998-207676/PT1 (may be released to U.S. persons with NASA approval).

[2] Bordignon, K., and Bessolo, J., "Control Allocation for the X-35B," AIAA Paper 2002-6020, 2002.

[3] Miller, C. J., "Nonlinear Dynamic Inversion Baseline Control Law: Flight Test Results," AIAA Paper 2011-6467, 2011.

[4] Bosworth, J., and Williams-Hayes, P., "Flight Test Results from the NF-15B Intelligent Flight Control System (IFCS) Project with Adaptation to a Simulated Stabilator Failure," NASA/TM-2007-214629, December 2007.

[5] Chambers, J., and Hall, R., "Historical Review of Uncommanded Lateral-Directional Motions at Transonic Conditions," *Journal of Aircraft*, Vol. 41, No. 3, May/June 2004.

[6] Bridges, M., "JSF Software Safety Process: Providing Developmental Assurance," Systems and Software Technology Conference, Tampa Bay, FL, 2007.

F-35 Propulsion System Integration, Development, and Verification

Steven P. Wurth* **and Mark S. Smith**†
Lockheed Martin Aeronautics Company, Fort Worth, TX

At the heart of every successful aircraft is a successful propulsion system design. The X-35 and the F-35 aircraft developments are at the top in the history of aviation. The concept of one aircraft design to meet the requirements of the Air Force, Marines, and Navy warfighter is not a new concept. What distinguishes the X-35 and F-35 programs is their success in developing a propulsion system design that meets performance, reliability, and affordability requirements. The success of the F-35 propulsion system can be attributed to developing verifiable requirements, conducting cohesive systems engineering, and quickly overcoming development challenges. The successful ground and flight testing of the unique propulsion system led to groundbreaking flight test programs. The F-35 propulsion system continues to successfully power the F-35 fleet well past its current 135,000 flight-hour mark.

I. INTRODUCTION

This paper presents the development of the F-35 propulsion system following a three-phase development life cycle. The development is similar to the design of any major engineering effort, with conceptual design, preliminary design, and detailed design phases. The conceptual design phase centered on the test and verification of the concept's feasibility. It included Advanced Short Takeoff and Vertical Landing (ASTOVL) and a Large-Scale Powered Model (LSPM). The preliminary design phase also entailed test and verification but was focused on assessing functional feasibility. For the X-35, this also included a Concept Demonstrator Phase (CDP). For the detailed design phase, test and verification reflected operational feasibility and included the System Development and Demonstration (SDD) Phase for the F-35. The phases proceeded chronologically from early ASTOVL through the F-35 SDD program.

*Lockheed Martin Technical Fellow, Integrated Flight Propulsion Control, AIAA Sr. Member.
†Lockheed Martin F-35 STOVL Chief Engineer (Ret.).

II. F-35 PROPULSION SYSTEM OVERVIEW

Developing a propulsion system that supports all three F-35 variants is a monumental task. The F-35A requires a Conventional Takeoff and Landing (CTOL) capability, the F-35B requires a Short Takeoff and Vertical Landing (STOVL) capability, and the F-35C Carrier Variant (CV) must withstand the rigors of naval carrier operation. Each of these supports a branch of the military: F-35A for the Air Force, F-35B for the Marines, and F-35C for the Navy. In order to test the concept of one propulsion system that could satisfy all propulsion requirements for the three services, the STOVL system needed a special focus. The propulsion system was designed as a tri-variant solution to maximize commonality and meet affordability goals.

The CTOL and CV propulsion systems are identical, whereas the STOVL propulsion system uses a CTOL/CV turbomachine coupled with STOVL-unique components. These components are a LiftFan®, clutch, driveshaft, roll post system, three-bearing swivel duct (3BSD), and STOVL exhaust nozzle. Previous efforts under the Defense Advanced Research Projects Agency (DARPA) allowed collaboration among Lockheed Martin, Pratt & Whitney, and Allison Advanced Development Co. (now LibertyWorks, a Rolls-Royce division). This team developed a STOVL propulsion system with a main engine that powers a dual-stage LiftFan via a shaft connected to the low-pressure spool of the main engine. Figure 1 shows the layout of the F-35 propulsion systems.

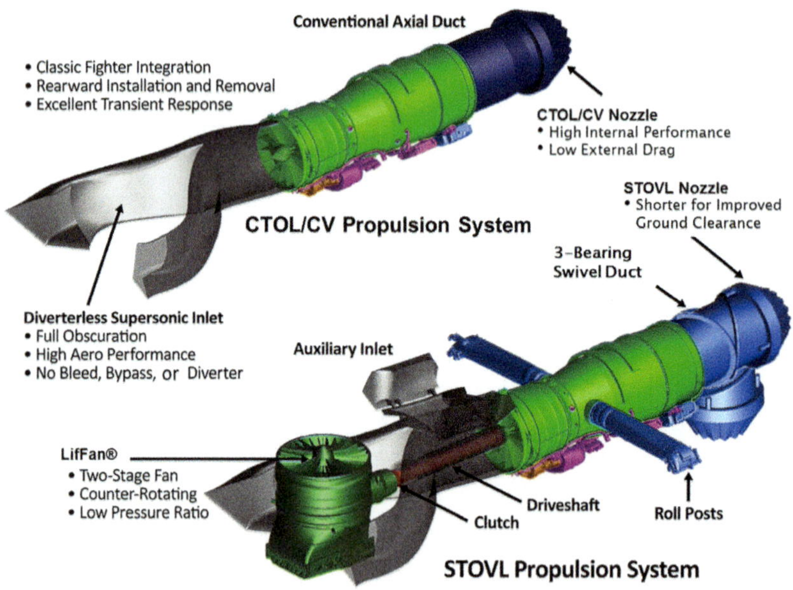

Fig. 1 F-35 CTOL/CV and STOVL propulsion system features.

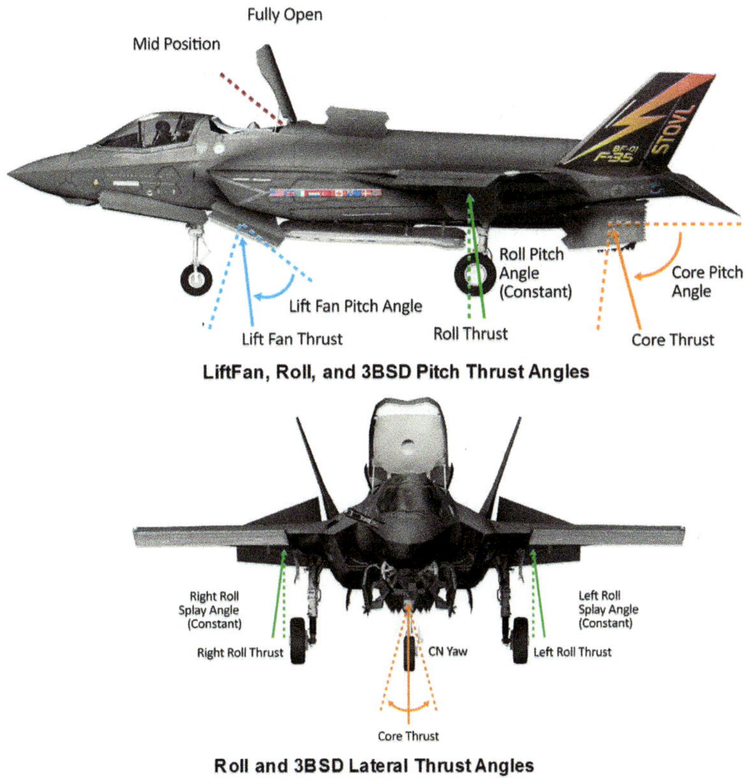

Fig. 2 F-35B STOVL mode effectors overview.

Figure 1 illustrates the basic differences between the CTOL/CV and STOVL propulsion systems. Among these is a shaft-driven LiftFan for thrust augmentation in STOVL mode, driven via a clutch/driveshaft from the main engine. The STOVL variant also has a unique LiftFan exhaust nozzle: the Variable Area Vane Box Nozzle (VAVBN). In addition, it has an auxiliary inlet for improved main engine airflow for STOVL operation and a pitch and yaw vectoring 3BSD. Its STOVL main engine nozzle is slightly shorter than that of the CTOL/CV variant. A similarity between the two propulsion systems is the main engine Diverterless Supersonic Inlet (DSI). In it, the diffuser section for the STOVL variant is modified by the presence of the auxiliary inlet.

A critical feature of the STOVL propulsion system is its control of the air vehicle during transition and conversion, low-speed operation, and hover. In transition, the aircraft transforms from conventional flight mode to STOVL mode, including flight down to a stationary hover. During conversion, the propulsion system is reconfigured from a conventional CTOL/CV propulsion system to a STOVL propulsion system with an operating LiftFan. Figure 2 shows the various STOVL effectors and their physical range of operation. STOVL control of the vehicle during conversion and transition, hover, and vertical landing was developed using Control Laws (CLAW). The CLAW govern the engine's Full-Authority Digital Engine Control (FADEC) and the aircraft's vehicle management computer [1].

Figure 2 also introduces an additional characteristic of the STOVL variant: its unique doors and inlets that allow airflow into the STOVL propulsion system. The STOVL variant has a three-position LiftFan inlet door that transitions based on vehicle speed, a LiftFan exhaust door for the VAVBN, roll post doors, and 3BSD doors. These doors were an integral part of the propulsion integration of the STOVL variant.

The following sections detail the basic design progression of the F-35 propulsion system. They illustrate challenges, discuss successes, and provide the history of the design, development, and verification of the F-35 propulsion system.

III. CONCEPTUAL DESIGN, TEST, AND VERIFICATION FOR CONCEPT FEASIBILITY: ASTOVL/LSPM

The ASTOVL program, described further in Ref. [2], included the construction of a 91%-scale LSPM. This model (Fig. 3) established concept feasibility by assessing such factors as:

- Low-speed and powered lift aerodynamics
- Hover and low-speed control power
- Hot Gas Ingestion (HGI) characteristics
- Propulsion system layout and general integration

The result was a very successful 160-hour ground test program executed at NASA's Ames Research Center. The LSPM had a representative propulsion system and was evaluated at both static and low-speed conditions [3].

A. KEY CAPABILITIES AND DESIGN FEATURES

One objective of the LSPM was to show the feasibility of harnessing the low spool of a main

Fig. 3 Lockheed Martin ASTOVL LSPM ground demonstrator.

engine to a bevel gearbox driving a fan. Another was to explore the STOVL aerodynamics of the air vehicle. To minimize costs, various components from F100 and legacy engines were used to develop this unique propulsion system. The LSPM had no clutch or actuating front nozzle, but had an actuating aft nozzle.

The LSPM LiftFan was the first-stage fan and Inlet Guide Vanes (IGVs) from a legacy engine. The LSPM LiftFan had only a single stage, compared to the X-35 and F-35B's two-stage LiftFan. The performance of the assembled LiftFan, gearbox, and driveshaft were at a power level representative of a two-stage design. The main engine combined the fan and high-pressure core of an F100-PW-220 and the low-pressure turbine stage from an F100-PW-229. The engine main fan rotor was modified to attach the driveshaft to the LiftFan. The engine case was modified so that the bypass air could be diverted to the ducts that supply the roll control jets [4].

A bifurcated F-22-like inlet was used, along with the LiftFan inlet and auxiliary door inlets that had been developed for the LSPM. A two-dimensional variable area thrust-deflecting nozzle was used for the rear main nozzle. The FADEC software was modified to control fuel flow and nozzle area as a function of the STOVL operating line of the turbine map [4].

The propulsion system was demonstrated at Pratt & Whitney's facility in West Palm Beach, Florida. The performance of the main engine's Air Induction System (AIS) was demonstrated under hover and low-speed conditions at NASA's Ames Research Center. Those demonstrations also showed the LiftFan AIS's performance under hover and low-speed conditions. Further, they investigated the propulsion effects on external aerodynamics, hover and low-speed aerodynamics, and HGI characteristics.

B. VERIFICATION TESTING

The demonstration of the first shaft-driven LiftFan and main engine operation occurred at Pratt & Whitney's outdoor engine test facility in West Palm Beach. More than 40 hours of static testing were accomplished with very few problems. The testing demonstrated the feasibility of changing the cycle of the cruise engine to provide shaft horsepower for the LiftFan [4]. It also demonstrated the capability of thrust split control. That refers to the ratio of main engine thrust to LiftFan thrust by modulation of the legacy first-stage fan and IGVs used as the LiftFan.

The next phase of testing involved suspending the LSPM at the Outdoor Aerodynamic Research Facility (OARF) at NASA's Ames Research Center. This testing investigated external aerodynamics under hover conditions, as shown in Fig. 4.

With the LSPM suspended on the OARF, free-air hover external measurements were taken while operating the STOVL propulsion system. These showed that the jet suckdown (i.e., the tendency of STOVL jet effects to lower the static pressure underneath the aircraft out-of-ground effects [OGE]) was less than 3% of the total lift. The model was also tested In Ground Effect (IGE),

Fig. 4 LSPM mounted at the OARF.

as shown in Fig. 5. This demonstrated that the jet fountain and lift improvement devices limited suckdown to less than 7% of the total lift [4].

As part of the IGE testing, HGI was also investigated and verified. Prior subscale HGI testing was questionable, so the LSPM was an ideal test platform to see how the shaft-driven LiftFan configuration would perform. The pressures and temperatures of the jet flow field around the aircraft showed that the jet footprint of the LiftFan system was more benign. This was applicable at the same thrust levels as those of a Harrier II AV-8B [4]. The relatively cool exhaust from the LiftFan blocked the hot engine exhaust from moving forward into the main engine inlets. This prevented detrimental HGI that would lower overall thrust due to higher temperatures. Figure 6 shows the use of ground flow visualization paint to understand the flow fields.

Another phase of the testing involved taking the LSPM to the National Full-scale Aerodynamics Complex (NFAC) at NASA's Ames Research Center. There, we used the wind tunnel at low speeds to investigate propulsion control effects on external aerodynamics, as shown in Fig. 7. Performance data were collected on Short Takeoff (STO) performance, transition (from jetborne to wingborne speeds) aerodynamic performance, and control power, including yaw control power for crosswinds [4]. Data showed excellent performance characteristics and proved the feasibility of a shaft-driven LiftFan and its integration into an ASTOVL airframe configuration.

The ASTOVL program established concept feasibility by assessing multiple factors, including low-speed and powered-lift aerodynamics, hover and low-speed control power, and the HGI of a shaft-driven LiftFan connected

Fig. 5 LSPM IGE testing.

Fig. 6 LSPM flow field investigation.

to the main engine of a traditional propulsion system. With the success of this program, the Lockheed Martin team was well-prepared to enter the CDP. This next section covers the progression of the design, development, and testing of the X-35 propulsion system.

IV. PRELIMINARY DESIGN, TEST, AND VERIFICATION OF FUNCTIONAL FEASIBILITY: CDP/X-35

In December 1996 the Joint Strike Fighter (JSF) Program Office awarded Lockheed Martin and Boeing CDP contracts. Our objectives were to conduct preliminary design studies for an operational aircraft and build a Concept Demonstrator Aircraft (CDA). Our CDA would demonstrate that similar airframes and propulsion systems could be designed and produced to meet Air Force CTOL, Navy CV, and Marine Corps STOVL requirements. Key requirements were to demonstrate vehicle characteristics that were either new (STOVL operations) or unsuitable for modeling or subscale demonstration (carrier approach handling qualities). We were not required to demonstrate legacy capabilities, such as stealth, avionics, and weapons carriage in the demonstrator; however, we were required to do so for the operational air vehicle design on which the demonstrator would be based.

The Lockheed Martin team was composed of Lockheed Martin, Northrop Grumman, British Aerospace (now BAE Systems), Pratt & Whitney, Rolls-Royce, and Allison Advanced Development Co. (now LibertyWorks). We were tasked with developing and integrating the propulsion system into the CDA, designated the X-35. The X-35 propulsion system consisted of a conventional afterburning gas turbine for the CTOL and CV variants. The STOVL propulsion system added the LiftFan for thrust augmentation for vertical lift while using the same basic gas turbine as the CTOL and CV variants.

Fig. 7 LSPM at the NFAC wind tunnel.

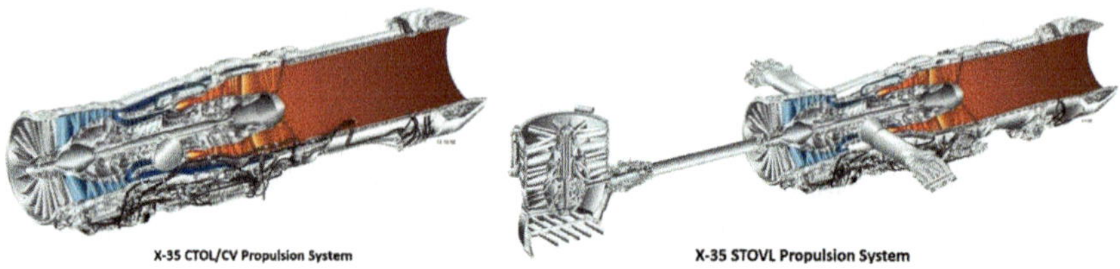

Fig. 8 X-35 CTOL/CV and STOVL propulsion systems.

Other components, such as roll post nozzles and a 3BSD, were also added to provide propulsive attitude control without needing a bleed-air reaction control system. This approach allowed us to use a common gas generator while mitigating STOVL impacts on the CTOL/CV propulsion system. Figure 8 illustrates the X-35 CTOL/CV and STOVL propulsion systems.

Lockheed Martin's demonstrator approach was to build two airframes capable of being configured as a Marine Corps STOVL variant. Aircraft 301 was to be initially configured as an Air Force CTOL variant to conduct CTOL flight tests. It would then be reconfigured as a Marine Corps STOVL variant for further flight testing. Aircraft 300 would be configured as a Navy CV variant with a larger wing area and control surfaces. This would demonstrate the low-speed handling qualities required for a carrier-based operation.

Pratt & Whitney proposed an engine based on the legacy design, designated the JSF SE611, with a larger fan and a redesigned turbine for all variants. In the STOVL variant, the engine control system used the unique turbine design of the JSF SE611. It provided the power to drive the LiftFan via a driveshaft from the engine's low spool. The development of the LiftFan, pitch vectoring LiftFan nozzle, clutch, and driveshaft was the responsibility of Allison Advanced Development Co. (now LibertyWorks). Pratt & Whitney and Rolls-Royce developed the roll post system and the 3BSD.

The STOVL propulsion system was designed to provide attitude control while in jetborne or semi-jetborne mode. The roll post nozzles provided roll control by regulating engine bypass air to nozzles located in the wings. The 3BSD provided both yaw and pitch control supplemented by a pitch vectoring nozzle on the LiftFan. Fore/aft balance of the aircraft, as well as pitch control, was provided by varying the thrust split (the ratio of 3BSD thrust to LiftFan thrust).

The development and integration of a traditional propulsion system for CTOL and CV aircraft is not a trivial task. The addition of a STOVL variant and the need to maximize commonality across all three variants increased the magnitude of the task. The primary propulsion focus for the X-35 was to demonstrate a tri-variant propulsion system that supported the two key objectives of the CDP effort: STOVL operations and carrier approach handling qualities. Propulsion carrier approach requirements were known from legacy operations, leaving STOVL operations as the primary propulsion challenge for the CDP.

The layout of the LiftFan STOVL propulsion system had been demonstrated in Lockheed Martin's LSPM under the DARPA ASTOVL program (1993 to 1995). However, it was not representative of an operational propulsion system. Although the LSPM gathered valuable data on propulsion-induced aerodynamics, the propulsion system featured a single-stage LiftFan without a clutch. This prevented conversion into a CTOL-configured propulsion system and lacked the control system necessary to execute a conversion. The engine exhaust system also featured a heavy two-dimensional nozzle rather than the 3BSD proposed for the X-35. Starting with the proven legacy core, all remaining elements of the STOVL propulsion system had to be developed under the CDP and flown in the X-35.

Developing the key STOVL propulsion system elements was a major technical challenge for the CDP. It required a STOVL demonstrator aircraft that could prove the viability of the operational aircraft design. To minimize cost and schedule risks, only the necessary elements were included. The advanced integrated systems that were part of the operational aircraft were not necessary to meet contract requirements. Conventional subsystems were to be employed whenever possible; however, for the STOVL propulsion system, virtually all elements that would be required in an operational aircraft were necessary for a successful demonstrator.

Other requirements were:

- Performance in both CTOL and STOVL modes
- Jetborne and semi-jetborne controllability
- The ability to convert between CTOL and STOVL modes

Other characteristics, such as system life, weight, and full envelope capability, were not as critical. This was due to the limited flight test program planned and the absence of full-up avionics and weapons carriage requirements.

As a target for the design team, Lockheed Martin defined a flight test mission, Mission X, that encompassed all contract requirements in a single flight. The mission made a strong case for the X-35 being a stepping stone for an operational aircraft. Mission X would accomplish what no STOVL aircraft had ever accomplished in a single flight:

- A short field STOVL takeoff
- Conversion into CTOL mode
- A level flight acceleration to supersonic speeds
- A conversion back into STOVL mode, followed by a vertical landing

Mission X truly represented what an operational aircraft would be called upon to accomplish. In order to accomplish it, the X-35 would be required to possess multiple capabilities. Its thrust and fuel burn characteristics had to meet the up-and-away requirements of an operational combat aircraft. Its performance had to meet the STOVL requirements of an operational combat aircraft. It needed to have a fault-tolerant Integrated Flight Propulsion Control (IFPC) system that seamlessly interfaced with the aircraft CLAW and minimized pilot workload. Also, it needed to be able to convert from STOVL into CTOL mode, and vice versa.

Developing and maturing these capabilities for the X-35 in the time available was the program's key challenge. Risk mitigation was integral and required identifying areas where contingency plans were required to allow a rapid response to development problems. Detailed schedules were prepared showing development milestones, presenting a methodical buildup approach leading to flight test.

A. PROPULSION SYSTEM REQUIREMENTS

1. PERFORMANCE

Performance of the X-35 propulsion system had to satisfy both CTOL/CV and STOVL requirements using the same basic gas generator. Extensive modeling was performed leading up to the CDP contract by all parties. We defined the propulsion cycle that would best meet the needs of the operational air vehicle. Pratt & Whitney generated engine models, and Allison Advanced Development Co. (now LibertyWorks) generated LiftFan performance models. Lockheed Martin combined these models with aircraft installation effects to predict operational aircraft performance considering range, combat performance, and payload. The result was the JSF SE611 engine cycle coupled with a LiftFan. This configuration balanced CTOL/CV requirements with STOVL requirements to produce a truly tri-variant propulsion system.

For the CTOL and CV variants, what remained was to provide an adequate engine response to ensure acceptable handling qualities. For the STOVL variant, having sufficient power to accomplish a mission was just the beginning. Controlling the aircraft while jetborne required not only thrust, but also thrust through the appropriate effectors. These had to be coupled with a control system that could keep the aircraft in balance and execute the pilot's commands. A new level of controls integration was required to couple the propulsion system with the aircraft flight controls, provided in the IFPC.

CTOL/CV performance requirements centered on mission range, dry power thrust (military power), and afterburner thrust levels (maximum power). The STOVL requirements were more extensive, including all the CTOL/CV requirements plus the need to provide STO thrust and propulsive lift for a vertical landing. Maintaining and controlling thrust during the transitions between wingborne and jetborne flight were also key requirements.

During the development period, the most stressing condition for the STOVL propulsion system was either STO thrust at deck-edge rotation or vertical thrust at the end of a mission. These two conditions were captured as part of the STOVL variant's overall mission requirements. STO thrust was dictated by mission payload and fuel load at the deck edge necessary to meet the design mission.

Vertical thrust was driven by the end of the mission where vertical lift bring-back dictated minimum landing weight. Vertical lift bring-back comprises a specified weapons load and a minimum fuel load that includes sufficient fuel for a go-around pass. These competing requirements confirmed that the correct engine cycle had been selected to balance performance requirements. The final result was a propulsion system in the 40,000 lbf class. STOVL thrust is distributed among the four lift posts: the LiftFan, two roll posts, and engine primary nozzle.

2. AIR INDUCTION SYSTEM AND HOT GAS INGESTION

Designing the AIS was a key element in meeting performance requirements. Lockheed Martin proposed a diverterless bump inlet, a design that had both weight and survivability benefits. Without any moving parts and no need for boundary layer bleed, the inlet minimized weight and provided good supersonic performance; however, like all supersonic inlets, performance was inadequate at the low speeds that would be encountered during STOVL operations. An auxiliary inlet was required to improve static inlet recovery and meet performance goals.

Similar performance goals were required for the LiftFan inlet. The inlet was shaped as much like a bell mouth as possible, subject to integration constraints. Multiple wind tunnel entries, Computational Fluid Dynamics (CFD) modeling, and static tests were conducted to maximize the performance of the AIS, including modeling the doors. Both the auxiliary inlet and the LiftFan required doors that would not impact inlet recovery while closing for up-and-away flight as part of a stealth configuration.

A critical factor in the performance of all STOVL aircraft is the potential for HGI. When a STOVL aircraft performs a vertical landing, hot exhaust from the aircraft itself strikes the ground and reflects upward. It can potentially enter the engine inlet and result in a dramatic decline in performance. The layout of the X-35 STOVL propulsion system relied on the relatively cool exhaust of the LiftFan. The fan provided an aerodynamic barrier that would prevent the engine exhaust gases from making their way to the inlet. Quantifying how effectively this prevented HGI was a critical step in defining the safe envelope for a vertical landing.

3. STOVL INTEGRATED FLIGHT PROPULSION CONTROL AND CONVERSION DEVELOPMENT

The development of the IFPC system required a rigorous and thorough application of an established Fly-By-Wire (FBW) flight control standard. This standard was modified to adapt a legacy twin-engine propulsion system into a single-engine, flight-critical application. The JSF SE611 engine was adapted from a twin-engine legacy application with a control system and redundancy architecture appropriate for a twin-engine application. There were clear deficiencies in the redundancy architecture when simply carried over to a single-engine airframe.

Addressing this issue with hardware and software redesign efforts was the major focus of IFPC activities.

The legacy engine used an active/standby hydraulic configuration, but an active/active electrical configuration for actuator control. The hydraulic active/standby feature posed the most concern due to the inherent transient that occurred as the system switched from active to standby after a failure. Considerable effort had been spent to minimize these transients to reduce the aircraft-level effects. The legacy engine lacked fail-safe positioning after second failures. This deficiency led to efforts implementing features that would drive some actuation systems to aircraft-safe positions. These concerns were the impetus to establish fault-tolerance requirements for the JSF X-35B lift system [5].

The X-35B's flight test success was critical to winning the program. As such, select members from the Flight Control Integrated Product Team (IPT) and the Propulsion IPT formed the IFPC IPT to apply flight control standards to the legacy propulsion system. The IFPC IPT was tasked with developing the integrated CLAW that integrated aircraft, main engine, and LiftFan hardware and software for STOVL mode flight. Additionally, the IFPC IPT had to apply the flight control standards and requirements for fault tolerance and failure mode handling to the development of the new hardware STOVL effectors. The IPT's objectives were to [5]:

- Rigorously and thoroughly apply an established FBW flight control standard.
- Develop the standard into an achievable process for the X-35B development.
- Ensure that the effects of considered failures were well-understood, simulated, and documented.
- Design the X-35B propulsion system such that, under the effects of these considered failures, the X-35B would complete a successful and safe flight test program.

Converting the aircraft from CTOL mode into STOVL mode and vice versa required the development of a clutch system that would be new to the CDP. This development required the integration of clutch CLAW with the engine and aircraft CLAW so that the transition between modes was seamless to the pilot. Pilot/vehicle interface design was also part of the IFPC development for this new STOVL X-35 development so that controlling the propulsion system was an easy task. The synergy of aircraft and engine CLAW development integrated with the STOVL effectors became the genesis of the IFPC IPT.

Requirements for the hardware were developed based on previous Lockheed Martin flight control FBW experience and applied to the new STOVL effectors. Propulsion requirements for redundancy and control characteristics were flowed to Pratt & Whitney and Allison Advanced Development Co. (now LibertyWorks)/Rolls-Royce for the main engine, LiftFan, and clutch system. These requirements were defined similarly to actuation requirements, but with an added complication: the actuation requirements had to be related to thrust and pitch control, roll control, and yaw control.

Pratt & Whitney provided the General Actuator Model (GAM) to help Lockheed Martin establish the effector requirements to provide the needed aircraft desired dynamics. The handling quality requirements were then decomposed to identify which effector requirements were necessary to meet those requirements. The redundancy requirements were established using a CDP-developed fault tolerance process [5]. This process involved reviewing legacy propulsion systems and adding redundancy to manage the X-35B STOVL flight test risk. The results exposed areas that needed added redundancy. For areas to which redundancy could not be added, minimizing exposure to these risks could be performed by flight procedures.

Conversion requirements drove redundancy and engagement characteristics while engaging and disengaging the LiftFan. This ensured that the aircraft CLAW could deal with the introduction of LiftFan thrust during inbound conversions and the removal of LiftFan thrust for outbound conversions. The conversion modes are shown in Fig. 9, along with a photograph of the X-35B completing a conversion.

Because engaging and disengaging the clutch was so critical to the success of the X-35B program, special considerations for the requirements of the clutch and lube system were established. No possible single or dual failure(s) of the clutch system were permitted to engage the LiftFan when not commanded to do so. Also prohibited were any possible single or dual failure(s) of the clutch system that disengaged the LiftFan when not commanded to do so. Failures of the clutch system during conversion were to be detected and safely accommodated.

These requirements ensured that the design of the clutch control system had mitigated any failures that would cause an inadvertent disengagement or engagement of the LiftFan. The establishment of a specialized built-in test was the result of these requirements. This test checked all conversion-related actuation and software systems immediately prior to engaging or disengaging the clutch. If that test failed, the pilot could land in CTOL or STOVL mode and prevent the aircraft from being in the precarious state of being in between modes.

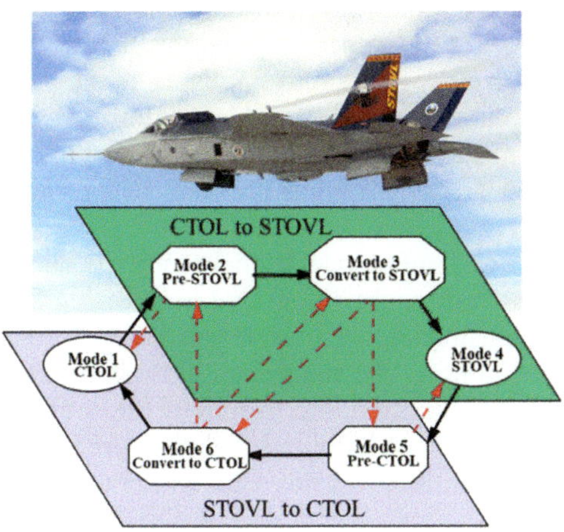

Fig. 9 X-35B conversion propulsion requirements and conversion modes.

The conversion process and the choreography of all the various propulsion systems, LiftFan, clutch, and main engine actuation were established by the IFPC IPT. This involved the systems integration of not only propulsion and flight controls, but also hydraulics and electrical power systems. This ensured that all of them operated in concert to convert the X-35B into/from conventional flight into/from jetborne flight.

The IFPC requirements for this supersonic STOVL prototype required a specialized effort. We defined flight control and fault tolerance requirements that would meet aircraft STOVL flight control handling quality and safety requirements to ensure a successful X-35B flight test program. Adapting a legacy propulsion main engine and developing new LiftFan and clutch hardware to meet these requirements would ensure this success. The IFPC IPT established these propulsion system STOVL effector requirements and incorporated changes to the system. This placed the aircraft in a fail-safe configuration in the event of a catastrophic failure of the propulsion system. As a result of establishing hardware and software requirements, extensive modeling and planning during the flight test program were performed. This ensured that pilots could safely recover the aircraft in the event of any improbable failures in the STOVL control system.

4. AIR VEHICLE INTEGRATION

Air vehicle integration of the propulsion system in the X-35 followed a conventional approach to minimize technical, cost, and schedule risk. Employing conventional subsystems meant that off-the-shelf components could be used where suitable. An Airframe-Mounted Accessory Drive (AMAD) was connected to the engine gearbox via a power takeoff shaft and used to drive the aircraft subsystems. The Auxiliary Power Unit (APU) was taken from the F-22, and the hydraulic pumps were taken from the YF-23. The air turbine starter, generators, and Environmental Control System (ECS) were all obtained from legacy aircraft. By using this approach, subsystem components were readily available to support development and flight testing with a supply of spares. This minimized cost and avoided Safety-of-Flight (SoF) component qualification testing. This philosophy was followed throughout the X-35 design, allowing the team to focus on the unique capabilities required.

Because the X-35 was a demonstrator and not an operational prototype, some capabilities required in an operational aircraft could be omitted, simplifying the design task. The capabilities to fly at night in clear or adverse weather, and to a full flight envelope, were not required. Limiting the maximum altitude relaxed emergency power requirements in terms of duration and simplified life-support system design. Similarly, although carrier approach handling qualities were a contract requirement, performing an arrested landing was not, eliminating the need for a carrier tail hook.

B. DEVELOPMENTAL, GROUND, AND FLIGHT TESTING

The X-35 propulsion system development effort followed a methodical approach. It began with component-level testing and proceeded through system-level ground tests, culminating in air vehicle ground tests of the installed propulsion system. Ground testing for the X-35A mimicked legacy practices, whereas extensive STOVL ground testing was required for the X-35B. This section provides an overview of that testing.

1. PROPULSION SYSTEM DEVELOPMENTAL TESTING

The development of the propulsion system followed multiple component-level paths. It culminated in an integrated STOVL propulsion system test at Pratt & Whitney's test facilities in West Palm Beach. Pratt & Whitney initially began testing the CTOL/CV propulsion system, proving out the JSF SE611 engine in a conventional environment. Parallel efforts were underway to develop STOVL components and construct test facilities. Conventional mode engine testing was accomplished in a run station before sending the engine to the Arnold Engineering Development Complex (AEDC) for altitude testing. The JSF SE611 was common between the CTOL/CV and STOVL systems, the so altitude testing applied to all variants and included performance, throttle transients, air starts, and inlet stability. Several engines were produced by P&W for engine system verification and validation testing. Allison Advanced Development Co. (now LibertyWorks) developed several LiftFans, clutches, and driveshafts for similar testing. Rolls-Royce developed the roll post system (nozzles, actuators, and ducting) and the 3BSD module for development testing. Allison used various rigs to exercise the gearbox and the clutch. A full-up test of the LiftFan would occur only when coupled with a JSF SE611 engine during the system-level test in West Palm Beach. The driveshaft was tested separately on rigs at the supplier's facility. In addition to the torque and speed requirements, key characteristics of the driveshaft were its ability to tolerate misalignment between the engine and LiftFan, aircraft deflections, and axial growth of the engine due to thermals. Due to the high rotational speed of the driveshaft, the rapid accumulation of load cycles also required it be designed for infinite life.

The 3BSD and roll post system were designed and underwent component testing at Rolls-Royce's facilities in Filton, United Kingdom, just north of Bristol. The 3BSD was a special challenge, featuring large-diameter flexible bearings that would allow the rotation of the exhaust duct segments to provide the desired vectoring capability. Actuators mounted on the duct would engage gears on the outside of the bearings to provide the motive force for vectoring. The 3BSD in both unvectored and vectored positions is shown in Fig. 10.

A key test objective was to collect data on the engine's Low-Pressure Turbine (LPT) during both CTOL and STOVL operation. The design of the LPT was the most challenging aspect of the JSF SE611 due to its wide operating range. The

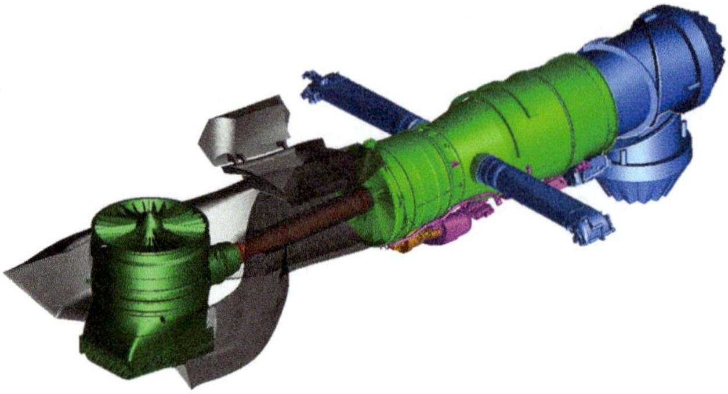

Fig. 10 3BSD in CTOL and STOVL positions.

turbine was designed to meet all up-and-away requirements while also being able to provide the horsepower necessary to drive the LiftFan. Pratt & Whitney modified two of its West Palm Beach test cells to accomplish these key objectives and demonstrate STOVL system operation. One test stand was reconfigured to allow the installation of the STOVL propulsion system. This included ducting to carry exhaust gases from the LiftFan and roll posts outside the test cell. The exhaust collector for the engine exhaust was modified to allow vectoring of the 3BSD. Initial testing occurred in this test cell with a highly instrumented engine and LiftFan. The second test stand (Fig. 11), which had been reconfigured by Pratt & Whitney for STOVL testing, was an outdoor commercial engine test stand. This allowed mounting the STOVL propulsion system at a height that permitted STOVL mode operation without using exhaust ducts or collectors. This thrust bed allowed the measurement of system performance to confirm modeling of the integrated system. It also enabled demonstrating that the system would meet the requirements for a successful flight test program. Other test facility modifications included placing high-temperature concrete beneath the test stand to permit extended STOVL test runs and deflectors to minimize any HGI.

This initial testing was the first time all STOVL propulsion components had been assembled as a system. It was also the first time the LiftFan was operated at high power. In order to reduce risk and allow for control system development, initial runs were accomplished without engaging or disengaging the clutch. Engine starts were accomplished with the clutch preset to the desired configuration based on the intended test program. Initial test runs were at low power as confidence in the system grew and the control system matured. High-power runs were first accomplished with the LiftFan disengaged, followed by engaged runs. Once high-power runs had been accomplished, testing proceeded with individually exercising each propulsion effector before proceeding with integrated system operation. The final step was the execution of clutch engagements, proving the capability of the system to convert into/from CTOL and STOVL modes.

Testing on outdoor STOVL test stands confirmed that system performance would meet the needs of the flight test program. It also verified that the propulsion effectors could produce the required control authority for acceptable handling qualities. The thrust measurement system provided steady-state measurements. From these, data from the test stand were used to calibrate system models to predict how the control system would perform under dynamic conditions. Rapid thrust modulation between control effectors also excited the structure of the test stand. This resulted in modifications during SDD to increase the overall stiffness of the test stand.

Once the STOVL propulsion system had sufficiently matured to fly a simulated mission on the test stand, an endurance test or Accelerated Mission Test (AMT) was designed. The AMT would demonstrate sufficient life and durability to meet the needs of the flight test program. It would also demonstrate twice the needs of the flight test program, exercising the propulsion system in both CTOL and STOVL modes. This would show that the system was safe and ready to fly. Developing the AMT program required condensing the planned flight test program and removing any nondamaging time (e.g., ground idle). It also required exercising the system through a series of missions that simulated the stressing portions of the flight test program. A single AMT would be performed to encompass the flight test programs of all three variants.

The CDP AMT program had some interesting differences from what an operational AMT program would require. An operational aircraft would spend a considerable portion of its life at a cruise condition. By contrast, the X-35 test program would stress the propulsion system constantly while executing test points. This was especially true for the lift system, because STOVL operations

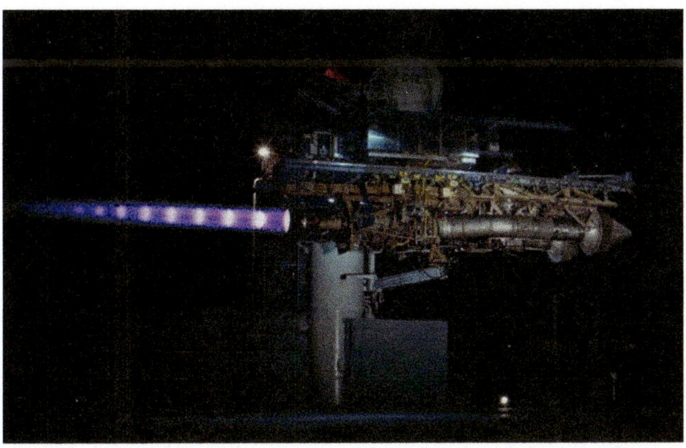

Fig. 11 STOVL propulsion system in full afterburner on test stand.

during the demonstrator flight test program were expected to approach the total design life of the lift system for an operational aircraft. Whereas the engine would have to demonstrate only a portion of a full-life operational system, the lift system was required to demonstrate a nearly full design life capability.

The objectives of an AMT are to demonstrate readiness for flight and highlight areas requiring special monitoring or components that might require replacement during the flight test program. The AMT of the X-35 propulsion system went smoothly, with only minor issues identified, such as fuel pump leaks partway through the program. No major issues were found with the propulsion system, clearing the way to execute the planned flight test program.

2. IFPC DEVELOPMENT TESTING

The IFPC system used an extensive process of testing and simulating all the various STOVL and CTOL systems to ensure that the X-35 had a very robust IFPC system. This included performing component testing for every actuation system, to include:

- Functional and environmental testing
- Integrated propulsion testing
- Final vehicle integration testing

Integrated propulsion testing included fault detection and accommodation testing, and the air vehicle and IFPC integrated verification of software- and hardware-represented models in vehicle integration laboratories. This approach to testing components, integrated systems, and vehicle integrated modeling of the IFPC system was very successful in mitigating X-35 flight test program risk. Simulation of IFPC failures established flight test procedures and allowed pilots to practice emergency procedures for these failures. The Lockheed Martin/Pratt & Whitney/BAE Systems/Rolls-Royce/Allison Advanced Development Co. (now LibertyWorks) team understood all failures and flight characteristics. This was as a result of this process, through which we acquired both a CTOL and STOVL flight clearance to proceed to first flights of all three variants.

3. AIR INDUCTION SYSTEM DEVELOPMENT TESTING

The development of the AIS for the engine and LiftFan occurred in parallel with that of the propulsion system. It proceeded from modeling and subscale testing (both statically and in wind tunnels) to the construction of the demonstrator aircraft. The X-35 incorporated the Lockheed Martin–developed and –patented DSI [6]. As Fig. 12 shows, the development of the DSI started in 1995 for the X-35. The DSI provides high aeroperformance across the flight envelope in a fixed-geometry design without using a boundary layer diverter, bleed system, or bypass system. Fixed-geometry, diverterless apertures are integrated on each side of

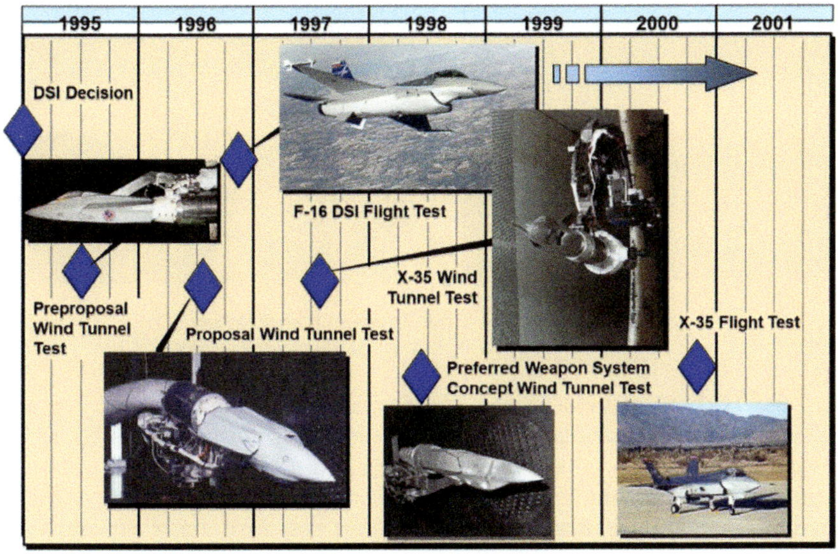

Fig. 12 DSI development timeline for the X-35.

the forward fuselage. The integration of the bump and the forward-swept aperture lips combine to force the boundary layer air out the aft close-off point of the aperture along the fuselage sidewalls.

Testing of the propulsion system with a representative AIS was not possible until the installation of the propulsion system in the X-35. All ground and altitude testing of the propulsion system was accomplished with ideal inlets (i.e., bell mouths). By contrast, distortion screens based on inlet modeling were used to demonstrate propulsion system operability. Designing the AIS for the engine and LiftFan began with CFD computer modeling, followed by testing wind tunnel models.

Both X-35 demonstrators had to be STOVL capable, requiring a single AIS design; however, this still required the development of three inlets as opposed to a single AIS for a conventional aircraft. These were the primary inlet in CTOL mode, primary inlet with the auxiliary inlet open, and the LiftFan inlet. The flight regime also required operation from static conditions all the way to supersonic speeds, with the ability to convert into/from CTOL and STOVL inlet configurations. Subscale testing through the speed regimes was conducted with both CTOL and STOVL inlets, as Fig. 13 shows. Subscale testing was also conducted to quantify the effects of HGI during a vertical landing.

A lesson learned while developing the LiftFan resulted in transient limitations on STOVL operations during the X-35 flight test program and a design change for SDD. Flow into the LiftFan exhibited higher than acceptable distortion at certain

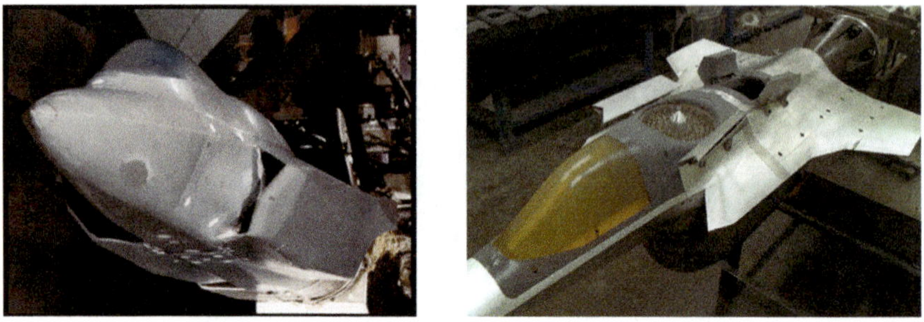

Fig. 13 CTOL and STOVL AIS wind tunnel testing.

airspeed/LiftFan rotational speed combinations. The result was higher and unacceptable aeromechanic stress levels in the blades of the LiftFan due to the distortion pattern shown in Fig. 14. This level of distortion limited the operating time at those conditions and thus required constant and real-time control room monitoring. Although manageable for a flight test program, a solution was required for an operational aircraft.

BAE Systems characterized HGI during a vertical landing at a unique test facility in Wharton, United Kingdom. The company used a subscale model of the X-35, complete with flowing inlets and temperature instrumentation, to perform dynamic testing. It did so by translating the model vertically into a ground plane, thereby simulating a vertical landing. More than 10,000 test runs were accomplished, investigating descent rates, aircraft attitude, ground slope, nozzle vector angles, and ground speed/headwinds. Analyses of the results provided a set of conditions under which the X-35 could land without risking HGI. Hover flow field IGE wind tunnel testing using oil flow indicated that, in a calm hover IGE, the jet of air from the LiftFan created a wall. This wall prevented

Fig. 14 X-35 LiftFan inlet distortion pattern due to side-hinged LiftFan inlet doors.

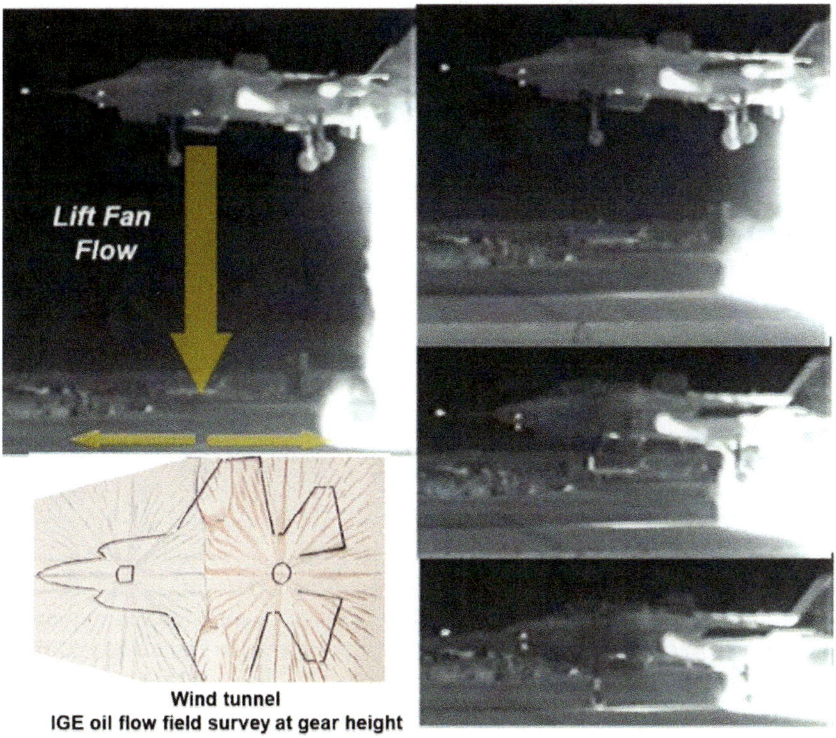

Fig. 15 Comparison of X-35B IGE wind tunnel oil flow and flight test demonstration.

the main engine exhaust from moving forward. During X-35B vertical landings, a comparison of the IGE oil flow results with flight test results, using night vision goggles (Fig. 15), shows that the LiftFan prevents main engine hot gases from moving forward as the X-35B approaches the ground during a vertical landing [7].

4. AIRCRAFT PROPULSION GROUND AND FLIGHT TESTING

The final verification of the X-35 propulsion system was accomplished with the ground testing of the installed propulsion system in the X-35A, X-35B, and X-35C aircraft. X-35A and X-35C propulsion testing involved restrained full-power engine runs comparable to those of legacy systems. These ensured that the aircraft and subsystems were ready for flight. These aircraft then completed very successful flight test programs, proving that the propulsion systems met X-35 requirements for the Air Force and Navy. X-35B STOVL propulsion system ground testing was considerably more involved and was the primary focus in preparation for the X-35B flight test program. After completing ground tests, the X-35B progressed to STOVL press-ups on the hover pit to demonstrate vertical capability and control. The X-35B then moved to Edwards Air Force Base (EAFB), where the flight test program exercised the STOVL propulsion system in flight. It then progressed from wingborne flight to semi-jetborne flight, to jetborne flight, and culminated in a vertical landing.

X-35A

As the first of the variants to enter flight test, the X-35A was charged with accomplishing all the basic SoF testing expected of any new aircraft configuration. Among other things, these tests included:

- Aircraft handling qualities checks
- Subsystem operations checks
- Takeoff and landing characteristics
- Engine throttle transients
- Turns at elevated *gs*

The flight test program was highly successful at meeting or exceeding all flight test objectives, clearing the way for the first flights of the X-35B and X-35C. The first flight of the X-35A occurred on 24 October 2000. A key milestone accomplished near the end of the flight test period was the first supersonic flight of the X-35 on 21 November 2000, less than one month from first flight. In order to increase redundancy, the X-35 was designed for continuous operation of the APU in flight, except for the supersonic portion of the flight envelope. This required demonstrating the ability to shut down and restart the APU in flight before performing a supersonic flight. Flight test missions were planned and executed that exercised the APU in flight. These determined which flight conditions would best support an APU restart.

The X-35A test program encompassed 27 flights, totaling 27.4 flight hours, with six pilots and an unprecedented flight rate of approximately seven flights per week.

X-35B

After the successful X-35A flight test program, Aircraft 301 was converted into an X-35B. The forward fuel tank was removed, and the LiftFan and roll post system were installed (Fig. 16). The 3BSD was part of the basic engine assembly, so no special installation was required. The process of converting Aircraft 301 from the CTOL X-35A configuration into the STOVL X-35B configuration spanned approximately four months.

Ground testing of the X-35B began with subsystem checks to verify the installation of the STOVL propulsion system. All engine-mounted STOVL components, such as the 3BSD, were driven by engine fueldraulics and could be tested during an engine ground run; however, the remainder of the STOVL

Fig. 16 First LiftFan installation in an X-35B.

propulsion effectors was powered by aircraft hydraulics. The variable geometry in the LiftFan, including the vectoring nozzle, clutch, and roll post system, was driven by the aircraft's hydraulic systems.

A purpose-built test facility was necessary to conduct all the necessary checkouts on the X-35B. To this end, Lockheed Martin constructed a hover pit at its Palmdale, California, facility specifically for testing the X-35. An additional hover pit was built in Fort Worth, Texas, for STOVL checkouts of the F-35B. The purpose of the hover pit was to allow restrained testing of the X-35B in STOVL mode. During testing, the exhaust from the STOVL system was directed away from the aircraft, aerodynamically simulating a free-air hover while avoiding HGI and exhaust gas impingement on the aircraft. This permitted extended test runs that were not representative of an operational environment.

Two systems were used for restraining the X-35B on the hover pit to support different test objectives. A force-and-moment hard-mounted restraint system was used to measure forces on the aircraft in STOVL mode, confirming performance predictions and control power. This was the first time the STOVL propulsion system was operated with representative inlets and other installation effects, as Fig. 17 shows.

All previous performance predictions had analytically combined STOVL propulsion system testing results with inlet performance predictions from CFD and subscale tests. Results confirmed predictions, with performance being as good as or better than predicted. An interesting aspect of the testing was the first demonstration of full thrust split range. Previous system testing had included ideal inlets, so the testing in the X-35 demonstrated the full thrust split range achievable with representative inlets. The testing highlighted the need for improvements in

Fig. 17 X-35B force-and-moment ground testing using hard support.

instrumentation, control sensor placement, and seals around STOVL door hinges that were implemented prior to flight. Control room experience for the flight test program was also gained by using a full control room during all ground testing. This reduced test risk while training personnel.

The second restraint system used on the hover pit was a soft-mounted support to complete STOVL mode structural coupling testing. Structural coupling tests are routine with today's FBW flight control systems. They allow tuning of the CLAW to avoid structural modes that could be driven by the flight control system. The soft system on the nose gear, shown in Fig. 18, was composed of air springs. These isolated the vehicle from the rigid ground while exercising the STOVL propulsion system with the CLAW loops closed and monitoring the aircraft structure with accelerometers. Results of the tests were used to adjust the

Fig. 18 X-35B nose gear soft support.

Fig. 19 X-35B structural coupling test soft support system.

STOVL CLAW to avoid any interactions between the flight control system and the aircraft's structure. Figure 19 shows the X-35B over the hover pit in preparation for engine-on STOVL structural coupling testing.

Completion of the STOVL ground tests on the X-35B cleared the way for flight operations. Lockheed Martin's plans for clearing the X-35B for STOVL operations called for a build-down approach from conventional flight to semi-jetborne flight, to jetborne operations. Before this could be accomplished, we needed to verify that the STOVL propulsion system could support the aircraft in a jetborne condition with sufficient control power. This verification would be accomplished in a series of short hops (press-ups) from the hover pit. The first hops would demonstrate performance necessary to support the aircraft, whereas later press-ups would investigate control authority in hover.

The first hop of the X-35B (Fig. 20) was piloted by Simon Hargreaves on 24 June 2000 and was a complete success. Simon found the aircraft rising higher than intended due to increasing thrust of the propulsion system; however, he did not want to move the throttle abruptly on his first flight. As such, he slowly retarded throttle as the aircraft climbed until the ascent was arrested and the aircraft descended back onto the pit. Required performance had been demonstrated, with the IFPC providing a stable and controllable platform. Posttest analyses showed that thrust had increased as engine thermals stabilized, requiring more throttle control than anticipated to retard the climb.

Press-up testing on the hover pit continued with accomplishing two short hops the next day and a stable hover lasting longer than two minutes on the third day of press-ups. Testing continued for a week with three test pilots accomplishing a total of 14 press-ups. The final press-up test included control system checks and was the longest hover accomplished, lasting more than three and a half minutes.

Completing the press-ups from the hover pit cleared the X-35B for the move to EAFB and the continuation of the flight test program. Over the next two and a half weeks, the X-35B exercised the STOVL propulsion system in flight, steadily

Fig. 20 Two views of X-35B's first press-up from hover pit.

progressing through the test program to lower and lower airspeeds. Testing culminated in a vertical landing on the EAFB hover pad.

The X-35B flight test program was extremely successful, conducting 39 flights totaling 21.5 flight hours with four pilots in an intensive six-week test program. With its unique propulsion system, the X-35B accomplished:

- 22 hovers
- 18 STOs
- 27 vertical landings
- 21 airborne conversions
- 6 supersonic flights

The aircraft also executed Mission X twice, reaching the milestones of STO, supersonic dash, and vertical landing in a single flight for the first time in history.

c. X-35C

Aircraft 300 commenced its flight test program configured as a Navy variant in late 2000, the same time when the X-35A completed its flight test program. The X-35C accomplished 73 flights totaling 58 flight hours with eight pilots, including the first transcontinental ferry by an X airplane. After initial flights at EAFB, the X-35C continued its flight test program at a Naval Air Station Patuxent River test facility. There it accomplished 250 field carrier landing practices to demonstrate carrier approach handling qualities. The propulsion system provided the necessary throttle response and performance required to meet all criteria with the flight CLAW. It performed well enough to prompt a pilot to state, "We are ready to go to the boat with this airplane."

The three variants, using two airframes and two propulsion systems (CTOL/CV and STOVL), successfully met all objectives. No flight control or engine software control changes were required, and the propulsion systems worked flawlessly through a very rigorous flight schedule. The flight test program concluded on 31 July 2001. The X-35's flight test program successes were key to Lockheed Martin being awarded the JSF F-35 SDD program on 26 October 2001. The success of the propulsion system during the flight test program in particular was instrumental in this.

V. DETAILED DESIGN, TEST, AND VERIFICATION FOR OPERATIONAL DEPLOYMENT: SDD/F-35

The X-35 flight test program was a success, proving the functionality of the propulsion system and meeting the demonstration requirements of STOVL operation and a carrier approach. The Lockheed Martin JSF team was then tasked with combining everything learned from the X-35 demonstrators with the operational aircraft preliminary design and producing an operational warfighting machine.

Christened the F-35, the aircraft incorporated the very successful propulsion system design into an airframe design that was lethal, stealthy, and survivable. Those requirements alone were difficult engineering challenges for the Lockheed Martin/BAE Systems/Northrop Grumman/Pratt & Whitney/Rolls-Royce F-35 team. Adding to them, the team had to make everything work with a propulsion system design that met Air Force, Marine Corps, and Navy tri-variant requirements for performance, reliability, maintainability, supportability, lethality, and survivability. The production propulsion system was designated the F135, departing from the demonstrator JSF SE611 nomenclature.

Beginning with the X-35 propulsion system, major changes were needed to address the CDP lessons learned and produce an operationally suitable full-life propulsion system. The most significant changes were made to the LiftFan and its integration into the air vehicle. An inlet redesign was necessary to address the aeromechanical problems encountered during the flight test program, and a complete LiftFan nozzle redesign made the nozzle part of the airframe structure. This reduced weight and provided area control that could be used to enhance operability. Other changes to the propulsion system and its integration are discussed in the following sections.

A. PROPULSION SYSTEM DESIGN REQUIREMENTS

1. PERFORMANCE

The X-35 flight test program demonstrated slightly better STOVL performance than had been expected due to the performance of the AIS and the propulsion system itself. This highlighted an opportunity to increase the up-and-away performance of the air vehicle and the vertical landing capability of the STOVL aircraft. Slightly increasing the size of the primary inlet of the operational aircraft took advantage of a new engine fan design being implemented for producibility. This allowed higher airflows with reduced distortion and promised even better inlet recoveries than had been experienced with the X-35.

By far the biggest challenge for the operational propulsion system was the requirement for full life. The STOVL system life requirements were also a challenge for the hot parts due to the more strenuous duty cycle of the STOVL aircraft. The goal was to use common parts across all variants and minimize any redesigns due to STOVL requirements. A key element in meeting full life was the specialized FADEC control system employed. Conventional engine control systems run to schedules, such as fan speed. Accordingly, as an engine deteriorates, thrust actually increases due to increasing turbine temperatures. With the new F135 control, engine performance remains constant throughout its life. Turbine temperatures are controlled to meet thrust targets rather than holding the engine on a control schedule as it deteriorates.

The major challenges for the engine, as might be expected, were in the turbine section. The high-pressure turbine was refined to better utilize cooling air, improving both life and efficiency over the JSF SE611. The LPT was a major challenge as well, with the JSF SE611 featuring a design that could not meet full-life requirements. New blade damping techniques were needed to meet operational life requirements.

Some of the changes to the propulsion system were refinements based on experience gained from the CDP. Variable geometry in the turbine exhaust case included for STOVL operation was found to be unnecessary and was removed from all variants. The aggregate vector angle of the 3BSD was found to be more than required. This allowed a reduction in bearing angles that produced benefits in overall tailpipe length, weight, and ground clearance. The offtake valves included as part of the roll post system were deleted because they did not add redundancy and were heavy due to their stiffness requirements. These types of changes were instrumental in maintaining performance targets while reducing weight and cost.

Operational design changes for the air system also drove changes to the propulsion system. The F-35 does not have an AMAD, which requires aircraft generators, hydraulic pumps, and lubrication pumps to be mounted directly on the engine gearbox. Heat rejection, a major problem for 5th-Generation aircraft, required a thermally efficient fuel pumping system on the engine and the ability to handle fuel temperatures as high as possible. Heat exchangers were mounted in the engine gas path to cool bleed air from the engine and provide a heat sink for the aircraft Power and Thermal Management System (PTMS). At the same time, every effort was made to improve maintainability by tailoring engine externals to line up with airframe access points as much as possible. Propulsion system access was more limited, relative to the X-35. This was due to the presence of weapons bays and the need to minimize the number of access panels in the aircraft outer surface.

2. AIR INDUCTION SYSTEM

The redesign of the primary inlet for higher airflows was only a portion of the inlet redesign. The inlet was also shortened as a weight-saving measure. It entailed moving the aperture aft and changing the cowl from the four-edge design on the X-35 to a simpler three-edge design for the F-35. The redesigned inlet aperture also provided improved inlet performance at high angle of attack. The aft portion of the inlet diffuser was made variant-unique, with the CTOL and CV variants sharing one design and the STOVL variant's auxiliary inlet having its own configuration. The STOVL inlet also featured a tunnel for the LiftFan driveshaft, extending forward from the engine face to the inlet bifurcation.

Common requirements across all variants included a source of nacelle ventilation air, the addition of an ice detector, and the introduction of an inlet debris monitoring system. The location of the ice detector was to be common across all variants, if possible, but was subject to the results of ice accretion tests that would be performed. The inlet debris monitoring system was a developmental prognostics and health monitoring system that promised to reduce or eliminate the need for inlet and engine face inspections between flights.

The refinement of the auxiliary inlet design was an evolutionary step from the system, which had performed well on the X-35B (Fig. 21). Recovery improved significantly, and the doors were lighter than the original X-35B doors. The combined thrust increase with the recovery improvement and the lighter weight added 500 lb of STOVL thrust over the X-35B configuration. Additional F-35B changes focused on lessons learned from the X-35, such as the following improved features:

- Seals on the door hinges
- Locations of inlet sensors on the LiftFan
- Smooth contours with any air vehicle outer mold line changes

The LiftFan inlet was the biggest challenge facing the Lockheed Martin internal aerodynamics team. The X-35B inlet had performed adequately, but the uneven flow field had unintended consequences on LiftFan aeromechanics.

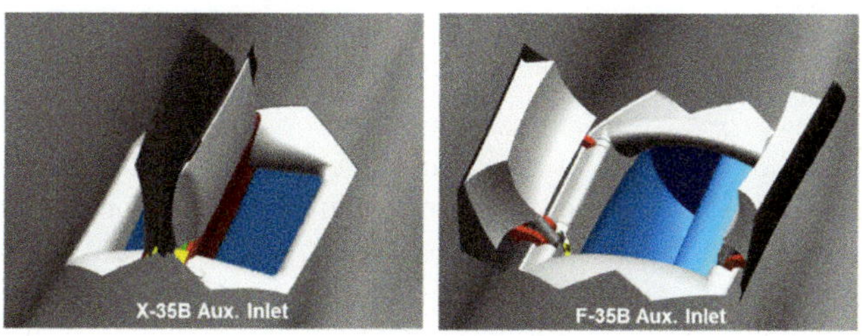

Fig. 21 F-35B improved auxiliary inlet.

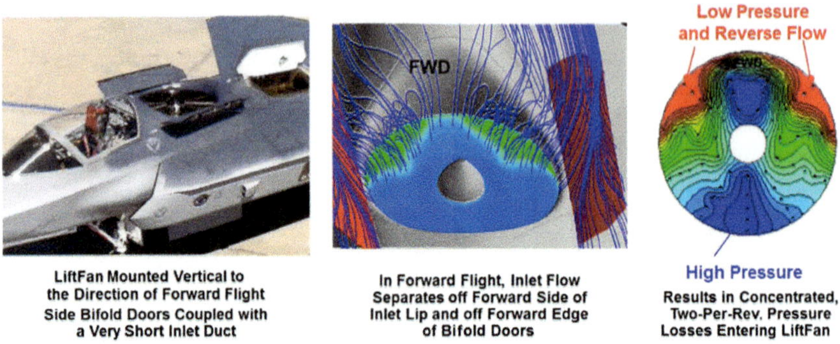

Fig. 22 X-35B LiftFan flow field causing LiftFan aeromechanic issues during forward flight.

Figure 22 shows the X-35B LiftFan inlet configuration, CFD streamlines into the inlet, and the resultant adverse pressure flow field that created the aeromechanics issue. The existence of transient-only zones was not acceptable for an operational aircraft.

The stream tube was a special inlet test fixture designed from CFD streamlines that replicated the LiftFan flow field for a given forward velocity while testing at static conditions. This test technique was developed during SDD and was validated through comparisons with X-35 flight test data and CFD. The technique proved invaluable and was used on subscale tests, full-scale LiftFan tests at Rolls-Royce, and Pratt & Whitney test stands in West Palm Beach. Figure 23 shows flight test total pressure (recovery) and a comparison of CFD and a test using a LiftFan stream tube to reproduce the adverse flow field [8]. The stream tube test technique proved its value by supporting the development of a LiftFan inlet that had acceptable levels of distortion [8]. With the technique, the inlet also eliminated the transient-only zones experienced during the X-35 flight test program [8].

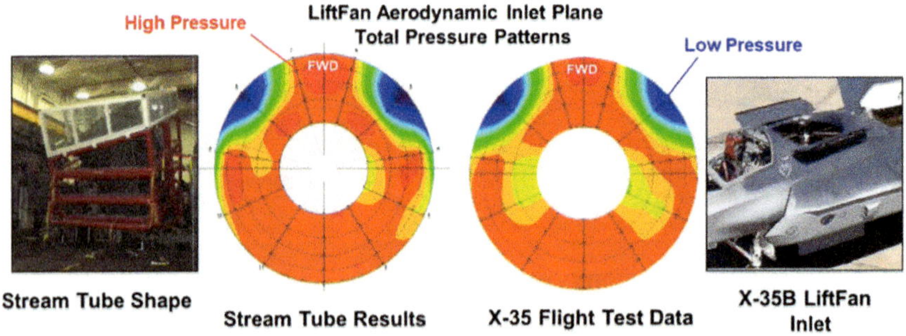

Fig. 23 X-35B LiftFan aeromechanic stream tube test rig results.

The level of inlet distortion was known and easily understood. The challenge was how to improve the inlet such that the LiftFan would tolerate the resulting flow field while having acceptable performance. The aft hinge configuration was subsequently chosen for future development, and work began on characterizing the resultant flow field. Distortion estimates were generated and provided to Rolls-Royce to assess the aeromechanics effects. It was also necessary to redefine the interface between the LiftFan and the inlet stream to more accurately model the performance of the system. Once a preferred configuration had been defined, testing was planned to verify that the system worked as predicted [8]. The aft hinge door had an inlet bell mouth lip modification and LiftFan case flare. It was the only concept to meet the distortion and high-cycle fatigue goals while maintaining an acceptable flight envelope. The F-35 propulsion team used CFD, subscale inlet testing, and stream tube testing to solve the X-35B LiftFan aeromechanics issue [8].

3. F-35 INTEGRATED FLIGHT PROPULSION CONTROL DEVELOPMENT AND FAILURE IMMUNITY

The F-35 presents one of the most complex integrated control problems for aircraft and propulsion to date through its tight interaction of numerous subsystems. These interactions span the obvious flight controls and propulsion systems and, as Fig. 24 shows, extend into utilities and subsystems, power and thermal management, and fuel systems. Utilities and subsystems include STOVL door control, hydraulics, fuel, electrical power systems, landing gear, weapons bay

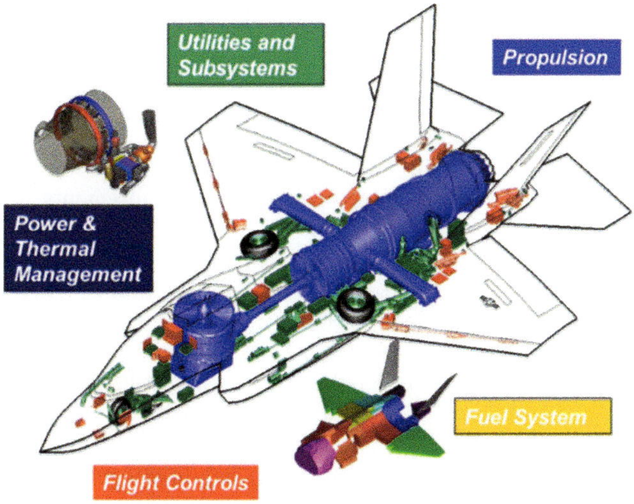

Fig. 24 F-35 integrated systems layout.

door systems, and pilot ejection systems. Power and thermal management includes an Integrated Power Package (IPP) and thermal cooling.

The synergy between aircraft CLAW and engine software was drastically improved over the X-35. IFPC requirements were inherent from the first day of the program, in contrast with the X-35 process of adapting IFPC requirements to existing propulsion control design. The IFPC CLAW architecture is a complex series of multivariable loop closures involving [1]:

- Six-degrees-of-freedom flight controls
- Five-degrees-of-freedom propulsion control
- Fuel system control
- Lift system door control

The Lockheed Martin nonlinear dynamic inversion flight control methodology has evolved from the early X-35 implementation to the present-day F-35 system. The architecture enables a single CLAW structure supporting all three aircraft variants. The STOVL design is the most complex, with the direct control of all six degrees of freedom. Higher levels of augmentation have been incorporated to reduce pilot workload and increase safety. The primary goal of the CLAW modes was to make STOVL takeoffs and landings as routine as they are for conventional aircraft. This also benefits fleet usage by reducing the training burden for STOVL mode-unique portions of the syllabus [1].

The F-35 CLAW improved from the X-35 in that more modes of augmentation were incorporated to ease pilot workload. This augmentation drove the need for the F135 engine control to drastically improve its capability with higher precision control and faster engine dynamics. These changes were necessary for the challenging STOVL and Navy requirements. From the beginning, the JSF program set out to develop performance-based specifications in which only the performance capability and desired operational mission requirements were put on contract. The specification levied against the aircraft contractor required Level 1 flying qualities for all three variants. This, in turn, caused the flow down of propulsion system dynamic performance requirements to constitute the most technically challenging requirement of the entire program. It required the STOVL propulsion system to deliver commanded total thrust (sum of all posts) to within 0.5% accuracy, with a bandwidth requirement that was up to several times faster than any legacy designs.

The most technically challenging of all requirements was either the thrust-to-pitch and pitch-to-thrust decoupling or the ability to stay wings level as the aircraft heaved in the vertical axis. Additionally, the aircraft could not have uncommanded movement in the pitch axis while performing a roll maneuver. To achieve these requirements, a quantum leap in integrated flight and propulsion control technology had to be realized. During the X-35B concept demonstrator phase, the operability of the integrated STOVL propulsion system and the variant's flying qualities from hover through supersonic flight were demonstrated. Several advanced flight and engine control concepts were being explored. Among them was a government-sponsored research program on the United Kingdom's Vectored thrust Aircraft Advanced Control (VAAC) Harrier. Many revolutionary areas of STOVL advanced control were explored, from pilot inceptor types to aircraft response mapping to those inceptors. This was demonstrated in large-amplitude, motion-based simulators and flight demonstrations, both ashore and during embarked operations at sea. Studies from VAAC changed the pilot control concept used in the X-35 to the F-35's unified flight control strategy, discussed later in this paper. This strategy was made possible, in part, with the successful development and deployment of Pratt & Whitney's F135 Advanced Multi-Variable Control (AMVC) design [9].

The AMVC is a multivariable engine control using a unique control technology that allows the engine software to decouple the highly coupled engine system composed of a main engine connected to the LiftFan via a driveshaft. The resultant aircraft performance was wings-level hover and vertical landing Level 1 flying qualities. The AMVC team met the immense challenge to develop and certify the control architecture concurrent with the JSF F135 development program. It did not have the benefit of a dedicated technology development program, however. Notwithstanding, the team successfully matured the state-of-the-art control system as part of the F135 development program, culminating in the operation of the F-35B in powered lift mode.

Pratt & Whitney's AMVC uncouples complex, cross-coupled, nonlinear propulsion system dynamics and makes them appear similar to a set of uncoupled virtual flight effectors. In contrast to normal propulsion system control, this one controls five degrees of freedom of thrust at four nozzles. These responses must be linear and consistent with relatively high bandwidths and with relatively little cross-coupling among the degrees of freedom. The virtual effector command response bandwidth required is greater than the propulsion feedback loop bandwidth and approaches the propulsion actuator bandwidths. Also, similar to most engine controls, the CLAW must run the engine close to and onto numerous limits without coupling into the virtual effector responses.

The failure management strategy for a single-engine FBW STOVL 5th-Generation aircraft has extensively evolved from its early implementation in the X-35B aircraft to the current F-35B system [5]. Based on modern triple- and quadruple-redundant FBW flight controls, the program requirements drove redundant channel control and feedback for propulsion effectors. In this way, they ensured a fail-operational/fail-degrade capability in the event of a failure of primary flight-critical flight control systems in STOVL flight. During STOVL mode, the propulsion system becomes the primary flight control system for control and, as such, it drove this redundancy into the design.

Achieving a fail-operational/fail-safe philosophy for dual mechanical failures for all the propulsion effectors was deemed impractical for a weight-sensitive STOVL aircraft. The performance-based requirements for the F-35B IFPC

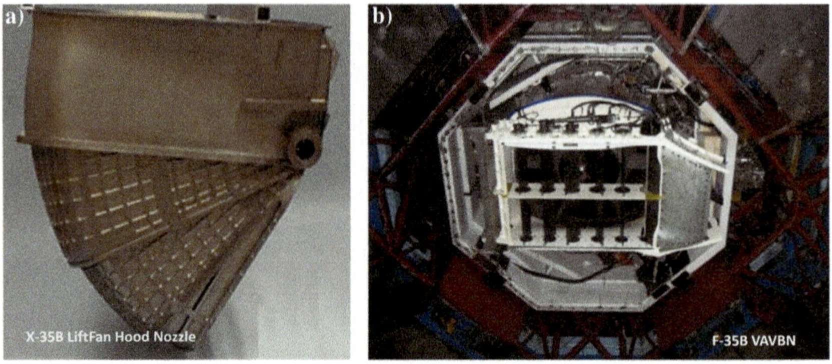

Fig. 25 (a) X-35B LiftFan hood nozzle and (b) F-35B VAVBN.

failure management drove a philosophy into the design that was different from that of the X-35B. This is not to say, however, that the aircraft could not achieve fail operational/fail safe for a wide range of failures. Nor was it the case that the aircraft could achieve fail operational in *all* cases for dual electrical failures. Rather, the basic difference in philosophy focused on how faults affected flying qualities and their probability of occurrence. Design success was achieved by adhering to a basic tenet giving a carrier pilot the utmost confidence of aircraft recovery under a wide range of failures and meteorological conditions. This tenet became known as "failure immunity in blue water ops." As a result, the F-35B IFPC system and its fault detection and accommodation strategy vastly improved over the X-35B and achieved a lighter-weight solution. IFPC redundancy has been designed into all the propulsion effectors and meets the failure immunity requirements. The FADEC and aircraft CLAW failure accommodation allow the pilot to achieve a safe recovery to a ship in blue water operations. During this, the pilot has at least Level 2 flying qualities for any single or dual combination of failures that have a greater than 10^{-7} probability of occurrence.

Many hardware redundancy improvements were incorporated into all engine and LiftFan systems as a result of the new failure immunity requirements. Specifically, the X-35B LiftFan hood nozzle was removed and replaced with the VAVBN. This greatly improved the capability to vector in the event of a failure. The VAVBN also provided nearly 40% variation in exit area, allowing for additional LiftFan vector and thrust control in the event of failures. The VAVBN system also represented a 35% reduction in weight over the X-35 LiftFan hood system [10]. Comparative pictures of both systems are shown in Figs. 25a and 25b.

The LiftFan clutch, gearbox, driveshaft, and lube system were vastly improved over those of the X-35. The F-35 clutch system used aircraft dual hydraulics to actuate and lock the clutch with improved control and feedback systems. A new lube pump was driven off the driveshaft via a gear drive that only supplied oil to the gearbox and clutch bearings. It did not have the task of engaging the clutch like the X-35B lube system. The driveshaft was an all-new design that improved producibility while maintaining an angular misalignment capability. It also addressed compression/tension tolerances between the engine and LiftFan commensurate with an operational F-35 system. The number of clutch plates was reduced, allowing thicker plates that increased the number of engagements before replacement.

To avoid using twice as many actuators to support a dual hydraulic clamp and lock system, an ingenious system was implemented to detect leaks in the clutch system. If the system detected a leak, it would alert the pilot and allow him or her to preserve the remaining hydraulic system to ensure a successful landing in STOVL mode if required. In 2015, the F-35B clutch system was incurring an inordinate amount of clutch failures attributed to nuisance issues and a nonrobust design. The Lockheed Martin/Pratt & Whitney/Rolls-Royce IFPC team redesigned the clutch, engine, and hydraulic redundancy algorithms. The team's modifications allowed the removal of certain clutch components while providing the necessary redundancy tolerance to meet failure immunity requirements.

F-35B conversion and aircraft flight control during conversion was also dramatically improved. In the X-35, the conversion was initiated when the pilot moved the Thrust Vector Lever (TVL) out of the detent, which initiated the conversion process. The TVL was a legacy device from AV-8B to control the vector angle of the nozzles while the pilot controlled height using the throttle. The F-35B had an improved pilot inceptor design, named Unified Control Strategy (Fig. 26). It used traditional stick, called the Right-Hand Inceptor (RHI), yaw pedals, and throttle controls, called the Left-Hand Inceptor (LHI), to reduce the amount of pilot workload significantly. The strategy was to decouple the aircraft control axis and provide the pilot with the most desired response for his or her inputs. It also simplified the cockpit by removing the third TVL inceptor, providing the same inceptor functionality whether the aircraft was stationary or flying at 500 kt.

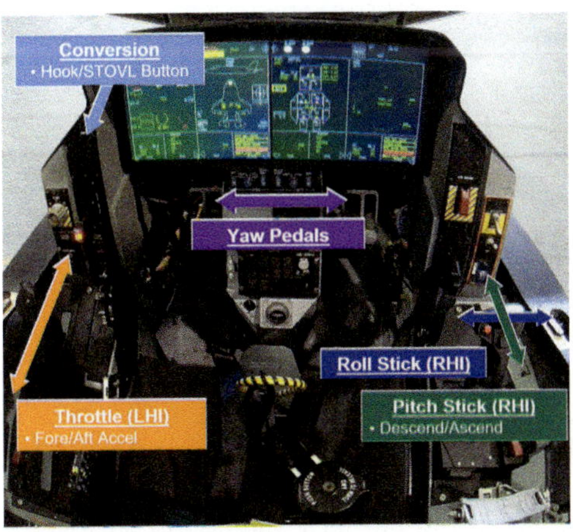

Fig. 26 F-35B Unified Control Strategy inceptor configuration.

Conversion was now handled by the identical method with which the F-35C (CV) would extend the carrier hook, a simple button in the upper-left region of the cockpit. The difference with the F-35B, however, was that it was labeled HOOK/STOVL. Conversion in the F-35B was designed to be as simple as the extension of landing gear or the carrier arresting hook. The pilot simply flies the aircraft below 250 kt and pushes the HOOK/STOVL conversion button, which initiates a complex transformation of hardware and software in the aircraft. The conversion is done completely automatically. The aircraft CLAW initiate commands to the aircraft to open doors, as shown in Fig. 27. Commands are also sent to the FADEC to engage the LiftFan and prepare the engine to drive the lift system. Commands then move the 3BSD and VAVBN vectors into position for STOVL flight—all while maintaining the pilot's desired flight path and speed.

The improved redundancy architecture and associated Lockheed Martin/Rolls-Royce/Pratt & Whitney fault detection logic provided acceptable fault tolerance for all failures meeting the failure immunity requirement. The fault detection logic was driven by the new failure immunity requirements to meet affordability goals. Improvements in all these STOVL effectors improved redundancy and capability to allow STOVL to be as simple as CTOL.

The success of this philosophy has been observed in flight test and fleet pilot confidence. If a failure were to occur when a pilot entered the slow STOVL regime, the aircraft would accommodate it. The aircraft's response would keep a transient from forcing a pilot to make the decision to egress (via ejection) the aircraft.

4. AIR VEHICLE INTEGRATION

The integration of the F135 into the F-35 was a complete rework, compared to the X-35, as Fig. 28 shows [11]. The entire suite of subsystems was changed from off-the-shelf systems pulled from legacy platforms to an innovative, highly integrated set of subsystems. One immediately obvious change is the lack of an air vehicle AMAD. Eliminating the AMAD reduced the overall length of the aircraft

Fig. 27 F-35B lift system doors.

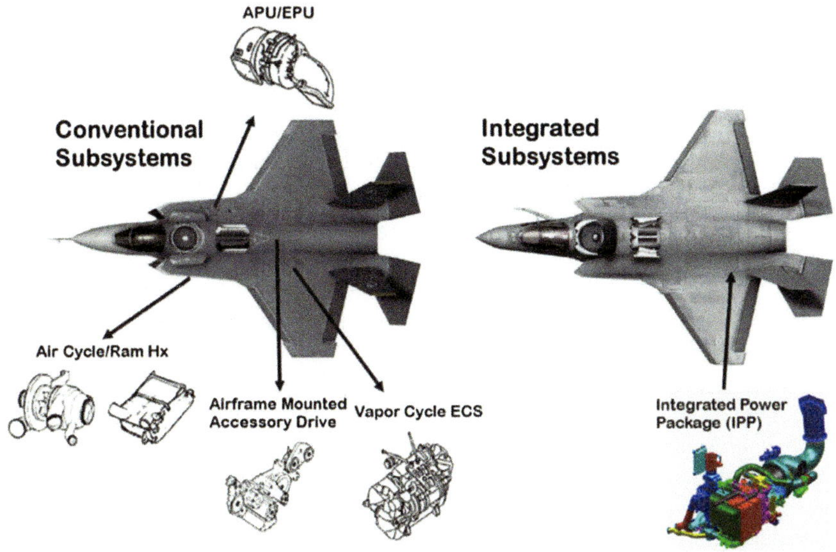

Fig. 28 Comparison of X-35 federated components with F-35 integrated PTMS.

by providing more flexibility in the placement of the weapons bays. In the case of the STOVL variant, it created more room for the LiftFan nozzle. Without an AMAD, this required all mechanically driven aircraft systems to be mounted on the engine gearbox. One of the systems that made this approach possible was the use of electrical power for flight controls rather than hydraulics.

Electrical power for flight controls reduced hydraulic demands to the level of a utility system. An Engine Starter/Generator (ESG) would be mounted on the gearbox, serving as both the engine starter and the prime generator for the entire aircraft. The internal architecture of the ESG made it functionally a triplex generator, allowing redundancy on a single gearbox pad. The electrical system architecture was configured to shed a load in the event of a failure, prioritizing power based on the criticality of the need. In the event of a total ESG failure, backup power would be provided by batteries and the IPP, another highly integrated subsystem providing multiple functions. The integration trade-off was that multiple power feeders would have to be connected to the generator once the engine was installed (generator mounts on engine prior to engine install). In addition, the lubrication system would be partially mounted on the gearbox (pump), with the remainder on the aircraft (heat exchanger and oil reservoir).

The use of electrically powered flight control actuators avoids much of the heat that would be generated by conventional hydraulic systems. Hydraulic systems operate continuously at full pressure, whereas electrically powered actuators operate on an on-demand basis. This reduction was a key enabler in meeting thermal requirements, as discussed later in this section.

The PTMS combined the functions of an APU and an ECS. The PTMS provides conditioned air for cockpit and equipment cooling, refrigerated liquid for equipment cooling, and electrical power for ground maintenance, main engine starting, and in-flight emergencies. The cooling function is performed by a closed-loop air refrigeration cycle powered by expanding engine bleed air across a power turbine. Waste heat from the refrigeration cycle is transferred to fuel and to engine fan bypass air. A polyalphaolephin liquid loop and two heat exchangers are used to transport heat between the air cycle and the fuel. Heat is transferred to the engine fan air through air-to-air HXs mounted in the engine Fan Duct HXs (FDHXs).

APU functions were facilitated by reconfiguring valves so that the PTMS turbomachine could operate like a traditional APU. In this way, it provided power for self-start, ground maintenance, main engine start, and in-flight emergency power. Effectively combining two functions into one system, the PTMS provided all necessary functionality in a much smaller package. The use of FDHXs required more ducts to be connected to the engine, but it reduced the need for ram air inlets and heat exchangers.

As previously mentioned, F-35 hydraulic heat loads were greatly reduced by the use of electric flight control actuators; however, hydraulic waste heat is just one of several heat sources transferred to fuel. The PTMS, electrical generator, engine lubrication system, and engine fuel system also dissipate waste heat to fuel. The mechanically driven, continuously running engine fuel pumps are a major source of engine waste heat. In addition to boosting fuel pressure to engine combustor pressure, the pumps also provide the power for engine nozzle and other actuation. Minimizing the heat generated by the engine fuel pumping system reduced the total cooling load on the aircraft. It was accomplished by using a thermally efficient fuel pump for the main engine while optimizing pump configuration when not required. When coupled with the electric flight control actuators, these changes made dramatic differences in the total heat load and allowed the air vehicle to meet its mission thermal requirements.

One major source of heat that could not be completely mitigated was the heat generated by the LiftFan during STOVL operations. When engaged, the gearbox generated large quantities of heat that were removed by the lubrication system. The most challenging condition was at the end of a mission, when the fuel heat sink was smallest (low fuel warmed during the mission) and the quantity of heat generated could not be entirely removed. Consequently, when a vertical landing was to be performed at the end of a mission, the aircraft would operate at a deficit, heating up during STOVL operations. Due to the short durations of vertical landings, coupled with the thermal mass of the system, this deficit proved to be manageable, as confirmed by flight test operations.

B. DEVELOPMENTAL INLET TESTING AND PROPULSION INTEGRATION GROUND TESTING

The development test program for the F-35 propulsion system was much more extensive than employed by the CDP. During the CDP, testing was accomplished using existing facilities with minimal modifications plus a few purpose-built facilities (e.g., the 3BSD rig in Bristol). For SDD, dedicated test facilities were constructed and existing facilities were modified to provide support for the life of the test program. Legacy test facilities already existed for CTOL/CV propulsion systems, so most facilities were targeted at STOVL testing. One example is the dedicated LiftFan test facility at Rolls-Royce Indianapolis. Another example is Pratt & Whitney's modified C-12 test stand in West Palm Beach (Fig. 29), which provided a STOVL test capability. A third example is Lockheed Martin's hover pit in Fort Worth, Texas, for force-and-moment testing and acceptance testing of production STOVL aircraft.

The ground test program started in 2004 and was completed with more than 10,000 engine test hours. Modeling of the propulsion system was critical to the success of the F-35. The program developed detailed models that were integrated with the aircraft models to develop the aircraft and propulsion CLAW. The engine and LiftFan ground hardware and software components underwent extensive testing [1]. This included:

- Bench and rig testing
- LiftFan testing
- Engine testing
- Integrated propulsion system testing
- AMT

Fig. 29 STOVL propulsion system on the C-12 test stand.

- Vehicle Integration Facility testing
- Vehicle Systems Integration Facility testing
- Air vehicle testing on the F-35 Hover Run Facility

1. DEVELOPMENTAL INLET TESTING

An expansive F-35 inlet test plan was launched at the program start in 2001. This plan called for using subscale wind tunnel models to improve the X-35 inlet system for an operational F-35 aircraft. CFD and subscale and full-scale inlet testing were added to the test program as the SDD propulsion testing phase continued [11]. This testing supported all milestones for the program and for lines freeze and validation.

Beginning with the main inlet, CFD and model testing were used to determine the performance of the primary inlet. Changes in aperture location, number of edges, and diffuser shape drove the need to treat the inlet development as a new configuration relative to the X-35. In addition, lessons learned from the CDP were incorporated. Subscale testing, from subsonic to supersonic, was conducted in multiple wind tunnels, including wind tunnel facilities at AEDC and NASA Glenn, shown in Figs. 30a and 30b, respectively. Test results were used to calibrate and refine CFD models, generating an iterative process that continued to refine the overall configuration while responding to integration considerations.

An ice accretion test program was included as part of the SDD program to investigate the buildup of ice and identify potential locations for an ice detector in the inlet. Testing was conducted at Italy's Centro Italiano Ricerche Aerospaziali (CIRA) facility in Capua, Italy north of Naples. The test featured a large half-model of the primary inlet that was subjected to various combinations of freezing moisture to determine where and how fast ice would accumulate. The data were also used to design the engine front-frame ice protection system.

Results from the test were used to calibrate an analytical ice prediction model for the inlet, which allowed later simulations for locating the ice detector in the

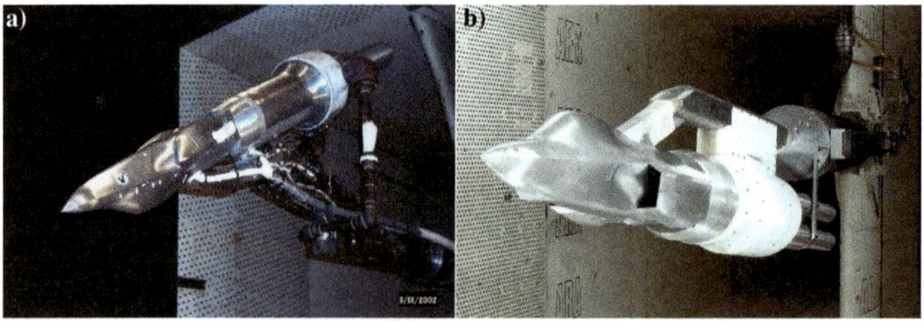

Fig. 30 (a) Main inlet development subscale testing at AEDC 16T and (b) NASA Glenn.

Fig. 31 Main inlet development full-scale testing.

inlet. Although the test identified the optimum location for an ice detector, integration considerations required compromising its placement. (Such considerations included the backup structure and the auxiliary inlet for the STOVL variant.) Later design changes made to the auxiliary inlet doors for structural reasons, discussed later in this paper, drove a subsequent relocation of the detector for the STOVL variant, requiring additional analyses.

The primary inlet development test program followed the same basic approach as that used for recent legacy systems. It relied on CFD and model testing to arrive at a final configuration. The addition of STOVL requirements drove additional testing, relative to legacy systems. This, in turn, required conducting full-scale tests at both Rolls-Royce's and Pratt & Whitney's test facilities. By contrast, STOVL requirements drove stream tube testing for the LiftFan at Rolls-Royce's test facility. Figures 31 and 32 show full-scale inlet testing at Pratt & Whitney's West Palm Beach test facilities [8].

Fig. 32 STOVL propulsion system inlet development full-scale testing.

The most unique and extensive inlet test program addressed the LiftFan inlet and the aeromechanics issues encountered during the CDP. The development test program included low-speed wind tunnel testing and static testing using stream tubes. Figure 33 shows multiple views of a fully instrumented 20%-scale model tested at German-Dutch Wind Tunnels' (DNW's) low-speed wind tunnel in Marknesse, Netherlands. The model was equipped with an instrumented LiftFan, including variable IGVs and an auxiliary inlet with adjustable doors. The simulated inlet face included a standard inlet rake to map distortion of the inlet with the auxiliary inlet open. Suction was applied to the LiftFan exhaust and the simulated engine to generate representative flows through the AIS [8].

Further testing of the LiftFan inlet was accomplished at static conditions through the use of stream tubes under both subscale and full-scale conditions. A drawback of the stream tube approach is that the shape of the stream tube is unique for a given LiftFan airflow/airspeed combination. Building multiple stream tubes at subscale is manageable, but constructing full-scale stream tubes is much more difficult and costly. This is due to the sheer size of the stream tube itself and the loads imposed by an operating LiftFan. A compromise was reached for the approach using multiple subscale assets. First, the most stressing conditions were identified, and then full-scale stream tubes were constructed to reflect these conditions.

Figure 34 shows the streamlines from a CFD simulation that define the shape of a stream tube. Combined with a model of the upper surface of the aircraft, the

Fig. 33 Twenty-percent STOVL inlet model at DNW's Marknesse facility.

Fig. 34 Representative stream tube lines for a given inlet flow/airspeed condition.

blue surface would be constructed to generate the flow field representative of the given flight condition [8].

The low loads generated during subscale static testing allowed the use of stereolithography to fabricate the stream tubes, reducing cost and minimizing turnaround time. CFD solutions were required to generate the lines to fabricate the stream tube. Accordingly, a CFD solution was available to directly compare against subscale test results, confirming that the desired flow field had been generated. Figure 35 shows a group of 20%-scale stream tubes used during the test program [8].

A full-scale stream tube was constructed for use at Rolls-Royce's LiftFan test facility in Indianapolis, Indiana. This was the final test of the LiftFan inlet design. With a flow field representing a stressing condition, a strain-gaged LiftFan was operated with the stream tube in place. This test collected the data that proved that the aeromechanics issue encountered during the CDP had been successfully mitigated. Figure 36 shows a full-scale stream tube used at the Rolls-Royce test facility [8].

Multiple STOVL jet effects aerodynamics wind tunnel test programs were conducted to characterize all aspects of the STOVL air vehicle during jetborne and semi-jetborne modes. These test programs were closely related to the characterization of the STOVL propulsion system. Many of the forces on the air vehicle

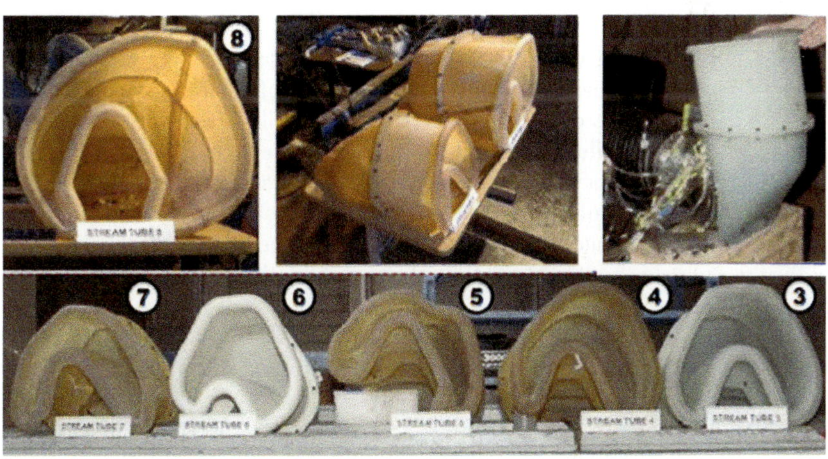

Fig. 35 Subscale LiftFan stream tube models (20% scale).

Fig. 36 Full-scale LiftFan stream tube prior to test stand installation.

during STOVL operations depended on system thrust and vector angles. It was necessary to characterize how the forces would be accounted for in analytical models of the aircraft. We also needed to identify how they fed into the overall air vehicle flight control system.

Accurate STOVL jet effects data are an integral part of the high-fidelity, STOVL-mode flight simulations required to execute the F-35 program. In jetborne mode, the STOVL jet effects can account for a reduction in thrust of approximately 8% OGE. This OGE value depends heavily on multiple variables, including wind speed and direction. When the F-35 is IGE, the OGE level can be magnified many times, depending on such variables as relative jet angles and thrust values, and height above the ground. For STO-critical conditions, the STOVL jet effects can reduce vertical thrust by approximately 20% and axial thrust by approximately 2%. One force affecting the pitching moment on the aircraft that must be accounted for is the inlet suction on the various inlets of the STOVL propulsion system. Another force comes from the suckdown effects from the various nozzles. Figure 37 shows an F-35 STOVL 12%-subscale jet effects model upon entry at DNW's Large Low-Speed Facility. Ten separate test entries were conducted, covering all STOVL flight regimes from pure hover, IGE, through transition to wingborne flight [12].

The particle image velocimetry technique is used to examine the flow field generated by the exhaust system installed in the aircraft. The technique requires the flow field to be seeded with a vegetable-based oil. A laser mounted aft of the model is used to illuminate and capture the particles at a given moment in time at a given frequency (approximately 3 Hz). Two cameras mounted to a traversing rig on the side of the wind tunnel are used to capture the image produced by the laser. These images are combined and used to generate a complete picture of the flow field. In the case of STOVL jet effects, both the LiftFan and core nozzle plumes are seeded.

Figure 38 shows a flow field center-line cut example with forward airspeed. A horseshoe vortex is formed as the wall jet spreads forward and is re-entrained into the LiftFan plume. The location of the horseshoe vortex is heavily dependent on the strength of the forward wall jet and forward aircraft velocity. At these higher forward velocities, the fountain is tilted aft and re-entrained into the core nozzle plume [12].

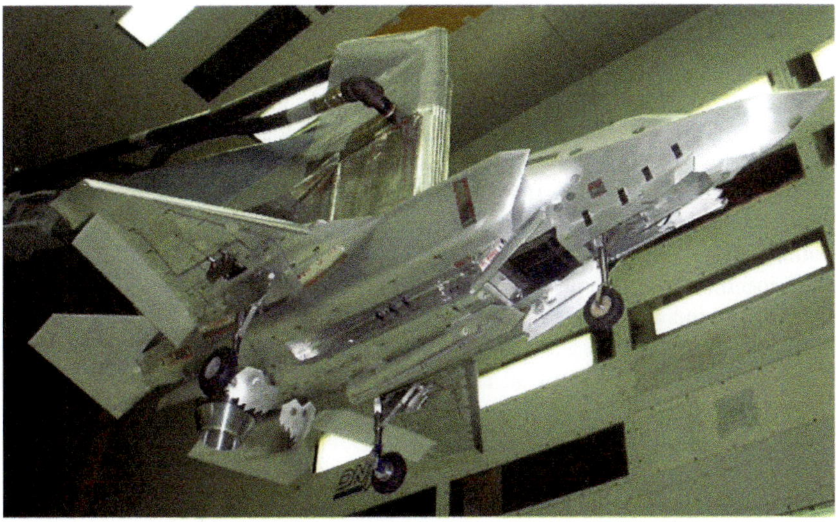

Fig. 37 F-35B STOVL jet effects subscale testing.

Similar to what was accomplished during the CDP, HGI testing was conducted at the BAE Systems facility in Wharton, United Kingdom. The test setup was updated to reflect the configuration of the F-35 and a similar but expanded test program conducted to establish a suitable operational envelope.

Fig. 38 Particle image velocimetry example (hover IGE).

2. PROPULSION INTEGRATION AND INTEGRATED FLIGHT PROPULSION CONTROL GROUND TESTING

The degree of integration between the air vehicle and the propulsion system resulted in test programs not found in legacy systems. Two key areas were investigated at Pratt & Whitney's West Palm Beach test facilities: the integrated subsystem architecture and the IFPC during STOVL operation.

The level of systems integration between the air vehicle and the propulsion system dictated that risk reduction testing be accomplished before all systems came together in the air vehicle. An enhanced integrated systems test was conducted that included installing major aircraft subsystems on a test stand in West Palm Beach. This test included the IPP, the ESG, hydraulic pumps, the ESG lubrication system, and all associated control hardware. The IPP was put through its paces, beginning with operation as an APU, then providing power to start the engine. It then transitioned to cooling mode, operating off engine bleed air and using the FDHX as a heat sink.

Failure scenarios were also simulated, such as engine flameouts in which the IPP would convert from ECS mode to an emergency power combustion mode. The ESG was used to both start the engine and provide power as if operating in the aircraft. The hydraulic pumps operated as if installed, generating minimal load on the engine gearbox during starts, then pressurizing once the engine had achieved a stable idle. Load banks to simulate aircraft systems were included on the test stand to represent aircraft operation. The testing verified the design architecture while gaining experience connecting all the systems and observing their interactions before an actual aircraft installation took place.

The early design of the AMVC was initially tested in support of a 2003 F135 ground test. This initial architecture was matured and qualified to support flight operations of the initial CTOL variant aircraft AA-1. First flight was achieved in December 2006, and the aircraft has been flying flawlessly since, earning the praise of the pilot community. F-35 Chief Test Pilot Jon Beesly stated [8]: "The engine performed flawlessly. The performance was actually better than I expected, which is interesting because I expected a lot."

Building on the initial architecture, improvements necessary to meet the strict pitch control and limit control requirements of STOVL operation were implemented in 2007. Following the completion of the ground test program for STOVL, powered-lift flight clearance was granted for the AMVC control architecture in January 2009. This was a major milestone for Pratt & Whitney and the F135 program. Meeting the pitch coupling and limit control requirements with this software represented a key milestone as the program headed toward powered-lift operation. A key area of STOVL propulsion system operation is engaging a physical rotor speed redline limit. In this situation, a traditional control system would automatically throttle back fuel flow to the engine, cutting back thrust. This would be an unacceptable and undesirable outcome for a STOVL aircraft operating in vertical mode. As described previously, the AMVC automatically takes the appropriate action to adjust not only fuel flow, but also nozzle areas and LiftFan actuators such that thrust and pitch control are maintained [8].

C. AIRCRAFT PROPULSION GROUND AND FLIGHT TESTING

Combined, the 18 flight test aircraft completed more than 16,000 flight hours and 64,000 test points in the program, which started in 2008. The validation and verification programs of the F-35 propulsion system had to ensure that the propulsion system could withstand the extreme rigors of warfighter usage at all maneuver, climatic, and battlefield conditions. On 15 December 2006, the first flight of the CTOL variant (F-35A), under Jon Beesly, was completed and initiated one of the largest flight test programs in the history of aviation. On 11 June 2008, the first F-35B (BF-1), piloted by Graham Tomlinson, took to the skies in CTOL mode flight for its first flight milestone. Shortly after its first flight, BF-1 was moved to the hover run facility to exercise the propulsion system in STOVL mode. No high-power runs were conducted initially, but the successful operation cleared the use of STOVL mode, if required, for an aircraft recovery. STOVL clearance would follow later, after full-power operations on the hover pit. By April 2009, the F-35 was ready for full-power STOVL testing on the hover pit.

Similar to the X-35 program, the aircraft was mounted on a force-and-moment system to measure performance and control power, as shown in Fig. 39. Unlike the X-35, no press-ups from the hover pit were to be performed, because an in-flight build-down approach was selected as a lower-risk approach than conducting press-ups. STOVL performance was confirmed, and control authority of the various effectors was verified. The aircraft completed an aggressive series of full-up powered-lift engine runs to exercise and evaluate the robustness of the F135 control systems and the performance of the AMVC. Hover pit

Fig. 39 Mounted on force-and-moment system at hover run facility.

ground tests validated the propulsion system and aircraft response, and the first F-35B was ready for its first STOVL flight.

With the successful completion of the hover pit test, the F-35 was cleared to begin powered-lift operations. On 17 October 2009, BF-1 completed its first non-tethered test of the STOVL system during taxi tests. On 15 November 2009, BF-1 arrived at Naval Air Station Patuxent River, Maryland, in preparation for hovers and the first vertical landing. Finally, on 7 January 2010, BF-1 engaged its STOVL propulsion system in flight for the first time.

On 18 March 2010, BF-1, under the control of Graham Tomlinson again, converted, slowed to a hover, and completed the program's first vertical landing. The STOVL flight test program continued, and challenges were uncovered. One was that the Auxiliary Air Inlet Door (AAID) design was not stiff enough to handle turbulence generated by the Upper Liftfan Door (ULFD) at certain speeds. During envelope expansion of the F-35B, it was discovered that under certain flight conditions during semi-jetborne flight with the ULFD opened to its mid-position, there were increased oscillatory loads acting on the AAID. The primary culprit for these increased loads was buffeting from the ULFD. However, the physical mechanism by which the buffeting was occurring was not fully understood. CFD was used to provide that understanding. A summary of the steady-state results is presented in Fig. 40 for the mid-position ULFD configuration. The streamlines tend to flow around the AAID at the higher Mach

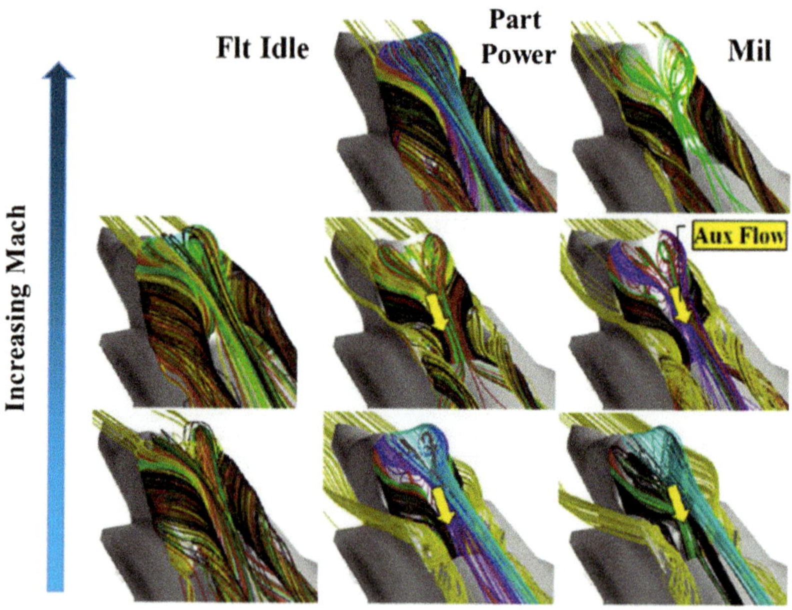

Fig. 40 Summary of steady-state AAID CFD solutions.

numbers, whereas at the lower Mach numbers the yellow streamlines are much lower and impinging on the AAID. The effect of the engine power setting can also be seen in the sequence of pictures. As the engine power setting increases, the streamlines are pulled into the AAID opening [13].

After extensive CFD and analysis of the stiffness of the door and its actuation system, a redesign of the actuation system and stiffening of doors was implemented in 2011. This returned the door design to a full-life capability [13]. The F135 propulsion system has performed acceptably throughout its flight test program and, as a result, has led the F-35A/B/C flight test programs to a successful SDD closure.

VI. CONCLUSION

The integration of the F135 propulsion system with the F-35 spans more than 20 years of propulsion design, development, and testing. Both STOVL and CTOL/CV propulsion systems meet all requirements. The LiftFan-based system, one of the higher-risk technologies for the program, works and integrates well with the F-35, with many challenges overcome during its development.

The ASTOVL program established the concept feasibility of a LiftFan-based STOVL system. The lift system architecture was proven using the LSPM and testing at facilities at NASA's Ames Research Center. These successfully showed that low-speed, powered-lift aerodynamics, hover control power, and HGI of a shaft-driven LiftFan were acceptable and an improvement over legacy STOVL systems.

The demonstration of a flightworthy propulsion system in the X-35 proved that a well-integrated team is required for success. The formation of a highly integrated IFPC IPT allowed for quick solutions to the formidable control challenges the team faced. The adaptation of a legacy conventional main engine with an unconventional LiftFan was a success. The IFPC IPT integrated X-35B aircraft and engine software and related hardware that led to a successful flight test program. The innovative DSI inlet design allowed for a short, lightweight inlet system that had excellent performance characteristics. Also, HGI, a concern for decades with legacy STOVL aircraft, was proven to be manageable and essentially a nonissue with the LiftFan arrangement.

The F-35's successful development of the propulsion system took the lessons learned from the X-35 and applied them across all the propulsion systems. The highly integrated cross-company team of Lockheed Martin, Northrop Grumman, BAE Systems, Pratt & Whitney, Rolls-Royce, and the F-35 Joint Program Office worked very well together. With a great combination of CFD, subscale wind tunnel testing, full-scale testing, stream tubes, and engine and LiftFan testing, the team used design techniques that maximized the benefit of each. The LiftFan inlet was improved to solve the serious LiftFan aeromechanical problem seen on X-35B with a new door design. The auxiliary inlet benefited from CFD analysis to solve an unforeseen door vibration issue. The full-scale testing, stream tubes, and extensive subscale inlet testing gave the propulsion system unprecedented performance for a single-engine 5th-Generation fighter. The VAVBN improvement over the X-35B's LiftFan hood provided much more flexibility in LiftFan operation. All these examples show how the team evolved these designs from concepts to operational reality. The degree of IFPC integration was absolutely unprecedented, with the propulsion and flight control systems in complete control from conversion to STOVL mode, slowing to a hover, and performing a vertical landing. The F-35 propulsion integration exemplified the importance of power and thermal management in the development of an integrated solution.

A testament to the F-35's propulsion design success was captured with a quote from a Royal Air Force F-35B test pilot in an interview on 26 January 2018. Sqn. Ldr. Andy Edgell stated:

> I will never forget my first hover in the Harrier and I will never forget my first hover in the F-35B. My first hover in the Harrier is akin to trying to stay alive on a unicycle and I don't think I made much of a conscious thought at the time in the hover. It was only once I managed to get her down on deck safely, and then retrospectively I thought, "How on earth did I manage that?" Whereas in my first hover in the F-35B I sat there at that moment, and looked around, realized, I was stationary in every axis, and thought, "Goodness gracious me, the guys who designed this are absolute geniuses." [14]

Fig. 41 Propulsion systems at work in the F-35A, F-35B (CTOL and STOVL modes), and F-35C.

The fleet continues to grow, with more than 290 F-35As, F-35Bs, and F-35Cs, totaling more than 135,000 flight hours in the field. As such, the F-35 and its unique and successful propulsion system will continue to meet the warfighter's every challenge, as shown in Fig. 41. In the final analysis, the F-35's propulsion integration, design, and development were a success.

ACKNOWLEDGMENTS

Steven P. Wurth and Mark S. Smith would like to thank their colleagues throughout the multicompany propulsion enterprise whose efforts brought this propulsion system to life. The ASTOVL, X-35, and F-35 propulsion teams constituted one of the best collections of subject-matter experts in the fields of propulsion aerodynamics, integration, and IFPC. The authors would also like to thank the executive review team for providing guidance in the development of this paper.

REFERENCES

[1] Wurth, S., Walker, G., and Fuller, Dr. J., "F-35B IFPC Development," AIAA Paper 2013-4243, August 2013.

[2] Maddock, I. A., and Hirschberg, M. J., "The Quest for Stable Jet Borne Vertical Lift: ASTOVL to F-35 STOVL," AIAA Centennial of Naval Aviation Forum, 11th AIAA ATIO Conference, Virginia Beach, VA, Sept. 2011.

[3] Bevilaqua, P. M., "Joint Strike Fighter Dual-Cycle Propulsion System," *Journal of Propulsion and Power*, Vol. 21, No. 5, Sept./Oct. 2005.

[4] Bevilaqua, P. M., "Future Applications of the JSF Variable Propulsion Cycle," AIAA Paper 2003-2614, July 2003.

[5] Wurth, S., Mahone, T., Hart, J., and Baxter, J., "X-35B Integrated Flight Propulsion Control Fault Tolerance Development," AIAA Paper 2002-6019, Nov. 2002.

[6] Weigand, C., et al., "F-35 Air Vehicle Technology Overview," AIAA Aviation 2018 Conference, Atlanta, GA, May 2018 (submitted for publication).

[7] Buchholz, M. D., "Highlights of the X-35 STOVL Jet Effects Test Effort," AIAA Paper 2002-59622002, Nov. 2002.

[8] Sylvester, T. G., Brown, R. J., and O'Conner, C. F., "F-35B LiftFan Inlet Development," AIAA Paper 2011-69402011, Sept. 2011.

[9] Fuller, J., "Advanced Multi Variable Control (AMVC) and Its Application in Turbomachinery," AIAA Paper 2010-01-1737, Nov. 2010.

[10] Lo Gatto, E., "Overview of the JSF LiftSystem™," International Powered Lift Conference, Hartford, CT, Sept. 2016.

[11] Smith, M. S., "Joint Strike Fighter: X-35 and F-35 Propulsion System Integration," 39th AIAA/ASME/SAE/ASEE Joint Propulsion Conference, Huntsville, AL, July 2003.

[12] Mange, R., and Hoggarth, R., "Highlights of the Lockheed Martin F-35 STOVL Jet Effects Program," 08IPLC-0022, Royal Aeronautical Society International Lift Conference, London, UK, July 2008.

[13] Cox, Dr. C. F., "F-35B Auxiliary Air Inlet Analysis and Design," AIAA Paper 2013-218, Jan. 2013.

[14] Robinson, T., "Inside F-35B Flight Test," Royal Aeronautical Society [interview with Sqn. Ldr. Andy Edgell], https://www.aerosociety.com/news/inside-f-35b-flight-test/ [retrieved 26 Jan. 2018].

F-35 Subsystems Design, Development, and Verification

Drew Robbins*, **John Bobalik**†, **David De Stena**‡, **Ken Plag**§, **Keith Rail**¶ **and Ken Wall****

Lockheed Martin Aeronautics Company, Fort Worth, TX

F-35 vehicle subsystems were developed, qualified, and fielded as part of an overall integrated systems architecture. Because of the unique requirements derived from the common tri-variant integrated systems architecture, several key technologies were developed as part of the overall subsystems development. These include electric primary and secondary flight control actuation, electrical starter/generator systems, lithium-ion batteries, integrated power and thermal management systems, and expanded ejection system performance envelopes. With any development and qualification of new technologies, several discoveries were observed and resolved during qualification. With the successful implementation of a fully integrated subsystem architecture, F-35 vehicle subsystems are fully qualified and operational, providing a high-performing, integrated architecture. These subsystems continue to successfully support the F-35 fleet well past its current 100,000 flight-hour mark.

I. INTRODUCTION

The vehicle subsystems on the F-35 were developed, qualified, and fielded as part of the F-35 System Development and Demonstration (SDD) program. Continuing from previous efforts in both configuration assessments and technology development [1], these systems were developed as part of an overall integrated air vehicle philosophy [2]. This paper discusses five systems: the Electro-Hydrostatic Actuation System (EHAS), the Power and Thermal Management System (PTMS), the Electrical Power System (EPS), the Hydraulic and Utility Actuation (HUA)

*Senior Manager, F-35 Utilities and Subsystems, MZ 6500.

†Aeronautical Engineer Principal, Advanced Development Programs, MZ 2353.

‡Principal Systems Engineer, F-35 Utilities and Subsystems, MZ 6500.

§Principal Systems Engineer, F-35 Electrical Power Systems, MZ 6500.

¶Deputy Manager, F-35 Vehicle Management Hardware, MZ 6468.

**Manager, F-35 Power and Thermal Management System and Pilot Systems, MZ 6500.

system, and the ejection system. As part of the development of these systems, the key technologies discussed are electric primary and secondary flight control actuation, electrical Starter/Generator (S/G) systems, lithium-ion batteries, integrated PTMSs, and expanded ejection system performance capabilities (Fig. 1). With any development program, integration challenges occur. The challenges this paper reviews are the increased priority of modeling and simulation, improvements implemented because of the early AA-1 inflight emergency, development of lithium-ion batteries, regenerative power impacts on both the EHAS and EPS, engine start functionality, emergency power transitions and qualification, and hydraulic system integration. Because of the integrated effort by design engineers across multiple companies, the F-35 flight controls and utilities and subsystem hardware are fully fielded and operational, supporting the growing F-35 fleet.

As discussed in Ref. [2], the final aircraft design required an integrated architecture vs a more traditional federated or distributed system. A federated system relies on several independently designed subsystems, whereas an integrated architecture attempts to achieve equivalent or greater capability with less hardware. As a result, these architectures require more integrated development and testing. Integrated architectures allow for increased performance, improved affordability, and improved weight due to dramatically reduced amounts of hardware. Multiple development programs formulated the final configuration, as discussed in Ref. [3]. Figure 1 compares a federated system to the F-35 integrated system. Of note, the integrated architecture drove new hardware development and had an impact on the overall system management (including software), qualification, and flight test approach. Rather than relying on individual systems to operate to a predefined set of criteria, an integrated architecture relied heavily on a robust system engineering approach. This applied throughout the process, from requirement derivation and allocation through integrated qualification and flight testing.

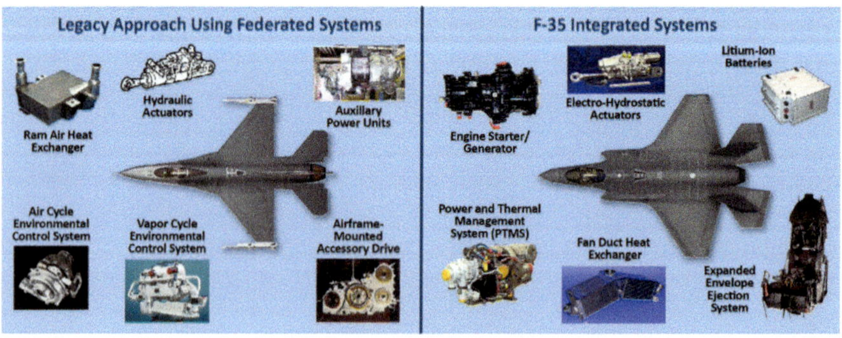

Fig. 1 Federated vs integrated vehicle subsystems.

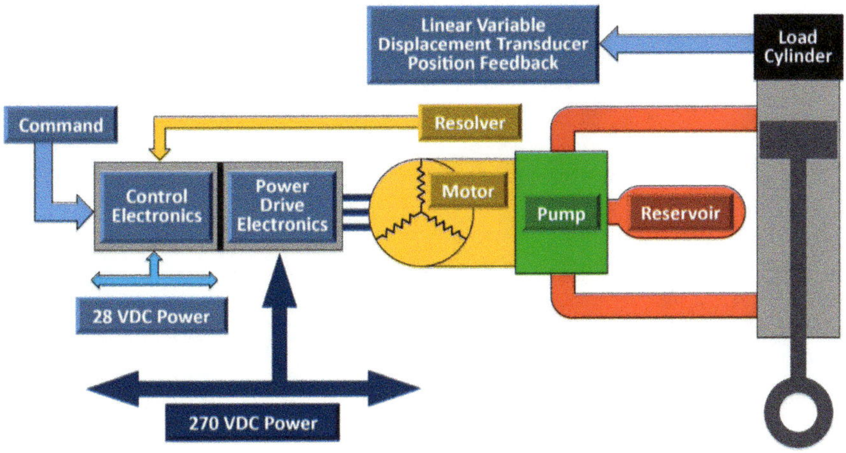

Fig. 2 Basic schematic of an F-35 EHAS.

II. SYSTEMS OVERVIEW

This section provides a brief overview of the five systems comprising the vehicle subsystems, including a brief history of how the systems evolved under the pressures to integrate functions into more efficient products. It provides a very basic understanding of how the federated functions combined into integrated systems. This is important in understanding how and why the aircraft-level requirements drove the flow down of requirements, the key technologies, and the integration and development issues discussed in later sections.

A. ELECTRO-HYDROSTATIC ACTUATION SYSTEM

The F-35 is the first manned aircraft and first fighter aircraft to use electric primary and secondary flight control actuation. As the primary actuation system, the EHAS (Fig. 2) incorporates a new type of self-contained actuator with an integral motor and pump driven solely by electric power (Fig. 3). These components, along with a reservoir, reside directly on the manifold of the actuator and replace the control valve used on a conventional actuator. The pump directly pushes fluid into the actuator, and the direction of actuator movement follows the direction of the pump rotation. Each Electro-Hydrostatic Actuation (EHA) motor is controlled by an Electronic Unit (EU) channel that contains a low-power (28 VDC) side and a high-power (270 VDC) side. The low-power side is responsible for digital loop closure and fault monitoring, and the high-power side is responsible for three-phase power switching.

Due to this new technology, the development effort included significant design maturation and challenges with scaling the design for the unique needs of the

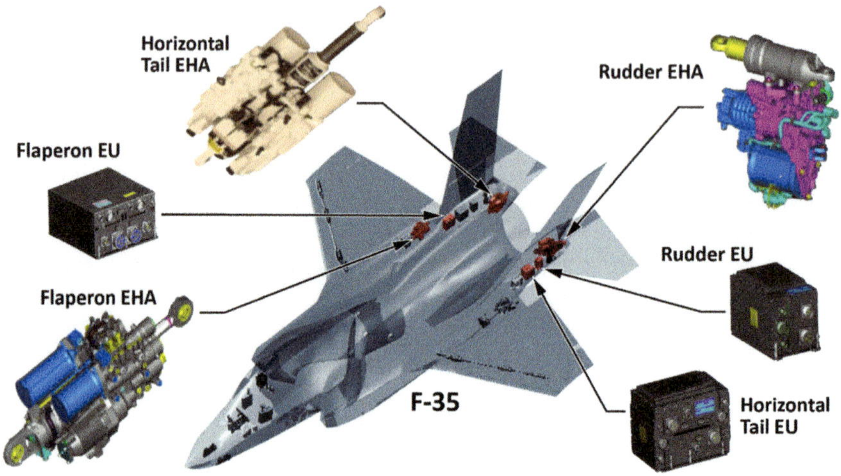

Fig. 3 EHAS installations.

three aircraft variants. The aircraft size, performance, and safety requirements drove the system to be larger and more integrated than previous electric actuators used in drone applications or for concept demonstration. One example of the latter case is the Joint Strike Fighter (JSF) Integrated Subsystems Technology (J/IST) program [2]. The interfaces between the EHAS and PTMS had to go beyond basic connections and performance criteria. They had to include modes of operation and how to collectively transition between modes in case of a failure in any of the systems. All these considerations played into iterative improvements as the EHAS matured through design, integration, and flight test.

B. POWER AND THERMAL MANAGEMENT SYSTEM

The PTMS is a multifunctional system designed by Honeywell International Inc. combining traditionally separate environmental control and auxiliary power functions into one integrated system. It provides conditioned air for cockpit and equipment cooling, refrigerated liquid for equipment cooling, and electrical power for ground maintenance, main engine starting, and inflight emergencies. Figure 4 provides a simplified schematic of the system.

The cooling function is performed by a closed-loop air cycle powered by expanding bleed air across a Power Turbine (PT). The closed-loop air cycle includes a compressor (*C* in the figure), Cooling Turbine (CT), and S/G mounted on a single shaft. High-temperature, high-pressure air exiting the compressor dissipates heat to a Fan Duct Heat Exchanger (FDHX) installed in the engine fan air duct. The engine FDHX is unique with respect to keeping exchanger waste heat output within the propulsion cycle rather than overboard. The air then passes through a polyalphaolephin (PAO) to air Heat Exchanger (HX) to supplement air cycle waste heat removal when fan duct temperatures are too high to be used as an effective heat sink. The air is cooled further in a recuperator HX before being expanded through a CT. The cold, low-pressure air exiting the CT passes through the load HX.

The load HX includes an air-to-air core to cool preconditioned bleed air to cool the cockpit, equipment, and Onboard Oxygen Generation System (OBOGS). It also has a liquid-to-air core to provide liquid cooling to the majority of the mission systems equipment. Air exiting the load HX passes through the recuperator HX before returning to the compressor inlet. In addition to providing preconditioned air to the load HX, the engine precooler supplies the Onboard Inert Gas Generation System (OBIGGS). The OBIGGS removes oxygen from air and provides nitrogen-enriched air to the fuel tanks for vulnerability and lightning protection.

In addition to the cooling and air conditioning mode, the PTMS has four other modes: ground maintenance, main engine start, and self-powered and bleed-driven emergency electrical power. Figure 5 summarizes the PTMS operating modes and depicts the turbomachine. This integrated approach to cooling relied on unprecedented integration among the propulsion, power, and environmental control system cycle.

Similar to the way in which it operates in the bleed-driven cooling mode, the PTMS can operate as an emergency generator in flight by reducing cooling and

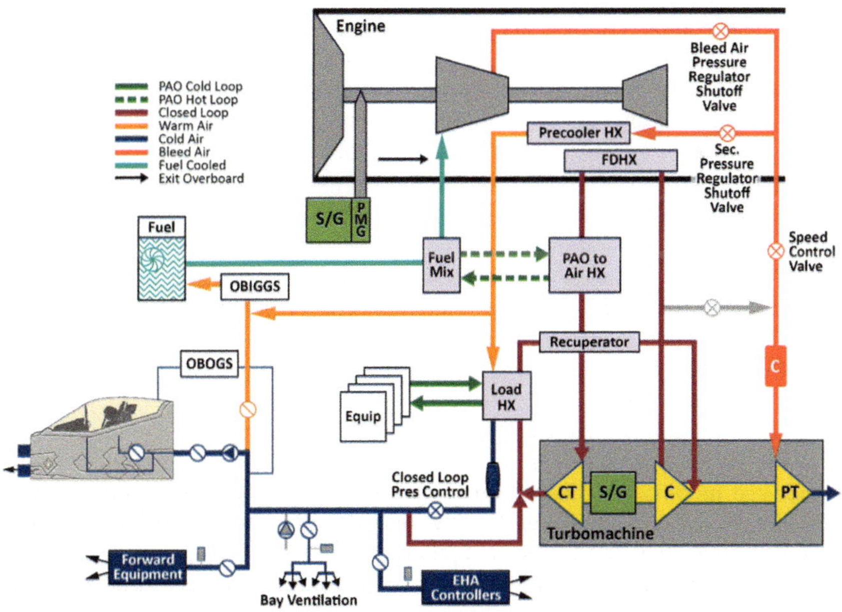

Fig. 4 Simplified PTMS schematic.

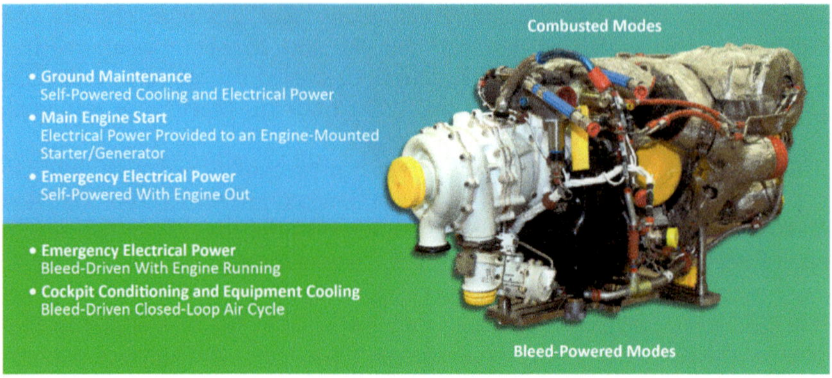

Fig. 5 The PTMS provides both environmental control and auxiliary power functions.

diverting shaft power to the S/G. The PTMS can provide cockpit conditioning, flight-critical equipment air cooling, and electrical power in this mode.

The PTMS can be reconfigured by valves to operate like a traditional Auxiliary Power Unit (APU) for self-start, ground maintenance, main engine start, and inflight emergency power. The compressor-combustor-turbine section of the PTMS turbomachine is effectively an APU that can produce shaft power to drive an integrated S/G and/or an open-loop air cycle. The S/G mounted on the same shaft system as the turbomachine provides electrical power for checkout of the aircraft's electrical, avionics, and flight control systems, as well as inflight emergency power. When provided with 270-VDC aircraft battery power, the S/G also acts as a start motor, providing the initial rotation and torque required to start turbomachine operation in the combusted mode.

Once the PTMS has been reconfigured to the combusted mode, the propulsion engine can be started electrically by providing PTMS-generated 270-VDC electrical power to an S/G mounted on the engine gearbox. In this same configuration, the PTMS can also provide air and liquid cooling and electrical power to perform ground maintenance without using ground support power or cooling. Because the FDHX is not available in this mode, cooling cycle waste heat is transferred to fuel from the PAO/air HX. An onboard fan is used to enhance heat transfer from the heated fuel to the atmosphere.

The PTMS also can reconfigure to the combusted mode in flight. In the event of an engine flameout, the PTMS transitions from the normal cooling mode to the combusted mode throughout the flight envelope. The 270-VDC aircraft battery provides flight-critical electrical power at high altitudes until the aircraft descends and the PTMS can provide sufficient power to the flight control actuation system. After mode transition is complete, the PTMS provides nonessential electrical power at high altitudes and flight-critical electrical power at lower altitudes, including enough power to facilitate a safe landing. Along with providing flight control power, the PTMS simultaneously provides power to assist with propulsion engine air-start within the engine start-assist flight envelope.

C. ELECTRICAL POWER SYSTEM

The EPS consists of two subsystems: the Electrical Power Generating System (EPGS) and the Electrical Power Management System (EPMS). The overall control and monitoring of the system is supplied by redundant software running in the Vehicle Management Computers (VMCs).

The EPGS consists of a single Engine Starter/Generator (ES/G) that provides two independent outputs of 80-kW power each during normal operation. During main engine start, the ES/G uses electrical power from the PTMS via a pair of Inverter/Converter/Controllers (ICCs) to create mechanical shaft power to motor the engine. After engine start, the ES/G automatically switches to generate mode, in which the ICCs convert the unconditioned power from the ES/G into regulated 270 VDC. Two voltage converters, each rated at 160A of power, provide 28 VDC for low-power, critical loads. The 270-VDC-to-115-VAC inverter provides the aircraft with 5.4 kVA of 115-VAC power for weapons stations and wing-fold on the F-35C aircraft. The three Converter/Regulators (C/Rs) provide redundant 28-VDC power to the flight-critical loads. These C/Rs accept power from the 28-V distribution system, 28-VDC battery system, and two sets of permanent magnet generators: one located within the ES/G and the other within the PTMS turbomachine. These inputs provide highly reliable, break-free power. Each C/R provides power to one branch of the triple-redundant flight control system.

The EPMS consists of the power distribution equipment, providing the on/off switch control of the loads and short-circuit protection for the wiring. The equipment uses a combination of traditional mechanical circuit breakers and relays, smart contactors, and solid-state switches to distribute power from the various sources to the using loads. (Smart contactors combine overcurrent monitoring and switching into a single unit.) The 28-VDC battery system consists of an eight-cell lithium-ion battery and charger/controller. The 270-VDC battery system is made up of an 84-cell lithium-ion battery and a charger/controller. This system provides fill-in power to 270-V flight-critical loads and provides electrical power for PTMS self-start.

The F-35 EPS has expanded on the level of integration incorporated compared to the F-22 aircraft. All load switching is now performed by the EPS via software commands from the using system. When the PTMS turbomachine is the power source, a hand-off between the PTMS and EPS provides a power-available signal. This depends on the instantaneous capacity of the turbomachine, but allows the EPS to actively load manage to meet critical needs within the available capacity. A similar relationship is present between the EPS and the stores management system. In it, the instantaneous power capacity is provided so they can manage which weapons are powered based on the available power and mission needs.

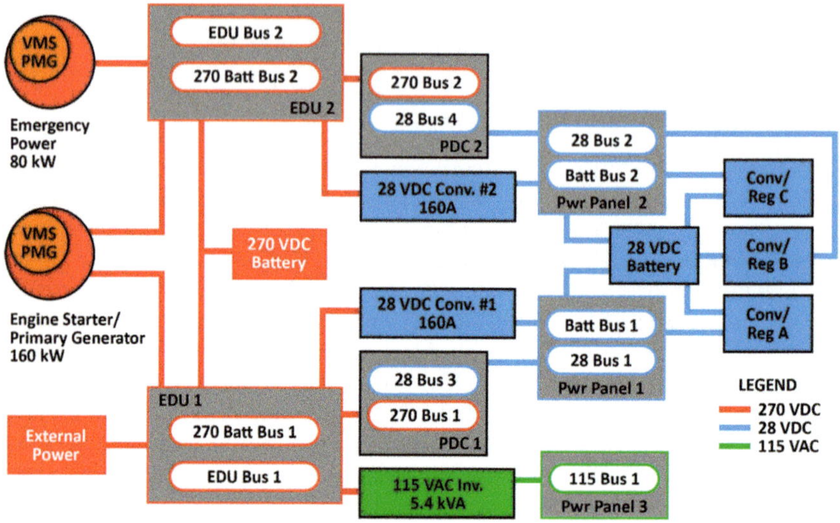

Fig. 6 EPS architecture.

An overview of the system architecture is shown in Fig. 6, with the part locations depicted in Fig. 7. The system is nearly common across all variants. The exception is the size of the output rating for the inverter, which is slightly larger to accommodate the wing-fold function of the F-35C aircraft.

D. HYDRAULIC AND UTILITY ACTUATION

The hydraulic system was designed to provide maximum power to F-35 utility actuation systems and F-35B lift system functions. It uses approximately one-fourth of the hydraulic power capacity that would have been required for conventional subsystems. The relatively small capacity of the hydraulic system and large power demands of the utility systems led to a unique balance of actuation technologies, power sources, and system architecture. Although the hydraulic power system architecture is common across the F-35 aircraft variants, the unique actuation systems across the aircraft variants resulted in approximately three times the number of utility actuation functions, compared to legacy aircraft. A block diagram of the F-35 utility actuation systems is shown in Fig. 8.

Characteristics of the hydraulic system include two independent 4000-psi hydraulic power systems that are common across all three variants. Each hydraulic system consists of:

- A 29.8-gallon per minute (gpm) engine-driven variable-displacement pump
- A boot-strap reservoir (F-35A/C at 640 in^3, F-35B at 945 in^3)

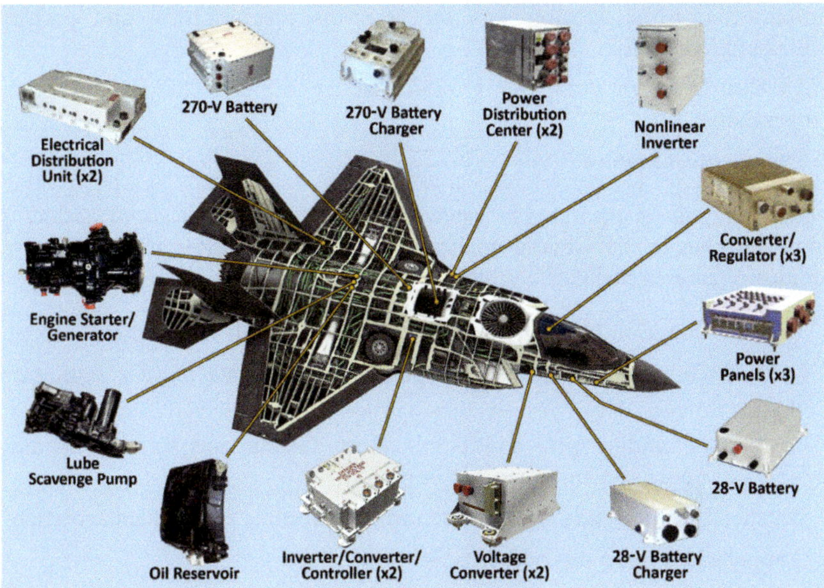

Fig. 7 EPS installations.

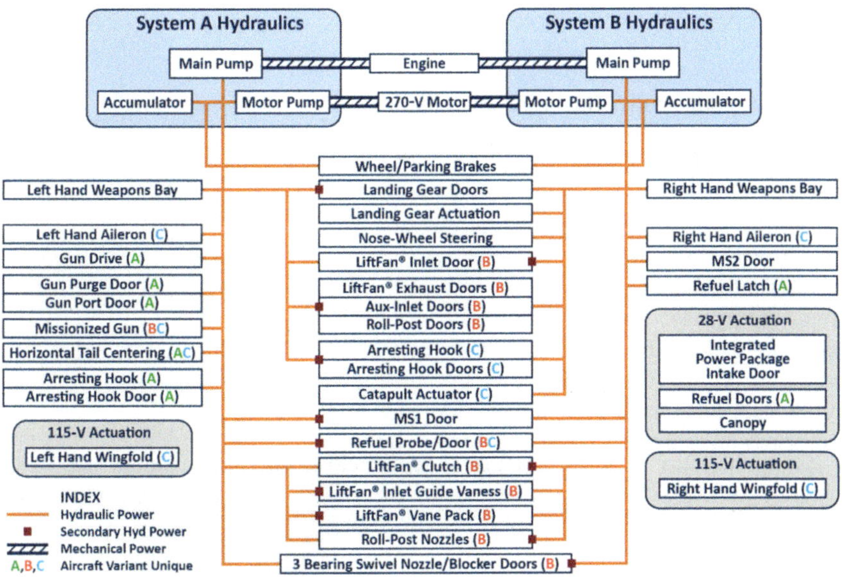

Fig. 8 HUA block diagram.

- Maintenance-free accumulators for reservoir pressurization and parking brake functionality
- An electric-driven pump for ground maintenance and inflight emergency operations
- Five-micron filtration

Redundancy was incorporated for safety-critical actuation applications, including redundant lines to the actuator applications and shuttle valves at the actuators. Additional features include:

- Inflight stowage and latching to support aircraft stealth requirements
- Quick disconnects on selected components to facilitate rapid aircraft turn times
- Isolation of landing gear and F-35B Short Takeoff and Vertical Landing (STOVL) doors systems when not in use
- Switched high/low-side electrical controls to preclude unintended actuation
- Computerized reservoir-level sensing
- An electronic filter replacement indication

E. EJECTION SYSTEM

The F-35 ejection seat (Fig. 9) provides a means for rapid egress from the cockpit under emergency conditions. To initiate ejection, the pilot pulls the ejection handle, which allows the transparency removal system to cut the canopy, fires the rocket motor, and activates the leg/arm restraints. When the seat separates from the aircraft, the drogue chute deploys to stabilize and decelerate the seat/occupant mass. The parachute deployment box fires to initiate seat/occupant separation and deploy the main parachute at the calculated delay provided by the sequencer to ensure deceleration to a non-injurious parachute opening load. The parachute harness features a head support panel to control the head from excessive aft extension.

Fig. 9 Production ejection seat.

The auto-escape system is configured in the F-35B only for the event of catastrophic failure of the LiftFan®. Such a failure can produce a rapid onset of forward pitch accelerations, perhaps exceeding the pilot's ability to react and manually eject. The system is designed to provide detection of inertial attitudes and rates in the pitch-down plane within tolerances to filter out normal flight control input and response. The detection occurs in the control law software application of each VMC during conditions in which the system is armed.

III. REQUIREMENTS DERIVATION

This section discusses the flow down of requirements driven by the contract specification and airframe design decisions. Several major design decisions drove the requirement for several key technologies to be discussed in the next section. These technologies were critical to the success of the overall platform. In addition, the weight and volume growth obstacle discussed in Ref. [1] introduced, in some cases, an additional design cycle midway through the development.

A. ELECTRO-HYDROSTATIC ACTUATION SYSTEM

The EHAS is much more complex than a conventional hydraulic actuation system and does not follow the classical growth trades well when compared to a stand-alone system. This is especially true because the EHAS is scaled up to the size required for the F-35C. The increased F-35C capability and, therefore, size are driven by the unique and stringent performance requirements for landing the F-35C on Navy carriers. The reasons for choosing the EHAS as the flight control actuation for the F-35 go beyond the individual system and look at the benefits to the overall aircraft configuration [2]. Large components like hydraulic pumps, which normally reside on the accessory gearbox in the body of the aircraft, can be removed and distributed in the wings and aft portions of the aircraft. This provides available fuel volume and improves the aircraft cross-section, benefitting aircraft performance. As a result, the EHAS became the system of choice for the F-35.

The architecture chosen for the EHAS and vehicle systems in general relies on the distributed processing of certain lower-level functions based on commands from the VMC. This design, along with greater interaction among the EHAS, EPS, and PTMS, increased the complexity of the interface requirements used to maintain coordination and the timing needed to safely control the systems. The additional requirements were derived and managed in the subsystem specifications and the Vehicle Systems Network interface control document.

B. POWER AND THERMAL MANAGEMENT SYSTEM

The F-35 PTMS requirements were derived from legacy federated subsystem architecture requirements for both auxiliary power and environmental control.

The PTMS turbomachine performs legacy air cycle machine functions to provide equipment cooling and cockpit environmental conditioning. It also performs legacy APU functions to start the propulsion engine, provide ground maintenance power and cooling, and provide inflight emergency power. The integration of these functions required turbomachinery with a wide operating range to cover open-loop, APU-like operation from sea level to 50,000 ft and closed-loop, refrigeration mode operation throughout the flight envelope. Propulsion engine start requirements sized the turbomachine PT and the compressor required to feed it. Sizing the compressor for engine start resulted in a compressor that was a good match for the F-35 environmental control functions, although not optimized for those functions. Propulsion engine start requirements also sized the generator integrated into the turbomachine. Sizing the generator for engine start provided ample power for inflight emergency modes and ground maintenance.

C. ELECTRICAL POWER SYSTEM

Following on from prior designs, the F-35 EPS was designed around a 270-VDC system that provides a significant weight savings and enables the system to provide essentially break-free power transfers under many conditions. Break-free power transfers ensure that the power-to-aircraft loads are not interrupted when a transfer to another power source occurs. The ability to provide break-free power at very high load levels enabled the flight control system to implement the EHAS, which benefitted the overall aircraft configuration extensively.

The advent of high-powered flight controls created the need for redundant sources, and these are provided by the pair of 80-kW generators housed in a single chassis and spinning on the same shaft. This provides sufficient electrical isolation and redundancy to meet the flight-critical power needs, but also minimizes weight. The original configuration of these generators was a switched-reluctance topology, with the idea that the units would be implemented deep within the engine. This dictated a unit capable of operating in that high-temperature environment. Other issues prevented the generators from being located within the engine, so the need for this specialized topology was never realized. As part of the continued evaluation of the system, it was determined that a weight savings of more than 100 lb could be realized by converting from a switched-reluctance system to a more conventional synchronous system. This system was implemented midway through the flight test program.

The next step aligned with the more-electric aircraft technology thrusts of the Joint Advanced Strike Technology and J/IST programs [4, 5] and the drive toward more integrated systems. The generating system was implemented as the electric start system to provide significant weight and complexity savings to the overall aircraft through the removal of the more traditional engine starting system. Additionally, and more important to the overall aircraft configuration, this allowed for the removal of the accessory drive gearbox. This allowed for a reduction in the cross-section of the airframe. Through the creation of the

ICCs, the generators are operated as motors during the start mode to provide torque to the engine.

To provide nearly instantaneous fill-in power to the EHAS while recovering from a generator or engine fault, a lithium-ion battery was tied to the 270-VDC generation bus. It is generally referred to as a 270-VDC battery because of the bus voltage, although its actual operating voltage exceeds 300 V.

D. HYDRAULIC AND UTILITY ACTUATION

The transition from legacy hydraulic flight-control actuation systems to electro-hydrostatic flight control actuation led to unique challenges for the actuation of the numerous utility-actuation functions. The quantity of utility functions and wide range of power demands led to a combination of hydraulic and electrical power sources for the utilities. Extensive studies with numerous hydraulic, pneumatic, and electric actuation suppliers with a focus on weight and cost efficiencies led to the utility system architecture shown in Fig. 7. Although there was extensive use of common utility functions across the F-35 aircraft variants, unique aspects of each aircraft resulted in a wide range of utility actuation systems (Table 1).

The electric architecture of the F-35 flight control actuation systems, high quantity of utility system functions, and stealth aircraft requirements resulted in several differences in the utility system architecture, compared to legacy utility actuation systems. These differences included the elimination of the airframe-mounted accessory drive gearbox because the hydraulic pumps were mounted directly to the main engine gearbox. A dual hydraulic system architecture was used to provide fault tolerance for the F-35B lift system actuation and

TABLE 1 HUA SYSTEM COMMONALITY

Common	Partial Commonality	F-35A Unique	F-35B Unique	F-35C Unique
• Hydraulic Power Generation • Landing Gear Architecture • PTMS and Utility Function Door	• Weapon Bay Door Drive • Refuel Probe (F-35B/C) • Missionized Gun (F-35B/C)	• Refuel Receptable • Arresting Gear • Internal Gun and Door	• Lift-Fan Inlet Guide Vanes, Vane-Box Nozzle and Roll-Post Nozzles • Lift-System Doors • LiftFan® Clutch	• Wing-fold • CV Arresting Hook • Ailerons • Cat-Launch Bar • Gear Shrink

backup actuation for essential landing, tail hook, and aerial refueling functions. Utilization of the redundant hydraulic system for safety-critical applications eliminated the legacy pneumatic bottles and associated servicing. In addition, advanced hydraulic motors were incorporated onto the Weapon Bay Door Drive (WBDD) systems based on their power efficiency and weight optimization. Electric motor pumps were implemented to support maintenance at austere sites, aircraft tow braking, and dead-stick landings. There was also extensive use of door sequencing functions for the F-35B STOVL doors, weapon bay doors, and landing gear doors. Intermediate door positions for the LiftFan inlet door actuation were incorporated to accommodate the LiftFan inlet recovery. The weapon bay doors are driven to a mid-stroke position and held to support the lift improvement device's F-35B vertical landing function. Finally, an advanced wing-fold system was implemented on the F-35C for weight optimization and volume reductions.

E. EJECTION SYSTEM

The F-35 Martin-Baker US16E ejection seat was built on legacy ejection seat designs. The design was modified to incorporate performance features to comply with the increasingly challenging anthropometric and pilot safety requirements. The ejection seat needed to be able to accommodate a wide range of pilot anthropometry, from Case 1 to Case 8, with a nude mass range of 103 lbm to 245 lbm. At the same time, it had to balance this with the advanced aircraft operating envelope and strict terrain clearance requirements defined for the F-35. The US16E design also integrates components of the F-35 life support systems, such as the breathing regulator and the backup oxygen supply. Further, it integrates into the unique helmet-mounted display system.

IV. KEY TECHNOLOGY DEVELOPMENT

Several key technologies were critical to the development program to enable the weight, volume, and performance promises of the integrated systems. These technologies included the EHAS, an FDHX, the integration of a permanent magnet S/G on the same high-speed shaft as the turbomachine, a 270-V battery, the ES/G, the WBDD system, specialized door sequencing, an advanced wing-fold system, and the integration of several features of the seat, along with a sensing of the environment and management of parachute deployment.

A. ELECTRO-HYDROSTATIC ACTUATION SYSTEM

A number of key technologies went into developing the EHAS. One area that most affected integration into the aircraft was power control and regenerative energy handling. New hardware and software were developed to safely manage the EHAS's power draw and dissipate the regenerative energy that occurs during motor deceleration so there are no effects on the power system. Another area affecting development was the management of cold temperature performance.

The EHAS is a large user of electrical power on the aircraft. How this power is managed is critical to the successful operation of the EHAS, as well as the other components that share the power buses. As the aircraft and EHAS matured, the losses and operating needs increased in fidelity and the required power draw went up. This increase was driven by the need to meet performance at cold temperatures. In addition to the power increase, design features were added to help mitigate temperature effects.

When the cold temperature issue was identified, it looked inevitable that the EHAS would need to live with added weight, cost, and complexity due to the addition of heaters in the manifolds to warm the system in cold environments. As the issue matured, it was determined that the software could be used to aid in actuator warming by allowing the EHAS to use losses in the motor as a heat source. The software was altered so that it could modify the commutation efficiency to increase the losses and, therefore, the heat generated in the motor. The heat then dissipates into the pump and manifold. This helps to keep the fluid warm and the fluid properties within the desired range.

The F-35 is a dynamic, naturally unstable aircraft, and controlling it requires high-bandwidth actuators that can stop and reverse directions quickly. To accomplish this, the EU uses regenerative energy to slow down the motor. The regenerative energy must be used or dissipated to prevent voltage transients that can damage electrical components. Early in the program the decision was made to keep the regenerative energy in the EU instead of allowing it back onto the power bus. This was intended to avoid transients on the bus but required additional provisions within each EU channel to dissipate the regenerative energy.

As the program developed, it was determined that the magnitude of this regenerative energy exceeded the capability of the EUs to dissipate it. The ceramic resistors would crack under thermal stress and the solder used to connect them would reflow and cause the circuits to open. Through multiple design iterations, additional resistor capacity was incorporated and the thermal bonding between the resistor and the lid of the EU was improved to better dissipate the heat to the environment. This included the use of steel plate resistors and high-temperature solder to increase the operating temperature range.

B. POWER AND THERMAL MANAGEMENT SYSTEM

Two key technologies were developed to maximize PTMS integration benefits at the aircraft level. The FDHX minimized the fuel displacement and ram drag of a typical ram air circuit and ram air/bleed air HX. Integrating a permanent magnet S/G on the same high-speed shaft as the PTMS turbomachine facilitated the elimination of a gearbox on the PTMS turbomachine. It also reduced the weight penalties of a more conventional low-speed generator.

The FDHX consists of three titanium cores and one INCONEL® core arranged in parallel in the F135 fan air duct. The FDHX is critical to dissipating heat generated by the aircraft electronics into the propulsion system. Hot side flow for the three closed-loop HXs is provided in parallel. The titanium cores transfer heat from the closed-loop air refrigeration cycle to fan air. The INCONEL core transfers heat from bleed air to fan air before it is conditioned by the closed-loop air refrigeration cycle and provided to the cockpit and forced air–cooled equipment.

A typical HX core is shown in Fig. 10. Each HX includes two separate cores welded together at an angle to allow for installation in the F135 annular fan duct area while maximizing the frontal HX fan air capture area. Hot closed-loop air enters the center header and splits between the left and right cores. The internal fin geometry turns the hot flow as shown in the figure to facilitate a counter-flow design, thereby maximizing performance and minimizing weight and volume. The cooled bleed or closed-loop air exits the cores through the left and right exit headers and is collected in a manifold installed on the outside of the F135 engine.

The PTMS turbomachine includes a compressor, a CT, a PT, and an S/G mounted on a single shaft. The turbomachine is shown in Fig. 11 with callouts indicating locations of the major components. To reduce weight and volume, the turbomachine was designed to operate at the highest speed possible. The high shaft speed, up to 59,000 rpm, drove the generator design to a permanent magnet rotor design. The S/G produces approximately 5 hp when used as a motor to start the turbomachine. The motor is powered by a 270-VDC battery that also provides inflight emergency power for the flight control system. In generator mode, the electrical power generation system is capable of producing 80

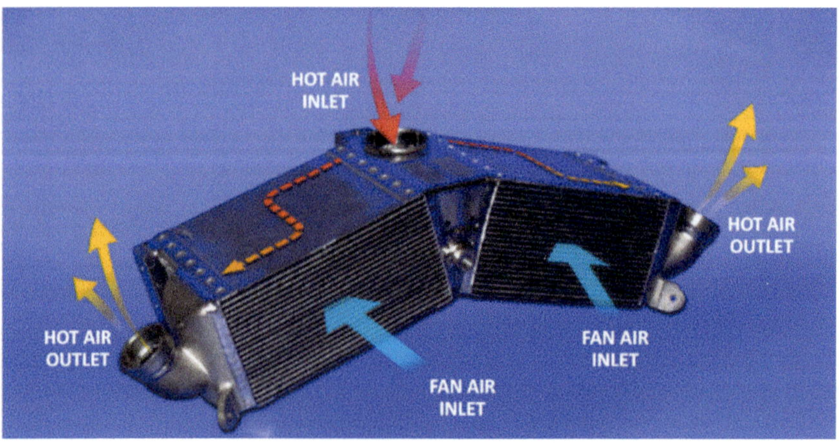

Fig. 10 Typical fan duct heat exchanger.

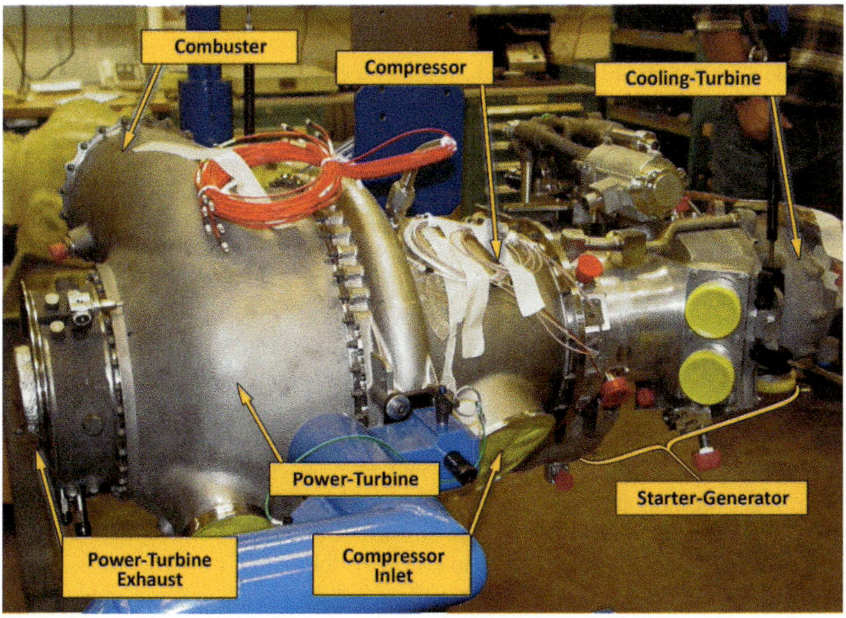

Fig. 11 PTMS turbomachine mounted on a single shaft.

kW continuous and 120 kW peak 270-VDC power. The generator provides power for main engine start and inflight emergency power. The F-35 flight controls are electrically actuated, so there is no need for an emergency hydraulic pump on the turbomachine. In addition, electrically driving the turbomachine fuel pump and lubrication system, along with the single-shaft design approach, facilitated removing the gearbox typically used on aircraft APUs.

C. ELECTRICAL POWER SYSTEM

Many of the technologies used in the F-35 EPS were developed during the development of the F-22, greatly reducing risk during the development phase. The overall bus architectures are very similar, and the solid-state power controllers are almost identical. The use of a central processing system to manage the overall control of the system demonstrated many benefits on the previous program. In addition to raising the level of integration among the various subsystems, two new key technologies were developed: a 270-VDC battery and the ES/G. These met the requirements that had been developed during the trade study phase.

To support the need for nearly instantaneous fill-in power to the EHAS, a battery was required. The battery provided fill-in power on the 270-VDC main bus while load and source management tasks were performed. The driving requirement for this battery was the need to provide 6.5-kW of power to start the PTMS turbomachine at 40°F. Then, following the start load, it would provide 8.6 kW with 40-kW peaks to support the EHAS load for an inflight emergency fill-in. The maximum charge voltage was limited to 350 VDC, and the voltage at the terminal was required to remain above 207 VDC throughout the event. Weight and volume are always a driver for aerospace applications, so an 84-cell lithium-ion battery system was defined. Saft Batteries was selected by General Electric as the supplier with the best chance of meeting this requirement, and its cells proved to have excellent performance after extensive development efforts.

The generating system consisting of the ES/G, ICCs, and a dedicated oil system for cooling the ES/G was developed by Hamilton Sundstrand (now UTC Aerospace Systems). In addition to providing the prime generator power, this system replaced the traditional engine-starting system (i.e., providing torque to the engine gearbox for engine starting). As mentioned previously, two complete systems were developed: the switched-reluctance system for the early flight test program, and its replacement, a more traditional synchronous system for later jets. The electric start system was the driving requirement for the ICC, while the continuous and transient loads drove the size of the generators.

Much of the technological development work had been accomplished during the Power Management and Distribution System for a More Electric Aircraft program. However, many issues were uncovered and surmounted through these development efforts. One feature of the More Electric Aircraft concept that was not obvious initially was the effect of having transient loads representing a very large percentage of the total load. With the F-35C configuration, 65% of the transient capability of the generating system is allocated to the EHAS. This drove a requirement for specifying the generating system's capability at 50 μs in addition to the more traditional steady-state, two-minute and five-second periods. Additionally, unique control functions are required to differentiate normal transient loads from short-circuit conditions.

D. HYDRAULIC AND UTILITY ACTUATION

Advanced developments were included in the weapon bay door drive system to meet the large hydraulic power demands. The F-35 WBDD system incorporated an advanced over-center motor that optimized the drive system performance to align with the hydraulic system's power capacity. Legacy pressure-compensated hydraulic motors experience hydraulic power losses across the control valves upstream of the motor. The use of the advanced over-center motor maximized the hydraulic pressure available to the motor, resulting in optimum power efficiency of the drive system. It also facilitated integration with the F-35 hydraulic power system. The decision to build a WBDD system integration facility at the supplier's site was instrumental in developing the drive system, controls, and software to facilitate successful and timely integration into the aircraft.

The utility actuation system also used door sequencing extensively to meet the needs of the air vehicle requirements. Overlapping doors prompted extensive use of left-to-right door sequencing. A variety of methods were used for the overlapping door sequencing, all tailored to the applications and software controlled to ensure reliability. The majority of the air vehicle doors incorporated door latches to support the stealth aircraft requirements, all requiring methods to integrate and sequence the latches with the door system applications.

The F-35C variant included an advanced wing-fold system for aircraft weight and volume efficiencies. The wing-fold system included new technologies developed and patented by Moog Aircraft Group that reduced the wing thickness requirements by approximately 1.5 in. These reduced weight reductions significantly, compared to legacy wing-fold systems. Extensive development, buildup testing, and system qualification led to the successful incorporation of this advanced technology into the F-35C aircraft. Figure 12 provides a view of the uninstalled wing-fold system.

E. EJECTION SYSTEM

The US16E ejection seat system incorporates several key technologies to support a safe ejection environment across a wide range of pilot body types (weight and height). Several features of the seat, such as head position control and arm and

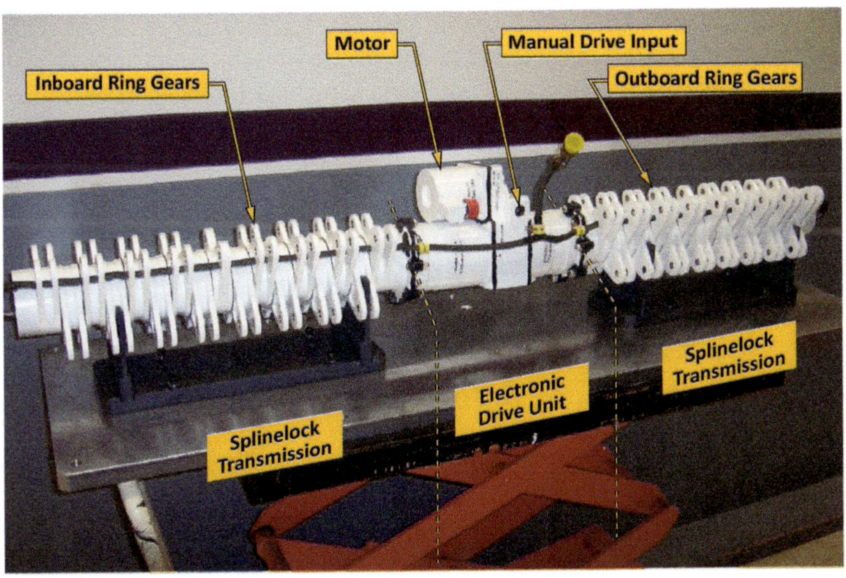

Fig. 12 Wing-fold actuation system.

leg restraints, are integrated. There is also sensing of environment and management of parachute deployment. With this combination, the seat successfully demonstrated a full envelope capability that minimizes the risk of injury.

The seat is qualified for safe ejection at up to 550 kt Calibrated Air Speed (KCAS) and provides superior terrain clearance performance over other fighter aircraft. Some of the design features that enable this performance are a five-mode electronic sequencer, active arm and leg restraint system, and neck protection device. The five-mode electronic sequencer energizes upon ejection and senses environmental data, such as pressure and acceleration. With this information, it determines the sequencing of drogue parachute deployment, main parachute deployment, and seat/pilot separation. A unique feature of the sequencer is its manual selection of main parachute deployment based on the pilot-weight range. To minimize accelerations during main parachute deployment, pilots weighing less than 135 lbm select a longer delay, minimizing acceleration and allowing the US16E to minimize head and neck loads. The selection is made by toggling a switch mounted on the left side of the seat, which is easily accomplished and verified during aircraft ingress. As a result of this design, the US16E ejection system achieves a greater terrain clearance and escape injury-protection envelope, compared to legacy systems (Fig. 13).

A second feature of the US16E ejection seat is the active arm and leg restraint. To minimize injury due to arm and leg flail during ejection, the US16E actively restrains the pilot's arms and legs during ejection. It uses fabric lines that are integrated into the pilot's flight ensemble. During ejection, the legs are drawn back and the arms are drawn to the lap and held until the lines have been severed. Figure 14 shows the US16E arm and leg restraint system. A third feature of the US16E is the neck protection device. This is an inflatable air beam that deploys during ejection to stabilize the pilot's head and reduce head and neck loading.

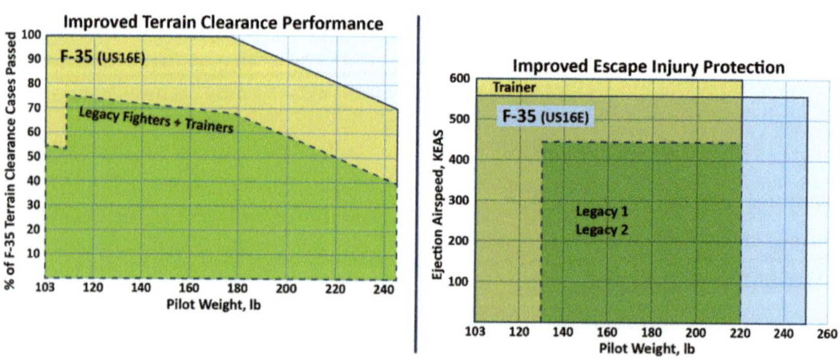

Fig. 13 F-35 US16E ejection envelope comparison.

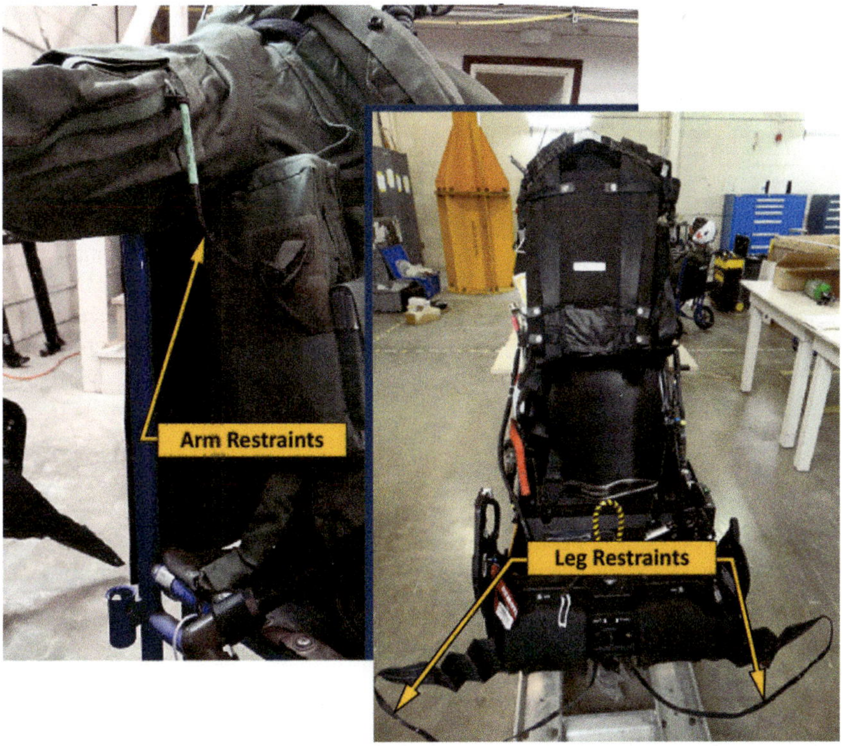

Fig. 14 US16E arm and leg restraints.

V. INTEGRATION QUALIFICATION EFFORT/CHALLENGES

There was a need for integrated systems and the development and qualification of several key technologies. Based on this, integrated testing and qualification resulted in several challenges during the development phase, including:

- The development of integrated modeling and testing
- Early flight test electrical system failures
- Lithium-ion battery development
- Integrated engine start and emergency power
- Hydraulic system integration
- Full envelope ejection seat qualification

Along with the individual system development at the suppliers, there was a need for a full-up integration facility. The facility had to be capable of testing the integrated system architecture not only for normal operation, but also—and perhaps more importantly—under failure conditions.

The Vehicle System Integration Facility (VSIF) supports both hardware-in-the-loop and pilot-in-the-loop capabilities for all three F-35 variants. It includes the following:

- EHAS
- EPS
- HUA systems
- PTMS and main engine drive stands
- Cockpit and visual display
- Bus architecture (including aircraft hardware)
- VMCs

Each of those systems can be integrated into the laboratory or emulated with software models, depending on the test needs. A six-degrees-of-freedom simulation provides the necessary plant models to simulate the aircraft environment. Figure 15 provides the layout for the VSIF.

A. ELECTRO-HYDROSTATIC ACTUATION SYSTEM

Aircraft-level integration played a significant role in the development of the EHAS. The complexity of the system and the time-critical interactions among the EHAS, VMC, EPS, and PTMS made integration a necessity. It was integral in verifying that the systems worked in coordination with each other both in

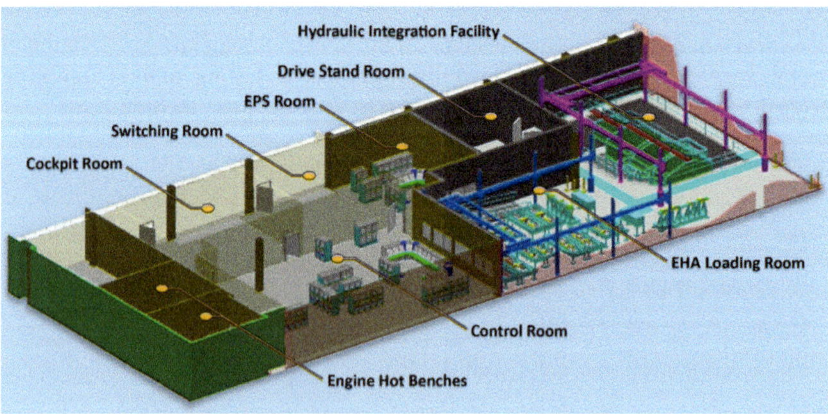

Fig. 15 Vehicle System Integration Facility layout.

Fig. 16 EHA loading fixtures.

normal conditions and as failures were inserted. Within the VSIF, an EHA loading fixture was developed to house the EHAS, along with loading fixtures (Fig. 16). Load cells were required to correctly mimic the EHAS environment specifically as it relates to thermal characteristics and electrical regenerative performance. The load cells' forces were computed by the aircraft simulation aerodynamic model. This allowed for varying hinge moments as a function of the aircraft envelope, rates, and accelerations, along with surface positions.

The PTMS provides the cooling air for the EHAS's EUs. The cooling air primarily cools the high-power switches that drive the EHA motors, but it also has indirect effects on all components in the box. The software solution to the cold temperature issue noted in the key technology development section was implemented. From this it was found that although increasing the motor losses could improve the performance of the EHA, it also increased the temperatures of the switches. To account for this, a new interface was created to request increased cooling flow from the PTMS. Additionally, the PTMS could not provide the increased flow under all conditions, which drove interlocks into the design to limit when the motor heating could be applied and prevent the switches from overheating.

Further complicating the cooling interface, the tubing carrying cooling air to the EUs on the right side of the aircraft was running around the engine and causing heat impingement. It was also elevating cooling air temperatures when compared to the left side. During normal operations the warmer airflow does not impact the ability of the EUs to operate. However, when the EUs are being stressed, the internal temperatures can rise and reach the component limits. This made it necessary to add a level of cooling flow that could be requested based on actual and predicted component temperatures.

The VMC is connected to the EPS, PTMS, and EHAS through a communications bus. Accordingly, commands and status messages must travel back and forth going from, to, and through the VMC. This allows for a more integrated aircraft and generates latency, timing issues, and the need for increased coordination among the systems.

The EUs control electric motors that commutate at a high frequency and can fail and burn out in milliseconds. As such, they must operate at internal frame rates that significantly exceed the bus. This means that the EU must be able to manage its state somewhat autonomously and address failures as they occur; however, it must do so predictably and within the expectations of the VMC running the aircraft control laws and providing position commands. The commands from the VMC had to be orchestrated such that it maintained ultimate control but did not limit the ability of the aircraft to self-protect during a failure. Reset processing and a built-in test capability also had to be sequenced to accommodate latencies and system interdependencies. This was most evident on the F-35B with the additional components required for STOVL operations.

All these interfaces were tested with hardware in the loop at the VSIF. That facility includes a complete triplex VMC and remote input/output setup, along with a full set of EHAs and EUs connected to a load system. It can also connect the EHAS to the EPS in the same manner as how it would be in an aircraft. This allowed interconnected failure modes to be tested so that the EHAS and EPS were both robust in responding to the failure events that can occur in a high-power system.

Early in flight test, the systems experienced one of these events. A high-power short in one channel of a dual-channel EU propagated to the opposite channel, causing a short across both power buses and a temporary loss of power. In spite of this, the aircraft was still able to recover, reconfigure, and land safely. Overcoming this problem illustrated that the design and integration testing performed had adequately prepared the systems for such an issue; however, it also brought into question other scenarios that required improvements. In part, the EUs needed to allow the EPS the opportunity to fully recover and address the variability of the timing of that recovery. As a result, new criteria were established to limit power draw and improve failure hold-off timing during a power loss event. The criteria also needed to allow for critical surfaces to automatically reset after a total power loss. The approach had to be balanced between the EPS response time and the time before the actuator would lose stiffness and cause surface flutter. New integration test cases showed improved handling of shorts and power losses and a significant increase in the robustness of the system.

B. POWER AND THERMAL MANAGEMENT SYSTEM

The PTMS normally operates in the closed-loop refrigeration mode, providing cockpit conditioning and equipment cooling during flight. In the event of an engine flameout, the PTMS transitions to an open-loop combusted power cycle and produces electrical power to drive flight control actuation and engine-start assist. Transition from the cooling mode to the emergency power mode requires deploying a ram air scoop to feed the compressor. In the process, this manipulates several valves to convert the closed-loop refrigeration cycle into an open power cycle, while also lighting the combustor.

The maturation of this transition started by performing a high-fidelity transient simulation of the process. A detailed transient model of the PTMS was built using MATLAB®/Simulink®. The model was used to develop the sequence of valve openings and closing and combustor light-off required for successful mode transition. Software was developed based on the transient modeling results, and both software and hardware were tested statically in a ground test stand and in an altitude chamber. The rigorous modeling and laboratory testing proved their value in the very successful flight test sequences for transition from bleed-cooling to self-powered emergency, in-flight propulsion engine start-assist, and in-flight PTMS self-start.

Legacy fighter aircraft programs mitigated the risk of propulsion system loss by running the APU during early flights. This was done until confidence in the propulsion system was gained through successful flight envelope expansion. The F-35 EPS contains a 270-VDC battery capable of providing emergency flight control power. However, actually landing the aircraft with only the available battery power is considered risky due to the unknown battery state of charge during the high current draw expected during landing. In addition, the PTMS's capability to produce electrical power depends on the recovery of the system's turbomachine inlet. Prior to flight, inlet recovery had only been evaluated by computational fluid dynamics analysis. In addition, the ability to transition from the closed-loop refrigeration mode to the open-loop power mode had only been analytically modeled and tested statically in an altitude chamber. Statistical failure analysis was used to show that the probability of the engine failure scenario, especially at benign flight conditions, was low enough to begin the flight test program with confidence.

Additional flight testing evaluated PTMS turbomachine inlet recovery by installing a total and static pressure rake in the turbomachine inlet throat. The PTMS was also operated in the open-loop, emergency power mode with the propulsion system operating. The process of transitioning from the closed-loop refrigeration mode to the open-loop emergency power mode was also demonstrated. It was tested in flight with the propulsion system operating while being facilitated by a flight test aid to simulate engine failure conditions.

Once the PTMS's inflight operation had successfully been demonstrated, engine inflight spool-down restarts were demonstrated. These had the PTMS preset in the open-loop emergency power mode, followed by engine-start assist testing with the PTMS preset in the open-loop emergency power mode. Ultimately the complete engine flameout scenario with the PTMS providing engine-start assist was demonstrated by cutting fuel flow to the engine. This simulated an engine flameout and allowed the PTMS to transition from the closed-loop refrigeration mode to the open-loop emergency power mode, followed by a successful engine-start assist.

The aircraft operating environment is critical to the PTMS and main engine, so the VSIF laboratory did not incorporate the PTMS or main engine systems. Instead, drive stands were incorporated to mimic the necessary inputs in support of the overall aircraft. The two drive stands are shown in Fig. 17.

C. ELECTRICAL POWER SYSTEM

As with any development program, changes were implemented as a result of discoveries from planned testing and unplanned events. Early in the design cycle, Northrop Grumman assembled a dedicated team to construct and run computer-based models. With these, the team analyzed various system interactions and helped to guide the requirements development and implementation. Models

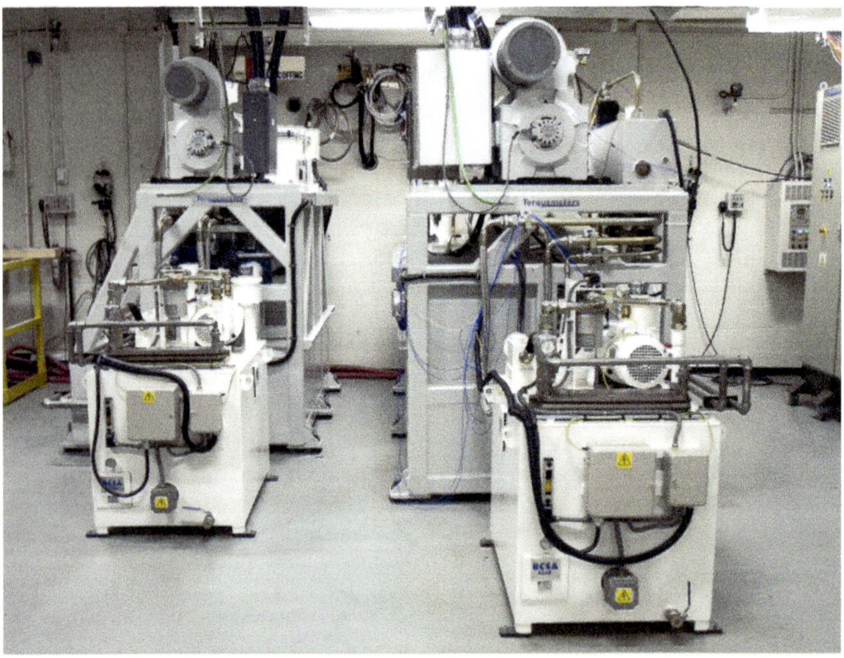

Fig. 17 Main engine and PTMS generator drive stands.

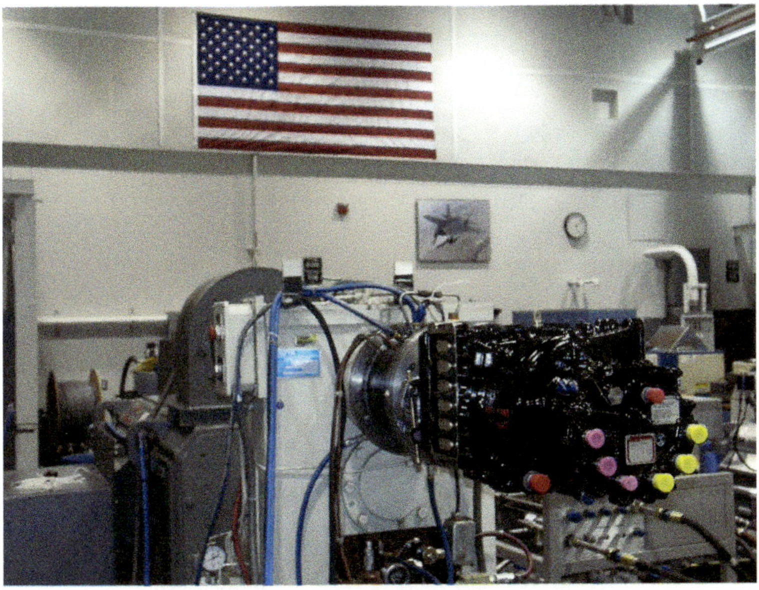

Fig. 18 ES/G testing at Northrop Grumman.

were used to predict power quality during main engine starts and high-power operations, and the effects of EHA in-rush currents during normal operations and failure modes. The models were extremely useful and were maintained throughout the entire qualification effort. Northrop Grumman also conducted extensive development testing in its facilities, particularly on the generating system and the batteries (Fig. 18).

Each component was subjected to extensive environmental, life, and performance testing as an individual component. In some cases, such as the battery systems, the components were tested as a sub-subsystem to ensure that the final products would be robust in the specified environments. As is typical, the electromagnetic environment and vibration environment were the most difficult.

Extensive testing was performed on the EPS as a stand-alone system and with several other key systems to ensure that the high level of systems-level integration was successful. A System Integration Laboratory (SIL) was constructed at a Hamilton Sundstrand (now UTC Aerospace Systems) facility where all the components of the EPS were assembled. The controlling software was also assembled there to ensure that the EPS components were well-integrated with each other. Lockheed Martin's VSIF (Fig. 19) was used extensively to expand on the efforts in the SIL. This included integration with the EHAS and key mission systems components, ultimately including a pilot in the loop as well. The latter was accomplished via a flight simulator in which end-to-end integration and an exhaustive list of failure modes were extensively examined.

Fig. 19 EPS test stand in Lockheed Martin's VSIF.

During laboratory testing of the overall EPS, an inadvertent short circuit was applied to the output of the 270-VDC battery. As expected, the extremely low impedance of the battery allowed extremely high currents into the fault. The distribution system was designed to tolerate the current, but the incident revealed a problem with the cell design: the internal connections were not sized to tolerate the extremely high currents, which resulted in the destruction of the battery. As a result, a fuse was added into the battery case to prevent external short circuits from cascading damage to the battery.

Similarly, an early failure during flight testing showed that the logic implemented to protect and isolate shorts in the electrical distribution unit loads was inadequate. Although protection was in place to ensure that both generators would not be susceptible to a short circuit, the battery system was not adequately considered. Based on the short circuit capability of the battery, it was decided that it would be prudent to postpone connecting the battery as long as possible. Waiting would allow the fault to clear before bringing online the remaining power source for flight control actuation.

In spite of our modelling and integration efforts, some problems were still beyond our ability to analytically duplicate even after they were discovered. One example was how the generating (starting) system reacts with the power source (PTMS turbomachine) during engine starting. An interaction between the load (ICC and ES/G), source (turbomachine), and interconnect wiring was discovered during the early production program. In it, the turbomachine would occasionally trip for instantaneous overcurrent conditions during specific frequency-dependent portions of the start profile. This anomaly was not discovered during development testing and, despite numerous attempts to duplicate the phenomenon through simulations, the team was unable to duplicate it; however, additional inductance between the turbomachine and ICC resolved the issue.

Because of the highly integrated systems implemented in the subsystems, the EPS was also influenced by failures in other systems. The EHAS inherently regenerates energy that must be dissipated. Early testing of the circuitry for consuming this energy within the EHAS controller showed that much more energy was being generated than could be dissipated by the circuitry. This was resulting in early failures of the circuitry.

Since volume in the EHA controller was limited, increasing the dissipation capacity was difficult, so alternate methods of consuming the regenerative energy were investigated. An early alternative was to allow the energy to flow back onto the electrical bus, where (in theory) it would either be consumed by other loads or back-drive the engine through the generating system.

Implementing a circuit to manage the regenerative energy within the E/SG ICC was not especially difficult; however, implementing the overall system was not as easy, particularly after trying to accommodate failure modes. It was finally determined that there was no feasible way to handle failure modes of the regenerative circuitry (open circuits) without risking the destruction of all the other equipment on the bus. This was due to the high voltage that would be created when a regenerative pulse was unable to be consumed by the regenerative circuitry. The first solution tested was the addition of burden resistors within the ICCs to ensure that there was always a sink for the regenerative energy. This then led to the final solution of localized regenerative consumption within the EHAS.

In spite of the team's efforts to integrate with the various stakeholders, a problem was discovered just prior to the start of the flight test program. This was the sharp rise-time of power demand from the EHAS, particularly during an F-35C carrier-based landing with numerous, rapid movements of the large horizontal tails. The affected power demand generated concern over how the electrical transients would be handled by the drive train through the generating system, gearbox, and engine. This was particularly concerning during the conversion of the electrical energy draw from the mechanical system. Extensive testing in the VSIF with multiple pilots in the loop established more precise usage profiles throughout multiple flight profiles, particularly the landing profiles of the F-35C. Analyses and modeling of this new data was beneficial, with no changes required to any of the units; however, the life of the ES/G stub shaft became a life-limited part as a result of the dynamic stresses imparted on the shear section.

As a final lesson learned, an incident during early flight testing revealed the need to ensure that the supplier base abides by the intricacies of high-voltage power. Proper isolation of 270-VDC power is absolutely necessary when designing units, particularly those containing both high- and low-voltage inputs. Such isolation ensures that the high voltage cannot short to the low-voltage inputs, or between isolated high-voltage inputs, during other catastrophic failure modes.

Through this development effort the EPS has proven to be a very robust and reliable system. The team spent considerable effort integrating the EPS with the PTMS and engine to minimize issues between/among these systems. Extensive testing during the Enhanced Integrated Systems Test [6] proved to be a worthwhile endeavor as well. No significant problems were discovered between/among these systems throughout the development and flight test programs.

Looking to the future, the EPS is well-situated for growth, with margin available on the 270-VDC system. By contrast, the 28-VDC system has little growth available as currently implemented. Implementing active, real-time load management (planned for a future software release) will allow the current system to meet all the projected needs for the foreseeable future. It is expected that future improvements in switching and conversion technologies will allow extending the capacity of the current 28-VDC system without increasing weight or volume.

D. HYDRAULIC AND UTILITY ACTUATION

The complexities of the hydraulic system requirements and associated design implementation led to the development of integration facilities to support the systems development. The development of a Hydraulic Integration Facility (HIF), or iron bird facility, was originally planned to take place at a supplier facility. Early planning transitioned this test facility to Fort Worth, Texas, to incorporate the test facility as part of the overall F-35 VSIF. Transitioning the HIF (Fig. 20) to the VSIF in Fort Worth offered cost efficiencies and supported a higher level of system integration. The HIF was instrumental in the HUA hardware and software development, and supported the rapid resolution of system anomalies during the development phase of the F-35 program.

Another test facility that was instrumental in the successful development of the F-35 utility actuation systems was the WBDD system test rig (Fig. 21). It was developed by UTC Aerospace Systems in Wolverhampton, United Kingdom. The weapon bay doors are designed to operate within the full flight envelope of the aircraft. The WBDD system includes various operational modes for weapon deployment, vertical landing, and ground operations. In addition, various safety features were designed into the system to protect maintenance personnel. The WBDD test rig included the complete actuation system, weapon doors, locks, and aircraft door lands. Overall system operation and performance were thoroughly tested throughout the operational loads and temperature extremes. The test facility was instrumental in developing the system software. In addition, aircraft rigging procedures were developed at the test facility, supporting successful aircraft integration and initial operations.

Fig. 20 Hydraulic integration facility.

E. EJECTION SYSTEM

The successful performance of the US16E ejection seat was proven through a rigorous qualification test program. The program included traditional component testing and incorporated a series of catapult tests, sled track ejection tests, and ejection testing. The latter was performed at altitudes requiring the seat to be ejected out of the cockpit of a Martin-Baker Aircraft Co. Ltd.-owned Gloster Meteor airplane. The catapult testing was carried out at Martin-Baker's facility in Chalgrove, United Kingdom. The catapult testing demonstrated the performance of the seat through the firing of the primary and secondary cartridges and the deployment of the neck protection device. Catapult testing is an efficient, cost-effective means to ensure that the impulse provided by the main cartridges meets the head and neck injury criteria without the complexities of a full-up sled test. Figure 22 shows a picture of a catapult test.

The US16E also incorporated an extensive sled test program to ensure that the entire escape system operated as designed across the required speed range. Two ground test tracks were used during the qualification program: Martin-Baker's

Fig. 21 F-35 WBDD system test facility.

Fig. 22 Catapult test at Martin-Baker's Chalgrove facility.

Fig. 23 Sled test at Martin-Baker's Langford Lodge facility.

facility in Langford Lodge, Ireland, and the test facility at Holloman Air Force Base, New Mexico. The Langford Lodge facility was used for most of the sled testing, and the Holloman facility was used for the 550-to-600 KCAS testing. Figure 23 shows a picture of a sled test at the Martin-Baker facility. Martin-Baker modified a Gloster Meteor aircraft to enable an ejection seat to be ejected at altitude. The altitude testing was executed at Cazaux Air Base, France (Fig. 24).

Fig. 24 US16E altitude ejection testing.

VI. CONCLUSION

The F-35 Vehicle Subsystems team successfully integrated previously federated systems into a cohesive system of systems that reliably and safely meets very stringent aircraft requirements. The team's rigorous, structured process of systems engineering included extensive modeling, simulation, integration, and testing. Through it, the resultant systems demonstrated significant reductions in weight, volume, and complexity across a wide range of operating conditions. The F-35 Vehicle Systems constitute the first production deployment of an integrated subsystems architecture for a combat aircraft. These systems and their designers overcame several developmental problems. The systems are well-positioned to meet future needs. Lessons learned from the team's efforts provide a firm grasp of the processes and techniques required to progress the level of integration to meet future requirements.

ACKNOWLEDGMENTS

The authors thank all their colleagues on the F-35 Vehicle Subsystems team who were instrumental in developing, qualifying, and fielding these systems on the F-35. Composing this paper was only made possible by their dedication and expertise. The authors would also like to thank graphic artists Daniel Buck and Dennis Soultaire for their assistance. In addition, we extend our gratitude to the executive review team whose members provided guidance in the development of this paper.

REFERENCES

[1] Sheridan, A., and Rapp, D., "F-35 Program History—From JAST to IOC," AIAA Aviation Forum, June 2018 (to be published).

[2] Counts, M., Kiger, B., Hoffschwelle, J., and Houtman, A., "F-35 Air Vehicle Configuration Development," AIAA Aviation Forum, June 2018 (to be published).

[3] Wiegand, C., et al., "F-35 Air Vehicle Technology Overview," AIAA Aviation Forum, June 2018 (to be published).

[4] Burkhard, A., and Deitrich, R., "Joint Strike Fighter Integrated Subsystems Technology, A Demonstration for Industry, by Industry," *Journal of Aircraft*, Vol. 40, No. 5, Sept./Oct. 2003.

[5] Weimer, J., "Past, Present & Future of Aircraft Electrical Power Systems," AIAA Paper 2001-1147, Jan. 2001.

[6] Wurth, S., Smith, M., Celberti, L., and Walters, J., "F-35 Propulsion System Design, Development, and Verification," AIAA Aviation Forum, June 2018 (to be published).

CHAPTER 11

F-35 Mission Systems Design, Development, and Verification

Greg Lemons*, Karen Carrington†, Thomas Frey‡ and John Ledyard§

Lockheed Martin Aeronautics Company, Fort Worth, TX

F-35 mission systems allow pilots to execute traditional advanced tactical missions critical to allied and partner nations. The systems include the most advanced sensor management and data fusion of any current fighter aircraft. These capabilities offer unmatched situational awareness to the pilot and provide decision aids that allow the pilot to make critical timely decisions. The development of these capabilities was accomplished utilizing use cases derived from mission vignettes to perform capability and model-based development. This approach was extended to the test and evaluation phase to develop mission-level scenarios used for validating the models and capabilities.

I. INTRODUCTION

The F-35 comprises a set of highly common aircraft for the Air Force, Marine Corps, and Navy, as well as the 12 current F-35 partner nations. Although the airframes themselves possess slight differences unique to each variant's operating environment requirements, the mission systems' hardware and software are common. In addition, product-line engineering tags have been integrated into the mission systems' software requirements baseline. This enables repeatable and affordable country-unique builds for production off of a U.S. baseline.

The F-35 sensor suite includes the following:

- AN/APG-81 Active Electronically Scanned Array (AESA) radar
- AN/ASQ-239 Electronic Warfare (EW)/Countermeasures (CM) system
- AN/AAQ-40 Electro-Optical Targeting System (EOTS)

*Systems Engineer Sr. Mgr., F-35 Mission Systems Design.
†Systems Engineer Dir., F-35 Mission Systems Design.
‡Senior Technical Fellow, Software Engineering.
§Technical Fellow, F-35 Mission Systems.

- AN/AAQ-37 Electro-Optical (EO) Distributed Aperture System (DAS)
- AN/ASQ-242 Communications, Navigation, and Identification (CNI) avionics suite

These five sensors provide F-35 fusion with object detection and measurements in the Radio Frequency (RF) and Infrared (IR) spectrums. This compilation of data gathering disseminates more information about the environment than what has ever been available on a fighter aircraft.

In addition to receiving information from the onboard sensors, the F-35 receives off-board tracks and measurements from the Link 16 datalink and the Multifunction Advanced Data Link (MADL). Designed for 5th-Generation aircraft, MADL provides fusion-quality data on all air and surface tracks to other members of the flight group. These data include the track state, track covariance, identification features, and passive RF data.

The amount and fidelity of the off-board information provided by MADL was one of the largest challenges for the fusion design. The capability of the sensors and information sharing across MADL presented a challenge for sensor fusion. The challenge was to ensure that the tracks displayed were real and not duplicated, which would result in display clutter. The last few software builds in the System Development and Demonstration (SDD) phase of the F-35 program were aimed at tackling display clutter problems. The objective was to ensure that the pilot had accurate and timely information to make real-time tactical decisions in the cockpit.

This paper discusses the design, development, and verification of each of these systems, as well as the system of systems integrated into the F-35 aircraft.

II. THE VISION

A. AIRCRAFT CONCEPT

The F-35 was born of the need for an affordable multirole, multiservice, multinational (U.S. Air Force, Marine Corps, and Navy, and international operators) fighter aircraft to replace an aging fleet of fighter and attack aircraft. The aircraft being replaced were battle tested by their operators. The challenge of this concept was rooted in the varied missions and operating environments that these platforms were satisfying. In order to replace their collective capabilities with a single fighter platform, a new way of thinking about single-seat fighter avionics design was needed.

The concept for the F-35 developed by Lockheed Martin centered on returning the pilot to the role of tactician. This principle was the driving force behind the avionics development plan. One of the Lockheed Martin Mission Systems team's design goals was to develop a set of sensors that could collect information across multiple spectrums. Another goal was to develop a sensor control scheme

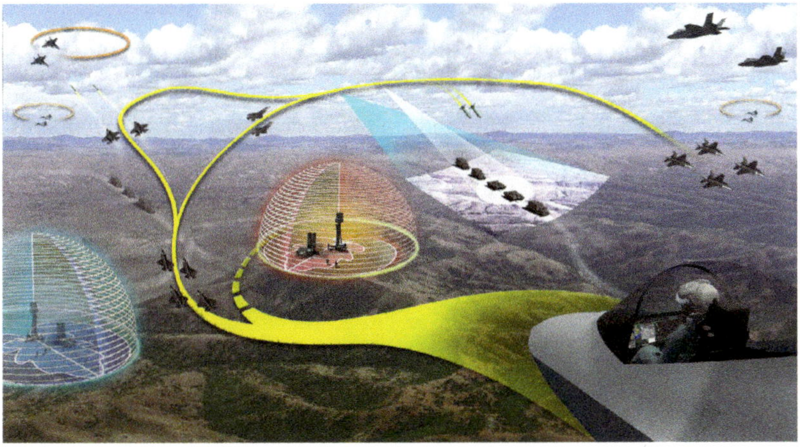

Fig. 1 Returning the pilot to the role of tactician.

of autonomous sensor management. This, along with a next-generation cockpit, would provide the pilot with an unprecedented amount of information distilled down to an easily consumable format. Prior to this, the pilot spent precious minutes setting up radar scans and adjusting tilt, gain, and refresh rates, while also monitoring multiple displays to run an intercept. With this new suite, the pilot is able to view a picture of the multispectral battlespace in a consolidated format, as depicted in Fig. 1.

The strategy for developing the avionics system was based on a block buildup strategy. It was founded on basic warfighting capabilities and then built into the most advanced fighter weapons system currently in service. The most fundamental elements were developed first, and then the design moved to higher levels of capability. In this way, the team reduced the risk of software development produced by having a single large software release. This strategy was adapted from previous tactical fighter programs, such as the F-16, and more recently, the F-22. Those programs demonstrated that it is essential to break the software development into manageable blocks to reduce the complexity and cost of testing. Incremental releases also provide more of an opportunity to manage requirements creep and incorporate technology changes that directly benefit the warfighting capability. This was evident in the F-35 program's ability to implement additional weapons (e.g., GBU-39) and capabilities (e.g., operational test support changes) to support emerging requirements. The approach is structured as three development blocks that establish the basic flight control systems and essential mission systems before building up mission capability. The approach to building up the block plan in detail is depicted in Fig. 2.

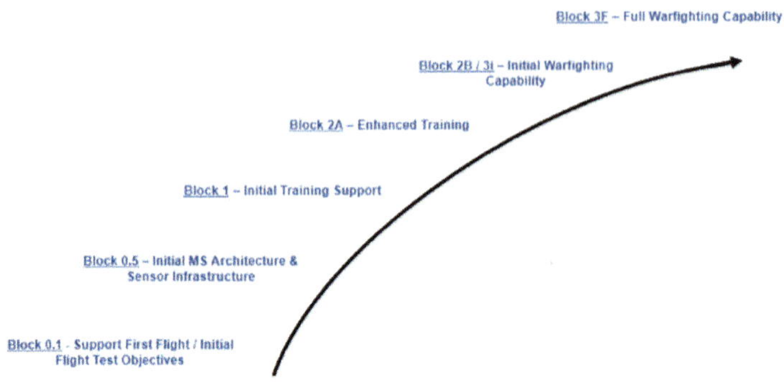

Fig. 2 Block development buildup.

B. SYSTEM ARCHITECTURE

It was recognized early in the concept development phase that the architecture of the mission systems would be key to the program's success. To succeed, many challenges had to be faced to develop the right architecture. One was that the computing resources needed for the full set of capabilities could not fit within the power, weight, volume, and thermal limitations of the air vehicle using available technology. Another was that the long development cycle and initial low-rate production was expected to result in Diminishing Manufacturing Source (DMS) problems. In addition, the aircraft needed to be easily adaptable to support the unique needs of multiple countries. Further, it needed to be unclassified on the production line and on the ramp to avoid increased production and sustainment costs. Beyond this, it also needed to operate in future battlespaces where the movement of data at multiple levels would be key to interoperability.

The plan for overcoming computing resource and DMS challenges was to execute multiple technology refreshes of the computers during development. The processing update would allow Moore's law to take effect, providing increased processing capability over time that would fit within the limitations of the air vehicle. The updates would also allow updates to mitigate DMS and validate that the application software was independent of the underlying processor.

To achieve the goal of making the application software independent of the processor changes, three design approaches were used. The first approach (Fig. 3) was to layer the software on top of Commercial, Off-the-Shelf (COTS) operating systems, under the assumption that the virtual platform would not change.

The second approach was to use rate-based processing for all threads when timing and latency were critical. This approach would achieve constant system-level timing, even with faster processing. Also, it would enable analyzing the system and proving that it was schedulable using rate-monotonic theory.

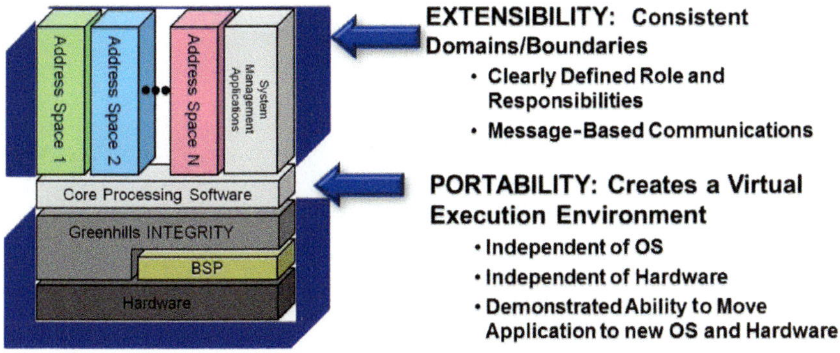

Fig. 3 Layered software design.

Both benefits supported easier integration, reduced regression testing, and supported airworthiness and safety certification.

The third approach was to use messages for communication among all application software components, as well as components and subsystems [1]. This created controlled interfaces among the components and enabled moving applications to different processors without impacting the software. The approach contributed to solving interoperability and adapting the software for multiple countries. With the clearly defined interfaces and communication paths in the system, it was possible to control data paths using the trusted computing base. This enabled isolating data access for specific address spaces and ensured that the application remained at the designed security level for a specific datalink. Further, combining the messaging and access control with a COTS operating system with a high assurance level made it possible to design write-down applications for each datalink.

The capability was then advanced to interoperate with multiple participants in the battlespace at different operational levels and message formats. From this, the messaging and access control was partitioned to the external communications domain. This provided a broker for data on and off the aircraft, ensuring the correct classification level and translating the external data into formats consistent with internal data. It also transformed internal data into the message formats and needs of the external links.

III. F-35 SENSOR SUITE

The F-35 sensor suite includes the AN/APG-81 AESA radar, AN/ASQ-239 EW/CM system, AN/AAQ-40 EOTS, AN/AAQ-37 EO DAS, and AN/ASQ-242 CNI system. This collection of advanced multispectral sensors, shown in Fig. 4, provides the F-35 with a next-generation ability to see the battlespace.

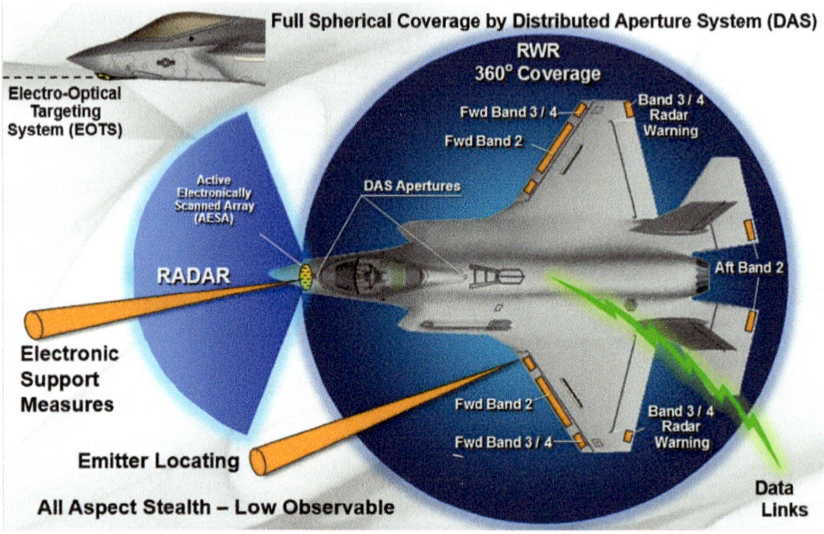

Fig. 4 F-35 installed sensor locations.

A. AN/APG-81 RADAR

Northrop Grumman's Electronic Systems sector's AN/APG-81 radar was developed as a next-generation version of its AN/APG-77 AESA radar, first fielded on the F-22A Raptor. The design was further refined with the AN/APG-80 fielded on the Block 60 F-16. This lineage of AESA radar designs allowed for a rapid development and insertion of previously fielded common waveforms in early software blocks. It also paved the way for more complex functions during later deliveries.

AN/APG-81 testing was approached in a buildup fashion of increasingly complex integration into the rest of the avionics system. The integration began with stand-alone laboratory testing isolated from the rest of the avionics system. It then progressed through the Northrup Grumman flying testbed, where dynamic stand-alone open-air testing was performed. From there the AN/APG-81 was integrated into the F-35 avionics suite to continue both laboratory testing and dynamic open-air testing on the Lockheed Martin Cooperative Avionics Test Bed (CATB). After the system was proven on the CATB it progressed to full airborne testing on the F-35. The buildup timeline for the AN/APG-81 is shown in Fig. 5.

The F-35 radar system has an active, electronically scanned Multifunction Array (MFA) and the RF support electronics necessary to support a fully functional radar. It also has integrated radar software modes that are hosted on the integrated core processor. The radar operates through the nose radome, which

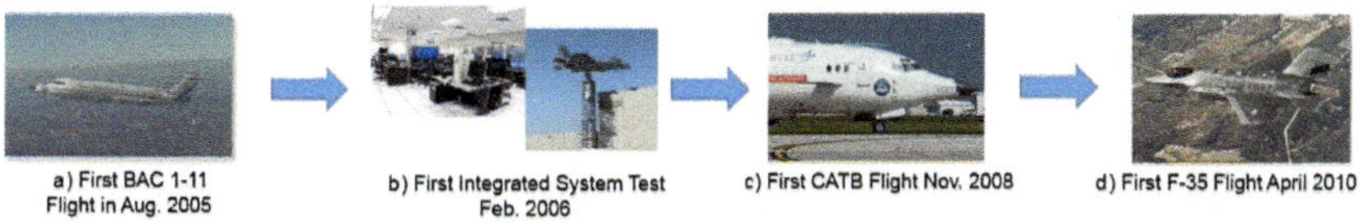

Fig. 5 AN/APG-81 integration buildup.

has a wide bandwidth, enabling high-power transmissions over a large frequency range (Fig. 6).

The AN/APG-81 is designed to operate as a radar, Electronics Support Measures (ESM) receiver, and jammer. It includes active and passive Air-to-Air (A/A) and Air-to-Surface (A/S) target detection, track, and identification capabilities. In addition, it allows many of these to be interleaved, providing both A/A and A/S functionality. The sensor also supports the Advanced Medium-Range Air-to-Air Missile (AMRAAM®) and synthetic aperture radar mapping, ground and sea moving target detection and track, and A/S ranging. Radar functions include electronic protection for operation in jamming environments and low probability of intercept features to minimize the likelihood of emissions being usefully detected by airborne or surface-based receivers. Radar functions also support system health determination and calibration. Figure 7 depicts the various radar functions.

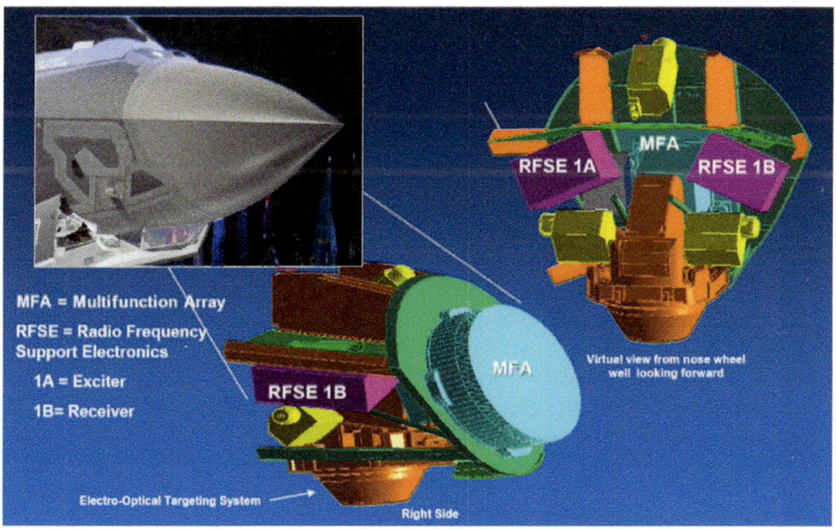

Fig. 6 Installed radome and radar locations.

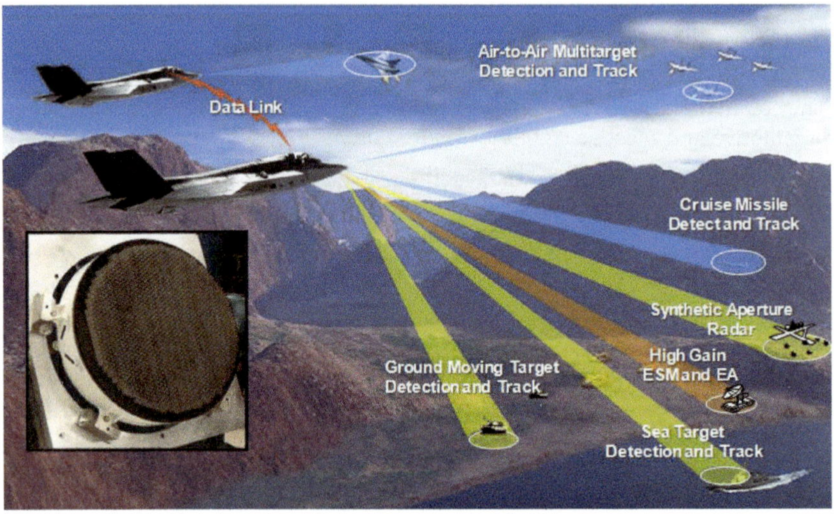

Fig. 7 Radar system air-to-air and air-to-surface operation.

B. AN/ASQ-239 ELECTRONIC WARFARE/COUNTERMEASURE SYSTEM

The AN/ASQ-239 EW/CM system is an integrated suite of hardware and software. It is optimized and designed to provide the F-35 with a high level of A/A and A/S threat detection and self-protection. It can search, detect, identify, locate, and counter RF and IR threats. The EW system supports the application of Electronic Support Measures (ESM) through such functions as:

- Radar warning
- Emitter geolocation
- Multiship emitter location (including high-sensitivity states)
- High-Gain (HG) ESM
- HG electronic CM
- HG Electronic Attack (EA) via radar MFA utilization

The EW functions are designed for:

- Wide frequency coverage
- Quick reaction time
- High sensitivity and probability of intercept
- Accurate direction finding and emitter geolocation

- Multiship geolocation
- Self-protection countermeasures and jamming

The countermeasure subsystem provides multiple self-defense responses, including preemptive and reactive techniques, based on available expendable payload and/or threat-specific self-protection plans. The EW/CM system provides emitter tracks to the sensor fusion function, which fuses EW track reports and other sensors (e.g., radar and DAS, off-board sensors) and displays the information to the pilot. The EW/CM system consists of the following primary elements:

- Band 3/4 apertures
- Band 3/4 aperture electronics
- Centralized EW electronics (Racks 2A and 2B)
- CM controller unit
- CM dispensers
- RF and digital interfaces with the MFA
- Digital clock reference interfaces with the CNI system

The installation locations of the EW/CM system-related equipment are depicted in Fig. 8. The EW system is common among the three F-35 variants, except for the forward Band 3/4 arrays, which employ longer elements for the F-35C

Fig. 8 EW equipment location.

Carrier Variant (CV). Also differing, the distance between the inboard and outboard arrays is less on the CV variant due to the wing fold. In addition to the EW Band 3/4 apertures, the radar MFA is employed to support EW functions. There are growth provisions allocated for Band 5 radar warning such that Band 5 apertures, aperture electronics, and the Band 5 switch can be incorporated into the EW subsystem architecture.

The EW apertures comprise six multielement antenna array sets covering portions of the Band 3 and Band 4 frequency spectrum, along with both vertical and horizontal polarization. All the arrays have Azimuth (AZ)-only designs that do not rely on the use of Elevation (EL) arrays. The passive array assemblies use a traveling wave-notch element approach designed to balance gain, polarization, Field of View (FOV), and radar cross-section features. Each Band 3/4 aperture feeds an aperture electronics module that amplifies and passes the detected RF signals from the apertures. It does this through a switch matrix and tuners that distribute the RF to a set of wideband EW Receivers (EWRs). The switch matrix also receives RF signals from the radar MFA when tasked to support HG modes.

The EWR suite consists of 12 wideband receivers grouped into three integrated sets of four channel receivers. The wideband receivers take in RF energy and convert the data into digital information via a set of high-speed analog-to-digital converters for processing. Each EWR performs initial data processing and generates pulse parameter reports that are sent to the EW controller/preprocessor. The processor then provides further signal processing and algorithms to support all EW activities. Various intelligence products combine to produce an Electronic Intelligence (ELINT) database in a preplanned mission data file. This data file provides the system with the necessary parametric descriptions of the emitters of interest for emitter identification and scan schedule operations.

EW system testing was performed in a buildup fashion of increasingly complex integration into the rest of the avionics system. The integration began with stand-alone laboratory testing isolated from the rest of the avionics system and progressed through a Sabreliner T-39-based flying testbed. From there the EW system was integrated into the F-35 avionics suite to continue both laboratory testing and dynamic open-air testing on the Lockheed Martin CATB. After the system was proven on the CATB it progressed to full airborne testing on the F-35. The buildup timeline for the EW system is shown in Fig. 9.

a) First T-39 Flight June 2005 b) First Integrated System Test May 2006 c) First CATB Flight Aug. 2007 d) First F-35 Flight April 2010

Fig. 9 Electronic warfare system integration buildup.

C. AN/AAQ-40 ELECTRO-OPTICAL TARGETING SYSTEM

The F-35 requirement for a Low Observable (LO) combat configuration did not allow for the traditional Targeting Forward-Looking Infrared (TFLIR) solution. The legacy pod systems could not be both missionized and easily concealed for LO operations. The solution was to integrate the targeting pod system into the outer mold line of the aircraft. The EOTS, built by Lockheed Martin Missiles and Fire Control, was built specifically for the F-35 to provide the jet with an LO IR targeting capability. Its integration was approached in a buildup fashion, as shown in Fig. 10. The initial open-air testing was performed with a modified Sabreliner T-39 jet to test the EOTS as a stand-alone sensor to verify sensor-level behavior. The EOTS was integrated into the rest of the avionics system on the Lockheed Martin CATB flying testbed to test the interactions between the sensor and the full avionics system with a pilot in the loop. The final testing and verification came with the full integration of the EOTS into the F-35.

The EOTS is an internally mounted advanced Midwave Infrared (MWIR) targeting system with a faceted window having LO characteristics, designed for A/A and A/S targeting support. By using the midwave portion of the IR spectrum the EOTS provides a sharper image and less susceptibility to target obscuration from smoke and haze. The EOTS may be used in the imaging mode in A/A and A/S, or in the IR Search and Track (IRST) mode in A/A. Design consideration has been given to achieving the following:

- A good receiver signal-to-noise ratio
- Effective FOVs for A/A and A/S performance
- A broad field of regard
- Auto-search pattern modes
- Low false alarm rates

The EOTS uses low-profile gimbals with an optical system that maintains boresight accuracy between the Forward-Looking Infrared (FLIR) and laser functions. Precise stabilization of the EOTS's line of sight is achieved by gyro-controlled AZ and EL gimbals, and fine stabilization is achieved through a fast-steering mirror. Equipped with a staring 1024-by-1024-element MWIR focal plane array, the

a) First Sabreliner Flight May 2007 b) First CATB Flight Nov. 2010 c) First F-35 Flight March 2011

Fig. 10 EOTS integration buildup.

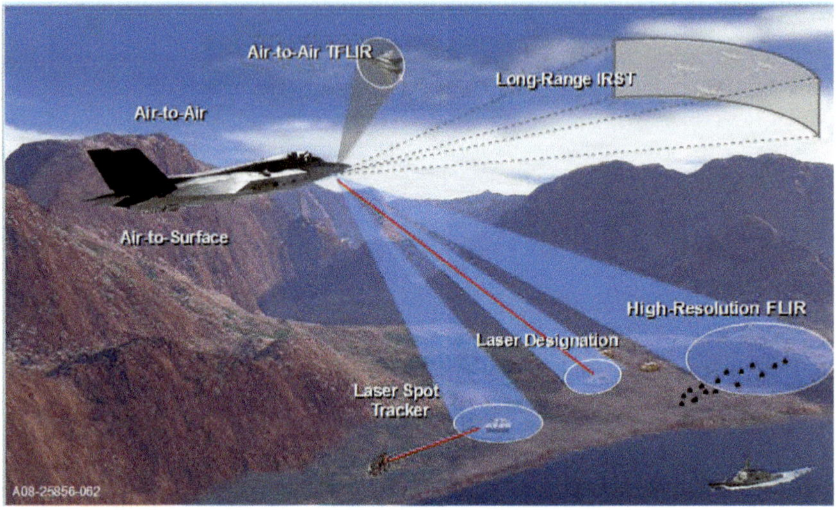

Fig. 11 EOTS capabilities.

EOTS is a dual-FOV system. The narrow FOV is optimized for targeting functions, and the wide FOV is developed to maximize search performance.

The EOTS's functionality consists of a TFLIR image, laser range finder/designator, laser spot tracker, and IRST, as shown in Fig. 11.

D. AN/AAQ-37 ELECTRO-OPTICAL DISTRIBUTED APERTURE SYSTEM

The program required a 360-deg spherical coverage missile warning system. The EO DAS consists of six identical MWIR sensors distributed on the aircraft, each with a corresponding airframe window panel. The sensors are installed such that their respective FOVs (95-deg AZ and EL) overlap to provide total spherical coverage. This EO DAS subsystem provides the pilot with both an MWIR tracking capability and FLIR visual scene, but its FLIR is more comprehensive. In legacy FLIR systems, the pilot's visual scene was limited to the forward sector. With the F-35's EO DAS, the pilot has a 360-deg spherical view of the environment. This allows for a true synthetic vision system, with the image displayed on the pilot's Helmet-Mounted Display (HMD).

The EO DAS integration began with a single sensor installation within a pod. This pod was mounted on an F-16 to support initial testing and data collection for image processing algorithm validation. This podded system was also mounted on a QF-4 drone to enable testing of the missile warning function. The next step in integration was to mount a sensor in an integration-representative fashion on the Northrop Grumman–owned BAC 1-11 flying testbed. The first introduction of

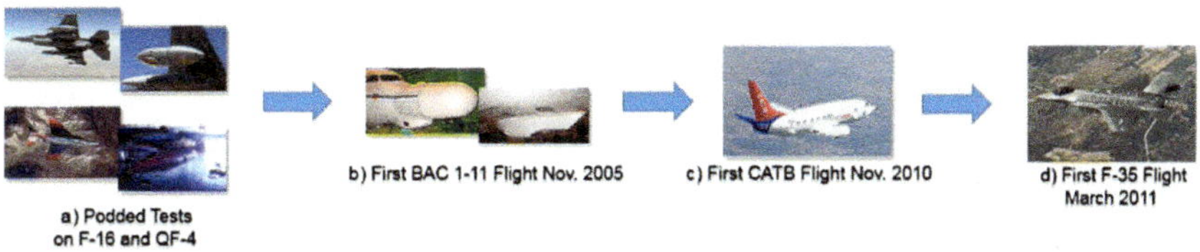

Fig. 12 DAS buildup integration.

multiple EO DAS cameras into the integrated avionics system was performed on the Lockheed Martin CATB platform. This marked the beginning of integrating the EO DAS sensors into the Lockheed Martin–developed fusion algorithms. The final step to fully incorporate the EO DAS into the integrated avionics system came in March 2011, with the first flight testing on an F-35. The EO DAS integration timeline is depicted in Fig. 12.

Key EO DAS operational functions in Block 1 of Flight Test Update B are Navigation Forward-Looking Infrared (NAVFLIR) and missile warning. Block 2 of Flight Test Update B added Surface-to-Air Missile (SAM) launch point reporting and situational awareness IRST. These EO DAS functions are available simultaneously and serve to enhance situational awareness and defensive response. Figure 13 illustrates these capabilities.

Fig. 13 DAS capabilities.

E. AN/ASQ-242 COMMUNICATIONS, NAVIGATION, AND IDENTIFICATION SYSTEM

The CNI system (Fig. 14) is an integrated subsystem designed to provide a broad spectrum of:

- Secure/anti-jam/covert voice and data communications
- Precise radio navigation and landing capability
- Self-identification, beyond-visual-range target identification
- Connectivity with off-board sources of information

In support of the stealthy operation and design goals of the F-35, the CNI subsystem includes techniques to reduce the probability of detection, interception, and exploitation, and can deploy electronic CM. These techniques include frequency agility, spread spectrum, emission control, antenna directivity, and low probability of intercept design capabilities. The CNI system provides interoperability with existing (legacy) military and civilian communication, RF navigation, and Identification Friend Foe (IFF)/surveillance systems. It is also interoperable with appropriate civilian systems for U.S. and European airspace operations. The CNI system provides an inherent growth capability and the flexibility to incorporate additional functionality through software upgrades. It also provides for hardware upgrades driven by parts obsolescence and enables manufacturing cost reduction and/or performance improvement.

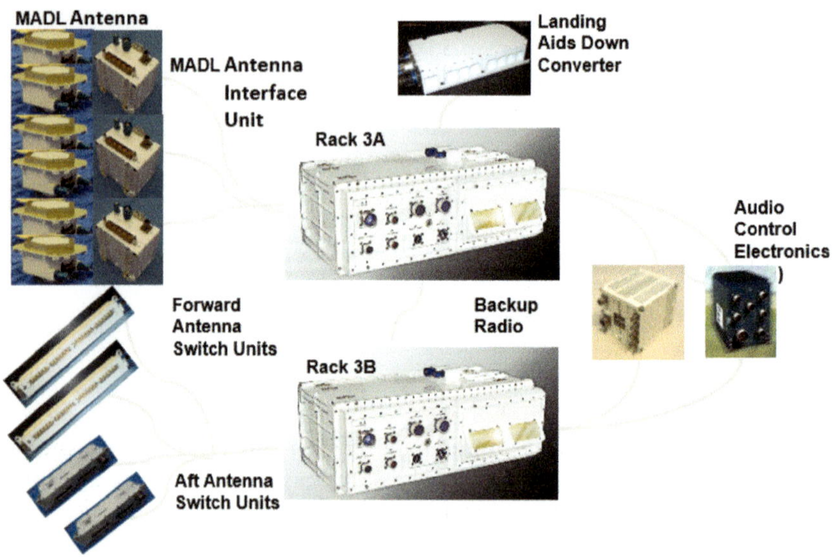

Fig. 14 CNI system components.

The CNI-specific data, signal, and cryptographic processing are performed in unique CNI processors, and the integrated core processor as required. The CNI system includes all functionality related to audio generation and distribution for the aircraft. This includes the pilot intercom; integrated caution, advisory, and warning messaging; pilot audio alerts; and support for the voice recognition function.

The CNI system includes an all-attitude Inertial Navigation System (INS) and anti-jam GPS. These provide outputs of linear and angular acceleration, velocity, body angular rates, position, attitude (roll, pitch, and platform AZ), magnetic and true heading, altitude, time tags, and time. The INS and GPS provide navigation data to the ownship kinematic model, which produces the navigation solution for the aircraft. The baseline system provides high-rate motion compensation data to the radar and EOTS.

IV. FUSING THE DATA INTO INFORMATION

The F-35 fusion engine is the software module at the heart of the integrated mission systems capability on the aircraft. Fusion involves constructing an integrated description and interpretation of the tactical situation surrounding ownship [2]. It draws from onboard, cooperative, and off-board data sources to enhance situational awareness, lethality, and survivability [2]. The fusion functionality is divided into two major subfunctions: Air Target Management (ATM) and Surface Target Management (STM). The purposes of these functions are to optimize the quality of air and surface target information, respectively. Their functionality is implemented in three primarily software modules: the A/A Tactical Situation Model (AATSM), the A/S Tactical Situation Model (ASTSM), and the Sensor Schedule (SS).

The AATSM software module receives data from onboard and off-board sources about air objects in the environment. It then integrates this information into kinematic and identification estimates for each air object. Similarly, the ASTSM software module receives data from onboard and off-board sources about surface objects in the environment. It then integrates this information into kinematic and identification estimates for each surface object.

Objects that are ambiguous between air and surface are sent to both Tactical Situation Models (TSMs). Each TSM assesses the quality of its tracks to identify any information needs. The System Track Information Needs (STINs) are sent from the TSMs to the SS software module. The SS prioritizes the information needs by track and selects the appropriate sensor mode command to issue in order to satisfy the information need. The SS provides the autonomous control of the tactical sensors to balance the track information need and the background volume search needs.

Measurement and track data are sent to fusion from the onboard sensors (e.g., radar, EW, CNI, EOTS, DAS) and off-board sources (e.g., MADL, Link 16).

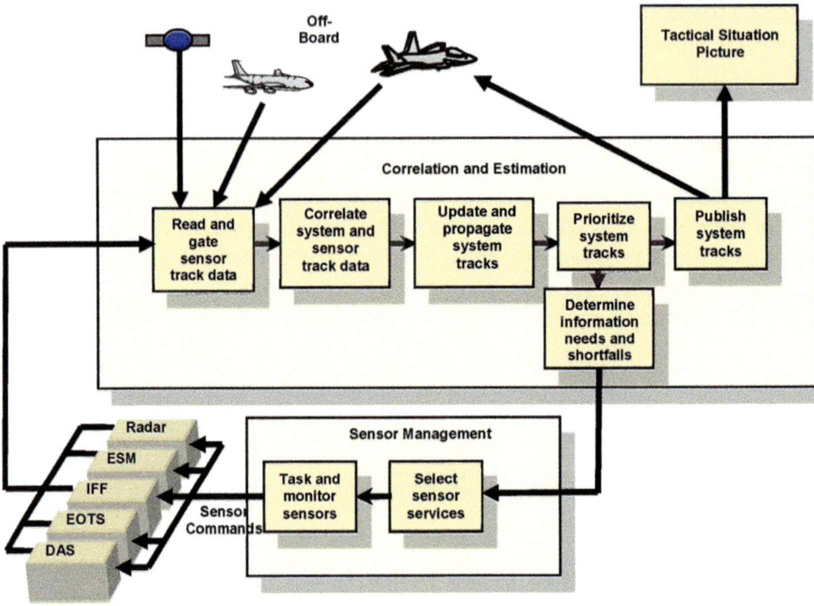

Fig. 15 Closed-loop sensor data fusion.

When this information is received at the TSM, the data enter the data association process. This process determines whether the new data constitute an update for an existing system (fusion) track or potentially new tracks. After being associated with a new or existing track, data are sent to the state estimation to update the kinematic, identification, and Rules of Engagement (ROE) states of the object.

Kinematic estimation refers to the position and velocity estimate of an object. It can also include an acceleration estimate for maneuvering air track. The kinematic estimate also includes the covariance for the track, an estimate of the track accuracy. Identification estimation provides an estimate and confidence of the affiliation, class, and type (platform) of the object. The identification process also evaluates the pilot-programmable ROE assistant rule to determine when the sensing states and confidences have been met for declaration. Estimation publishes the updated track state (kinematic, identification, and ROE statuses) to the system track file. At a periodic rate (about once a second), each track is prioritized and then evaluated to determine whether the kinematic and identification content meets the required accuracy and completeness. Any shortfall for a given track becomes STINs. The STIN message for the air and surface tracks are sent to the SS to make future tasking decisions for the onboard sensor resource. The process continues in a closed-loop fashion with new pieces of data from the sensors or datalinks. Figure 15 illustrates this process.

The results of this fusion of information are provided to the other elements in mission systems. They are provided to the Pilot/Vehicle Interface (PVI) for display, fire control and stores for weapon support, and EW for CM support. This allows these elements to perform their related mission functions to provide:

- A clearer tactical picture
- Improved spatial and temporal coverage
- Improved kinematic accuracy and identification confidence
- Enhanced operational robustness.

For a clearer tactical picture, multiple detections of an entity are combined into a single track instead of multiple tracks. For improved spatial and temporal coverage, a target can be continuously tracked across multiple sensors and FOVs. This is made possible by the extended spatial and temporal coverage of the onboard sensors, as well as the off-board contributor. Improved kinematic accuracy and identification confidence requires the effective integration of independent measurements of the track from multiple sensors or aircraft. This integration is what improves the detection, tracking, positional accuracy, and identification confidence. Enhanced operational robustness requires the abilities to fuse observations from different sensors and hand off targets between sensors. This leads to increased track resilience to sensor outages or countermeasures. Increased dimensionality of the measurement space (i.e., different sensors measuring various portions of the electromagnetic spectrum) then reduces vulnerability to denial of any single portion of the measurement space.

V. NEXT-GENERATION COCKPIT

Fig. 16 F-35 cockpit.

The visible product of this information gathering is ultimately the fighter pilot's office. The F-35 cockpit (Fig. 16) was designed to accommodate the unprecedented amount of data available to the pilot of a single-seat fighter. The F-35 display suite and associated PVI were

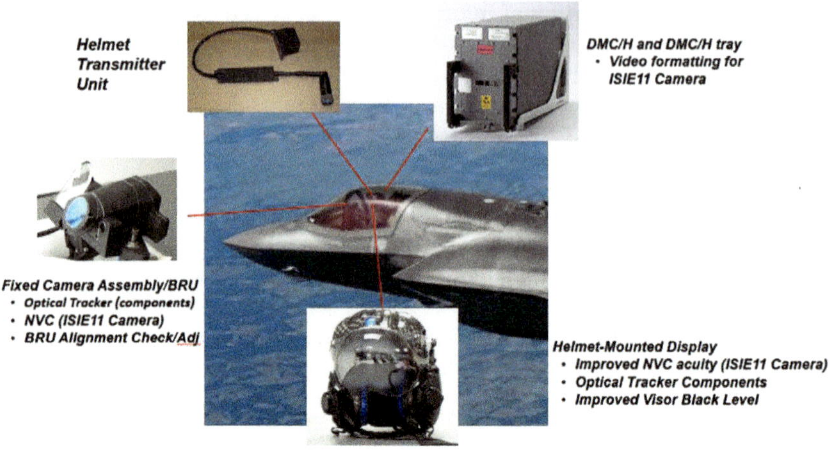

Fig. 17 Gen III helmet-mounted display system.

conceptualized through a multiyear immersive simulation and evaluation, along with flight demonstrations on surrogate aircraft.

A. HELMET-MOUNTED DISPLAY

The HMD, included in Fig. 17, was developed to provide the pilot with both existing and evolutionary tactical advantages. In it, legacy systems, such as the Joint Helmet-Mounted Cueing System (JHMCS), were supplemented by an integrated digital Night Vision Camera (NVC). The HMD also provides integrated flight reference information that allowed for the removal of the traditional Head-Up Display (HUD). This improved upon the legacy tactical and navigation display, with an integration surpassing that of the HUD/JHMCS combination. The HMD design evolved from concepts demonstrated using the Viper-II HMD on the Variable-stability In-flight Simulator Test Aircraft (VISTA) F-16 program. This was the basis for proving the concept of a HUD-less cockpit. The HMD design and integration then matured through several iterations during the SDD phase. Ultimately, it met ejection safety limits (Gen I to Gen II), program requirements for line-of-sight accuracy (Gen II to Gen III), and (later) full ejection envelope (Gen III to Gen III Lite).

The HMD is a monochromatic 1280-by-1024-resolution bi-ocular display with a 30-by-40-deg FOV. The helmet position is determined by a hybrid (magnetic, inertial, and optical) tracker system that allows for low-latency symbol positioning. The HMD can display either the MWIR image provided by the DAS or a near-IR image provided by the embedded NVC. When the pilot is looking forward, the NVC image is blended with the image provided by a fixed camera mounted on the glare shield. This is done to eliminate the obstruction caused by the canopy bow frame.

B. PANORAMIC COCKPIT DISPLAY

A striking difference between the F-35 and other legacy fighters is the incorporation of a large-format touchscreen display. The Panoramic Cockpit Display (PCD) provides the pilot with a programmable and reconfigurable display that allows for customized views to satisfy mission needs. The PCD was the replacement selected for the original rear-projection Multifunction Display Suite (MFDS) [3]. The MFDS rear-projection display technology was abandoned due to several factors. The lack of available contrast and insufficient resolution for tactical displays within the F-35 cockpit environment caused the symbols to be difficult to read. Pilots required the system to be at full brightness for all daytime flights, which shortened the projection lamp lifespan. This resulted in frequent maintenance being needed for the developmental test aircraft. Due to these and other factors, the program chose to recompete the display and associated processing elements early in the SDD contract.

The PCD consists of an 8-by-20 active matrix liquid crystal display and an Electronics Unit (EU), as shown in Fig. 18. The display resolution is 2560-by-1024 color pixels. The PCD uses a portal window concept to support multiple formats on a single display surface. A touchscreen interface and/or hands-on throttle and stick provide pilot input to the PCD. The EU has two separate power supplies and independent display management computers for each half (left and right) of the display. This, along with the ability to display any format on either display, provides redundancy to support mission operations and airworthiness. The EU accepts multiple video inputs for display on the PCD and is also responsible for outputting an MPEG-2 standard compressed video stream to a recording function to facilitate pilot debrief. The PCD unit is compatible with the night vision imagery system.

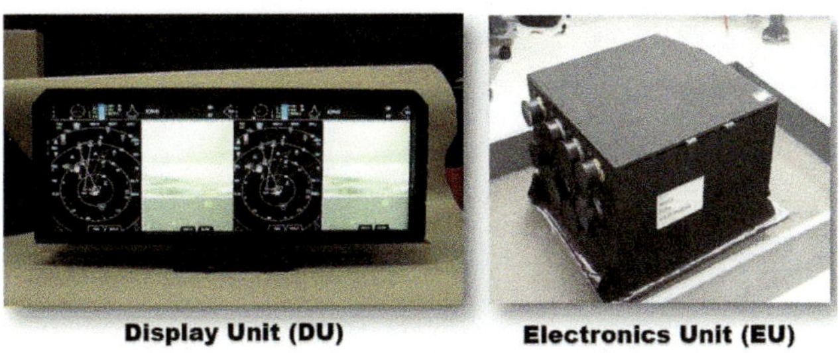

Fig. 18 PCD display unit and electronics unit.

VI. VERIFICATION

A challenge for the F-35 program was to reduce the cost of integration and verification. The concept developed was based on verification of performance to specific missions depicted in Fig. 19. The focus of the plan was to break down the missions into verifiable pieces that could be developed into a buildup plan for test and verification. This allowed a view of the problem space that did not center on any one specific scenario. Instead, it took a broader look at how the aircraft performed in representative test scenarios.

To rely less on flight testing for developmental buildup, the F-35 Mission Systems team leveraged mission-representative vignettes that had been developed for verification. The team applied those same scenarios and criteria to the lower-level testing. The subsystem and laboratory testing venues allowed for a rapid turn of changes once a testable product was provided. The laboratories constructed, along with their associated capabilities listed in Table 1, provided unique venues to add or subtract levels of complexity to grow the system and isolate issues.

The F-35 Mission Systems team used mission-based scenarios and a pyramid-type test approach (Fig. 20). With these, the team could progressively add complexity to the integrated system and quickly compare system performance to uncover problems early in integration. This also allowed for releases to flight test that may have a limitation in one area but verification-capable performance in another. By including the developmental test pilots in the final laboratory testing sessions, pilots were able to get an idea of performance and learn where they may experience shortfalls. This allowed them to more efficiently perform

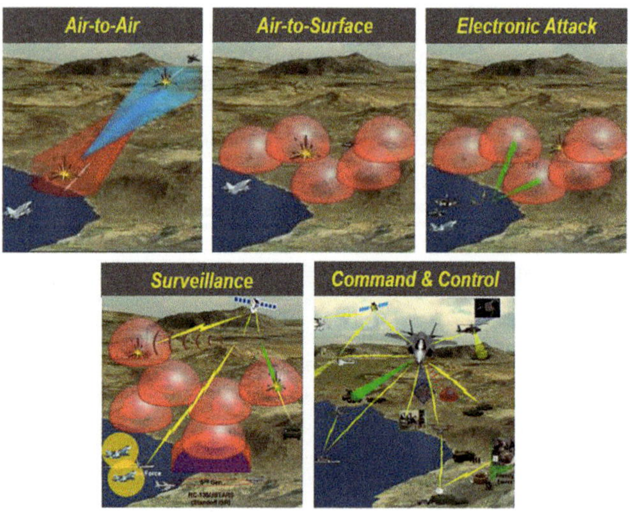

Fig. 19 F-35 mission types.

TABLE 1 VERIFICATION LABORATORIES

Laboratories	Capabilities
Simulated Systems Integration Station	1. Core avionics hardware and PVI 2. Supplier-provided models for sensors and weapons 3. Ability to connect to vehicle system and other integrated avionics laboratories
Open-Air Systems Integration Station	1. Open-air environment with representative apertures and associated hardware 2. Full avionics hardware suite 3. Capable of CATB cooperative testing
Stimulator-Based Systems Integration Station	1. RF and IR stimulation capability 2. Representative sensor hardware minus apertures 3. Multiaircraft cooperative simulation utilizing using connections to multiple laboratories

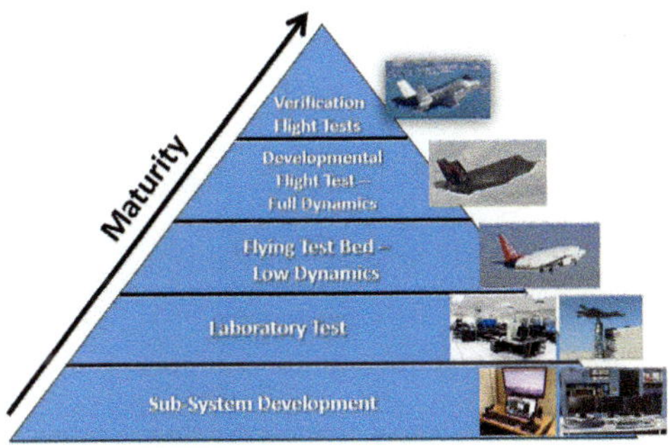

Fig. 20 Test buildup methodology.

flight tests to gather the required data. The partnership and communication between the developmental test pilots and the development team proved essential to maintaining the rapid pace required to achieve full system verification.

VII. CONCLUSION

The F-35 mission systems suite of sensors, displays, and advanced fusion algorithms was developed to satisfy the need to provide tomorrow's fighter pilot with an unprecedented level of information. At the same time, the suite allows pilots to still perform the tactics mandated by the mission. The F-35's performance has been demonstrated in more than 100,000 flight hours covering developmental test, operational test buildup, and training flights. The F-35 also showcased its capability during Red Flag 17-1, where the platform achieved a 20:1 kill ratio in simulated combat exercises [4].

F-35 mission systems design, development, and verification were achieved with an object-oriented, multilevel COTS-based architecture and a suite of powerful multispectral sensors. As successful products of the F-35 Mission Systems team's efforts, the advanced fusion algorithms and next-generation cockpit enable the pilot to return to the role of tactician.

REFERENCES

[1] Levis, J., Sutterfield, B., and Stevens, R., "Fiber Optic Communication within the F-35 Mission Systems," *IEEE Conference Avionics Fiber-Optics and Photonics*, Annapolis, MD, 2006, pp. 12–13. doi: 10.1109/AVFOP.2006.1707476

[2] Frey, T., "F-35 Information Fusion," AIAA Paper, June 2018 (to be published).

[3] Kalmanash, M. H., "Status of Development of LCOS Projection Displays for F-22A, F/A-18E/F, and JSF Cockpits," *Proc. SPIE 4362, Cockpit Displays VIII: Displays for Defense Applications*, 7 Sept. 2001. doi: 10.1117/12.439116

[4] Demerly, T., "'Red Flag Confirmed F-35 Dominance with a 20:1 Kill Ratio' U.S. Air Force Says," *The Aviationist*, https://theaviationist.com/2017/02/28/red-flag-confirmed-f-35-dominance-with-a-201-kill-ratio-u-s-air-force-says/ [retrieved 3 May 2018].

F-35 Information Fusion

Thomas L. Frey*, J. Chris Aguilar†, Kent R. Engebretson‡, David K. Faulk§ and Layne G. Lenning¶

Lockheed Martin Aeronautics Company, Fort Worth, TX

Information fusion is a set of algorithms that combines data from all sources to create an integrated view of the environment to provide situational awareness. Fusion is a core attribute of the F-35, designed into the mission systems from initial conception. The F-35 Information Fusion development leveraged experience from past fusion projects across the corporation and industry; however, there were some fundamental architectural decisions and algorithmic solutions that are unique to the F-35 concept of operation. This paper discusses some of the key design decisions and features that shaped the final F-35 Information Fusion solution.

I. NOMENCLATURE

$\vec{D}$	vector difference between Observer 1 and Observer 2
K_k	Kalman filter gain
κ	Dempster-Shafer measure of conflict between two masses
$m(A)$	Dempster-Shafer probability mass of proposition A
$P_{k\|k-1}$	previous covariance estimate propagated to time k
R_k	measurement covariance for Z_k
$\hat{R}_1$	range estimate from Observer 1 to the target
$\hat{R}_2$	range estimate from Observer 2 to the target
Σ_1	covariance matrix for the target from Observer 1
Σ_2	covariance matrix for the target from Observer 2
Σ_{int}	covariance matrix for the intersection point

*Senior Fellow, F-35 Information Fusion, AIAA Associate Fellow.
†Software Engineer Sr. Staff, F-35 Information Fusion.
‡Fellow, F-35 Information Fusion, AIAA Senior Member.
§Embedded S/W Engineer Staff, F-35 Information Fusion, AIAA Senior Member.
¶Principal Research Scientist, F-35 Information Fusion.

$\hat{X}_{k\|k}$	resultant state estimate at time k
$\hat{X}_{k\|k-1}$	previous state estimate propagated to time k
$\hat{u}_1$	unit direction vector from Observer 1 to the target
$\hat{u}_2$	unit direction vector from Observer 2 to the target
Z_k	measurement vector at time k

II. INTRODUCTION

In his 1982 book *Megatrends*, John Naisbitt predicted that in the Information Age we would find ourselves "... drowning in information but starved for knowledge" [1]. For many aviation or military applications, as the amount of information increases, the sheer volume of data becomes overwhelming and results in the loss of situational awareness [2]. As the amount of data increases and control choices multiply, the pilot workload increases exponentially to a point where eventually even the most able pilots begin to miss important information or fail to recognize critical situations [3]. The loss of situational awareness degrades the reaction time of the user [4, 5]. Shenk coined the term *data smog* to describe the data overload caused by the dramatic increase in information without a means to readily incorporate this information or its significance [6]. In *Consilience: The Unity of Knowledge*, E. O. Wilson predicted the need for a synthesizer that could combine relevant information at the right time to support critical decisions [7]. Kline identified fusion algorithms as this synthesizer to reduce information overload, improve situational awareness, and reduce user reaction time [8].

Fifth-Generation fighters, with their diverse sensor suites and multirole mission sets, require some form of information fusion to support the pilot's situational awareness (Fig. 1). The F-35 avionics suite comprises several complementary sensors and off-board datalinks but is a single-seat fighter with no weapon systems operator. Without some form of information fusion, the pilot would be left to manually correlate sensor and datalink tracks together, while executing tactical air, land, and sea missions while also trying to fly the aircraft. This combination of factors can increase pilot workload and quickly lead to an overwhelming amount of displayed information. Information fusion algorithms aggregate the onboard and off-board sensor information to provide a complete and accurate representation of the environment, resulting in an increase in situational awareness, which is the ultimate goal of fusion [9, 10].

The terms *data fusion*, *sensor fusion*, and *information fusion* are often used interchangeably, and yet these terms have subtle distinctive connotations within the community. The Joint Directors of Laboratories (JDL) Data Fusion Model defines a useful categorization of fusion algorithms and techniques used in the solution of many general fusion problems [11]. It defines *data fusion* as the combining of information to estimate or predict the current or future state of the environment. Level 1 fusion is focused on object assessment. Level 1 fusion algorithms include: 1) data association algorithms, which determine whether

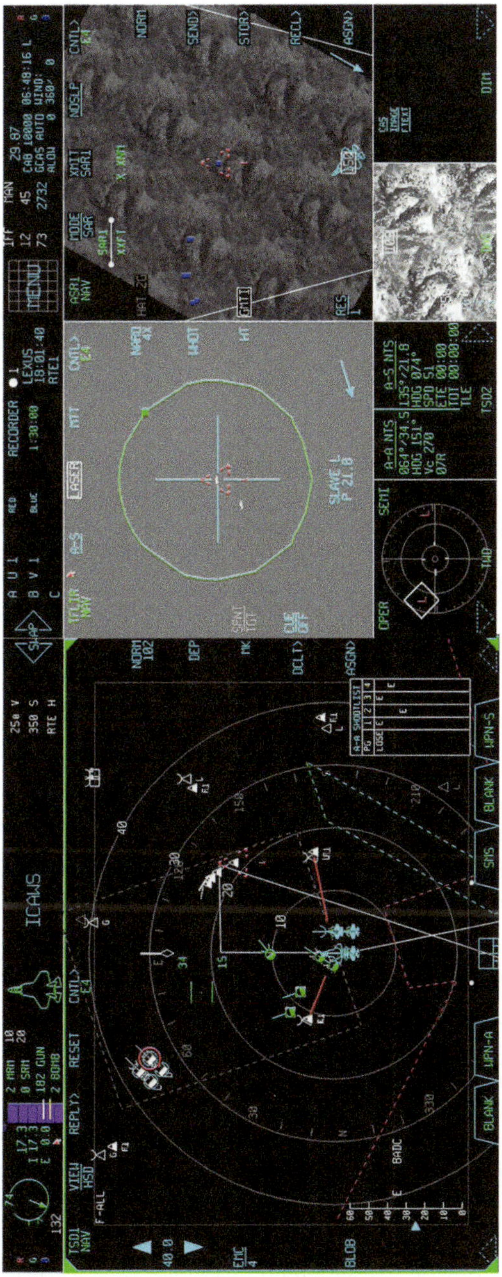

Fig. 1 Information fusion is a set of algorithms that combines data from all sources to create an integrated view of the environment.

information from multiple sources describes the same object; and 2) state estimation algorithms, which estimate the current (and, in some cases, future) state of the physical object in the environment. The estimate includes both the kinematic state (e.g., position, velocity) and an estimate of the object's Identification (ID). Level 2 fusion focuses on aggregating the Level 1 objects, inferring relationships between/among the objects and corresponding events, and assessing the unfolding situation. Level 3 fusion assesses the impact of perceived, anticipated, or planned actions in the context of the unfolding situation, for instance, in terms of lethality and survivability. Level 4 fusion is focused on process refinement, including sensor resource management or sensor feedback to modify sensor actions and refine the overall situational picture.

There are many published algorithms that offer methods to address each of these fusion functions; however, in the field the imperfect nature of data and the varying fidelity of disparate data sources make the fusion problem much more challenging to solve. To quote Yogi Berra, "In theory there is no difference between theory and practice. In practice, there is" [12]. For example, object refinement is relatively easy when there are a few well-separated objects, but as the number of objects increases and their spacing decreases, the data association problem can become much more difficult or even unsolvable. The potential for the object to perform maneuvers, coupled with a highly dynamic environment involving multipath, interference, signal blockage, and weak signals, can complicate the ability of fusion to make sense of the environment. Information fusion must be robust enough to provide reliable situational awareness, even in this challenging environment.

A. THE EVOLUTION OF FUSION TECHNOLOGY

The first radar was introduced to a fighter in the mid-1940s. As confidence grew in this technology, pilots came to rely on the radar to provide situational awareness in the environment. A radar warning receiver was later added to give a coarse indication of the general direction of a hostile emitter. In the 1970s and 1980s, new sensors and datalinks added secondary sources of ranged tracks. This ushered in the initial generation of fusion technology, which used data association or correlation algorithms to identify tracks that most likely described the same object in space and then suppressed all except the most accurate copy of the track from the display (Fig. 2). This is sometimes referred to as correlation or display fusion. The accuracy of this technique was equal to the accuracy of the best track.

The next step in fusion technology was to combine the output of multiple sensor tracks into a blended system solution. By blending the tracks from two or more sensors, the resultant system track accuracy approached the accuracy of the best parameter of the contributing sensors. For example, blending tracks from a radar and an Infrared Search and Track (IRST) sensor could have the range and range rate accuracy of the radar along with the angle and angle rate

Fig. 2 The evolution of fusion technology.

accuracy of the IRST; however, the accuracy of the resultant track remains limited by the track's update rate. If the track's update rate (fusion rate) is larger than the measurement's rate, then there is a loss of accuracy, even with optimal algorithms [13].

Fifth-Generation aircraft are designed to process the sensor measurements rather than the sensor tracks, resulting in an integrated system track containing the most precise track accuracy and enabling cooperative sensing across aircraft. Measurement-level processing can provide earlier discovery of objects in the environment that are hard to detect. By processing the measurement-level data, the system can use detections from any sensor (or aircraft) to confirm a track before any single sensor can make the declaration. The focus on the measurement data rather than track data also means that combat ID information from a sensor is retained by the system track, even when the track is no longer in the sensor's field of view because the system track can be maintained by other sensors or aircraft.

In addition to improved accuracy and detection performance, the introduction of an Autonomous Sensor Management capability provided the ability to react and refine objects in the environment much faster than any human could respond [14]. The addition of the Autonomous Sensor Manager is referred to as *closed loop fusion*. This capability provides the fusion process a feedback loop to coordinate the actions of the sensors in a complementary way to detect, refine, and maintain tracks based on system priorities [15]. The sensor management capability evaluates each system track, determines any kinematic or ID needs, assesses those needs according to system track prioritization, and cues the sensors to collect the required information. Analogous to John Boyd's Observe, Orient, Decide, and Act (OODA) loop [16], which expressed the engagement advantage related to the pilot's ability to understand and react to an adversary, closed loop fusion accelerates the ability of the pilot to understand and respond to an object in space faster and often at a much greater range than legacy systems.

III. FUSION ARCHITECTURE

The F-35 is not one but *three* highly common aircraft for the U.S. Air Force, U.S. Marine Corps, and U.S. Navy, designed to avoid the higher costs of developing,

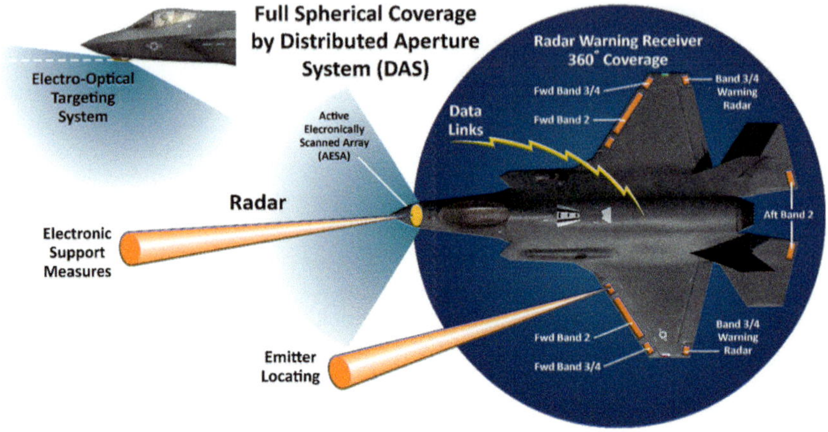

Fig. 3 The F-35 sensor suite.

procuring, operating, and supporting three separate tactical aircraft designs to meet the services' similar but not identical operational needs. Numerous partner countries are also involved, and their needs are incorporated into one or more of these three basic variants. The mission system sensors and software, on the other hand, are common among all three variants. The F-35 sensor suite (Fig. 3) includes the APG-81 Active Electronically Scanned Array (AESA) radar, the ASQ-239 Electronic Warfare (EW)/Counter Measures (CM) suite, the AAQ-40 Electro-Optical Targeting System (EOTS), the AAQ-37 Electro-Optical Distributed Aperture System (DAS) system, and the ASQ-242 Communication, Navigation, and Identification (CNI) system. These five sensors provide object detection and measurements in the Radiofrequency (RF) and Electro-Optical/Infrared (EO/IR) spectrum to F-35 Information Fusion, resulting in more information about the objects in the environment around the aircraft than has been available to a fighter aircraft before.

In addition to the onboard sensors, the F-35 transmits and receives tracks via the Link 16 datalink and the Multifunction Advanced Data Link (MADL). Link 16 provides connectivity with legacy aircraft and command and control systems, which allows the aircraft to transmit information about selected tracks and receive surveillance information from command and control centers. Designed primarily for cueing, information about the quality and timeliness of the track is limited.

The F-35 MADL communication link was designed explicitly for the F-35 to support the sharing of information among the flight group. Unlike legacy datalinks, MADL bandwidth supports passing fusion-quality information on all air and surface tracks to other members of the flight group. These data include a locally derived track state, track covariance, ID measurement history, and RF history, as well as other metadata associated with each track. The amount of potentially redundant off-board information provided by MADL was one of the largest challenges for fusion design. It was now possible to have many copies (in some cases greater than 10) of the same track from onboard and off-board sources simultaneously, creating a high potential for clutter. The fusion of this spatially diverse data offered a significant improvement in situational awareness and cooperative sensing.

IV. THE F-35 INFORMATION FUSION APPROACH

Prior to the introduction of the 5th-Generation fusion systems, fusion historically referred to only the data association and estimation processes. The earliest partitioning of the F-35 fusion capability envisioned the sensor management capability to be independent of the fusion process; however, there was already strong evidence that the autonomous sensor manager was fundamental to efficient fusion performance and sensor optimization. During the early stages of design, the sensor manager was repartitioned to the F-35 fusion design. Figure 4 shows the top-level functional architecture of the F-35 fusion design, highlighting the data association, estimation (both kinematic and ID), and sensor management functionality.

The F-35 Information Fusion design isolates fusion algorithms from both the sensor and datalink inputs, as well as any consumers of fused data. Essentially, the fusion algorithms comprise a black box, known internally as the fusion engine, and sensor inputs and data consumers are encapsulated in external software objects known as Virtual Interface Models (VIMs). For incoming data, the sensor-specific or datalink-specific VIMs fill in missing data (e.g., navigation state, sensor bias values), preprocess the information, and translate it into a standard form for the fusion process. For data leaving fusion, the outgoing VIM, known internally as

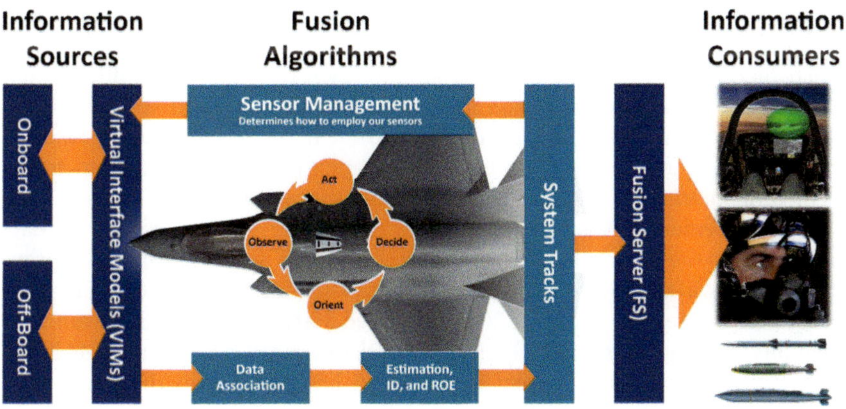

Fig. 4 F-35 fusion architecture.

the fusion server, provides data to the various consumers of fused information, both onboard and off-board. The fusion server isolates users of the fused information from both the fusion process and data sources. Legacy fusion implementations reported fusion tracks as a monolithic block (i.e., one size fits all) where all data consumers received the same message. Any propagation of the data or conversion was the responsibility of the recipient. This created a coupled interface between fusion and the data consumers. When a new data source was introduced to fusion, the interface changes to make these data available impacted all consumers of that message, whether the data were used or not, making changes to fusion very costly. The fusion server sends each information consumer a tailored message that contains only the information required to support that consumer. This isolates that consumer from changes to any data source or to the fusion algorithm. The use of VIMs enables the fusion architecture to be extensible to new sensors and data sources, as well as new data consumers, over its lifetime.

V. INFORMATION TIERS

Sensor fusion can result in poor performance if incorrect information about sensor performance is used:

> A common failure in data fusion is to characterize the sensor performance in an ad hoc or convenient way. Failure to accurately model sensor performance will result in corruption of the fused results. [17]

One of the key architecture decisions for F-35 fusion is how to share information among aircraft. Independent data can be incorporated optimally into a filter for the highest accuracy; however, if dependent data are incorporated under the assumption of independence, the result will be track instability and, eventually, track loss [18]. Data consumers on the F-35, including the pilot, receive the kinematic and ID estimate of each track based on all available data sources, both onboard and off-board. This is referred to as the Tier 3 solution. However, when sharing information with other aircraft, each F-35 shares the information describing a track based solely on measurements from onboard sensors. This is referred to as the Tier 1 solution. By ensuring that the information received from MADL is independent, the track information can be converted into equivalent measurements [19] by the recipient supporting both track-to-track and measurement-to-track of the information. The sharing of Tier 1 data ensures that the information is not coupled to any specific fusion algorithm and provides a method for dissimilar fusion platforms to share optimal fusion data in the future (Fig. 5). In late 2016, Lockheed Martin and the U.S. government used this technique to share an F-35 fused track of a target drone across MADL to a surface-based weapons system that had no line of sight to the drone. The surface-based weapons system converted the F-35 MADL Tier 1 information into equivalent measurements that were consumed by the native engagement

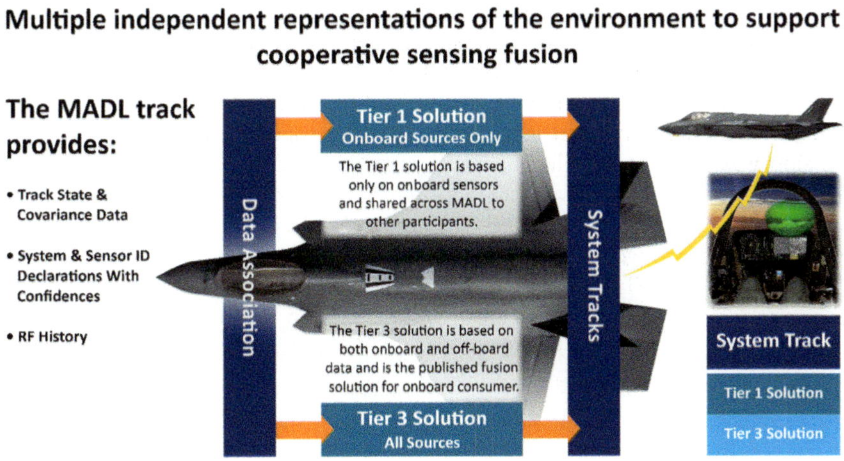

Fig. 5 Information tiers.

tracker. Together, the networked systems achieved a successful acquisition, guidance, and kinematic intercept of the track using a surface-to-air missile.

To incorporate dependent information or unknown pedigree information into the F-35 estimate, fusion includes a covariance intersection update that is more tolerant of rebroadcasted data [20]. The covariance intersection algorithm reduces the error of the estimate when novel information is introduced but does not improve the error with redundant data. This technique provides flexibility for incorporating data from many different datalinks in the future, where the pedigree of the data is not known.

The integration of legacy datalinks uncovered other challenges to optimal fusion. When system tracks are sent off-board over a datalink, it is important to send the most precise error characterization available for the system track. The covariance matrix provides this sensor characterization to other users of the data; however, in many legacy datalinks, quality factors are used to describe track errors rather than multidimensional covariance matrices. These quality factors denote a maximum area or volume uncertainty value for the data. This lack of specificity in dimensional accuracy requires the recipient to assume the worst possible accuracy for the data in all dimensions to ensure stable tracking behavior under all conditions. From an algorithmic perspective, the pessimistic error characterization leads to a higher incidence of false correlation and a deweighting of the off-board data [21].

Given the off-board data have successfully correlated to the correct system track, the reported error of the remote track is used to weight the influence of the data. Recall from the Kalman filter that the optimal gain K_k is used to determine the blending of the new measurement [22, 23]. Equation (1) is an alternative form of the Kalman filter, showing how the optimal filter gain blends the propagated system estimate and the observation data. As the gain increases, the observation data Z_k has a larger influence on the new estimate.

$$\hat{X}_{k|k} = K_k \, Z_k + (1 - K_k) \, \hat{X}_{k|k-1} \tag{1}$$

The gain for a Kalman filter is determined by the estimated system track errors and the measurement errors, as shown in Eq. (2).

$$K_k = P_{k|k-1} \left[P_{k|k-1} + R_k \right]^{-1} \tag{2}$$

If the measurement errors are pessimistic, the filter gain decreases and results in a natural deweighting of the measurement influence on the new estimate. Conversely, if the measurement errors are optimistic, the filter gain increases, and the measurement has too large an effect on the track. The use of quality factors on legacy datalinks leads to smaller gain values and reduces the potential accuracy of the resultant track. The F-35 MADL datalink reports the track covariance for each track, allowing for accurate weighting of the off-board contributions.

VI. EVIDENCE-BASED COMBAT IDENTIFICATION

Combat ID is the process of attaining an accurate characterization of detected objects in the joint battlespace [24]. Characterizations may include affiliation, class, type, nationality, and mission configuration. The F-35 employs a probabilistic, evidential reasoning approach to ID in a fundamental departure from heuristic algorithms employed on other tactical platforms. The decision to switch to a probabilistic formulation was driven by lessons learned from those platforms to eliminate misidentification due to forced declarations and the inability to handle ambiguous information sources. The transition to a probabilistic framework forced design changes throughout the rest of the information fusion design, including sensors, pilot displays, and sensor tasking.

Early in the F-35 design, a trade was performed between Bayesian and Dempster-Shafer inferencing algorithms. Ultimately, Dempster-Shafer was selected due to greater robustness to conflicting data, less reliance on a priori target distributions, and a more natural formulation of relationships between sensor attributes and corresponding platform sets. Algorithms based on Bayes's rule provide intuitive probabilistic confidence values; however, Bayesian inference places a higher dependence on a priori probabilities, the $P(A)$ term in Eq. (3), to seed the probability of a given target ID declaration. Accurate a priori information can be difficult to obtain in practice.

$$P(A|B) = \frac{P(B|A)P(A)}{P(B)} \tag{3}$$

In contrast, Dempster-Shafer belief theory does not require explicit a priori information. Further, Fixsen and Mahler have shown that Dempster-Shafer probability masses can approach Bayesian probabilities [25] when probability masses are chosen appropriately. In our Dempster-Shafer formulation, each sensor declaration maps to a set of platform disjunctions, or a list of disjunctions in the case of ambiguities. These sensor disjunctions are combined using Dempster's rule of combination, shown in Eq. (4). Inconsistent propositions, designated by κ in Eq. (5), are removed and their confidence is distributed across the consistent terms through normalization.

$$m(A) = \frac{1}{1-\kappa} \sum_{B \cap C \neq \emptyset} m(B)m(C) \tag{4}$$

$$\kappa = \sum_{B \cap C = \emptyset} m(B)m(C) \tag{5}$$

Implementation of the Dempster-Shafer algorithm requires first defining the frame of discernment: the set of objects over which the algorithm reasons. Previous approaches to combat ID have been forced to limit their platform set due to processing and memory constraints. By contrast, the F-35 design took a holistic approach to include all tactically relevant platforms in the theater, minimizing misidentification of platforms that are not in library. Platforms are interconnected in a strict hierarchical taxonomy with levels in the taxonomy labeled subtype, type, class, and affiliation, as shown in Fig. 6. The algorithm creates ambiguity lists and associated confidences for each level of the taxonomy. The type ambiguity list is inferred from the subtype ambiguity list; the class ambiguity list is inferred from the type ambiguity list; and the affiliation ambiguity list is inferred from the class ambiguity list. The strictly enforced hierarchy allows for processing simplifications and ensures that the confidence at a higher level in the taxonomy is always greater than or equal to the confidence at lower levels [26].

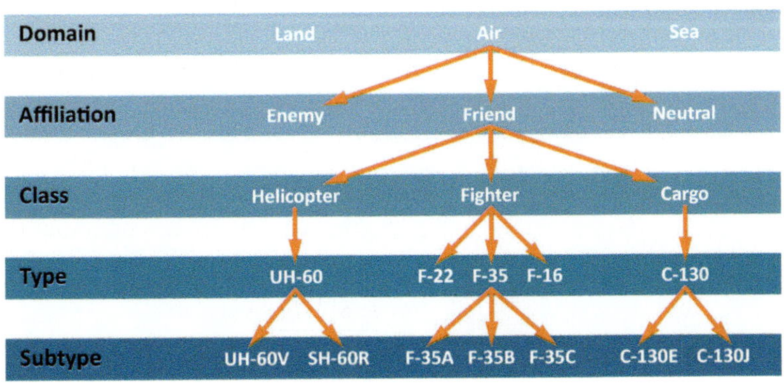

Fig. 6 Platform taxonomy levels.

The platform taxonomy also provides the framework for mapping sensor attributes to a platform set. This mapping, contained in an onboard relational threat database, encodes the transitional probability of observing a feature given a particular platform subtype. Each sensor declaration must also have an associated probability or confidence so the input to fusion is weighted based on its relative merit in the same way a kinematic report is weighted based on its covariance matrix. At the beginning of the F-35 program, none of the sensor modes produced confidence values, so these needed to be generated. Called the Basic Probability Assignment (BPA), the generation of confidence values proved to be one of the more challenging aspects of the design. It should go without saying, the generated error must also correctly bound the true error distribution. Continuing with the kinematic tracking analogy, it does no good to report a track range error as having a standard deviation of 1 m when the true error statistics are 100 m. Similarly, the ID probabilities must accurately reflect the true target uncertainty.

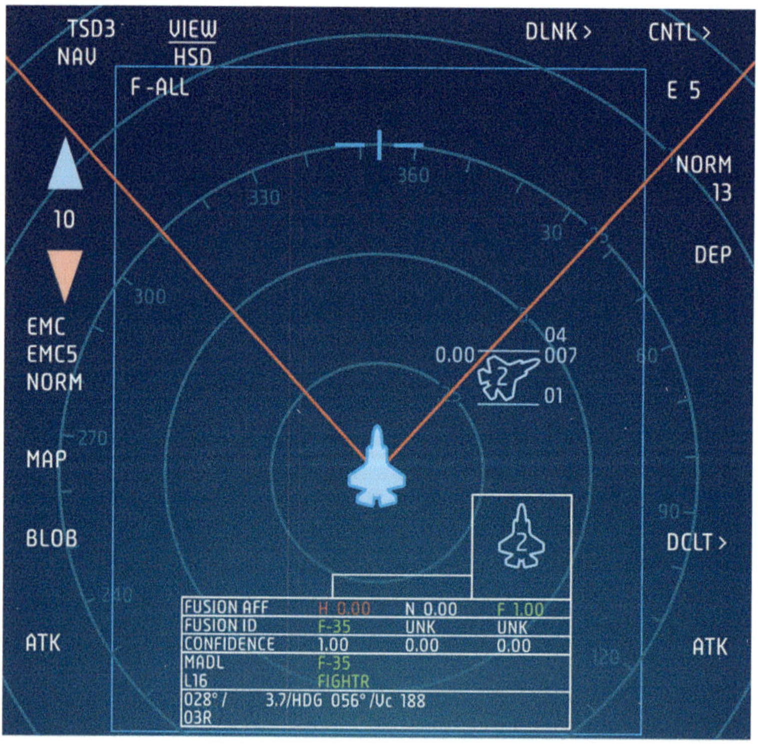

Fig. 7 F-35 expanded data window.

Although the soft decisions developed by fusion provide a more accurate representation of what is known and not known about a given target's ID state, the pilot requires actionable information, which requires that the soft decision be converted into a hard declaration. The Dempster-Shafer algorithm produces support-plausibility intervals that bound the estimate of probability but do not directly provide platform declarations. The F-35 design converts the probability masses into pignistic probabilities that effectively distribute the platform disjunction confidence across all elements of the disjunction [27]. The system makes a hard ID declaration when the translated confidence exceeds a user-defined threshold. The taxonomy is traversed from lower nodes to higher nodes until the confidence threshold is exceeded. The F-35 combat ID output, shown in the second row of Fig. 7, is flexible enough to allow a display of information from any level in the taxonomy. The output also contains ID declarations from other sensors and off-board sources, which helped to develop trust in the fused outputs and transition from legacy platforms. In this example, a Link 16 declaration of fighter was combined with the MADL declaration of F-35 to produce a high-confidence type declaration of F-35 and a friendly confidence of 1.

VII. AUTONOMOUS SENSOR MANAGEMENT

Technical advances in the capability of modern sensors, increases in the number of multiple sensor systems, and the migration to increased connectivity have led to a sensor network that has exceeded the ability of a human to efficiently control them [28]. This performance gap led to the development of an automatic feedback mechanism in the closed-loop fusion model that autonomously modifies the actions of the sensor suite to achieve a mission-level objective or behavior. The use of an algorithm to task or modify the behavior of the information sources is referred to as process refinement and is categorized as JDL Fusion Level 4 [29]. The goals of the sensor manager are to: 1) reduce pilot workload by automating sensor actions and selection; 2) prioritize information requests (including pilot requests, background volume searches, and fusion information needs); and 3) reconfigure sensor assets to compensate for individual sensor loss or unavailability [30].

Sensor management strategies focus generally on the control of heterogeneous collocated or spatially diverse sensors to achieve a desired goal (e.g., track accuracy). An autonomous sensor management system requires an objective function to optimize sensor utilization to select a preferred action over many possible alternatives. Many early strategies [31–33] were ad hoc or rule-based, having a strong coupling to sensor requirements. Others [34–38] explored information-theoretic approaches that sought to minimize uncertainty in the track covariance, thus decoupling the objective function from a specific sensor suite. The advantage of information-theoretic solutions is their adaptability to many problems; however, they are focused on optimizing the information of a given track while neglecting the value of that information to the mission objectives [39]. For a fighter aircraft, the primary objective is to provide situational awareness to the operator during a complex and changing environment to support critical mission decisions [40].

Endsley [41] defines situational awareness from the perspective of the fusion customer (pilot), which could then be decomposed further into times or distances where key decisions must be made. The necessary information to support these key decisions are independent of the sensor's ability to meet these needs. The information boundaries and associated information needs define a dimensional space that can be used to derive a global objective function to support autonomous sensor management [42] and can be used to define the sensor and fusion capabilities necessary to support these decisions. A benefit from this mapping is that the mission goals can be directly related to the sensor capabilities in terms of range, accuracy, and latency. This allows the designer to trace system-level fusion situational awareness requirements to individual sensor performance requirements.

For the F-35, the automated sensor management function is responsible for efficiently managing the sensor suite to provide critical information about the objects in the environment to the pilot, supporting critical decisions and actions. It does this by prioritizing the systems tracks, autonomously directing system resources to maintain existing tracks, gathering mission ID and Rules of Engagement (ROE) information, and balancing track maintenance with new track discovery through searching. System priorities, track information needs, and track accuracies are based on the track type, pilot emphasis (if any), and information boundaries around the aircraft. The autonomous sensor manager also provides methods for the pilot to collaborate with the fusion system, both for refinement or reprioritization of existing tracks and to cue the system to search for new objects in the environment. The pilot can designate a line of sight (air) or area (surface) and command a cued search. The sensor manager will direct a series of active and passive scans focused on the detection of new air or surface objects. If the pilot selects an existing track, sensor management will raise the priority of the track and will cue sensors to meet the information needs associated with the tactical zone (full state).

The autonomous sensor management algorithm focuses on providing the information needs for every track based on priority. The goal is not to drive each track to the best accuracy, but to instead drive it to sufficient accuracy and information content. In practice, for situational awareness, there is a level of component accuracy (e.g., range, angle) where the information is sufficient to support the pilot's understanding of the environment to decide. Additional accuracy beyond this point does not significantly improve the pilot's awareness or decision-making ability. Therefore, information gain above a sufficient level does not support the mission objective and should be directed toward other objectives in the environment, such as searching for new objects. This concept of sufficiency can be used to define a new constrained objective function that incorporates the concept of sufficient information and targets specific information needs. The addition of the sufficiency constraint removes the dependency of the objective function on the accuracy of the sensor making the update, as long as the sensor can provide the required accuracy. This objective function encapsulates the mission goals in terms of situational awareness while incorporating the constraint of sufficiency. The definition of sufficiency changes at each boundary based on the pilot's needs to make critical decisions. The closed-loop nature of the autonomous sensor manager enables the system to respond more rapidly to changes in the environment and to optimize the sensor behavior, freeing the pilot from the role of sensor manager and returning him or her to the role of tactician.

VIII. COOPERATIVE SENSING

The F-35 MADL was designed to support full sharing of information among aircraft. MADL bandwidth supports the exchange of all air and surface tracks between/among participants within the flight group. Given that each F-35 has multiple sensors detecting multiple targets—and sometimes spurious signals—this can lead to the exchange of numerous, potentially duplicative tracks over MADL. Therefore, the F-35 places limits on the kinds of tracks and associated information that can be transmitted over the link.

For MADL distribution, a single F-35 system track is divided into three messages: the basic MADL surveillance track, extended combat ID (XID), and RF parametric extensions. The basic surveillance track provides the independent kinematic state estimate and track covariance at the time of the last measurement update. It is important to note that the kinematic estimate for a sent track can be either ranged or angle-only (no observed range). This distinction becomes important for advanced multiship tracking techniques, such as angle/angle ranging or Time Difference of Arrival (TDOA), described later. The MADL surveillance track also includes a list of sensors contributing to this track, as well as ID summary data. The XID message contains a higher-fidelity ID ambiguity list, in addition to ID measurements (e.g., IFF). The RF parametric message contains the Electronic Signal Measurement (ESM) data correlated to this track. The sharing of this detailed information allows each aircraft to leverage the spatial diversity of the flight group.

One of the initial multiship capabilities of the F-35 was the ability to cooperatively range airborne emitters by finding the intersection (or point of nearest approach) for angle-only tracks on two or more different aircraft (Fig. 8).

On the receipt of a MADL angle-only track, the receiving fusion system determines possible intersection points with its own onboard angle-only tracks. It is possible for a given MADL angle-only track to intersect with multiple ownship tracks. In truth, only one of the intersections is correct. These alternate angle/angle candidates are referred to colloquially as ghosts. Once all ghosts have been eliminated, the range of the track from each participant can be calculated

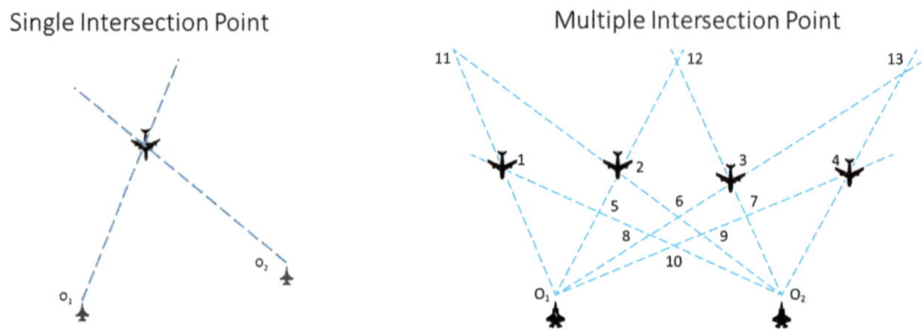

Fig. 8 Passive angle/angle ranging of airborne emitters.

in Eqs. (6) and (7) as follow [43]:

$$\hat{R}_1 = \left\|\vec{R}_1\right\| = \frac{\left|\vec{D} \times \vec{u}_2\right|}{\left|\vec{u}_1 \times \vec{u}_2\right|} \tag{6}$$

$$\hat{R}_2 = \left\|\vec{R}_2\right\| = \frac{\left|\vec{D} \times \vec{u}_1\right|}{\left|\vec{u}_1 \times \vec{u}_2\right|} \tag{7}$$

Rotating the covariance of each observer into a common reference frame, the range error in the common reference frame is approximately the intersection of

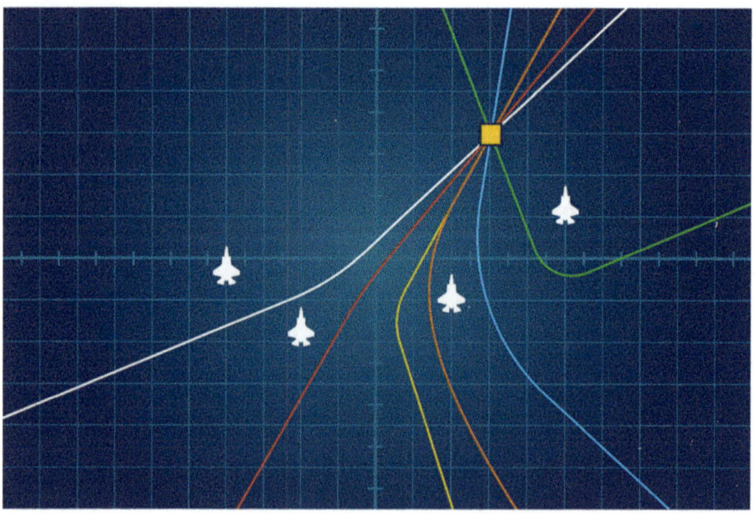

Fig. 9 Passive ranging of surface emitters using TDOA techniques.

the two error covariance matrices, which can be expressed in Eq. (8) as

$$\Sigma_{\text{int}} = \left(\Sigma_1^{-1} + \Sigma_2^{-1}\right)^{-1} \tag{8}$$

For surface emitters, fusion incorporates a TDOA capability for precision location. The TDOA capability provides the ability for multiple aircraft to synchronize ESM dwells in time and frequency. Upon initiation, the autonomous sensor manager configures the dwells across the network, and then all aircraft send the time of arrival of any received pulses to the initiating aircraft. Fusion processes these pulse streams from the cooperating participants to detect common pulse pairs. Common pulse pairs form a surface of constant delta-time that, when intersected with the Earth, produce a hyperbola of constant time difference called an isochrone. Multiple pulse pairs represent the intersection of one or more isochrones and produce range estimate (Fig. 9).

IX. SUMMARY

The F-35 Information Fusion software combines information from both onboard and off-board data sources, providing the pilot with advanced capabilities not available on legacy aircraft. Further, this extensible approach to information fusion leverages the spatial and spectral diversity among multiple F-35 wingmen, creating an innovative tactical network where data are shared instantaneously with other F-35s and legacy aircraft. The F-35 Information Fusion implementation of data association, state estimation, and combat ID ensures that the pilot has accurate situational awareness, allowing for advanced target detection, tracking, and tactical employment. The autonomous sensor manager provides timely reaction to a changing environment and ensures that all tracks are refined to a prespecified quality based on priority, allowing the pilot to return to the role of tactician. The F-35 MADL provides sufficient bandwidth for complete sharing of detailed fusion solutions and accuracies for all air and surface targets, resulting in improved situational awareness for all pilots in the MADL network. Using data-sharing methods to ensure that the data pedigree is maintained, the MADL information can be processed like a remote sensor, resulting in improved accuracies and new capabilities.

REFERENCES

[1] Naisbitt, J., *Megatrends: Ten New Directions Transforming Our Lives*, Warner Books, New York, 1982.

[2] Chalmers, B. A., "On the Design of a Decision Support System for Data Fusion and Resource Management in a Modern Frigate," NATO System Concepts and Integration Panel Symposium, 1998, pp. 2-1–2-13.

[3] Musick, S., and Malhorta, R., "Chasing the Elusive Sensor Manager," *Proceedings of the IEEE 1994 National Aerospace and Electronics Conference, NAECON 1994* (23–27 May 1994, Dayton, OH), Vol. 1, IEEE, New York, 1994, pp. 606–613.

[4] Endsley, M. R., "Situation Awareness in an Advanced Strategic Mission," NOR DOC 89-32, Northrop Corporation, Hawthorne, CA, 1989.

[5] Endsley, M. R., Farley, T. C., Jones, W. M., Midkiff, A. H., and Hansman, R. J., "Situation Awareness Information Requirements for Commercial Airline Pilots," ICAT-98-1, International Center for Air Transportation, Massachusetts Institute of Technology, Cambridge, MA, 1998.

[6] Shenk, D., *Data Smog: Surviving the Information Glut*, Harper, San Francisco, 1997.

[7] Wilson, E. O., *Consilience: The Unity of Knowledge*, Vantage Books, New York, 1998.

[8] Kline, G. B., Moon, B., and Hoffman, R. R., "Making Sense of Sensemaking 1: Alternative Perspectives," *IEEE Intelligent Systems*, Vol. 21, No. 4, July/Aug. 2006, pp. 70–73.

[9] Canavan, R., Ganesh, C., and Matuszak, B., "Evolution of Fusion in Navy Tactical Systems," *17th International Conference on Information Fusion (FUSION)*, 2014, pp. 1–7.

[10] Endsley, M. R., and Garland D. J (eds.), *Situation Awareness Analysis and Measurement*, Lawrence Erlbaum, Mahwah, NJ, 2000.

[11] Steinberg, A., Bowman, C., and White, F., "Revisions to the JDF Data Fusion Model," *Proceedings of AIAA Missile Sciences Conference*, Naval Postgraduate School, Monterey, CA, 17–19 Nov. 1998.

[12] Yogi Berra Quotes, BrainyQuote.com, Xplore Inc. [online database], https://www.brainyquote.com/quotes/yogi_berra_141506 [retrieved 22 Jan. 2018].

[13] Chong, C., Mori, S., Govaers, F., and Koch, W., "Comparison of Tracklet Fusion and Distributed Kalman Filter for Track Fusion," *Proceedings of 17th International Conference on Information Fusion*, Salamanca, Spain, 2014.

[14] Steinberg, A. N., and Bowman, C. L., "Revisions to the JDL Data Fusion Model," Chapter 3 in *Handbook of Multisensor Data Fusion*, Liggins, M. E., Hall, D. L., and Llinas, J. (eds.), CRC Press, London, 2009.

[15] Steinberg, A., and Snidaro, L., "Levels?," *Proceedings of the 18th International Conference on Information Fusion*, Washington, DC, July 2015.

[16] Schechtman, G. M., "Manipulating the OODA Loop: The Overlooked Role of Information Resource Management in Information Warfare," master's thesis, AFIT/GIR/LAL/96D-10, Air Force Institute of Technology, Wright-Patterson Air Force Base, OH, 1996, p. 28.

[17] Hall, D. L., and Steinberg, A. N., "Dirty Secrets in Multisensor Data Fusion," Chapter 21 in *Handbook of Multisensor Data Fusion*, Hall, D. L., and Llinas, J. (eds.), CRC Press, London, 2001.

[18] Blasch, E., and Israel, S., "Situation/Threat Context Assessment," *Proceedings of the 18th International Conference on Information Fusion*, Washington, DC, 6–9 July 2015.

[19] Chong, C., Mori, S., Barker, W. H., and Chang, K., "Architectures and Algorithms for Track Association and Fusion," *IEEE Aerospace and Electronic Systems Magazine*, 2000, Vol. 15, No. 1, pp. 5–13.

[20] Julier, S. J., and Uhlmann, J. K., "A Non-divergent Estimation Algorithm in the Presence of Unknown Correlations," *Proceeding of the American Control Conference*, 1997, pp. 2369–2373.

[21] Frey, T., Rasmussen, N., and Engebretson, K., "Information Exchange Considerations for Effective Fusion among Heterogeneous Network Participants," *Proceedings of the AIAA Infotech@Aerospace Conference*, Jan. 2018.

[22] Kalman, R. E., " A New Approach to Linear Filtering and Prediction Problems," *Journal of Basic Engineering*, Vol. 82, 1960, p. 35.

[23] Kalman, R. E., and Bucy, R. S., "New Results in Linear Filtering and Prediction Theory," *Journal of Basic Engineering*, Vol. 83, No. 1, 1961, pp. 95–108.

[24] Capstone Requirements Document for Combat Identification, Joint Requirements Oversight Council, 19 Mar. 2001, pp. I–1.

[25] Fixsen, D., and Mahler, R. P. S., "The Modified Dempster-Shafer Approach to Classification," *IEEE Transactions on Systems, Man, and Cybernetics—Part A: Systems and Humans*, Vol. 27, No. 1, Jan. 1997, pp. 96–104.

[26] Shafer, G., and Logan, R., "Implementing Dempster's Rule for Hierarchical Evidence," *Artificial Intelligence*, Vol. 33, 1987, pp. 271–298.

[27] Smets, P., "Data Fusion in the Transferable Belief Model," *Proceedings of the 3rd International Conference on Information Fusion*, Paris, France, 2000, pp. 21–33.

[28] Freedberg, S. "F-22, F-35 Outsmart Test Ranges, AWACS," *Breaking Defense*, 7 Nov. 2016, https://breakingdefense.com/2016/11/f-22-f-35-outsmart-test-ranges-awacs/ [retrieved 23 Jan. 2018].

[29] McIntyre, G., "A Comprehensive Approach to Sensor Management and Scheduling," Ph.D. dissertation, George Mason University, 1998.

[30] Steinberg, A., and Bowman, C. L., "Revisions to the JDL Data Fusion Model," in *Handbook of Multisensor Data Fusion*, Hall, D. L. and Llinas, J. (eds.), Boca Raton, FL, CRC Press, 2001, pp. 2-1–2-18.

[31] Adrian, R., "Sensor Management," *Proceedings of the 1993 AIAA/IEEE Digital Avionics System Conference*, Fort Worth, TX, 25–28 Oct. 1993, pp. 32–37.

[32] Bier, P., Rothman, P., and Manske, R., "Intelligent Sensor Management for Beyond Visual Range Air-to-Air Combat," *Proceedings of the IEEE 1988 National Aerospace and Electronics Conference NAECON 1988*, Vol. 1, Dayton, OH, May 1990.

[33] Rothman, P., and Bier, S., "Intelligent Sensor Management Systems for Tactical Aircraft," *Second National Symposium on Sensors and Sensor Fusion*, Chicago, IL, March 1989.

[34] Buede, D., Martin, J., and Sipos, J., "AIFSARA: A Test Bed for Fusion, Situation Assessment and Response," *Proceedings of the 4th AAAIC 1988 Conference*, Dayton, OH, Oct. 1988.

[35] Greenway, P., and Deaves, R., "Sensor Management Using the Decentralized Kalman Filter," Sensor Fusion VII, *Proceedings of the SPIE—The International Society for Optical Engineering*, Vol. 2355, Boston, MA, 31 Oct–1 Nov. 2004, pp. 216–225.

[36] Greenway, P., and Daves, R., "An Information Filter for Decentralized Sensor Management," Signal Processing, Sensor Fusion, and Target Recognition III, *Proceedings of the SPIE—The International Society for Optical Engineering*, Vol. 2232, Orlando, FL, 4–6 April 1994, pp. 70–78.

[37] Jenison, T., "Improved Data Structures, Multiple Targets/Target Types and Multiple Sensors in Discrimination Based Sensor Management," master's thesis, University of Minnesota, July 1996.

[38] Hintz, K., "A Measure of the Information Gain Attributable to Cueing," *IEEE Transactions on Systems, Man, and Cybernetics*, Vol. 21, No. 2, March/April 1991, pp. 237–244.

[39] McIntyre, G. A., and Hintz, K., "Sensor Management Simulation and Comparative Study," Signal Processing, Sensor Fusion, and Target Recognition VI, *Proceedings of the SPIE—The International Society for Optical Engineering*, Vol. 3068, Orlando, FL, 20–25 April 1997.

[40] Avasarala, V., Mullen, T., and Hall, D., "A Market-Based Approach to Sensor Management," *Journal of Advances in Information Fusion*, Vol. 4, No. 1, June 2009, pp. 52–53.

[41] Endsley, M. R., "Designing for Situation Awareness in Complex Systems," *Proceeding of the Second International Workshop on Symbiosis of Humans*, Artifacts and Environment, Kyoto, Japan, 2001.

[42] Frey, T., "A Constrained, Information Needs Based Sensor Fusion Resource Management Design Methodology," *Proceedings of AIAA Infotech@Aerospace Conference*, St. Louis, MO, March 2011.

[43] Wadley, J., "Important Considerations for Cooperative Passive Ranging of Aircraft," *Proceedings of the AIAA Infotech@Aerospace Conference*, Jan. 2018.

CHAPTER 13

F-35 Carrier Suitability Testing

Tony Wilson*
Lockheed Martin Aeronautics Company, Fort Worth, TX

Carrier suitability is a multidiscipline specialization of aircraft test and evaluation. The discipline combines theories from aircraft loads, flying qualities, and performance in a system of systems approach to assess the suitability of an aircraft to operate from ships and austere sites. Additionally, navigation and guidance, sensor integration, data link interoperability, pilot–vehicle interface, supportability, and maintainability are evaluated to ensure the aircraft is capable of operating as a system within a system. This paper will provide an introduction to carrier suitability flight test of the F-35C, a carrier-based, multirole, 5th-Generation stealth fighter to be used by the U.S. Navy and U.S. Marine Corps. Topics will include a discussion of the shore-based testing prerequisites and results, an overview of the challenges experienced with the original tailhook and its redesign, and the use of autopilot functionality in shore-based testing. This will be followed by an examination of the necessity to test in the shipboard environment, a discussion of shipboard testing, shipboard catapult and arrested landing methods of test and results. Throughout, an investigation on how advanced approach mode control laws used for shipboard landings were implemented, the results from three ship trials, and the implications to future operations will be analyzed.

I. NOMENCLATURE

α	angle of attack
d	off-center distance
H/E	hook to pilot eye distance
i	initial condition
L	pilot eye to hook point length

*Test Pilot Senior Staff, Fighter Flight Operations.

P_T total probability
t_x time increments
V_A approach speed
V_E engaging speed
V_V sink speed
δ difference between flight path angle and glideslope angle
γ flight path angle
Φ pilot eye to fuselage reference line angle
φ roll angle
ψ yaw angle
Θ glideslope angle
θ pitch angle

II. INTRODUCTION

The idea of naval aviation, the launching and recovering of aircraft from a seagoing vessel, is neither new nor novel; its development started less than a decade after the Wright brothers' first flight. U.S. naval aviation can trace its inception back to 14 November 1910, when Eugene Ely took off in a Curtiss Pusher from a temporary platform erected over the bow of the light cruiser USS *Birmingham*. Upon rolling off the deck, the airplane sank 83 ft downward and did not start to climb until the aircraft's wheels had dipped into the water. Just two months later, Ely was at it again, this time landing his aircraft on a 120-ft platform attached to the deck of the USS *Pennsylvania*, an armored cruiser anchored in San Francisco Bay. A series of ropes attached to sandbags were stretched across the landing platform designed to be snatched by a hook attached to the aircraft and stop it prior to rolling into a canvas awning attached to the ship [1]. With these historic events, Eugene Ely not only proved the concept of naval aviation, but also started the discipline of ship suitability flight test.

Ship suitability flight test is a multidiscipline approach to aircraft test and evaluation used to determine how well an aircraft system either integrates within a ship system or operates from an austere site. Multiple nations operate varied aircraft (fixed wing, rotary wing, and powered lift aircraft) from the decks of military and civilian ships. To try to address all combinations would be a herculean effort. Therefore, this paper will provide an introduction to ship suitability flight test of the F-35C, shown in Fig. 1, a carrier-based, multirole, 5th-Generation stealth fighter to be used by the U.S. Navy and U.S. Marine Corps both shore-based and aboard U.S. aircraft carriers (CVN). (This acronym is the U.S. Navy's designation for nuclear-powered aircraft carriers. The C designates the ship as a carrier, the V is a designation for fixed wing operations, and the N represents the nuclear power aspect of the ship).

Carrier Suitability (CVS), a subset of the ship suitability discipline, is composed of both shore-based testing and ship-based trials during which a

Fig. 1 The author conducting the first shipboard landing of the F-35C.

system-of-systems approach is used to assess the suitability of an aircraft to operate from the deck of an aircraft carrier. This assessment begins ashore with an assessment of deck handling evaluations, catapult spotting, steam ingestion, Jet Blast Deflector (JBD) compatibility, approach speed survey, Powered Approach (PA) throttle transients, bolter and wave-off performance, approach handling qualities, and a structural survey or demonstration. The difference between a survey and a demonstration is the degree to which envelope exploration or expansion is required: a demonstration requires a higher degree of expansion whereas a survey is used to validate models and predictions. For the F-35C, a structural survey was completed; this is a series of shore-based catapults and arrestments designed to test the structural integrity of the landing gear, arresting hook, aircraft structure, weapons, and weapon suspension equipment.

Shore-based testing of the F-35C commenced in the spring of 2011. It was quickly realized that the original tailhook design would need alterations due to geometric constraints. This resulted in a prolonged pause in testing while the tailhook was redesigned. The new tailhook was installed in December 2013 and testing recommenced. The delay in testing acted as a forcing function to develop new techniques to increase testing efficiency. Prior to the F-35, structural survey test points were flown mostly, if not completely, with the pilot in the loop, making test point completion highly dependent upon pilot experience and technique, resulting in multiple attempts at a single test point. The advanced control laws of the F-35C were used in a manner never attempted before to attain test point parameters and improve repeatability. This resulted in a rapid completion of the clean wing aircraft structural survey, a prerequisite for shipboard testing, and paved the way for F-35C ship trials. Benefits to future carrier aviation programs will be realized through the examination of how the advanced control laws were used, the advantages and disadvantages to each mode, and how F-35C flight test benefited from this breakthrough.

Once shore-based prerequisites are complete, ship trials commence. Shipboard testing is a necessity due to the ship's flight deck being approximately 60 ft above the ocean's surface. This elevation results in the aircraft rotating about its Center of Gravity (CG) as it leaves the deck during a catapult launch vs rotating about its main gear during a conventional takeoff. Additionally, the wind interacts with the structure of the ship and produces disturbances and perturbations both behind and in front of the ship that cannot be reproduced or simulated on shore. The F-35C conducted its initial ship trial (DT-I) in November 2014, aboard USS *Nimitz* (CVN 68) where initial evaluation of the clean wing aircraft catapult launch and recoveries began. DT-I consisted of 124 catapult launches, 124 arrested landings, and 222 Touch-and-Go (T&G) landings with two test aircraft during day and night Visual Meteorological Conditions (VMC). DT-II was completed in October 2015 aboard USS *Dwight D. Eisenhower* (CVN 69), where increased weight and forward CG were evaluated with 66 catapult launches, 66 arrested landings, and 40 T&G landings during day and night VMC. The third ship trial (DT-III) was conducted aboard USS *George Washington* (CVN 73) in August 2016, consisting of 121 catapult launches, 121 arrested landing, and 67 T&G in which shipboard operations with full internal weapons bays and external weapons were evaluated. With respect to catapult testing, the largest technical challenge and most dangerous portion of the evolution is Minimum End Airspeed (MEAS) catapults. During these particular launches, the fly-away airspeed of the aircraft is slowly decreased, intentionally allowing the aircraft to sink below deck level to a targeted height above the sea surface, allowing the minimum safe fly-away airspeed to be anchored for a given aircraft gross weight. This section will provide a detailed discussion on the theory of MEAS testing and other catapult launches, the results of the shipboard evaluation, and lessons learned while deployed during the three ship trials.

Once an aircraft is launched, it must be recovered via an arrested landing. Almost 600 approaches were made during ship trials to various aircraft carriers in a multitude of wind conditions ranging from minimum recovery headwinds to winds in excess of 40 kt. During these approaches, three different approach modes were tested. The basic mode, also known as manual, is flown on the backside of the power curve in which the pilot controls rate of descent with thrust and airspeed with aircraft attitude. The next mode is Approach Power Compensation (APC), in which the pilot commands the flight path rate via longitudinal control stick inputs while the control laws contained in the vehicle management computers maintain Angle of Attack (AOA) via engine thrust request and throttle backdrive. The final mode, Delta Flight Path (DFP), allows precision control of flight path (gamma) by integrating APC with Integrated Direct Lift Control (IDLC). These advanced approach modes are synthesized via the nonlinear dynamic inversion control paradigm that is shared will all other F-35 variants. The remainder of the paper will focus on the design and implementation of these advanced control modes, the methods used to evaluate the approach handling qualities in the various wind cells using all three approach modes, and the results of these tests.

In addition to catapults and arrested landings, several other areas were evaluated including navigation and guidance, sensor integration, data link interoperability, pilot–vehicle interface, supportability, and maintainability—all of which will be touched upon in this paper. In the end, the reader will learn about the theories and applications of the multidisciplinary field of carrier suitability flight test, how they were applied to the F-35C, and the implications to future operations and test programs.

III. BACKGROUND

This section is intended to provide the reader the necessary information to understand shore-based and ship-based carrier suitability testing. It will begin with an overview of the F-35C with a focus on the Control Laws (CLAW) and Arresting Hook System (AHS). This will be followed by a description of the landing aids, arresting gear, and catapult. The background section will be concluded with a brief discussion of the pilot's tasks while working with a Landing Signal Officer (LSO). If the reader is already familiar with carrier suitability, this section may be skipped or serve as a refresher on the basics.

A. F-35C LIGHTNING II

The F-35C, shown in Fig. 2, is a carrier based, multirole, 5th-Generation stealth fighter intended for use by the U.S. Navy as a replacement for its aging 4th-Generation fighters. The F-35 is a single-seat, single-engine, all-weather fighter with low observable features designed for ground attack and air superiority missions. It is powered by the Pratt & Whitney F135 engine that generates approximately 40,000 lb of thrust. The F-35C shares its basic structure and primary load paths with the F-35A and F-35B variants but incorporates specialized design features and strengthened internal structure to perform catapult launches and arrested landings. The F-35C was designed to be compatible with and operationally supported from the Nimitz class (CVN 68) aircraft carrier.

The F-35C design incorporates 45% more wing area, larger horizontal stabilators, increased control surface size, and outer wing ailerons, which contribute to precise slow-speed handling characteristics required for carrier approaches

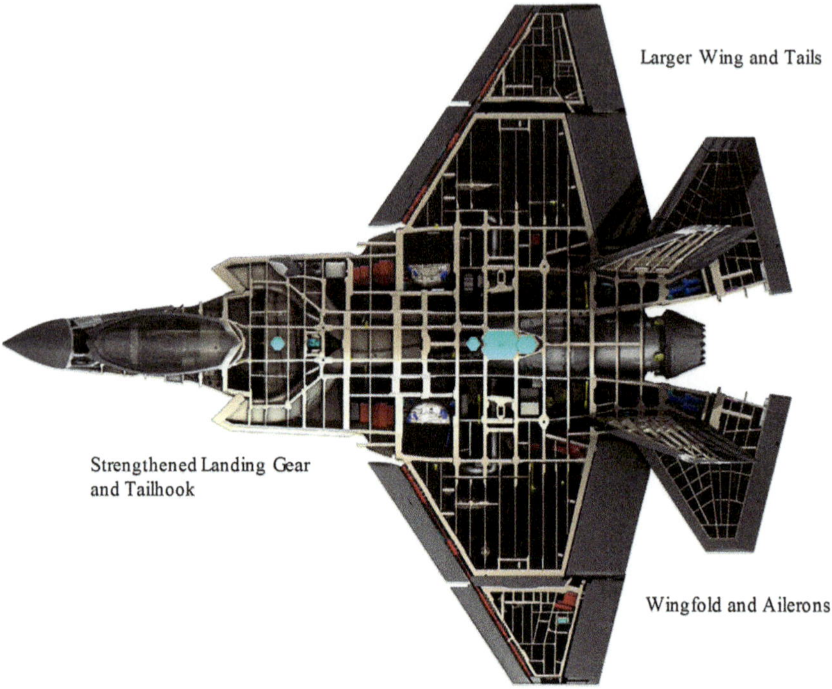

Fig. 2 Unique features of the F-35C Lightning II.

and landings. Unique features to the F-35C include folding wings to reduce the amount of deck space needed while aboard the carrier, a launch bar for catapult operations, and a recessed stowable tailhook for shipboard arrestments. The aircraft has 11 weapons stations, 4 internal and 7 external. The internal weapon bay stations are designed for a combination of Advanced Medium-Range Air-to-Air Missiles (AMRAAMs) and two Air-to-Surface (A/S) weapons up to 2000 lb each. The F-35C has three weapons stations under each wing, allowing it to carry a combination of Air-to-Air (A/A) and A/S weapons externally. The centerline fuselage station is designed for carriage of a low-observable gun pod. Table 1 depicts all the configurations that were tested shipboard [2].

B. F-35C CONTROL LAWS

The F-35 uses nonlinear dynamic inversion control architecture that is operated in three redundant Vehicle Management Computers (VMCs) to provide fly-by-wire control. The CLAW that augment aircraft dynamics and provide stability are common among the three variants; however, each variant's CLAW has been optimized and contains advanced features to reduce pilot workload and

TABLE 1 SHIPBOARD FLIGHT TEST WEAPONS CONFIGURATIONS

Loading	Station										
	1	2	3	4	5	6	7	8	9	10	11
Clean Wing Empty Bays				DART-1k	Suspension Equip		Suspension Equip	Suspension Equip			
Clean Wing (Most Fwd CG)				DART-1k	AIM-120		AIM-120	GBU-31			
Pylons Only	LAU	Pylon	Pylon	DART-1k	Suspension Equip		Suspension Equip	Suspension Equip	Pylon	Pylon	LAU
Heavy Symmetric	LAU	GBU-12	GBU-12	DART-1k	AIM-120	GUN	AIM-120	GBU-31	GBU-12	GBU-12	LAU
Heavy Symmetric No Gun	LAU	GBU-12	GBU-12	DART-1k	Suspension Equip		Suspension Equip	GBU-31	GBU-12	GBU-12	LAU
Symmetric (Most Aft CG)	LAU	GBU-12	GBU-12	DART-1k	Suspension Equip		Suspension Equip	Suspension Equip	GBU-12	GBU-12	LAU
Light Symmetric	LAU	GBU-12	Pylon	DART-1k	Suspension Equip		Suspension Equip	Suspension Equip	Pylon	GBU-12	LAU
Asymmetric (10,000 ft/lb)	LAU	GBU-12	Pylon	DART-1k	Suspension Equip		Suspension Equip	GBU-12	Pylon	Pylon	LAU

provide carefree handling for specific tasks (e.g., hovering in an F-35B). For the F-35C, these optimizations include, but are not limited to, shipboard launches and recoveries.

A shipboard catapult launch can be divided into three distinct phases: acceleration, rotation, and flyaway. The launch starts with an acceleration. The aircraft is attached to a shuttle on the deck by a launch bar on the nose strut. The shuttle, attached to a steam-powered piston under the deck, provides the motive force to accelerate the aircraft to a safe flyaway speed. During this phase, the CLAW pre-positions the control surfaces to provide aircraft rotation upon shuttle release. The horizontal tail is scheduled to achieve roughly 12.5 deg/second of pitch rate at operationally representative aircraft weights, Wind Over the Deck (WOD), and energy imparted on the aircraft by the catapult. Next, the launch bar is released from the shuttle at the end of the catapult stroke; stored energy in the nose strut and aerodynamic moments generated by pro-rotation control surfaces (symmetric horizontal tail trim, symmetric ailerons, and toe-in [trailing edge inboard] rudders) combine to provide a smooth and quick aircraft rotation. Finally, during the flyaway phase, the CLAW transitions from pro-rotation to pro-lift to optimize lift and minimize sink once airborne. The pro-lift control surfaces are symmetric Trailing Edge Flaps (TEFs) and symmetric ailerons.

The catapult launch control law logic is evoked when the launch bar is commanded down while the aircraft is on deck, positioning the control surfaces for pro-rotation based on gross weight and the aircraft's CG. The pro-rotation configuration is designed to achieve a positive flight path as quickly as possible following catapult launch in order to minimize sink. Additionally, catapult launch logic causes a 30-lb throttle detent to engage at the 100 Engine Thrust Request (ETR) MIL gate, Nose Wheel Steering (NWS) is disengaged, and engine AB Limit (ABLIM) is activated. The ABLIM feature is designed to reduce thermal heating of the tail boom, empennage structure, JBD panels, and flight deck. If throttle position is set for ETR $>$ 122% during the catapult launch sequence, ABLIM control logic will automatically limit ETR to 122% power prior to launch. When a catapult launch is declared or longitudinal acceleration (Nx) $>$ 0.5 g, the ABLIM feature will clear, allowing thrust to increase up to the commanded power setting during aircraft acceleration down the catapult track. The Flight Control System (FLCS) uses multiple sensor to provide feedback to the CLAW to declare a catapult launch:

- ETR $\geq$ 95% and longitudinal acceleration (Nx) $>$ 2 g or calibrated airspeed (VCAS) $>$ 100 kt *or*
- ETR $<$ 95%, wheel speed $>$ 20 kt, and Nx $>$ 2 g or VCAS $>$ 100 kt

As the launch bar leaves the shuttle, the pro-rotation control surfaces begin to be removed as Angle of Attack (AOA) or pitch rate increase to protect from overrotation. The control system commands 20 deg AOA initially but backs off to

15 deg as AOA passes through 14 deg AOA. This command is then blended back to zero once a 1-deg flight path (gamma) is achieved. If AOA and pitch rate do not increase prior to the aircraft rolling off the deck edge, the pro-rotation aids are removed quickly once weight-off-wheels is sensed and replaced with pro-lift aids. Once airborne, catapult launch mode is disabled. The control law reschedules surfaces from pro-rotation to pro-lift positions in order to minimize sink and to enhance fly-away characteristics.

The CLAW also augments the roll axis during catapult launches. Roll rates and bank angles induced by asymmetries and crosswinds at the deck edge are removed via a launch roll trim mode that commands differential flaps and ailerons during a catapult launch. Launch roll trim activates when VCAS exceeds 100 kt and is reset as the aircraft leaves the deck. Additionally, a bank assist was added to hold wings level on flyaway. Bank assist activates at weight off wheels (WoffW) and attempts to maintain a wings-level attitude (0 deg bank angle) during flyaway for 10 seconds. The pilot may override bank assist at any time with a lateral stick input.

The F-35C has three approach modes: manual, Approach Power Compensation (APC), and DFP. All three modes are designed to maintain a precise AOA throughout the approach, providing the pilot fine control of touchdown location on the deck at the optimum pitch angle to facilitate a successful arrestment. Manual mode allows the pilot to fly a traditional backside approach to a ship, using thrust via the throttle to control rate of descent and longitudinal stick inputs to control airspeed. APC automates engine response and AOA such that the pilot controls flight path, and thus rate of descent, with longitudinal stick inputs. DFP uses APC logic and automates flight path maintenance by capturing and maintaining a designated flight path.

All approach modes use IDLC, a pilot command TEF/aileron interconnect that provides improved wave-off, bolter, and approach flying qualities by enhancing the flight path control of the nominal PA control system. This dynamic symmetric deflection command is added to the nominal PA schedules. Pilot input is passed through a first-order high pass filter to create a dynamic IDLC command. This dynamic command deflects the symmetric TEFs/ailerons for rapid direct lift capability to offset the slow response of the engine and provide a more immediate flight path response. IDLC is only used while in the AOA portion of the PA system where the aircraft is designed to be "back-sided" (i.e., throttle changes control rate of descent). IDLC response is initiated via the throttle or longitudinal stick inputs when in manual or APC, respectively; however, TEF/aileron command gains are increased 300% in APC over that of manual.

Delta Flight Path engages APC and commands an automated gamma capture and maintenance. As perturbations move the aircraft away from the desired glideslope, the pilot inputs and holds a longitudinal stick displacement to generate a change in gamma. As the aircraft approaches the desired glideslope, the pilot leads the correction and releases the stick, thus commanding the aircraft to recapture the programmed gamma.

C. F-35C ARRESTING HOOK SYSTEM

The Arresting Hook System (AHS) consists of a tailhook, hydraulic actuators and dampers, and a set of clamshell doors to conceal the tailhook in a bay for Low Observability (LO) considerations. It is designed to achieve rapid deceleration during shipboard landings by engaging an arresting wire on the deck of an aircraft carrier, or during shore-based aborted takeoffs or emergency landings. The arresting hook is electrically controlled, hydraulically actuated, and attached to the aircraft structure via a pitch pivot pin. Figure 3 shows the redesigned arresting hook, and Fig. 4 shows the AHS components.

Testing on the original AHS began in 2011, and it was quickly realized that the AHS was not functioning per design: the tailhook engagement rate (number of times the hook engaged and maintained contact with the arresting wire) was significantly low and the hook shank did not have enough load carrying capacity. An engineering investigation identified three root causes for the low engagement rate:

1. *Aircraft geometry:* The distance between the tailhook point and the main landing gear is short, 7.1 ft, as shown in Fig. 5 [3]. The short distance was the result of geometric constraints imposed by low observability requirements. To reduce radar cross-section, the AHS is contained in a recessed bay that is covered with doors similar to the weapons bay. The size of the bay, limited by aircraft size, dictates the allowable length of the AHS.
2. *Tailhook point design:* The hook point's geometry was too rounded, thus lowering its scooping efficiency, as shown in Fig. 6 [3].
3. *Tailhook hold-down performance:* The hold-down damper provided an insufficient hold-down force, thus allowing excessive tailhook bounce [3].

These three factors culminated in the main landing gear tires rolling over the arresting

Fig. 3 Redesigned F-35C arresting hook.

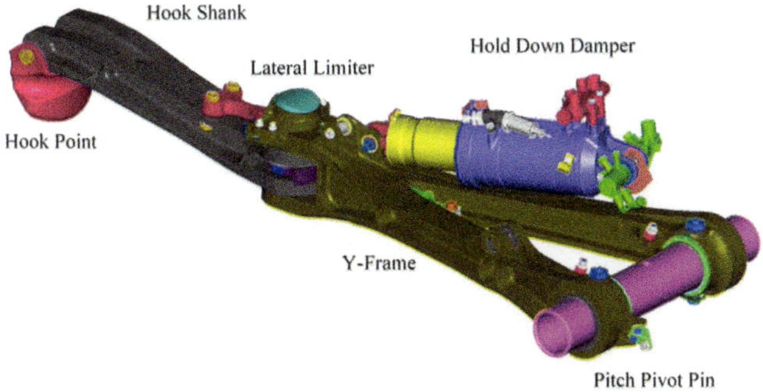

Fig. 4 F-35C arresting hook components.

wire, initiating a transverse or "kink" wave. This wave propagated away from the point of tire rollover, both outwards and inwards. The wave propagation speed and aircraft forward velocity were such that the arresting wire was lying flat on the ground as the rounded tailhook point arrived at the wire, causing the hook to skip over the wire. These dynamics were further exacerbated by the low hold-down force. When the hook point initially contacted the deck or was perturbed by an imperfection on the landing surface, the hook would bounce [4], leaving the surface and either glancing off or missing the arresting wire completely. These dynamics can be seen in the series of photos shown in Figs. 7–10.

In 2013, the redesigned AHS was completed, delivered, installed, and tested. The new design offered several improvements. First, the hook point geometry was altered from a rounded design to a wedge design (Fig. 11), ensuring that the apex of the wedge was lower than the centroid of the arresting wire, as shown in Fig. 12.

Additional improvements included increased structure to the y-frame and hook shank, thereby increasing the load carrying capacity; the hold-down

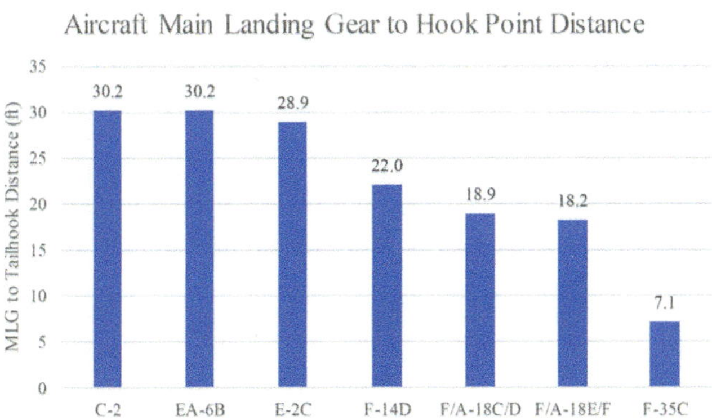

Fig. 5 U.S. Navy aircraft main landing gear to hook point geometry comparison.

Fig. 6 Original hook point with rounded geometry.

damper size also was increased to increase the hold-down force and reduce hook bounce. Furthermore, an end-of-stroke damper and lateral limiter were installed. The end-of-stroke damper was added to compensate for the increased down force of the hold-down damper. When the aircraft is on the deck and the hook is commanded down, the hook point will strike the deck prior to the hook shank achieving full range of motion; however, when airborne, the hook shank will travel its fully allowable arclength, striking the end-of-stroke damper. The end-of-stroke damper absorbs the shock downwards force imparted on the hook by the hold-down damper, preventing overload conditions. The lateral limiter acts to prevent excessive movement of the hook shank. It is attached to the hinge point where the shank meets the y-frame. If the aircraft rolls or touches down in a bank angle and imparts an angular velocity on the hook shank, the lateral limiter acts as a bumper, absorbing the energy. If too much energy is imparted,

Fig. 7 During roll-in testing, nosewheel initiates kink wave.

Fig. 8 Kink wave propagates away from the source of excitation.

the elastic deformation region of the material is exceeded, resulting in plastic deformation of the lateral limiter, providing a telltale sign to maintainers and pilots that an overload condition has been prevented. The resulting maintenance requires the removal and replacement of the lateral limiter, but provides a significant savings over damaging the hook shank.

The result of the redesigned AHS was tested for the first time in December 2013, where it successfully engaged the arresting wire on the first attempt, as

Fig. 9 Tailhook arrives at arresting wire flush with ground.

Fig. 10 Tailhook bounces over flush arresting wire.

seen in Fig. 13. Subsequent flight tests have provided a statistically significant sample size to validate the effectiveness of the F-35C's AHS.

D. NIMITZ-CLASS AIRCRAFT CARRIER

Testing was conducted aboard USS *Nimitz* (CVN 68), USS *Dwight D. Eisenhower* (CVN 69), and USS *George Washington* (CVN 73), all Nimitz-class aircraft carriers. Nimitz-class aircraft carriers are 1092 ft long, 135 ft wide at the waterline, 252 ft wide at the widest point on the flight deck, have a 37-ft draft, and displace approximately 100,000 long tons fully loaded. Each Nimitz-class ship has an angled flight deck, four C-13 steam catapults, four Mk 7 arresting gear (three Mk 7 arresting gear CVN 76 and later), an IFLOLS, drop lights, laser line-up system, flight deck lighting, a large below-decks aircraft hangar, and four aircraft elevators. Each ship has an approximately 788-ft angled deck (9 deg to port from

Fig. 11 Wedge hook point geometry.

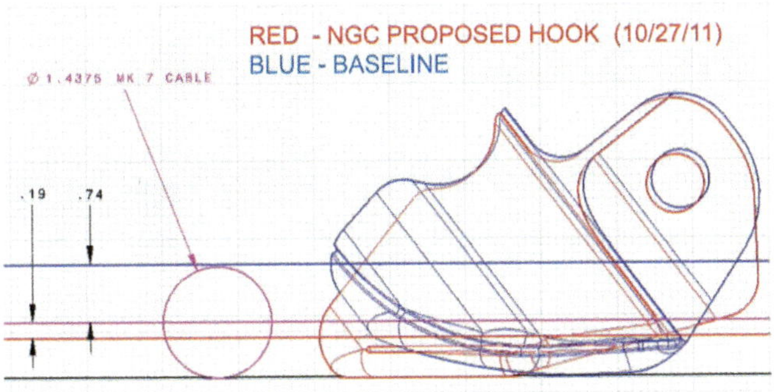

Fig. 12 Geometric comparison of rounded to wedge hook point.

centerline) to facilitate simultaneous launch and recovery of aircraft. Aircraft are lowered to and raised from the hangar deck via the four elevators: three located on the starboard side and the other located to port near the stern. A flight deck diagram is presented in Fig. 14.

E. CATAPULT

The catapult provides the motive force for shipboard launches of aircraft. The main components of a catapult are the power cylinders, piston assemblies, shuttle assembly, and a water brake, illustrated in Fig. 15. Steam, provided by the ship's boilers, enters the aft end of the cylinders through the launching valve. This steam pressure then acts on the piston assemblies and accelerates

Fig. 13 The first successful arrested landing with the redesigned arresting hook system.

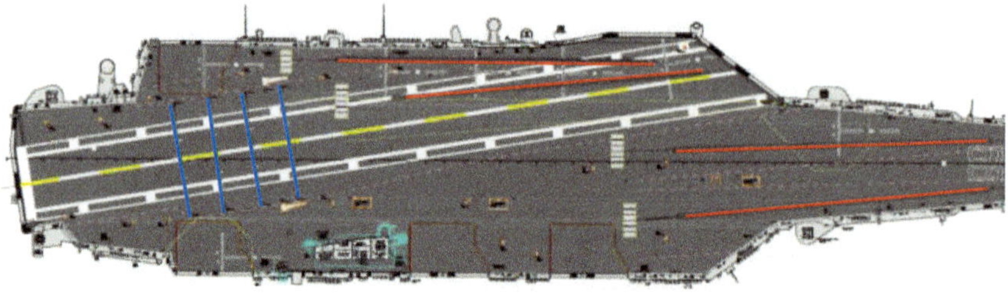

Fig. 14 Nimitz-class aircraft carrier deck diagram with catapults in red and arresting wires in blue.

the shuttle assembly to a safe launching velocity. The aircraft is connected to the shuttle assembly via the launch bar attached to the aircraft's nosewheel, shown in Fig. 16. At the end of the catapult stroke (which is the forward end of the ship), the tapered spears, one mounted to the front of each piston, enters the water brake assembly, which absorbs the energy of the piston assemblies and rapidly brings them to a stop. At this time the aircraft is released from the shuttle with sufficient speed to become airborne.

Proper heating and temperature maintenance are a critical factor for the safe operation of the catapult. Finned tub heat exchangers installed around the power cylinders, also supplied with steam from the ship's boilers, provide heat to maintain the cylinders at a near constant temperature. The catapults are preheated to achieve the proper thermal expansion. Once the proper temperature is reached, the flow of steam is varied to maintain a constant elongation of both power cylinders. The maintenance of proper cylinder elongation is critical in that all launch bulletins (documents that provide the settings for the Capacity Selector Valve [CSV]) are based on a constant elongation, thus providing for reliable and repeatable catapult performance between launches.

The water brake assembly provides the catapults braking capability via a 9-ft spear on the forward end of the piston assembly. At the end of the catapult stroke, the spear is guided into the water brake cylinder, which is kept full of water by a centrifugal vortex induced by jets of water [5].

Two types of catapults were used during testing: a shore-based C-7 catapult and a C-13 aboard the Nimitz-class aircraft carriers. The primary difference between the two catapults is length. The C-7 was designed for an earlier class

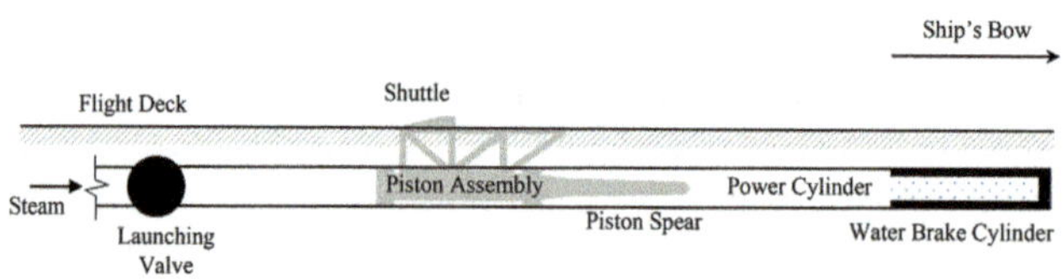

Fig. 15 Simplified diagram of steam catapult components.

Fig. 16 F-35C launch bar connects to shuttle forward of nose strut.

aircraft carrier, thus limiting its length to 276 ft with a 253-ft stroke. Conversely, the C-13 has a length of 325 ft with a 310-ft stroke [6]. Due to this difference, the jerk (time rate of change of acceleration), or jolt, is greater on the C-7 catapult. Despite this difference, the C-7 provides operationally representative launches, mirroring accelerations and both aircraft and catapult tow loads experienced during a shipboard launch. A typical longitudinal acceleration and launch bar axial loading as a function of time during a catapult launch are shown in Fig. 17.

F. VISUAL LANDING AIDS

The U.S. Navy has two shore-based visual landing aids: the Mk 8 Fresnel Lens Optical Landing System (FLOLS) and the Mk 14 Improved Fresnel Lens Optical Landing System (IFLOLS) to display glideslope and trend information to pilots approaching the runway. Each system consists of a yellow source light assembly, green datum lights, red wave-off lights, and green "cut" lights. The lights are useable at a range of 1.0 n mile. The yellow source light is projected through a series of linear Fresnel lenses to focus the light into narrow horizontal bands. A lenticular lens is attached in front of the Fresnel lenses to provide the desired source light color, increase the azimuth visibility arc to ± 20 deg, and disperse sun reflection from the lens face. The result is a light source that appears to the pilot as the "meatball" or a "ball" that moves vertically across the face of the lens as the pilot's eye moves through the beam's arc length. The ball provides the pilot a visual means to determine the aircraft's position relative to the designated glideslope. A jackscrew and hand-crank on the front of the trailer base are used to adjust the system to achieve desired glideslope. The task of the pilot is to capture and maintain the ball between the datum lights. If the aircraft is flying on the desired glideslope, the pilot will see the ball in line with the green datum lights.

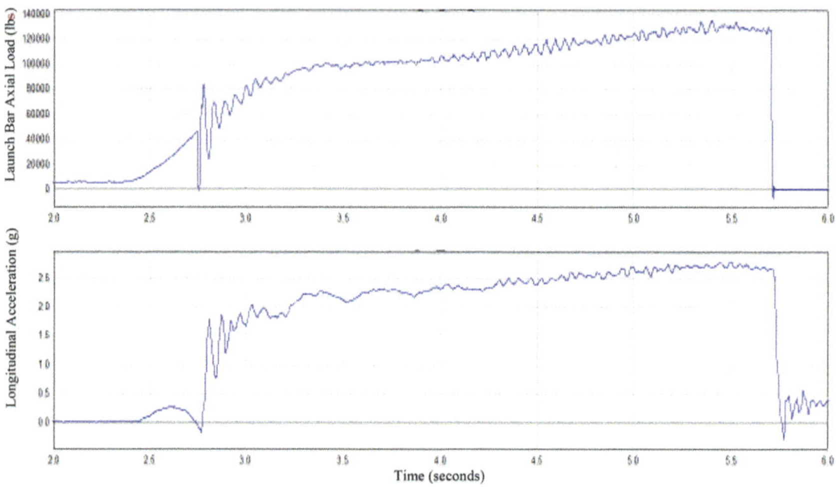

Fig. 17 Launch bar axial loading and longitudinal acceleration of the F-35C as a function of time.

If the aircraft is high, the pilot will see the amber ball above the datum arms; conversely, if the aircraft is below glideslope, the pilot will see the ball below the datum lights. If the pilot proceeds too far below glideslope, the amber ball will change to a red ball. The red wave-off lights are manually activated by the LSO when dangerous landing conditions exist. The green cut lights were historically used to inform the pilot of a propeller-driven aircraft when to "cut" the engines; however, today cut lights are used to indicate a landing clearance in the event of lost communications [7].

The Mk 8's source box, shown in Fig. 18, contains fewer cells than the Mk 14, shown in Fig. 19, and is consequently a shorter display—a 4-ft-tall source box vs the 6-ft-tall source box of the Mk 14. This results in the fidelity of the Mk 14 display being approximately double that of the Mk 8, as can be seen when comparing Tables 2 and 3. The Mk 8 FLOLS provides optical glideslope information within approximately $\pm\frac{3}{4}$ deg of the desired glideslope. The source light unit is composed of five cells (each cell is designed to illuminate $20.45'$ of arc) for a total of approximately 1.7 deg of vertical coverage. A subsequent consequence of the Mk 14's larger size is that it is transversely mounted on its trailer as compared to the mounting orientation of the Mk 8. Although

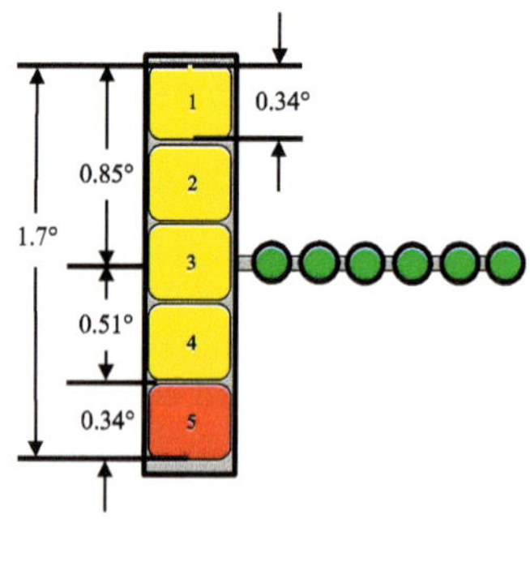

Fig. 18 Mk 8 FLOLS light box geometry.

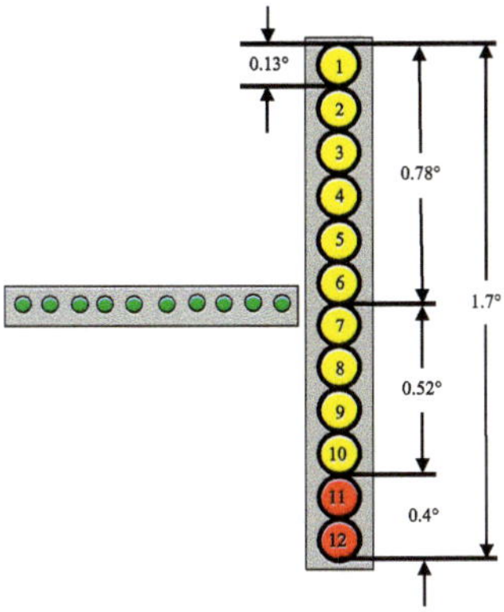

Fig. 19 Mk 14 IFLOLS light box geometry.

the Mk 14 offers higher fidelity with respect to changes in glideslope, it is not easily or quickly moved when a change in touchdown point is desired. To conduct carrier suitability testing efficiently, there is a need to rapidly move the desired touchdown location. The orientation of the Mk 8 on its trailer provides this flexibility over the Mk 14, thus making it the lens of choice for carrier suitability testing.

Mk 13 Mod 0, shown in Fig. 20, is used for shipboard operations. It provides glideslope and trend information to the pilot in the same exact manner as the Mk 14 shore-based model. It is set to a base angle of 3.5 deg to ensure a Hook-to-Ramp (H/R) distance of at least 14 ft, as shown in Fig. 21. This will be discussed in more detail in the "Glideslope Geometry" section. The largest difference between the two is that the Mk 13 is dynamically stabilized to compensate for ship's motion in pitch, roll, and heave [5].

G. MK 7 ARRESTING ENGINE

The Mk 7 constant runout arresting gear, illustrated in Fig. 22, is a hydro-pneumatic system composed of the engine structure, a cylinder and ram assembly, a crosshead and fixed sheaves, a control valve system, an accumulator system, air flasks, and a sheave and cable arrangement. An Mk 7 arresting engine is installed in the approach end of runway 32 at Naval Air Station Patuxent River, Maryland. The shore-based Mk 7 site is similar to shipboard arresting gear and considered operationally representative.

TABLE 2 MK 8 FLOLS GEOMETRY BY CELL

Cell	Basic Angle (deg)	Hook-to-Ramp (ft)
Top	4.4	23.3
1	4.2	20.0
2	3.8	17.7
3	3.5	14.0
4	3.2	10.3
5	2.8	6.6
Bottom	2.7	4.7

TABLE 3 MK 14 FLOLS GEOMETRY BY CELL

Cell	Basic Angle (deg)	Hook-to-Ramp (ft)
Top	4.3	23.6
1	4.2	22.8
2	4.1	21.2
3	4.0	19.6
4	3.8	18.0
5	3.7	16.4
6	3.6	14.8
Center	3.5	14.0
7	3.4	13.2
8	3.3	11.6
9	3.2	10.0
10	3.0	8.4
11	2.9	7.2
12	2.7	4.7
Bottom	2.6	3.4

During an arrested landing, the Mk 7 brings the aircraft to a smooth, controlled stop in the following manner: The aircraft's arresting hook engages the cross-deck pendant (also called the "wire" or arresting cable) spanning the landing area. The cross-deck pendant is a 1-3/8-in.-diameter cable with a polyester core, weighing approximately 3 lb/ft and having a minimum static breaking strength of 205,000 lb [5]. The forces generated by the aircraft's forward motion are transferred via the cross-deck pendant to purchase cables that are reeved around a movable crosshead of sheaves and the fixed sheave assembly of the arresting engine. The movable crosshead is moved toward the fixed sheave assembly as the aircraft pulls the purchase cables off the arresting engine, forcing a ram into the cylinder holding pressurized hydraulic fluid (ethylene glycol). The flow of fluid is controlled by the constant runout control valve, which in turn controls the pressure in the cylinder thereby providing the restraining force to absorb the aircraft's energy.

H. LANDING SIGNAL OFFICER

The Landing Signal Officer, pictured in Fig. 23, is a pivotal member of the test team. The LSO is a pilot who, through training and experience, is capable of quickly correlating the effects of wind, weather, aircraft capabilities, and pilot

Fig. 20 Shipboard Mk 13 IFLOLS.

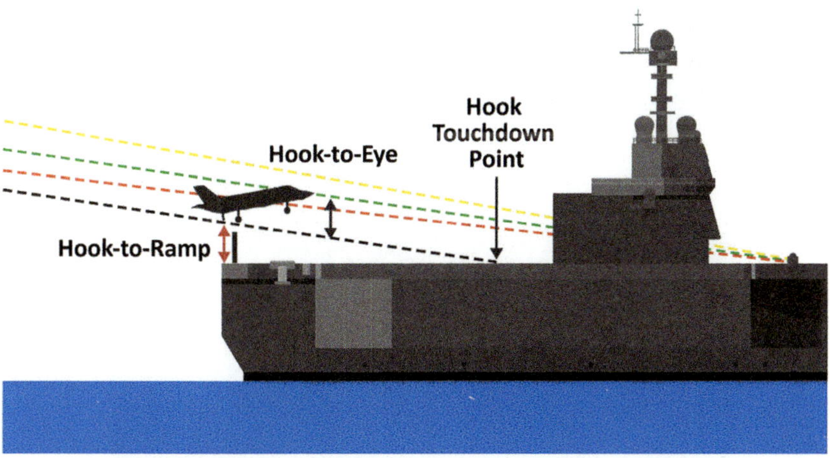

Fig. 21 Glideslope geometry relationships: hook-to-ramp, hook-to-eye, hook touchdown point, and IFLOLS glideslope projection.

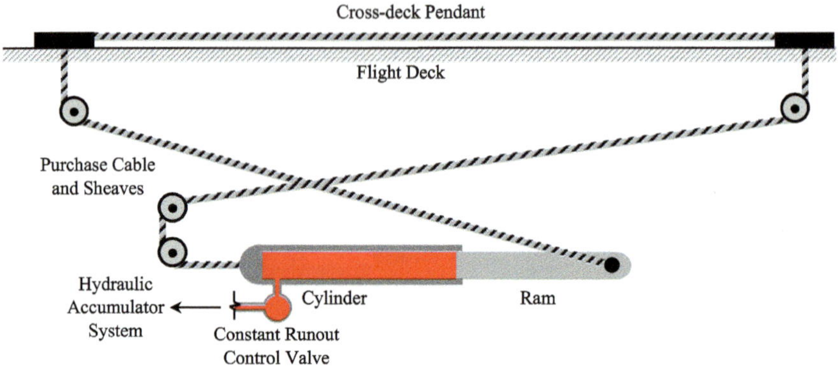

Fig. 22 Simplified diagram of arresting gear components.

performance to provide optimum control and assistance in achieving the test point. The LSO's primary responsibility is safety. Additional duties include serving in a capacity similar to that of a test conductor, directing the testing efforts of the day. The test pilot and LSO form a professional and disciplined team founded on mutual respect and the highest degree of trust.

Shore-based, the LSO is stationed in a mobile structure on the side of the runway abeam the intended point of landing and is able to observe the test aircraft throughout the landing pattern. Ship-based, the LSO and team are stationed on the port (left) side of the flight deck, approximately 50 ft from the aft end of

Fig. 23 The Landing Signal Officer (LSO).

the ship. From the moment of the "ball call," signifying that the light source on the Fresnel lens is visible, to the arrestment, touch and go, or wave-off, the pilot is under the control of the LSO. Analyzing trends in both pilot and aircraft performance, assessing external environmental factors, and having a thorough understanding of what techniques the pilot is using, the LSO transmits informative advisory and directive calls to aid the pilot in safely achieving the test point.

I. PILOT'S TASK

A test pilot flying a structural survey arrested landing has a different task than that of the fleet aviator executing a carrier-style landing. The fleet aviator is expected to capture and maintain glideslope, centerline, and angle of attack from the start of the approach all the way to touchdown. Although there are tolerances for deviations, if the appropriate correction is not made in a timely manner, the fleet aviator will be waved off. However, a carrier suitability structural survey is a specialized loads test, thus in addition to glideslope, line up, and angle of attack, the test pilot has the additional tasks, depending on the specific test point, of capturing and maintaining a ground speed; setting a pitch, roll, and/or yaw attitude prior to touchdown; or attaining a target sink speed. Another difference is hook touchdown location. The fleet aviator, while expected to maintain a centered ball, has an acceptable touchdown window of 120 ft whereas the test pilot's acceptable touchdown window varies from 40 ft to less than a foot, again depending on the test point.

The base glideslope is a 3.25-deg (above the horizon) fixed path determined by the angle of the Fresnel lens. The rate of descent necessary to maintain this glideslope depends on the aircraft ground speed (and therefore changes slightly with wind conditions). Proper execution of the approach requires an accurate, rapid scan to allow the pilot to identify deviations and make corrections as soon as one is required; if an appropriate correction is not made, the deviation will only increase in magnitude. This is due to the constant arc length of the light the Fresnel lens is displaying. As the pilot approaches the light source, the lens fidelity increases, resulting in ball movement increasing at an exponential rate; for example, if the pilot captures the desired 3.25-deg glideslope but is one ball high and does not center the light source between the datum, the pilot will initially see the ball moving upward at a slow rate. As the pilot's eye approaches the source, the rate of ball movement increases exponentially, resulting in the ball appearing to shoot off the top of the lens.

Lineup for shipboard landings is critical; the relatively small size of the landing area makes it imperative that the pilot lands on the centerline with no drift. What adds to the difficulty of shipboard lineup is that the landing area is misaligned with the ship's centerline by 9 deg, resulting in the landing area having a sideward velocity or drift. Conversely, shore-based testing is completed to a runway with no drift; however, lineup is no less critical for the test pilot. For shore-based testing, an extended landing pattern is used with the pilot capturing runway centerline somewhere between 2.0 and 1.5 n miles from touchdown, thus requiring the pilot to perform the centerline maintenance task for up to one minute. Depending on the wind gradient and magnitude of gust, this task can be challenging, requiring the pilot to identify and correct lineup deviations as early as possible. Failure to make lineup corrections in a timely manner will typically cause the test pilot's scan to break down. This becomes evident to both the pilot and LSO with a manifestation of a deviation in glideslope or another critical test parameter. Lineup deviations as little as 3 ft from centerline can have a profound impact on hook loading. Additionally, one of the test points that will be discussed later requires the test pilot to engage the cross-deck pendant within 2 ft. The test point calls for an off-center engagement of at least 18 ft from centerline; however, if the pilot exceeds 20 ft, the arresting engine requires an immediate inspection because damage may have occurred. For both the fleet aviator and the test pilot, lineup deviations that are not corrected in a timely manner will typically cause scan breakdown, resulting in deviations in other parameters. Shipboard, a camera is mounted on the centerline flush with the deck facing aft. This image is monitored by the LSO and allows him or her to see if the pilot is failing to maintain the landing area centerline. Unfortunately, the shore-based LSO does not have this capability. This results in the pilot having to assess crab angle and make lateral and/or directional inputs just prior to touchdown to precisely place the hook point within the lateral tolerances of the test point.

Historically, the pilot was also required to maintain the aircraft within a specific AOA, or alpha, range. The range is specific to each aircraft and is determined by two factors: speed and geometry. The lower end of the acceptable alpha range is determined by maximum arresting gear engagement speed. If the aircraft's velocity is too high (low AOA), the aircraft structure may yield or the arresting gear engine may be damaged. Conversely, if the aircraft is too slow (high AOA), the airspeed margin above aerodynamic stall is unacceptable or the aircraft's tailhook point is sufficiently lowered that the threat of an in-flight engagement exist. Due to the carefree handling offered by the F-35's FLCS, the CLAW attempts to maintain AOA tightly within the acceptable range, removing the pilot from controlling AOA directly; although it is important for the pilot to mind the alpha and avoid large AOA transients, this is now a secondary concern for the pilot.

J. GLIDESLOPE GEOMETRY

The task of executing carrier suitability testing, be it on land or sea, requires the capability to precisely have the hook point touchdown on the targeted position. To facilitate this, glideslope geometry, shipboard landing dynamics, and the environmental effects must be thoroughly understood. The following are definitions of commonly used terms with respect to glideslope geometry also illustrated in Figs. 24 and 25:

- *Natural wind*: Existing surface wind caused by the moving air mass independent of ship's motion

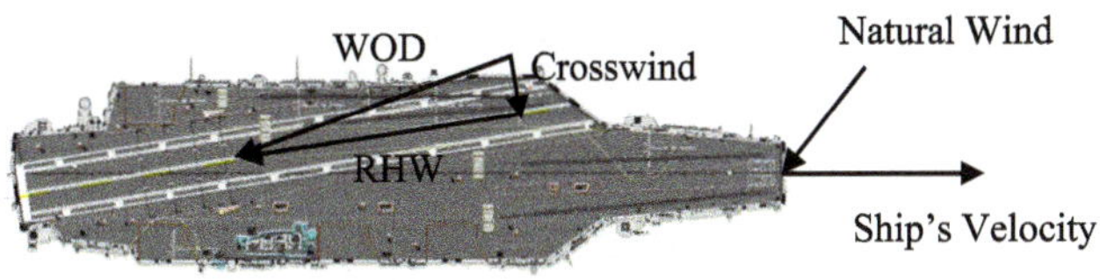

Fig. 24 Vector relationships of shipboard winds.

- *Ship's wind*: Wind created by ship's velocity, equal in magnitude and opposite in direction
- *Relative wind or Wind Over Deck (WOD)*: The vector sum of natural wind and ship's wind
- *Recovery Headwind (RHW)*: That component of WOD parallel to the centerline of the landing area
- *Recovery crosswind*: That component of WOD perpendicular to the centerline of the landing area
- *Angle of attack (α)*: Angle between the aircraft Fuselage Reference Line (FRL) and flight path
- *Flight path (γ)*: Aircraft's velocity vector through the air mass measured from the FRL
- *Glideslope angle (Θ)*: Angle between the optical landing aid light source and the horizon

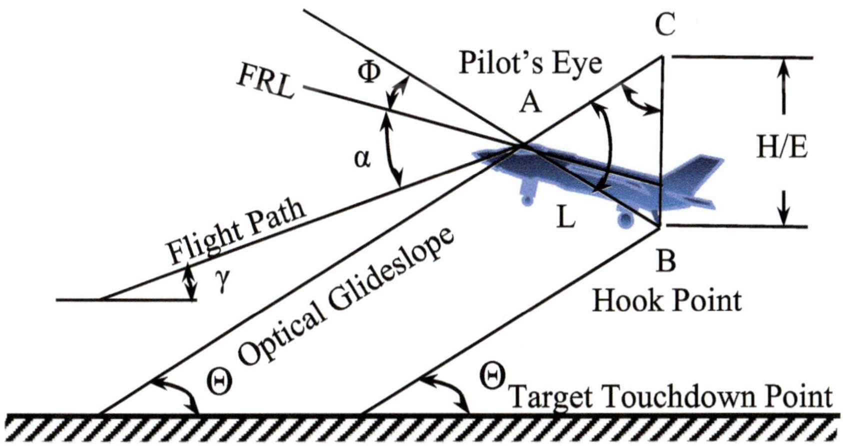

Fig. 25 Glideslope geometry.

- *Pilot eye to FRL angle (Φ)*: Angle between FRL and a line connecting the pilot's eye to the hook point
- *Pilot eye to hook point length (L)*: Straight line distance between the pilot's eye and the hook point

The Hook-to-Eye (H/E) distance of an aircraft is of critical importance to place the aircraft touchdown at the proper point in the landing zone of the aircraft carrier. The H/E distance needs to be precisely determined; any error will result in either an early touchdown that reduces Hook-to-Ramp (H/R) clearance or a late touchdown that increases bolter rates. A stadiametric view of glideslope geometry is presented in Fig. 25 [6].

Applying the law of sines to triangle *ABC*,

$$> A = (\Theta - \gamma) + \alpha + \Phi \tag{1}$$

$$> C = 90 - \Theta \tag{2}$$

$$\frac{H/E}{\sin A} = \frac{L}{\sin C} \tag{3}$$

$$H/E = \frac{L \sin ((\Theta - \gamma) + \alpha + \Phi)}{\sin (90 - \Theta)} \tag{4}$$

$$\sin (90 - \Theta) = \cos \Theta \tag{5}$$

$$H/E = \frac{L \sin ((\Theta - \gamma) + \alpha + \Phi)}{\cos \Theta} \tag{6}$$

Small angle theorem, $\cos \Theta = 1$; therefore,

$$H/E = L \sin ((\Theta - \gamma) + \alpha + \Phi) \tag{7}$$

To determine the difference between the glideslope angle and the flight path angle, the relative velocity between the aircraft's true airspeed and the landing point's velocity relationship is evaluated. The landing point's relative velocity is a function of ship's speed and natural winds, as shown in Fig. 26.

$$\delta = \Theta - \gamma \tag{8}$$

$$H/E = L \sin (\delta + \alpha + \Phi) \tag{9}$$

Applying the law of sines to Fig. 26,

$$\frac{RHW}{\sin \delta} = \frac{TAS}{\sin (180 - \Theta)} \tag{10}$$

$$\sin (180 - \Theta) = \sin \Theta \tag{11}$$

$$\sin \delta = \frac{\sin \Theta RHW}{TAS} \tag{12}$$

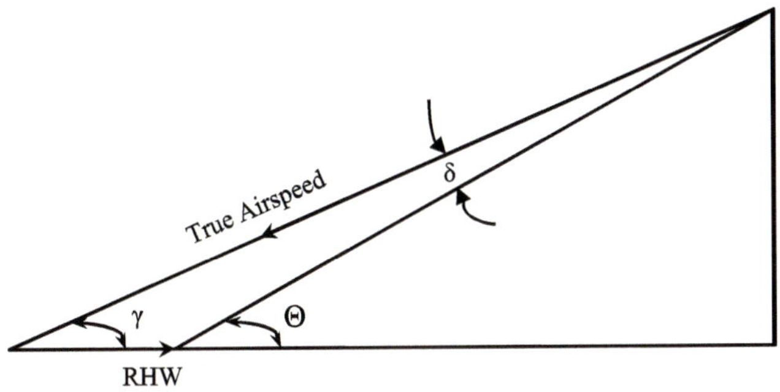

Fig. 26 Vector relationship of true airspeed, recovery head wind, and glideslope angle.

Small angle theorem, sin $\Theta = \Theta$; therefore,

$$\delta = \frac{\Theta RHW}{TAS} \tag{13}$$

It is seen that the largest value of δ will be obtained with increasing glideslope angle and RHW and decreasing approach airspeed. Another way of looking at Eq. (13) is to manipulate glideslope angle based on changes in RHW to maintain δ near constant for pilot considerations. To accomplish this, the shipboard glideslope is routinely increased during high recovery headwind conditions.

The flight path angle can be determined by

$$\gamma = \Theta - \delta = \Theta - \frac{\Theta RHW}{TAS} \tag{14}$$

$$\gamma = \Theta\left(1 - \frac{RHW}{TAS}\right) \tag{15}$$

An alternative way to observe effective glideslope is presented in Fig. 27. At time one t_1, the pilot's eye is on the optical glideslope. A few moments later, as

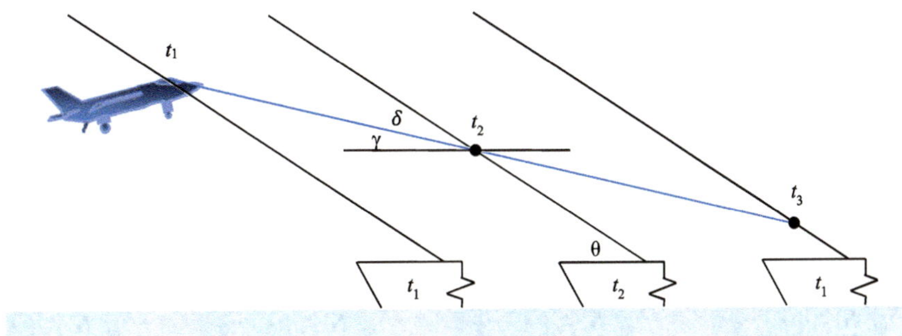

Fig. 27 Effect of the ship's forward motion on the effective glideslope.

the pilot continues to approach the ship, the ship's position is displaced by its velocity through the water. This dynamic continues all the way to touchdown at t_3. Under ideal operational conditions, the glideslope is set to 3.5 deg, but the pilot maintains a 3-deg flight path, the effective glideslope.

IV. SHORE-BASED TESTING

Shore-based carrier suitability testing of the F-35C started on 17 March 2011. Historically, new seagoing aircraft were required to demonstrate compliance with MIL-D-8708; however, for the F-35C, requirements that outlined the shore-based tests required were derived from the Joint Contract Specification (JCS) and JSF Operation Requirements Document (ORD). Specifically, a structural survey vs a structural demonstration would be executed. The difference between a survey and demonstration is the degree to which envelope exploration or expansion is required: a demonstration requires a higher degree of expansion whereas a survey is used to validate models and predictions. To meet the requirements of the JCS and ORD, shore-based testing was broken into three phases: prerequisites to support CVS (phase one), the structural survey (phase two), and initial ship trial prerequisites (phase three).

During the first phase, prerequisites to support CVS testing were completed to include static loading, drop test of a ground article (CG-1), and PA flying qualities. The second phase was designed to support shore-based catapult launches and arrested landings. To facilitate shore-based launches, JBD compatibility and environmental testing was completed to evaluate deck heating, JBD panel cooling, and vibro-acoustic, thermal, and Hot Gas Ingestion (HGI) environments on and near the aircraft, shown in Figs. 28 and 29. The C13-2 catapult at Naval Air Engineering Site (NAES) Lakehurst, New Jersey, was intentionally degraded to represent a worst-case steam leakage condition to evaluate the effect on engine performance and tolerance to steam ingestion. Additionally, ground and deck handling tasks were conducted to ensure aircraft compatibility with the confined space taxi requirements, catapult spotting procedures, and physical geometry of the above deck equipment (launch bar and repeatable release hold back [RRHB] bar compatibility). The final task with respect to shore-based catapult testing was to subject the aircraft to design limit load launches from the TC-7 catapult at Naval Air Station Patuxent River, Maryland (PAX). Prior to starting shore-based arrested landings, bolter and wave-off performance test points were flown and roll-in arrestments were completed. Roll-in arrestments, completed at NAES Lakehurst, were a ground-based test where the aircraft was positioned at a predetermined distance on the runway from the arresting gear so as to reach a targeted groundspeed by the arresting gear. When the aircraft was within the allowable groundspeed band, the tailhook was lowered, and the F-35C "rolled" into the arresting gear. This line of testing was performed both on and off runway centerline (to a maximum of 20 ft off centerline) and is

Fig. 28 F-35C conducting deck heating and JBD panel cooling evaluations.

designed to measure both arresting gear and tailhook loads prior to executing fly-in arrestments. The third, and final, phase consisted of testing required to support the initial ship trials: reliability and maintenance (deck tie-down and towing), electromagnetic spectrum capability, navigation and guidance, sensor integration, data link interoperability, pilot–vehicle interface, and supportability

Fig. 29 JBD panel cooling evaluation with the F-35C engine in afterburner limited by RATS.

Fig. 30 First shore-based catapult launch of the F-35C.

were evaluated to ensure the aircraft was capable of operating as a system within a system (aircraft aboard a seagoing vessel).

A. SHORE-BASED CATAPULTS

Shore-based catapult launches of the F-35C conducted during the structural survey, shown in Fig. 30, were used to define and/or validate engineering models of the structural components to withstand the forces and accelerations the aircraft would be subjected to during a shipboard launch. The extent to which the aircraft was stressed was based on the maximum shipboard takeoff weight and the performance characteristics of the launching catapult. An additional objective of shore-based catapults was to ensure adequate clearance between the F-35C (including all external stores) and the catapult launch components (i.e., catapult shuttle, RRHB, etc.) throughout the launch cycle (hookup, launch, shuttle release, and rotation).

The results of shore-based catapult testing are used to predict the limit conditions and establish pilot technique that will be used for shipboard operations. For the F-35C, the pilot technique used was open-loop, or hands-off, because the CLAW is designed to capture and maintain the optimum flyaway attitude and maintain wings level. Testing consisted of a series of launches within an allowable weight band that built up in Nx. The Nx was incrementally increased until the endpoint, maximum longitudinal acceleration (Nx_{max}) was achieved or the launch bar horizontal tow load limit was reached. After Nx_{max} was achieved, a series of off-center catapult launches were conducted in which the longitudinal axis of the aircraft was misaligned with the longitudinal axis of the catapult to excite directional oscillations. The off-center testing was conducted within a constant weight band and at a constant *Nx* while incrementally increasing the off-center angle until the maximum delta had been achieved [6].

Most airplanes will experience directional oscillations during catapult launches when excited by crosswinds, asymmetric loadings, or inadvertent lateral or directional inputs. When directional oscillations are excited, aerodynamic forces and tire cornering forces combine as a forcing function to realign the aircraft with the catapult; however, inertia will cause an overshoot. This cycle continues until the forcing functions act to dampen out the oscillations. Off-center catapult testing is used to excite directional oscillations for two reasons:

1. Confirm that forcing function dynamics will dampen oscillations during the power stroke prior to shuttle release.
2. Measure side forces and bending moment loads on the launch bar and nosewheel strut.

Catapult launch test points at high gross weights resulted in conditions at which the catapult end airspeed was less than that required to become airborne. Testing was conducted at PAX, where the centerline of the C-7 catapult intersects runway 32 at an angle of 8 deg approximately 1000 ft from the end of the catapult power stroke, as seen in Fig. 31. For these points, significant performance calculations, modeling, and simulation were conducted to ensure the aircraft would be airborne

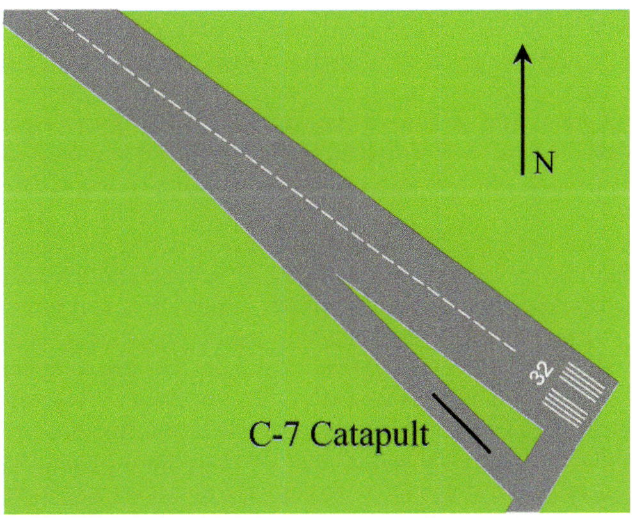

Fig. 31 Relative geometry of NAS Patuxent River catapult site intersecting runway 32 at an 8-deg angle.

prior to reaching the centerline of runway 32. The results of this effort were used in training the test pilots in simulators where aircraft system failures were input during the catapult power stroke to ensure pilot recognition and appropriate actions were taken to align the aircraft with the runway for continued takeoff or abort. If the pilot would be unable to safely realign the aircraft or was required to eject, the test point was deferred until it could be completed at NAES Lakehurst where the catapult is aligned with the runway.

Shore-based catapults were completed in each of the configurations shown in Table 1. The data obtained from shore-based testing were used to construct the allowable launch envelope and initial Aircraft Launch Bulletins (ALBs), tables that use aircraft weight and WOD as inputs to determine the CSV setting for launch.

B. SHORE-BASED ARRESTMENTS

In carrying out the carrier suitability testing, a majority of the time, effort and budget are spent on the shore-based arrestment portion of the structural survey. This phase of testing can be thought of as specialized loads testing in which the aircraft executes a series of arrested landings at various sink rates, attitudes, and airspeeds to evaluate the loads on the landing gear, arresting hook, and other critical load nodes. Prior to the F-35, limitations in computer processing power constrained the test team's ability to use aircraft capabilities to increase testing efficiency; however, with computer processor power ever-increasing, advances in flight control computers and Full Authority Digital Engine (or Electronics) Control (FADEC) systems have lessened the need to keep the pilot in the loop, thus providing an avenue to increase testing efficiency.

The first F-35C structural survey arrestment was accomplished on 4 February 2014. This testing was filled with discoveries and engineering challenges that resulted in delays; however, using a "blank canvas" philosophy to challenge legacy assumptions and techniques, new methods were developed to complete testing and recover lost time. To achieve this end, a methodical five-step iterative process was created. The first step was to evaluate current test equipment in use and conduct a historic review of carrier suitability testing to understand the reasons for assumptions under which shore-based testing was conducted, thus giving insight into the legacy techniques. The second step consisted of completing a comprehensive examination of the geometry, control laws, and flying qualities of the F-35C with respect to approach handling qualities and landing and arrestment characteristics. Third, the results of the first two phases were compared and contrasted, the outcome of which was used to identify every aircraft capability that could be used or exploited to achieve test points in the structural survey. Next, as new techniques were identified, a rigid developmental process was followed consisting of extensive simulator utilization for maneuver development, flying qualities evaluation and maneuver safety assessment. Finally, the maneuvers and techniques were used airborne to achieve test points during which their merits were evaluated. The methods that worked were adopted and the ones that didn't were refined by repeating the process. The "blank canvas" philosophical approach consisted of a systematic iterative approach that resulted in the development of new test techniques that will be used to advance the field of carrier suitability flight test.

A structural survey is essentially a loads envelope expansion program. Ten specific maneuvers are flown to endpoints, to achieve 80–100% of the components allowable load limit. For each maneuver, multiple components' loads are being monitored via instrumentation, such as main gear compression and bending moments, nose strut compression and bending moments, hook side forces, and ordnance suspension forces. To replicate the landing attitudes the aircraft may experience during a shipboard landing, 10 maneuvers were flown to reach the limits of the landing design envelope. The landing design envelope for U.S. Navy aircraft is expressed as a mathematical probability in terms of the multivariate distribution of landing impact conditions [Eq. (16)] [6]. The total probability of occurrence of any landing is expressed in terms of the probability of eight variables that can be expected during landing. Sink speed is defined by a Pearson Type III distribution function, and the other seven variables are expressed in terms of a normal or Gaussian distribution function [6].

$$P_T = \{V_A >< V_{A_i}\}\{V_E >< V_{E_i}\}\{V_V >< V_{V_i}\}\{\theta >< \theta_i\} \times \{\varphi >< \varphi_i\}\{\dot{\varphi} >< \dot{\varphi}_i\}\{\psi >< \psi_i\}\{d >< d_i\} \tag{16}$$

P_T = Total probability

V_A = Approach speed

V_E = Engaging speed

V_V = Sink speed

θ = Pitch attitude

φ = Roll attitude

$\dot{\varphi}$ = Roll rate

ψ = Yaw attitude

d = Off-center distance

i = Initial condition

$><$ = Greater than and, alternately, less than

All three approach modes of the F-35C were used in pursuit of structural survey test points. Through the iterative process of the blank canvas philosophy, each was found to be best suited for certain test points and environmental conditions. Manual with IDLC was used either as pilot preference or in cases where the aircraft was being flown beyond the CLAW limits for APC or DFP (i.e., high or low AOA). The challenge with manual was IDLC. Although IDLC provides fine glideslope control in an operational sense, the "heave" response to power inputs complicated the pilot tasks for some test points, requiring a simultaneous longitudinal stick input to counter the heave or sink to maintain the desired sink rate; however, this, in turn, would change the pitch attitude at landing. As will be discussed, both sink rate and pitch attitude had tight tolerance on most test points, making manual approaches a viable option only when outside of PA control AOA bands.

APC with 300% IDLC was also used either to establish the pilot on the desired glideslope for the test point or for high sink test points when environmental conditions allowed. The same heave challenges faced with manual were also faced with APC, but due to mechanization, the pilot was unable to counter changes in sink rate with pitch inputs.

DFP was found to be the mode of choice for most test points, provided the aircraft was inside of the PA AOA band. It provided a very stable sink rate even when the pilot would make lineup corrections with lateral stick inputs. DFP CLAW maintains tight control of gamma, and thus sink rate, while allowing AOA to vary but still be maintained within tolerances. With the CLAW maintaining gamma, sink rate, and AOA, this provided spare capacity for the pilot to maintain other parameters for the test point.

In addition to DFP, the pilots were provided flight test scales on their displays to fine tune test parameters such as AOA, sink rate, and yaw angle. With the flight test scales and DFP, the pilot was able to set sink rates accurately to 1 ft per second (fps). The ability for the pilot to capture and maintain the glideslope to within 1 fps resulted in the realization that the FLOLS no longer provided truth data. Historically, the pilot did not have the tools within the cockpit to precisely set glideslope parameters (sink rate and gamma), and therefore relied on the FLOLS for truth data. The challenge with the FLOLS at the test site is that it does not sit on a surveyed pad; it resides on asphalt on the side of a crowned runway. The imperfections in both the asphalt surface and the crowned slope that cause the lens to have a roll angle introduce variances that change the desired glideslope. The lens would be set in increments of 0.25 deg; however, it was noticeable to the pilots using information in the cockpit that the lens was not set to the proper angle. This meant that the only useful information the lens provided was touchdown location; as long as the pilot touched down with a centered ball, the hook point would be properly positioned for the arrestment. The takeaway at this point was that the FLOLS no longer presented truth data.

To complicate the task, for valid loads measurements the aircraft needed to land on the nonskid, outlined in red in Fig. 32, which extends 20 ft longitudinally in either direction from the wire. Nonskid is a surface treatment that increases the coefficient of friction on the deck of aircraft carriers; there is a patch of nonskid at the test site's arresting gear to replicate the shipboard landing environment. The landing gear is required to touch down on the nonskid for proper wheel spin-up and landing gear spring-back. With respect to tailhook aircraft, all AHSs exhibit a bounce on touchdown. When the aircraft and hook hit the deck, the hook bounces

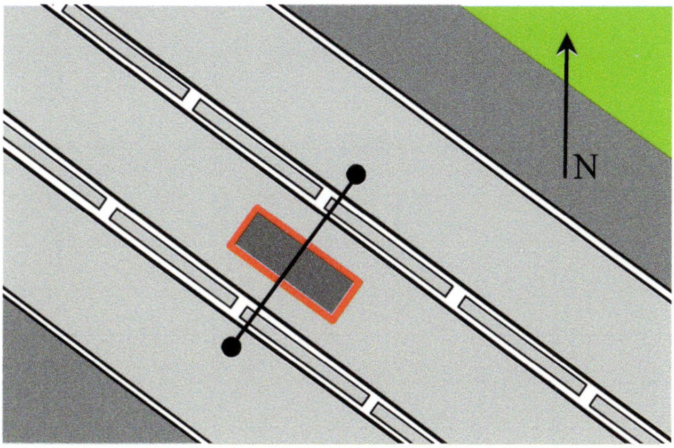

Fig. 32 Overhead view of NAS Patuxent River arresting site on runway 32 with nonskid outlined in red.

off the deck for a finite distance as the main mounts begin to compress. For the F-35C, this initial bounce is approximately 5 ft. When added to the nonskid constrain, that meant that the pilot had to precisely place the hook between 5 and 20 ft in front of the wire for a successful test point.

The final aircraft characteristic that will be discussed in this section is ground effect, the aerodynamic phenomenon in which lift increases and drag decreases when the aircraft is within approximately a half of a wingspan off the deck. This effect results in a shallowing of the flight path and reduction in sink rate. This presented a challenge initially as the test team observed a difference between the calculated and actual sink rates, presented in Fig. 33. There was a hesitance in targeting high sink rates by increasing the lens angle due to concerns that the ground effect reduction would be reduced or negated at higher gammas, resulting in an overshoot in targeted sink rates, thus damaging the aircraft. Eventually, the decision was made to progress and monitor the ground effect. There is a point at glideslopes in excess of 5 deg where a reduction in ground effect was observed.

High sink test points are designed to achieve 80% or more of the landing gear allowable load limit. They are conducted to verify structural integrity of an aircraft that has developed a high sink rate landing aboard the deck of a pitching aircraft carrier in various sea states. The test point emulation scheme is mean pitch correlates to a level deck, tail down mimics a ship bow down attitude, and nose down equates to the ship's bow being up.

1. MEAN PITCH HIGH SINK

Despite the title of this landing being self-explanatory, the objective was to achieve a landing at or above 21.1 fps at a targeted pitch angle. This endpoint was required

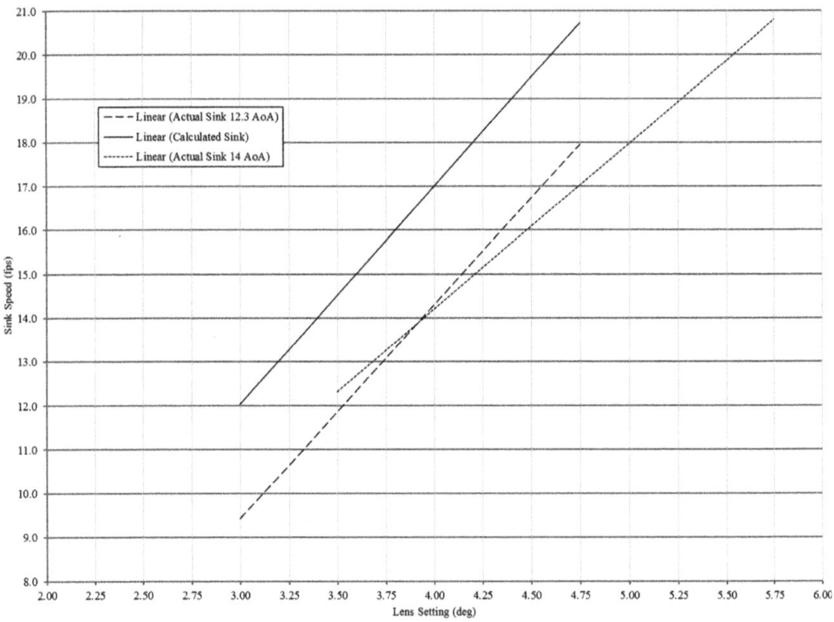

Fig. 33 Calculated and observed aircraft sink speed as a function of lens angle.

to be hit a minimum of three times with at least one deck obstruction (e.g., cross-deck pendant), with rollovers at 75% of the maximum main landing gear compression load or greater. One of the reasons that the shore-based structural survey is time consuming is that regardless of what was achieved the day before, each day began with a "nominal" arrestment using a 3.5-deg glideslope to ensure the aircraft, instrumentation, and arresting gear were operating properly. From there, the lens angle would be increased at 0.5-deg increments until 5.0 deg, at which point the increments were decreased by 0.25 deg until the endpoint was achieved. This meant that for high sink test points (mean pitch, tail down, and nose down), on average a minimum of approximately six arrestments were needed to achieve an endpoint, barring any complications such as delays with the airfield, instrumentation issues, aircraft maintenance challenges, or limitations on working hours.

Delta Flight Path was the mode of choice for this test point, providing a high degree of consistency and repeatability regardless of pilot techniques or environmental conditions. The pilot would be briefed on the lens angle and target sink rate, and then using the flight test scales, would set the sink rate and engage DFP. The team decided that the pilots would not compensate for ground effect, again removing variability due to pilot technique and increasing consistency and repeatability. The importance and utility of DFP cannot be overemphasized;

its use allowed the test team to achieve these test points in minimum time, making up for delays incurred due to weather, engineering discoveries, or programmatic issues.

2. TAIL DOWN HIGH SINK

The goal of this test point was to achieve a landing at or above 21.1 fps at the targeted pitch angle plus 2.2 deg, putting the aircraft in a tail down attitude. The buildup for this test point was the same as discussed in the mean pitch high sink section. The largest variable with this and the nose down test point was wind. If the aircraft maintains a constant AOA as the lens angle is increased, the aircraft's pitch attitude decreases. To achieve the required high pitch (tail down) attitude, the aircraft was flown at a high AOA and thus slower TAS. A slower TAS means a reduction in vertical velocity or sink rate. One solution to this dilemma is to continue to increase the glideslope angle; however, this has two adverse safety ramifications:

1. *Field of view*: At high AOAs, the pilot's field of view over the aircraft's nose is restricted and he or she loses sight of the lens or even the runway.
2. *Margin to stall and handling qualities*: At high AOAs, the aircraft stall margin is reduced and becomes a safety concern. Additionally, the handling qualities change; the aircraft's response is asymmetric. If the pilot is high on the glideslope, a forward longitudinal input results in a rapid decrease in AOA and pitch attitude (perhaps out of the test band) and a sharp increase in rate of descent. Conversely, a low on glideslope correction, aft longitudinal stick, may initially result in nothing more than increased AOA and pitch attitude—the aircraft squats, before shallowing the gamma and returning to the desired glideslope. It is a disharmonious control scheme.

Therefore, the optimal solution is wind direction. For tail down high sink, the best environmental days were either no wind or a tailwind. This would allow the aircraft to fly at the required AOA to achieve the target pitch attitude while still allowing the sink rate to be achieved without giving up sight of the landing area or increasing the risk of the test point.

Again, DFP was the mode of choice for this test point, provided the AOA was within the CLAW limit, for all of the same reasons discussed previously. At AOAs higher than the CLAW limit, either the point was flown manually with little success or the team targeted another type of landing.

3. NOSE DOWN HIGH SINK

The goal of this test point was to achieve a landing at or above 21.1 fps at the targeted pitch angle minus 4.5 deg, putting the aircraft in a nose down attitude. The buildup for this test point was the same as discussed in the previous high sink sections. Again, the largest variable with this and the nose down test point was wind.

Despite the pitch attitude decreasing as the glideslope was increased, theta was still high and out of the desired test band. To reduce theta more, AOA was reduced, thereby increasing airspeed. The challenge faced was that at a low enough AOA to put theta in the test band, airspeed was too high (above maximum engaging speed) with zero to little headwind. Achieving the endpoints safely required a headwind in excess of 5 to 10 kt, depending on the aircraft's weight.

As with the other high sink test points, DFP was the mode of choice, provided the AOA was within the CLAW limit, for all of the same reasons discussed previously. Alternatively, manual or APC could be used for these points, but they introduced variations due to pilot technique, reducing consistency and repeatability.

4. ROLL

The roll only test point, shown in Fig. 34, simulates either a late lineup correction initiated by the pilot or the ship having a roll attitude caused by sea motion, a slight turn, or not being properly ballasted. It was conducted in both directions, left roll and right roll. The objective of this test point was to achieve a sink rate in excess of 17 fps, with the aircraft's roll attitude statically set (zero roll rate)

Fig. 34 F-35C conducting a roll-only arrestment during the structural survey.

at 6 deg or more. The test point was set up based on the wind direction such that the maneuver was performed into the crosswind component. This was done to reduce the magnitude of the lateral drift that would develop as a result of the aircraft turning prior to touchdown. The test team was fortunate in that the environmental conditions supported all of the roll arrestment testing into the wind without incurring delays waiting on favorable wind directions.

To achieve the roll arrestment, the pilot would offset laterally from the runway centerline approximately 5 to 10 ft opposite the side of the roll (i.e., offset left for right roll), depending on the direction and magnitude of crosswind feet. The LSO provided a "roll now" call over the radio at an appropriate height above the runway (just prior to touchdown). With the aid of the LSO, the pilot would capture and maintain the desired bank angle. The safety concerns were twofold. First, if the LSO or pilot initiated the roll too early, the difference between the runway's longitudinal axis and the aircraft's longitudinal axis would be significant, resulting in a dangerous condition as the aircraft's directional vector at touchdown was biased towards the side of the runway vs aligned with the runway. Second, despite having adequate displays for setting other aircraft parameters, the Pilot-to-Vehicle Interface (PVI) is not optimized for setting precise roll angles required in flight test. The pilot has a small bank angle indicator on the bottom of the display consisting of an inverted triangle and an arced scale out to 30 deg in 10-degincrements. The size of the PVI requires the pilot to divert attention from glideslope maintenance to set a bank angle with any degree of accuracy. Additionally, the aircraft's roll mode time constant is relatively high in PA, complicating the pilot's task of capturing the required bank angle. To capture bank angle, the typical pilot's control inputs resulted in an initial target overshoot followed by a reduction in bank angle. The bank angle reduction was necessary for safety; as bank angle increased, the separation between the wingtip and runway surface decreases. Similarly, due to the relatively high roll mode time constant, when the pilot would attempt to reduce the roll angle, an overshoot would occur resulting in a bank angle of less than 6 deg and an invalid test point.

Again, DFP was the mode of choice to achieve this test point for its capability of maintaining a constant sink rate. Another benefit of DFP is the response to lateral stick inputs. DFP is designed to maintain a constant gamma; therefore, as the pilot commands a roll rate with lateral stick inputs, the control surfaces respond not only to provide the requested roll rate, but also to maintain the commanded gamma. This allows the pilot to divert attention away from glideslope maintenance, and set the bank angle precisely with little to no roll rate (lateral stick inputs removed) prior to touchdown.

5. YAW

The yaw arrestment point is a buildup to the roll/yaw arrestments and is only conducted in one direction—into the wind. The objective of this test point was to achieve a sink rate in excess of 17 fps, with the aircraft's yaw attitude statically set (zero yaw rate) at 5 deg or more with bank angle as close to zero as possible. The yaw angle was defined by the difference in the aircraft's longitudinal axis and directional velocity vector. Again, the arrested landing was accomplished primarily for safety. To achieve the yaw test point, the setup was the same as the rolled arrestment—the pilot would offset laterally from the runway centerline approximately 5 ft opposite the side of the yaw input. The offset side was chosen based on the wind direction such that the maneuver was performed into the crosswind component to limit lateral drift induced by the yaw. The LSO provided a "yaw now" call over the radio at an appropriate height above the runway (just prior to touchdown). Prior to executing the test point, the pilot would engage a Flight Test Aid (FTA) that enabled a beta limiter. This allowed the pilot to use a single step or ramp input to full rudder pedal deflection to command the target beta. The safety concern was similar to those associated with the rolled arrestment: If the LSO or pilot initiated the yaw too early, the difference between the runway's longitudinal axis and the aircraft's longitudinal axis would be significant, resulting in a dangerous condition as the aircraft's directional vector at touchdown was biased towards the side of the runway vice aligned with the runway.

A combination of advanced approach mode CLAW and FTA logic resulted in this maneuver being consistent and repeatable independent of pilot technique and with low pilot workload. APC was used to capture the desired glideslope followed by DFP to maintain the desired sink rate. The FTA beta limiter greatly increased the consistency and repeatability by removing all variability of pilot technique—full rudder pedal deflection. Another design point of the CLAW that reduced pilot workload for this test point was the decoupling of the lateral and directional axes. As the rudders deflect, the roll induced (roll due to yaw) is near zero, meaning that the pilot was able to accept any roll experienced because it would be minimal and inside of the test point tolerance.

6. ROLL/YAW OPPOSITE

Roll/yaw test points simulate a late lineup correction in a crosswind landing. The objective of this test point was to achieve a sink rate in excess of 17 fps and the aircraft established in a steady-heading sideslip with the flight path perpendicular to the cross-deck pendant. The test point was valid provided both the roll angle and yaw angle were equal to or greater than 5 deg. As a general rule, when the environmental conditions allowed, the yaw was into the relative crosswind; this decreased the pilot's centerline maintenance workload.

The employment of DFP and the FTA beta limiter resulted in the completion of this line of testing in one day. DFP was the optimum approach mode for this test point. The pilot would engage the beta limiter while setting up for the test point. Once aligned with the runway centerline at a range of approximately 1.5 miles with the desired glideslope captured, DFP was engaged, the appropriate rudder pedal was fully deflected, and the pilot used lateral stick inputs to set an angle of bank that maintained the desired ground track. The test pilot scans glideslope (the ball), line up, bank angle, yaw angle, and rate of descent; however, using this technique, the pilot's workload was low and repeatability was high. DFP maintained the desired rate of descent while the pilot used primarily lateral stick inputs to maintain centerline. Just prior to touchdown, the LSO would give a "roll now" call as a preemptive command to have the pilot check the roll angle. This call was made with enough time to allow the pilot to make one to two more lateral stick inputs to ensure the bank angle was within test tolerance. All of the difficulties and safety concerns discussed in the roll section are also applicable to roll/yaw opposite with the exception of runway misalignment; despite the unusual landing attitude, on touchdown, the aircraft was aligned with the runway.

7. ROLL/YAW SAME

The objective of this test point was to achieve a sink rate in excess of 17 fps and the aircraft's attitude statically set with both a roll angle and a yaw angle equal to or greater than 5 deg in the same direction. The maneuver was performed into the relative crosswind to limit the lateral component of the velocity vector while allowing the yaw angle to reach the desired target. As with the roll only and yaw only test points, the pilot needed to be prepared for potentially high drift rates and runway misalignment on touchdown should a bolter occur.

To achieve the test point, the pilot would engage the beta limiter while setting up for the test point. Once aligned with the runway centerline at a range of approximately 1.5 miles with the desired glideslope captured, DFP was engaged to maintain the desired sink rate. As the aircraft approached the runway, the pilot would establish a 5-ft lateral offset from the runway centerline opposite the direction of the maneuver. Two techniques were employed in the achievement of this test point: a steady-heading sideslip reversal and a dynamic maneuver. In the dynamic method, the pilot would simultaneously set the roll angle with lateral stick and input a full pedal deflection when the LSO called "roll now." The pilot relied on the beta limiter both to establish a yaw angle greater than 5 deg and not to exceed yaw angle limits where high main gear side loads were predicted. With the steady-heading sideslip reversal method, the pilot would establish the aircraft in a steady-heading sideslip using the roll/yaw opposite technique. Upon the LSO's "roll now" call, the pilot would reverse the bank angle with lateral stick while relying upon DFP to maintain the sink speed and the beta limiter to maintain the yaw angle. Although the dynamic maneuver was successful in completing the test point, it was deemed too susceptible to pilot technique with insufficient repeatability; therefore, the steady-heading sideslip reversal method was preferred. Both techniques utilized DFP and FTA beta limiters extensively to reduce pilot workload and increase repeatability, thereby increasing testing efficiency and effectiveness.

8. MAXIMUM ENGAGING SPEED ON-CENTER

The data from the maximum engaging speed arrestment were used to measure and verify the design limit arresting hook axial load. An initial maximum engaging speed had been established during the ground-based roll-in testing at NAES Lakehurst. To evaluate the load, an arrested landing was accomplished at a sink rate of at least 17 fps while flying a nominal approach AOA. This allowed the pilot, control room team, and LSO to evaluate the headwind gradient during the approach, descending from an approximate altitude of 1000 ft to sea level. After the hook loading was evaluated, the pilot was given a ground speed to target for the next arrestment.

The pilot used the speed hold autopilot functionality of the CLAW that captures and holds a constant calibrated airspeed. This mode was intended for up-and-away flight. It was found that when engaged in PA, it easily excited the phugoid mode, and the open loop response resulted in a gentle longitudinal undulation about the desired glideslope. Although the magnitude of this undulation was relatively small, if unchecked, it could have resulted in the aircraft landing before the nonskid or boltering, both of which result in an invalid test point. During the approach, the pilot was able to accept the undulations provided the mean was about the desired glideslope; however, starting at approximately 0.5 n mile from touchdown, the pilot would dampen out the undulations and maintain the desired glideslope with longitudinal stick inputs. Not using DFP further increased pilot workload. When making a lateral correction to maintain alignment with the runway centerline, the flight control system does not automatically compensate for loss of lift or the increase in drag, nor is the engine response coupled with control surface movement. Because the parameter of primary concern was ground speed, if the aircraft developed a lateral drift, the pilot would initiate a correction towards centerline with roll and then use longitudinal stick to increase AOA momentarily to maintain the desired glideslope. A similar series of control inputs were used to recapture runway centerline. Both the pilot and LSO have instruments that provide the aircraft's ground speed. As the F-35C approaches the arresting gear, the LSO advises the pilot of required speed changes. As target groundspeeds were increased, longitudinal stick forces changed due to the CLAW trimming to a nominal approach AOA. The pilot would engage Speed Hold at a calibrated airspeed that equated to the targeted ground speed for the environmental conditions of that particular day. The pilot manually trimmed the longitudinal stick forces to zero. Then, using a switch on the throttle, the pilot adjusted the captured calibrated airspeed in 1-kt increments as required to maintain the desired ground speed. The pilot and LSO worked as a team to ensure that the aircraft engaged the arresting gear at the targeted ground speed.

This technique of the pilot maintaining glideslope with longitudinal stick inputs, lineup with a combination of lateral and longitudinal stick inputs, and ground speed with multiple switch actuations results in a higher pilot workload but was very repeatable, resulting in accurate engaging speeds. The endpoint of this series of testing was achieving an allowable hook axial load limit of 100% while engaging the cross-deck pendant within 5 ft of the runway centerline and the aircraft's groundspeed within 1 kt—1 kt too slow resulted in a hook loading less than 100% and 1 kt too fast resulted in an overstress condition requiring testing to stop while both the aircraft and arresting engine were inspected for damage.

9. MAXIMUM ENGAGING SPEED OFF-CENTER

This series of testing is designed to stress both the arresting hook and arresting gear loading. Maximum engaging speed endpoints were required to be executed to both the left and right side of the cross-deck pendant. This is due to the path the arresting gear cables travel from the arresting engine to the cross-deck pendant—one length is shorter than the other, resulting in slightly higher loads on the short side. Testing was conducted using the same techniques and with the same safety concerns discussed in "Maximum Engaging Speed On-Center"; however, the buildup sequence was slightly different. After an on-center engagement at a 17-fps sink rate had been achieved at a nominal approach AOA, the off-center distances of 10 ft, 15 ft, and 18 ft were targeted. Once reaching the maximum off-center distance (18 ft with a tolerance of −1 and +2 ft), the engaging speed was incrementally increased until the arresting hook loads or the arresting gear limits were reached.

Asymmetric loading of the aircraft added another level of complexity to this series of test points as well as the yaw and roll/yaw series. Due to the lateral shift in CG, the aircraft's dynamics during the arrestment were closely monitored for wing tip-to-runway clearances, arresting gear-to-aircraft interactions, weapon bay clearances, or adverse stores response during inertially induced rolls.

10. FREE FLIGHT OR HIGH PITCH/LOW ANGLE

This test point duplicates an approach where the pilot initiates an aggressive nose-up pitch maneuver close to the deck, causing the arresting hook to engage the cross-deck pendant prior to main gear touchdown or with the aircraft's velocity vector above the horizon. The objective of this point was to achieve at least an 85% arresting hook limit load while generating an as high as possible nose gear limit load. This type of landing, also called an "inflight engagement," results in high nose-down pitch rate that causes peak loads in the nose gear. Although not impossible, the F-35C's geometry makes a free flight or inflight engagement extremely unlikely because at optimum approach AOA, the bottoms of the main gear tires are lower than the hook point. This results in the main gear touching down on the deck or runway prior to the hook point.

The setup for this test point was to set the FLOLS to a minimum safe glideslope of 2.25 deg. The aircraft was flown at 15 deg AOA. Under these conditions, this places the hook point and the main gear in the same horizontal plane, resulting in simultaneous touchdown of the main gear and hook point. Although the pilot was able to utilize DFP, at the higher AOA the response was asymmetric and disharmonious. If the pilot was above the desired glideslope, a forward longitudinal input resulted in a rapid decrease in AOA and pitch attitude, raising the hook point and increasing the rate of descent. Conversely, to correct for being below the desired glideslope, an aft longitudinal stick input initially resulted in an increased AOA and pitch attitude, a lowering of the hook point, and an aircraft squat before shallowing the gamma and returning to the desired glideslope. Additionally, flying at high AOAs limited the pilot's field of view over the aircraft's nose. Finally, due to the desire for the hook to engage the cross-deck pendant simultaneously as the main gear touched down, hook point touchdown placement needed to be precise. These factors combined to make this series of testing one of the most difficult for the pilot to achieve with a marginal success rate.

As a result of aircraft geometry and the test point's lack of repeatability due to the sensitivity in pilot technique, it was decided that the test point would be redesigned for future testing. The new test, high pitch/low angle, uses a low-angle glideslope being flown with the aircraft at an elevated AOA. The touchdown requirements are loosened with the objective being to generate pitch rates and nose gear loading above calculated values that still allow the models and predictions to be validated.

Shore-based ship suitability testing of the F-35C began on 17 March 2011 and culminated in the first shipboard arrestment on 3 November 2014. It took approximately three and a half years to complete the prerequisite testing to

Fig. 35 The F-35C conducting a structural survey arrestment with GBU-12s and a gun pod.

safely conduct the initial ship trials. During that time, multiple discoveries were made and engineering challenges faced, primarily the AHS being redesigned, but that is to be expected in developmental flight test. The structural survey started on 4 February 2014, after the installation of the new AHS; however, the program continued to make discoveries that complicated the completion of the structural survey. Between the start of the structural survey and 19 August 2014, 51 arrested landings, of which 15 were endpoint, were achieved. That is an average of 0.26 arrestments per day. On August 19 the last remaining restriction (no more than six arrested landings per day) was removed. And on September 22, the test team completed the structural survey, executing an additional 102 arrestments, of which 23 were endpoints, in 34 days. See Fig. 35 for an example of a structural survey arrestment. This pace increased the average daily arrestment-per-day over 11-fold from 0.26 to 3.0. This order of magnitude increase in testing efficiency has been shown to be linked to three factors:

1. Challenging the assumptions that previous carrier suitability projects were conducted under, the F-35C was approached with a blank canvas philosophy that paved the way for the use of automation in the pursuit of CVS testing that increased efficiency.
2. The advanced approach modes and IDLC provide the pilot with an unprecedented ability to control the aircraft during shipboard approaches. This technology, specifically DFP, and FTAs provided tools that allowed challenging test points to be achieved with a high level of repeatability without relying on a test pilot's skill (or lack thereof) to finesse aircraft performance.
3. The blank canvas philosophy approach enabled the use of the advanced approach modes and FTAs in ways that had been previously thought unusable or unsafe.

V. SHIP-BASED TESTING

Flying qualities and performance testing are first conducted shore-based, then evaluated at the ship. The shore-based tests are used to define the minimum acceptable approach airspeed and give the testers an initial look at airplane flying qualities in the more benign shore-based environment. Once satisfied with the shore-based results from a structural suitability, flying qualities, and performance standpoint, the airplane is ready to proceed to the ship. Some of the highest risk shipboard tests are the minimum catapult end airspeed tests where the slowest safe speed for catapult flyaway is determined. Other shipboard launch tests include establishing crosswind limits, evaluating launches off waist catapults, and evaluating the effects of low energy launches (low gross weights, high winds) on flyaway. Recovery testing includes establishing crosswind limits, evaluating bolter and wave-off performance, and evaluating handling qualities at various wind-over-deck conditions. Additional testing includes evaluating

Fig. 36 Example of ship attitudes routinely attainable while at sea.

compatibility with shipboard facilities and support equipment: heavy weather tie-down (as seen in Fig. 36), canopy opening under high wind conditions, dynamic tipback following an arrestment, and hangar bay towing and spotting.

Over the course of two years, three ship trials were completed: the initial ship trial, DT-I, aboard the USS *Nimitz* (CVN 68) in November 2014, DT-II aboard the USS *Eisenhower* (CVN 69) in October 2015, and DT-III aboard the USS *George Washington* (CVN 73) in August 2016. While embarked, the F-35C flew in several different configurations, shown in Table 1, building up in weight, shifting center of gravity, and increasing asymmetry.

A. WHY TEST SHIPBOARD?

Shipboard testing is conducted for two reasons: first and foremost is to evaluate the aircraft in the environment in which it will be operated, and second, it is not currently possible to re-create or simulate the environmental conditions experienced at sea. The environmental conditions include launching from a deck approximately 60 ft above sea level, the aircraft carrier air wake or "burble," and deck motion.

The dynamics of a shipboard launch are very different than taking off from a runway. When an aircraft takes off from a runway, the pilot applies power and the aircraft accelerates down the runway. When the aircraft's velocity has reached rotation speed, the pilot applies an aft longitudinal input on the control stick to deflect the horizontal stabilators, thus producing an aerodynamic pitching moment. The aircraft rotates about the main wheels, increasing the AOA of the wing and thereby lift. As the aircraft continues to accelerate down the runway, lift increases and the aircraft flies off the runway. This process is illustrated in Fig. 37. Conversely, for a shipboard launch, the aircraft is accelerated to flyaway speed in just under three seconds with the horizontal stabilators prepositioned to generate a pitching moment. At shuttle release, stored energy in the nose strut is released and aided by the aerodynamic pitching moment, a pitch rate is started; however, a moment later (under one second) the aircraft leaves the deck. The aircraft now rotates about its CG vs the main wheels. This process is illustrated in Fig. 38. The dynamics and flyaway characteristics of a shipboard launch are more easily affected by crosswinds, pitch angle of the aircraft carrier's deck, and the carrier wake turbulence, and therefore preclude shore-based re-creation.

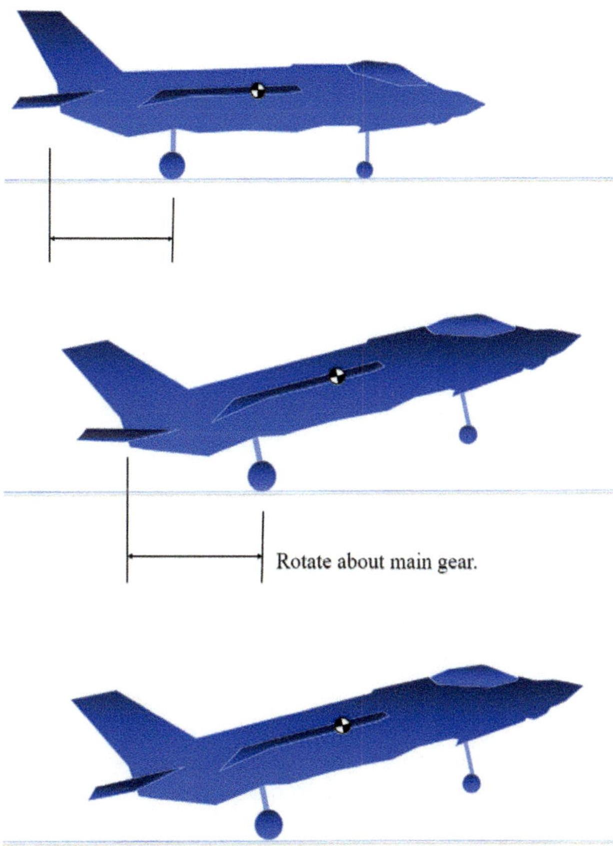

Fig. 37 Conventional takeoff dynamics rotating about the aircraft's main landing gear.

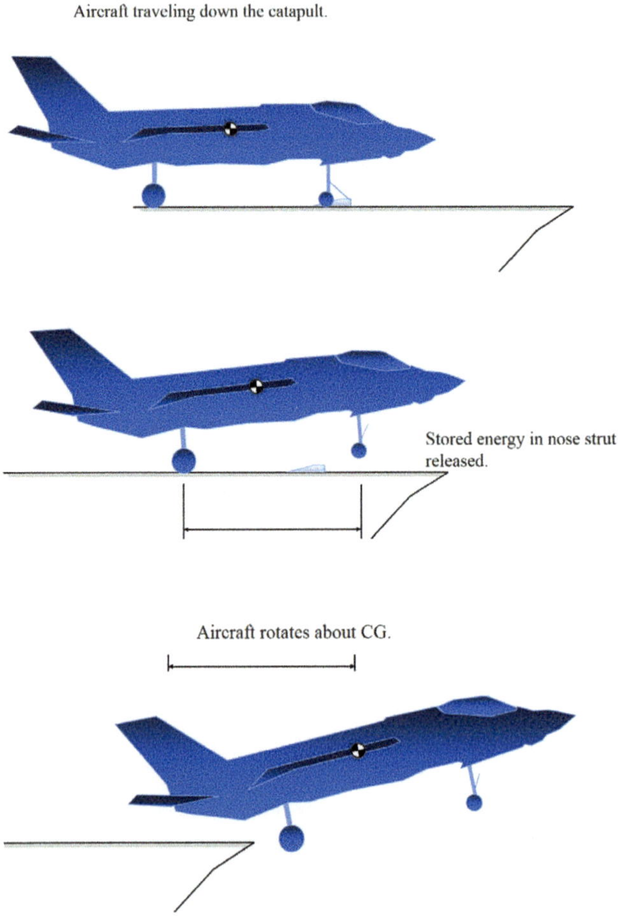

Fig. 38 Catapult launch dynamics rotating about the aircraft's center of gravity.

As a U.S. aircraft carrier moves through the water, it has a profound effect on the air mass through which it moves. Starting at the bow of the ship, two vortices are generated aligned with the longitudinal axis of the ship, similar to wingtip vortices. The vortex generated on the starboard side rotates counterclockwise as viewed from the deck of the ship looking forward but has negligible effects on the launch and recovery of aircraft. The vortices of concern with respect to recovering aircraft are generated off of the port bow and the outboard leading edge of the landing area; both are aligned with the longitudinal axis of the ship, rotate counterclockwise, and flow through the landing area. Next, the aircraft carrier's superstructure creates an area of turbulent air. Due to the angled landing area,

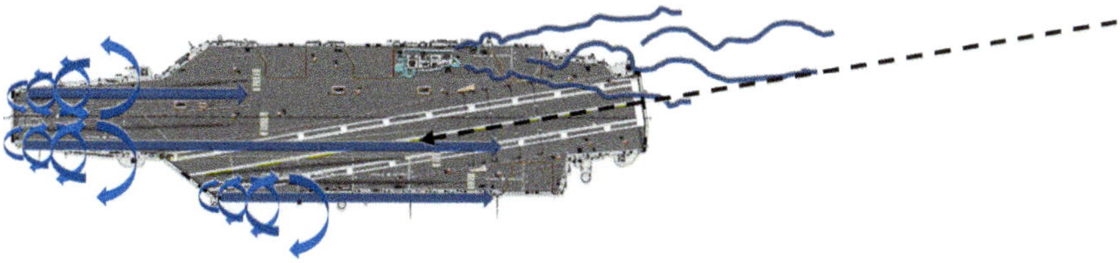

Fig. 39 Overhead view of ship airwake effects on the deck of an aircraft carrier.

aircraft transit through the turbulent air during the approach. All of these effects are shown in Fig. 39 with the approach path of the aircraft depicted by the dashed black arrow.

The airwake off the end of the aircraft carrier, called the "burble," is characterized by a downdraft as the air flows off of the deck followed by an area of updraft aft of the downdraft, as seen in Fig. 40. The magnitude and position of the downdraft and updraft are a function of natural wind and the ship's velocity. As an aircraft on approach interacts with the burble it will first experience an uplift; the open looped response would be a reduction in rate of descent followed by a greater than normal increase in rate of descent. With computational fluid dynamics, the fidelity of computer-modeled airwake effects is increasing; however, outside of a simulator, the shore-based re-creation to date is impossible.

The deck of the aircraft carrier is subject to the motions of the ship: pitch, roll, heave, drift, and Dutch roll. These dynamics result in complicating launches and recoveries. For launches, any bow-down attitude results in an initial negative gamma (downward flight path) for the aircraft. With respect to recoveries, ship's motion results in landing area and optical lens motion, increasing the difficulty of a shipboard landing. In addition to complicating launches and recoveries, the ship's dynamics also induce airwake effects. As the ship pitches and heaves up and down, a vortex aligned with the ship's lateral axis is formed at the bow. The magnitude of this vortex is a function of pitch and/or heave rates, growing as the bow pitches down or the ship settles after a heave. This lateral vortex has been found to affect catapult launches. Finally, the ship's dynamics prevent deterministic loads calculations, requiring stochastic analysis of the multivariate distribution of landing impact conditions.

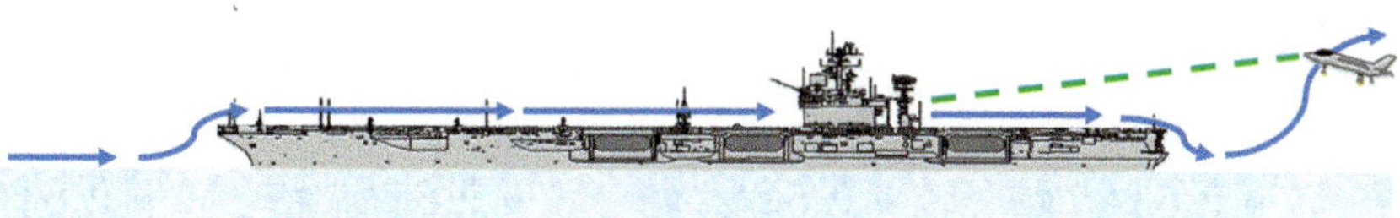

Fig. 40 Airflow profile over the flight deck and aircraft's approach path.

B. CATAPULT LAUNCHES

Shipboard catapult launches provided data used to verify the launch envelope, similar to that seen in Fig. 41 [6], construct ALBs, and develop sections of the flight manual. The launch envelope is bounded by the aircraft's maximum allowable gross weight, launch bar axial load limit, and longitudinal acceleration limit. Other envelope-bounding considerations include catapult limits and the minimum airspeed required to safely fly away. The launch envelope and ALBs provide limitations and procedures to ship personnel to safely catapult the F-35C; these documents contain information such as wind envelopes and catapult settings. Conversely, the flight manual provides the pilot limitations such as maximum gross launch weight, maximum asymmetric loading, and maximum gross arrestment weight (also known as "max trap").

To collect the required data, four distinct types of catapult launches were conducted from both the bow and waist catapults aboard the ship: Minimum End Airspeed (MEAS), high excess, low energy, and crosswind. To better understand shipboard catapult testing, the following terms and phrases are defined:

- *Digital End Speed Indicator (DESI)*: A chronograph that is integrated with the catapult and uses a series of magnetic sensors to provide a catapult end speed.
- *Catapult end speed*: The speed of the aircraft relative to the deck (catapult shuttle speed) at the end of the catapult power stroke, normally measured by the DESI.

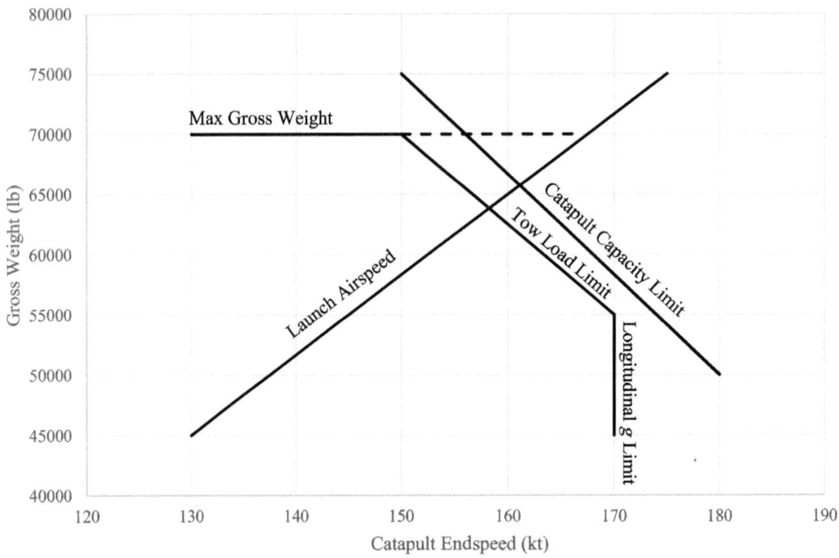

Fig. 41 Example catapult envelope illustrating the factors that are evaluated and verified during testing.

- *Minimum End Airspeed (MEAS)*: The end airspeed at which the aircraft can safely fly away at a given weight but not without the aircraft's CG sinking a predetermined amount below the deck of the aircraft carrier.
- *Recommended launch airspeed*: A launching airspeed in excess of MEAS that provides a margin of safety for variations in launch parameters. It is normally expressed in terms of excess airspeed above MEAS (e.g., an aircraft launched at an end speed 20 kt above MEAS would be phrased "min plus twenty" and would be written min +20). Under normal operational conditions, aircraft are launched at min +15.

1. MINIMUM END AIRSPEED

All catapult launches are based on the minimum end airspeed values. Through examination of the figure and intuition, it can be deduced that if the catapult launches the aircraft at higher speeds or the ship produces a higher WOD, then margins are increased; however, launching aircraft at the lowest safe airspeed possible has many advantages, including:

- Lower WOD requirements reduce the speeds at which the ship must move through the water to produce winds, thereby reducing the rate at which the nuclear fuel is consumed.
- Lower WOD requirements reduce the environmental stress on the deck crew. (Exposure to sustained high winds results in fatigue.)
- Lower launch airspeeds result in lower catapult acceleration profiles, thereby decreasing loads imposed on the airframe and increasing fatigue life.
- Lower launch airspeeds reduce the amount of energy that the catapult must impart to the aircraft, thereby conserving the ship's fresh water supply (the catapults are steam-driven) and nuclear fuel.
- Lower launch airspeeds increase the operational capability of both the ship and the aircraft.

Minimum end airspeed launches were conducted at various weights, CGs, and power settings (military power [MIL], which is 100% of rated thrust, and maximum afterburner [MAX] at 150% of rated thrust) (see Fig. 42). The MEAS is calculated based on the aircraft's CG sink below the flight deck or "sink off bow." Several factors are considered when determining the targeted sink off bow including thrust required, thrust available, proximity to stall warning, aircraft flying qualities, longitudinal acceleration over vertical acceleration (a/g), flight control response, and pilot comfort [6].

The build-down process began at the predicted minimum end airspeed plus 15 kt excess using CSV, then 10 kt of excess, and finally 6 kt excess, all while keeping WOD constant. CSV was then held constant to remove catapult variability, and

Fig. 42 Minimum end airspeed testing was conducted in multiple configurations, including heavy weight external loadings, as shown here.

WOD was reduced by slowing the ship's speed for the next two launches targeting 3 kt and 0 kt excess (minimum end airspeed). Environmental conditions needed to be ideal each time MEAS testing was conducted. Winds needed to be steady with little to no gust for consistency and safety. Additionally, the winds needed to be high enough such that testing could continue once the ship's speed reached minimum steerage (the minimum ship's speed to maintain rudder effectiveness). The targeted CG sink off bow for MIL and MAX was 13 ft and 18 ft, respectively.

2. HIGH EXCESS END AIRSPEED

Launching with high excess end airspeeds is an operational capability requirement in which the aircraft is catapulted with up to 40 kt of excess. During these launches, the aircraft may "kite" or become airborne prior to the end of the catapult power stroke [6] or overrotate upon shuttle release. Although these undesirable flying qualities were not encountered, had they been, the testing would have been halted and a maximum excess end airspeed limit would have been defined, bounding the catapult launch envelope. High excess end airspeed launches utilized a buildup from minimum end airspeed plus 30 kt to plus 45 kt. CSV was the primary driver to provide the excess; however, a minimum of 15 kt of WOD was required. The F-35C exhibited neither kiting nor overrotation, but instead, high excess end airspeeds resulted in the aircraft flying off of the deck in a nonobjectionable flat attitude, shown in Fig. 43, vs the typical rotation about the CG.

3. LOW CATAPULT ENERGY

During pilot carrier qualification, aircraft are routinely launched at lighter weights, slightly above max trap, to facilitate an immediate recovery. When lighter weight aircraft are launched in high WOD conditions, CSV settings are lowered to reduce fatigue stresses on the airframe, resulting in lower catapult end speeds and longitudinal accelerations. Historically, low catapult energy launches reduced the inertial loads action of control surfaces, resulting in small control surface deflections during open-looped launches; unintentional sink off bows were a potential due to degraded rotation characteristics [6]. However, the F-35C is designed to be launched stick free; the CLAW schedules the control surfaces as a function of air data, longitudinal acceleration, and wheel speed. Low catapult energy launches were conducted to assess the flyaway performance and rotation characteristics in high WOD conditions; low launch bar loads result in lower nose strut squat at RRHB release, imparting less stored energy at shuttle release. This series of catapult launches started at minimum plus 15 kt excess. Using ship's speed, WOD was then incrementally increased above 45 kt and until a minimum CSV setting or $\leq$ 5% launch bar axial tow force was achieved. The safety concern with low axial tow forces was that the shuttle could spit the launch bar.

4. CROSSWIND CATAPULT

During crosswind catapult launches, an up-wash condition is created on the windward side of the ship's hull, effectively increasing the AOA of the upwind wing, resulting in high torque loads on the launch bar. Upon shuttle release, the aircraft's open loop response is to yaw into and roll away from the crosswind

Fig. 43 F-35C flying away in a flat attitude after a high excess end airspeed launch.

while developing a lateral drift aligned with the crosswind. The objective of the stick-free response was to characterize the aircraft's stability and evaluate the flying qualities and FLCS to return the F-35C to balanced flight.

Catapults in crosswinds were conducted to evaluate and establish the aircraft's operational limits. Launches were accomplished from the bow and waist catapults with the aircraft in two configurations: the maximum allowable shipboard asymmetry and the most forward CG operationally possible. There is a correlation between excess end speed and crosswind effects: crosswinds under 10 kt have negligible effects on the launching characteristics of an aircraft [6]; therefore, testing began at 10 kt and expanded outwards to a maximum of 15-kt crosswind. Testing was achieved first by building up in crosswind from 10 to 15 kt, then by building down in excess end airspeed from 15 to 10 kt. Additionally, the crosswinds were evaluated from both the port and starboard side of the ship. During testing, launch bar torqueing loads were well within structural tolerances. Additionally, aircraft dynamic response in yaw, roll, and drift at shuttle release were deemed nonobjectionable by the test pilots. Both of these factors allowed the crosswind envelope to be cleared out to 15 kt in either direction down to a minimum excess of 10 kt. (With the exception of catapult 4, 15 kt excess is the minimum due to a longer deck roll than catapults 1, 2, and 3).

5. AIRCRAFT LAUNCH BULLETINS

The results of shipboard catapult testing were used to develop a series of ALBs, an example of which is shown in Table 4 [8], that contain the catapult settings and WOD required to launch the aircraft throughout the full range of gross weights as a function of aircraft configuration and power setting. The bulletins also contain information on WOD limitations, corrections for atmospheric conditions, corrections for catapult expansion due to heat, JBD limitations, and CSV settings.

TABLE 4 EXAMPLE OF AN ALB THAT PROVIDES THE UNITLESS CSV SETTING FOR CATAPULT LAUNCHES AS A FUNCTION OF AIRCRAFT WEIGHT AND HEADWIND

	Headwind for 15 Kt Excess				
	20	22	24	26	28
Aircraft Weight (1000 lb)	**CSV Setting**				
45	114	110	106	102	98
46	118	114	110	106	102
47	124	120	116	112	108
48	128	124	120	118	116

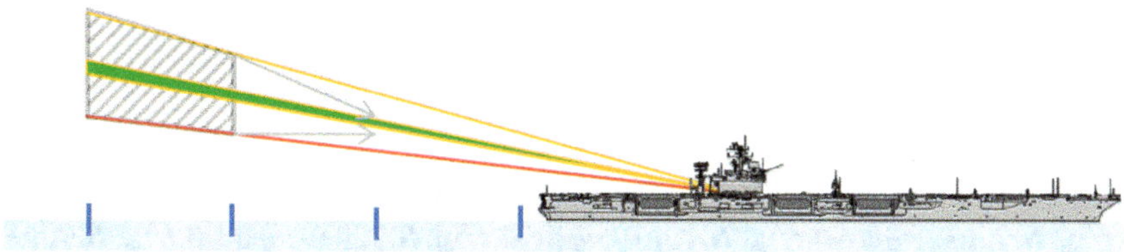

Fig. 44 Carrier approach segment nomenclature and approach handling qualities testing envelope bounded by the limits of the IFLOLS (bottom red and top yellow lines) and proximity to the ship (from the start [X] to in the middle [IM]).

C. ARRESTED LANDINGS

Shipboard recovery handling qualities were evaluated using both nominal and off-nominal (intentional error) approaches in environmental conditions emulating those to be expected by fleet pilots. In the course of the three ship trials, evaluations were conducted in multiple configurations including forward CG, maximum allowable lateral asymmetries, and full external stores. In addition to the many configurations, all three approach modes were evaluated in four different wind cells. Although handling qualities were not evaluated with the aircraft in simulated degraded states (control surface failures or degraded modes) during ship trials, degraded and failure state shipboard recovery handling qualities were extensively evaluated in the simulator to develop emergency procedures and recommended pilot recovery techniques. The results of the simulator trials and shipboard testing were used to establish procedures and publish pilot techniques in the flight manual.

The final approach to an aircraft carrier starts with the "ball call" and ends with an arrestment, touch and go, bolter, or wave-off. The ball call is a radio transmission made at $\frac{3}{4}$ n mile from the aft end of the ship by the pilot to inform the LSO that the lights of the OLS and ships landing area are in sight. The approach is segmented into five parts, four of which are shown in Fig. 44, and are divided by distance in nautical miles from the aft of the ship. The ball call coincides with the start. The times shown below the distances are based on a fighter aircraft flying the optimum approach airspeed and correlate to time in seconds from touchdown.

The test pilots would intentionally set up a deviation in glideslope, lineup, or airspeed to evaluate the handling qualities. Testing began with single deviations at the start such as a "high start" or lined up left start, also known as an angling approach (see Fig. 45). The pilot would then correct to be on glideslope, on centerline, and on-speed no later than the in close position. As a safety measure, if the pilot had failed to make the appropriate corrections by the in close position, he or she would have been waved off.

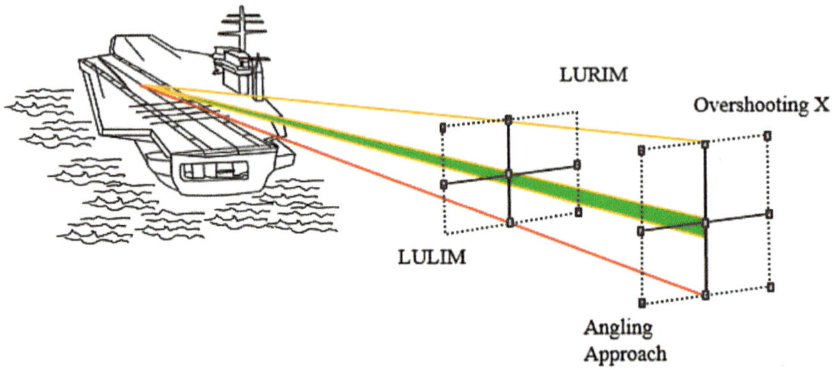

Fig. 45 Carrier approach nomenclature for lateral deviations.

After single deviations were evaluated, testing progressed to multiple deviations, ultimately evaluating multiaxis and speed deviations such as low on glideslope, slow airspeed, high bank angle, and laterally displaced to the right of the landing area centerline at the start—known to naval aviators as a low slow wrapped up overshooting start. All of these deviations were evaluated in the three approach modes: manual, APC, and DFP. After start deviations were evaluated, the test pilots would hold the deviation to the in the middle position with a requirement to correct the deviation by the in close position.

The initial rounds of evaluations were completed in the nominal wind cell (25 kt $\pm$ 5 kt WOD). The burble effects aft of the ship increase as a function of WOD and may result in special pilot techniques due to increased ship airwake disturbances. After the nominal wind cell was complete, low wind (16 kt $+4/-0$ kt WOD), medium wind (35 kt $\pm$ 5 kt WOD), and high wind (40 kt $+0/-5$ kt WOD) cells were evaluated, again using all three approach modes. Handling Qualities Ratings (HQRs), using the Cooper-Harper rating scale, were assigned for both the capture and maintenance tasks associated with glideslope, centerline, and airspeed. The results of all three ship trials are shown in Figs. 46 and 47. Airspeed, or AOA, ratings were not evaluated for APC and DFP because the CLAW is controlling AOA and the pilot has no direct control.

All three approach modes exhibited level one handling qualities, but DFP was the preferred approach mode for both test pilots and operational pilots. As a result of the handling qualities, the tasks associated with shipboard landings not only were deemed satisfactory in achieving desired performance, but DFP also reduced pilot workload. The general control scheme/philosophy is that the pilot engages DFP as the glideslope is captured. As perturbations move the aircraft away from the desired glideslope, the pilot inputs and holds a longitudinal stick displacement to generate a change in gamma. As the aircraft returns to the desired glideslope, the pilot leads the correction and releases the stick, thus commanding the aircraft to recapture the programmed gamma and desired glideslope.

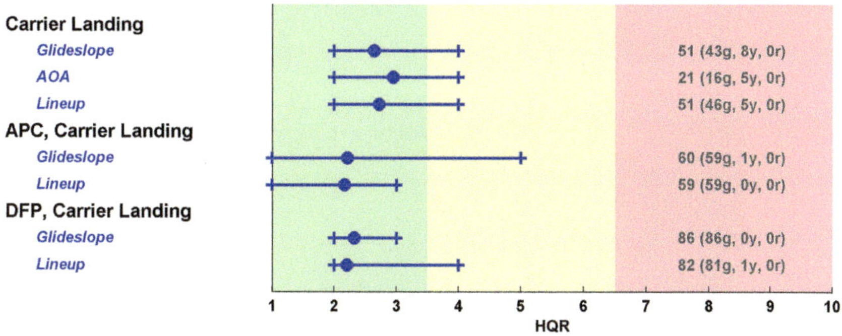

Fig. 46 DT-I and II approach handling qualities results for the clean wing configuration (no external stores) evaluation. (Numbers to the right in red correspond to the number of deviations and approaches attempted for that specific task with breakouts according to handling quality levels.)

Additionally, as pilots are making lateral inputs to capture and maintain centerline, the DFP functionality attempts to maintain the commanded gamma such that once the pilot's scan returns to the OLS, the ball is still between the datums.

The F-35C was designed to conduct carrier approaches with the TEFs positioned at 30 deg down. This was done to minimize the approach airspeed. However, due to aircraft geometry and hook shank length, the driving design objective was to minimize hook touchdown point dispersion to ensure successful arresting gear engagement. It was determined during testing that the task of glideslope capture and maintenance was easier when the TEFs were positioned to 15 deg and IDLC was increased 300%, as seen in Fig. 48. A flight path (gamma) hold feature was added to the CLAW, in addition to Approach Power

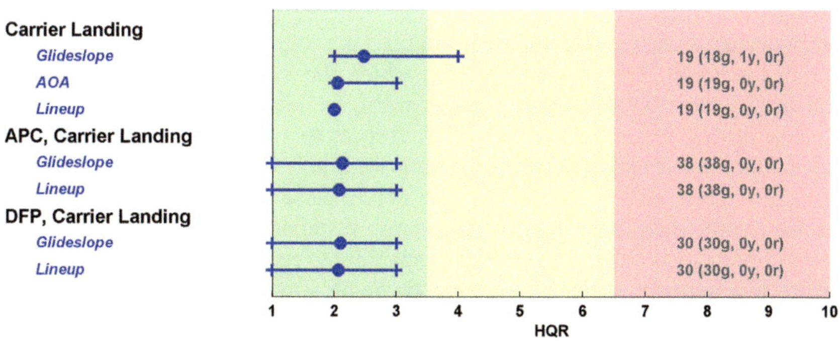

Fig. 47 DT-III approach handling quality evaluation results with external stores, both symmetrically and asymmetrically, loaded.

Fig. 48 F-35C approaching the ship with 15-degTEFs and 300% IDLC.

Compensation (APC), that resulted in a pilot workload reduction. The combination of improved handling qualities and reduced pilot workload culminated in a reduced hook touchdown dispersion, thus increasing the probability of a successful hook-to-wire engagement, as seen in Figs. 49, 50, and 51.

The potential benefits that DFP offers are a reduction in training requirements (both shore-based and shipboard), an increase in safety margins, and increased operational capability. As a result of the level one handling qualities, reduced pilot workload, and small hook point touchdown dispersion, the initial cadre of fleet aviators (all of whom had previous carrier experience in other aircraft) to conduct F-35C carrier qualifications used DFP successfully. During that evolution, the aircraft neither boltered (landing long, missing all of the cross-deck pendants) nor engaged the one wire (minimum safety margin tolerated). The consistency with which the aviators were able to safely land on the deck on the target location with ease leads to the logical question of reducing training requirements. With respect to safety, the consistency and repeatability with which the fleet aviators were able to execute approaches to the deck and precisely place the hook point on or close to the targeted touchdown point increases safety margins (hook-to-ramp distance consistently greater than the minimum allowable). The aircraft's capability to reliably land aboard the ship on the first attempt without boltering or being waved off serves to improve boarding rates. Operational capability will be increased as a result of reliable boarding rates, which affect the time the ship is required to steam into the wind and allow more time for maintenance personnel to prepare the aircraft for the next launch.

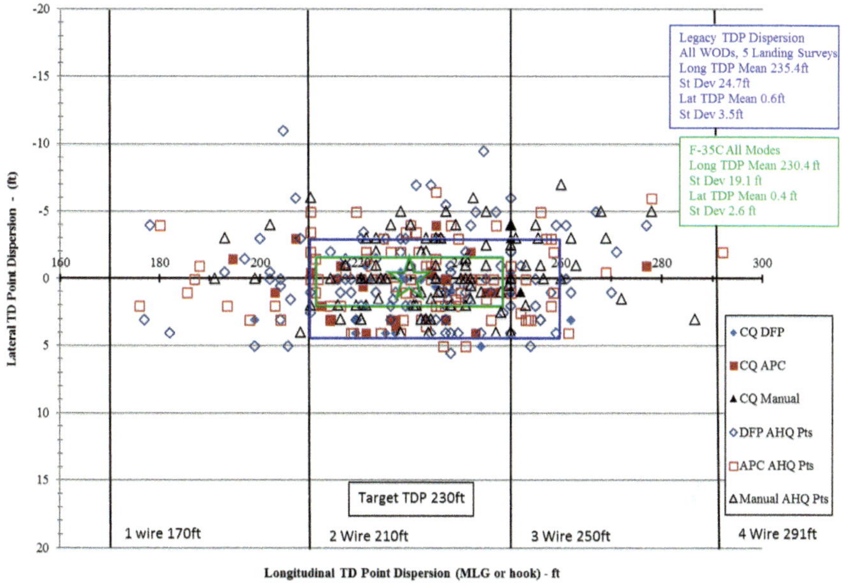

Fig. 49 DT-I hook point touchdown dispersion.

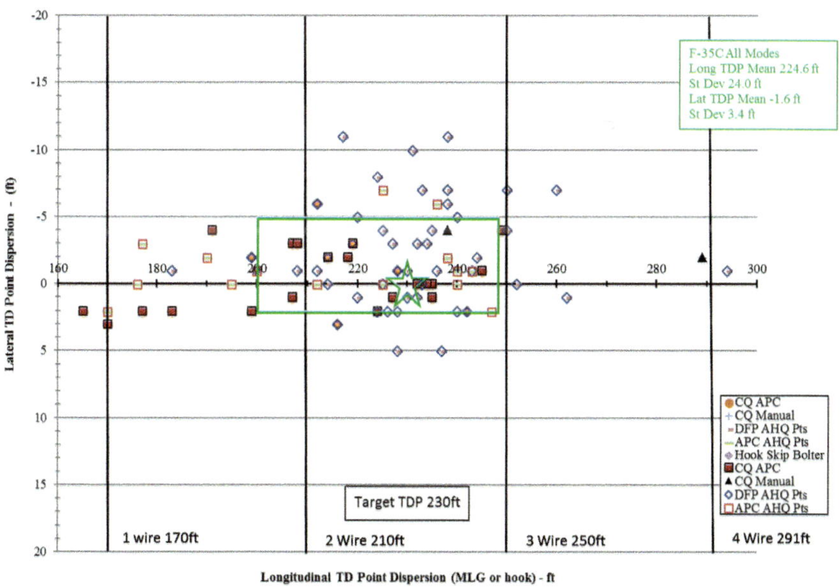

Fig. 50 DT-II hook point touchdown dispersion.

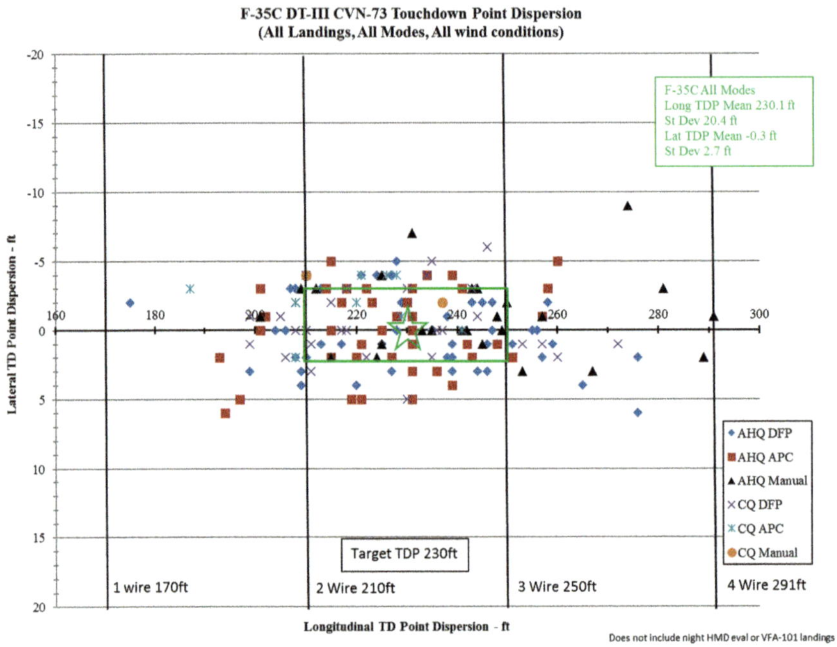

Fig. 51 DT-III hook point touchdown dispersion.

VI. CONCLUSION

Designing an aircraft that will operate from the decks of seagoing vessels is not easy. The design space is filled with a myriad of constraints and intricacies such as sizing for deck space considerations, wing fold capabilities to reduce that size factor, and the proper sizing of control surfaces for the tasks of launching and recovering. The F-35C couples those baseline constraints with the additional requirements of low observability and commonality with two other variants, which affected such things as arresting hook geometry that had second-order effects on arresting gear dynamics (perturbations of kink waves). These engineering challenges were faced by the multidiscipline specialization team of carrier suitability that combines theories from aircraft loads, flying qualities, and performance in a system of systems approach to assess the suitability of an aircraft to operate from ships and austere sites. In addition to flight sciences, carrier suitability includes many other nuances and facets such as navigation and guidance, sensor integration, data link interoperability, pilot–vehicle interface, supportability, maintainability, and night evaluations, shown in Fig. 52, that were all beyond the scope of this paper.

Aircraft that are launched from and perform arrested landings to the deck of a ship require enhanced structure to withstand the loads to which they are subjected. As such, a majority of the carrier suitability process was spent on conducting specialized loads testing in the form of shore-based catapults and arrested landings. During the structural survey, the F-35C faced and overcame many challenges such as the AHS. The original tailhook is proof that although designs may work on paper, they do not always go right the first time. The AHS redesign process was validation of the engineering problem-solving process.

Additionally, strides towards the improvement of the carrier suitability discipline were also made by challenging the current assumptions and philosophies. The F-35C was approached with a blank canvas mentality that paved the way for the use of automation in the pursuit of CVS testing that increased efficiency. The advanced approach modes and IDLC provided the pilot with an unprecedented ability to control the aircraft. This technology, specifically DFP, and FTAs provided tools that allowed challenging test points to be achieved with a high level of repeatability without relying on a test pilot's skill to finesse aircraft performance.

Undeniably, the CLAW, which provides the basis for the advanced approach modes and FTAs, demonstrated many benefits to both the F-35C and the entire F-35 program. Delta Flight Path is a game changer. It was an instrumental tool in allowing the test team to complete the structural survey efficiently, effectively, and safely. DFP, which has been attainable only with the implementation of fly-by-wire controlled aircraft and advances in computing power, will be the basis for changes in how the discipline of CVS flight test is approached in the future. DFP has not only been beneficial to flight test, but also has provided the fleet aviator with an unprecedented ability to control the aircraft during shipboard approaches, which will increase operational capabilities and make naval aviation safer.

Fig. 52 First night shipboard evaluation.

REFERENCES

[1] Jakab, P., "Eugene Ely and the Birth of Naval Aviation—January 18, 1911," Smithsonian National Air and Space Museum, 18 Jan. 2011, https://airandspace.si.edu/stories/editorial/eugene-ely-and-birth-naval-aviation—january-18-1911?lipi=urn%3Ali%3Apage%3Ad_flagship3_feed%3BB3eSYDI7RcKZ3oim3aV9Kg%3D%3D [retrieved 22 Aug. 2018].

[2] Kandler, E., and Pope, J., *F-35C Ship Suitability Third Sea Trials (DT-III) Joint Test Plan*, F-35 Integrated Test Force, Patuxent River, MD, 2016 (unpublished).

[3] Cenciotti, D., "'F-35 Unable to Land Aboard Aircraft Carriers' Report Says. U.S. Navy and Royal Navy Have Something to Be Worried About," *Aviationist*, 9 Jan. 2012, https://theaviationist.com/2012/01/09/f-35c-hook-problems/ [retrieved 22 Aug. 2018].

[4] Thomlinson, J., "A Study of the Aircraft Arresting-Hook Bounce Problem," Ministry of Supply, Aeronautical Research Council Reports and Memoranda, London, England, 1957.

[5] Wedertz, R., Hess, E., Taylor, B., Stockton, K., and Antel, J., *The Landing Signal Officer Reference Manual*, U.S. Navy Landing Signal Officer School, Virginia Beach, VA, 2010 (unpublished).

[6] *Carrier Suitability Testing Manual*, SA FTM-01, Naval Air Warfare Center Aircraft Division, Patuxent River, MD, 1994.

[7] "NATOPS Landing Signal Officer Manual," NAVAIR 00-80T-104, Naval Air Technical Data and Engineering Service Command, San Diego, CA, 2001.

[8] "Aircraft Launching Bulletin No. 33-76," Steam Catapult Performance Engineering, Naval Air Technical Data and Engineering Service Command, San Diego, CA, 2016.

F-35 Aerodynamic Performance Verification

David G. Parsons*, Austin G. Eckstein† and Jeff J. Azevedo‡
Lockheed Martin Aeronautics Company, Fort Worth, TX

The combined efforts of the airframe contractor, propulsion contractor, and government program office culminated in verifying the primary conventional performance requirements of the F-35 aircraft. The joint group collaborated to develop a cost-effective and credible modeling and simulation-based approach to validate the engineering databases used to calculate aircraft performance. Instrumental to the group's success was a rigorous weight management process, supported by an incremental burndown of conservative factors applied to preflight test performance predictions. These approaches ensured that the F-35 met the key contractual performance requirements. Throughout our efforts, we maintained attention to detail when performing focused analyses of test data, from individual sensor measurements to calculated performance parameters. These efforts allowed a minimal flight test matrix to be sufficient in resolving even minor adjustments to preflight test aerodynamics. Ultimately, the flight test results proved that the design of the F-35 aircraft exceeded requirements. Further, the results formed the basis of the operational performance capability provided to the operator throughout the flight envelope.

I. NOMENCLATURE

α	angle of attack
a_X	longitudinal acceleration
a_Z	normal acceleration
C_D	drag coefficient
C_{Df}	friction drag coefficient
$C_{D\min}$	minimum drag coefficient

*Aeronautical Engineer, Senior Staff, Flight Sciences.
†Aeronautical Engineering Manager, Flight Sciences.
‡Aeronautical Engineer, Senior Staff, Flight Sciences.

C_{DT}	test drag coefficient
C_f	skin friction coefficient
C_L	lift coefficient
C_{LT}	test lift coefficient
δ	ambient pressure ratio
D	drag
D_E	engine drag
F_G	gross thrust
g	acceleration due to gravity
γ	flight path angle
K	equivalent sand grain roughness
L	lift
n_Z	normal load factor
n_X	axial load factor
q	dynamic pressure
S	reference wing area
θ	pitch attitude
W	gross weight
V	velocity

II. INTRODUCTION

This paper provides a top-level understanding of the approach taken and analytical techniques used on the F-35 during our performance validation. With these we verified the conventional performance requirements of the F-35, with particular emphasis on the Key Performance Parameters (KPPs) of the Joint Contract Specification (JCS). Our modeling and simulation-based verification process successfully validated the aerodynamics and performance databases used to calculate performance with a minimal flight test matrix. Ultimately, the process indicated that the mission performance of all three variants of the F-35 exceeded requirements. This paper presents the conventional aircraft performance management approach and highlights some of the challenges, issues, and successes involved in the verification process.

III. BACKGROUND

The F-35 is a fighter aircraft designed for use by the U.S. Air Force, Marine Corps, and Navy, as well as the current 12 partner nations. The aircraft comes in three largely common configurations/variants: F-35A Conventional Takeoff and Landing (CTOL), F-35B Short Takeoff and Vertical Landing (STOVL), and F-35C Carrier Variant (CV). Figure 1 provides an overview of the variant planforms and basic configuration details.

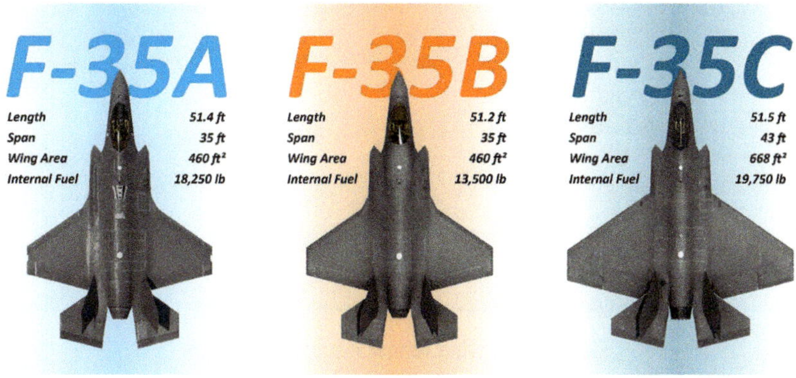

Fig. 1 F-35 variants.

Performance requirements for the F-35 were defined in the JCS. The primary conventional aircraft performance requirement for each variant was the KPP design mission radius. The F-35B (STOVL) had additional KPP requirements for short takeoff distance and vertical landing bring-back to emphasize operations aboard Navy LH-class amphibious assault ships. The approach to verifying each of those requirements followed processes similar to those used for the conventional performance requirement approach presented herein [1]. Vertical landing bring-back is a measure of the aircraft's ability to recover to the ship with unexpended munitions and fuel. The F-35C (CV) had a separate KPP requirement for approach speed to emphasize operations on Navy aircraft carriers [2].

The F-35 program used a modeling and simulation-based approach to determine and verify aircraft aerodynamics and performance. The wind tunnel testing component of the program consisted of approximately 50,000 hours of testing using 15 models in 23 tunnel facilities in seven states and four countries.

We used two force and moment models: a one-twelfth-scale model to baseline for all aerodynamic databases, and a one-fifteenth-scale model for aerodynamic store increments and most stability and control databases. We validated both models to ensure the accurate representation of the aircraft's outer mold line. Figure 2 shows two of the installations.

We incorporated several innovations into the larger one-twelfth-scale model to improve data quality. The model had two aft ends, which allowed for different balance/sting installations. The smaller aft end accommodated a 2-in. balance and sting, allowing it to be completely accurate to the airframe/nozzle interface. The 2-in. balance was also gauged to a lower capacity that resulted in higher-sensitivity data with greater accuracy. We used the smaller aft end installation to gather data across the Mach range to 15 deg α. The larger aft end had a 2.5-in. higher-capacity/lower-sensitivity balance/sting and was used to gather data from 15 to 40 deg α.

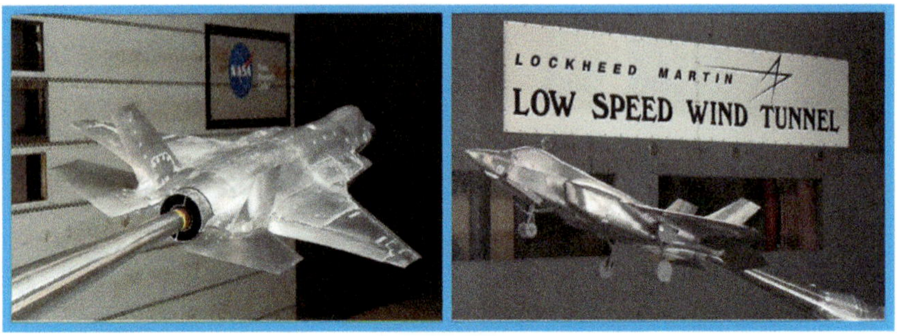

Fig. 2 F-35 wind tunnel model installations.

The one-twelfth-scale model also incorporated remote actuation of the horizontal tails. This permitted us to change the configuration without opening the tunnel. That, in turn, enabled us to gather far more data within the budget and time constraints imposed on any wind tunnel program. The increased amount of trailing-edge flap/horizontal tail interaction data greatly improved the accuracy of the final aerodynamic database.

We used Computational Fluid Dynamics (CFD) extensively to complement the wind tunnel testing. This included evaluating dozens of trade study options with CFD to determine the most effective options to test in the wind tunnel [3]. We also used CFD to account for items that could not be tested in the tunnel, such as nacelle ventilation and the propulsion thermal heat exchanger inlet. In addition, we used it to account for any minor configuration changes that occurred after the wind tunnel testing had concluded. This ensured that the aerodynamic database accurately represented the flight test aircraft before entering the flight test phase.

F-35 wind tunnel testing also designed a secure network for communications and data exchange between the wind tunnel test facilities and Fort Worth, Texas, and between different tunnels. The wind tunnel data were automatically synced among the source tunnel facility, other specified tunnel facilities, and Lockheed Martin in Fort Worth. This allowed a limited number of engineers to run and monitor multiple tests worldwide. At one point in the test program, 11 tests were being run simultaneously.

Table 1 provides the total wind tunnel testing hours from the F-35 System Development and Demonstration (SDD) program across all disciplines.

Figure 3 illustrates the F-35 three-degrees-of-freedom (3DOF) force and moment accounting system. The data obtained during the wind tunnel testing were supplemented with CFD analyses [3–5]. The idea behind this approach was to invest in accurate modeling of the configurations early in the program. Doing so minimized, to the greatest extent possible, the relatively expensive flight testing used to validate the databases.

TABLE 1 F-35 SDD PROGRAM WIND TUNNEL TESTING HOURS

User Occupancy Hour Summary (Actuals thru Mar. 30, 2010)							
	Actual Hours per Test Discipline						
Year	Aero Analysis	Stability & Control	External Environment	Store Separation	Propulsion Aero	Flutter	Total
2002	3975	1435	868	1175	2269	192	**9914**
2003	3464	3168	677	578	2060	–	**9947**
2004	1991	2582	810	373	2892	–	**8648**
2005	4054	3319	556	732	2935	–	**11,596**
2006	2677	2315	468	342	1253	–	**7055**
2007	780	–	308	445	250	–	**1783**
2008	–	–	–	264	–	–	**264**
2009	–	–	–	–	–	–	**-**
2010	–	–	777	–	–	–	**777**
Actual thru Mar. 30, 2010	**16,941**	**12,819**	**4464**	**3909**	**11,659**	**192**	**49,984**
Plan	**16,941**	**12,819**	**4464**	**3909**	**11,659**	**192**	**49,984**
Complete	**100.0%**	**100.0%**	**100.0%**	**100.0%**	**100.0%**	**100.0%**	**100.0%**

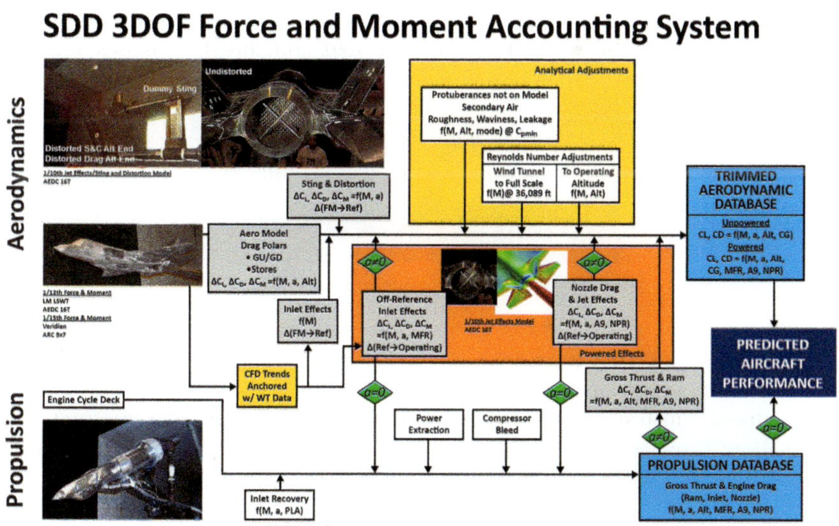

Fig. 3 F-35 3DOF force and moment accounting system.

IV. PERFORMANCE MANAGEMENT

At contract award, the F-35 program mandated the use of a conservative factor on aircraft performance calculations to cover risks in several aspects of the new design. The factor was implemented as a 5% increase in predicted fuel flow to cover immaturity in the design, among other things. It also covered the uncertainty in predicted aerodynamics and propulsion databases, as well as the possibility of weight growth during the maturation of the design. Further, it covered the potential for configuration changes resulting from discoveries during flight testing that could adversely affect performance. As the aircraft design matured and flight testing was accomplished, uncertainty in each of these areas was to be retired. In tandem, the factor was to be progressively reduced until no conservatism was to be applied for the final calculation of specification performance. Two of the largest areas of uncertainty at contract award were the accuracy of the predicted aerodynamics database and the process of adjusting subscale-based testing results to a full-scale aircraft.

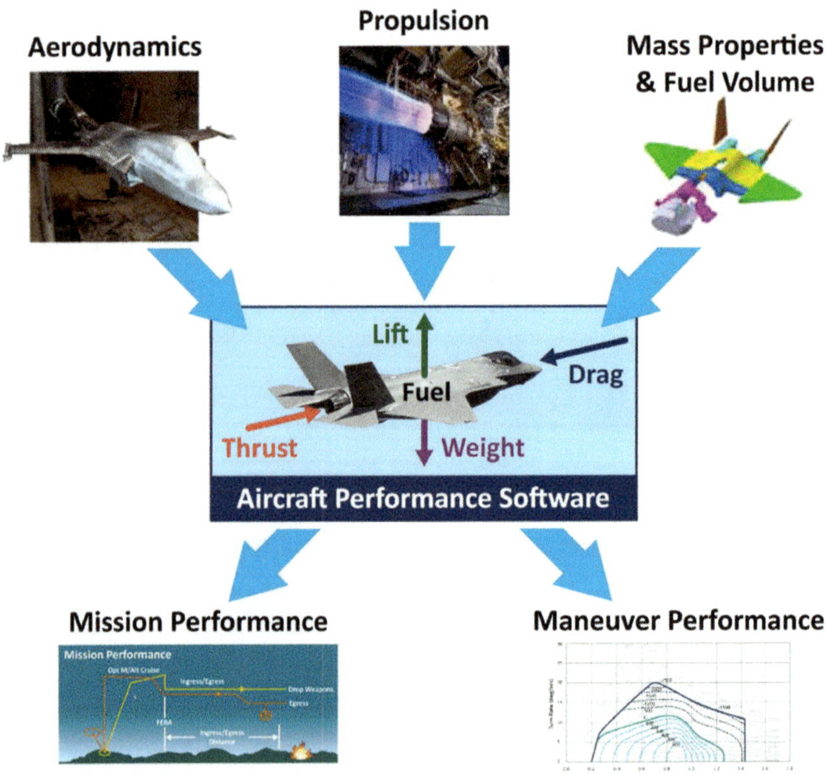

Fig. 4 F-35 aircraft performance calculation process.

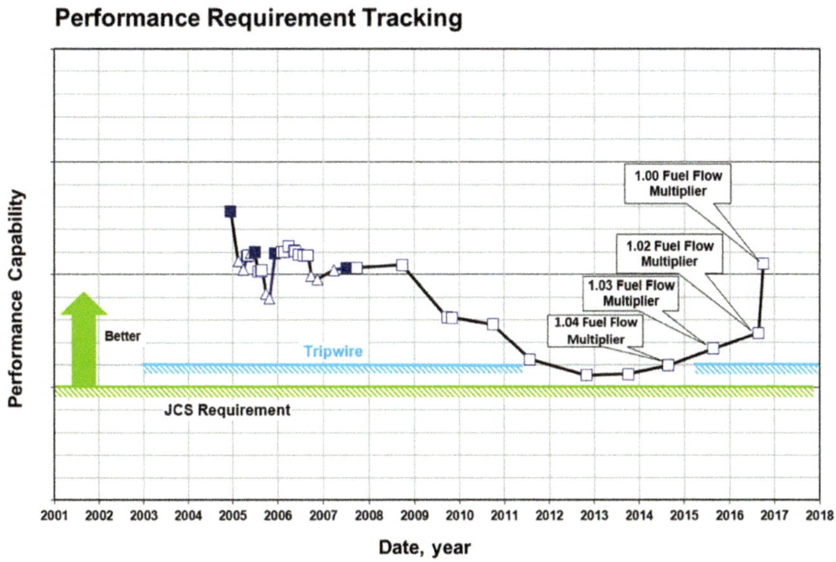

Fig. 5 F-35 performance capability tracking against tripwire and requirement.

From a simplified perspective, aircraft performance calculations involve combining inputs from aerodynamics, propulsion, and mass properties databases, as illustrated in Fig. 4. These are used to calculate mission and maneuver capabilities, and changes in any of the databases can either improve or degrade calculated aircraft performance. The F-35 program implemented an aircraft performance management process early in the development program to ensure that the final aircraft design would meet the KPP requirements. We calculated the F-35 capability against the JCS performance requirements and tracked the levels of performance as the databases matured through wind tunnel testing and analyses.

For each requirement we developed tripwires to indicate when the margin to the requirement had eroded enough to compromise meeting the requirement. If predicted performance fell below a tripwire, we would initiate a recovery plan to recover margin and improve performance. Figure 5 provides an illustration of how the performance for a particular requirement changed over time. Each symbol in it indicates a formal calculation of the requirement. For the early phases of the program, aerodynamics, propulsion, and mass properties databases were being updated frequently. As part of this, performance calculations for the specification requirements were executed roughly once a month. As the design matured and stabilized, changes that affected performance calculations became less common. Consequently, the frequency of calculations of specification performance was reduced.

For the performance requirement in Fig. 5, calculated performance dipped below the tripwire level in late 2012. A subset of maneuvers from the dedicated performance flight test matrix was executed to evaluate the relative accuracy of the preflight test aerodynamics database. Results of this dataset analysis indicated that the pretest databases provided a good match of measured performance, and that uncertainty in the final level of aerodynamics was not high. From this analysis, and in agreement with the F-35 Joint Program Office (JPO), we reduced the fuel flow conservatism factor used for specification performance calculations from 5% to 4%. The change in conservatism was enough to allow the calculated performance to exceed the tripwire level. The conservatism in the fuel flow factor was then incrementally reduced as the remaining dedicated performance flight test matrix was executed and analyzed. Incremental reductions were also the product of the flight test program maturing and not indicating the need for configuration changes that would adversely affect performance.

In Fig. 5, the final point on the plot indicates the official JCS performance verification calculation where the fuel flow conservatism factor had been completely removed. This represented the final validated flight test–level aerodynamic databases, validated propulsion installation effects, and actual measured production aircraft weight and fuel volume.

A key component of managing aircraft performance was to control weight growth during the program's design and development phase. Aircraft gross weight typically grows as the design matures, so F-35 weight was actively managed, and weight growth was limited to a projected increase with time [6]. The calculations of aircraft performance against program requirements used the projected weight at

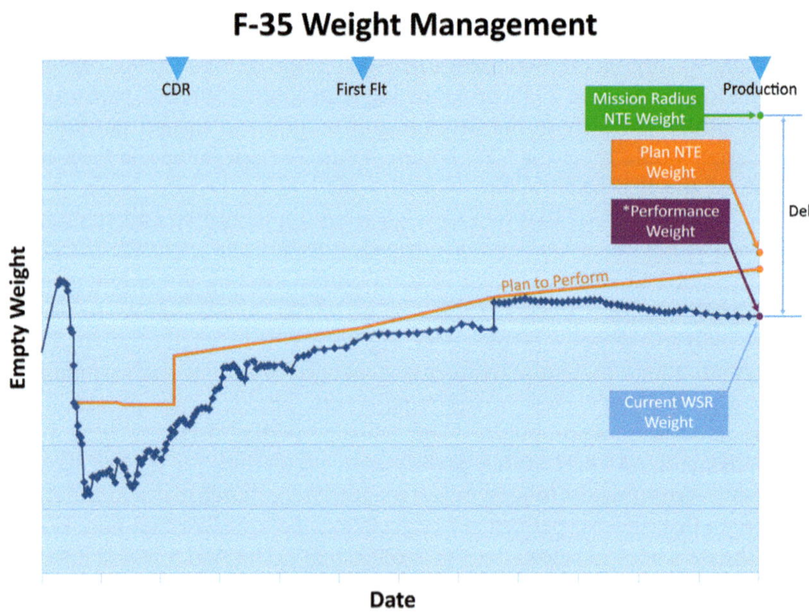

Fig. 6 F-35 weight management.

the end of the SDD program. This was in contrast to the status weight at that particular point in time. Figure 6 illustrates F-35 weight growth with time and shows the plan-to-perform line that the program used as a ceiling on weight increase with time. We waited to implement design changes that might cause the weight to cross the plan-to-perform line. Only after we could implement offsetting changes that reduced weight to keep status weight below the target did we proceed with design changes.

V. FLIGHT TEST APPROACH

The overarching strategy chosen by the F-35 aerodynamics and performance verification team was one of database validation rather than performance demonstration. Aircraft performance predictions were calculated by combining databases for aerodynamics, propulsion, mass properties, and landing gear (for basing performance) through the laws of physics. The predictions' calculations also used the equations of motion defining the mission or maneuver in question. Each variant had a unique design mission profile, but the approach to verify each of those mission capabilities was common among the variants. The team's strategy was to acquire the data and complete the analysis necessary to validate the databases used in those calculations. It then used the validated databases to calculate verification performance.

The team's approach provided several tangible and cost-effective benefits. Among these, the amount of dedicated flight testing necessary to validate the databases was considerably less than that for demonstrating performance for all aircraft configurations, in all ambient operating conditions. Minimizing flight testing provided significant cost savings. As another benefit, the contributions of aerodynamics and propulsion to the overall performance of the aircraft were able to accurately be allocated and validated. As a result, the amount of testing for future aircraft or propulsion system modifications can be minimized because testing can be initiated from a well-established baseline. Although this effect does not generally provide immediate cost benefits at this phase of the program, it will provide significant cost avoidance for future F-35 modernization/enhancement contracts.

A. IN-FLIGHT THRUST AND PERFORMANCE WORKING GROUP

A key contributor to the success of the aerodynamics and performance verification team's approach was the early establishment of the In-Flight Thrust and Performance Working Group (IFTPWG). Shortly after contract award, the IFTPWG was established with representatives from Lockheed Martin, BAE Systems, Pratt & Whitney, and Air Force and Navy personnel at the F-35 JPO. The group's goal was to collaborate on the testing and analysis necessary to credibly validate the databases and successfully verify aircraft performance specification requirements.

The IFTPWG leveraged its combined expertise in aerodynamics, performance, and propulsion testing to define and agree upon the analyses and testing necessary to achieve database validation. By agreeing on the approach prior to initiating testing, scope growth in expensive flight testing could be minimized or prevented. The IFTPWG served as a forum for vetting and discussing all problems and results, and was key to fostering a cooperative working relationship among all parties.

A critical component of the working group's efforts was the In-Flight Thrust Calculation Deck (IFTCD). The IFTCD was a Pratt & Whitney–supplied tool for calculating thrust at any discrete point from flight or ground testing of the aircraft, based on measured engine and aircraft parameters. Two independent methods of calculating thrust were included as options for the IFTCD: airflow square root of temperature and flow area pressure [7]. Each method relied on different parameters to determine propulsive thrust for inflight conditions. One of the benefits of having two independent methods was obtaining the ability to better identify any systematic bias within one method.

Thrust and aerodynamics cannot be independently measured in flight. As such, the IFTCD was crucial for determining aircraft aerodynamics from flight test maneuvers. In Fig. 7, values for W, a_X and a_Z, V, and α can be directly measured; however, the aerodynamics values of L and D and the propulsive terms of F_G and D_E cannot be directly measured. To determine the aerodynamics parameters, the propulsive terms are required to be known.

The IFTCD was calibrated against the results of engine development testing in altitude chambers with measured load cells to improve the fidelity of the two calculation methods. The IFTPWG continually evaluated options for increasing

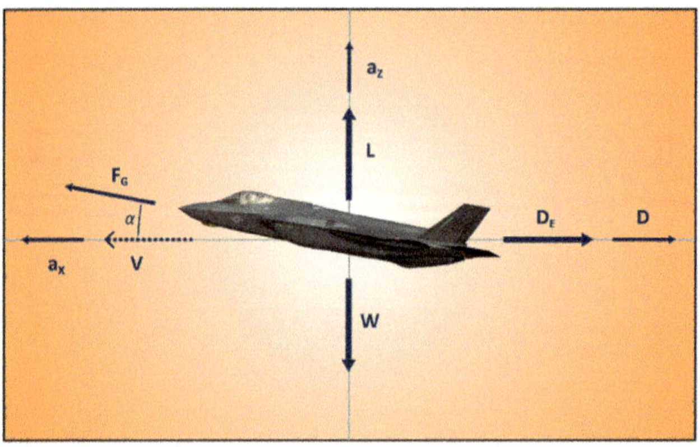

$$F_G \cos \alpha - D_E - D = W/g\ a_x \longrightarrow C_{D_T} = (F_G \cos \alpha - D_E - Wn_x) / q\ S$$

$$L + F_G \sin \alpha - W = W/g\ a_x \longrightarrow C_{L_T} = (Wn_z - F_G \sin \alpha) / q\ S$$

Fig. 7 Conventional forces acting on an aircraft in level flight.

fidelity in the flight test results. It used improved methods for individual parameter measurement, along with specific ground and flight test techniques, to reduce uncertainty in the end product. Each option was evaluated for its technical feasibility and to assess its cost/benefit tradeoff. Any additional cost to the program had to provide realistic, obtainable, and significant benefits to be included as part of the verification process.

B. TEST MATRIX

The F-35 Aerodynamics and Propulsion teams entered the flight test phase of the SDD program with a high-fidelity wind tunnel analytical database. Extensive work was accomplished to ensure that all forces and moments identified in the force and moment bookkeeping system were well-defined from wind tunnel testing and CFD. This was particularly important when objectives were difficult to achieve from testing. The comprehensive nature of the preflight databases was key to successfully verifying aircraft performance requirements with a minimal matrix of flight test maneuvers.

The matrix of dedicated aircraft performance flight test maneuvers was designed to minimize required test flights. At the same time, it continued to provide the data necessary to verify the KPP performance requirements and validate the performance databases for all phases of flight. The goal was to validate a credible collection of the databases that formed the basis of the performance products provided to the operator (e.g., flight manual, pilot checklist, performance on the glass). The IFTPWG balanced fidelity in the final databases with the cost of testing and analysis. Its members were constantly evaluating and refining the test matrices due to programmatic pressure to reduce flight test costs and shorten the schedule. In the end, the number of dedicated test points flown for any one variant was roughly half the number used for previous fighter aircraft. Table 2 summarizes the test points for clean configuration.

One element of the dedicated flight test matrix that supported verifying the KPP design missions was the focus on the critical portions of each mission. The Aerodynamics and Performance Validation team analyzed each design mission to determine which segments consumed the most fuel. One or more dedicated flight test point(s) were defined to anchor the validation of the databases. This was done for each flight condition that represented a significant portion of the design mission fuel usage (e.g., optimum altitude/optimum Mach cruise). Figure 8 illustrates the flight envelope coverage of dedicated performance flight test maneuvers.

The validation team created predictions for each flight test maneuver. It then spotted the predictions onto the predicted drag polars and lift curves to ensure sufficient coverage and avoid unnecessary duplication of conditions. It used a combination of steady-state and quasi-steady-state maneuvers to evaluate the lift and drag of the F-35 at a full range of sustainable α. Examples of such maneuvers were cruise points, fixed-power sustained turns, Roller Coaster Maneuvers

TABLE 2 CLEAN-AIRCRAFT DEDICATED PERFORMANCE MANEUVERS

Unique Dedicated Performance Test Points for Clean Aircraft Configuration			
Maneuver	**F-35A**	**F-35B**	**F-35C**
Speed Powers (cruise)	51	56	54
Powered Approach (gear down) Speed Powers	7	12	28
Roller Coaster Maneuvers	8	9	9
Accelerations	9	9	9
Decelerations	6	6	6
Sustained Turns	14	14	14
Climbs	8	8	8
Descents	4	4	4
Landings	8	8	8
Takeoffs	0*	0*	0*

*Maneuver analyzed from ride-along data for other disciplines.

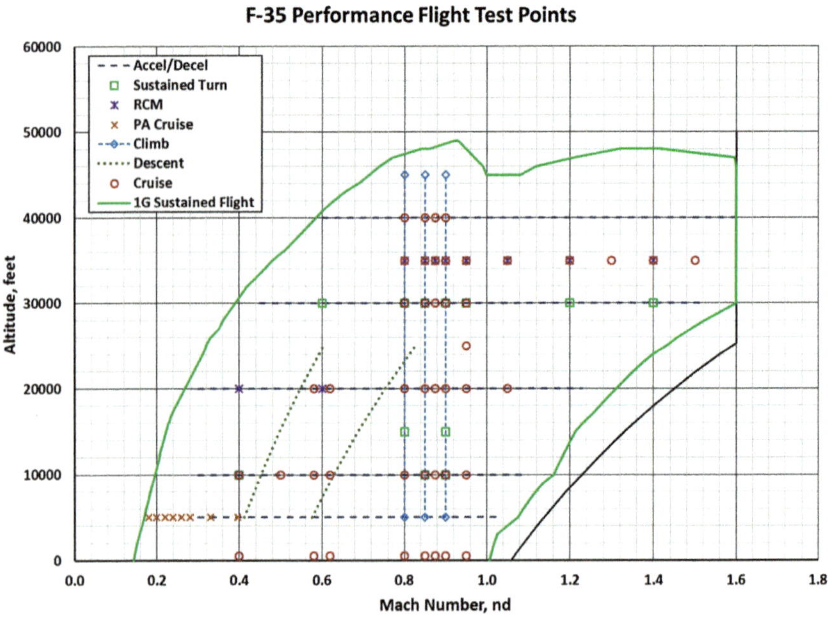

Fig. 8 Dedicated performance flight test maneuvers.

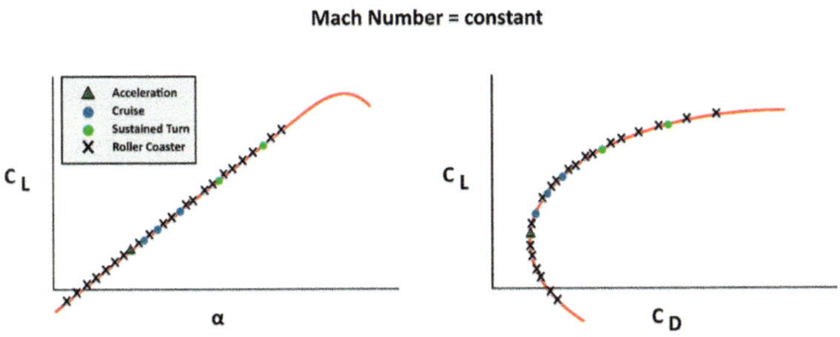

Fig. 9 Example lift curve and drag polar.

(RCMs), and fixed-power accelerations. The Flying Qualities and Loads teams' maneuvers were used to assess higher α aerodynamics and performance, rather than executing dedicated maneuvers, for cost savings and greater efficiency. Figure 9 illustrates the relative locations of multiple types of maneuvers on a predicted lift curve and drag polar. In this scenario, cruise maneuvers would have been executed at multiple altitudes to provide the spread of C_L; higher-altitude maneuvers require higher C_L values to sustain level flight at a given weight as a result of lower dynamic pressure.

The team also utilized cruise maneuver repeats flown at identical W divided by δ conditions to evaluate the repeatability of aerodynamics results from flight test. Flying the same Mach condition at a different weight and determining the test altitude from the W/δ equation drives both test points to the same C_L. Therefore, it drives them to the same point on the drag polar as well. For instance, if the original test point was flown at a W of 45,000 lb and an altitude of 35,000 ft, the W/δ would be approximately 190,500. Then, a repeat test point could be flown at 42,000 lb, and the test altitude would need to be approximately 36,400 ft. Consistency in drag levels from the two maneuvers, once standardized for altitude, center of gravity, and engine operating conditions, would demonstrate the repeatability of the flight test data.

C. UNCERTAINTY ANALYSIS

To make a minimal test matrix work, accurate and redundant sources of critical instrumentation were required. An uncertainty analysis was accomplished early in the program. With it, we identified the parameters that drive accuracy in the aerodynamics and performance data calculated from flight test maneuvers. The results of the analysis provided insight into how best to utilize limited resources to improve the final product.

The uncertainty analysis process involved creating influence coefficients for each sensor measurement that was an input to the aerodynamics and performance data reduction process. The influence coefficients varied with test condition and maneuver type and needed to be evaluated across the expected test operating range. These influence coefficients were then combined with values for sensor accuracy to determine the uncertainty in the calculated aerodynamics and performance parameters. Realistic sensor measurement accuracies are required for a realistic uncertainty analysis. The results of the analysis allowed the team to focus its efforts on the parameters that were significant drivers in the uncertainty of the results. In so doing, the team avoided wasting resources on measurements that had little or no effect on the results.

The detailed uncertainty analysis also identifies where one method of inflight thrust calculation would have less uncertainty than another. Just as the influence coefficients varied with condition and maneuver type, so did the analytical accuracy of the two inflight thrust methods.

VI. FLIGHT TEST ANALYSIS RESULTS

The effort to validate aerodynamics and performance databases overcame several technical challenges during the course of analysis. Aircraft α is a critical measurement that influences the accuracy of the derived aerodynamics data from flight test. The Aerodynamics and Performance Verification team worked with the Air Data team to refine the accuracy of the flight test nose boom data beyond what was required for the other disciplines. The teams used all of the dedicated performance cruise data acquired for each variant to develop a secondary α calibration. They analyzed the difference between the primary corrected nose boom α and an inertially derived α determined from the inertial navigation sensors on the aircraft. The primary nose boom correction mostly adjusts for up wash and pitch rate. Equation (1) shows the formula used in the teams' calculation for inertial α.

$$\alpha_{\mathrm{INERTIAL}} = \theta - \gamma \tag{1}$$

Equation (1) assumes there are no ambient vertical winds, and this is generally a reasonable assumption; however, a detailed analysis conducted by the Validation team revealed that certain maneuvers exhibited higher levels of vertical wind than others. This applied to maneuvers conducted near the mountains around Edwards Air Force Base, compared to maneuvers conducted away from the mountains or over the water at Naval Air Station Patuxent River. These results are consistent with ambient winds traveling up and around the mountains. The winds can also be seen generating increased levels of turbulence during the maneuvers. High-rate aircraft acceleration and α data were also filtered to reduce spurious noise in the signals. All these refinements led to more accurate lift and drag data with reduced scatter.

Aircraft W is another parameter for which the team looked to reduce uncertainty. The test team employed a process to increase the fidelity of aircraft W at any point in time. The test aircraft was weighed with a set of portable scales following on-ground fueling and prior to any test flight. The aircraft was weighed again following the test flight. The pretest and posttest weighings provided anchor points used to determine aircraft W at any point during the flight. The amount of fuel burned from each anchor point was integrated for the entire flight. This determined the amount of fuel and the weight of the jet at any discrete point during the flight. This process reduced uncertainty from the fuel gauging system due to fuel slosh or aircraft attitude.

Fuel flow is a primary driver in aerodynamics and performance calculations. As mentioned previously, fuel flow is used to help determine aircraft weight at any discrete point in time. It is also a primary input for the inflight thrust calculation methods and a primary driver in inflight thrust calculation accuracy. The team used test flow meters for both gas-generator and afterburner fuel flow. This increased the accuracy in the fuel flow measurement over the production source of fuel flow for the engine control. The team intended to use the production fuel flow measurement as a backup in case of sensor failure of the test flow meter; however, the installation of the test flow meter caused an unintended consequence for the Initial Service Release (ISR) engine standard used in F-35A and F-35C testing.

The software for the ISR production engine control included several corrections to improve the accuracy of the production fuel flow measurement. With the test flow meter installed, one of the corrections was no longer applicable, introducing an error of up to 3% to the production measure of fuel flow. The team did not discover this problem until the difference was revealed by an engine stand run to calibrate the test flow meter. Correcting the error would have required a modification to the production engine control software, which was deemed unacceptable due to its effects on cost and schedule. As a result, for the ISR engine standard, the production fuel flow measurement could not be used as a backup for the test flow meter. This problem was not present for testing with the first flight release engine standard used for F-35B conventional performance testing.

After having acquired all test data for a particular variant, we processed the raw test data through a set of data reduction routines. With these, we calculated test-day aerodynamics and performance parameters, and then standardized those parameters to a set of reference conditions to compare them to the pretest predicted databases. The standardization process involved analytically adjusting data from test-day conditions to the reference conditions via predicted increments. Standardization allowed multiple maneuvers to be evaluated against pretest predictions, rather than evaluating each point against a test-day prediction. This process allowed us to readily identify trends between/among prediction and flight test results.

Aerodynamics data for the aircraft in an up-and-away configuration (i.e., gear up) were standardized to the nearest database breakpoint Mach number, 36,089-ft pressure altitude, a constant center of gravity, and a set of engine operating parameters generally consistent with cruise conditions. The altitude serves as the baseline condition for the aerodynamics database. Aerodynamics data for the aircraft in a powered-approach configuration (i.e., gear down) followed that process but were modified due to the difference in typical operating altitudes between an up-and-away and a powered-approach configuration; the latter occurs nearer to sea level so that sea level serves as the baseline for powered-approach configuration aerodynamics. All aerodynamics standardization was completed holding test α constant because the databases are a function of α, rather than of C_L.

Performance data standardization varied slightly by maneuver type. All performance data were standardized to flight design gross weight, center of gravity, and standard day temperature. Cruise maneuvers were standardized to the nearest 50-foot pressure altitude and wings level, constant altitude, and unaccelerated flight conditions holding test Mach number constant. Acceleration and deceleration maneuvers were standardized to the chosen power setting (idle, military, or maximum afterburner), nearest 5000-ft pressure altitude, and wings-level attitude. We held the test Mach number constant for each discrete point in the maneuver. Climb maneuvers were standardized to the chosen power setting and desired climb schedule, holding the test altitude constant for each discrete point in the maneuver.

The standardized aerodynamics data were evaluated against predicted lift curves, drag polars, and trimmed horizontal tail deflections at each database Mach number. We used this approach, with the α and acceleration filtering discussed previously implemented to reduce scatter. In our assessment, the flight-derived aerodynamics data were generally found to reconcile among maneuver types with varying power settings and test altitudes. As expected, hysteresis was noted in some maneuvers (mainly RCMs) near the lift curve break. This occurred where the flow separations differ when pitching up and pitching back down, being attached to separated when pitching up and separated to attached when pitching down. The resultant data quality was very good, and the final flight test-basis aerodynamics levels were readily defined using standard regression curve fitting techniques.

By evaluating the data as a whole, the team was able to identify a trend with altitude that had not been present in the predictions. The difference was attributed to the predicted skin friction roughness used to adjust the subscale pretest wind tunnel–based database to full-scale conditions. Figure 10 illustrates the predicted variation of minimum coefficient of drag with altitude for different levels of roughness. The pretest aerodynamics database estimated a roughness of 250 μ in., whereas the flight test data indicated the roughness to be closer to 450 μ in.

The skin friction roughness was updated to be more representative of the results of flight test and the comparison process accomplished again. Further aerodynamic investigations revealed transonic drag rise differences from the wind tunnel–predicted aerodynamics data. These were attributed to differences in shock locations in flight, as well as complex interactions among the close-coupled wing/empennage, including power influences. Wind tunnel and flight data from the F-35 Loads and Flying Qualities teams were leveraged for these analyses.

Fairings of the flight test data were created to facilitate the development of untrimmed flight test increments to the pretest databases. The implementation

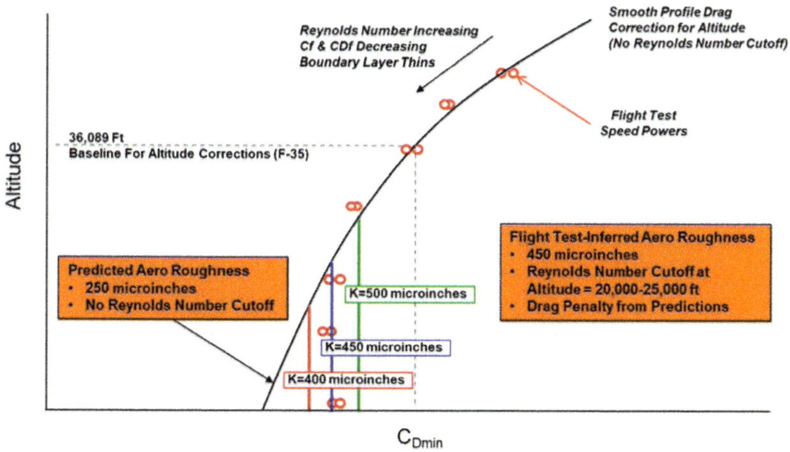

Fig. 10 Effect of roughness on variation of $C_{D\min}$ with altitude.

of the increments provided a flight test–basis aerodynamics database across the full operating envelope.

In addition to conducting the aerodynamics analysis, we calculated residual differences for each flight test maneuver from the adjusted database. A thorough evaluation of these residuals allowed the team to identify adjustments to the predicted jet effects that further improved the fidelity of the aerodynamics databases. Figure 11 illustrates a residual drag variation with nozzle exit area present in the data. After correcting the aerodynamics databases for the observed flight test jet effects, the resolution among all flight test maneuvers was approximately five counts of drag (0.0005 C_D) for cruise conditions. Samples of a lift curve and drag polar illustrating the final flight test results against original and flight test–based predictions are provided in Figs. 12 and 13, respectively.

After the aerodynamics analyses were completed, the adjusted database was used to create predictions for the standardized performance data. As had been done with the aerodynamics analysis, the performance data were evaluated as a whole rather than maneuver by maneuver. The team compared the performance predictions and the standardized test data. This allowed its members to validate that the adjustments made to the aerodynamics databases produced performance data that matched flight test results. Where differences still existed, further inspection led to identifying differences between the predicted engine model and the actual test engine. Some of these differences were due to different levels of engine bleed flow between the predicted model and the test engine. By contrast, others could be attributed to the number and severity of operating hours on the test engines. Although the inflight thrust methods typically accounted for these differences, the predicted steady-state engine model used to create predicted performance did not.

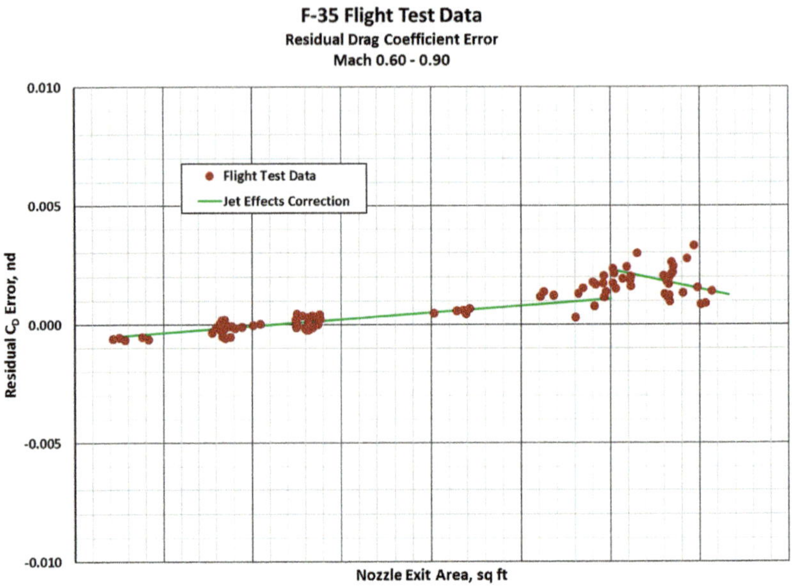

Fig. 11 Residual drag difference with nozzle area.

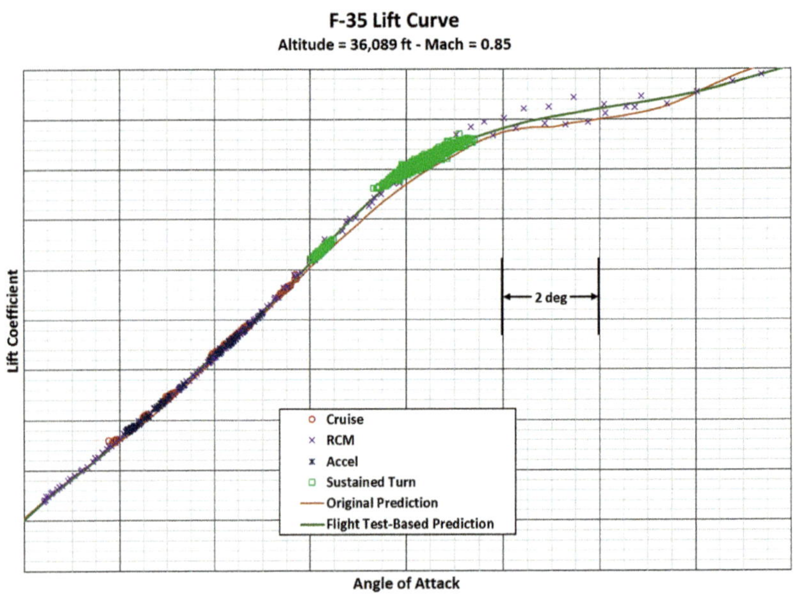

Fig. 12 F-35 flight test lift curve.

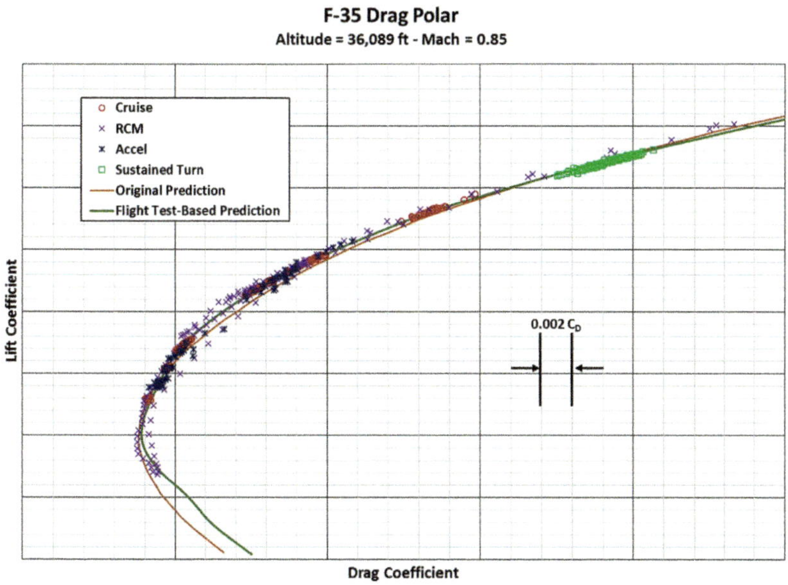

Fig. 13 F-35 flight test drag polar.

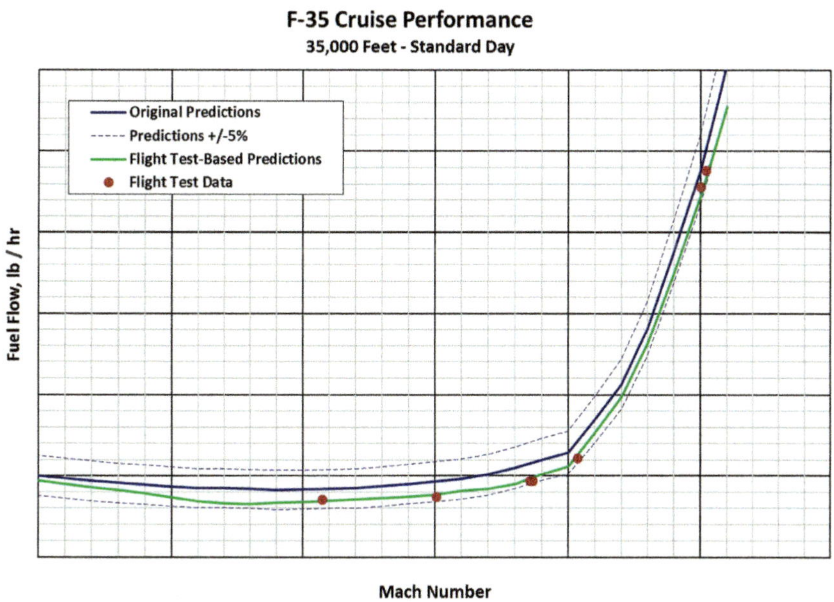

Fig. 14 Flight test cruise fuel flow vs predictions at 35,000 ft.

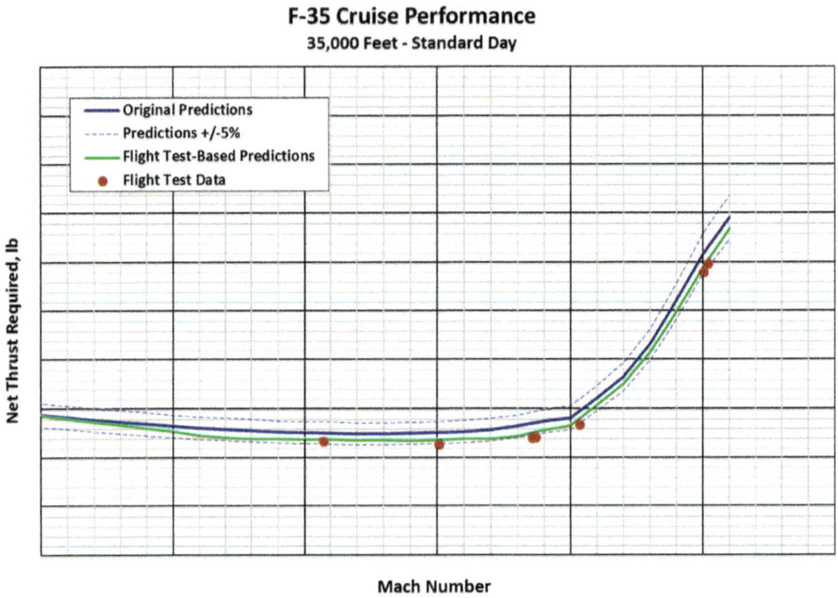

Fig. 15 Flight test cruise net thrust required vs predictions at 35,000 ft.

Figures 14 and 15 compare original predictions, flight test–based predictions, and standardized flight test data for cruise conditions at 35,000 ft. Figure 14 depicts the fuel flow, and Fig. 15 depicts the net thrust required. The original preflight test predictions are represented by the solid blue line, with dashed lines representing a deviation from that level of 5%. The flight test–based predictions are represented by the solid green line, and the standardized flight test data are represented by the symbols, one for each flight test maneuver. These plots are representative of the comparison of flight test data to predictions for all maneuvers in all conditions.

After completing all flight test analysis and obtaining the F-35 JPO's concurrence, we used the validated aerodynamics databases to calculate the KPP mission performance for each variant. The results showed that each of the variants exceeded the JCS requirement for mission range by more than 10%, as illustrated in Fig. 16.

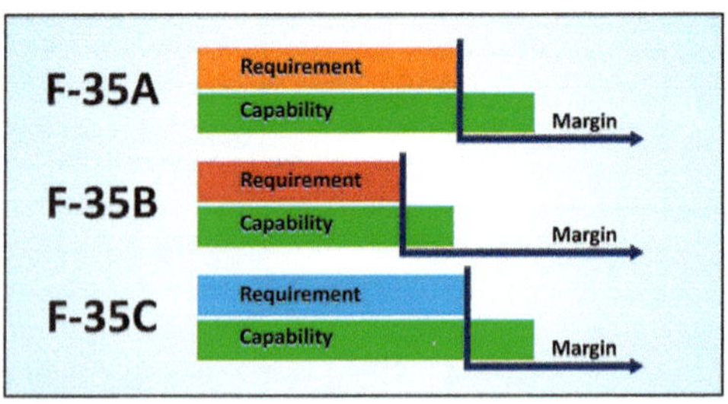

Fig. 16 Mission radius capabilities vs JCS requirements per F-35 variant.

VII. CONCLUSION

On the F-35 program, we successfully implemented a modeling and simulation-based approach to aerodynamic performance verification. Applying conservatism to performance calculations early in the program protected against potential uncertainties in configuration, weight, or aerodynamics levels. Our rigorous process controlled aircraft weight growth and helped to ensure that the performance of the final F-35 design met the KPP requirements of the program specification. The efforts of a government/contractor team culminated in delivering a credible, flight test–based aerodynamics and performance database that accurately represents the performance of the F-35. This will be applicable for not only specification verification, but also the operational performance products used by the fleet.

ACKNOWLEDGMENTS

D. G. Parsons thanks Brian Losos, Scott Rudin, and Jason Westly for their contributions to this paper and the analysis of the F-35 performance flight test data.

REFERENCES

[1] Levin, D. A., Parsons, D. G., Panteny, D. J., Rask, M. R., and Wilson, P., "F-35 STOVL Performance Requirements Verification," 2018 (to be published).

[2] Wilson, M. A., "F-35 Carrier Suitability (CVS) Testing," 2018 (to be published).

[3] Wooden, P. A., and Azevedo, J. J., "Use of CFD in Developing the JSF F-35 Outer Mold Lines," AIAA Paper 2006-3663, 2006.

[4] Wooden, P. A., Smith, B. R., and Azevedo, J. J., "CFD Predictions of Wing Pressure Distributions on the F-35 at Angles-of-Attack for Transonic Maneuvers," AIAA Paper 2007-4433, 2007.

[5] Karman, S. L., Jr., and Wooden, P. A., "CFD Modeling of F-35 Using Hybrid Unstructured Meshes," AIAA Paper 2009-3662, 2009.

[6] Counts, M. A., Kiger, B. A., Hoffschwelle, J. E., and Houtman, A. M., "F-35 Air Vehicle Configuration Development," 2018 (to be published).

[7] Vorwerk, A. V., and Ciszek, R. S., "Use of In-Flight Thrust on JSF Program," International Powered Lift Conference, Oct. 2010.

CHAPTER 15

F-35 High Angle of Attack Flight Control Development and Flight Test Results

Daniel G. Canin* Jeffrey K. McConnell† and Paul W. James‡

Lockheed Martin Aeronautics Company, Fort Worth, TX

The defining requirements of the F-35 relate to its range, performance, and 5th Generation combat capabilities, including all-aspect stealth, sensor fusion and network-enabled operations. Meeting signature and performance requirements while providing a robust high Angle-of-Attack (AOA) maneuvering capability presented challenges in configuration and flight control design. The selection of Nonlinear Dynamic Inversion (NDI) as the flight control methodology, with its dependence on highly accurate aircraft state information and onboard aerodynamic modeling, presented additional challenges when applied to the high AOA regime. This paper describes the F-35 high AOA flight control system development and testing conducted during the System Development and Demonstration (SDD) phase, with an emphasis on control law development, technical challenges, and key flight test results (Fig. 1).

Fig. 1 Author flying F-35C test aircraft CF-5 through a high AOA maneuver (8 January 2014).

---

*Experimental Test Pilot (High AOA Lead, F-35B/C), Lockheed Martin.

†F-35 Stability and Control Lead Engineer, Lockheed Martin.

‡F-35 Control Law Design Engineer (High AOA Lead), Lockheed Martin.

I. NOMENCLATURE

α	angle of attack
α-*dot*	rate of change of angle of attack
β	sideslip angle
β-*dot*	rate of change of sideslip angle
φ, θ, ψ	Euler roll, pitch and yaw angles
C_l, C_m, C_n	body-axis aerodynamic rolling, pitching, and yawing moment coefficients
$C_{l\beta}$, $C_{n\beta}$	body-axis aerodynamic rolling and yawing moment due to sideslip
$C_{n\beta, \text{dynamic}}$	departure susceptibility parameter
p, q, r	body-axis pitch, roll and yaw rates
Pb	body-axis roll rate
Pb-*dot*	body-axis roll acceleration
$\bar{q}$, *qbar*	dynamic pressure
Nz	longitudinal acceleration

II. INTRODUCTION

The unique tactical capabilities of the F-35, including all-aspect stealth, advanced sensor fusion, and network-enabled operations, support a strong argument that Within-Visual-Range (WVR), high Angle-of-Attack (AOA) maneuvering (i.e., dogfighting) may be a thing of the past. Indeed, the advent of high-off-boresight air-to-air missiles alone makes the close-in, low-speed engagement something to be avoided in *any* fighter. That said, there are cases in which a WVR engagement may be unavoidable, such as when rules of engagement require a visual identification; when onboard missiles have been depleted; or when the aircraft is required to dwell in a threat area to defend assets on the ground. Even if these scenarios are rarely encountered in combat, training for them results in routine exposure to the high AOA regime, with the attendant risk of Out-of-Control Flight (OCF).

With that understanding, the F-35 was required to provide high AOA maneuvering capabilities comparable to that of legacy fighters without compromising its more critical 5th Generation tactical capabilities. Specifically, the F-35 was required to:

- Be capable of performing air-to-air tracking tasks up to the stall AOA.[§]
- Provide positive and predictable pitch, yaw, and roll control response in the post-stall flight regime.

[§]For the purpose of these requirements, the F-35's stall AOA was defined as the AOA for maximum coefficient of lift (CL_{max}).

- Be departure resistant.[¶]
- Recover from all out-of-control modes using aerodynamic control with minimum pilot input.

The challenge for the F-35 design team was to extract the required high AOA maneuvering capability from an aircraft whose configuration was driven almost exclusively by other requirements.

III. AIRCRAFT DESCRIPTION

The F-35 is a family of single-seat, single-engine, all-weather stealth fighters designed for ground attack and air superiority missions. The family consists of the three variants shown in Fig. 2: the F-35A Conventional Takeoff and Landing (CTOL) variant, the F-35B Short Takeoff and Vertical Landing (STOVL) variant, and the F-35C Carrier-based Variant (CV).

A. FLIGHT CONTROL SURFACES

The flight control surfaces of the F-35 are conventional, consisting of Trailing Edge Flaps (TEFs), Horizontal Tails (HTs), dual rudders, and full-span Leading Edge Flaps (LEFs). Ailerons are incorporated on the CV variant only. The LEFs are scheduled symmetrically with Mach number and AOA to optimize performance and to improve lateral/directional stability at elevated AOAs but are not actively used for control.

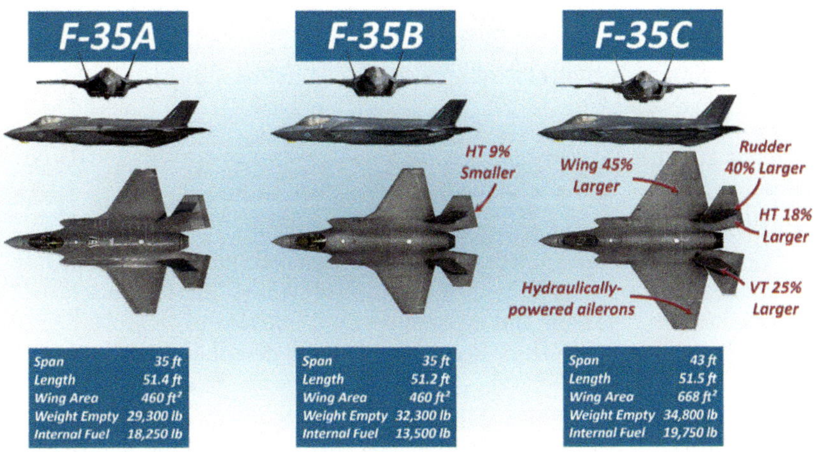

Fig. 2 F-35 variants.

[¶]By *departure resistant*, the intent was not necessarily to be carefree or departure-proof, if doing so would unnecessarily restrict capability. The desire was to incorporate appropriate flight control features that would protect the aircraft from loss of control during reasonably expected maneuvers.

The design features that affect high AOA aerodynamics differ significantly among the three variants:

- The STOVL variant has a similar planform to CTOL, but the upper fuselage contours associated with its powered lift system have an important influence on vortex behavior at elevated AOAs. More significantly, the lower maximum speed and *Nz* requirements of the STOVL aircraft allowed the area of the HT to be reduced by nine percent relative to CTOL's. While the STOVL aircraft has a vectoring nozzle for powered lift operations, it is not used at high AOA.
- The CV planform is the most obviously different from the other two. The requirement for shipboard launch and recovery resulted in a wing that is 45 percent larger than that of the other two variants, and the addition of ailerons to enhance roll and flight path control at low speeds. Additionally, the CV's horizontal and vertical tails are much larger due to the aircraft's lower approach speed and greater inertia.

B. CONTROL SURFACE MONITORING AND ACTUATION

The F-35 is the first manned fighter to use electric power to drive its primary flight control surfaces. The TEFs, HTs, and rudders are driven by Electro-Hydrostatic Actuators (EHAs), each of which is comprised of an integral 270V motor, hydraulic pump, and reservoir [1]. Relative to legacy hydraulic systems, EHAs improve the maintainability, reliability, and survivability of the flight control system, but introduce challenges of their own. Since the direction of the pump rotation determines the direction of surface movement, the motor and pump must reverse direction with each control surface reversal. The motor is assisted in this reversal by regenerative braking during the deceleration cycle. At low airspeed and high AOA, the control surfaces are often operating at their rate limits, with frequent reversals, resulting in a significant amount of regenerative energy. Dissipating this energy presented a significant challenge in the design and the cooling of the EHAs and associated power switching equipment.

Control surface rates and deflections, the limits of which are shown in Table 1, are monitored continuously by the flight control system. This feedback is used not only for failure detection and accommodation, but also to improve high AOA flying qualities and Departure Resistance (DR), as described later.

C. FUEL SYSTEM AND STORES MANAGEMENT SYSTEM

The F-35's use of Nonlinear Dynamic Inversion (NDI) requires the aircraft's mass properties to be known with sufficient accuracy and redundancy to be used as primary inputs to the Control Laws (CLAW). This dependency places stringent requirements on the accuracy, reliability, and failure detection capabilities of the aircraft's fuel quantity measurement and stores management systems. The

TABLE 1. CONTROL SURFACE DEFLECTION LIMITS (IN DEGREES)

Surface	F-35A/B	F-35C
LEF	↑3 ↓40	↑3 ↓36
Flaps	↑30 ↓30	↑30 ↓30
HT	↑30 ↓25	↑24 ↓27
Rudder	±30	±30
Aileron		↑30 ↓30

payoff for this investment is the ability to optimize handling qualities over a wide range of loadings and fuel states, without pilot interaction. This capability is particularly valuable at high AOA, where maneuvering performance and command limits can be closely tailored to mass properties, obviating the need for overly conservative (worst case) limits as are typically invoked on legacy fighters.

D. AIR DATA SYSTEM AND TACTICAL NAVIGATION SYSTEM

The F-35 pneumatic Air Data System (ADS) uses two dog-leg probes and two flush-mounted ports that are located on the forebody below the chine, just aft of the radome (Fig. 3). These sensors are used to measure AOA, angle of sideslip, static pressure (P_s), and total pressure (P_{total}). At very low speeds or extreme AOA or sideslip angles, where the pneumatic measurements become unreliable, the system transitions to inertially derived AOA, sideslip, and P_{total}. To support the calculation of these inertial parameters, the velocity of the air mass (wind) is continuously computed whenever the aircraft is in the pneumatic air data envelope. When outside the pneumatic air data envelope, the stored wind information is

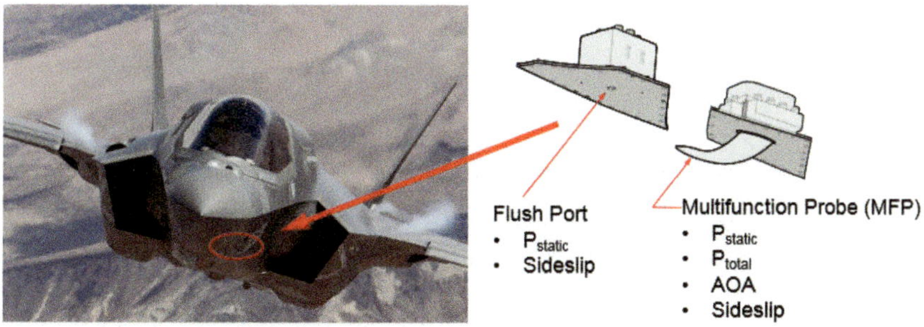

Fig. 3 Air data system flush port and multifunction probe.

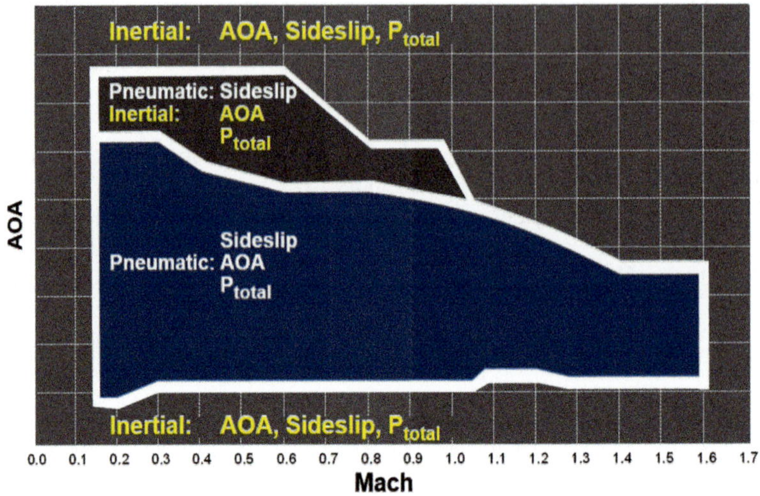

Fig. 4 Pneumatic and inertial air data envelopes.

used along with the aircraft's inertial velocity** to derive the required air data parameters.

The AOA thresholds for transitioning between pneumatic and inertial sources, illustrated in Fig. 4, evolved significantly during the high AOA test program. Although a flight test nose boom was available, it was not installed during high AOA testing because of its potential aerodynamic impact. At the beginning of each flight for which excursions outside the pneumatic envelope were anticipated, a "wind calibration" climb or descent through the test altitude band was performed. Wind data collected in this manner combined with aircraft inertial attitude and velocity provided an independent source of speed, AOA, and sideslip for real-time monitoring and post-flight analysis.

E. STABILITY AND CONTROL

The following section provides a brief overview of the aerodynamic characteristics that most impact the F-35's ability to maneuver at high AOA and to recover from OCF.

LONGITUDINAL STABILITY AND PITCH CONTROL POWER

The low-speed pitching moment characteristics of the STOVL variant at aft Center of Gravity (CG) are presented in Fig. 5. The STOVL variant was selected

**Aircraft attitude and inertial velocity are provided with very high reliability in the F-35 by the Tactical Navigation System (TNS). This system provides a redundant, six-degree-of-freedom (6-DOF) inertial solution, required to support hover in STOVL mode, and is an integral part of the flight control system in all variants of the F-35.

for this illustration because it is the most pitch-challenged of the three variants as a result of its smaller HT. The solid lines represent the pitching moments associated with maximum nose-up, maximum nose-down, and zero HT positions, while the dashed line shows the reduction in nose-down capability with the Weapon Bay Doors (WBDs) open.

All three variants are neutrally or negatively stable at low AOA, and strongly stable at high positive and negative AOA. The reduced size of the STOVL aircraft's horizontal tail makes it the most susceptible to deep stall (also known as a pitch hang-up) at AOAs beyond the CLAW limiter. The CTOL variant exhibits the potential for a weak deep stall only at extremely aft CGs achievable at very low fuel states. The CV variant, which has the largest HT of the three, does not exhibit deep stalls even in worst-case conditions. Inverted deep stalls are not a concern in the F-35 since all variants can generate positive pitching moment for all loadings across the full range of negative AOAs.

Each of the F-35's two weapon bays are equipped with two doors, a small door near the aircraft centerline and a larger outer door. The pitch effect of the inner doors is negligible, but the outer doors generate significant nose-up moment when open, shown by the dotted line in Fig. 5. Normally, the WBDs close automatically after internal weapons employment, so a deep stall due to their effect would occur only if they failed open while at high AOA with an aft CG. In keeping with the objective of ensuring recovery from the worst-case departures, recoveries were demonstrated for these conditions despite their remote probability.

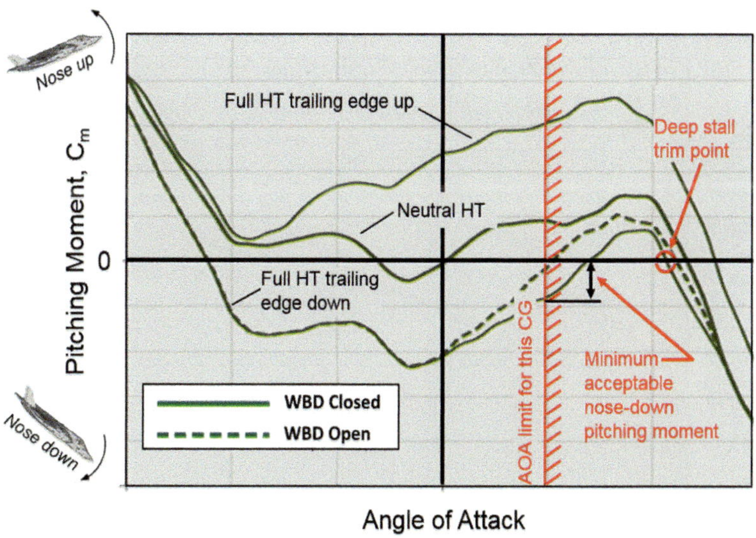

Fig. 5 Longitudinal stability.

LATERAL/DIRECTIONAL STABILITY

The lateral/directional stability characteristics of the F-35 are typical of twin-tailed, swept-wing fighters. Fig. 6 provides a representative illustration of the variation in static lateral/directional stability with AOA for a low-speed, aft-CG condition. Directional stability ($C_{n\beta}$) is slightly stable at low AOA but becomes unstable at higher AOA as the vertical tails are increasingly immersed in the wing/forebody wake. Dihedral effect is stable ($C_{l\beta}$ negative) at all AOAs, with a minimum at moderate AOA followed by a steady increase beyond that. Overall, the bare airframe provides some level of departure resistance (positive $C_{n\beta,\text{dynamic}}$) throughout the positive AOA range, with a notable reduction in the medium α range.

ROLL/YAW CONTROL POWER

For any fighter, high AOA agility is determined primarily by the amount of control authority available to generate roll and yaw moments. The effects of AOA on roll/yaw surface effectiveness in the F-35 are shown in Figs. 7 and 8. In each figure, the vertical scales are common for the three effectors to show their relative capability. The red, green, and blue lines represent increasing deflection of the control surfaces.

Plots of roll control (Fig. 7) show that the HTs and TEFs produce similar levels of rolling moment across the full AOA range. As AOA increases, the change in roll moment per degree of deflection reduces notably for the TEF and HT near the maximum trailing-edge-down deflection limits.

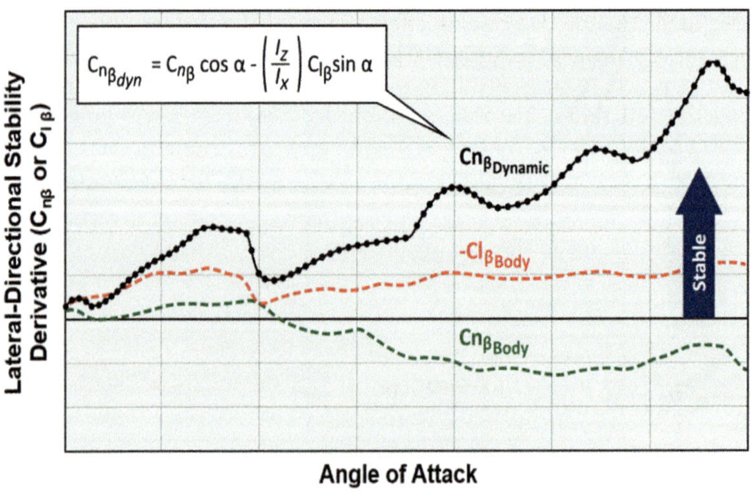

Fig. 6 Lateral/directional stability.

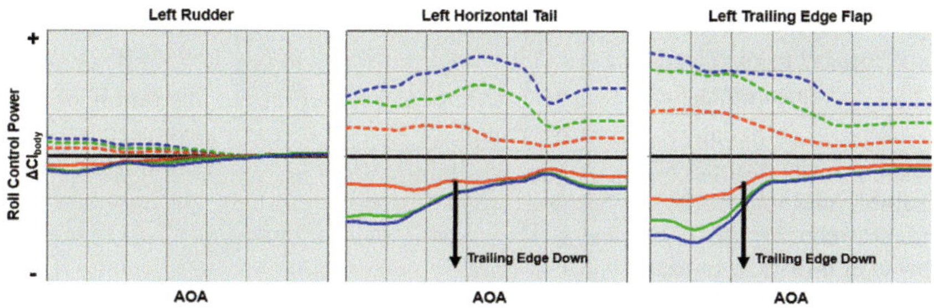

Fig. 7 Roll control power of the left rudder, HT, and flap.

In the yaw axis (Fig. 8), the most obvious control challenge is the near-complete loss of rudder effectiveness as AOA increases in the medium AOA range. By contrast, differential HT produces little yaw at low AOA but becomes the primary yaw effector at higher AOAs. Note that at high AOAs, the incremental increase in yaw control power of differential HT (i.e., the change in yaw effect with a change in HT deflection) decreases at large deflections. This has important consequences when the HTs are commanded to large deflections (symmetrically) to meet pitch axis demands, as their ability to generate yaw is greatly reduced.

To illustrate how limitations in rudder control power impact maneuverability at high AOA, consider a 1 g roll produced by 15 degrees of differential flap. Figure 9 compares the amounts of rudder and differential HT required to coordinate such a roll when each is used independently. As AOA increases the yawing moment required increases, with the rudder being the preferred surface since it produces relatively pure yaw with minimal opposing roll. With further increases in AOA, the decrease in effectiveness of the rudder ultimately result in it becoming saturated, requiring differential HT to coordinate the roll. Differential HT, however, generates rolling moment opposing that of the flaps, thereby reducing the roll rate achievable. Additionally, the amount of differential HT available for yaw may be limited by the amount of symmetric HT needed for

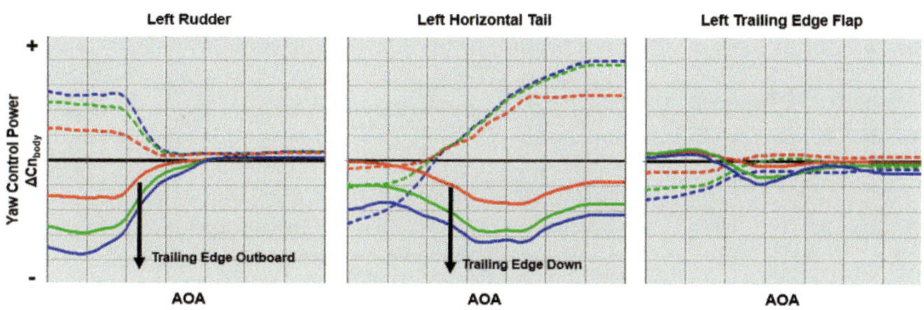

Fig. 8 Yaw control power of the left rudder, HT, and flap.

Fig. 9 Rudder vs. differential HT needed to coordinate a representative roll.

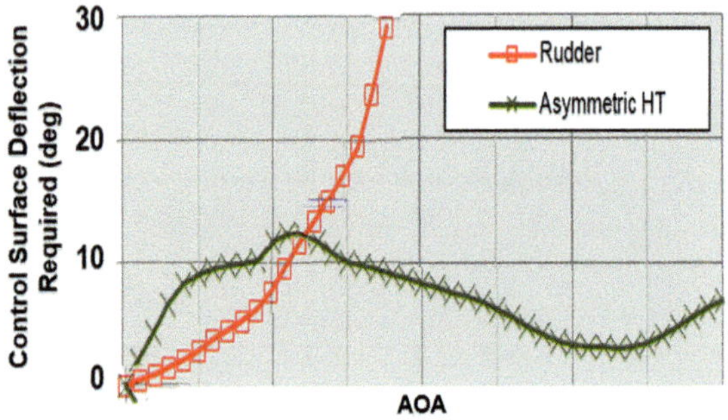

pitch, which has priority. Proper allocation of the HT between the pitch and yaw axes is critical to maximizing maneuverability at high AOA while avoiding departure.

IV. CONTROL LAW DESIGN

The F-35 is the first production fighter to incorporate NDI as the control allocation scheme for its Fly-By-Wire (FBW) control system [2, 3]. As distinguished from legacy FBW schemes, which use preprogrammed (scheduled) gains, the NDI approach allows the aircraft to determine a control solution on the fly, based on a detailed Onboard Model (OBM) of the aircraft's mass properties and stability and control characteristics.

In the NDI implementation (Fig. 10), the desired flying qualities (aircraft responses to stick and pedal inputs) are contained in the front end, the command and regulator modules. The outputs of these modules are the pitch, roll, and yaw accelerations needed to achieve those responses. The incremental effector commands needed to provide these accelerations are computed in the back end by the Effector Blender (EB) based on information received from the OBM.

To support this task, the OBM provides two types of information as a function of surface positions, configuration, mass properties, and flight condition: The aircraft's linear and angular accelerations, and the effectiveness of each control surface pair. Using this information, the EB determines the additional moments

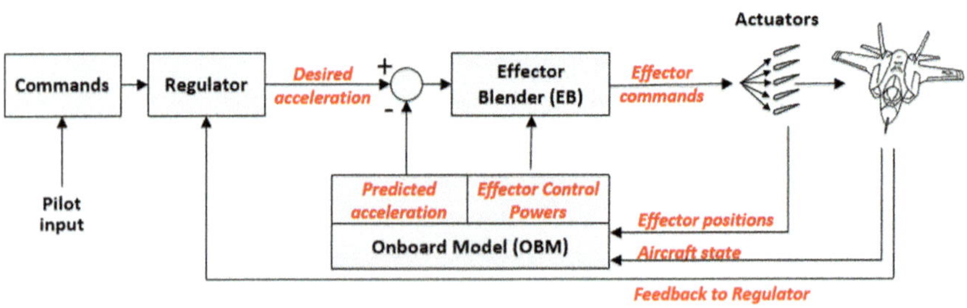

Fig. 10 Control law structure.

required to produce the desired changes in angular acceleration, and the changes in surface deflections needed to produce those moments. An error in the OBM (i.e., a difference between the aerodynamic model and the actual aircraft aerodynamics) for either the accelerations or the control effectiveness estimates will result in persistent errors in the control solution.

Of the many considerations that led to the selection of NDI for the F-35, a primary factor was NDI's ability to manage the complex propulsive and aerodynamic control allocation challenge in STOVL. The applicability of NDI to the high AOA flight regime, however, was not seen as an NDI strongpoint. Specific concerns included the ability to accurately model highly nonlinear and asymmetric stability trends, complex control interactions, and significantly reduced control surface effectiveness. Prior to initiating detailed design, therefore, it was necessary to evaluate the potential benefits and disadvantages of using NDI at high AOA. The conclusion of that analysis was that, although there were areas of concern, there were no issues that would preclude the use of NDI throughout the high AOA envelope or would require fundamental changes to the basic NDI approach. Indeed, the extensive information contained in the OBM had the potential to provide significant control benefits at high AOA.

A. HIGH AOA CLAW MODES

The Up-and-Away (UA) control laws of the F-35 at low-to-moderate AOA are typical of most FBW fighters: Longitudinal stick commands pitch rate at low speed and *Nz* at high speed; lateral stick commands roll rate around the flight path; and pedals command sideslip.

As AOA increases, however, the directional response to large pedal inputs blends from sideslip command to yaw rate command (Fig. 11). In this mode, the pilot has no direct control over the roll axis. CLAW controls roll rate and sideslip, exploiting the natural stability of the aircraft (via Cn_β and Cl_β) to drive roll and yaw. This strategy significantly enhances roll and yaw performance beyond that which could be achieved using control surfaces alone.

In the high AOA mode, the pilot commands yaw rate primarily with the pedals. Lateral stick input increases or decreases the

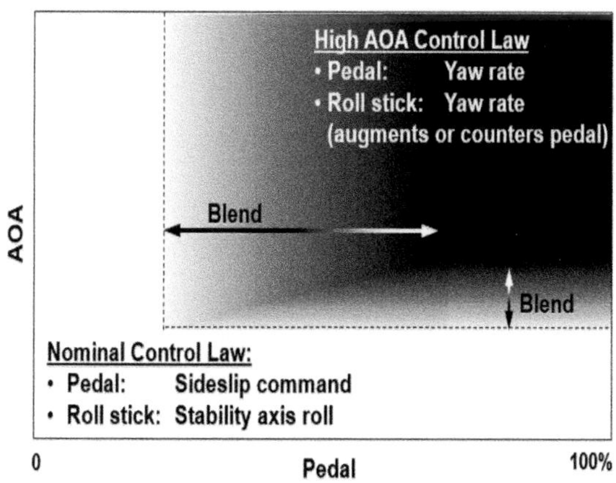

Fig. 11 Up-and-away control laws.

commanded rate, depending on whether the stick is displaced into or opposite the pedals. The intent of this approach is to allow the pedals to be used for gross maneuvering, with the stick used for fine tuning of that command.

Figure 12 and Fig. 13 compare bank-to-bank maneuvers using the two lateral-directional command modes: roll rate command using lateral stick only, and yaw rate command using stick and pedal.

- For the stick-only case (Fig. 12), the commanded parameter is stability axis roll rate, which exhibits a well-controlled first-order response. Yaw rate follows the roll rate, resulting in near-zero sideslip.
- When pedal is used (Fig. 13), the primary command parameter is body axis yaw rate, which is well-controlled with only minor overshoots. Rather than attempting to coordinate the roll, CLAW commands proverse sideslip, using the natural stability of the aircraft (Cn_β and Cl_β) to generate roll and yaw moments.

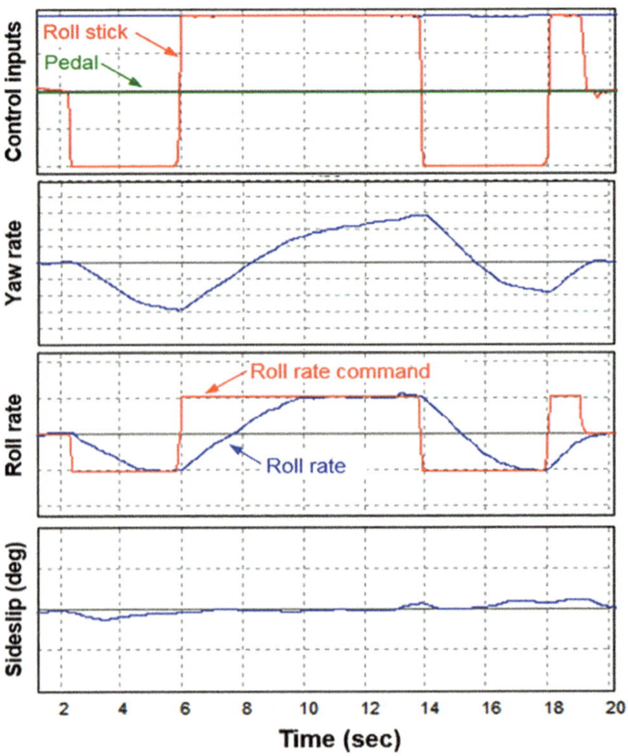

Fig. 12 Bank-to-bank roll using roll stick only (roll rate command).

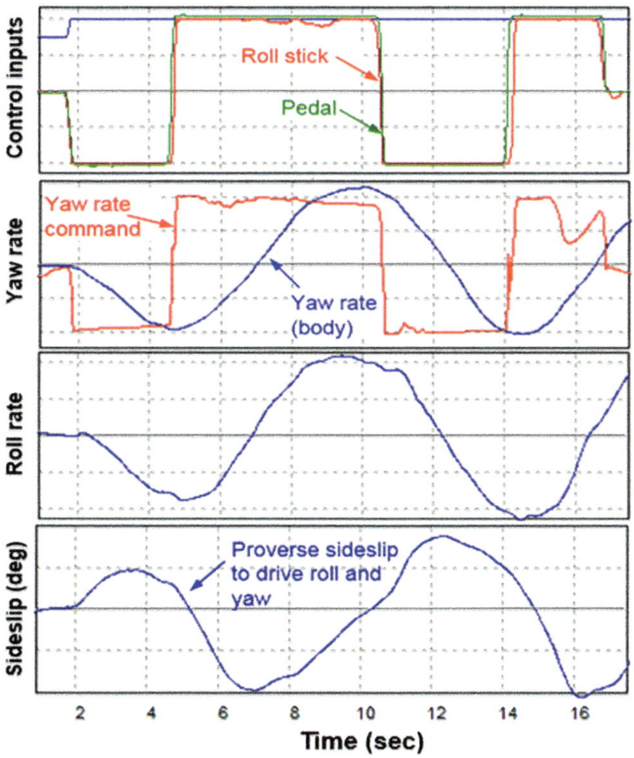

Fig. 13 Bank-to-bank roll using full pedal and roll stick (yaw rate command)

B. DEPARTURE PREVENTION

One of the most commonly held perceptions among pilots of highly augmented fighters is that CLAW features incorporated to prevent departures (most notably AOA and roll/yaw rate command limiters) are unnecessarily intrusive and prevent pilots from getting the maximum maneuverability from the airplane. Historically, there is some validity to this perception. Less sophisticated designs required more conservatism to provide the required levels of departure protection. The complexity of the original F-16 CLAW, for example, was constrained by limits in computing power and aircraft state information available to the CLAW, resulting in compromises and simplifications in the design of its DR features. The F-16 pitch axis limiter, for example, was based on a simple schedule of AOA and G. The consequence of that simplification was that as AOA increased, the available G decreased (with AOA still below the absolute AOA limit), leaving some amount of maneuvering capability unavailable. By contrast, the F-22 was able to exploit its greater computing power and aircraft state information to incorporate more complex and less intrusive DR features. The F-35 continued this trend, using its NDI control structure and significantly increased state information to support a DR logic that extends well beyond traditional command limiters. By continuously monitoring measured and projected aircraft states (Mach, dynamic pressure, angular rates, mass properties, effector positions, and OBM-predicted accelerations), the F-35 CLAW anticipate and respond to potential departure conditions in the least intrusive manner possible.

Brief descriptions of the F-35 AOA and command limiters are presented in the following subsections. Although the limiters are described individually, it is important to note that they do not act in isolation. Whereas legacy systems set limits in each axis based on the worst-case inputs in other axes, the F-35 takes a more adaptive approach, establishing baseline limits for each axis and then modifying them in real time based on the rates and commands in the other axes. Designing and tuning these features was one of the more challenging aspects of the high AOA CLAW development. The result is an aircraft that is extremely resistant to departure while wringing as much capability as possible out of its controls.

AOA LIMITERS

The AOA limiter in the F-35 represents a significant leap in complexity and performance relative to legacy aircraft. Inputs to the limiter computation include longitudinal CG, lateral asymmetry, Mach, external stores configuration, WBD position, control surface failures, and failures of the air data, fuel gauging, electrical, and tactical navigation systems. These various AOA limiter schedules run in parallel, with the most restrictive being applied.

At low speed, the primary focus of the AOA limiter design was the preservation of nose-down pitch acceleration capability (pitch margin). There is a rich history of research in this area, and a number of guidelines were referenced in selecting pitch margin criteria that would be acceptable across all variants, loadings, and fuel states [4–6]. The result of this study was the selection, as the low-speed criteria, of $-0.20\ \mathrm{rad/sec^2}$ in UA mode and $-0.15\ \mathrm{rad/sec^2}$ in Powered Approach (PA) mode. Based on these criteria, AOA limiter schedules were developed as a function of CG, WBD position and other factors affecting pitching moment. Secondary effects on pitching moment, such as inlet momentum, jet effects, and non-zero sideslip, were also accounted for to ensure that the aircraft was robust to these variations.

The F-35 fuel system helps to maximize the AOA capability by keeping the CG as far forward as possible. Figure 14 shows the CG movement as a function of fuel burn and internal store loading for the CTOL variant.

Regardless of the specific AOA limiter criteria in effect, the goal of the limiter mechanization was to allow the pilot to attain the limit as rapidly as possible, with acceptable overshoot, following a full-aft-stick pull. In some cases, extensive logic and gain scheduling were required to achieve this goal. One particular challenge

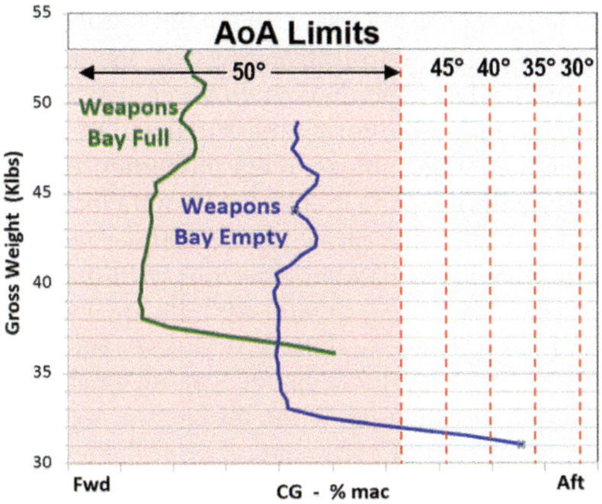

Fig. 14 Variation of the AOA limiter with fuel (CTOL clean-wing configuration).

was presented in the case of a maximum slowdown turn from high speed, where the rapidly changing relationship between *Nz* and AOA results in a switch from an *Nz* limit at high speed to an AOA limit at lower speed. This required the inclusion of airspeed and airspeed rate-of-change parameters in the limiter design structure to ensure that both the AOA and the *Nz* limits were attained.

ROLL/YAW LIMITERS

Command limits in the lateral and directional axes are scheduled as a function of AOA, dynamic pressure, Mach, stores loading, and CG, and then adjusted based on feedback of actual aircraft dynamics. The feedback-based adjustments are selective and temporary, reducing the limits in appropriate axes based on observed aircraft rate and acceleration parameters (α*-dot*, β*-dot*, *Pb-dot*, etc.). These adjustments are essential to maximizing maneuverability in that they preclude the need for overly restrictive, preemptive command limits. While important for any fighter design, adaptive limiters of this sort are especially important for an NDI-based high AOA CLAW, as they provide robustness to aerodynamic modeling errors that are common in this regime.

At low speed, where control authority is limited, the design of the roll/yaw command limits was guided by handling qualities considerations rather than DR concerns. The limits on roll rate, for example, were defined such that a roll initiated in one direction can be arrested and reversed without unacceptable bank angle overshoot (roll coast). In the absence of guidance defining acceptable levels of coast (MIL-STD-1797 is silent on this point [7]), command limits were based on piloted simulation and flight experience, including lessons learned from the F-16 and F-22.

At high speed, the lateral/directional control limits were determined by control surface loads limitations and by departure considerations related to gyroscopic (inertial) coupling, since the roll rates in this regime can be high. For illustration, a typical inertia-coupled departure might begin with an aggressive roll at low AOA combined with an abrupt pitch command. As AOA increases, the yaw controls may be unable to balance the inertial and aerodynamic moments generated in this maneuver, resulting in sideslip increase and potential departure. A simple way to prevent the coupled roll-pitch departures would be to limit roll rate based on the yaw control available and the maximum pitch rate achievable at that condition. While effective, this solution would be overly restrictive for cases in which the pilot has no intention of applying simultaneous roll and pitch inputs. By contrast, the F-35 limits roll rate based on the pilot's actual control inputs. With accurate knowledge of the yaw control available, and accurate predictions of the yaw control required as a function of mass properties and roll rate, CLAW will reduce the roll command only by the amount required to prevent a departure. If the aircraft is already rolling at high rate when a pitch input is made, CLAW will temporarily reduce the pitch rate command until the roll rate is reduced, at which point the pitch rate command is restored to full authority. The combination of command and feedback-based limits provides enhanced agility and predictability at high AOA, while maximizing maneuvering capability.

C. AUTO RECOVERY MODES

While the F-35 CLAW is extremely effective at preventing departures from the normal flight envelope, the aircraft is not completely carefree. CLAW will not, for example, prevent departures that result from sustained high-pitch-attitude climbs, where AOA may stay within the nominal range until airspeed drops below that required for aerodynamic control. The most extreme example of a departure of this sort is the entry into a tailslide, where the aircraft remains at low AOA right up to the point where its speed goes to zero. CLAW functions could be incorporated that intercede and maneuver the aircraft away from such conditions preemptively, as does the Eurofighter Typhoon's Automatic Low Speed Recovery (ALSR) system [8]. The F-35 design philosophy, however, was to give the pilot the greatest freedom to maneuver possible, as long as controllability cliffs – if they exist – are recognizable and avoidable. An enabling element of that philosophy was the incorporation of automatic recovery modes that recognize a departure from controlled flight and affect a quick recovery.

There are two auto-recovery modes in the F-35: an anti-spin mode to counter uncommanded yaw rate, and an Automatic Pitch Rocker (APR) to respond to locked-in deep stalls. Given the criticality of these modes, it was essential that they function in a robust and predictable manner and not be sensitive to modeling errors or sensor failures.

AUTOMATIC PITCH ROCKER MODE

APR is designed to recognize and recover from deep stalls without pilot input. The recovery strategy used by APR is derived from the Manual Pitch Override (MPO) technique developed for deep stall recovery in the F-16, in which the pilot takes direct control of the tails using MPO and manually pitch-rocks the aircraft out of the deep stall. Recovery from deep stall in the F-16 relies on the pilot to apply well-timed inputs, the success of which varies with the pilot's familiarity with the technique. Lessons learned from the F-16, combined with reliable aircraft state data in the F-35, allowed the development of a highly effective, robust, and autonomous F-35 APR.

APR declares a deep stall when AOA is stabilized above CLAW limits and pitch rate has dropped below a threshold value, indicating that the nose is hung up. APR then pitch-rocks the aircraft, using pitch rate feedback, until AOA is returned to within the basic CLAW envelope.

The initial APR design was extremely effective, but refinements were made to the entry criteria and to the mode itself, several of which are discussed in the Flight Test Results section. The result of this development was an APR system that consistently and autonomously recovers the aircraft from the worst-case deep stall conditions, including the most aft CG combined with the WBDs failed open.

ANTI-SPIN MODES

The anti-spin functions include a spin recovery mode and a yaw rate suppression mode. As their names indicate, the primary functions of these modes are (a) to arrest the large, self-propelling yaw rates associated with incipient or fully developed spins and (b) to suppress yaw rate during high AOA conditions where sustained spins are possible. These two functions, described below, work in a complementary fashion in conjunction with APR to recover automatically from any type of departure, regardless of loading or fuel state.

- Spin recovery logic is engaged if body-axis yaw rate exceeds a defined threshold while at high (positive or negative) AOAs where sustained spins are possible. The yaw rate threshold is a function of speed and is beyond that which would be expected during the most aggressive maneuvers. When spin recovery logic is activated, the relationship (gain) between the HT and yaw rate is scheduled based on lateral CG and uses yaw rate and yaw acceleration feedbacks to ensure a prompt, smooth recovery for all loadings.
- Yaw rate suppression logic is invoked when the positive AOA limiter has been exceeded by a specified amount. The AOA schedule is designed so as not to cause engagement in response to minor AOA limiter overshoots. Yaw rate suppression is not invoked at negative AOA, as the aircraft has no inverted deep stall and the rudders and vertical tails are effective at negative AOA.

A key element of the anti-spin CLAW is a change in the prioritization of the HTs. During normal flight, CLAW prioritizes pitch control (versus yaw control) in its use of the HT. This priority ensures that AOA or *Nz* is maintained as precisely as possible, even at the expense of lateral/directional control. In a high AOA departure, however, minimizing yaw rate is essential since even a low sustained yaw rate can significantly delay—or completely prevent—a successful recovery. Accordingly, when a spin mode is activated, HT priority is shifted to the roll/yaw axes and pitch control is sacrificed (if necessary), even if APR is active.

Once yaw rate is reduced to a low value, the magnitude of differential HT needed to keep it low is fairly small, leaving a substantial proportion of symmetric HT available for APR or a direct nose-down recovery.

In addition to axis prioritization, another key to the effectiveness of the anti-spin recovery mode is an advanced system of adaptive filters applied to body-axis yaw rate. The original F-16 design was overly aggressive in its attempts to control yaw rate in a deep stall, resulting in ineffective pitch rocking and, in one case, an extremely delayed recovery. This event led to a filter design that reacts to steady-state yaw rate but avoids over-responding to high-frequency yaw rate oscillations. The F-35 variable filter design is a direct result of these lessons learned.

AUTO RECOVERY SWITCH

To minimize nuisance engagement, the entry criteria for the auto recovery modes require a relatively high certainty that a sustained departure has occurred. The result of this conservatism is that there are cases where the pilot might recognize the departure before the automatic system engages. An AUTO RECOVERY switch (Fig. 15) allows the pilot to expedite auto recovery engagement by relaxing or eliminating some of the entry criteria (AOA, yaw rate, pitch rate, dynamic pressure, etc.). A beeping aural tone indicates to the pilot that activating the switch will engage one of the recovery modes.

MANUAL PITCH ROCKER MODE

A Manual Pitch Rocker (MPR) mode is incorporated into the F-35 CLAW, activated by pressing and holding a button on the throttle. This mode gives the pilot direct control of the HT in a manner similar to that of the F-16 MPO mode. MPR was intended primarily as a backup to APR for deep stall recovery

Fig. 15 AUTO RECOVERY switch.

during flight test. It was used extensively during initial pitching moment verification, as it allowed the pilot to set the HT manually at various positions when outside the normal CLAW AOA envelope.

D. NONLINEAR DYNAMIC INVERSION IMPLEMENTATION AT HIGH AOA

LATERAL CG OFFSETS

One of the first issues encountered during high AOA analysis related to the way NDI handles lateral CG offsets.

In conventional aircraft, a lateral CG offset results in a rolling moment that the pilot naturally counters with a roll stick or trim input. The roll input creates a non-zero sideslip that the pilot either ignores or removes with yaw trim.

By comparison, the F-35's NDI CLAW is aware of the mass asymmetry and automatically adjusts control surfaces to zero the sideslip and roll rate. This auto-trim capability works well as long as control power is abundant. However, at low speed and high AOA, the reduced control surface effectiveness can result in very large differential control deflections. While an acceptable static trim may be achieved, this control solution results in highly asymmetric roll performance, with high rates being achievable in one direction and very low rates in the other.

Pilots of conventionally controlled swept-wing fighters learn that the solution to the roll control problem at high AOA is to use pedal (versus lateral stick) to roll the aircraft; the pedal input creates proverse sideslip, and dihedral effect (Cl_{β}) produces the roll. By contrast, a baseline NDI design counters the roll-off using rolling surfaces only, while seeking to maintain zero sideslip, resulting in large asymmetries in trimmed surface deflection and roll response.

In the F-35, the remedy to the control asymmetry involved incorporating features that command non-zero sideslip to offload the roll control requirement when trimmed. The result is an augmentation of the classic NDI control strategy that provides symmetric roll performance for very large lateral CG offsets across the high AOA envelope.

ONBOARD MODEL DEVELOPMENT

Providing an accurate aerodynamic OBM, which is fundamental to the success of NDI-based CLAW, is particularly challenging at high AOA. The aircraft's aerodynamics at high AOA are dominated by a complex system of vortices and transitions between attached and separated flow that result in nonlinear control effectiveness and strong control interactions. Extensive wind tunnel data were collected in this regime to develop a truth model and the OBM. These data were also used to establish control surface deflection limits and optimum control surface trim positions, with the objective of eliminating aerodynamic deadband and ensuring the effective use of the limited control power.

One particularly challenging modeling issue involved asymmetric flow separation resulting in uncommanded roll. This behavior, observed early in the flight test program at moderate AOAs and transonic speeds (referred to as transonic roll-off, or TRO), was also observed in the high AOA regime. The conditions under which the roll-offs occurred followed a trend of increasing AOA with decreasing Mach. The precise trigger point, however, varied considerably as a function of surface positions, aircraft rates, AOA, Mach rate, and other parameters. Since precise modeling of these aerodynamic anomalies was not practical, logic was added to augment the predicted OBM moments using aircraft feedback. This implementation, discussed in Ref. [2], was similar to that developed to handle TRO and was used in other areas of the flight envelope where modeling errors were expected.

EFFECTOR BLENDER LIMITATIONS

A central element of the NDI control scheme is the EB, which uses high-fidelity OBM data to determine the optimal effector solution as a function of aircraft state. In the lateral/directional axes, the EB performs the function of a traditional control surface mixer and, at low AOA, nominally arrives at the same surface allocation as would a human designer. In the high AOA regime, however, cases arise (e.g. when control power is limited) in which the standard NDI approach may not result in the optimal effector solution. In these cases, the human designer would invoke a variable (adaptive) control surface strategy. Unfortunately, the basic NDI architecture restricts the designer's ability to do so directly, requiring modifications to the EB structure.

An example may help to clarify this point. Rudder effectiveness is extremely limited at high AOA, leading to rudder saturation during maneuvers such as full-command rolls and roll reversals. When the rudder saturates, the EB automatically overflows the unmet yaw requirement to the next best source of yaw: differential HT. If differential HT cannot satisfy the yaw demand (e.g., if the HTs are symmetrically deflected for pitch), the EB overflows the remaining yaw requirement to the next best effector: differential flap. Fundamentally, this solution is no different from what a human designer would implement with a traditional mixer. The human designer, however, can identify consequences of this solution that the nominal EB design will not recognize. For example, since the EB prioritizes yaw control over roll, it will do so using as much differential flap as needed, even if the flaps are weak yaw effectors, with the undesirable effect of producing an opposing moment in roll. Knowing this, the human designer might decide that, where rudder effectiveness is inadequate, the optimal solution would be to conditionally relax the priority on yaw. The designer could implement this by setting limits on the amounts of differential HT and differential flap used, allowing some amount of sideslip to occur. Incorporating functionality of this sort in a traditional mixer architecture is straightforward. In an OBM/EB structure, however, where the designer does not have direct control of the effector solution, incorporating such a function is a significant challenge. For the F-35, the resolution of this type of problem involves actively varying control surface and axis weightings using a combination of aircraft feedbacks. The logic evaluates the current state of the aircraft, including control surface usage, and determines whether full yaw axis priority will result in the best control surface utilization or a relaxed yaw priority would be preferred.

V. FLIGHT TEST PROGRAM

This section provides a general overview of the high AOA test program, including scope, schedule, and a description of prerequisite testing and the major test phases. This provides a backdrop for the next section, which discusses key test results.

A. OVERVIEW

The primary objective during the early stages of the System Development and Demonstration (SDD) program was to certify an initial operational envelope for the fleet. The AOA envelope for that initial capability, $-10^\circ\alpha$ to $+20^\circ\alpha$, became the *de facto* definition of "low AOA" for the F-35 program, and testing outside that range was deferred until more than four years after the beginning of SDD flight test. High AOA testing commenced in late 2012 on CTOL, with the other two variants starting approximately one year later.

The high AOA test program followed a phased approach that was similar for all three variants. Figure 16 presents a high-level description and list of the objectives of each phase, along with a schedule and test-point breakdown. Approximately 320 flights and 3,500 test points were flown, the majority of which were dedicated to DR testing.

The rest of this section presents a more detailed description of the scope and objectives of each phase, and how each fit in the overall high AOA clearance process. A discussion of flight test results is presented in Section VI.

B. PREPARATION

Several aircraft modifications and system checks were performed as prerequisites to high AOA testing. In addition to the risks directly associated with high AOA flight, the potential for dwelling at zero or negative g during tail slides and other nose-high departures raised concerns related to the engine oil system, the fuel system, the electrical power system, and the Power and Thermal Management System (PTMS).

- To mitigate the risk of engine flameout at high AOA, engine airstart testing was completed in the CTOL and STOVL variants prior to starting high AOA testing. (The CV variant was cleared by its similarity to CTOL.)
- Waivers were received from the engine manufacturer for a limited number of excursions beyond the normal negative G (low oil pressure) time limit.

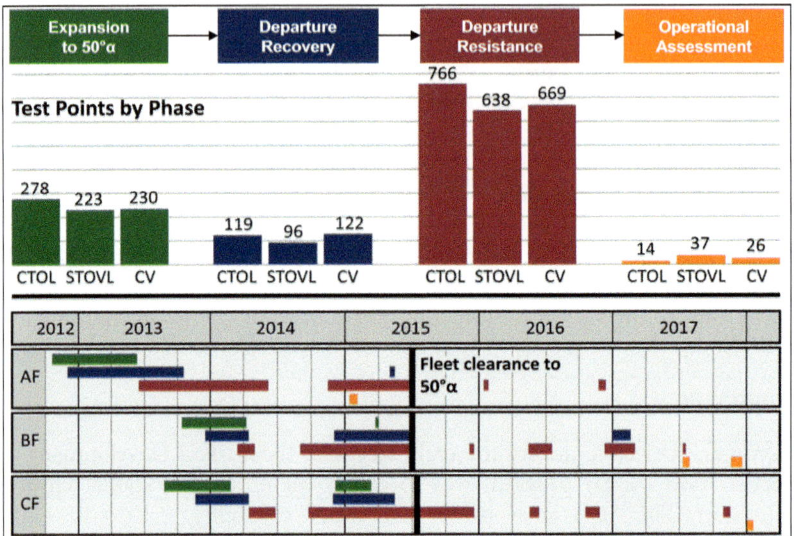

Phase	Description	Objectives
Preparation	• Installation and checkout of flight test modifications: – Spin recovery chute – Fuel system modifications for extended inverted flight – Supplemental cabin pressurization (CTOL only) • Prerequisite flights, including air-start and SRC effects	• Verify air-start capability • Functional check of flight test modifications • Check aerodynamic effects of SRC installation at low AOA
Envelope Expansion	• Low-speed 1g integrated test blocks • Expansion from 20°α to the CLAW limit (50°α)	• Engine and subsystem functional checks • Verify aero model within CLAW limits • Validate air data accuracy
Departure Recovery	• Increasing climb angles until AOA limit is exceeded • Stable trim points beyond CLAW limit using MPR • Anti-spin demonstration • Tailslides with static and dynamic entry	• Verify aero model beyond CLAW limits • Demonstrate manual and automatic recovery capabilities • Justify removing the spin recovery chute for subsequent testing
Departure Resistance	• Worst-case combinations of pilot inputs over all flight conditions, loadings, and CGs	• Demonstrate departure resistance • Update OBM and CLAW
Operational Assessment	• Mission-related tasks involving flight at high AOA	• Fleet recommendations and flight manual guidance

Fig. 16 High AOA test phases.

- The fuel system in the CTOL and CV variants was modified to allow greater time at negative g, and fuel-boost-pump pressures were monitored closely during all zero and negative g operations.
- Hardware-in-the-loop (laboratory) testing was performed to verify that the electrical system could withstand the increased demand of sustained periods of large amplitude surface motion during worst case departures.
- A Flight Test Aid (FTA) was used to extend the amount of negative g time available on the integrated power package (the central component of the PTMS) before it would automatically perform a self-protective shutdown.

- A supplemental cabin pressurization system, installed in CTOL for air-starts, was retained for the initial high AOA testing to mitigate the risk of injury to the pilot in the event of a flameout at high altitude. This system was not incorporated into the F-35B and F-35C variants, but cabin leak checks were performed in those aircraft to ensure that adequate cockpit pressure would be retained in the event of a flameout at high altitude.

Provisions were made for the installation of a Spin Recovery Chute (SRC) on one aircraft of each variant. The SRC installation consisted of a chute canister mounted to a four-legged support (shown in white in Fig. 17), a cockpit control panel, and four brackets to carry the chute loads into the aircraft structure (shown in orange in Fig. 18). Although this design allowed for the rapid removal/installation of the support structure and canister, the same cannot be said for the brackets, which required several days to remove or install. This reconfiguration time provided a strong incentive to group the SRC-required testing into blocks in order to minimize the number of on/off cycles.

After careful consideration, the decision was made to forgo an in-air checkout of the chute. An end-to-end check of the SRC system was conducted during a 65-knot taxi on the CTOL aircraft (Fig. 19). This test went smoothly and was the only time the chute was ever deployed on the F-35 program. The final check before testing above 20° α was to investigate the effects of the SRC installation on flying qualities at low AOAs. A one-flight checkout was flown with each variant, which included standard maneuver blocks (doublets, sideslips, rolls, and windup turns) at a range of conditions from 0.95 Mach down to 1 g/20° α in UA mode, and across the full PA envelope. Minor aerodynamic effects due to the SRC installation were noted, but they were consistent with wind tunnel predictions and had only minor effects on aircraft handling.

Fig. 17 SRC canister.

Pilot-selectable FTAs, many of which were

Fig. 18 SRC installation.

developed specifically for high AOA testing, were used extensively and contributed immensely to the efficiency of these tests. The FTAs most commonly used during high AOA testing were those that

- modified the fuel burn sequence (fuel ballasting) to set a desired longitudinal CG;
- set reduced *Nz* or AOA limits;
- enabled hands-on-stick-and-throttle control of the WBDs;
- disabled the DANG COND tone for CLAW AOA limit exceedance (to facilitate communication),
- disabled automatic APR engagement (to allow MPR testing);
- disabled anti-spin logic and allowed the pilot to command specified spin rates using pedal inputs; and
- biased the CG reported to CLAW to allow maximum AOA to be achieved over a wider fuel range.

Fig. 19 SRC ground deployment test.

C. EXPANSION TO 50-DEGREE AOA

The objective of this phase was to validate the aerodynamic model to the limits of the CLAW envelope. Since departure recovery capability had not yet been established, all testing in this phase was done at the most conservative CG (forward) with the SRC installed. The expansion was performed in increments of $5°\alpha$, building up from $20°\alpha$ to the CLAW limit of $50°\alpha$. The test cards for a typical envelope expansion integrated test block are shown in Fig. 20.

The first maneuvers conducted at each new AOA were a trim, a doublet series and a pushover, illustrated on the left card in Fig. 20. These were performed at an AOA five degrees beyond that to be used for the lateral directional tests, illustrated on the right card in Fig. 20. The pushover was performed to ensure that adequate nose-down pitch acceleration was available prior to proceeding to the more challenging lateral/directional maneuvers. At each AOA, a stringent set of flight test continuation criteria was applied before proceeding to the next increment in AOA.

Maneuvers in this phase were limited to modest bank angles and did not involve simultaneous pitch and yaw/roll inputs. The AOA limit FTA was used extensively to assist the pilot in maintaining constant AOA (using full aft stick) and minimizing cross-axis contamination. Testing was initially done with the clean aircraft, followed by various air-to-air external stores and WBD configurations.

13.0 *1g ITB at 45° AoA*			2CF:0005	Card 23
ALT	AoA	CG	ETR	GEAR
$35000_{\pm 5k}$	$45_{\pm 1.0}$	$26_{\pm 0.5\%}$	≤MIL	UP

RUN #	TEST DESCRIPTION	LIMITS
13.1	FTA sw – ON Set AoA limiter to 45° Set Burn Curve __ ☐ SRC – ON / ARM ☐ Trim Reset	450 / 1.0M
13.2	Climb a. Setup: 28-30k/ 0.9M b. Pullup: MAX / 35°θ, then MIL c. As γ reduces: Full AFT stick	
13.3	TRIM / Stabilize	
13.4	Yaw / Roll / Pitch Doublets	
13.5	Pushover to 0° AOA	
13.6	(CR) Call "Recover" FTA sw – OFF	

14.0 *1g ITB at 40° AoA*			2CF:0005	Card 24
ALT	AoA	CG	ETR	GEAR
$35000_{\pm 5k}$	$40_{\pm 1.0}$	$26_{\pm 0.5\%}$	≤MIL	UP

RUN #	TEST DESCRIPTION	LIMITS
14.1	FTA sw – ON Set AoA limiter to 40° Set Burn Curve __ ☐ SRC – ON / ARM ☐ Trim Reset	450 / 1.0M
14.2	Climb a. Setup: 28-30k/ 0.9M b. Pullup: MAX / 35°θ, then MIL c. As γ reduces: Full AFT stick	
14.3	BANK-TO-BANK (0° to 30° to 30°) w/Check	
14.4	CROSS (Release on CR Call or 10s) a. Left stick / Right pedal b. Right stick / Left pedal	
14.5	RUDDER STOMP (Release on CR Call or 10s) a. Left b. Right	
14.6	COORDINATED (Release on 60° ΔHDG or CR Call) a. Left stick / Left pedal b. Right stick / Right pedal	
14.7	Yaw Trim Survey	
14.8	(CR) Call "Recover" FTA sw – OFF *Flameout landing capability required, 15kts cross wind and 10 tail wind max at designated RNY	

Fig. 20 Test card showing envelope expansion integrated test block maneuvers and buildup.

D. DEPARTURE RECOVERY

The goal of this phase was to demonstrate that the F-35 recovers from OCF without pilot input in all air-to-air configurations. The departure recovery phase consisted of two distinct sub-phases: (1) pitch-axis-only departures executed using a controlled, wings-level buildup in AOA beyond the $50°\alpha$ CLAW limit; and (2) dynamic departures using multi-axis inputs to fully check the auto-recovery capability of the aircraft. The primary objective of these demonstrations was to gain confidence that we could remove the SRC for the next phase, DR testing, so the DR tests could be performed in a production-representative configuration.

The departure recovery demonstrations began with a series of climbs using progressively higher pitch attitudes to produce low-speed, ballistic recoveries that resulted in momentary exceedances of the CLAW AOA limit. These initial tests were conducted at forward CGs, where recovery was guaranteed, in order to assess any tendencies to yaw or roll at extreme AOAs and to provide a check of air data and engine behavior. Next, the MPR feature was used to command various symmetric HT positions to establish stabilized trim points beyond CLAW limits. These trims were used to verify pitching moment predictions, to investigate deep stall susceptibility, and (on recovery) to verify nose-down control power. The initial excursions beyond the CLAW limit were conducted in the clean configuration at nominal CG. Subsequent tests progressed to more aft CGs and more critical configurations, including WBDs open. (The WBDs normally close automatically after stores release, but the outer doors produce a significant nose-up pitching moment, so it was important to test this failure mode.)

Once the pitching moment predictions had been verified, a comprehensive evaluation of APR recovery logic began. Having established adequate control power for recovery using MPR, the APR testing was done almost exclusively at the most aft CGs. Initial APR engagements were done in a controlled fashion by using MPR to establish a stable trim beyond the CLAW AOA limit, then releasing it to allow APR to engage. Since the F-35 does not have a negative-AOA deep stall, the majority of the MPR-induced APR checks were performed upright, with a handful initiated inverted to validate the negative-AOA logic.

Preflight analysis indicated that all variants of the F-35 were highly resistant to spins, an attribute credited both to aerodynamics and the anti-spin logic in CLAW. The initial concept, therefore, was to validate the spin recovery logic using simulation only. Based on past experience, however, the Navy insisted that a spin recovery be demonstrated in flight before proceeding with more aggressive departures. Since no means of creating a spin could be devised using the normal CLAW, the demonstration required the development of a dedicated pro-spin FTA. The FTA disabled the nominal anti-spin logic and allowed a high-rate, spin-like condition to be created using pedal commands, after which the FTA would be turned off, allowing the spin recovery mode to engage.

The graduation exercise for the departure recovery phase was the performance of dynamic departures. With the confidence gained from the more controlled testing described above, and the successful correlation of those results with analytic models, a buildup in configuration and CG was not required for this testing. Dynamic departure testing was therefore done almost exclusively at the most aft CG and in the worst case aerodynamic configuration, including WBDs open. Departures were generated by conducting high-pitch attitude climbs, including tail slides, applying stick and pedal inputs to generate as much rate in all three axes as possible at the point of departure. The results from the first round of maneuvers were used to devise even more diabolical combinations of pitch attitudes and control inputs in an attempt to create sustained departures. This no-holds-barred approach to demonstrating recoveries from any conceivable situation was deliberate, in order to build a convincing body of evidence to justify removing the SRC for DR testing. As a final verification of the recovery capability, a set of the most critical configurations and maneuvers was repeated with the SRC removed to ensure that the SRC did not have a notable effect on recovery.

E. DEPARTURE RESISTANCE

The majority of the high AOA test program was dedicated to DR testing, i.e., verifying that the aircraft was highly resistant to departures for the full range of configurations and pilot control inputs. Having fully verified the recovery capability of the aircraft in the previous phase, almost all DR testing was done in a production-representative configuration with the SRC off.

Removing the SRC for the majority of this phase was somewhat controversial from a safety perspective but was crucial to the efficiency of the F-35 high AOA program. DR testing represented about two-thirds of the high AOA test program. Performing these tests in a production-representative configuration eliminated the potential for SRC effects to impact test results, thereby avoiding the need for costly SRC-off regression. Just as importantly, it allowed the data collected in this phase to be used to refine the aerodynamic OBM and associated truth model, without concerns about its validity. Had testing a production-representative configuration been deferred until the end of the program, the data would have been severely limited and would have been acquired too late to be integrated during SDD.

DR tests involved the application of aggressive, maximum, single- and multi-axis commands, reversals, and assaults on both the positive and negative AOA limiters. To illustrate the types of maneuvers performed, test card examples are shown in Fig. 21. Because some of the control inputs and timing were relatively complex, graphic illustrations of stick and pedal inputs reduced pilot error and added significantly to test efficiency.

A standard buildup was done in altitude (high to low), speed (low to high), and configuration (clean, then air-to-air loadings, then air-to-ground). The primary test altitudes for gear-up testing were 35,000 and 20,000 feet, although,

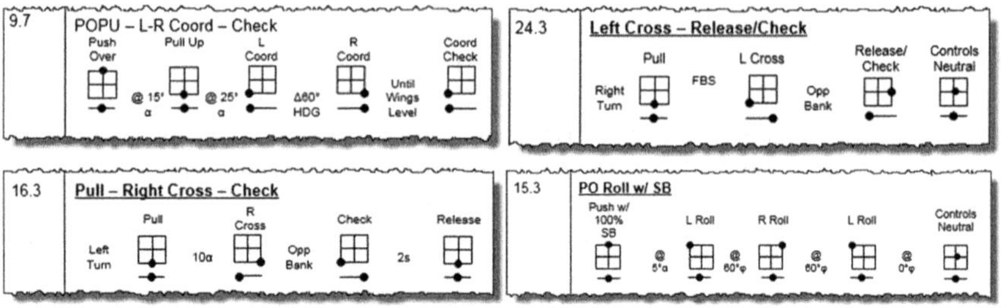

Fig. 21 Sample departure resistance test cards.

as discussed in the Flight Test Results section, some testing at lower altitudes was conducted when simulation or flight test results indicated it was necessary.

Longitudinal CG was held at target condition with an FTA, and lateral CG variations were established using internal and external stores. The majority of DR testing was performed with air-to-air loadings since CLAW limits are close to controllability boundaries for these configurations. Critical external air-to-ground loadings were also tested to their (more restrictive) limits. Since departure recovery had not been demonstrated (or required) for air-to-ground loadings, the SRC was installed for the initial DR testing of those configurations. Gear down DR testing was also performed for critical air-to-air and air-to-ground loadings.

Offline simulation was used as the primary means of verifying that each CLAW update performed as intended and was safe to operate over the range of conditions in the test plan. For each software release, roughly 30,000 simulation runs were conducted, among them a routine analysis set known as the Matrix of Death. Automated tools were used to scan the results to identify cases that, due to flying qualities and/or loads concerns, required further scrutiny. The simulation results were also used to identify potential trouble spots that required investigation in flight, and to design critical stressing maneuvers.

LOW-SPEED DEPARTURE RESISTANCE TESTING

The first objective of DR testing was to clear the low-speed envelope. While this testing covered the same flight conditions explored during the expansion to $50^\circ \alpha$, DR tests were done with aggravated, multi-axis inputs over a range of CG and AOA combinations. An FTA was used to change CG expeditiously during the flight, greatly enhancing test efficiency.

HIGH-SPEED DEPARTURE RESISTANCE TESTING

The approach to DR testing at higher speeds was similar to that used for low speeds (aggressive inputs, worst-case loadings, etc.), with a few unique considerations. First, since the effect of CG on the AOA and sideslip command limits blends out at high Mach, testing in this regime was done almost exclusively at a single, most aft CG. Second, since many of these points would be performed at high dynamic pressure, there was a potential for exceeding airframe load limits. Accordingly, while the initial testing at 35,000 feet (and relatively low dynamic pressure) was done with a non-loads-instrumented jet, the testing at 20,000 feet and below required close coordination with loads engineers and real-time monitoring with the instrumented aircraft.

Prior to testing, the AOA limits at high speed had been established based on inlet compatibility alone, although it was acknowledged that there were several issues that had the potential to require more restrictive limits:

- Unlike at low speed, high-speed DR would reach combinations of Mach and AOA that fell outside the envelope where wind tunnel data were acquired. This meant relying on OBM data that were based on extrapolated trends.
- Testing below $20^{\circ}\alpha$ had confirmed predictions that the F-35 was subject to abrupt flow field changes at transonic speeds and elevated AOAs, resulting in unsteady rolling moments on the aircraft. The expectation was that these effects would subside at higher AOA, but only testing would tell.

The approach was to start with the AOA limiter set by inlet compatibility requirements and reduce it only if necessary. It was fully understood that departures were possible—even likely—with this test approach. However, this was considered to be a fully managed risk based on the confidence gained in our recovery capability.

F. OPERATIONAL ASSESSMENT

The final phase of high AOA testing for each variant was a limited evaluation of aircraft performance and handling qualities while conducting operational tasks above $20^{\circ}\alpha$. Unlike the previous phases, which consisted of highly scripted, open-loop maneuvering, this testing allowed the pilot to operate the aircraft to achieve an operational task. Testing included an evaluation of high AOA task elements (tracking and reversals, ditches, minimum speed overhead maneuvering, nose-high recoveries, etc.), as well as engagements with an adversary aircraft starting from offensive, defensive, and neutral setups. The data gathered from this testing were qualitative in nature and used primarily to demonstrate specification compliance and to support describing the high AOA characteristics in the flight manual.

VI. FLIGHT TEST RESULTS

This section provides an overview of the key findings from the SDD high AOA test program. In general, the aircraft performed as expected and designed. However, since the positive results are less interesting than the anomalies, this section focuses primarily on the latter. Correcting these anomalies involved combinations of air data refinements, CLAW logic updates, and OBM corrections.

A. AIR DATA SYSTEM

Several issues with the ADS were discovered during the initial expansion to 50° α and during departure recovery testing. These findings resulted in improvements to the calibration of pneumatic measurements and the logic used to transition between pneumatic and inertial sources.

INERTIAL SIDESLIP IN CHANGING WIND

At AOAs outside the pneumatic envelope of the ADS, estimated winds and inertial velocities are used to derive AOA and sideslip, referred to as inertial α and β. This scheme works well as long as the aircraft remains in an air mass of relatively uniform velocity.

The example in Fig. 22 shows a deceleration to high AOA while in a descent. During this maneuver, the aircraft descends through a wind gradient, causing the

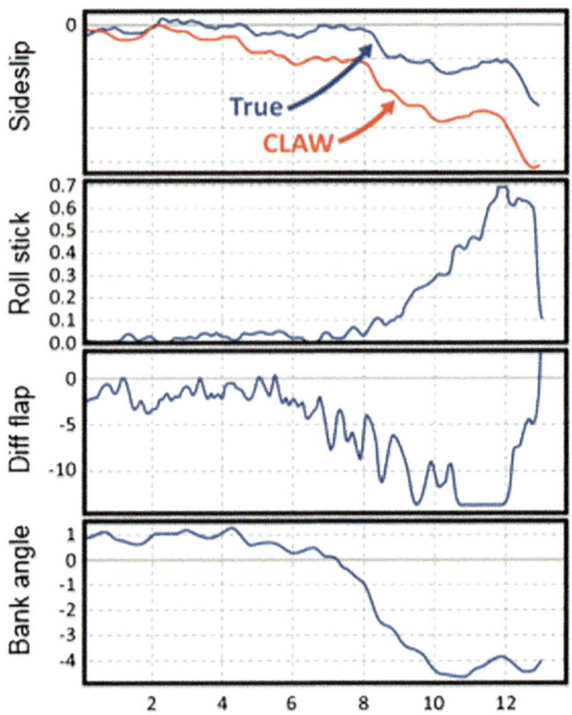

Fig. 22 Effect of inertial air data (sideslip) error during descent at high AOA.

sideslip being reported to CLAW to grow at about twice the rate of the actual value.

Starting at around six seconds, a slight uncommanded roll to the left begins. The CLAW respond with differential flap to create a right rolling moment, which should stop the roll and drive sideslip back to zero. However, the overestimated sideslip being reported to the OBM causes it to over-predict the amount of rolling moment it will get from dihedral effect ($C_{l\beta}$). The roll input by CLAW is therefore less than it needs to be, which the pilot corrects manually using right stick. The aircraft eventually stabilizes in a left bank with differential flap at its deflection limit. At this point, all the available roll control power from the flaps is being used just to hold the current bank angle, which precludes the aircraft from rolling right.

The solution to this issue involved both improving the air data accuracy and making the CLAW more robust to air data and aerodynamic modeling errors. The updates included:

- Raising the AOA for transitioning to inertial sideslip, as pneumatic sideslip was found to be accurate to higher AOAs than predicted.
- While the AOA transition point was not changed, robustness was improved by improving the calibration of pneumatic AOA and by using pneumatic AOA for wind computations even when the CLAW AOA had transitioned to inertial.

PNEUMATIC AOA ERROR AT LARGE NEGATIVE AOAS

Departure recovery testing identified the need for additional refinements to the air data algorithm. The most significant issue from this phase emerged during inverted MPR testing on CTOL. At large negative AOAs, separation off the fore-body chine created a low static pressure condition that impacted all primary air data outputs. The static pressure error resulted in a positive AOA reading that, since it was within the pneumatic envelope, caused the system to accept the erroneous pneumatic air data while the aircraft was still in an inverted departure. The aircraft was easily recovered using MPR, but post-flight analysis showed that this AOA sign error would adversely impact spin recovery. Departure recovery testing was suspended until logic was added to ensure that the aircraft's normal load factor (Nz) was positive before allowing the ADS to accept a positive pneumatic AOA.

B. ENGINE AND SUBSYSTEMS

A critical objective of the envelope expansion phase was to ensure that the engine and related emergency backup systems functioned properly at elevated AOAs and low speeds. During the entire high AOA test program, all critical subsystems (propulsion, fuel, electrical, environmental control, etc.) functioned flawlessly at all AOAs, at low speed, and during extended low-g conditions.

C. AERODYNAMIC MODEL VALIDATION

The most valuable flight data for high AOA aerodynamic model validation were acquired during the envelope expansion phase. In general, the aircraft response compared well with predictions for all maneuvers and configurations flown, including asymmetric air-to-air stores, gun pod installed, and WBDs open. Most importantly, there was very good agreement between flight test results and predictions for longitudinal stability and HT control power, as exemplified by the max push from a high AOA trim condition shown in Fig. 23. Although minor adjustments to the OBM were made as a result of data collected in this phase, none were consequential enough to require updates before proceeding to the DR testing. This result, which was pivotal in keeping the high AOA test

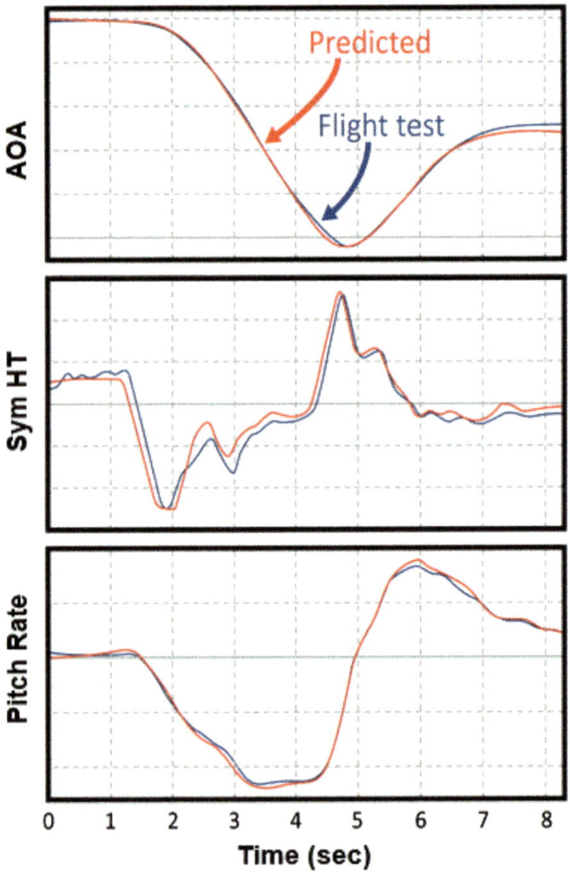

Fig. 23 Nose-down pitch maneuver from 50°α.

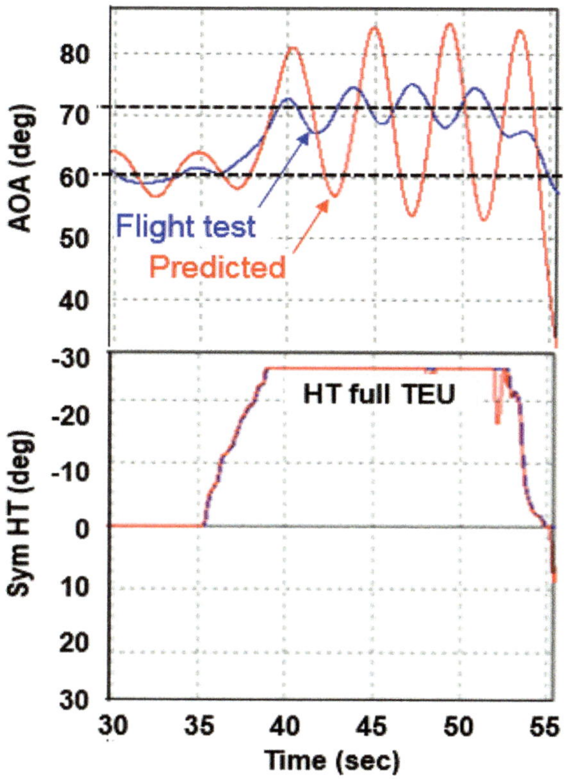

Fig. 24 MPR trim point showing lower-than-expected pitch damping.

program on schedule, was the payoff for the significant upfront investment made in collecting low speed wind tunnel data and assembling a high-fidelity aerodynamic model.

Figure 24 shows an example of how MPR was used to collect data for stabilized trims beyond the CLAW limit. The pilot initially let the aircraft stabilize near $60°\alpha$ with neutral HT (no stick input), then slowly pulled aft, causing the HT to go full trailing-edge-up. After the pull, AOA increased to around 72 degrees but exhibited a persistent oscillatory behavior. In this case, the predicted trim conditions were well-modeled for both HT settings, but the simulation predicted a much more dynamic response than that seen in flight. This damping mismatch was observed for most upright trim cases in all three variants. Based on these data, models were updated to include a sizable increase in pitch damping to match the more sedate behavior exhibited by the aircraft.

Similar testing was conducted for a range of configurations, both upright and inverted. Where necessary, models for basic pitch stability, symmetric tail control power, and pitch damping were adjusted. A small subset of the MPR test points were repeated with the SRC off. As expected based on wind tunnel testing, the impact of the SRC on trim AOAs was negligible.

Dedicated parameter identification testing was not included in the initial planning, but preparations (FTAs, analysis tools, etc.) were put in place so this capability could be used on an as-needed basis. Relatively late in the program, three dedicated high AOA parameter identification flights were performed with the STOVL and CV aircraft to verify control effectiveness in localized Mach and AOA regions. These data were used to adjust several terms in the aerodynamic truth models used for offline analysis, but the changes were not significant enough to warrant an update to the OBM, which would have required flight test regression.

D. DEPARTURE RECOVERY

Of the more than 220 dynamic departures performed, the vast majority self-recovered without engaging any auto-recovery mode. While no significant issues with the CLAW were encountered during this phase, test results were used to fine tune the recovery logic as described below. The end result was a system that, in all demonstrated and simulated cases, recovered without any pilot involvement whatsoever, clearly meeting the requirement to recover with "minimum pilot input." This result provided high confidence in the departure recovery logic and was critical to allowing DR testing to be conducted without the SRC installed.

PITCH ROCKER

Initial APR engagements were done in a controlled fashion by stabilizing beyond the CLAW AOA limit using MPR, then releasing the MPR switch to allow APR logic to engage. In total, 78 APR engagements were performed in this manner, all of which recovered the aircraft in one or two pitch-rock cycles and less than 4,000 feet of altitude loss. Of those that required two cycles to recover, most occurred with WBDs open (a failure mode of the doors).

During recoveries from dynamic departures (steep climbs and tailslides), APR engaged 45 times, all but one of which required one or two pitch-rock cycles to recover. The single event that required three cycles, while not operationally relevant since the WBDs were open, highlighted areas where further improvements could be made.

The three-cycle APR event, shown in Fig. 25 occurred with the CV variant following a tailslide at an extremely aft CG with all WBDs open. The maneuver begins with the aircraft in a vertical climb with a full right roll input applied at 120 knots. As speed drops to near zero, AOA peaks at 90 degrees and the HTs go to nearly full trailing-edge-down deflection, backing off as needed for yaw rate suppression. After five seconds of low pitch rate, APR engages automatically, but the initial nose-up kick is only marginally effective due to the extremely low speed (less than 40 KCAS). From 22 seconds until the recovery, the amplitude of

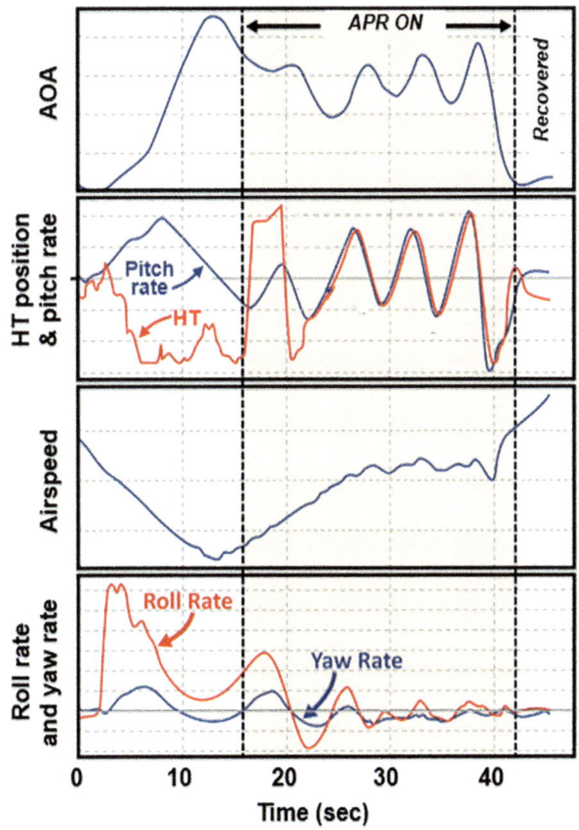

Fig. 25 Three cycle APR recovery.

the pitch rocking gradually increases until the deep stall is broken. Note that while APR was engaged, the roll and yaw rates subsided to near zero as a result of the spin suppression logic and increasing airspeed.

In this event, the CLAW functioned as designed and recovered the aircraft without pilot intervention. The delayed recovery, however, led to improvements to the APR entry criteria and function. Since the improvements were minor, inflight regression was not warranted. Offline analysis, however, indicates that all deep stalls will recover with a single pitch-rock cycle with WBDs closed, and no more than two cycles with WBDs open.

SPIN RECOVERY

There were 25 engagements of the spin recovery logic during dynamic departure testing, all of which occurred during post-departure gyrations at low dynamic

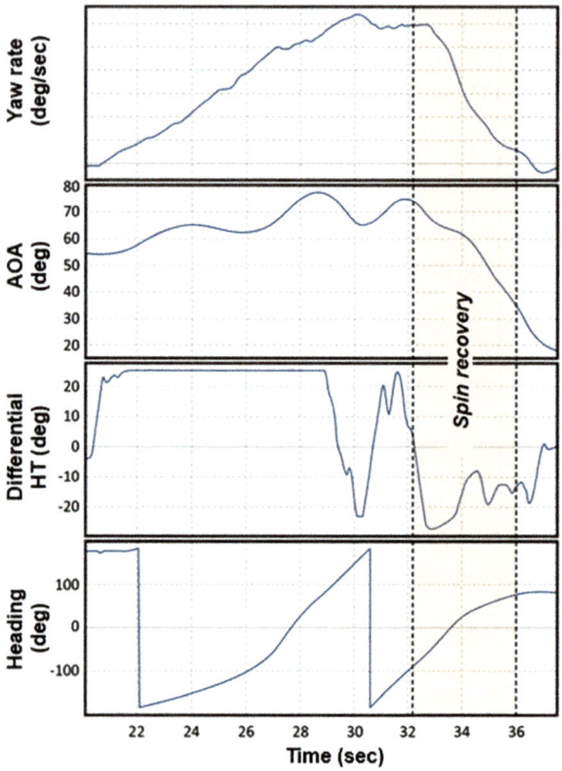

Fig. 26 Spin recovery demonstration.

pressure. The yaw rate suppression logic was effective, and the aircraft exhibited no propensity to spin, regardless of its dynamic state at departure entry.

The only sustained spins encountered were those that were staged using the spin mode FTA described earlier, a typical example of which is shown in Fig. 26. MPR was engaged at the start of the maneuver, allowing the pilot to establish the aircraft beyond the CLAW AOA limit. With the spin FTA engaged, the pilot applied full right pedal, and yaw rate built up to the high target value. (Note in the figure that AOA begins at 55 degrees but increases as the yaw rate grows and the spin flattens out.) The pilot then disengaged MPR and the FTA simultaneously, allowing immediate engagement of the CLAW spin recovery logic, which drove the differential tail to arrest the spin. As the yaw rate subsided, differential HT backed off to around half of its initial value, freeing up symmetric HT for pitch control. As the yaw rate reduced further, the associated decrease in nose-up coupling resulted in a steady drop in AOA, with the recovery completed around four seconds after spin recovery mode had been enabled.

TAILSLIDE EVENT

An interesting occurrence during departure recovery testing was the discovery of the perfect tailslide. Whereas the recoveries from the first 71 tailslides were relatively benign, the 72nd – which was performed with a loading and entry conditions not significantly different from its predecessors – provided a valuable lesson regarding the statistical nature of departure recovery testing. Figure 27 shows the pitch and yaw rates for all the tailslides flown, with event No. 72 highlighted in red. Unlike prior events, where the aircraft began pitching nose-down before achieving a significant negative velocity, airspeed in this case reached approximately 60 KCAS before the nose-down movement began. The higher aerodynamic moment resulting from this higher speed drove the initial nose-down pitch rate to 85 degrees per second, which was much higher than seen in any prior event, causing the aircraft to swing through the nose-down attitude back to near 90-degrees nose-high. With the aircraft now flying backward at more than 110 KCAS, an extremely dynamic recovery occurred, driving both pitch and yaw rates beyond the engine design limits.

Since sufficient tailslide data had been collected at this point to validate the recovery logic, tailslide entries were discontinued because of the risk of engine damage. No CLAW changes were made in response to this event, but a special aural tone was added to alert the pilot to an impending tailslide while airspeed was high enough to allow preventative action to be taken.

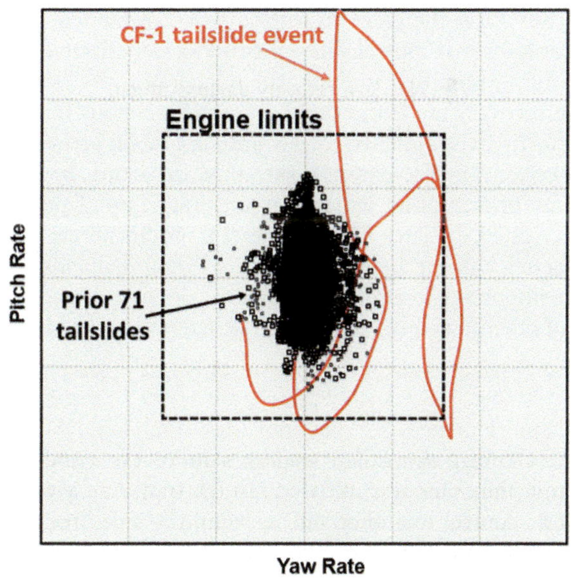

Fig. 27 Engine yaw and pitch rate limits exceedance in tailslide.

E. DEPARTURE RESISTANCE

This section provides examples of some of the more interesting findings from DR testing, along with a description of the actions taken to resolve them. Improvements in the CLAW logic and aero models derived from data acquired during this phase resulted in an aircraft that clearly met the requirement to be departure resistant and would more accurately be categorized as "extremely resistant to departure" based on the guidance in MIL-F-83691B [9].

CTOL DEPARTURE RESISTANCE

Since the CTOL aircraft was the first to start DR testing, much of the early discovery occurred during these tests, and several of the resultant CLAW updates were applied preemptively to the other two variants.

Low speed

The example in Fig. 28 illustrates a case in which competing pitch and yaw axis demands on limited HT control power led to a spin recovery engagement. This case occurred during one of the first bank-to-bank maneuvers done on the $50^\circ \alpha$ limiter using maximum (roll stick and pedal) yaw rate inputs. For this maneuver, the yaw rate tracked the command flawlessly on the initial bank to the right and had a small but controlled overshoot after reversing to the left. A similar overshoot occurred in response to the next reversal back to the right, but in this case the pilot released the pitch stick, creating an additional demand on the HTs to reduce AOA. The CLAW therefore prioritized nose-down pitch, leaving no differential HT available for yaw control. Since the rudder is ineffective at this AOA and the HTs were fully committed to pitch, the unchecked yaw rate was enough to exceed the spin recovery mode threshold. As noted previously, anti-spin engagement causes priority to be switched to yaw until the rate is arrested. Although the CLAW was successful in preventing a sustained departure, switching from full pitch to full yaw prioritization led to a momentary loss of pilot control while yaw rate was arrested. In later CLAW releases, modifications were introduced to achieve a better balance between pitch and yaw demands at high AOAs, where both are important to aircraft control.

Examples of other low-speed issues that were found and addressed are shown in Table 2.

High speed

Early in the CTOL high-speed DR testing, sideslip excursions were observed during a roll-and-pull maneuver initiated at high transonic Mach and high altitude. While basic control was never in question, the sideslip excursions were a concern due to the potential to exceed airframe load limits at lower altitude. Accordingly, an additional series of tests was conducted to focus on combined roll and pull maneuvers in this regime.

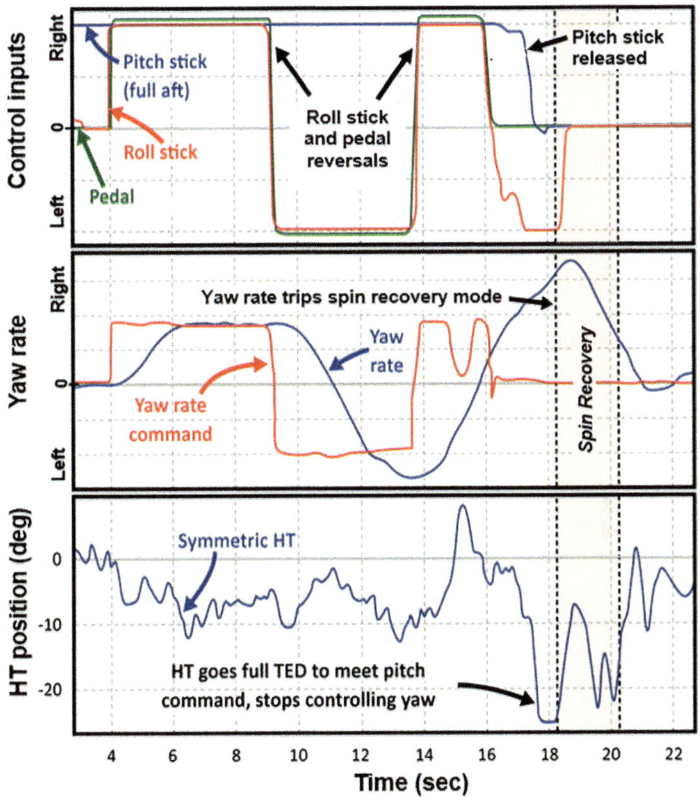

Fig. 28 Nose-down pitch command during bank-to-bank maneuver results in loss of yaw control.

More than 20 maneuvers were completed in the narrow Mach-AOA region of interest using a variety of pull-and-roll inputs. Although yaw control saturation and loose sideslip were observed, no major handling issues or departures were noted after completing all planned runs. However, with fuel remaining, the test team elected to repeat a few worst-case runs with minor variations in entry conditions and roll input timing. On the last of these bonus runs, the full-blown departure shown in Fig. 29 occurred. The maneuver was a full-aft-stick decelerating turn, with a max roll over the top initiated at $15^{\circ}\alpha$. The initial roll response was in the correct direction, but yaw control was inadequate to coordinate the roll, resulting in a buildup of five degrees of adverse sideslip. As AOA increased through 19 degrees, a large un-commanded left roll occurred due to a combination of adverse sideslip and sudden asymmetric shock movement on the wings. This uncoordinated left roll, along with the still-uncontrolled positive

TABLE 2 CTOL LOW-SPEED DEPARTURE ISSUES AND RESOLUTION.

Issue: HT ringing during assaults on the negative AOA limiter using a full-stick push followed by roll stick and opposite pedal. The large amplitude, high-rate HT motions were very lightly damped and resulted in sustained pitch axis oscillations.	**Issue:** After stabilizing in a maximum yaw rate maneuver, a max push caused a seven second saturation of the symmetric HT and a transient excursion above the AOA limiter. This issue was predicted by simulation and was confirmed in flight.	**Issue:** Excessive overshoots of yaw rate commanded trigger spin recovery during high yaw rate maneuvers. Yaw rate grew steadily to a value well past the commanded value, causing spin recovery logic to intervene and drive the yaw rate back to a controllable range before returning control to the pilot.
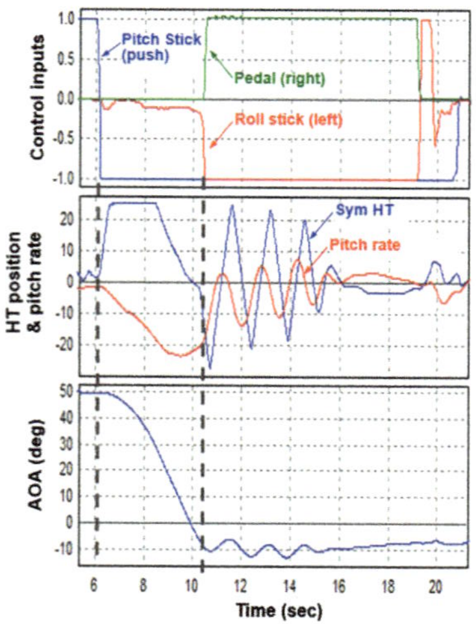	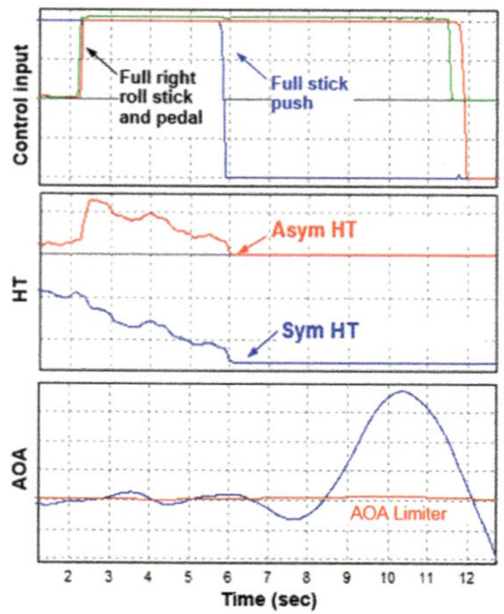	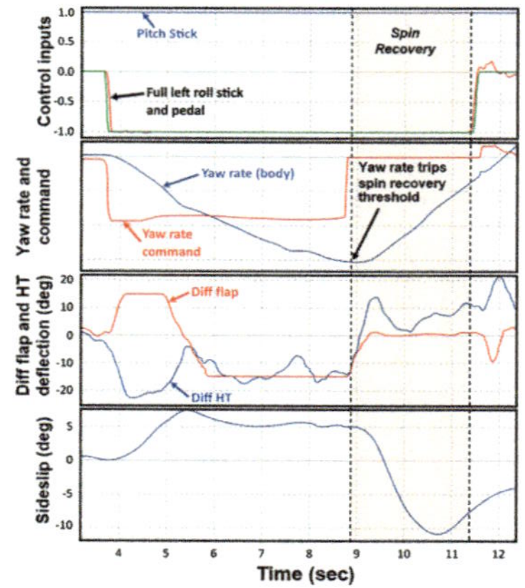
Solution (AOA limiter gain): The negative AOA limiter was softened by allowing it to begin intervening sooner, with lower gains. This fix was regressed successfully.	**Solution (CLAW and OBM):** Desired dynamics command models were adjusted when large errors between the requested and actual pitching moment occurred while maneuvering near the AOA limiter.	**Solution (CLAW and OBM):** CLAW logic and aero models were improved incrementally over several software releases to resolve issues such as this. Solutions were customized by variant and involved a fly-fix-fly approach to ensure that maneuverability limits were maximized while remaining within control power constraints.

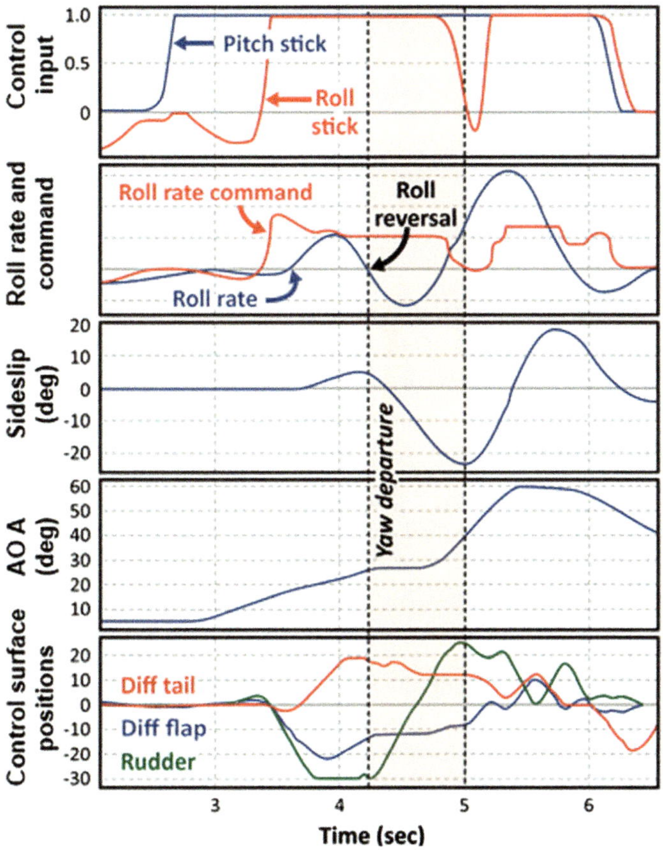

Fig. 29 CTOL high-speed departure.

yaw rate, caused sideslip to reach -24 degrees. The large roll moment due to sideslip led to a snap roll back to the right, which coupled into an AOA excursion to $60^{\circ}\alpha$ before the aircraft self-recovered.

This event led to incorporating several significant changes to the CLAW:

- Extensive modifications were made to the aerodynamic models (truth and OBM) in the transonic elevated-AOA regime. The changes included updates to roll and yaw stability, damping, and control power.
- DR logic was bolstered using aircraft states, surface positions, and pilot inputs to preemptively reduce command limits during aggressive maneuvering.
- The AOA limiter, which had previously been established by inlet compatibility considerations only, was reduced for high transonic and supersonic Mach numbers.

Once the updated software became available, the changes were successfully regressed and the remainder of the CTOL high-speed DR testing was completed without event.

STOVL DEPARTURE RESISTANCE

With the benefit of experience gained on the CTOL variant, the STOVL DR testing was largely uneventful. In fact, the similarity of the two planforms allowed the overall scope of the STOVL DR testing to be reduced. The most significant savings were realized by eliminating the buildup loadings for air-to-ground DR testing in both UA and PA modes, based on commonality assumptions and the success of the CTOL testing. Despite having a later start, the STOVL variant was cleared for full AOA capability concurrently with CTOL.

CV DEPARTURE RESISTANCE

The CV variant proved the most challenging from a DR perspective. Due to its distinct planform, lessons learned during DR testing of the other variants were generally not applicable, and several CV-unique problems were identified that required customized solutions.

Roll reversal

CV had the distinction of being the only variant to experience a departure prior to formal high AOA testing. The event, shown in Fig. 30, was flown relatively early in the test program for loads evaluation. The maneuver was a constant $20^{\circ}\alpha$ slowdown turn to the right, with a left roll commanded at 0.93 Mach via an auto-maneuver FTA. At roll initiation, the rudders and differential flaps saturated for nearly two seconds as adverse sideslip grew to 20 degrees, precipitating a rapid, un-commanded roll reversal.

As noted previously, this Mach/AOA regime is subject to un-commanded roll and pitch motions resulting from rapid, often asymmetric, shock movement on the wing. On CV, the resultant flow separations can be further aggravated by several CV-unique factors, including the influence of aileron deflection, increased wing span, and close proximity of the wing and HT. Addressing this issue required CLAW modifications that were similar to but more extensive than those made following the CTOL transonic departure described earlier. An extensive rework of the aerodynamic OBM was also required, including updates to roll damping and the effectiveness of all the lateral/directional control surfaces.

Sideslip control

The CV aircraft exhibited poor sideslip control while maneuvering near stall AOA around 0.75 Mach. Figure 31 shows one of the more extreme examples that occurred while reversing a maximum yaw rate command at high altitude. The initial input was applied at low AOA and generated the expected response.

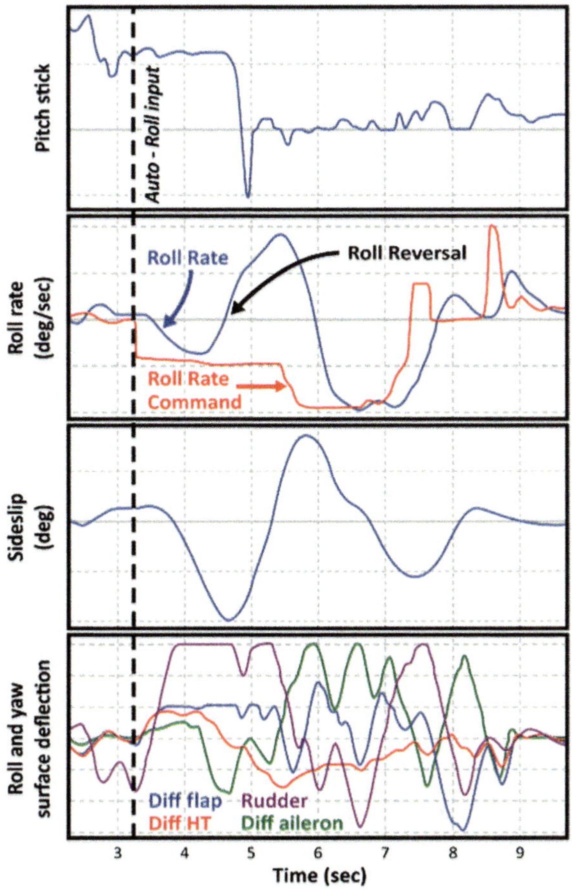

Fig. 30 CV roll reversal.

When the input was reversed just below the limit AOA, the yaw rate responded immediately but the roll rate continued to increase, leading to a larger-than-desired sideslip.

While the roll response and sideslip control were less than ideal, basic control was maintained. However, the large sideslips had to be addressed, as they had been predicted to create excessive loads at lower altitudes if left uncorrected. Several CLAW updates were developed to correct the issue, including modifications to desired dynamics and the addition of logic to improve sideslip control. To ensure that potential structural concerns were fully mitigated, the test plan was modified to include a progressive altitude build down to lower values than originally planned, using the loads-instrumented aircraft.

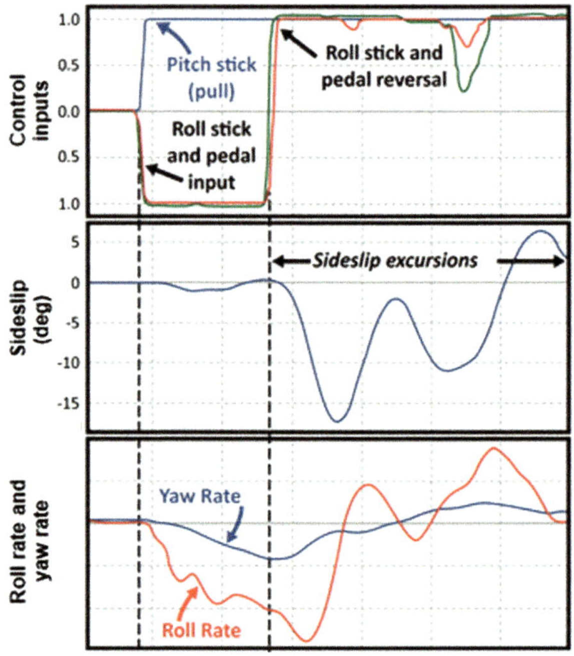

Fig. 31 CV excessive sideslip during control reversal.

Anti-spin mode engagements

Another CV-specific issue was the intervention of the spin recovery logic during stick-only rolls. Figure 32 shows a case that occurred as AOA increased through 25 degrees during a level pull with a full-stick roll input (pedal-free) at 15,000 feet and 0.6 Mach. Roll and yaw rates tracked command initially, but yaw rate continued to grow and eventually exceeded the spin recovery engagement threshold. Once the anti-spin logic engaged, CLAW took control of the aircraft and reduced yaw rate, leading to an uncommanded increase in the roll rate and some fluctuation in AOA. This issue was determined to be the result of small errors in Mach-dependent OBM coefficients that were only observable at low altitude, as discussed in the following paragraphs.

ALTITUDE EFFECTS

The initial plan for DR testing called for flight testing at 35,000 and 20,000 feet only, relying on simulation to clear lower altitudes. The rationale for this approach, used in past high AOA programs, was that the primary aerodynamic drivers are Mach and AOA, not altitude or dynamic pressure. However, a number of anomalies observed during low altitude testing on CV called this strategy into question.

Poor sideslip control was initially seen at 35,000 feet, but the issue did not pose a threat to aircraft control and was not considered to be operationally significant At lower altitude, however, stick-only rolls initiated at around 0.6 Mach and between 20° α and 30° α resulted in significant yaw excursions and occasional spin recovery engagements, as described above.

Analysis showed that the altitude dependence was not the result of changes in aerodynamic coefficients with altitude but to the increased effect of errors in Mach-dependent coefficients at low altitude. The increased effect was

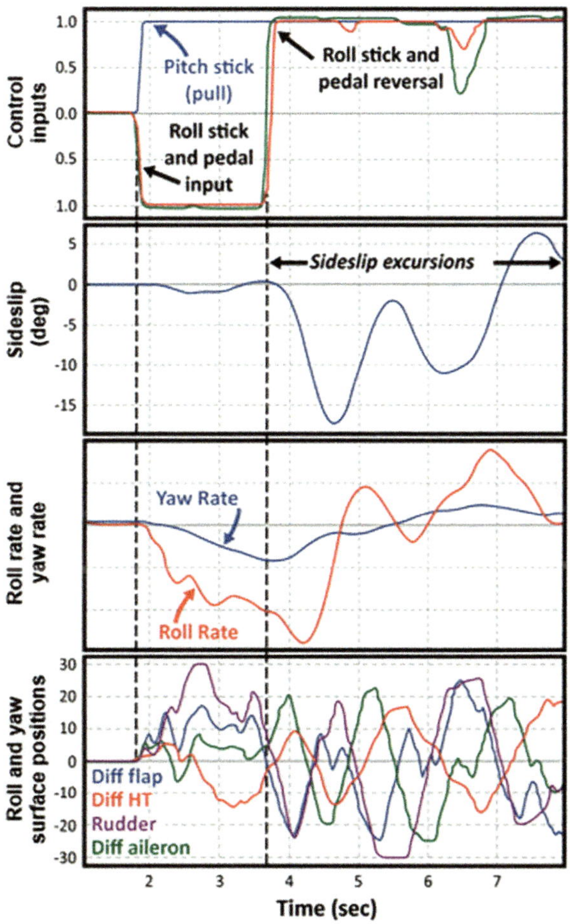

Fig. 32 CV anti-spin engagement during maneuvering.

attributable to (1) the higher dynamic pressure (for a given Mach) generating greater physical moments, making the effects of the errors more observable, and (2) the aircraft's performance being greater at low altitude, causing it to decelerate more slowly and thus dwell longer in the troublesome Mach-AOA regions.

Once the impact of these errors was understood, additional testing was conducted with the CV aircraft at lower altitude (down to 7,000 feet MSL), which provided the data to make the necessary corrections. Based on the CV testing, lower-altitude DR testing was then performed on the CTOL and STOVL variants, although no similar anomalies were discovered. As a result of this additional unplanned effort, all three variants were ultimately cleared for unrestricted high AOA maneuvering at all altitudes.

VII. SUMMARY

The F-35 high AOA development and flight test program proceeded with extraordinary efficiency and resulted in a maneuvering capability that met or exceeded all high AOA requirements. This success can be attributed to a number of factors, among them:

- **Highly accurate modeling and simulation.** Fundamental to NDI is the requirement for a highly accurate aerodynamic model, a requirement that is inherently difficult to achieve in the high AOA regime. Numerous refinements were made to the truth model and OBM as a result of flight test discoveries, several of which are described in this paper. In general, the accuracy of these models provided extremely accurate predictions prior to flight, allowing the high AOA test program to proceed quickly with minimal downtime for analysis.
- **CLAW design.** In addition to accurate models, NDI requires extensive real-time information to be provided to CLAW regarding the aerodynamic state and mass properties of the aircraft. Exploiting this state information in the high AOA regime allowed limiters to be precisely tailored to the aircraft state, the result being a design that is highly departure-resistant while extracting the maximum capability from the aircraft.
- **Robust OCF recovery capability.** Implementing a proven (gain-scheduled) approach for the automatic recovery modes leveraged lessons learned from past programs and eliminated dependence on NDI for these critical modes. This design decision, and the demonstration of consistent recovery from worst case conditions, gave the team and leadership confidence to remove the SRC for the vast majority of the DR phase. Doing so allowed the DR phase, which comprised nearly two-thirds of the high AOA flight test program, to be conducted in a production-representative configuration, precluding the need for costly regression testing.

- **Flight test aids.** FTAs (particularly the ability to set AOA limiters and to modify fuel burn to control CG) were critical to the efficiency and precision with which data were collected throughout the high AOA program. Specialized FTAs, such as those used to support the spin recovery demonstration, enabled testing that would not have been possible otherwise.
- **Flight test preparation and training.** Although not discussed in this paper, extensive planning and rehearsal preceded every test flight, resulting in an extremely low (less than 2 percent) test point re-fly rate.

As can be expected of any test program of this scope and complexity, many lessons can be learned that, even if impractical for implementation on the F-35 program, could potentially be used in future programs. These are:

- **Start high AOA testing earlier.** The programmatic need to provide an initial fleet operational capability (below 20 degrees AOA) delayed the start of high AOA testing until nearly four years after the SDD flight test program began. The delayed start, and the requirement to clear all variants to 50 degrees AOA by mid-2015, compressed the calendar time and reduced the number of Operational Flight Program (OFP) release cycles available to resolve high AOA issues. Additionally, the schedule compression forced testing to be done on all three variants concurrently, limiting the extent to which lessons learned on the lead aircraft (CTOL) could be exploited by the others. An earlier look would have provided greater efficiency and more fly-fix-fly opportunities (OFP cycles) for refining the OBM and high AOA CLAW.
- **Streamline the OFP release cycle process.** Other factors limiting the number of OFP releases possible during the SDD program were the personnel and laboratory time required to develop, test, and verify each software update. Each OFP was a comprehensive air vehicle release that combined updates for all disciplines. This software release process, appropriate for more deterministic disciplines and systems, was a pacing item in the development of the high AOA CLAW, which is intrinsically a fly-fix-fly process. Since the CLAW (in particular, the OBM) functions independently of all other vehicle functions, the use of an agile process for turning CLAW-unique software updates during high AOA testing is worthy of consideration in order to maximize the number of iterations available for optimization.
- **If implementing NDI at high AOA, expect to need innovative solutions.** The unsteady, nonlinear nature of aerodynamics at high AOA makes accurate modeling, which is intrinsic to NDI, a challenge. Errors in the OBM in this regime are inevitable and cannot always be remedied with model updates alone. Adaptive, customized solutions that fall outside the normal NDI structure may be required.
- **Allow freeform testing within the cleared envelope.** Perhaps to a greater extent here than in any other area of flight test, modeling uncertainty gives

rise to the potential to discover unknown unknowns (i.e., issues that the designers had not thought of or were not evident in simulation) during high AOA testing. Accordingly, once the basic envelope has been cleared and a robust recovery capability has been established, some amount of time should be allowed for unscripted exploration. Allowing the test pilot to explore the handling characteristics of the airplane and to fly in an operationally representative manner would allow issues to be discovered that may not have presented themselves during structured test point execution.

VIII. CONCLUSION

Although high AOA was not a primary consideration in the aircraft's design, the F-35's advanced flight control system provides a high AOA maneuvering capability that meets or exceeds all program requirements, including post-stall maneuvering, extreme DR, and automatic recovery from OCF. In providing this capability, the F-35 not only pioneered the use of NDI on a production fighter but also demonstrated its application in the inherently challenging environment of high AOA.

ACRONYMS

AOA	Angle of Attack
APR	Automatic Pitch Rocker
CG	Center of Gravity
CLAW	Control Laws
CTOL	Conventional Takeoff and Landing
CV	Carrier Variant
6-DOF	Six Degree of Freedom
DR	Departure Resistance
EB	Effector Blender
EHA	Electro-hydrostatic Actuators
FBW	Fly-By-Wire
FTA	Flight Test Aid
HT	Horizontal Tail
LEF	Leading Edge Flap
MPR	Manual Pitch Rocker
NDI	Nonlinear Dynamic Inversion
OBM	Onboard Model
OCF	Out-of-Control Flight
OFP	Operational Flight Program
PA	Powered Approach
PTMS	Power and Thermal Management System
SDD	System Development and Demonstration

SRC	Spin Recovery Chute
STOVL	Short Takeoff and Vertical Landing
TEF	Trailing Edge Flap
TNS	Tactical Navigation System
TRO	Transonic Roll-Off
WBD	Weapons Bay Door
UA	Up-and-Away

REFERENCES

[1] Robbins, A., Bobalik, J., De Stena, D., Plag, K., Rail, K., and Wall, K., "F-35 Subsystems Design, Development, and Verification," AIAA Aviation Forum, Atlanta, GA, June 2018 (unpublished).

[2] Harris, J., and Stanford, J., "F-35 Flight Control System Design, Development and Verification," AIAA Aviation Forum, Atlanta, GA, June 2018 (unpublished).

[3] Bordignon, K., and Bessolo, J., "Control Allocation for the X-35B," 2002 Biennial International Powered Lift Conference and Exhibit, AIAA 2002-6020, November 2002.

[4] Ogburn, M. E., et al., "High Angle of Attack Nose Down Pitch Control Requirements for Relaxed Static Stability Combat Aircraft," NASA High Angle of Attack Technology Conference, NASA Langley Research Center, Hampton, Virginia, 30 October to 1 November 1990, NASA CP-3149, Part 2, Paper No. 24, 1992.

[5] McNamara, W. G., et al., "Navy High Angle of Attack Pitch Control Margin Requirements for Class IV Aircraft," NAVAIRWARCENACDIV Technical Memorandum 91-167 SA, 25 June 1992.

[6] Nguyen, L. T., and Foster, J. V., "Development of a Preliminary High Angle of Attack Nose Down Pitch Control Requirement for High Performance Aircraft," NASA Technical Memorandum 101684, February 1990.

[7] MIL-STD-1797B, Flying Qualities of Piloted Aircraft, 2006.

[8] Bowman, M., and Bemridge, A., "The Automatic Low Speed Recovery Function of the Eurofighter Typhoon Aircraft and how it was Flight Tested," Society of Experimental Test Pilots, 2004.

[9] MIL-F-83691B, Military Specification: Flight Test Demonstration Requirements for Departure Resistance and Post-Departure Characteristics of Piloted Airplanes, March 1991.

CHAPTER 16

F-35 Weapons Separation Test and Verification

Christopher F. Hetreed*, Matthew D. Carroll† and Joe E. Collard‡

Lockheed Martin Aeronautics Company, Fort Worth, TX

Richard C. Snyder§

Naval Air Warfare Center, Naval Air Station Patuxent River, MD

Weapons separation flight testing is the most visible phase (Fig. 1) of a military aircraft program's store certification process. Store separation flight tests, commonly video-recorded from a chase aircraft, are generally regarded as both verification that the weapon release requirements have been satisfied and validation that the system will perform as expected during operational use by the customer. During the System Development and Demonstration (SDD) phase of the F-35 Lightning II program, Lockheed Martin has provided the F-35 Joint Program Office (JPO) with a "certification recommendation" to load, carry, and release 11 weapons in 20 different loadings across the F-35A, F-35B, and F35C variants. The basis for both Lockheed Martin's certification recommendation and verification that the weapon release requirements have been satisfied is flight test–validated Modeling and Simulation (M&S) that has been adjusted based on data gathered during the store separation flight test events, so that simulated store separation events replicate the actual store separation events. During the test program, the flight test team rapidly and successfully performed 183 safe store separation events, without the need for any reflown points.

I. NOMENCLATURE

6DOF	six degrees of freedom.
F-35A/B/C	the 3 F-35 Joint Strike Fighter (JSF) variants.
IMU	inertial measurement unit; measures store rates and accelerations.

*Aeronautical Engineer Senior Staff, F-35 Weapons.

†Aeronautical Engineer Staff, F-35 Weapons.

‡Aeronautical Engineer Manager, F-35 Weapons.

§F-35 Store Separation Lead and Task Area Expert, F-35 Lightning II Joint Program Office/NAVAIR-4321.

Fig. 1 F-35 Lightning II weapon separation flight test.

S&RE/AME aircraft devices used for carriage, suspension, employment, and jettison; Suspension and Release Equipment (S&RE) includes missile and bomb carriage/ejector racks, and pneumatic power systems; Alternate Mission Equipment (AME) includes adapters, missile rail launchers, pylons, etc. In this paper, S&RE includes AME, for convenience.

store any device intended for internal or external carriage and mounted on aircraft suspension and release equipment, whether or not the item is intended to be separated in flight from the aircraft. Stores include missiles, rockets, bombs, nuclear weapons, mines, torpedoes, pyrotechnic devices, detachable fuel and spray tanks, line-source disseminators, dispensers, pods (refueling, thrust augmentation, gun, electronic-counter measures, etc.), targets, cargo-drop containers, and drones.

TM kit telemetry kit; for transmitting 6DOF/IMU measured store rates and accelerations.

II. INTRODUCTION

The F-35 Lightning II Joint Strike Fighter (JSF) is a 5th-Generation fighter attack aircraft designed for survivability, lethality, supportability, and affordability, a culmination of the JSF program [1]. In order to meet the needs of multiple military services and international customers, while still maintaining affordability, there are three unique F-35 variants [2] with many common aircraft and systems components; for example, the F-35A and F-35C share a common weapon bay, capable of carrying larger air-to-ground weapons than the vertically landing F-35B. But unlike the F-35C variant, which requires larger wings to enable slower approach speeds for landing on naval aircraft carriers, the F-35A and F-35B variants share a weight-saving, similarly shaped smaller wing. During the SDD phase of the JSF program, 11 weapons were certified for carriage and release in 20 different loadings across these three aircraft variants (Fig. 2)—a challenging process that includes design, modeling, simulation, and testing across multiple engineering disciplines.

Many engineering disciplines follow the same general path (Fig. 3), starting from system requirements and proceeding to design, then to Modeling and Simulation (M&S), and then to testing for the purposes of M&S adjustment/validation and verification that requirements have been met. In the end, requirements verification is accomplished using both testing and test-validated M&S.

The left-to-right process for the weapons (store) separation discipline is similar, with additional detail depicted in Fig. 4 for the store separation modeling, simulation, and flight test path to store certification, and summarized within this paper. Maintaining affordability is critical in both the development and sustainment of the F-35 Lightning II, and it is shown in Ref. [3] that several techniques, such as Computational Fluid Dynamics (CFD) studies, were used to minimize preflight wind tunnel aerodynamics testing costs and reduce risk, prior to flight testing. But the most significant cost in the assessment of safe store separation, on the path to store certification, is usually flight testing. So with attention to

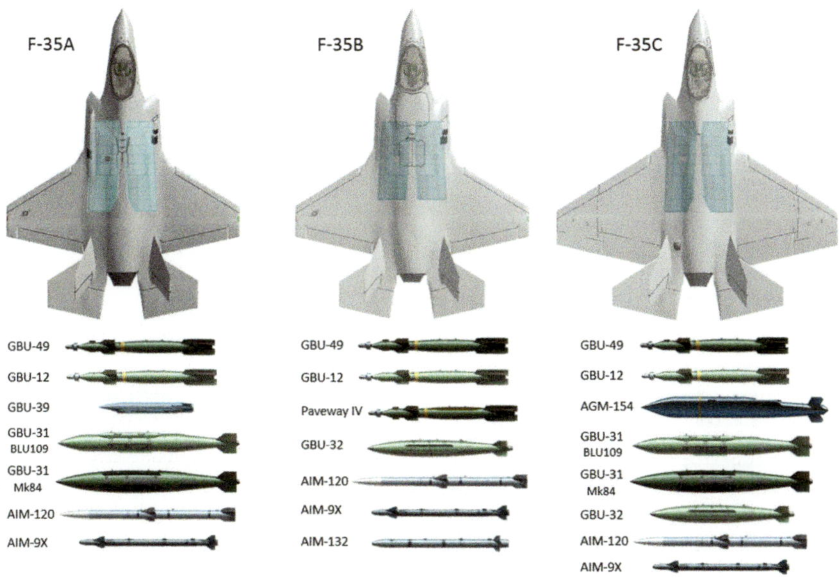

Fig. 2 F-35 Lightning II SDD weapons, certified for carriage and release.

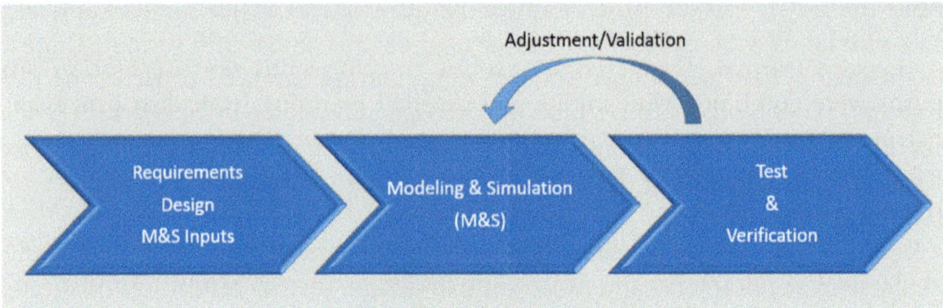

Fig. 3 Requirements to verification.

affordability, the philosophy on the F-35 Lightning II program was to extract as much accurate information during the fewest number of weapon separation flight test events, to confidently validate modeling and simulation in order to support this high-confidence safe store separation assessment and certification recommendation to the JPO.

Due to challenging requirements, 183 store separation test points were required for M&S validation over the weapons release flight envelope. In order to maintain an aggressive schedule, rapid progression through these test points was necessary, with no tolerance for delay—the Edwards Air Force Base (EDW) and Patuxent River Naval Air Station (PAX) flight test teams needed the ability to execute store separation events on back-to-back days. Within hours of receipt of flight test data, it was necessary for Lockheed Martin to understand

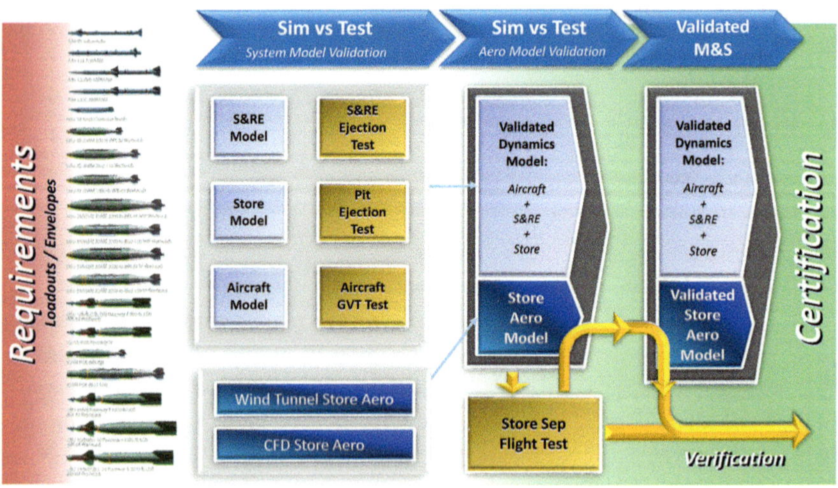

Fig. 4 Store separation path to certification.

and communicate not only what happened during a separation event, but also how it happened, especially if there was an unexpected store trajectory.

Lessons learned and recommendations from previous store separation programs were combined with inputs and detailed planning from data processing, instrumentation, survey/metrology, mass properties, store separation, flight test, and weapons experts within Lockheed Martin, EDW, PAX, U.S. Naval Air Systems Command (NAVAIR), U.S Air Force SEEK EAGLE Office (AFSEO), U.K. Ministry of Defense (UK MoD), Arnold Engineering Development Center (AEDC), Boeing, Raytheon Missile Systems, MBDA Missile Systems, and Raytheon Systems Limited. This planning, analysis, cooperation, and execution resulted in the F-35 SDD weapon separation test program, with the following highlights:

1. 183 safe store separation flight test events, across three aircraft variants, at several ranges at two test sites
2. 121 in-bay weapon releases:
   a. 58 large and/or bay-defining air-to-ground stores
   b. 41 air-to-air missiles
   c. 22 air-to-ground stores from four-place ejector racks
3. 62 external weapon releases:
   a. 32 air-to-air rail-launched missiles
   b. 30 air-to-ground pylon-mounted stores
4. One- to two-day typical approval/clearance to next test event:
   a. On-board high-speed camera and 6DOF store telemetry data analysis, computation of flight test-based aerodynamics increments, and adjusted predictions for the next store separation test event were usually accomplished within one day.
5. 17 flights with multiple store releases; "in-control-room" clear-to-next-point:
   a. Three flights with four store releases
6. One weapon fin-activation lanyard redesign/reconfiguration
7. Zero changes in ejector rack pitch settings
8. Zero unsafe store trajectories; zero store separation–related test program delay/pauses
9. Zero reflown test points

III. F-35 LIGHTNING II AND VARIOUS SDD WEAPONS

In the SDD phase of the F-35 program, there was a requirement for 11 weapons to be loaded, carried, and employed from three aircraft variants, from internal weapon bays and underwing; detailed descriptions of design considerations and hardware are presented in *F-35 Weapons Design Integration* [4]. In summary,

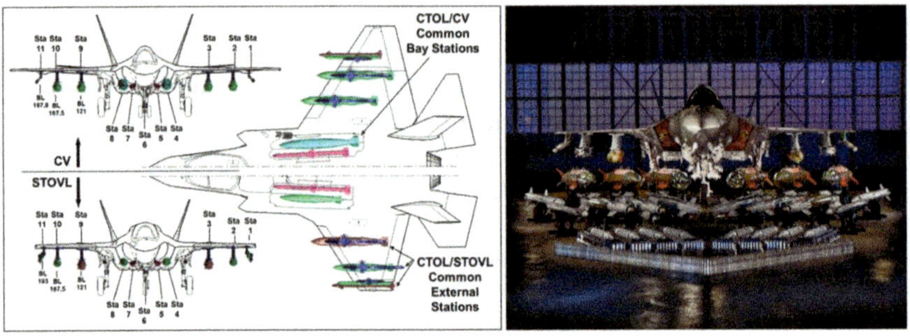

Fig. 5 F-35A SDD weapons carriage.

the F-35 Lightning II has 11 weapon stations (Fig. 5) and carries and releases 250-, 500-, 1000-, and 2000-lb-class air-to-ground bombs from selectable-performance pneumatic ejector racks mounted within space-constraining weapons bays and underwing pylons. Air-to-air missiles are carried and released from pneumatic ejector racks within the bay, and from underwing rail launchers.

The pneumatically actuated, two-piston BRU-68 ejector rack (Fig. 6) is capable of releasing 500-, 1000-, and 2000-lb-class bombs from the F-35A/C internal weapons bay, or from the F-35A/B/C external weapons stations on the wing when installed within an air-to-ground pylon. Similar to the BRU-68, the

Fig. 6 BRU-67/BRU-68 ejector rack (top), in bay (bottom left), and in pylon (bottom right).

smaller pneumatically actuated BRU-67 ejector rack is capable of releasing 500- and 1000-lb-class bombs but is designed for use within the F-35B internal weapons bay. Both the BRU-68 and BRU-67 have multiple pitch valve settings that allow for selectable pitch performance based on which weapon is installed. Inside the weapons bay, the Pneumatic Power System (PPS) pressurizes the bomb rack to its operating pressure. Air-to-ground pylons each contain a separate PPS for pressurizing bomb racks at external/wing weapons stations.

The LAU-147 (Fig. 7) is a pneumatically actuated, two-piston ejection launcher that can be installed on the F-35A/B/C at either internal weapons bay stations. LAU-147s are capable of launching medium range air-to-air missiles such as the AIM-120 AMRAAM. Along with pressurizing any BRU-68 or BRU-67 in the weapons bay, the same in-bay PPS pressurizes any LAU-147 to its operating pressure.

The BRU-61 (Fig. 8) is a four-station pneumatically actuated ejector rack that is capable of releasing up to four 250-lb-class bombs including the GBU-39 Small Diameter Bomb (SDB) and the GBU-53 SDB-II. The BRU-61 can be installed in the F-35 weapons bay on a fixed adapter or on a BRU-68 connected to an air-to-ground pylon on the aircraft wing. The BRU-61 has a self-contained compressor, independent of the PPS in both the weapons bay and the air-to-ground pylons. The four stations on the rack include an aft-inboard, forward-inboard, aft-outboard, and forward-outboard rack station, each with selectable-performance pitch valves and two pistons. The release order on the BRU-61 bomb rack is customizable.

During the internal weapon bay design phase, weapons were rigged in Computer-Aided Design (CAD) software, such that the stores would all be

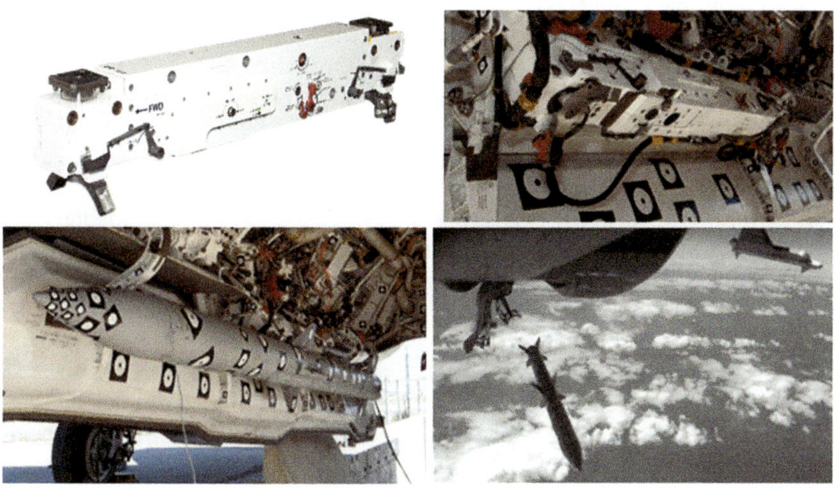

Fig. 7 LAU-147 missile ejector rack.

Fig. 8 BRU-61 with four-pack of GBU-39 small diameter bombs.

sufficiently far away from in-bay hardware during carriage and release from the F-35 weapon bays. Reference [4] contains detailed descriptions of internal weapon bay design considerations, along with descriptions of on-aircraft weapons-mounting hardware, such as the Ordnance Quick Latch System (OQLS). Stores have different sizes, shapes, and carriage lug locations, so it was not feasible to have a single ejector rack location that would be sufficient for carriage and employment of all stores. Some stores required a more-forward ejector rack location, relative to the aircraft OQLS, whereas others needed to be carried higher within the weapon bay, for instance. Therefore, various adapters are used (Fig. 9) to ensure adequate aircraft hardware miss distances during store carriage and release from the F-35. And because some of these adapters are attached to the in-bay OQLS differently, carriage and ejection loads are transferred differently to the aircraft structure.

The LAU-148 and LAU-151 (Fig. 10) are rail launchers that are installed on air-to-air pylons at the most outboard external F-3A/B/C weapons stations, and are capable of launching rail-launched missiles, such as the AIM-9X Sidewinder and the AIM-132 ASRAAM. The LAU-148/151 restricts the missile's vertical and lateral motion, while the missile's thrust propels the missile forward on the rail.

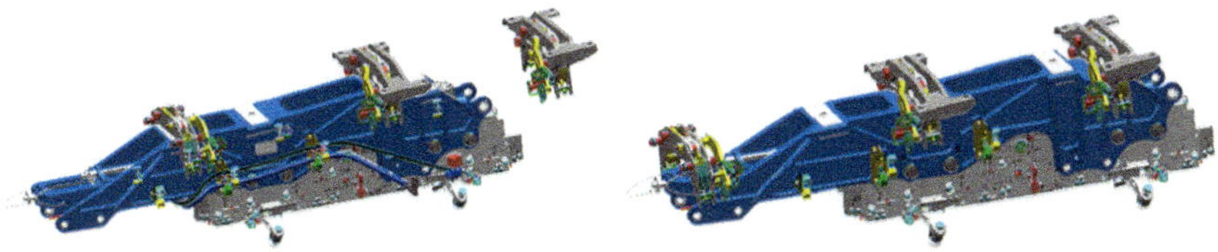

Fig. 9 BRU-67/68 within two different adapters (blue), each with different mounting to in-bay OQLS.

In addition to the differences in aircraft variants, weapon stations, and previously described S&RE from which stores are released from the F-35, each store has different features that result in unique release behavior for each store type, with respect to miss distance to aircraft hardware, aerodynamics, trajectory motion, and stability and control while in the near-field vicinity of the aircraft. For instance, the GBU-31 JDAM and AGM-154 JSOW stores are large, fill a majority of the internal weapon bay, and are in close proximity to aircraft hardware, such as weapon bay doors, when released from the internal bay. The Paveway II (GBU-12) and Paveway IV stores include forward canards and aft tailfins, the latter of which are deployed with the use of a lanyard device. The GBU-39 SDB is a lightweight store and also features deployment of stabilizing fins during the store separation

Fig. 10 AIM-132 and AIM-9X air-to-air rail-launched missiles from LAU-148/151.

event. The rail-launched AIM-9X Sidewinder missile includes forward canards, whereas the AIM-132 ASRAAM is absent of such forward aerodynamic surfaces. Several stores have relatively low mass moments of inertia, which can result in fast response to aircraft flow field aerodynamic forces, compared to stores with higher mass and inertia. Some stores feature active controlled flight, using moving fin control surfaces, when in close proximity to the F-35, whereas other stores delay such control or propulsion activation—the ejected AIM-120 AMRAAM, for instance, ignites its rocket motor at a delay sufficient to ensure that forward motion starts when the store is at a safe distance below the aircraft.

Therefore, modeling, simulation, and testing must all be conducted in consideration of these unique store features, behaviors, and manners in which they are released or separated from the F-35. The store separation event is a dynamic event, influenced by its own aerodynamic, propulsion, and inertia forces, and by interaction with the S&RE and aircraft.

IV. MODELING

Modeling and simulation are both on the store separation path to certification, but before considering models of the aircraft, S&RE, and stores previously described, it is useful to understand details of typical real store separation events, followed by consideration of how and whether such details could be modeled and simulated. Store separation problem descriptions, hardware requirements, compatibility guides, and weapons flight testing information are found in Ref. [5–10, 23]. Although some of the techniques have been replaced by modern equipment, tools, and methods, the store separation problem and general testing techniques and goals still apply. These particular references are provided because they served as the foundation documentation for understanding store separation fundamentals, aiding the implementation of newer modeling techniques, and reinforcing the importance of adequate flight test instrumentation for properly validated models and simulations. Reference [11] includes "lessons learned," providing recommendations for focusing resources in modeling, test, and analysis, which were generally adhered to during the F-35 store separation preflight, flight test, and postflight activities. But this was somewhat unintentional and coincidental, yet reassuring—the F-35 store separation team's grassroots approach was to 1) try to understand actual typical events; 2) consider available general-purpose modeling tools and techniques used in the aerospace and automotive industries; 3) try to reasonably model detailed mechanical, structural, and aerodynamic behavior; and 4) consider how to objectively judge and adjust the simulated model's behavior vs the real hardware's behavior, based on available instrumentation and typical testing procedures. As with other disciplines such as loads, dynamics, flutter, aero/performance, and stability and control, modeling and simulation adjustments were expected as part of the validation process. Store separation component models were built, simulated, and validated during and after component ground testing. Store separation full-aircraft models were built, simulated, and validated during and after full-aircraft ground and flight testing. And like those other disciplines, validated models and simulations are the basis for verification that store separation requirements have been met.

A. THE ACTUAL EVENT

The actual event is described in a manner to help think about construction of a virtual functional prototype—a simulation model that will behave like the real system. During captive carriage and prior to a store separation event, the store is constrained or restrained to the aircraft by S&RE. The store separation event begins at the moment the store carriage state changes, and the store is released, partially restrained, and/or pushed away from the carriage position. During this event, the aircraft and store each respond in a rigid and flexible manner due to mechanical, structural, inertial, gravitational, aerodynamic, and/or propulsive dynamic forces. Consider two examples.

1. RELEASED/EJECTED STORE EXAMPLE

An ejector S&RE system retracts its store restraint mechanism (hooks or clamps) and the store becomes free to fall. Almost immediately at this point, the ejector S&RE system extends pressurized pistons against the store, pushing the store away (Fig. 11). The interactive response of the aircraft, S&RE mechanism, and store may be complex.

The aircraft response depends on the aircraft mass and inertia, S&RE location relative to the aircraft CG, mass of the released store, S&RE ejection force, aircraft structural flexibility near the S&RE location, pilot aircraft control inputs, and flight control system response, for instance. The aircraft response may be important during the store separation event, or it may be negligible.

The S&RE behavior is influenced by the attachment constraint configuration to the aircraft or wing pylons, local aircraft and/or multiple-store carriage system

Fig. 11 Ejected store.

structural flexibility and response, contact constraint with the ejecting store, store mass and inertia, store CG, behavior of the extending ejection pistons from the pneumatic or pyrotechnic charging inputs, and aerodynamic forces on the store during the ejection process.

The store response is influenced by its structural flexibility, mass properties, contact constraint and ejection forces from the S&RE ejector system, store control laws inputs, and aerodynamic forces acting on the store.

2. RAIL-LAUNCHED STORE EXAMPLE

During captive carriage, two or more of a missile's hanger brackets are constrained within a long inner or outer rail channel on a rail-launcher S&RE system. When the missile's rocket motor ignites, the missile starts to move forward, and its hanger brackets push against and retract the launcher's forward-motion restraint/detent mechanism. The rocket motor then continues to push the missile forward, down the rail (Fig. 12), while the missile is partially restrained during this launch event by its hanger brackets within the rail channel. After tip-off, or the moment the last missile hanger bracket exits the rail channel, the missile is no longer restrained by the launcher on the aircraft and continues to be propelled forward. And like the ejected store example, the response of the aircraft, launcher S&RE, and missile may be complex.

The aircraft response during a normal rail-launched missile separation event is usually much less pronounced than during an ejected store separation event and is generally observed in the localized structural response near the S&RE location.

The launcher S&RE behavior is influenced by the attachment constraint configuration to the aircraft or wing pylons, launcher structural flexibility and response, rail constraint contact with the missile's hanger brackets, store mass and inertia, aerodynamic forces on the missile and launcher, and missile plume impingement forces, for instance.

The store response may be influenced by its own structural flexibility, mass properties, launcher flexibility, rail channel constraint contact, propulsion force,

Fig. 12 Rail-launched store.

aerodynamic forces, and fin or canard surface movements due to missile flight control law system inputs, for example.

B. THE MODEL AND SIMULATION EVENT

The prior two actual event examples illustrate the physical system (aircraft, S&RE, and separating store) behaving dynamically, in an interrelated manner—structurally, mechanically, and aerodynamically. In M&S of the dynamic behavior of such systems, the *simulation* portion can be regarded as the time-dependent solution of the set of mathematical dynamics equations representing the *model* of the actual physical system. Such models can range from simple to complex, depending on the desired fidelity and intent of the simulation. For example, in some M&S instances, it may be desirable to consider mechanical S&RE behavior during the separation event, beginning at the moment of store first motion, whereas in other instances it may be beneficial to exclude these interactions from the model and simply start simulations after End-of-Stroke (EOS) or after End-of-Rail (EOR), after the store would normally lose contact with the S&RE.

Typical store separation software involves the solution of mathematical equations of motion representing a rigid aircraft and a 6DOF store system, and includes store aerodynamics, propulsion, controls, S&RE-to-store interactions, and other provisions [12, 13]. In such software environments, the model, or mathematical equations of motion, are customized to include specific behaviors of specific systems; when additional components, constraints, or other mechanical effects are to be added, the equations of motion may need to be modified or customized. Manually customizing mathematical equations of motion, to capture modeling changes, is a task that requires tremendous skill and experience, arguably shifting valuable resources away from the important analysis task.

C. LOCKHEED MARTIN'S DYNAMIC SYSTEM MODEL

Lockheed Martin's store separation M&S software (Fig. 13), ASEP, is a plug-in to the commercially available VI-Aircraft and ADAMS (Automatic Dynamic Analysis of Mechanical Systems) mechanical system simulation software. In the *modeling* phase, virtual models of multibody rigid and flexible mechanical systems are built up, assembled, and archived in a CAD-like environment. So, when considering physical details of an actual event, such as friction, flexible components, surface contact, complicated restraint mechanisms, and so forth, these physical effects are easily "built" into the virtual model. The model can include detailed structural and mechanical behavior, as necessary. Similarly, "plugged-in" store aerodynamics, controls, and propulsion forces can be simple or complex, and may be derived from various sources, such as wind tunnel tests, CFD, or proprietary "black box" modules furnished from ejector rack and store suppliers.

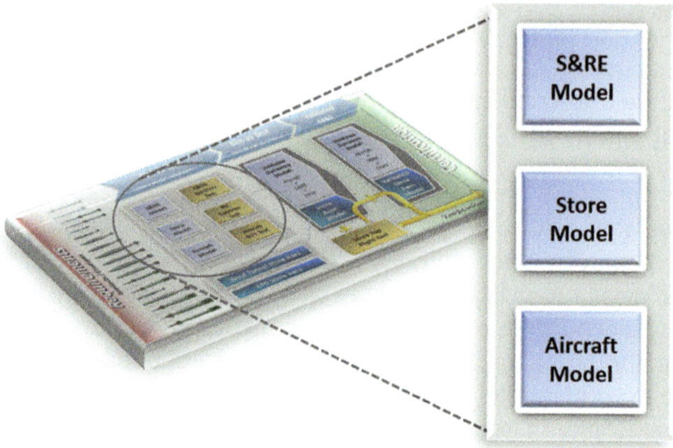

Fig. 13 Dynamic system model.

During ASEP's *simulation* phase, each unique assembled system model's mathematical equations of motion are automatically generated and solved. So, the inclusion of additional hardware components and physical behavior becomes an act of simply adding them to the model (Fig. 14), without the need to manually reformulate mathematical equations of motion.

Lockheed Martin's ASEP M&S software features two modeling approaches—a simpler system dynamics model to focus mainly on the store, and a more complex system dynamics model to include complex interaction among the aircraft, S&RE, and store. The modeling approaches are summarized as follows.

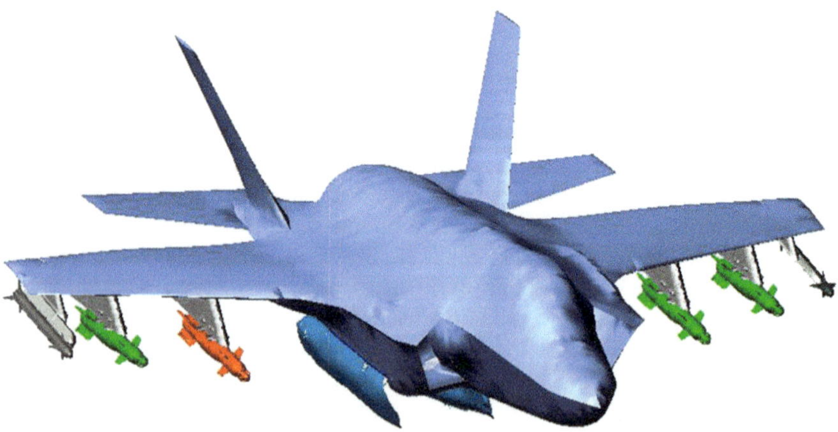

Fig. 14 ASEP full model.

1. SIMPLE DYNAMICS SYSTEM: INITIAL CONDITIONS MOTION M&S

This system modeling approach, involving two or three 6DOF bodies, includes a maneuvering rigid aircraft model, a rigid store model, and an optional 6DOF/IMU TM Kit rigid body. This approach is used when intentionally focusing only on the less-complicated unrestrained store free-flight region, excluding the direct S&RE mechanical connection between the store and aircraft. The primary focus is usually on a store's trajectory after EOS or tip-off. The rigid bodies are as follows.

The rigid aircraft body is "flown" in space, according to either a prescribed steady maneuver or an exact flight-test aircraft maneuver response. In both types of aircraft maneuvering, the aircraft is flown to match the desired aircraft translational accelerations and rotational rates.

The rigid store body motion can be simulated in two ways. Store trajectory motion begins from prescribed initial store displacement and velocity conditions. Alternatively, store is constrained to follow a prescribed store motion trajectory time-history, relative to the moving aircraft, and then optionally becomes detached from the aircraft at any point in the store trajectory time-history at the instantaneous displacement and velocity conditions. In this latter scenario, the source of the prescribed store trajectory time-history may be either from a store separation wind tunnel trajectory simulation, from a flight test event's photogrammetric solution store trajectory, or from a store trajectory solution derived from flight test measured rates and accelerations.

In both of these store model motion scenarios, only after the store rigid part becomes completely detached from constrained motion with the aircraft part and becomes its own independent 6DOF body (Fig. 15) can its dynamic trajectory be influenced by the forces acting on it: aerodynamics, propulsion, gravitational, and so forth. So, such force models are also included in the system model, usually through software links to aerodynamics databases or black box modules.

Like the rigid aircraft body and rigid store body, the optional 6DOF/IMU TM Kit rigid body is a moving part and is used to: 1) automatically replicate a flight test store's actual separation trajectory motion, and 2) automatically compute the applied aerodynamic forces that influenced this store trajectory motion. During a posttest simulation, this 6DOF/IMU TM Kit part is initially attached to the rigid store. At a selected point in the rigid store's trajectory (whether at carriage or when moving according to a prescribed photogrammetry-based motion relative to the aircraft), the 6DOF/IMU TM Kit part becomes completely detached from the store rigid body and "flies" according to the measured rates and accelerations.

2. FULL DYNAMICS M&S

This system modeling approach includes a maneuvering flexible or rigid aircraft model, a model of the S&RE components, and a flexible or rigid store model

Fig. 15 Simple dynamics.

(Fig. 16). This approach is used when simulating structural, mechanical, and aerodynamic behaviors of the combined aircraft/S&RE/store system—when attempting to achieve a detailed virtual model of the physical system. The simulation starts at store carriage; continues through the release, launch, or ejection phase; and includes the system mechanical dynamic response during the time the store and S&RE are in contact during the separation event. The simulation continues after the store loses contact with the S&RE, after ejector piston EOS, after launcher rail/missile tip-off, or after release of a pylon or fuel tank's hook/ball/socket restraint mechanism.

The aircraft model may be flexible or rigid (Fig. 17). It may be attached to a landing gear mechanical model resting on the ground, to simulate aircraft dynamic response during a store static ejection pit test. Or it may be "flown," or moved in space during simulations, according to either a prescribed steady maneuver or an exact flight-test aircraft maneuver response.

The flexible or rigid S&RE system model is attached to the aircraft model, in a rigid or flexible manner similar to that of the actual system, to ensure proper interdependent motion and intended structural load paths between the S&RE and the aircraft. The S&RE model (Fig. 18) could be a mechanically actuated model of an ejector rack, including rack supplier–provided gas or pyrotechnic piston pressure modules. Or the S&RE model could be of a flexible rail missile launcher, obtained from Finite Element Modeling (FEM) software. The model may also include flexible or rigid adapter and pylon structural models. And the S&RE is built to model

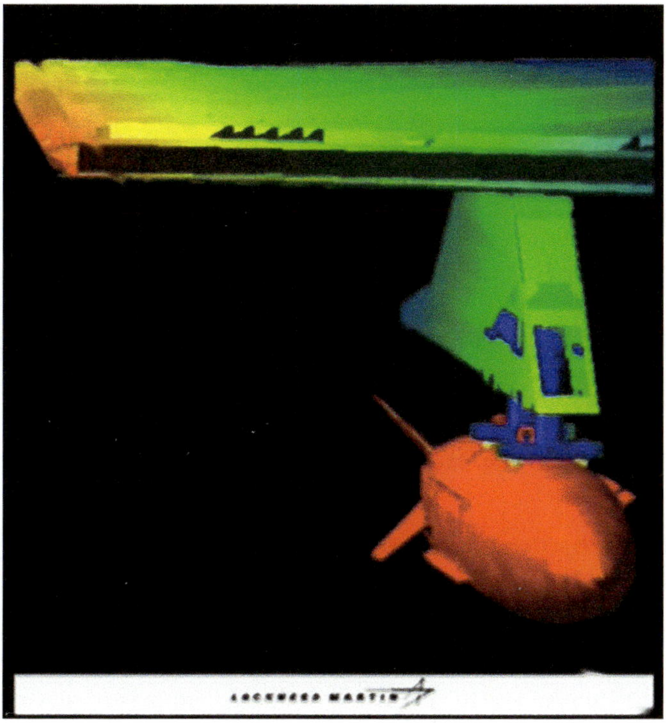

Fig. 16 Full dynamics.

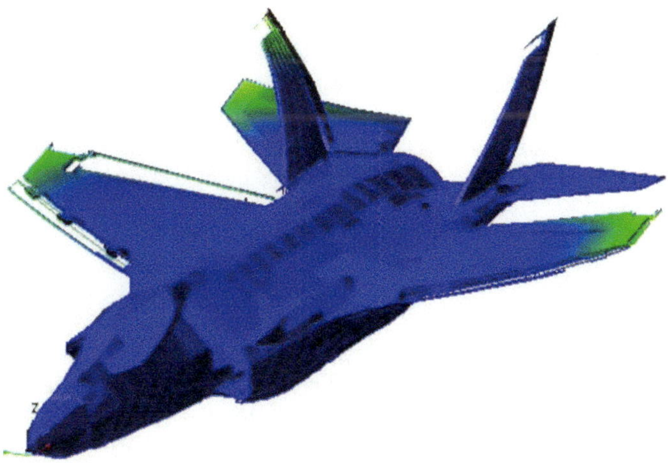

Fig. 17 Flexible aircraft.

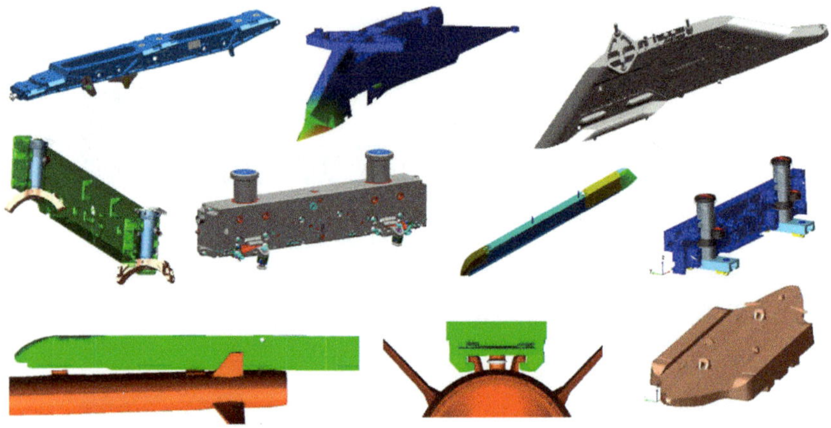

Fig. 18 S&RE system models in ASEP.

the physical interaction with the store. The S&RE subsystem model may have as much or as little detail as desired.

The flexible or rigid store model is fully or partially constrained or restrained by the S&RE mechanical model during the store separation simulation. In order to simulate the interaction forces between the store and S&RE, contact models between the store and S&RE are usually included, and may or may not include friction. This contact could be between a store body and S&RE ejector pistons and/or piston sway brace pads. Or this contact could be between store hangers and a launcher's rail channel. This contact could even be a complicated hook/ball/socket constraint typically seen in a rear-pivot mechanism connecting a fuel tank to a pylon, or a pylon to an aircraft wing. And these contact models may be as simple or as complex as necessary to replicate the physical interaction between the store and S&RE. In the simulated separation event, during and after constraint by the S&RE, other forces also act on the store, such as aerodynamics, propulsion, gravitational, and so forth. So, such force models are also included, usually through software links to aerodynamics databases or black box modules.

D. AERODYNAMICS MODELS

The store aerodynamics models (Fig. 19) are usually derived from wind tunnel testing [14], but can be replaced by, or combined with, CFD-generated store aerodynamics [15]. In building store separation aerodynamics models for released stores from all SDD loadings on the three F-35 aircraft variants' internal and external weapon stations, 11 total wind tunnel tests were conducted, with over 3900 User-Occupancy Hours (UOH) and over 1700 air-on hours. One-fifteenth-scale aircraft and store models were used in order to minimize wind

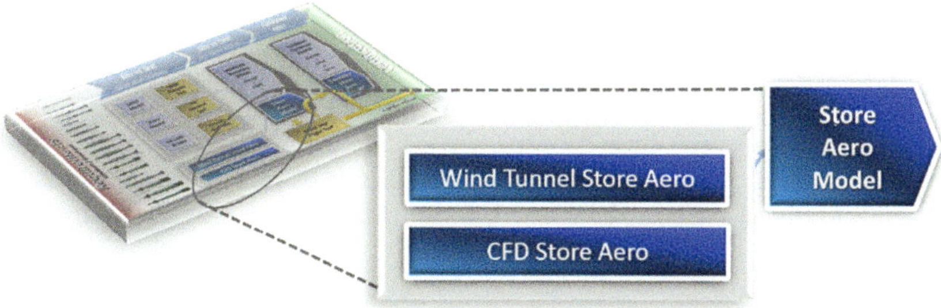

Fig. 19 Store aerodynamics model.

tunnel test costs by using the smaller 4-ft transonic/supersonic wind tunnel at AEDC, instead of AEDC's 16-ft transonic/supersonic wind tunnel. Details of store separation wind tunnel testing are best-described in wind tunnel literature and test reports, so only a summary is provided herein. During these tests, store aerodynamic forces were obtained from a load-measuring balance, to which the store was attached, during up to four test scenarios.

1. ATTACHED LOADS

The store and balance are attached to a strut or other mounting hardware directly on the aircraft model, usually within the weapon bay (Fig. 20). In this scenario, aerodynamics forces are obtained for the store, usually in a "carriage" configuration for two purposes: computing store-applied airloads onto the aircraft and obtaining aerodynamic forces in regions, such as a weapon bay cavity, where a sting-mounted store model cannot reach.

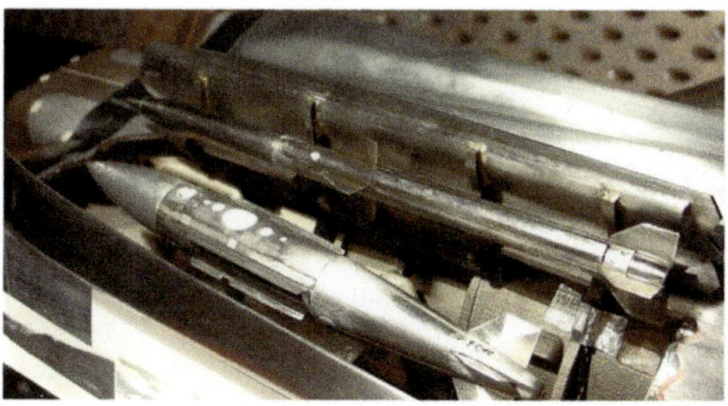

Fig. 20 Attached loads.

2. STORE-ONLY FREESTREAM

The store and balance are attached to a sting and swept through various angles relative to the freestream airflow direction (Fig. 21). Comparisons are usually made to larger-scale (full-scale, 1/2-scale, 1/4-scale, for instance) freestream aerodynamics models, obtained from weapon suppliers, to quantify uncertainty associated with the 1/15-scale model's "small-scale freestream" aerodynamics model.

3. FLOWFIELD GRID

The store and balance are attached to a sting, identically to that in the freestream store-and-balance sting arrangement, but the aircraft model is also present in the wind tunnel test section, mounted on an independent sting. In this arrangement, the aircraft and store can be moved independently, enabling the store to be positioned relative to the aircraft, while the aircraft is positioned relative to the freestream air flow to simulate various aircraft angles of attack and angles of sideslip (Fig. 22). This type of testing is known as the Captive Trajectory System (CTS) at AEDC, or the Two-Sting-Rig (TSR) system at Aircraft Research Association (ARA). The main goal in this arrangement is to measure store aerodynamics forces at various positions in the vicinity of the aircraft that populate the store flowfield aerodynamics database.

4. FLOWFIELD TRAJECTORY SIMULATION

In the same arrangement as in the flowfield grid scenario, the aircraft is positioned at the desired angle of attack and angle of sideslip, relative to the wind tunnel freestream air flow, and the store is positioned, relative to the aircraft, as close to

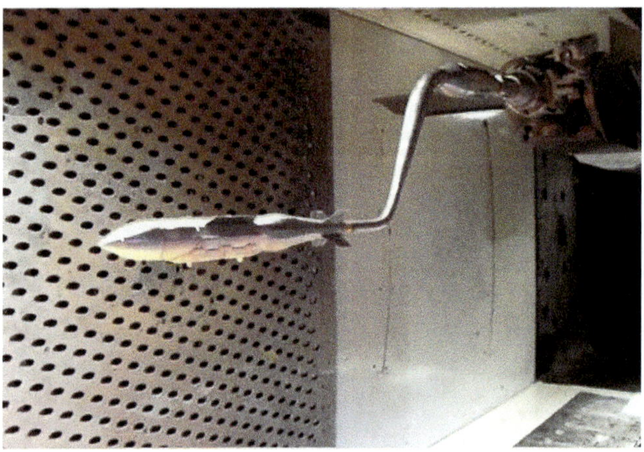

Fig. 21 Freestream.

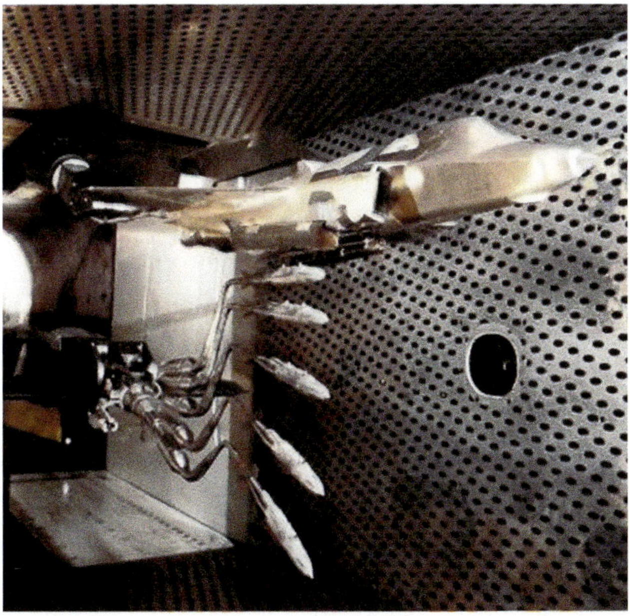

Fig. 22 Flowfield CTS.

carriage as practical. Measured store aerodynamics forces are input into the wind tunnel facility's trajectory simulation software, where 6DOF equations of dynamic rigid-body motion are solved in time, for the aircraft flight condition of interest, and the resulting "next time step" store position is output to the store sting controller, resulting in a store movement to this next position. Store ejection forces and constraints are usually included in the online trajectory simulation model, as well as weapon supplier-provided black-box autopilot control and propulsion forces. This mutual-feedback measure/simulate process results in a simulated store trajectory, which is used in several ways. The collection of trajectories results in a range of combined store positions and attitudes, which can influence the test's aerodynamic flowfield database grid density and ranges, while at the wind tunnel. Online trajectory simulations at the wind tunnel are usually a quick source for comparing trajectories of different aircraft weapons loadouts. During and after the wind tunnel test, comparisons of 1) simulations involving the same exact trajectory aerodynamics, and 2) simulations involving the grid-based aerodynamics database are conducted—"wind tunnel emulation" simulations can provide additional insight into grid-based aerodynamic model contributions to uncertainties in aerodynamics forces. More importantly, these wind tunnel emulations are used to ensure that offline aerodynamic grid-based simulations adequately replicate wind tunnel online trajectory simulations.

5. COMPUTATIONAL FLUID DYNAMICS

Although the industry-standard source for store aerodynamics models has usually been wind tunnel tests, CFD is gradually gaining acceptance. During the F-35 SDD phase, CFD has not been a primary source for store separation aerodynamics database models, primarily because a rigorous comparison with flight test–derived aerodynamics has not yet been completed. Many comparisons between CFD and wind tunnel have been made, with more-favorable comparisons taking place for underwing positions vs in-bay positions. Ultimately, however, the more important CFD comparisons will take place between CFD-based and flight test–derived "truth" aerodynamics, instead of between CFD-based and wind tunnel–based aerodynamics. The derivation of flight test "truth," or flight test–derived aerodynamics, is presented in the flight testing section of this paper.

During F-35 SDD, store separation CFD analysis has been used as a low-cost alternative to wind tunnel testing [16]. For instance, CFD was used to identify similar flowfield regions among the three different F-35 aircraft variants, in order to more efficiently perform wind tunnel testing. During such CFD studies, it was discovered that underwing store separation testing could effectively be performed with either the F-35A or F-35B variant (Fig. 23), and the wind tunnel store separation aerodynamics were applicable for both variants. Furthermore, it was also discovered that transonic in-bay store separation wind tunnel testing on either the F-35A or F-35C was equally applicable to both variants (Fig. 24). It was not necessary to perform duplicative store separation wind tunnel testing using all three F-35 aircraft variant models.

Store configuration effects on the aircraft flowfield were also analyzed. In the early stages of GBU-39 Small Diameter Bomb (SDB) integration on the F-35, the GBU-39 fins were set to deploy at a time after first motion such that the store was out of the region of influence of the aircraft flowfield. F-35/GBU-39 wind tunnel testing and preliminary store separation analysis were therefore conducted in a fins-stowed store configuration. Following this F-35/GBU-39 wind tunnel testing, the default fin deployment timeline was shortened during GBU-39

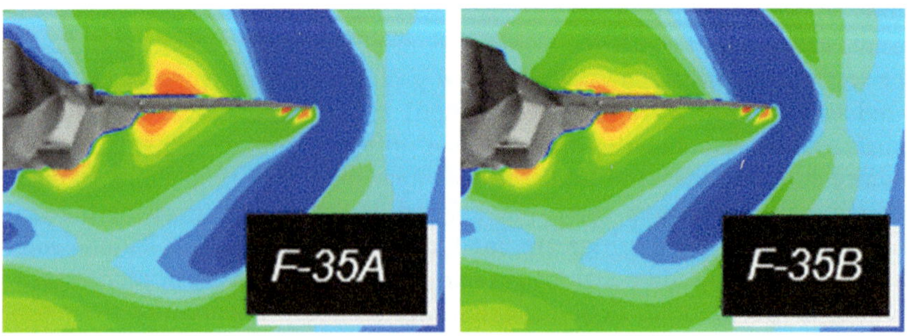

Fig. 23 Flowfield comparison.

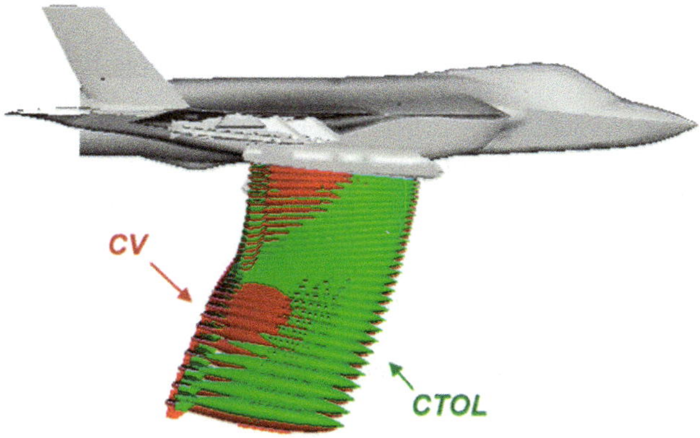

Fig. 24 Variant comparison.

development testing on non-F-35 store separation programs. As a result of this store change, when released from the F-35, the GBU-39 fins would then deploy while the store would still be in the region of influence of the F-35 aircraft flowfield. There were concerns that the underbay flowfield might be different for the GBU-39 with fins stowed and the GBU-39 with fins deployed. With F-35/GBU-39 flight testing on the near-term schedule, CFD analysis was chosen as an efficient, cost-effective method of quantifying the flowfield differences with GBU-39 in the fins-deployed configuration, possibly without the need for additional wind tunnel testing.

Simulated trajectories with a purely fins-stowed configuration (Fig. 25) still applied for the GBU-39 jettison case (cyan store). For the GBU-39 employment scenario with fins deploying early (green, yellow, and red stores), simulated store trajectories that used the supplemental CFD aircraft flowfield coefficients from the GBU-39 in the fins-deployed configuration were similar to simulated trajectories using wind tunnel aircraft flowfield coefficients derived from the GBU-39 in the fins-stowed configuration. Because of the availability of CFD analysis and the aforementioned trajectory similarities, store separation analysts were able to

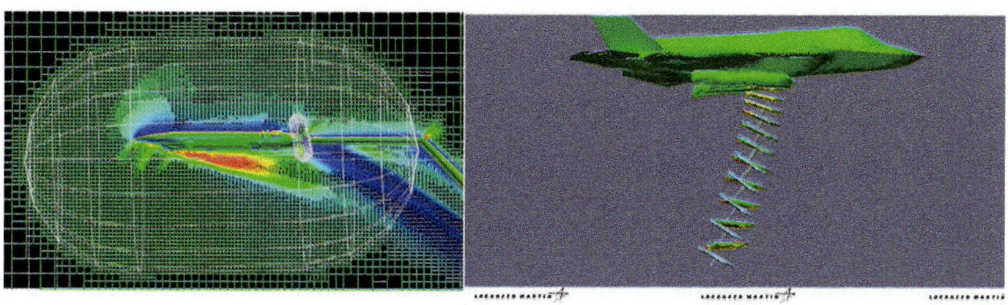

Fig. 25 GBU-39 CFD with fins deployed (left); simulated trajectories (right).

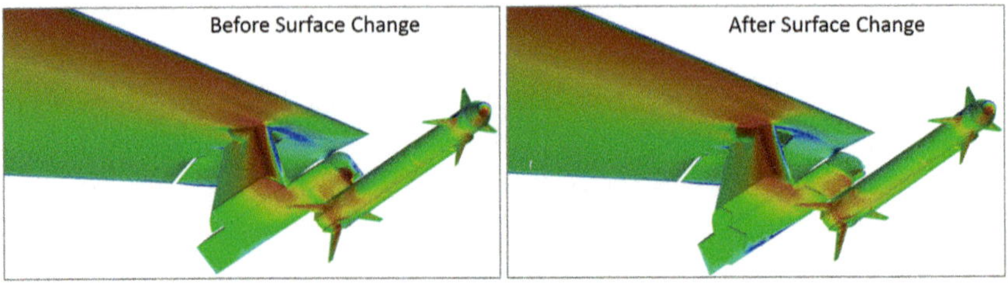

Fig. 26 AIM-9X surface pressure comparison LAU-148 vs LAU-151.

conclude that the fins-deployed effect on the aircraft flowfield was negligible. Unfortunately, aerodynamics modeling schemes, such as combining store freestream and flowfield aerodynamics to obtain total aerodynamics, are beyond the scope of this paper. However, it was concluded that for simulations with the fins-deployed configuration, complete GBU-39 store aerodynamics, as a combination of fins-deployed GBU-39 freestream aerodynamics plus fins-stowed GBU-39/F-35 flowfield aerodynamics from the original F-35/GBU-39 wind tunnel test, were applicable; the adverse budget and schedule effects of an additional wind tunnel test were avoided.

CFD has also been used to quickly assess changes in the air-to-air launcher surface geometry, with respect to quantifying differences in aerodynamic forces on the launching AIM-9X missile (Fig. 26). In performing such a comparison, the goal was to demonstrate that it was reasonable to conclude that the AIM-9X aerodynamics model and store separation flight test trajectory results applied equally for two different launchers.

Postflight CFD studies have also been used, for example, to confirm that differences between the GBU-32 JDAM wind tunnel–based aerodynamics model and flight test–derived aerodynamic forces were likely due to store rearward movement into a high pressure gradient region underneath the aircraft aft end (Fig. 27). This particular study highlighted the importance of obtaining store aerodynamics grid data, beyond the limits of the fore–aft grid that were included in the relevant wind tunnel test.

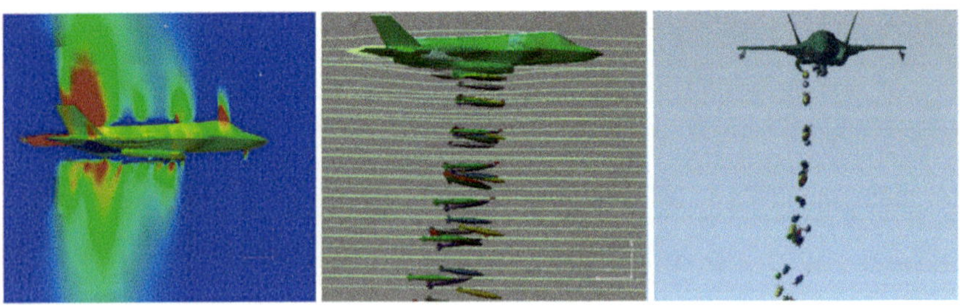

Fig. 27 CFD to investigate flight test store aero differences.

V. GROUND TESTS: VALIDATION OF THE DYNAMIC SYSTEM MODEL

The usefulness and value of both detailed dynamic systems models and aerodynamics models diminish if such models are not judged and validated, or appropriately adjusted, to be representative of the real dynamic system and real aerodynamics. There are a series of ground-based validation tests (Fig. 28) in which information is gathered and used for preflight *system model validation*, whereas store *aerodynamics model validation* can only be performed using information gathered in flight test.

Recall that the dynamic system includes the store, S&RE, and aircraft. Sometimes it is impractical or difficult to successfully validate the full system that includes the aircraft. Even though individual component behavior may change when it becomes part of a larger system of components, it remains good practice to validate isolated component models before attempting to validate the entire system, especially because it can sometimes be difficult to isolate areas of concern, in the event that full-system validation becomes unsatisfactory.

A. STATIC S&RE EJECTION TEST

Hundreds of BRU-67, BRU-68, and LAU-147 static rack ejection tests were conducted by the rack supplier, EDO LLC, in support of required performance, environmental, durability, and qualification testing. These ejections were conducted at room temperatures, at hot and cold temperatures, at various starting piston accumulator pressures, and with various ejector pitch performance settings

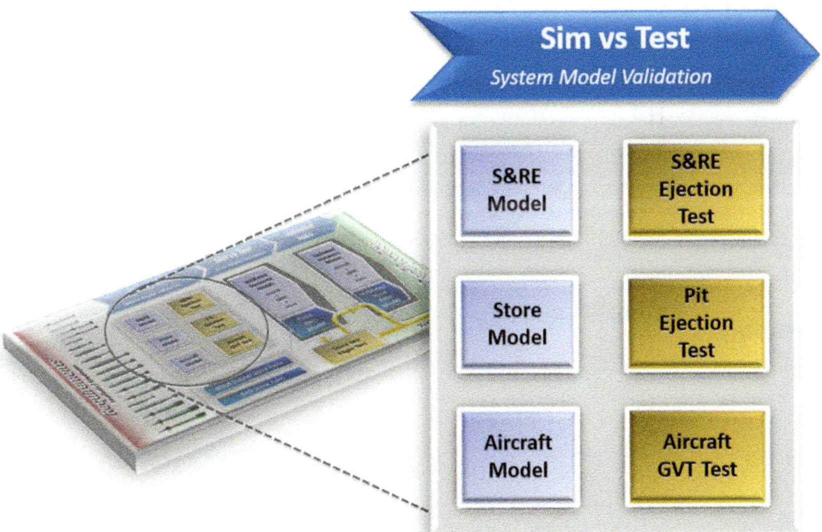

Fig. 28 Dynamic system model validation.

(if applicable). Because the BRU-67 and BRU-68 racks were designed to carry and release different store weight classes, 500-lb, 1000-lb, and 2000-lb stores were ejected. In addition, different racks with different production serial numbers were used. Furthermore, acceptance testing requirements are such that every ejector rack must be ejection-tested at least once, prior to customer delivery. In all of these tests for the F-35's BRU-67, BRU-68, and LAU-147 ejector racks, store EOS vertical velocity and pitch rate were measured or computed. This wealth of data was useful for quantifying rack-to-rack performance variation for uncertainty simulations/analysis.

Lockheed Martin's BRU-67, BRU-68, and LAU-147 S&RE ejector system models are based on EDO's mechanism model and include EDO's proprietary black box pneumatic piston force gas model, so the ejector rack S&RE model validation consisted of comparing Lockheed Martin's ASEP ejection simulations of the BRU-67 and BRU-68, using 500-lb, 1000-lb, and 2000-lb store models with mass properties identical to those in EDO's tests (Fig. 29). The same model validation was performed for the LAU-147 S&RE ejector system model, using a store model with the same mass properties as EDO's AIM-120-like test store. Lockheed Martin's test-validated BRU-67, BRU-68, and LAU-147 S&RE system models were the basis for store separation preflight predictions.

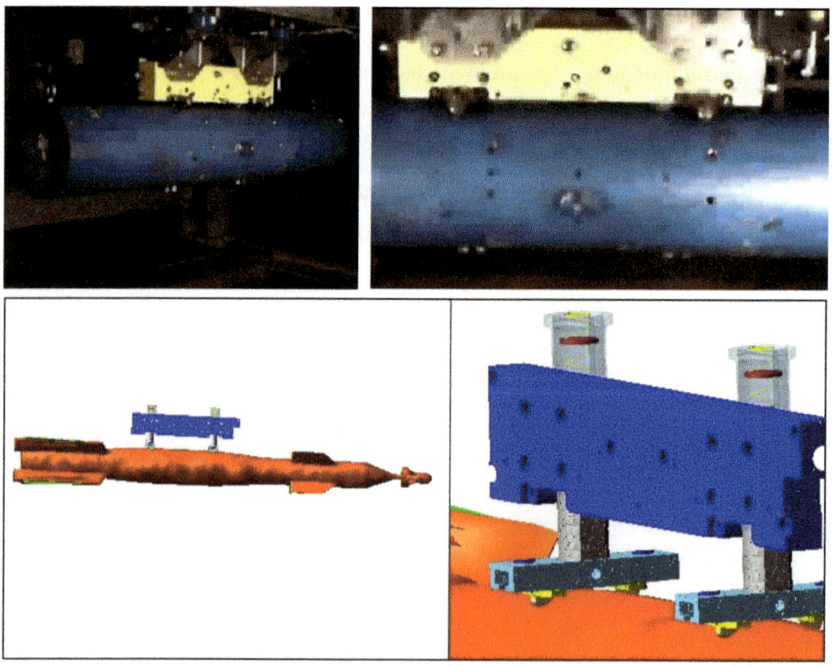

Fig. 29 Rack static ejection test.

Unfortunately, in the alternate dynamics modeling scenario of using a simple S&RE model, where only EOS displacement and velocity initial conditions are specified for specific store types and mass properties, these EDO test data are useful only for establishing nominal simulation store initial condition vertical velocity and pitch rates if the store separation simulations are performed with store mass properties identical to those tested by EDO.

B. STORE VIBRATION TEST

When the dynamic response of flexible dynamic stores was important, such as during AIM-9X and AIM-132 air-to-air missile launches, the store and pylon system simulation models included a flexible representation, usually from the store supplier's finite element model. The store supplier typically performs its own finite element model validation using store vibration tests. Lockheed Martin's flexible missile store models used test-validated AIM-9X and AIM-132 missile FEM models from Raytheon Missile Systems and MBDA Missile Systems, respectively.

C. GROUND VIBRATION TESTS (GVT)

The Lockheed Martin dynamic loads integrated product team conducted on-aircraft ground vibration tests to validate finite element models, which included flexible aircraft, pylon, launcher, and missile models, in addition to contact stiffness or freeplay in various degrees of freedom at every wing-to-pylon attachment bushing, every pylon-to-launcher attachment point, and every launcher rail channel-to-missile hanger/lug contact point. After adjusting their own flexible dynamics systems models, the dynamic loads team also similarly adjusted the flexible ASEP models shared with the store separation team. These GVT test-validated flexible AMR/S&RE system models were used in store separation preflight predictions for rail-launched flight test events.

D. ON-AIRCRAFT STATIC EJECTION PIT TEST

Some 158 on-aircraft SDD store ejection events were conducted from three F-35 test aircraft that were used for a majority of store separation flight testing, with each of these aircraft variants parked (and resting on the landing gear) over a pit filled with foam (Fig. 30). The main purposes were to 1) validate the aircraft-installed S&RE ejector rack model performance, 2) validate S&RE-to-aircraft loads models, 3) verify on-aircraft stores management system communication and function prior to flight testing, and 4) verify on-aircraft instrumentation function prior to flight testing. The on-aircraft static ejection pit test was the only opportunity to perform store + S&RE + aircraft dynamic system model validation, prior to flight testing, using the same aircraft and weapon stations, store types, and instrumentation that would ultimately be used during subsequent store separation in-flight testing.

Fig. 30 On-aircraft static ejection pit test.

During system model validation, aircraft landing gear models were also included in the system model to replicate the as-tested system, where during the downward ejection event, the aircraft always reacted with an upward motion with some aircraft roll. The general effect was a slight reduction in ejector forces and performance, relative to a nearly rigid S&RE-only pit ejection test, while also imparting some store EOS roll motion during ejections taking place at external wing stations. At that time, there was an expectation that the in-flight aircraft would also exhibit an upward and roll response during store ejection, which later proved to be true. After on-aircraft static ejection pit tests, the validated store + S&RE + aircraft system models were used in final preflight test store separation trajectory predictions for all of the planned flight test events, during which validation of the store aerodynamics model was expected to occur.

VI. TYPICAL STORE SEPARATION FLIGHT TEST INSTRUMENTATION

A brief description of typical store separation flight test instrumentation analysis is included in this section, in recognition of its existence prior to the F-35 flight test program. These same types of instrumentation would eventually be used during the F-35 store separation test program, so an understanding of the basic instrumentation was necessary. F-35-specific planning, considerations, and additional uses are described in the next section.

The act of judging the extent to which store separation simulation matches flight test has typically been focused on "what happened"—comparing simulated and actual store displacement trajectories and rotational rates. The determination of the actual store displacement trajectory and store rotational rates has evolved from estimation to high-confidence calculations and direct measurement, using one or both of the following: photogrammetric analysis of on-aircraft weapon separation camera video, and numerical integration of measured translational accelerations and rotational rates from either an in-store IMU or a 6DOF telemetry kit, collectively called "6DOF/IMU TM Kit" herein [17–19].

A. PHOTOGRAMMETRIC ANALYSIS

Video-dependent high-quality photogrammetric analysis results in an accurate computation of store translational and rotational displacement, relative to the aircraft. Such accurate computations are usually completed within 8–16 hours, and careful numerical differentiation of this store position solution results in a store velocity solution, and even a store acceleration solution. Photogrammetric analysis includes using high-capture-rate video imagery (Fig. 31) of the store and aircraft, and uses reference locations, such as high-contrast photogrammetry (photoG) targets and store geometric features.

1. PHOTOGRAMMETRY PROS

The direct product of the photogrammetry solution is position, and is therefore considered the position "truth," which is especially useful in computing store miss distance to aircraft hardware.

The photogrammetric position solution is usually a better initial condition starting point for complementary 6DOF/IMU TM Kit–based trajectory solutions, especially when the store, S&RE, and/or airframe maneuver-dependent flexibility and free-play results in mismatch between ideal CAD position and actual inflight carriage position.

Even if targets become obscured, a successful position solution is still possible by overlaying CAD wireframe onto video imagery, if the store outline is still visible.

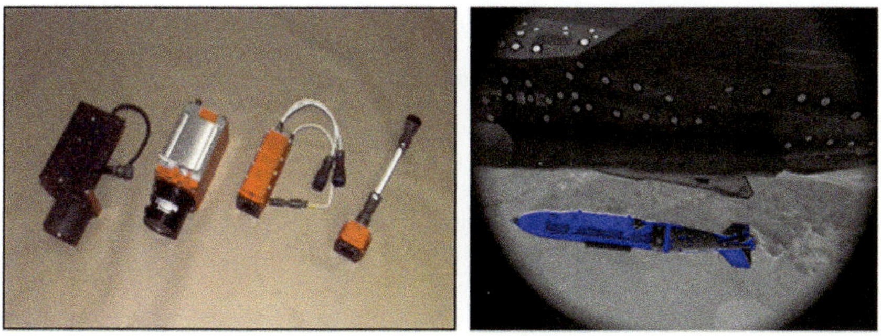

Fig. 31 Store separation video cameras [17] and typical overlaid solution.

The cameras required for photogrammetric analysis are also useful for capturing video evidence of other events, such as store, S&RE, or lanyard function.

2. PHOTOGRAMMETRY CONS

Too much or too little light may render imagery unusable for photogrammetric analysis. In addition, vapor condensation may obscure store and/or reference geometry and targets. Also, unless properly accounted for, vibrating cameras may decrease solution accuracy. Further, unless properly smoothed, numerically differentiated positions may produce noisy velocity solutions, and especially noisy acceleration solutions. Compounding this, typical video frame rates of 200–400 frames per second may not provide enough resolution for computing store aerodynamics from acceleration solutions.

B. 6DOF/IMU TM KIT ANALYSIS

The main strength of the in-store 6DOF/IMU TM Kit (Fig. 32) is that it provides a direct measurement of translational acceleration and rotational rate relative to the ground; when combined with numerical integration of the translational acceleration, the result is regarded as the best source of store 6DOF "velocity truth." An added strength is that 6DOF/IMU TM Kit measurements are usually available immediately to engineers in the flight test control room, as the store separation flight test event takes place. Careful numerical integration of the store velocity

Fig. 32 6DOF TM Kit with store nose installation [19].

results in a store position solution, relative to the aircraft, as long as the aircraft motion is accounted for.

1. 6DOF/IMU TM KIT PROS

The output is a telemetered fast and direct measurement of store rates and accelerations, so little to no data manipulation is needed to interpret basic store behavior, thereby enabling in-control-room "continue to next point" decisions for missions with multiple separation events per flight.

High sample rate ($>$ 4500 samples per second) enables many filtering strategies, which is useful when differentiating rotational rates to obtain rotational accelerations.

Modern 6DOF TM Kits have 1) delayed turn-on, to save battery during long missions; and 2) delayed transmission, which increase the opportunity to resolve regions of data loss by "stitching" recorded real-time and delayed transmission data streams.

2. 6DOF/IMU TM KIT CONS

Measured data are telemetered, sometimes resulting in data dropout and/or noise. Also, computing accurate store positions, relative to the maneuvering aircraft, requires accurate initial conditions and knowledge of aircraft maneuver response. In addition, data sometimes require alignment/nulling with aircraft rates and accelerations. Further, sometimes data from all 6DOF are not properly time-synchronized.

3. FURTHER ANALYSIS

It may initially appear as if redundancy exists when both of these flight test data sources are available, because they can indeed be used to compute similar information; however, these sources should instead be considered as complementary [20]. For instance, during several F-35 in-bay store separation events, the 6DOF/IMU TM Kit data stream had excessive or unresolved data noise while the store was still in contact with the S&RE, which resulted in a "TM Kit-solution" drift in store position and velocity, resulting in inaccurate late-trajectory solutions; however, when the photogrammetry solution was substituted for the TM Kit solution in the early separation phase, the late-trajectory solution accuracy was drastically increased. Furthermore, during approximately 10% of the missions during F-35 SDD store separation flight testing, one data source failed or became unusable, but the availability of the other data source resulted in data sufficient for posttest analysis, without the need to refly the test point. Many times, the more accurate photogrammetry position "truth" was needed in the earlier part of the separation trajectory where store miss distance to aircraft hardware was critical, whereas the more accurate (and higher sample-rate) 6DOF TM Kit store velocity and acceleration was needed after the store became free from the S&RE, so that the critical store aerodynamics computations could be performed.

C. F-35: USE THE SAME INSTRUMENTATION FOR AERODYNAMICS MODEL VALIDATION

Prior to the F-35 program, typical store aerodynamics models were adjusted during a trial and error process where postflight simulations were iteratively performed with simple modifications, such as scaling or shifting/biasing, to aerodynamic coefficients in one or more degrees of freedom. The criteria for successful aerodynamics model validation was subjective, and usually based on comparing successive postflight simulated trajectories or rates with those computed from the available instrumentation, such as on-aircraft high-speed video or store 6DOF/IMU TM Kits. Favorable comparisons were not always achieved in all six degrees of freedom.

During the F-35 store separation program, Lockheed Martin implemented a new strategy of also including a focus on where and how it happened—comparing simulated and actual store applied aerodynamics, in which the quality of the simulation aerodynamics model is directly judged in all six degrees of freedom, as a more-detailed function of distance from carriage. Using the same type of instrumentation that was used during prior store separation test programs, Lockheed Martin has created a rapid and repeatable process for computing accurate store 6DOF aerodynamic forces, and adjusting the aerodynamics model, such that the validated postflight simulations reproduce the exact flight test store trajectory in all six degrees of freedom. So, iterative trial and error adjustments and subjective interpretation have been replaced with a standard repeatable process. However, successful store aerodynamics model validation requires extra attention to the quality and detail of the flight test instrumentation, which is the subject of the following section.

VII. F-35 STORE SEPARATION FLIGHT TESTING: VALIDATING STORE AERODYNAMICS MODELS

Recall that the basis for both Lockheed Martin's certification recommendation and verification that the weapon release requirements have been satisfied is flight test–validated M&S—modeling and simulation that has been adjusted based on data gathered during the store separation flight test events, so that simulated store separation events replicate the actual store separation events. It is generally reasoned that if the store trajectory and rotational rates are similar between simulation and flight test, then the total forces and moments acting on the store during the simulation are representative of those acting on the store during the actual event, and the simulation is, therefore, validated. Ideally, the store separation preflight predictive simulation should match that of the flight test event. In reality, a nominal prediction doesn't always match its corresponding actual flight test trajectory. When such discoveries occur during flight testing, it is important to understand and explain such differences as quickly as possible, in order to adjust the model and prediction simulation for the next store separation flight test event, without adversely affecting the flight testing schedule.

In *both* the actual and simulated store separation events, the following occur:

1. Six degrees of freedom total forces and moments are applied to the specific store, which results in
2. Store translational and rotational acceleration, which then results in
3. Changes in store translational and rotational velocity, which then results in
4. Changes in store translational and rotational displacements.

The actual aircraft maneuver response, which may be different than that during a nominal simulation, must also be considered. So, *the simulated store separation event is only identical to the real event when all of the following are true*:

- Simulated and real store mass properties are identical.
- Simulated and real aircraft response are identical.
- Simulated and real store 6DOF applied forces and moments are identical.

Measurements (or calculations from measurements) of these real items are the basis for postflight model adjustments. The simulation and model, which includes the aerodynamics model, become validated (Fig. 33) only after verification that simulated and real store 6DOF response (displacement trajectory, velocity, and acceleration) are identical. Measurement sources were identified as follows:

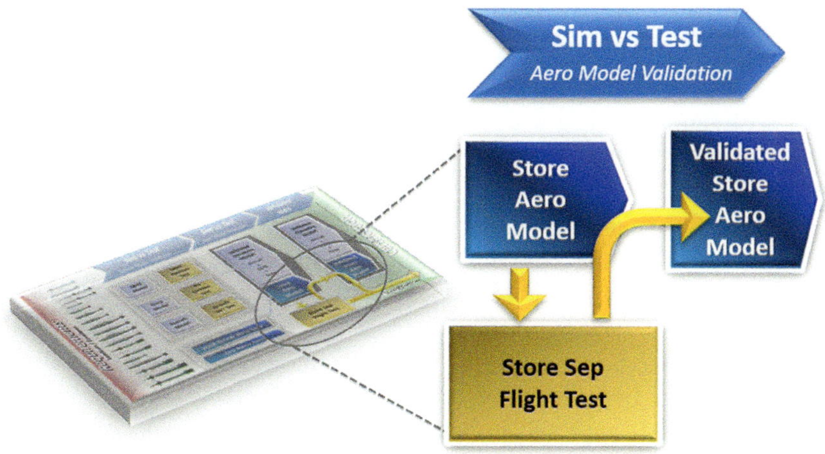

Fig. 33 Flight test and aerodynamics model validation.

- Store mass properties: Preflight measurements
- Aircraft response: On-aircraft instrumentation
- Store displacement, velocity, and acceleration: 6DOF/IMU TM Kit and on-aircraft camera photogrammetry
- Store aerodynamics: 6DOF/IMU TM Kit

The requirement of performing mass properties measurements for each store test asset prior to flight testing ensured the availability of store mass properties. Availability of aircraft maneuver response measurements was reasonably assured, because this information was telemetered to the control room and also recorded by the on-aircraft data acquisition system; however, to ensure high-confidence photogrammetric and 6DOF/IMU TM Kit postflight analyses, which were critically important to successful aerodynamics model validation, Lockheed Martin store separation engineers coordinated with data processing, instrumentation, survey/metrology, mass properties, flight test, and weapons experts on the F-35 test team to plan and enact standard processes in preflight measurement, in-flight data acquisition, and postflight computation and comparison/validation.

With M&S validation as the primary reason for store separation flight testing, and because of the importance of the quality of these measurements, an understanding of planning and processes are relevant and are described herein.

A. PLANNING PLAYBACK OF THE REAL FLIGHT TEST EVENT

In consideration of the moving parts and associated instrumentation during store separation flight test events, reviewing the store separation event would involve the playback of three items: the maneuvering aircraft, the store trajectory, and the 6DOF/IMU TM Kit response. All three would be treated as unique rigid bodies in a single-simulation playback of the event within the Lockheed Martin store separation modeling and simulation environment, ASEP. Using ASEP's simple dynamicsm model, described earlier, these three moving rigid bodies would be forced or constrained to move:

- The aircraft body would be "flown" according to on-aircraft response measurements, using customized artificial applied forces that yield the same aircraft rates and accelerations from the on-aircraft instrumentation measurements.
- The store would be constrained to move according to the trajectory displacements solution (provided by photogrammetry engineers).
- The 6DOF/IMU TM Kit body would be "flown" according to the 6DOF/IMU TM Kit response measurements, using customized artificial applied forces that yield the same 6DOF/IMU TM Kit rates and accelerations.

The common ingredient in the movement of each of these bodies within the playback simulation is the set of simplified rigid body equations of motion, Eqs. (1–6):

$$\Sigma F_x = ma_x \tag{1}$$

$$\Sigma F_y = ma_y \tag{2}$$

$$\Sigma F_z = ma_z \tag{3}$$

$$\Sigma M_x = I_{xx}\dot{\omega}_x - (I_{yy} - I_{zz})\omega_y\omega_z \tag{4}$$

$$\Sigma M_y = I_{yy}\dot{\omega}_y - (I_{zz} - I_{xx})\omega_z\omega_x \tag{5}$$

$$\Sigma M_z = I_{zz}\dot{\omega}_z - (I_{xx} - I_{yy})\omega_x\omega_y \tag{6}$$

- This form of the common $F = ma$ equations of motion was used, simply as a result of the available measurements.
- Because products of inertia were unavailable, the rigid store body axes were conveniently assumed to also be the principal axes.
- All three parts (aircraft, store, and 6DOF/IMU TM Kit) were treated as rigid, without rotors, liquid sloshing, and so forth.

The ASEP-specific application of these simplified equations of motion to each of the three bodies is described in relevant sections that follow.

B. STORE MASS PROPERTIES: PREFLIGHT MEASUREMENTS

As observed in Eqs. (1–6), the store mass and inertia must be known, in order to eventually compute the total forces and moments. During 6DOF/IMU TM Kit posttest processing, the relative positions of the actual store's center of gravity and the actual 6DOF/IMU TM Kit must be known. Furthermore, store features, such as photogrammetry reference targets, fin and strake corners, S&RE attachment locations (such as bomb lugs or missile hangers), nose, tail, and other body references must be known well before the store disappears into the flight test range's desert floor or ocean bottom at the end of the store separation test event.

Nominal mass properties are usually provided by the store supplier, but the nominal store center of gravity sometimes differs from tested stores, which typically results in simulation vs test EOS pitch rate performance differences. So, when performing validation exercises, or comparisons between test results and pretest/posttest simulations, it was necessary to be confident that the as-tested store mass properties were also the ones used in simulations. In order to provide measurement confidence and to quantify measurement uncertainty, several types of stores were mass-propertied, repeatedly, by different personnel at the same and different test sites. Part of the store separation preflight process was to ensure that all stores used during store separation on-aircraft pit ejection testing and flight testing were mass-propertied by the ordnance team at PAX, EDW, or

Fig. 34 GBU-12 mass properties and model.

AFSEO, where the store mass, center of gravity, and mass moments of inertia were quantified from a combination of measurements and calculations. The store was required to be in its as-tested configuration, complete with fins, strakes, and even installed instrumentation, such as the 6DOF/IMU TM Kit. As a result, every unique store separation test asset was represented by its equivalent ASEP model (Fig. 34).

C. AIRCRAFT RESPONSE: ON-AIRCRAFT INSTRUMENTATION

During store separation events with an air-to-ground store, the S&RE ejector forcibly pushes the store away from the aircraft; however, because the S&RE, store, and aircraft are all part of the dynamic system, it was reasonable to expect that the aircraft would also have a reaction to this S&RE ejector impulse. Furthermore, with the reality of occasional atmospheric turbulence, and a potential for aircraft transient reaction to opening weapon bay doors, there was sufficient reason to believe F-35 store separation flight test aircraft would inevitably experience transient responses during store separation events.

Considering reference frames, the 6DOF/IMU TM Kit provides measurements of store rates and accelerations relative to an inertial (Earth/ground) reference frame, independent from the separately moving aircraft; however, photogrammetric store trajectory solutions are relative to the moving aircraft, because the store separation cameras are attached to the aircraft. So, 1) transforming 6DOF/IMU TM Kit solutions into the moving aircraft's reference frame, and 2) transforming photogrammetric store trajectory solutions into an inertial reference frame both require consideration of the aircraft motion.

Furthermore, basic aircraft flight condition information is necessary to transform the computed aerodynamic forces and moments into nondimensional aerodynamic coefficients familiar to aerodynamicists, because the main goal in aerodynamics model validation is to compare these flight test–derived store aerodynamic coefficients to those used in the simulation.

So, as a standard process for every store separation flight test event, when post-processing store 6DOF/IMU TM Kit data for the purposes of computing a store's trajectory relative to the aircraft (and ground), Lockheed Martin has also included the aircraft response. The final motivating reason for this approach is simply to include all relevant data and avoid shortcuts based on assumptions, especially if these assumptions have the potential for adding confusion when troubleshooting potentially problematic postflight solutions, which could cause unnecessary delays in continuing to the next test point.

During store separation event playback simulations within ASEP, the aircraft rigid body was "flown" by applying customized artificial forces at the aircraft body's center of gravity, according to Eqs. (1 – 6), with the following considerations:

- The applied forces onto the aircraft body are dependent on measured aircraft response, where a_x, a_y, and a_z are the measured/transferred translational aircraft accelerations, and ω_x, ω_y, and ω_z are the measured aircraft rotational rates (Fig. 35). Simple differentiation of these aircraft rotational rates yields aircraft rotational accelerations: $\dot{\omega}_x$, $\dot{\omega}_y$, and $\dot{\omega}_z$.
- These applied forces are also dependent on the aircraft model's mass m and moments of inertia I_{xx}, I_{yy}, and I_{zz}; however, in order to ensure that the ASEP playback aircraft model response is not influenced by other forces (which are beyond the scope of this paper), the aircraft body's mass and moments of inertia were artificially increased by several orders of magnitude, in both the equations of motion and the aircraft rigid body ASEP model to which the aircraft "flying" forces are applied.
- The initial condition is coincident with store first motion, or the moment the store separation event begins. When examining aircraft measurement parameters, this moment is identified by certain behavior in the aircraft response, such as a sudden change in the aircraft roll rate and vertical acceleration, in

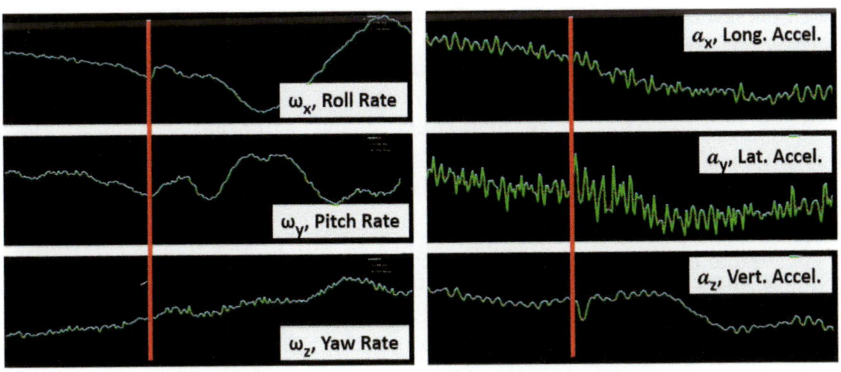

Fig. 35 Aircraft rate and acceleration response.

response to strong store ejection forces, for instance. The red vertical line in Fig. 35 marks this point in a typical flight test aircraft response during an air-to-ground store separation event.

- The aircraft body's initial playback flying conditions are set by other aircraft measurements, such as Mach, angle of attack, angle of sideslip, altitude, bank angle, pitch angle, yaw rate, pitch rate, and roll rate.

The PAX and EDW aircraft maneuver response data were usually available to the store separation engineers in Fort Worth within four hours after the flight test event.

D. STORE DISPLACEMENT: ON-AIRCRAFT VIDEO PHOTOGRAMMETRY

When validating the store aerodynamics model database, the flight test–derived aerodynamic forces are ultimately compared to the store aerodynamics model. The main input into the store's aerodynamics model is the store position and orientation. Furthermore, the photogrammetry-based trajectory solution is the best store position "truth" with which to sanity-check 6DOF/IMU TM Kit–based trajectory solutions. In addition, the primary consideration in safe store separation is store clearance to aircraft hardware and adjacent stores. For these aforementioned reasons, it is critically important to confidently know the real store trajectory positions and orientations.

Photogrammetric analysis includes using video imagery of the store and aircraft, and is dependent upon reference locations, such as high-contrast photoG targets and store geometric features; the key requirement for such video imagery is that these store and aircraft reference locations are visible and identifiable. So, prior to F-35 flight testing, several tasks were undertaken to ensure that reliable and accurate photogrammetric analyses could be performed for every SDD store, at aircraft station, on each of the three F-35 variants.

1. CAMERAS: LOCATION, ORIENTATION, AND FIELD OF VIEW

Well before flight testing, the store separation test aircraft were not available, so Lockheed Martin's Flight Test Instrumentation and Store Separation groups worked together to solidify high-speed digital Airborne Separation Video System (ASVS) camera locations and orientations on the three F-35 variants. This process involved assessing simulated CAD image views, with identical fields of view as the real ASVS cameras (Fig. 36).

For determining store position relative to the aircraft, Lockheed Martin's photogrammetric software, Digital Store Separation Analysis System (DSSAS) relies on visibility of at least three or four store location references from each camera, and store position computation accuracy is increased with additional location references and multiple camera views. Even though DSSAS is capable of producing reasonably accurate photoG solutions from as few as one camera's

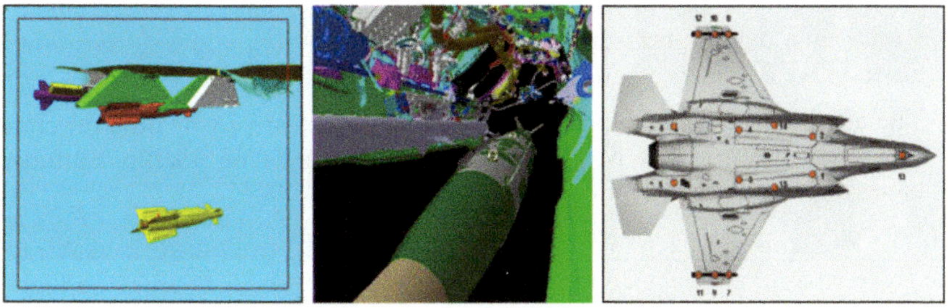

Fig. 36 Virtual camera views and flight test camera layout.

imagery, there is a need for achieving more than reasonably accurate photoG solutions, due to expected low store clearance distance from in-bay hardware. Although multiple cameras are used to capture imagery during a store separation flight test event, it is sometimes not known whether one or more cameras will malfunction or provide store imagery that is washed out due to bright sunlight, darkened due to poor ambient lighting, or obscured by water vapor condensation (Fig. 37). Although there was consideration of attempting prediction of likelihood of flowfield-induced water vapor condensation, based on forecasted atmospheric conditions [21], it was believed that flight test missions would not be cancelled due to the possibility of store-obscuring water vapor condensation, if other cameras could capture unobstructed store views.

Therefore, a collection of 15 ASVS camera locations were selected for F-35 store separation flight testing, in order to capture store imagery during separation events from the left-side and right-side weapons bays and external wing stations.

2. STORE TARGET LAYOUT

Because it was not practical to select specific store test assets and assign them to separation testing at specific aircraft weapon stations, each store would require sufficient coverage of high-contrast photoG targets so that a sufficient number

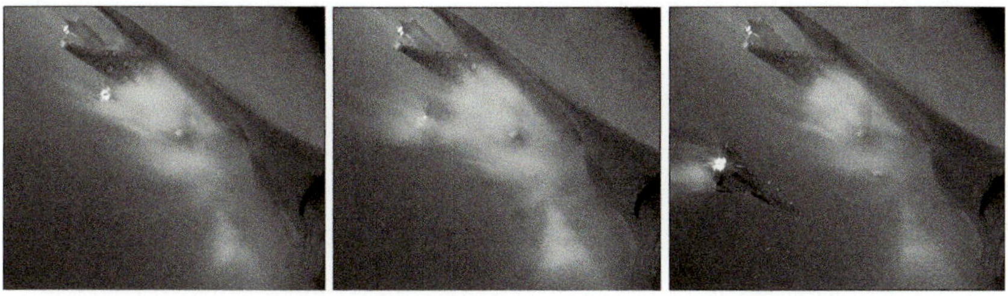

Fig. 37 Water vapor condensation temporarily obscuring separating store.

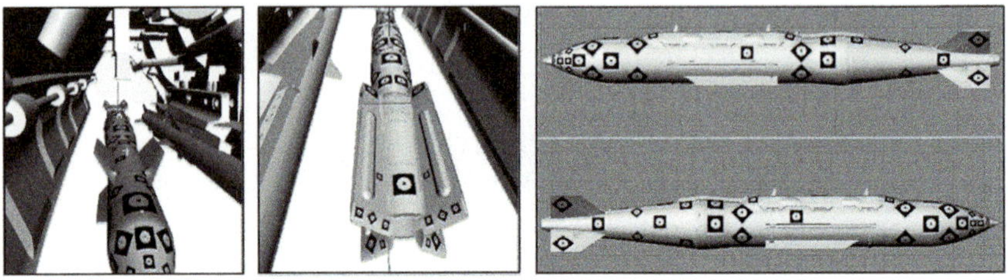

Fig. 38 Virtual store target layout.

of these targets could be identified (seen) within any of the 15 ASVS camera fields of view. It was also impractical to use the actual flight test aircraft to load weapons in order to view ASVS imagery of all SDD store types at all weapons stations and interactively assess photoG target schemes. So, with recognition of the balance between 1) too many photoG targets, which would result in time-consuming store metrology survey exercises for each store; and 2) too few photoG targets, which could result in inaccurate photoG solutions, Lockheed Martin relied on CAD and AEDC's store trajectory visualization software, TVIS, to place virtual photoG targets on each store (Fig. 38) and aircraft (Fig. 40), and to assess the resulting target schemes as seen by each camera's field of view.

For every store that was to be used in store separation flight testing, the PAX and EDW metrology experts applied photoG target stickers onto the store, per the virtual photoG target scheme. Finally, metrology surveys (Fig. 39) were performed twice, for confidence, to measure 1) all of these target locations, and 2) key features of the store, such as lug locations, nose, tail, fin/strake root and corner points, and even installed 6DOF/IMU TM Kit location and orientation. Because the 6DOF/IMU TM Kit's measured rate and acceleration data would be used in post-test computations at other store reference locations, it was necessary to confidently know the position and orientation of the 6DOF/IMU TM Kit (and its sensors) in the store reference frame.

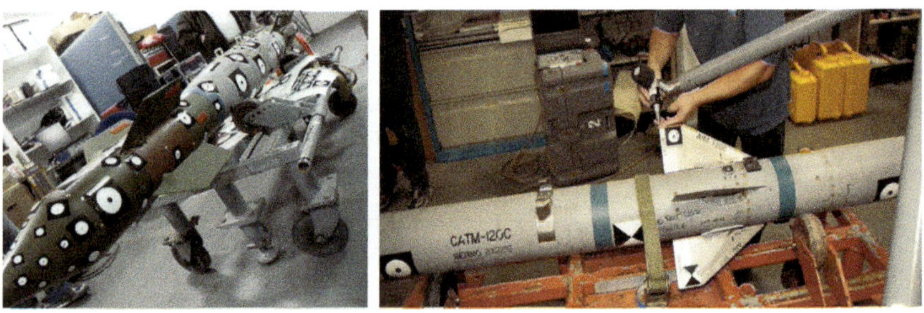

Fig. 39 Store metrology survey.

Fig. 40 Virtual and real aircraft photogrammetry targets.

3. AIRCRAFT TARGET LAYOUT

Aircraft-mounted ASVS cameras can move or vibrate in flight, which means that the ASVS camera focal point's location and field-of-view orientation can move from its nominal installed location, relative to the aircraft reference frame; however, with visibility of known references, it is possible to know where the camera moved. DSSAS relies on seeing known and stationary location targets on the aircraft, within the weapon bay and on the exterior surface. This enables the instantaneous computation, within DSSAS, of the instantaneous location and orientation of each ASVS camera, so that the final DSSAS photoG solution for the store position is relative to the fixed aircraft reference system, instead of the moving camera system's reference frame.

Using the virtual aircraft photoG target scheme as a guide, PAX and EDW metrology experts applied high-durability painted photoG targets to the store separation aircraft external surfaces. During this time, photoG target stickers were also applied within the left and right weapon bays, interactively with the real ASVS camera views. Finally, metrology surveys were performed twice (for confidence), to measure all of these target locations, in what would become the "known locations" of all the aircraft photoG targets in the aircraft reference frame.

4. ONE-TIME ACCURACY ASSESSMENT

The photogrammetry-based store trajectory solution has been regarded as the store position "truth," but this confidence was only realized after a measurement exercise at Northrup Grumman, well before flight testing commenced. Actual ASVS cameras were set up within an F-35 mockup, with similar fields of view as those that were being considered for F-35 store separation flight testing. PhotoG targets were attached to the aircraft mockup, an air-to-ground store mockup, and an air-to-air store mockup (Fig. 41); these photoG target locations were surveyed, just as they would be on real flight test aircraft and stores, to establish the known target locations in the aircraft's and stores' coordinate systems, respectively.

The following steps were then accomplished:

1. *ASVS image capture:* The store was placed at the first location, and the image was captured on each ASVS camera, simultaneously. This image (along with separate camera and lens calibration images) was to be used in the normal DSSAS photogrammetry position solution.
2. *Real store position:* With the metrology laser tracker system established within the coordinate system of the aircraft mockup, several store photoG targets were surveyed, and the metrology software subsequently revealed the store mockup's true location and orientation in the aircraft axis system. This real store position was locked up and hidden from the photogrammetry engineers until after the DSSAS store position solutions, from the ASVS camera images, were eventually computed.
3. *Repeat:* These two previous steps were repeated for five more real store positions below carriage, and then repeated again with the other store mockup.
4. *DSSAS photogrammetry store position:* Using the ASVS imagery, along with the known store and aircraft photoG target positions in their respective reference systems (as would identically be done during real flight test photogrammetry), the DSSAS displacement solution was computed for each of the six store mockup positions below carriage, for both stores.

The resultant DSSAS photogrammetry store positions were finally compared to the previously "locked-up and hidden" real store positions, revealing photogrammetry-based store displacement solutions 1) within 0.2 in. of the real store

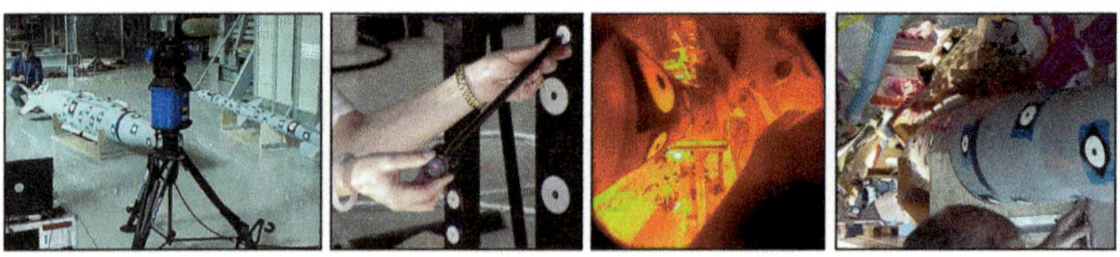

Fig. 41 DSSAS photogrammetry accuracy assessment.

three-dimensional translation position, 2) within 0.2 deg of the real store's yaw and pitch rotation angle, and 3) within 1 deg of the real store roll position. These differences were within the resolution needed for confidently determining store clearance (or miss distance) to aircraft hardware during real store separation flight test events. Because high accuracy was demonstrated for these solutions, the DSSAS photogrammetry solutions have routinely been considered the store displacement and rotation position truth.

5. FLIGHT TEST TRAJECTORY SOLUTIONS

Following all store separation events where there was little clearance margin to aircraft hardware, or where the photogrammetry solutions were needed to confirm, enhance, or substitute for 6DOF/IMU TM Kit solutions, DSSAS photogrammetry store position solutions were computed using the ASVS video imagery and identifiable store photoG targets. This digitization can be seen from the green points in Fig. 42. These store position solutions were ultimately provided to the store separation engineers within one to two days of the flight test event.

Occasionally, wireframe overlay representations of the final trajectory were produced (Fig. 43), but in all instances, the solution was in the form of tabular store translational and rotational displacements, expressed in the aircraft reference frame. These tabular photogrammetry solutions were used by store separation engineers during postflight simulations in ASEP, as the time-dependent store body motion constraint to the aircraft, resulting in the store trajectory playback in ASEP.

E. STORE DISPLACEMENT, VELOCITY, AND ACCELERATION: 6DOF/IMU TM KIT

As observed in Eqs. (1–6), the rates and accelerations must be known in order to compute the total forces and moments. The store rotational rate and translational acceleration "truth" is best-obtained from direct measurement, using an IMU and/or 6DOF TM Kit. Fundamental to the computation of the total

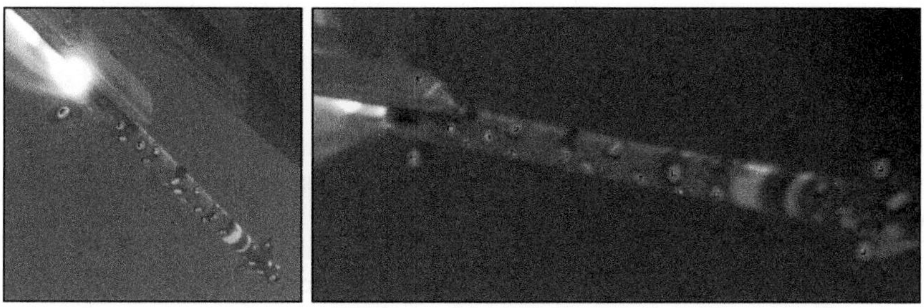

Fig. 42 PhotoG target digitization during DSSAS photogrammetry analysis.

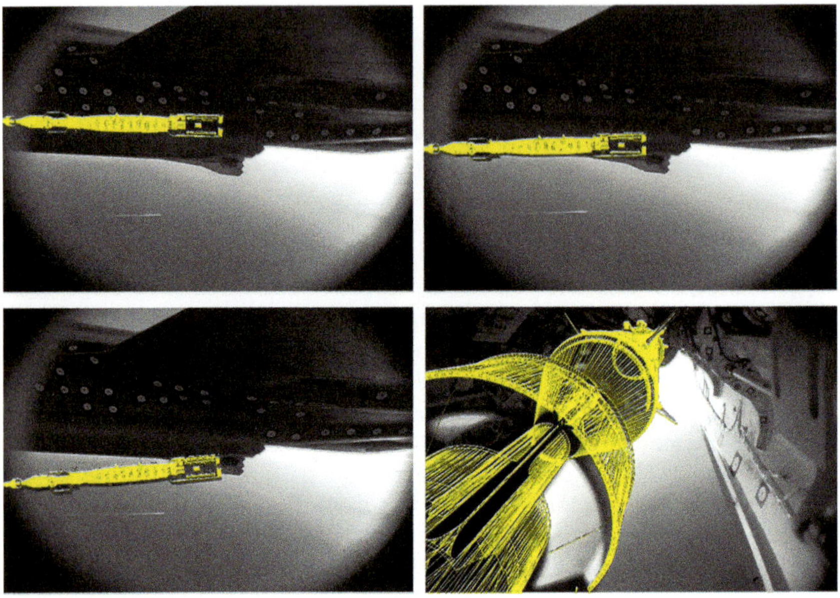

Fig. 43 DSSAS photogrammetry solution, wireframe overlay on ASVS video.

applied moment are the rotational accelerations, $\dot{\omega}_x$, $\dot{\omega}_y$, and $\dot{\omega}_z$, which are obtained simply by numerical differentiation of the measured rotational rates, ω_x, ω_y, and ω_z. Although it is possible to alternatively obtain store rate and acceleration information from photogrammetry solutions, the comparatively low sample size and numerical differentiation of positions to obtain store velocities and accelerations require extra time and skill. Store aerodynamics are more quickly and reliably obtained using directly measured store rate and acceleration "truth." Nevertheless, overall confidence is further increased when accurate photogrammetry *and* 6DOF/IMU TM Kit solutions are available, with one source's solution confirming or even improving the accuracy of the other.

1. STORE IMU AND 6DOF TM KIT

Certified weapons in the F-35 SDD phase are all actively controlled, guided stores. Such guided stores usually have an IMU to measure store rotational rates and translational accelerations, which are passed to the store's guidance control system. The store separation flight test versions of these guided stores also transmit the IMU's measured data to nearby antennas.

There are also unguided versions of these stores, which are used for various captive-carriage tests and for store separation jettison testing. A majority of F-35 SDD store separation testing involved these unguided jettison stores 1) because of their lower cost, and 2) because at similar in-flight release conditions, jettisoned stores' trajectories are generally not as safe and well-behaved as guided stores' trajectories. So, when unguided stores are used during separation testing, a 6DOF TM Kit (Fig. 44) is usually installed within these stores, enabling the measurement and telemetered transmittal of similar rotational rate and translation acceleration parameters as a guided store's IMU.

When these telemetered data are used only to compare rotational rates and translational acceleration between simulation and flight test, it is sometimes possible to use telemetered data that have localized loss, or dropouts, provided that the available rate time history at least reveals peak values, ejector end-of-stroke rates, or rail-launched end-of-rail tip-off rates, depending on the region of interest in the trajectory.

However, when these telemetered data are used primarily for computing the time-dependent store position and orientation trajectory and especially the store aerodynamic forces, data loss (dropout) becomes much less tolerable. Dropouts and inadequately filtered rotational rates will result in inaccurate flight test–derived aerodynamics forces and moments, and therefore, an incomplete or inaccurate aerodynamics model validation.

2. PREPARING FOR THE INEVITABILITY OF DATA LOSS

For the F-35 SDD store separation flight test program, several steps were taken to minimize the chances for untimely data loss.

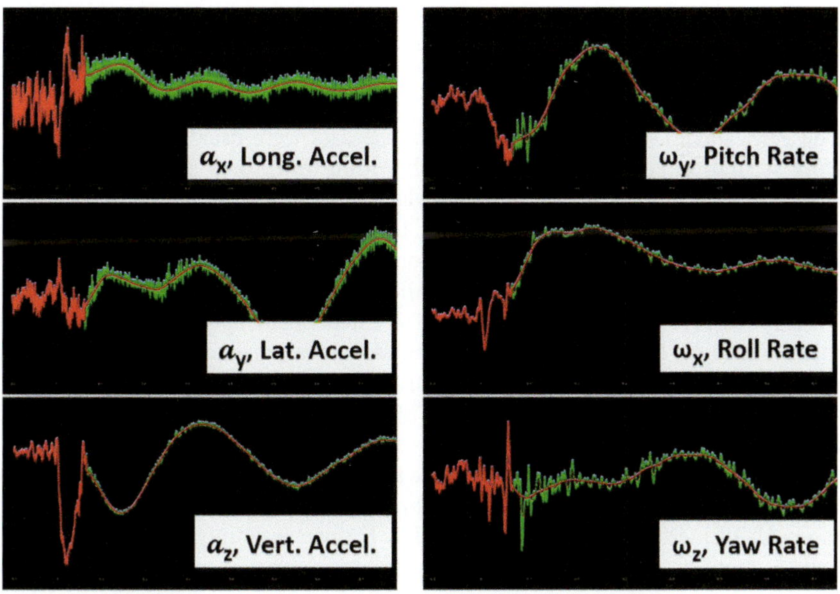

Fig. 44 Store 6DOF/IMU TM Kit acceleration and rate data.

For unguided/jettison stores, most of the installed 6DOF/IMU TM Kits included a TM-delay feature, enabling a delayed retransmittal of the sensed rates and accelerations. A delay duration of approximately 0.5 seconds was chosen, to ensure that the retransmittal occurred when the store was sufficiently far from the aircraft, without obstruction to the ground or support aircraft antennas.

For in-bay store separation events, when the store hasn't yet fallen completely out of the bay and past the weapon bay doors, ground-based antennas may not be able to receive the 6DOF/IMU TM Kit telemetered signal. Therefore, for F-35 in-bay store separation tests, Lockheed Martin used in-bay telemetry reception/retransmittal equipment, or Re-Rad Kit, to independently receive the close proximity store 6DOF/IMU TM Kit's telemetered signal and retransmit it without bay and door obstruction.

With both the F-35 Re-Rad Kit hardware and 6DOF/IMU TM Kit TM-delay feature, up to four separate signal transmission sources would be available for postflight use. In many cases, the best of the four sources was selected for postprocessing, without need of enhancement; however, in some cases where the best source still had dropouts in critical areas of the trajectory, one or more of the remaining sources were used to fill in the dropouts with better data.

For external underwing store separation events involving a guided weapon, only one data stream from the store's IMU was telemetered from the store, without the possibility of using the F-35 in-bay Re-Rad Kit. For such events, in order to ensure successful IMU data acquisition, multiple ground-based reception antennas were used, along with occasional use of support-aircraft data acquisition antennas.

3. REPLAYING THE 6DOF/IMU TRAJECTORY

The PAX and EDW 6DOF TM Kit response data were usually available to the store separation engineers in Fort Worth within four hours after the flight test event. During store separation event playback simulations within ASEP, the 6DOF/IMU TM Kit rigid body was "flown," separately and independently from the store model, by applying customized artificial forces at the TM Kit's center of gravity, according to Eqs. (1–6), with the following considerations.

Applied forces onto the TM Kit part are dependent upon TM Kit measurement, where a_x, a_y, and a_z are the measured translational accelerations, and ω_x, ω_y, and ω_z are the measured rotational rates. Simple differentiation of these 6DOF TM Kit rotational rates yields TM Kit rotational accelerations: $\dot{\omega}_X$, $\dot{\omega}_Y$, and $\dot{\omega}_Z$.

In ASEP, because the TM Kit part is a moving rigid body, this part's response to the applied forces is also dependent upon the TM Kit part's model's mass m and moments of inertia I_{xx}, I_{yy}, and I_{zz}. The TM Kit part is simply a convenient rigid body to move, from which model-based playback measurements of displacement, velocity, and acceleration measurements are requested. Conveniently, massless store geometry is attached to this rigid body for playback visualization. So, the

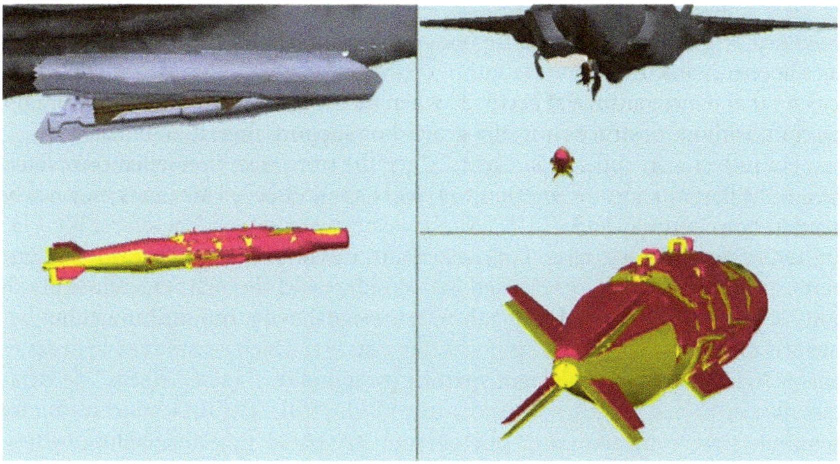

Fig. 45 Store 6DOF/IMU TM Kit and DSSAS photogrammetry trajectory, visualized.

TM Kit part's mass and moments of inertia were artificially set to a conveniently low value of one unit, in both the equations of motion and in the ASEP TM Kit part rigid body model to which the TM Kit part "flying" forces are applied.

The ASEP TM Kit part's initial condition is coincident with store at the CAD carriage position, or the initial condition could be at any point in a photogrammetry-produced trajectory, for instance. Because most store separation test events yielded high-quality 6DOF/IMU TM Kit data, almost all trajectory computations were begun at first motion, corresponding to the moment immediately before the rapid increase in vertical acceleration (Fig. 44), and also included the timespan when the actual store was still in contact with the ejector.

As often as was necessary or practical, the photogrammetry-based and 6DOF/IMU TM Kit–based trajectory solutions were both plotted or visualized (Fig. 45), to ensure favorable comparisons.

F. STORE AERODYNAMICS

Lockheed Martin's additional focus on actual store aerodynamics involves no additional measurement sources, relying on the measured 6DOF/IMU TM Kit rates and accelerations, measured store mass properties, measured aircraft flight conditions, and a standardized posttest process, usually within one day. The store total applied forces and moments in all six degrees of freedom are computed from the simplified rigid body equations of motion, Eqs. (1–6). This form of the common $F = ma$ equations of motion was used, simply as a result of the available measurements.

From the existing store mass property information, m is the measured actual store mass, and I_{xx}, I_{yy}, and I_{zz} are the measured store 3D rigid body axis moments of inertia. Because actual store mass properties measurements did not include products of inertia, the rigid store body axes were conveniently assumed to also be the principal axes.

After the in-flight 6DOF/IMU TM Kit measurements are transferred to the store's center of gravity, a_x, a_y, and a_z are the measured/transferred translational store accelerations, and ω_x, ω_y, and ω_z are the measured store rotational rates. Simple differentiation of these store rotational rates yields store rotational accelerations: $\dot{\omega}_x$, $\dot{\omega}_y$, and $\dot{\omega}_z$.

Other simplifying assumptions were made, for convenience. The store was presumed to be rigid, without rotors, liquid sloshing, and so forth. The 6DOF/IMU TM Kits were never located near stores' centers of gravity, and measurements frequently revealed flexible store response, for instance, which was filtered in order to enable rigid body–like store aerodynamics computation. "Sanity checks" were frequently performed during postprocessing: 6DOF/IMU TM Kit–based trajectory solutions were compared with photogrammetry-based trajectory solutions. For the air-to-air and air-to-ground stores in the F-35 SDD store separation test program, these frequently matching solutions provided confidence that the simplified $F = ma$ equations were sufficient for replicating photogrammetry-based actual store trajectories and for extracting valid aerodynamics.

Although the idea is simple, and is certainly not new, several considerations must be addressed in order to successfully implement this simple idea to understand where and how it happened—to achieve confident aerodynamics model validation. Although it might be convenient to further simplify Eqs. (4–6) as was described in Ref. [22], such a simplification did not yield flight test–computed store aerodynamics forces and moments sufficient for posttest simulation replication of the flight test trajectories in all six degrees of freedom.

During the store separation phase between carriage and EOR or EOS, the total forces and moments applied to the store include the interaction between the store and S&RE. Distinguishing between store aerodynamics and S&RE-applied forces relies on knowledge of the S&RE performance. Although this behavior can generally be characterized from other testing and roughly subtracted from the total forces and moments, the Lockheed Martin approach, when strictly performing aerodynamics (and propulsion) model validation, is to remove the S&RE portion from consideration and focus only on the phase beyond EOR/EOS (Fig. 46), where the store is no longer influenced by the mechanical and structural interaction with the aircraft and S&RE.

Additional forces may act on the store beyond EOR/EOS, but it should be noted that the stores involved in F-35 SDD store separation flight testing were sufficiently large, such that the forces associated with sudden mechanical deployment or rapid actuation of moving control surfaces, for instance, were extremely small, compared to applied aerodynamic and propulsion forces.

Considering an example of computing flight test–based aerodynamic pitching moment, and recalling Eq. (5), the applied pitching moment forces depend mainly on the store pitch acceleration $\dot{\omega}_y$. However, measured 6DOF TM Kit pitch rate

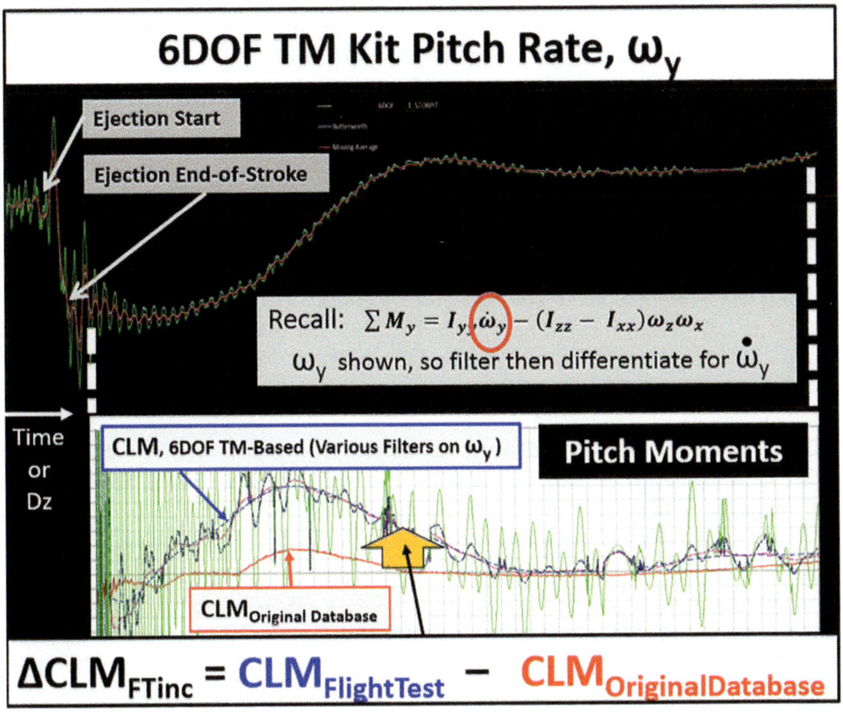

Fig. 46 Post-EOS store pitching moment flight test aero increments, derived from filtered 6DOF TM Kit pitch rates.

ω_y is not usually smooth (Fig. 46, green curve in top plot), and differentiation typically results in noisy pitch acceleration $\dot{\omega}_y$ producing an equally noisy pitch moment computation, observed in the green curve in the lower plot of Fig. 46. However, filtering the 6DOF TM Kit's pitch rate helps produce smoother pitch accelerations, yielding smoother 6DOF TM Kit–based aero pitch moment computations, observed in the successively filtered black, magenta, and blue curves in the lower plot of Fig. 46.

The 6DOF TM Kit–based aero pitch moment is the flight test "truth" against which the original store aerodynamics model will be compared and adjusted (validated). For this exact flight test trajectory, the photogrammetry-produced or 6DOF TM Kit–produced store trajectories are useful for obtaining the aerodynamics database model's table look-up pitching moment, observed in the red curve in the lower plot of Fig. 46. For this flight test example, the difference between the 6DOF TM Kit–derived aerodynamics pitching moment and the aerodynamics model's pitching moment is the aero flight test increment.

For several disciplines such as Loads, Dynamics, Flutter, Aero/Performance, and Stability and Control, modeling and simulation adjustments were expected

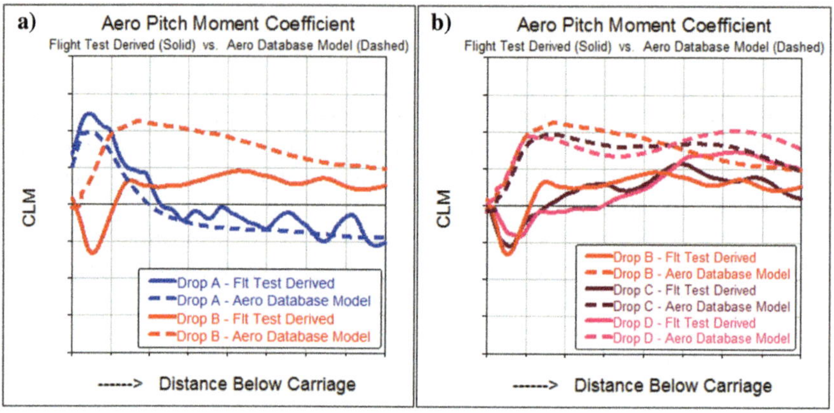

Fig. 47 Flight test–derived and aero database model pitch moment coefficient (CLM): a) two separation events at different Mach numbers, and b) three separation events at same Mach numbers.

as part of the flight test–based validation process, especially because wind tunnel models frequently exhibit transonic aerodynamics differences from full-scale aircraft in flight. The same is true for the Store Separation discipline. One of the goals during store aerodynamics model validation was to strive for adjustments as a function of Mach and/or aircraft angle of attack, for instance, instead of a flight test event-specific adjustment. Although various store separation events, even at nearly identical flight conditions, can sometimes produce slightly different store trajectories, it was reassuring to group similar events (when available) and observe similar store aerodynamics. In Fig. 47, the left plot's blue and red curves reflect the store aerodynamics pitch moment for test events at two different Mach numbers, whereas all of the right plot's curves reflect the store pitch moment during three different separation events at the same Mach number.

The Flight Test aero increment (FTinc), or the difference between the flight test–derived store aerodynamics and store aerodynamics models, is a function of store distance below the aircraft. Such increments are obtained for the forces and moments in all six degrees of freedom, producing the complete 6DOF aero adjustment, contributing to a flight test–validated aerodynamics model for the specific flight test trajectory. Finally, grouping 6DOF flight test aero increments from different flight test events produces a more general set of flight test aero increments, perhaps as a function or aircraft Mach or angle of attack, for instance. The lower middle plot in Fig. 48 represents the store aerodynamics pitch moment FTinc for the same events in Fig. 47. The three red-tinted FTinc curves in Fig. 48, associated with events with the same Mach numbers, reflect nearly identical aero model adjustments and are conveniently averaged (white dotted central curves in Fig. 48) to represent the nominal FTinc for that particular Mach number. The blue

FTinc curve, associated with a store separation event with a different Mach number, clearly reflects a different (and nearly zero) adjustment to the store aerodynamics model.

Explanation of reasons for differences between scale model wind tunnel–based aerodynamics and actual in-flight aerodynamics are beyond the scope of this paper, but it is clearly observed in Fig. 48 that such aerodynamics differences can be dependent upon store distance and aircraft (and wind tunnel) Mach, at the very least. So, separate studies (if deemed necessary) to investigate reasons for such differences can be focused directly on store aerodynamics, instead of indirectly on the trajectories and rates produced by the store aerodynamics.

But more importantly, it becomes possible to focus on aerodynamics model validation, separately from the mechanical, dynamic system model. Furthermore, analysis of such FTinc summaries enables a flight test–based estimation of aerodynamics coefficient uncertainties, represented by the white dashed curves/boundaries on either side from the white dotted central nominal FTinc curves. Such uncertainties are later used in Monte Carlo uncertainty simulations during analyses supporting certification recommendation.

The final verification, that both the aerodynamics and system models are validated, is conducted following posttest simulations, comparing the photogrammetry-based and 6DOF TM Kit–based playbacks (solutions) trajectories of the flight test event, the preflight (prior to validation) simulation trajectory, and the postflight validated simulation trajectory. Well-validated aerodynamics models produce favorable trajectory comparisons in all six degrees of freedom, as shown in a sample air-to-ground store separation flight test event in Fig. 49, where the validated M&S trajectory (gold curve) identically matches the trajectory

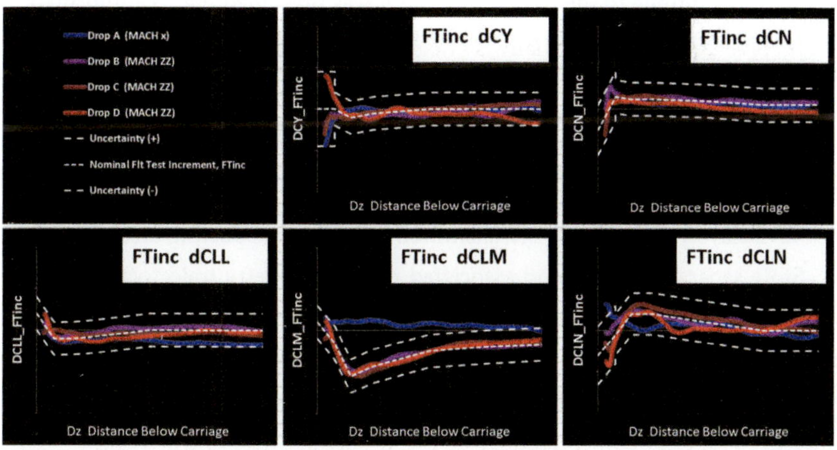

Fig. 48 Flight test–derived aerodynamics increments in five degrees of freedom (axial not shown).

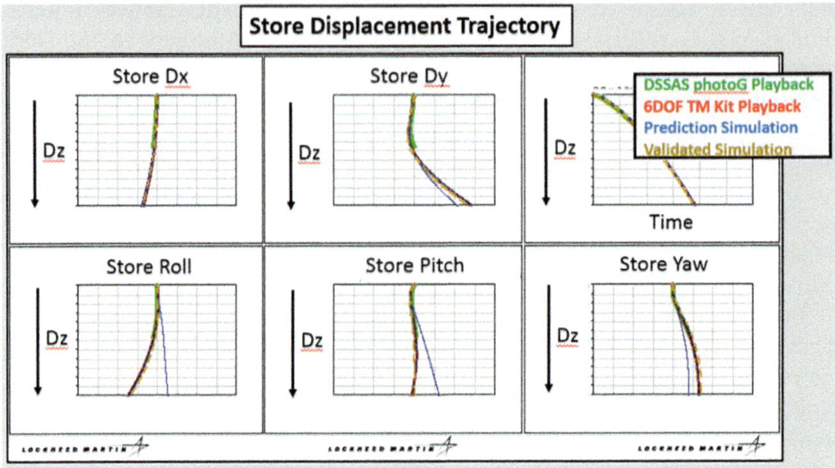

Fig. 49 Store trajectories compared: DSSAS photogrammetry–based (green), 6DOF/IMU TM Kit–based (red), preflight prediction (blue), postflight validated-aero simulation (gold).

playback from the DSSAS photogrammetry (green curve) and 6DOF TM Kit solution (red curve).

G. TIME CONSIDERATIONS

When planning for the rapid posttest analysis required for "next-day" clearance/approval to continue to the next test point, it was presumed that the information from the following data sources was *never* going to be time-synchronized when the Lockheed Martin store separation team received them:

1. 6DOF/IMU TM Kit data streams:
   a. Real-time data stream, directly from store
   b. Delayed data stream, directly from store
   c. Real-time data stream, retransmitted by test or supporting aircraft
   d. Delayed data stream, retransmitted by test or supporting aircraft
2. Store separation video and associated photogrammetry trajectory solution
3. Aircraft maneuver response data

Furthermore, posttest data streams that were telemetered to the control room were not time-synchronized with the same data streams that were recorded on the test aircraft. Yet all of these data were to be merged and used together. Because of aircraft system delays associated with electronic triggers, such as the cockpit pickle switch, or insufficient sample-rate resolution in S&RE-associated electronic messages, such as "hooks open," the search for reliable store first-motion evidence was biased toward evidence of physical response. The F-35 store separation posttest analysis process included an exercise of identifying the time at store first motion, independently in each data source, and then synchronizing the independent data sources with this common start time:

1. *6DOF/IMU TM Kit data:* Vivid store vertical or longitudinal acceleration reflected the store response to the S&RE impulse or store rocket motor ignition.
2. *Store separation video:* Video includes time-stamped video frames, but sometimes the identification of store first motion was subjective, with not-so-obvious small first movements; however, during review of processed photogrammetry trajectory solutions, the start of store translational displacement changes was easily identified.
3. *Aircraft maneuver response data:* The aircraft response to the S&RE store ejections was usually observed in sudden aircraft vertical acceleration and roll rate response. Selecting the start time within aircraft maneuver response data from rail-launched missiles events was slightly more difficult, requiring a less-accurate time selection from low-sample-rate Stores Management System (SMS) and flight controls information, such as a drop in total aircraft weight due to "store release" electronic messages, followed by a more-refined examination of aircraft maneuver response, such as yaw rate and longitudinal acceleration.

VIII. F-35 STORE SEPARATION FLIGHT TEST VIGNETTES

Even with commonality among aircraft variants and similarity amongst stor, there still exist differences unique to aircraft variants as well as the various stores, resulting in a significant amount of unique cases to be considered for store separation modeling, analysis, ground testing, and flight testing. Store separation analysis and flight testing has been conducted over the last several years by the industry partner/government integrated test forces at EDW and PAX, where 183 stores were released. This number of test points would have been higher if aircraft variant commonality wasn't proven in wind tunnel testing and CFD analyses and leveraged during flight test planning. For example, simulation model validation for GBU-31 JDAM Mk-84 weapon releases from the internal bays of both the F-35A and F-35C variants only required seven store separation test events.

Because F-35 store separation flight testing was conducted from aircraft that also performed testing required by other disciplines, such as loads, flutter, environment, and flying qualities, the store separation test flights would frequently be scheduled to occur in small blocks of time requiring expeditious clearance to subsequent test points. However, there was also a safety constraint on proceeding to the next test point: the flight clearance authority and flight test teams at PAX and EDW required that the previous test point was proven to be safe and that there was sufficient confidence that the next test point would also be safe. Several store separation flight vignettes are presented, and include instances of failures, discoveries, confirmations, and a process to increase test efficiency.

A. AERODYNAMICS EFFECTS ON S&RE PERFORMANCE AND INITIAL CONDITIONS

Actual and simulated store trajectories are highly dependent on the end-of-stroke conditions of the store as it leaves the ejector rack or launcher. Predicting end-of-stroke conditions for any given flight condition is a complex problem because the store end-of-stroke conditions are dependent on the flight condition, especially for stores released from a pylon on the wing or from any other station that is exposed to external flow when the store is at the carriage position. End-of-stroke conditions from static ejection testing can provide some insight into ejector rack performance variability as well as indicate expected end-of-stroke conditions at low-dynamic-pressure flight conditions. However, the end-of-stroke conditions observed in static ejection testing do not adequately predict the end-of-stroke conditions of stores subject to aerodynamic influence during the ejector stroke.

An enhanced simulation prediction can be accomplished by modeling the forces the store experiences on the ejector as soon as the store begins moving from the carriage position. Modeling the sway braces at the end of the pistons ensures that the modeled store is constrained laterally in the same manner as the actual store. Modeling the friction of the sway brace pads ensures that the store can be held in place with the same frictional force as an actual store on the ejector. Modeling the ejector piston forces using the EDO-provided black box gas model ensures that the S&RE ejector model accounts for the variable inertial force of the store under flight-condition-dependent aerodynamic loading. Simulation trajectories conducted with the aforementioned modeling considerations, together with aerodynamics from wind tunnel testing, match flight test trajectories better than simulation trajectories initialized using end-of-stroke conditions from static ejection testing and wind tunnel aerodynamics.

As previously discussed, validation of the S&RE-only models and the complete dynamic system models (including the aircraft, S&RE, and store) were accomplished following both store/S&RE ejection testing and on-aircraft ejection pit testing. Also, from previous discussion, store aerodynamics models were created primarily from grid aerodynamics data from store separation wind tunnel testing. In such tests, store aerodynamic forces and moments were obtained with the store positioned at many store locations and angles relative to the aircraft. For stores that are ejected, one of these important store positions is at or near carriage, because the store aerodynamics at this location are representative of those that the store will see during the mechanical ejection phase. So, per the normal preflight prediction process, the validated dynamic system model was combined with the store aerodynamics model to perform simulations of all anticipated flight test events. Certain simulation cases revealed different store/S&RE performance than that observed during the ground-based ejection testing, which was fully expected.

1. AIM-120 AMRAAM MISSILE

Numerous AIM-120 (Fig. 50) ejections from the LAU-147 during on-aircraft pit ejection tests were performed, with EOS pitch rates nearly identical to those observed during AIM-120/LAU-147 S&RE-only ejection stand tests. Lockheed Martin's fully dynamic (aircraft + S&RE + store) "no airload" AIM-120 store separation simulations yielded identical EOS pitch rates; however, flight test simulations performed with the same dynamic system model yielded substantially more EOS nose-down pitch rate, when including wind tunnel–based aerodynamic applied store loads during the ejection phase.

Store separation flight testing included AIM-120 separation events from the in-bay near-door position, where the AIM-120 was exposed to the air flow, instead of being hidden deep within the weapons bay. During these test events, EOS store pitch rate performance was proven to be nearly identical to that from the preflight prediction simulations. The significance of this favorable comparison with flight test behavior was that the dynamic system model was validated, not only in the no store aero state, but more importantly in the aero-loaded in-flight state—store separation simulations using the fully dynamic aircraft + S&RE + store + aerodynamic model could now be performed more confidently for many different aircraft flight conditions and maneuvers in the AIM-120 employment flight envelope.

Fig. 50 AIM-120 store separation: static ejection pit test and flight test.

2. EXTERNAL GBU-12

Store separation simulations were performed using ASEP's fully-dynamic (aircraft + S&RE + store) model for underwing GBU-12 (Fig. 51) ejections from the in-pylon BRU-68 ejector rack, matching on-aircraft pit ejection test results for the same configurations. Similar to the aforementioned AIM-120 case, GBU-12 store separation flight test simulations performed with the full dynamic system model yielded different EOS nose-down pitch rate, when including external GBU-12 wind tunnel–based applied store aerodynamic loads during the ejection phase. Such "with airload" simulations matched flight test ejector performance very well.

But pitch wasn't the only relevant degree of freedom. Lockheed Martin's fully dynamic system model included a frictional contact model for BRU-68 swaybrace-to-store interaction. So, like the actual GBU-12 interaction with the BRU-68 ejector rack, the simulated GBU-12 store was restrained (not constrained), such that the store could yaw relative to the BRU-68 and pylon during the ejection phase. At each swaybrace point, the store surface was able to frictionally slide fore–aft and sideways, while in contact with the four swaybraces during ejection. Simulated external GBU-12 store separation EOS yaw rates were consistent with those from actual GBU-12 external store separation flight test events. The fully dynamic system model was validated in the aeroloaded in-flight state, so like the AIM-120 case, GBU-12 store separation simulations using the fully dynamic aircraft + S&RE + store + aerodynamic model could now be performed confidently for many different aircraft flight conditions and maneuvers in the GBU-12 external release flight envelope.

B. MULTIPLE STORE SEPARATION EVENTS PER FLIGHT

Before F-35A GBU-39 small-diameter bomb store separation flight testing began at EDW, F-35 store separation flight test events were limited to one separation

Fig. 51 Flight test: GBU-12 store separation from BRU-68 rack in underwing pylon.

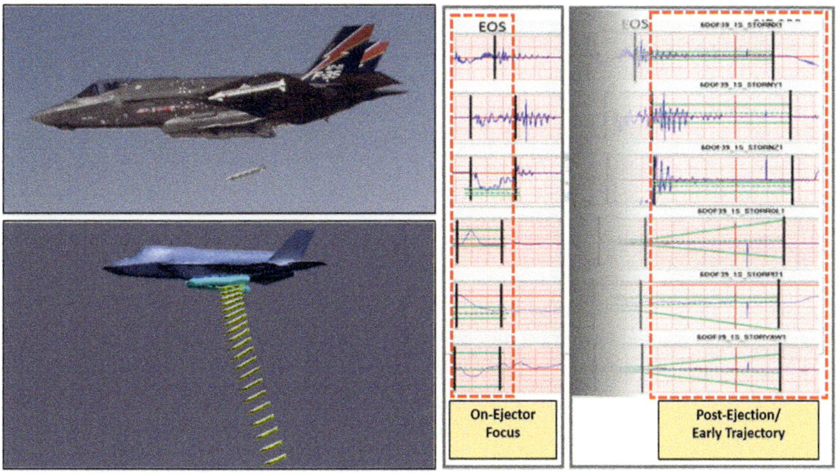

Fig. 52 GBU-39 store separation multistep criteria: chase pilot and 6DOF TM Kit assessment.

event per flight, partly in order to allow store separation engineers enough time to download, process, and analyze the data before making a clear-to-next-point call that would allow the test site Integrated Test Force to continue to the next store separation flight test point. For in-bay store separation events, this approach was in line with the strategy of primarily performing separations from the left-hand bay, because telemetry retransmission equipment, required to increase likelihood of successful 6DOF/IMU TM Kit data acquisition, was installed where stores would normally be carried in the right-hand bay. Furthermore, in-bay air-to-ground store separation events typically required a different weapons test range than the in-bay air-to-air AIM-120 AMRAAM store, further contributing to the necessity for single-release-per-flight missions.

With the four-place GBU-39 carriage arrangement from the BRU-61 within the weapon bay, however, there was an opportunity to plan and execute store separation missions involving the release of up to four of these stores from the weapons bay, thereby increasing test efficiency by reducing the total number of required store separation test flights. Fully realizing this opportunity required a clearly understood process, with unambiguous criteria for in-control-room approval for continuation to the next GBU-39 separation event. After gaining acceptance by the Integrated Test Force team, a three-step process (Fig. 52) was implemented, which led to a continuation decision within five minutes.

1. CHASE PILOT ASSESSMENT

The first rapid decision was based on the visual observation from the chase pilot. Before the mission, the predicted trajectories for all planned store separation events were briefed to the test team, including the test and chase pilots. This briefing includes animated movies from the prediction simulations, along with a description of important behaviors to consider and communicate back to the control room, such as pitch attitude (down vs flat or nose-up) near the aircraft and further away from the aircraft. The chase pilot's description of the visually observed event, relative to the preflight briefed store trajectory behavior, was audible to all engineers in the control room.

2. ON-EJECTOR PERFORMANCE

The second rapid communication was based on an in-control-room assessment of whether the ejector performed as expected. Store 6DOF/IMU TM Kit telemetry data were examined, with primary focus on store vertical acceleration, pitch rate, and roll rate. On-ejector parameter plots from prior static pit ejection and flight test events served as a guide for the in-control-room store separation engineer. For each of the four GBU-39 positions, there was an expectation of specific peak vertical acceleration during the ejection, in addition to specific store pitch and roll rates at the completion of the ejection phase, EOS. A rigid, documented tolerance band was allowed; if these tolerances were exceeded, continuation to the next GBU-39 was not allowed. On the other hand, verification of expected on-ejector performance allowed the store separation engineer to move to the next criterion.

3. OFF-EJECTOR EARLY TRAJECTORY

The final rapid decision was based on the off-ejector store trajectory between EOS and the time that the GBU-39 moved below the open weapon bay doors. Preflight predictions included Monte Carlo uncertainty analyses, accounting for variations in store mass properties, aircraft flight conditions, ejector rack performance, and store aerodynamic forces. Safe separation trajectories were gathered from these simulations and formed the basis for documenting acceptable ranges, or variations, in store linear acceleration and rotational rates during this brief phase in the store separation event. The criteria for acceptable early trajectory store behavior, as observed from the store 6DOF/IMU TM Kit data in the control room, were that these accelerations and rates must remain within the documented acceptable ranges. Exceedance of the allowed variation, with any one of the 6DOF parameters, automatically resulted in ending the mission—continuation to the flight's next release event was not allowed.

4. EVENTS

To minimize the in-control-room store separation engineer's comparison assessment subjectivity and decision time, the referenced criteria parameter plots were presented identically to those plotted in the control room playback of the telemetered 6DOF/IMU parameters—reference plots included the same time and parameter plot scales, and parameter variation tolerances were clearly marked. Complete confidence in all three of these assessments was required. With attention to safety, continuation to the next separation event in the flight mission was allowed only after a firm, clearly communicated favorable assessment in all three steps in the process. As a result, the EDW F-35 Integrated Test Force (ITF) team eliminated 12 GBU-39 store separation flights by efficiently conducting the following:

- Two flight missions, each including four store separation events
- Two flight missions, each including three store separation events
- Two flight missions, each including two store separation events

This same in-control-room process of analyzing 6DOF/IMU telemetered parameters within the control room was utilized during flights with multiple separation events with other stores:

- The PAX F-35 ITF team eliminated seven store separation flights by conducting multiple separations per flight for the GBU-12, GBU-32, AIM-132 ASRAAM, and Paveway IV weapons, including a flight during which all four external GBU-12 stores were released.
- The EDW F-35 ITF team eliminated four more store separation flights by conducting multiple GBU-12 separations in three flights.

C. GBU-12 TAILKIT LANYARD RECONFIGURATION

The GBU-12 store is an air-to-ground weapon that uses deploying tail fins for increased aerodynamic stability; the opening of these tail fins is mechanically activated by a wire lanyard arrangement. With rapid and unsteady air flow around the separating store, intricate lanyard arrangements are sometimes susceptible to improper function, especially when arrangements differ from typical externally carried, underwing arrangements on many prior aircraft programs. Because weapon bay and door geometry constraints are present, the GBU-12 tail fins could not be opened until the store was past the doors during an ejection event, which necessitated a longer and more complex lanyard arrangement.

Based on prior AFSEO experience with lanyards and GBU-12-like stores (Fig. 53), it was recommended that a failed lanyard condition, where fins do not open, should not be considered a statistically insignificant event—store separation simulation events should include both a normally functioning tail fin arrangement and a nonopening tail fin arrangement.

The unfavorable store stability associated with the latter case, with completely stowed fins, led the store separation team to mandate a BRU-67/BRU-68 rack ejector split setting with nose-down pitch rate bias in the GBU-12 store loading procedures, which ensured that the stowed-fins GBU-12 would not point up and fly back to the aircraft during the store separation event.

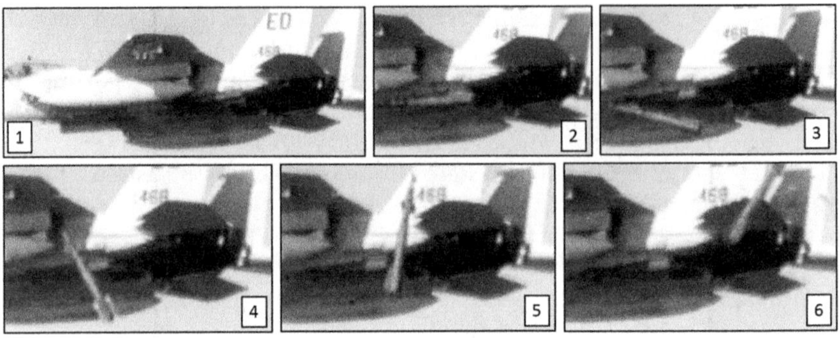

Fig. 53 F-15 aircraft with flyback of GBU-12-like store with stowed tail fins [23].

During one of the first four GBU-12 store separation events from the F-35, the tail fins did not deploy, remaining stowed. As a result, the store pitched down safely and as expected, due to the mandated nose-down pitch rate bias on the BRU-67/68 rack. In addition to reaffirming the prudence of including the lanyard failure scenario when selecting the appropriate ejector split setting, this store trajectory associated with this fins-stowed event proved to be beneficial because the store was in one of the exact store model arrangements during preflight wind tunnel testing. Preflight store separation wind tunnel testing for the GBU-12 included the model in various states of fins and canards, and aerodynamic forces and moments were measured for these states, including one state with fully stowed fins (Fig. 54). As it turned out, the flight test trajectory for this event identically matched the preflight simulation's trajectory (with fins stowed), effectively validating the GBU-12 store aerodynamics model in this fins-stowed configuration.

However, the demonstrated failure of the lanyard to properly function (Fig. 55) resulted in a brief root cause investigation. Because air flow between

Fig. 54 F-35 GBU-12 fins stowed and fins opening: flight test events and wind tunnel models.

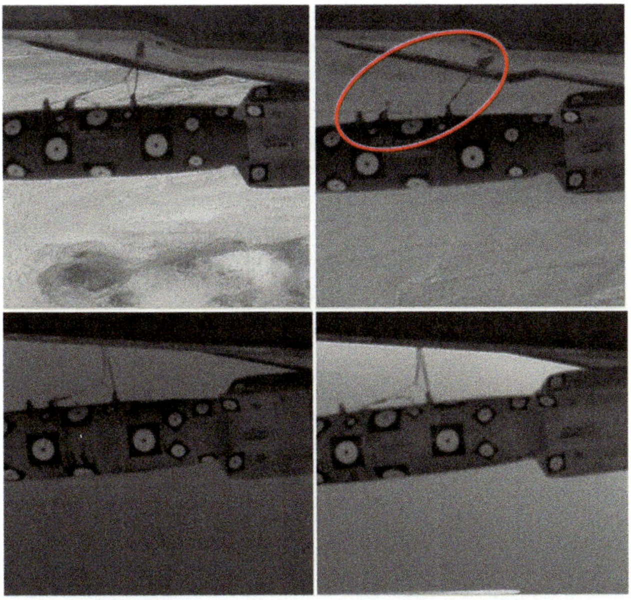

Fig. 55 Proper and improper (circled) GBU-12 lanyard function.

and near the weapon bay doors is both rapid and unsteady, the initial lanyard routing arrangement resulted in portions of the lanyard pulling prematurely under various conditions. A simpler yet more reliable arrangement was implemented after this event, resulting in increased consistency in lanyard behavior within the unsteady underbay aircraft region.

Several noteworthy items were associated with this event:

- The importance of ASVS camera video evidence was emphasized during the root cause investigation.
- The GBU-12 lanyard arrangement redesign, for in-bay releases, happened early in the test program, with enough remaining in-bay separation events to provide statistical lanyard function reliability metrics.
- Valuable GBU-12 aerodynamics model validation data were still obtained.
- Flight testing was not delayed; because of the intentionally biased nose-down fins stowed store trajectory behavior, flight testing could continue safely without the prospect of future unsafe trajectories.

D. IMPORTANCE OF INCLUDING AIRCRAFT MANEUVER IN POSTTEST ANALYSIS

During a GBU-31 store separation flight test event, test engineers in the control room noticed that a replay of the ASVS in-bay camera video revealed store lateral movement that was different than that typically observed in prior GBU-31 store separation events. Postprocessed 6DOF/IMU TM Kit–based and photogrammetry-based store trajectory solutions confirmed this additional lateral store movement relative to the aircraft. An ASVS in-bay camera image of this event, superimposed with an image from this same camera, but for a different event, visually reveals this difference (Fig. 56).

However, the raw, unprocessed store 6DOF/IMU TM Kit data for these two different events were nearly identical, even in the lateral acceleration measurement parameter. Air-to-ground store separation simulations typically reveal similar store trajectories, relative to the ground, or Earth, following release from the aircraft, even if the aircraft has different acceleration response, provided

Fig. 56 GBU-31: Two separation events, superimposed.

that the aircraft initial forward velocities are the same. So, the focus of the lateral store movement needed to include consideration of another culprit.

Because the Lockheed Martin posttest analysis process includes the maneuvering aircraft response data for each flight test condition, the combined store separation posttest simulation replays of the store and aircraft trajectories, for the two different flight test events, revealed the reason for the lateral store movement, relative to the aircraft: the two aircraft were moving differently. In fact, the aircraft flight control system, during the store separation event with the appearance of lateral movement, had responded to sudden pilot control stick inputs at the time of the separation event, briefly taking the aircraft out of the specified release acceleration limits. Although the image appears to show the store moving laterally during the separation event, it was instead the aircraft that was moving laterally.

Based on visual observation of the replay of the store separation event, the team thought that this might be considered an unexpected event, requiring a pause in the GBU-31 store separation test program; however, because of the rapid posttest analysis process, which includes combined store and aircraft response, continuation to the next flight test point was not delayed, with the full explanation provided to the test team within two days. So, the main lesson from this example is that when the aircraft response is not included in posttest replay simulations, incomplete or incorrect explanations of events similar to this one might have had cascading consequences. These may have included possible delays in the test program or attributing the perceived unexpected store motion to increased uncertainty or general factors of nonsimulation—unknowns that weren't directly included in the nominal or Monte Carlo uncertainty simulations.

IX. CONCLUSION

Many variables can prevent a store separation flight test event from taking place: bad weather, excessive clouds, chase aircraft issues, refueling aircraft issues, range-clearing aircraft or boat issues, instrumentation issues, test aircraft issues, telemetry issues, relatively short test range times, and even boats or whales in an off-shore test range. Difficult maneuvers required additional preparation to increase the probability of success—stressing high-g and high-angle-of-attack maneuvers were rehearsed many times within flight simulators, to minimize time-consuming (and fuel-consuming) in-flight cold-pass rehearsals. Store separation flight test events are expensive and challenging to achieve, requiring no barriers to stand in the way of a successfully executed store separation mission.

The basis for both Lockheed Martin's certification recommendation and verification that the weapon release requirements had been satisfied is flight test-validated M&S—modeling and simulation that has been adjusted based on data gathered during the store separation flight test events, so that simulated store separation events replicate the actual store separation events. Failure to obtain the necessary instrumentation measurement data for verification of safe store separation and proper M&S validation would mean that the test point would need to be reflown—an expensive, schedule-stressing situation, considering the aforementioned difficulty associated with achieving successful store separation flight test missions.

Overall confidence in the final M&S and certification is directly dependent on the quality of the dynamic system models, the quality of the aerodynamics models, and the quality of the adjustment and validation of these models. Increasing the quality of baseline dynamic system models was accomplished by using functional mechanical models of the aircraft, S&RE, and store in the ASEP modeling and simulation environment. The quality of the aerodynamics models, based on industry-standard scale-model wind tunnel testing, was sufficient for preflight simulations required to obtain flight clearance, and was adjusted and validated during the progression of flight testing. The high quality of the aerodynamics model validation and the rapid flight test pace was a direct result of extensive planning, with attention to ensuring that the instrumentation critical to confident validation was obtained reliably.

During F-35 weapons testing, the pieces were in place to support high-quality photogrammetry-based trajectory solutions, 6DOF/IMU TM Kit–based trajectory solutions, and store aerodynamics model validation. The body of knowledge gained in these ground-based and in-flight tests, combined with adjusted and validated model simulations, support the confident safe store separation assessments required for a certification recommendation to release weapons from the F-35.

ACKNOWLEDGMENTS

The authors would like to recognize several individuals for their contributions, specifically in regards to understanding or increasing confidence. Rob Crandall, from the PAX F-35 Integrated Test Force, is recognized for his persistence and insistence in quantifying 1) accuracy in DSSAS photogrammetry–based store position solutions, and 2) uncertainties in store mass property measurements. Paul Sasenbury, from the EDW F-35 Integrated Test Force, is recognized for his diligence and commitment to ensuring that the store separation team thoroughly understand detailed, real-hardware store-to-S&RE mechanical interaction, so that models reflect actual hardware behavior. Greg Harding, from the Air Force SEEK EAGLE Office, is recognized for his continual reinforcement of the importance of objectively assessing predictive confidence. The mere existence of numerous vivid flight test videos contributes to initial impressions that store separation flight test events were successful; however, the success of the store separation flight test program was dependent primarily upon behind-the-scenes efforts by the PAX, EDW, and Fort Worth F-35 Integrated Test Force teams, assuring high-quality instrumentation and measurements at the heart of high-confidence modeling and simulation validation leading to the rapid flight testing pace and certification of 11 stores across the F-35A, F-35B, and F35C JSF variants.

REFERENCES

[1] Sheridan, A. E., Rapp, D. C., and Burnes, R., "F-35 Program History—From JAST to IOC," June 2018 (to be published).

[2] Counts, M. A., Kiger, B. A., Hoffschwelle, J. E., Houtman, A. M., and Henderson, G., "F-35 Air Vehicle Configuration Development," June 2018 (to be published).

[3] Purdon, M. L., Hetreed, C. F., and Hudson, M. L., "F-35 Pre-Flight Store Separation Analyses: Innovative Techniques for Affordability," AIAA Paper 2009-102, Jan. 2009.

[4] Hayward, D. M., Duff, A. K., and Wagner, C., "F-35 Weapons Design Integration," June 2018 (to be published).

[5] Schindel, L. H., "Store Separation," AGARD-AG-202, June 1975.

[6] Arnold, R. J., Epstein, C. S., and Bogue, R. K., "Store Separation Flight Testing," AGARD-AG-300-Vol-5, 1986.

[7] Nadar, O., "Aircraft/Stores Compatibility, Integration and Separation Testing," AGARD-AG-300-Vol-29, Sept. 2014.

[8] MIL-HDBK-1763, "Aircraft/Stores Compatibility: Systems Engineering Data Requirements and Test Procedures," 1998.

[9] MIL-STD-2088A, "Bomb Rack Unit (BRU), Aircraft, General Design Criteria for," 1997.

[10] MIL-HDBK-244A, "Guide to Aircraft/Stores Compatibility," 1990.

[11] Keen, K. S., Morgret, C. H., Langham, T. F., and Baker, W. B., Jr., "Trajectory Simulations Should Match Flight Tests and Other Lessons Learned in 30 Years of Store-Separation Analysis," AIAA Paper 2009-99, Jan. 2009.

[12] Morgret, C. H., Moore, D. A., and Smith, M. E., "The FLIP 4 Store-Separation Trajectory Simulation Code," AIAA Paper 2009-100, Jan. 2009.

[13] Keen, K. S., "Equations for Store Separation Motion Simulations and Instrumented Model Data Reduction," AEDC TR-95-12, Aug. 1996.

[14] Veazey, D. T., "Current AEDC Weapons Separation Testing and Analysis to Support Flight Testing," AIAA Paper 2004-6847, Nov. 2004.

[15] Davis, M. B., Yagle, P., Smith, B. R., Chankaya, K. M., and Johnson, R. A., "Store Trajectory Response to Unsteady Weapons Bay Flowfields," AIAA Paper 2009-547, Jan. 2009.

[16] Hudson, M. L., and Charlton, E. F., "Many Uses of CFD in JSF Store Separation," International Aircraft-Stores Compatibility Symposium XIV, Fort Walton Beach, FL, 11–13 April 2006.

[17] Crandall, R., "Airborne Separation Video System," presented at Aircraft-Store Compatibility Symposium, Destin, FL, March 2001.

[18] Forsman, E., Getson, E. S., Schug, D., and Urtz, G., "Improved Analysis Techniques for More Efficient Weapon Separation Testing," *Proceedings of the 19th Annual*

Symposium of the Society of Flight Test Engineers European Chapter, Manching, Germany, September 2008.

[19] Getson, E. S., "Telemetry Solutions for Weapons Separation Testing," presented at ITEA Aircraft-Stores Compatibility Symposium and Workshop, Destin, FL, Feb. 2003.

[20] Forsman, E., and Schug, D., "Estimating Store 6DOF Trajectories Using Sensor Fusion between Photogrammetry and 6DOF Telemetry," presented at ITEA Test Instrumentation Workshop, 2011.

[21] Harding, G. C., and Barton, K. M.., "Predicting Aircraft Flowfield Induced Water Vapor Condensation for Store Separation Flight Tests," AIAA Paper 2007-4169, June 2007.

[22] Tutty, M., Akroyd, G., Cenko, A., and Piranian, A., "Stores Separation from Weapons Bays," presented at 30th Congress of the International Council of the Aeronautical Sciences, Daejeon, Korea, Sep. 2016.

[23] Roberts, E., "Lessons Learned: Limitations of Modern Tools and Applications for Store Separation Prediction," presented at Aircraft-Store Compatibility Symposium, Destin, FL, March 2001.

CHAPTER 17

F-35 STOVL Performance Requirements Verification

David G. Parsons* **and Daniel E. Levin†**
Lockheed Martin Aeronautics Company, Fort Worth, TX

David J. Panteny‡ and Peter N. Wilson§
BAE Systems, Warton, United Kingdom

Michael R. Rask¶
Cummings Aerospace, Niceville, FL

Brad L. Morris**
Naval Air Systems Command, Naval Air Station Patuxent River, MD

The F-35B is the Short Takeoff and Vertical Landing (STOVL) variant of the F-35, with two challenging and unique requirements. First, the aircraft must take off from both LHA/LHD-class flat-deck ships and ski-jump–equipped Queen Elizabeth–class ships while carrying the contractually specified fuel and payload. The second requirement is to return and perform vertical landings on the same ships or at austere sites while carrying unexpended ordnance. Flat-deck, short-takeoff distance and vertical landing bring-back capability requirements are two STOVL aircraft Key Performance Parameters (KPPs) that must be met for the aircraft to be successful. The verification of achieving these KPPs spanned more than a decade. It involved sub- and full-scale ground-based testing and model development, and culminated in flight testing in the System Development and Demonstration phase of the F-35 program. This paper focuses on the flight test element of the F-35 program's STOVL performance requirements verification effort. Flight test verification methods and results, as well as operational observations, are presented for both shore- and ship-based testing.

*Aeronautical Engineer, Senior Staff, Flight Sciences.
†Lockheed Martin Site Lead Test Pilot, F-35 Integrated Test Force, NAS Patuxent River.
‡BAE Lead STOVL Propulsion Engineer, F-35 Integrated Test Force, NAS Patuxent River.
§STOVL Lead Test Pilot, F-35 Integrated Test Force, NAS Patuxent River.
¶Principal Aerospace Engineer.
**Tactical Fixed Wing Performance Lead Engineer, NAWCAD, Patuxent River.

I. INTRODUCTION

The F-35 Joint Program Office required the Short Takeoff and Vertical Landing (STOVL) F-35 variant to have substantially improved range, payload, and bring-back capabilities over the legacy Harrier. With 5th-Generation stealth capabilities and supersonic speeds, the STOVL variant provides the operator with enhanced lethality, survivability, maintainability, and affordability.

The Joint Contract Specification (JCS) identifies two challenging Key Performance Parameters (KPPs) for STOVL operations. The first specifies that the aircraft must be able to launch from a flat-deck LHD/LHA-class ship with a full complement of internal stores and the fuel required to support a designated combat radius. The second specification establishes the aircraft's Vertical Landing Bring-Back (VLBB) requirement, which is based on recovery with those same internal stores. VLBB also requires an appropriate amount of fuel on board to conduct operations in the pattern. Figure 1 shows an F-35B hovering abeam an LHD prior to a Vertical Landing (VL).

Fig. 1 F-35B hovering abeam an LHD.

These requirements provide a metric for comparing competing designs and design changes. Although generally representative of the expected aircraft capabilities, the requirements are specified for ideal operating conditions that are not necessarily consistent with real-world operations. The actual capabilities of the aircraft that are provided to the operator must be validated across the expected range of operating conditions and configurations.

This paper describes the process used to verify the achievement of the two KPPs. It details how additional testing and analysis led to the operational performance capabilities of the F-35B. The discussion is limited to flat-deck Short Takeoff (STO) and VL. As such, it does not cover the additional launch and recovery capabilities that were developed and validated, such as ski jump STOs, vertical takeoffs, and slow landings. In addition, it discusses several operationally significant issues and lessons learned that were discovered during flight test.

II. BACKGROUND

The STOVL variant of the F-35 is intended for use by the U.S. Marine Corps and the U.K. and Italian ministries of defense as a replacement for the AV-8B Harrier. The F-35B flight regime is divided into conventional and STOVL modes. Conventional mode is used for nominal, fixed-wing-type flight operations. STOVL mode provides the operator with capabilities essential to the F-35B mission. This includes the abilities to take off from naval vessels or austere runways without using a catapult, and to return to land on the same surfaces.

In STOVL mode the propulsion system generates thrust that is exhausted through four nozzles: the main or core nozzle, the LiftFan® nozzle, and the left and right roll-post nozzles [1]. The core nozzle can pitch downward and yaw left and right through the three-bearing swivel module. The LiftFan is powered by diverting horsepower through a driveshaft from the main engine's low-pressure turbine via a clutch and gearbox. The LiftFan exhaust can be vectored forward and aft via a Variable Area Vane Box Nozzle (VAVBN). The roll-post nozzles are located under the wings just outboard of the main landing gear. The STOVL configuration is shown in Fig. 2.

To meet other requirements of the JCS, including a supersonic capability and low observability, the high-drag, nonstealth components of STOVL mode are hidden behind doors. STOVL operation is incorporated into the design as a unique mode that is used for only a small portion of a mission and is enabled through a process called conversion. STO is a submode of STOVL in which the Control Laws (CLAW) automatically manage thrust, Thrust Split (TS), and control effector positioning for optimal STO performance. TS is the ratio of core nozzle thrust to LiftFan nozzle thrust. For a STO, the pilot selects STOVL mode prior to takeoff (Fig. 3) and converts back to the Conventional Takeoff and Landing (CTOL) mode once airborne. In CTOL mode, the aircraft is prepared to use its supersonic, stealthy, 5th-Generation capability. At the end of the

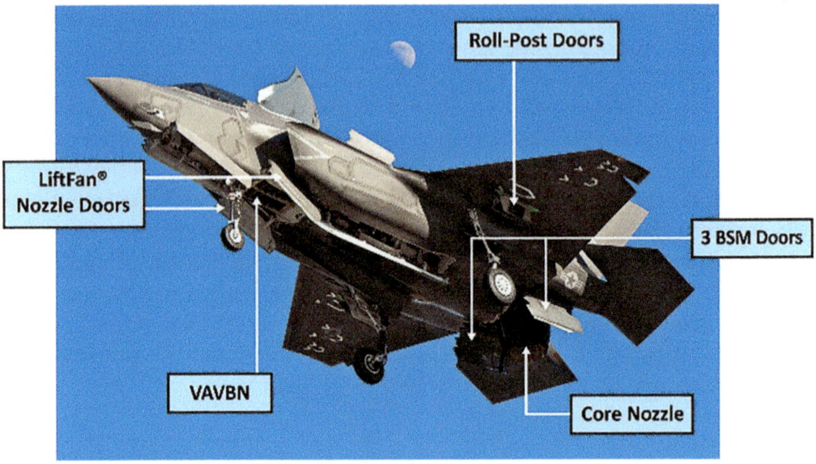

Fig. 2 STOVL features.

Fig. 3 F-35B STO from an LHA.

mission, the pilot converts back to STOVL mode in preparation for a slow landing or VL.

III. F-35 STOVL PERFORMANCE REQUIREMENTS VERIFICATION

The verification of the two KPPs was the culmination of years of development, as Fig. 4 shows. All performance verification work on the F-35 program followed the same approach. Aerodynamics models were developed from subscale wind tunnel and full-scale force and moment evaluations, and uninstalled propulsion performance was measured in indoor and outdoor test facilities. Flight testing was accomplished to validate the models used to calculate performance. Once validated, those models were then used to calculate performance against the specification requirements. Flight testing to verify the JCS requirements discussed in this paper is a subset of the larger performance test matrix. It was designed to substantiate the performance of the aircraft throughout the flight envelope to validate operational performance mission-planning products.

Through years of development testing on the F-35 program, the test team completed numerous maneuvers. Table 1 indicates the experience and knowledge the team developed for STOVL-related tasks, along with the robust capability of the F-35B. The data in the Unique column under Vertical Landings and Short Takeoffs indicate events completed to satisfy the requirements of specific test points.

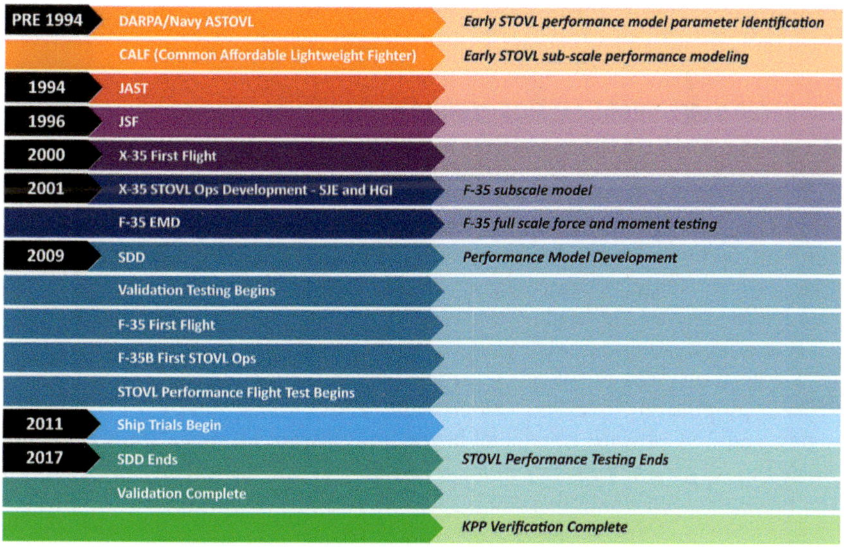

Fig. 4 F-35 STOVL performance verification timeline.

TABLE 1 TOTAL VLS AND STOS AS OF 14 FEBRUARY 2018

Vertical Landings			Short Takeoffs		
Aircraft	**Unique**	**Total**	**Aircraft**	**Unique**	**Total**
BF-1	360	782	BF-1	271	1407
BF-2	187	286	BF-2	92	358
BF-3	18	25	BF-3	20	29
BF-4	89	181	BF-4	158	458
BF-5	104	228	BF-5	105	470
Grand Total	**758**	**1502**	**Grand Total**	**646**	**2722**

IV. SHORT TAKEOFF PERFORMANCE VERIFICATION

The JCS flat-deck STO requirement is to take off within a specified distance from an LHD/LHA-class ship. The aircraft taking off must have internal weapons and only the fuel necessary to complete a specified mission profile. The total aircraft weight, including the fuel required to complete the mission and stores, affects the spotting distance required to launch the aircraft. Verifying the requirement entails validating both the conventional performance databases (for the fuel required to complete the mission) and STOVL-mode databases (for the distance required to launch the aircraft). The verification of the fuel required to complete the specified mission was part of the conventional performance requirements verification process [2]. Validating the STOVL-mode performance databases for STO calculations required both shore- and ship-based testing.

A. SHORE-BASED SHORT TAKEOFF TESTING

Shore- and ship-based STO performance validation was accomplished in a complementary manner. Shore-based testing was conducted as a prerequisite for ship-based testing, but it provided data for use in validating areas of the databases common to both. These databases were updated with and validated through these results, and the validated databases provided the foundation for ship-based STO predictions and testing.

Pretest analysis and modeling provided recommended rotation speeds, which are a critical element of the STO, for the beginning of shore-based testing. The STOVL propulsion effectors (core nozzle and LiftFan nozzle thrust vector angles, and TS) provide considerable rotation power. As such, these alone are capable of rotating the aircraft at a speed below which the aircraft can fly away. The recommended rotation speed needed to satisfy several criteria:

- The aircraft must reach 45 knots Calibrated Airspeed (KCAS) before the control system will allow rotation.

- The rotation and liftoff must provide sufficient clearance to avoid a nozzle or STOVL door strike.
- The aircraft must exhibit a positive climb rate and flight path acceleration throughout the climb out.

The STO predictions surveyed rotation speeds for each mass properties/power setting/ambient atmospheric condition. These determined the optimum speed to achieve a height of 50 ft above ground level in the shortest ground distance while meeting the three criteria above.

From a performance analysis standpoint, the STO maneuver can be broken into three major elements [3]. These are the ground acceleration to rotation speed, the rotation to flyaway attitude, and the climb out to a height consistent with Out-of-Ground Effects (OGE) conditions.

The ground acceleration is straightforward and dominated by thrust from the propulsion system. Aircraft control surface positions are relatively constant throughout the acceleration. The contributions from unpowered aerodynamics (mainly drag) are minimized due to the slow rotation speeds (low dynamic pressure generates low drag force). Test data validated the predicted levels of ground acceleration performance, and no adjustments were necessary for the models to match flight test. This was a crucial aspect of model validation that was required to be completed prior to moving to ship-based STOs.

STO rotation, shown in Fig. 5, is a dynamic event. Multiple control effectors (conventional control surface positions and STOVL propulsion system effectors) move in unison to increase pitch attitude on the aircraft. This, in turn, increases conventional aerodynamic lift while simultaneously providing thrust in the

Fig. 5 STO rotation.

vertical and longitudinal axes. Use of the effectors is managed by the flight control system and is a function of many variables [4].

Owing to the dynamic nature of the rotation event, model comparisons to flight test data are generated by feeding actual flight test control effector usage into the model. This is done as a function of groundspeed, and the model ground track, pitch path, and flight path are integrated. Ideally, true airspeed would be used instead of groundspeed to ensure a match of dynamic pressure; however, the true airspeed signal is occasionally noisy and subject to wind gusts that drive the true airspeed trace to be nonmonotonic with either time or distance. Because groundspeed is single-valued with both time and distance, it is used to determine the correct effector positions from flight test data. Time is not used to determine control effector usage from test because the entire maneuver is analyzed at once. If there were any deviation from the test airspeed and model airspeed at rotation, the causes of differences between test and model flyaway characteristics would be more difficult to ascertain. (Such a deviation would be due to differences in ground acceleration.) This approach forces rotation modeling to be done at airspeeds consistent with flight testing to achieve representative conventional lift and drag forces.

The modeling of the rotation control effectors is conducted in an open-loop fashion, for performance analysis purposes, until the model achieves nose gear liftoff. After model nose gear liftoff is achieved, the remaining rotation is conducted in a closed-loop fashion using horizontal tail deflection to close the loop on the flight test pitch attitude. For this portion of the prediction, the model uses whatever predicted tail deflection is necessary to match the flight test pitch vs the groundspeed characteristic. Modeling in this manner effectively uses the horizontal tail to account for any pitching moment errors in the predicted databases.

Analyzing the rotations of multiple STO maneuvers across a range of mass properties and ambient conditions showed that the model was a good match for the test aircraft. Minor adjustments to the model pitching moment and lift coefficient in ground effect at compressed gear height were included as well. These enabled us to better match test aircraft rotation characteristics. No other adjustments were required.

Flight test STO climb out to a height of 50 ft was conducted with the pilot out of the loop (as much as possible) for performance purposes. As a result, the aircraft control surfaces remained in nearly constant positions once flyaway attitude was attained. The F-35 flight control system schedules the pitch attitude of the aircraft for STO flyaway as a function of calibrated airspeed, with attitude gradually decreasing as airspeed increases. The modeling of the flyaway was conducted in the same manner as the last portion of the rotation. A model horizontal tail position was used to close the loop on aircraft pitch attitude as a function of groundspeed. Initial comparisons of STO climb out showed the predicted model to be conservative in terms of climb rate; however, conventional Powered Approach (PA) configuration performance flight test data indicated a lift benefit for the test aircraft versus predictions. Once this increment for conventional lift was included in the STO modeling, predictions for climb out to a height of 50 ft matched test data within acceptable tolerances. Some residual pitching moment differences existed during climb out, evidenced by differences in the horizontal tail deflections between the model and test; however, the differences in tail deflection had only minor effects on the climb out characteristic and were considered acceptable for performance modeling purposes.

Figure 6 provides an example comparison of the predicted model to the results of a complete shore-based STO. Climb out and ground acceleration matched the test data well. In the figure, the lower two plots illustrate how the comparisons were developed. Prior to liftoff, the prediction was calculated in an open-loop fashion, with control effectors in the model forced into the flight test positions as a function of groundspeed. This is demonstrated in the figure by the tail position in the Tail Behavior plot and was done prior to separating prediction and test data. Once the nose wheel lifted off, the pitch attitude in the model was forced to match the pitch attitude from flight test as a function of groundspeed. In the figure, this is depicted in the Pitch Attitude plot after approximately 83 kt. The residual pitching moment difference in the model is evidenced in the Tail Behavior plot as a difference between the flight test and model predicted tail positions.

Figure 7 summarizes the results of the shore-based STO analysis, obtained after all adjustments noted above had been included in the model. Data are presented as flight test distance to a height of 50 ft vs the predicted distance to a

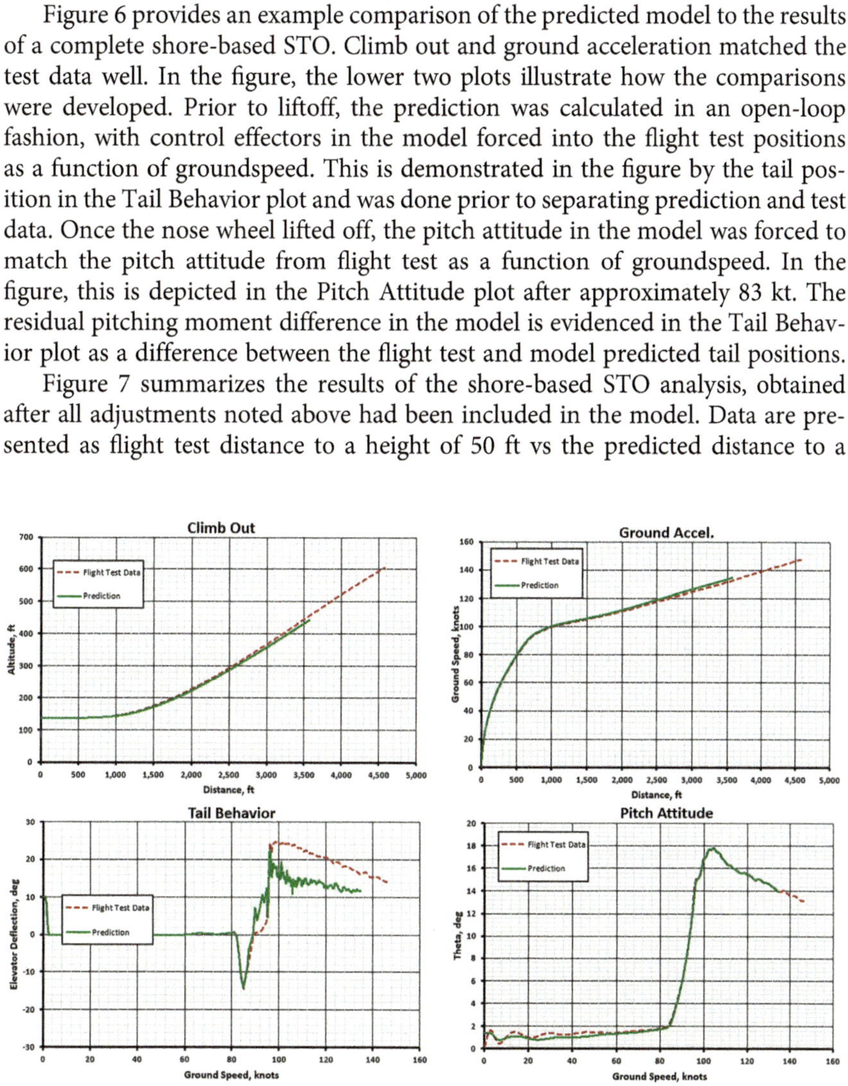

Fig. 6 STO performance characteristics.

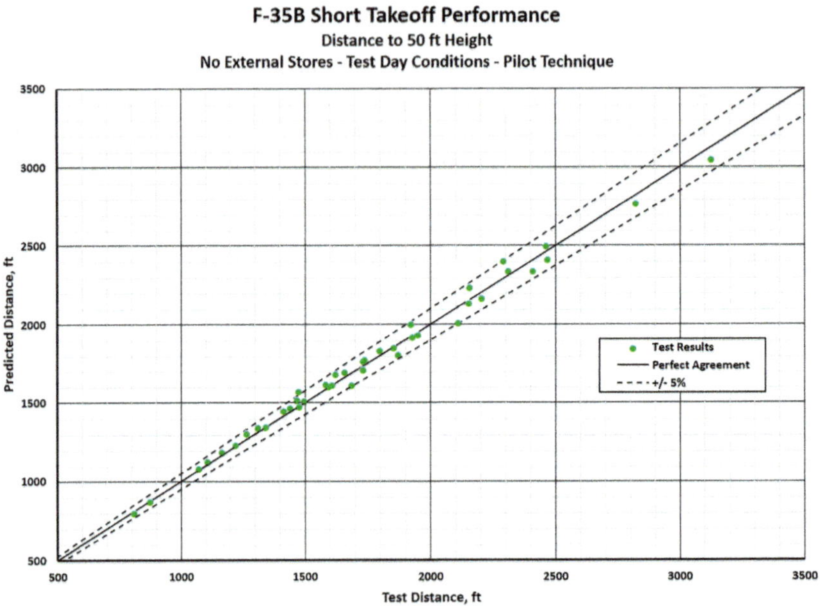

Fig. 7 Shore-based STO model comparison to flight test.

height of 50 ft. In general, all predicted data are within 5% of the test values for all conditions, configurations, and power settings tested.

B. SHIP-BASED SHORT TAKEOFF TESTING

Successful shore-based testing resulted in the development of a high-quality STO/STOVL model that permitted the team to enter the ship-based testing phase with confidence. It also allowed our team to focus mainly on the characteristics unique to ship-based operations. For ship STOs, the pilot does not initiate rotation at a target airspeed, as is done for shore-based STOs. On the ship, a reference rotation line is placed 225 ft from the ship's bow, and all rotations are initiated at this point on the deck. Predictions are generated to determine the proper starting distance (spot) from the bow of the ship from which to initiate the STO. The STO is initiated in a manner such that the aircraft will reach the rotation line at the correct speed. Spots are markings on the carrier deck in 50-ft increments from the bow, with the largest usable distance being 750 ft. The successful correlation of predictions to the shore-based ground acceleration data provided a solid foundation for this portion of the ship-based STO predictions.

An element of ship operations that could not be evaluated during shore-based testing was the variation of ambient wind over the deck of the ship. The amount of headwind and crosswind the aircraft experiences during a STO is a function of the magnitude and direction of the ambient wind and the speed, pitch, and roll of the ship. It is also affected by the position of the aircraft along the STO launch tram line and its location relative to the ship's superstructure. The environment can be very unsteady and unpredictable.

The aircraft's ability to successfully fly away from a ship STO launch is a function of airspeed and aircraft attitude at the deck edge. Figure 8 shows the aircraft approaching the deck edge as the flyaway attitude is established. A deck spot is used to target the airspeed and attitude for a given combination of aircraft weight and ambient conditions. The airspeed can be generated through a combination of ambient wind, ship speed, and aircraft longitudinal acceleration potential along the deck.

Similar to the shore-based STO, the preferred launch spot for a given ship-based STO is a balancing act of multiple criteria:

- The aircraft must achieve 45 KCAS prior to the rotation line for the CLAW to enable STO mode. Without STO mode, the aircraft cannot rotate to the attitude necessary to achieve a successful launch.
- The aircraft must have sufficient clearance to prevent a nozzle or STOVL door strike during rotation prior to leaving the deck.
- The aircraft must have margin to the launch conditions that yields 10 ft of aircraft Center of Gravity (CG) sink once the aircraft leaves the deck.

The 10-ft-of-sink criterion is the only one of the three that is mandated by the contract specification. The other criteria have been developed for use only with the F-35B.

Fig. 8 STO from the USS *Wasp*.

The JCS-based sink criterion is the focus of dedicated performance STO testing on the ship. To ensure that the aircraft has 15 kt of margin to 10 ft of sink off bow, the 10-ft-of-sink condition must be validated. Obviously, these criteria cannot be evaluated during shore-based testing. The complex environment experienced by the aircraft during operations at the ship is difficult to predict. That drives the focus in this area of testing and is mostly responsible for the requirement for operations to have 15 kt of margin to this criterion.

Predictions were generated prior to the ship trials to understand the variation of aircraft sink with wind over deck. At the ship, testing progressed in a controlled fashion by reducing the amount of wind over deck in measured steps. In this way, we quantified the wind over deck vs aircraft sink characteristic. For an F-35B flat-deck STO, the wind over deck must decrease from the operationally planned value by as much as 8 to 10 kt before the aircraft CG will experience any appreciable sink (less than 2 ft) relative to the ship deck. As wind over deck is reduced beyond the 8-to-10-kt level, the amount of sink off bow increases fairly rapidly. At the 10-ft-of-sink point, the slope of the airspeed vs the sink curve is very steep, which makes minimum-end airspeed testing very challenging.

Due to the sensitivity of aircraft sink to the maneuver technique, all performance ship-based STOs were conducted as auto STOs. An auto STO is a technique with which the pilot enters the deck spot through the cockpit displays. The flight control system determines the distance traveled from throttle up and brake release to the rotation line. Then, it initiates and executes rotation to capture flyaway attitude automatically. This STO type was designed to eliminate possible pilot rotation errors and variability, and it is the primary STO technique used at the ship.

The ship-based testing approach included mitigation procedures that could be implemented if the actual sink was expected to be excessive. This was based on the real-time monitoring of aircraft conditions in the control room by performance engineers. Shore-based training leading up to the ship trials was used to help the control room engineers recognize the characteristics of a poor launch. They did this by monitoring the aircraft height rate trace and the absolute level of sink below deck level. The mitigation procedure for a poor launch included advancing the throttle to request the maximum thrust available and increasing the pitch attitude to arrest the sink rate. The procedure proved effective because the F-35B flight characteristics provided sufficient time to recognize and respond to excessive sink.

Ship-based performance STO testing involved two series of build-downs to 10 ft of sink. The first series was conducted with no external stores on the aircraft during the second development test ship trial (DT-2). The flight test team established an approach to incrementally reduce wind over deck margin to gradually increase sink toward 10 ft. For DT-2, the build-down steps involved reducing spot distance and/or wind over deck to generate sink. The STOVL propulsion system performed strongly, so the team planned for tailwind conditions to generate any amount of sink at the maximum gross weight possible without external stores carriage. The other launch criteria, such as airspeed at rotation line and nozzle clearance, were still respected in the process. As a result, the aircraft could not simply be spotted as far forward as possible to generate sink.

The tailwind requirement proved to be problematic for successful testing. With ambient wind coming from behind the ship, the aircraft entered a somewhat unknown wind condition as it left the deck and sank below deck level. The ambient wind was not as constant in this airspace as it had been along the deck. This was likely due to the turbulence effects of the wind swirling off the front of the boat. The final test point conducted for performance during DT-2 exhibited more sink than desired, and the mitigation procedures were executed. As expected, once the pilot had increased the pitch attitude, the sink rate was arrested and the aircraft flew away without incident.

Figure 9 shows the aircraft height rate response to the pilot recovery action during the high-sink event. The blue line shows the raw Inertial Navigation System (INS) vertical velocity. The orange line shows the vertical velocity of the aircraft's CG, which is the reference parameter for aircraft sink off the bow. As the pilot increased the pitch attitude of the aircraft, the inertial velocity showed an immediate positive increase due to its location forward on the aircraft. Approximately one second from pilot input, the vertical velocity of the CG became positive. This indicated that the sink rate had been arrested and that the aircraft had begun to climb.

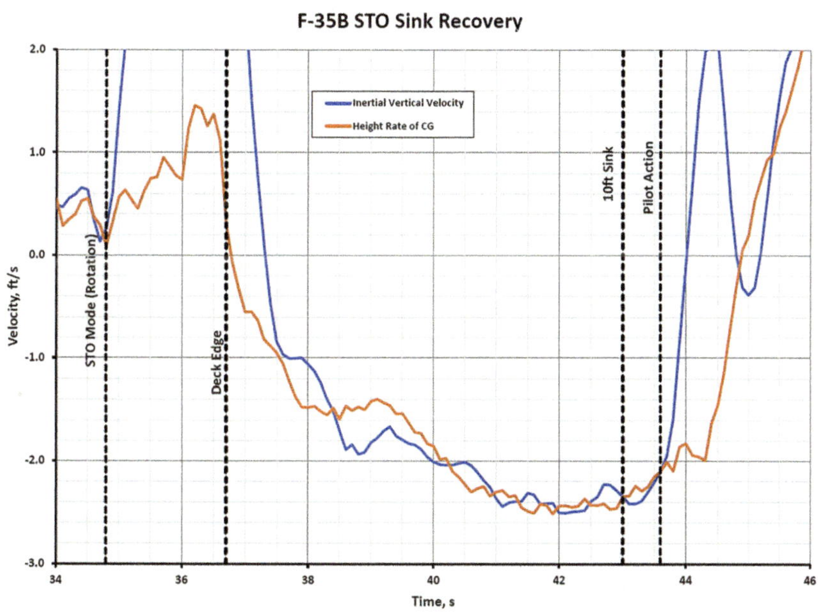

Fig. 9 Height rate change with pilot response action.

Figure 10 illustrates the characteristic of aircraft sink off bow with aircraft speed at the rotation line for DT-2. The solid line represents the predicted minimum height of the aircraft's CG. This is relative to deck edge vs speed at the rotation line for the target test weight and a nominal wind over deck condition. The symbols represent the flight test sink off bow adjusted to the reference conditions, such as weight, CG, pressure altitude, ambient temperature, and wind over deck. The sink is consistent with the predicted line. The final test point is circled, indicating that the maximum sink was achieved after pilot action.

The testing that was accomplished during DT-2 indicated that the predicted model was generally conservative; however, there was uncertainty in measuring the tailwind magnitude and direction in front of the ship. Owing to this, the validation team was hesitant to adjust models to reflect the results. Another element of the testing that proved problematic to understanding the results was varying both the lineup spot and wind over deck to progress from test point to test point. By changing more than one variable at a time, the team struggled to understand the complex combination of variables that determines aircraft sink off bow. A posttest detailed analysis of each performance launch provided evidence that the predicted model was conservative, relative to the flight test results (i.e., had less real sink than predicted). Further, it demonstrated that the tailwind conditions tested had not been particularly operationally representative. As such,

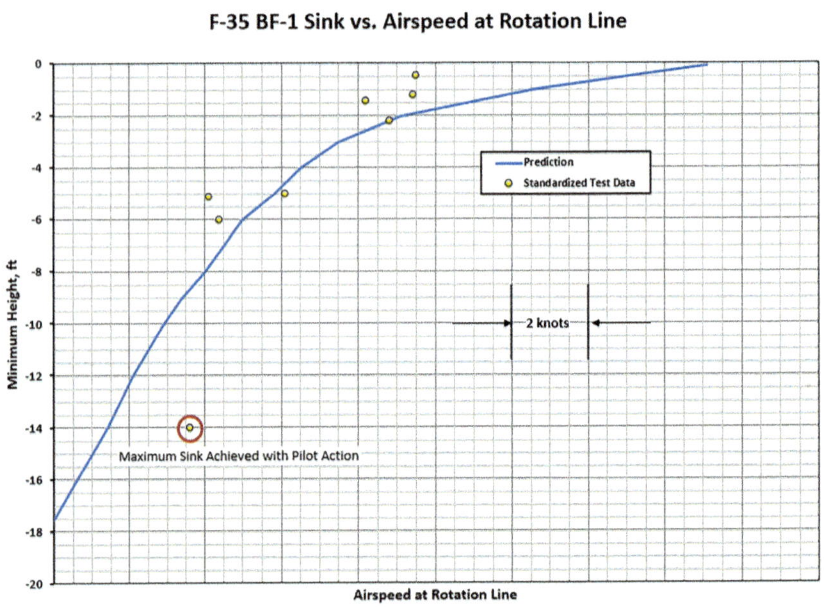

Fig. 10 Aircraft sink vs deck edge airspeed for DT-2.

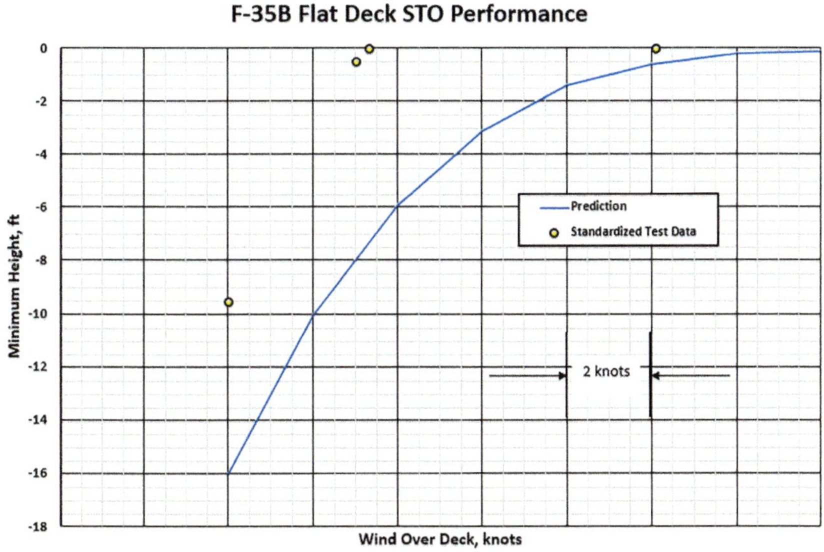

Fig. 11 Sink vs wind over deck for DT-3.

the team chose to leave the model unchanged prior to testing for the third development test ship trial (DT-3).

Testing during DT-3 utilized external stores to increase the weight of the jet so that minimum-end airspeed testing could be accomplished in positive headwind conditions. For this series of tests, aircraft weight and spot were held constant, leaving wind over deck as the only means to affect wind over deck margin changes. By holding weight and spot constant, the aircraft would be at nominally consistent groundspeed (relative to the deck) and attitude at deck edge for each test point. Again, the team took a measured approach to gradually increase sink off bow toward 10 ft. The increased weight, positive headwind conditions, and controlled spot provided higher-quality data than those from DT-2. DT-3 test data confirmed that the predicted sink characteristic was conservative (i.e., had less real sink than predicted sink) for a given wind over deck. The data also showed that the sink vs wind over deck curve characteristic was steeper than predicted. Figure 11 shows the results from DT-3. As with DT-2 data, the flight test data have all been adjusted to the same reference conditions: weight, CG, altitude, and ambient temperature; however, they have been held at a constant deck spot rather than wind over deck to illustrate the test results against a single predicted line. The final test point in Fig. 11 shows that the predicted wind over deck characteristic had approximately 2 kt of conservatism at the 10-ft-of-sink condition.

Deviations in ship pitch attitude and heave can impart initial vertical velocity to the aircraft as it exits the deck. This is difficult to account for accurately, and minor changes in initial vertical velocity can make measurable differences in the achieved sink at these minimal performance conditions. The test team made every effort to gather performance minimums test data in sea conditions that were as calm as possible; however, some deviations always exist and must be accounted for when assessing results.

Prior to DT-3, the verification team had not updated the models for the results of conventional aircraft testing in the PA or gear-down configurations. They also had not made any adjustments for the conservatisms observed during DT-2 in tailwind conditions. The conventional PA configuration testing showed increased flight test lift, relative to pretest predictions. After adjusting models for the results of this testing, each of the minimum-end airspeed performance test points was compared to the revised model predictions. When detailed model comparisons were made to the as-flown conditions, the predicted flight path matched the flight test results within acceptable tolerances. Figure 12 compares the flight test results for the largest sink on DT-3 to the detailed model. The model is an excellent representation of the flight test results for a very dynamic maneuver.

C. SHORT TAKEOFF PERFORMANCE VERIFICATION RESULTS

Shore-based and ship-based STO testing provided the performance verification team with the data necessary to update and validate the models used to predict

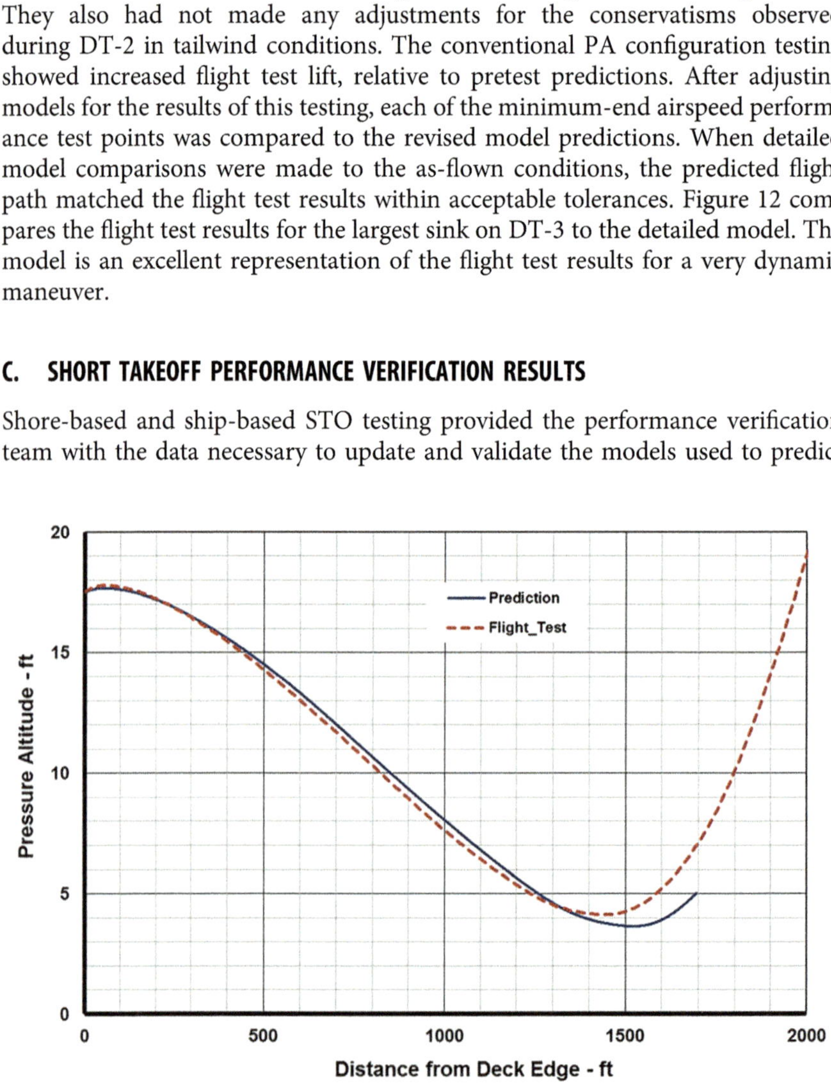

Fig. 12 Comparison of flight test and detailed modeling for sink.

Fig. 13 Flat-deck STO verification results.

STO performance. There were only a few changes necessary to have the pretest-predicted models match the flight test results. These consisted of minor adjustments to ground effect pitching moment and lift, and an update to the conventional PA flight test results.

The STO flight testing showed the performance of the F-35B to be better than originally predicted. The results of the F-35B conventional performance validation testing also reduced the amount of fuel required to complete the mission associated with the flat-deck STO requirement [2]. The combination of improvements in both conventional mission performance and STO performance allowed the F-35B to meet the JCS requirement for flat-deck STO. Moreover, it did so with a margin greater than 20%, which will greatly benefit future growth initiatives for the weapons system platform. Figure 13 illustrates this achievement.

D. OPERATIONAL CONSIDERATIONS

Performance testing focused on the capability of the aircraft in controlled and repeatable conditions; however, the aircraft must be operationally robust in a far wider range of operating conditions and techniques. The problem of sink, and the pilot's reaction to it, was discussed in the preceding subsection. Robustness testing also highlighted a potential problem that will eventually affect the operational capability of the aircraft.

Winds experienced by the aircraft during a STO deck run were anything but constant. Wind can vary in magnitude and direction throughout the STO event. With strong ambient winds from starboard during ship STO, the aircraft was in the lee of the island for the majority of the ground roll; however, it experienced a large change in crosswind as it cleared the island on the way to the bow. Figure 14 illustrates the difference in wind magnitude and direction as measured at the bow of the ship and farther aft near landing Spot 7. This was assessed during a wind calibration event conducted during DT-2.

An interesting result occurred from a wind gust during rotation and unstick during the first ship trial. The integrated flight and propulsion controls prioritized sideslip control over performance and used differential horizontal tail for that purpose. Planned performance was based on the tails being in their maximum lift position. A deviation resulted from the expected performance and dynamic nature of the rotation in the presence of a crosswind gust. This, in turn, resulted in a low nozzle clearance event that exceeded established flight test continuation

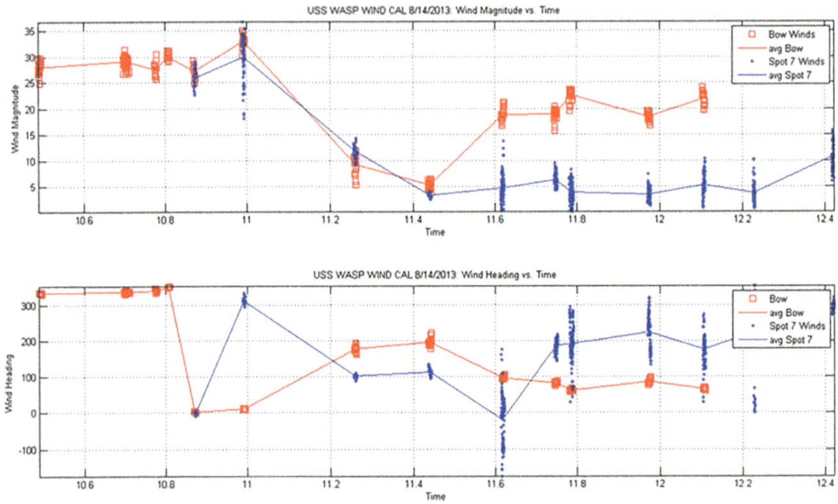

Fig. 14 Comparison of wind magnitude and direction for bow and Spot 7.

criteria. This event is illustrated in Fig. 15. In it, a nominal STO appears shortly before unstick on the top-left, and the CLAW reaction to a starboard crosswind is shown on the bottom-left. The nozzle clearance during the actual event is

Fig. 15 Nozzle clearance comparison.

shown on the right. This was barely noticed in the cockpit. However, the magnitude of the exceedance and the risk of a nozzle strike in gusty crosswind conditions ultimately led to adjustments in the aircraft CLAW to modify the prioritization scheme.

The net result of ship trials testing in crosswind led to the inclusion of additional margin for STOs in the presence of higher values of crosswind. This margin is carried in the form of additional speed at the rotation line over that which would be determined for nominal wind conditions down the deck. It is implemented as a farther-aft deck spot for these conditions over nominal winds down the deck.

V. VERTICAL LANDING BRING-BACK PERFORMANCE VERIFICATION

In the air, STOVL-mode operation has two major flight regimes: Semi-Jetborne (SJ) and Jetborne (JB). SJ covers speeds at which the lift required for flight is provided by both the aerodynamic surfaces and propulsion system thrust (approximately 45 KCAS and higher). JB is optimized for flight below 45 KCAS, for which nearly all lift is generated by propulsion system thrust. The CLAW automatically configures the propulsion system and effectors to optimize performance while ensuring sufficient control power. For a VL, the jet is decelerated into JB flight as it is positioned for landing.

The VLBB KPP requirement defines the combination of reserve fuel and unexpended stores that must be recovered at the end of a mission. The VLBB task is described in Fig. 16, and its analysis consists of elements in both the JB and SJ

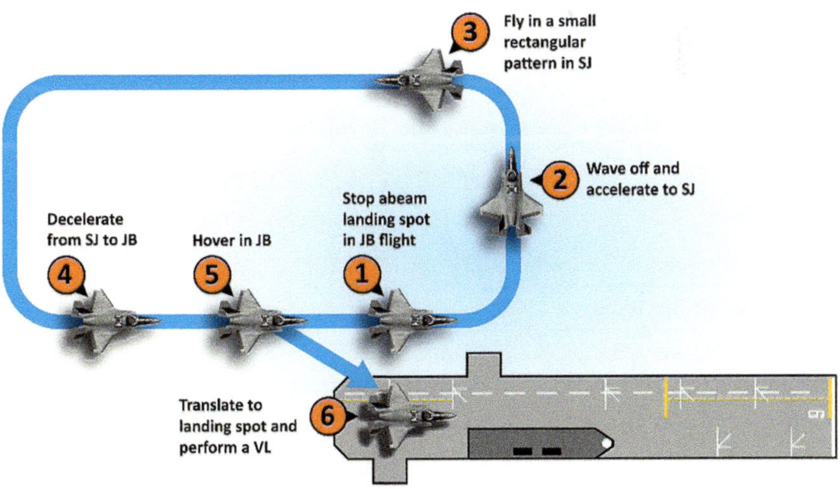

Fig. 16 VLBB pattern profile.

regimes. The maximum hover capability of the aircraft must be determined (JB), and the weight required for the specification hover must be determined. The latter is determined from a pattern flown around the ship during a recovery (SJ). In addition, the ability to land vertically at the required weight must also be validated (JB). The performance team validated each of these regimes separately.

Forces affecting aircraft performance in STOVL mode flight are:

- Propulsive forces, including thrusts from the four nozzles and inlet and nozzle drags
- Unpowered aerodynamics, or classical aircraft lift and drag
- STOVL Jet Effects (SJEs), or the effects of entrained and reflected flow, the component of SJEs in the vertical axis commonly referred to as suck-down because the result is an effective reduction in propulsive lift

The preflight test SJE database was created based from the results of extensive wind tunnel testing [5]. Figure 17 shows the relative magnitudes of the three

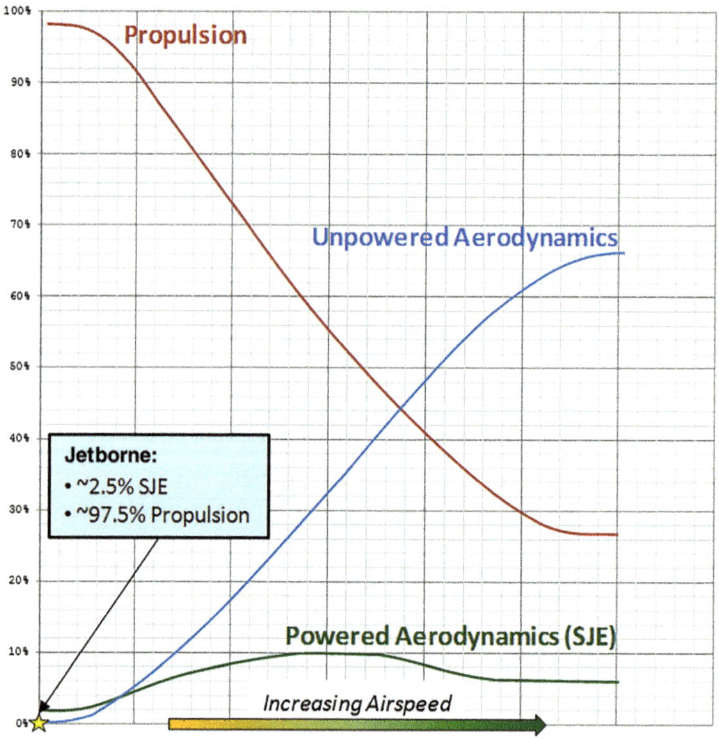

Fig. 17 Relative magnitude of forces on the STOVL aircraft.

categories of forces as a function of airspeed. During a stabilized hover at OGE heights, SJE effectively offset propulsive total vertical thrust by approximately 2.5%. Otherwise put, the actual usable vertical thrust is 97.5% of the net vertical thrust.

Prior to flight test, Pratt & Whitney and Lockheed Martin concentrated considerable resources on validating propulsion system performance characteristics in the JB flight regime [6]. As a result, it was one of the most accurately modeled regions of the entire flight envelope in conventional or STOVL mode. For the inflight thrust methodology [7], data from an instrumented propulsion system fed a propulsion cycle model to provide an accurate estimate of thrust. This methodology was a critical element in validating aircraft SJE for low-speed, JB flight. Pratt & Whitney developed the In-Flight Thrust Calculation Deck (IFTCD) for use by the performance validation team. The IFTCD was calibrated for propulsion system performance at hover conditions based on extensive thrust stand testing at Pratt & Whitney facilities in West Palm Beach, Florida.

Reconciliation of aircraft performance in hover was also improved by using data from the force and moment hover pit testing [8, 9]. The force and moment testing on the hover pit in Fort Worth, Texas, provided the first opportunity to evaluate multiple factors. These included the consistency of IFTCD calculations, wind tunnel–based SJE databases, and measured forces and moments from load cells installed in the pit restraint system for simulated OGE conditions. The results of the pit testing confirmed that the combination of IFTCD-calculated thrust and the preflight test SJE database matched the measured forces within defined goals. These results allowed the validation team to enter the flight test phase with confidence in the hover performance prediction model.

Having an aircraft gross weight capability in hover in OGE conditions does not necessarily directly indicate having a VL capability. Hot Gas Ingestion (HGI) is a significant factor affecting performance and operability during VL. It reduces total propulsion system thrust, potentially influencing aircraft descent rate and handling qualities and reducing engine and LiftFan stall margins. When executing a VL, the aircraft must not experience excessive uncommanded acceleration or adverse handling qualities during the landing as the jet enters ground effects.

HGI has two main mechanisms of action: ground sheet roll-up ingestion and fountain ingestion. These two mechanisms are visually depicted in Fig. 18.

The aircraft exhaust plumes impinging on the ground form a ground sheet that flows away from the aircraft in all directions. Ground sheet roll-up occurs when a strong environmental wind picks up this buoyant layer of gas and turns it back toward the aircraft. Due to the well-mixed nature of the flow returning to the aircraft, the HGI experienced from this mechanism is of a low magnitude. Usually, it is only seen near the ground in the last stages of a VL. The flight-test-measured levels of HGI from this mechanism have not been significant and do not require mitigation through performance limitations.

There are certain heights above ground where the propulsive jets do not merge before impacting the ground. There, the ground sheets from the individual

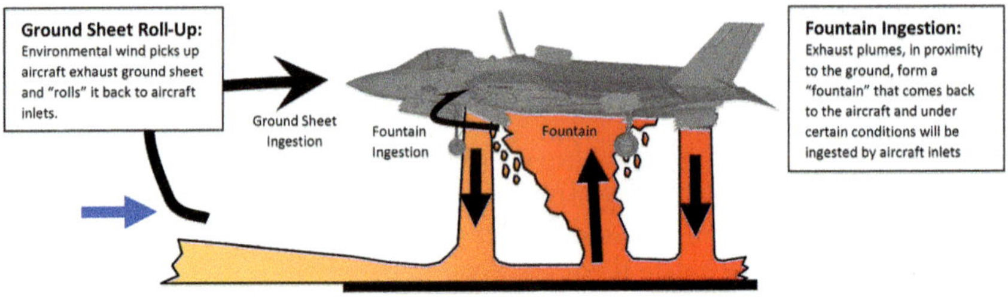

Fig. 18 HGI mechanisms.

exhaust plumes of the LiftFan, core nozzle, and roll-posts flow inward toward one another under the aircraft. An upward fountain is formed where they collide, and it then impacts the undersurface of the aircraft. Under certain conditions, exhaust gas from the fountain can enter the main engine inlets and LiftFan inlet. HGI in this scenario can result in a high temperature rise and high-temperature inlet distortion. This leads to concerns for both aircraft performance and engine/LiftFan operability that must then be mitigated.

For the F-35B, the amount of HGI is a function of height above ground, aircraft relative wind, aircraft Fuselage Station Center of Gravity (FSCG), and ground slope. Fountain HGI is typically only seen with tailwind or crosswind conditions relative to the aircraft and not with headwind conditions relative to the aircraft. For the aircraft to remain stationary in the hover in the environmental wind, the exhaust plumes must be directed aft for a headwind and forward for a tailwind. Additionally, the environmental wind acts on both exhaust plumes as they travel toward the ground and on the fountain as it returns. Therefore, a headwind tends to angle the fountain aft away from the inlets, reducing HGI levels. By contrast, a tailwind tends to angle the fountain forward toward the inlets, resulting in potentially higher levels of HGI.

An aft aircraft FSCG increases the trim TS required. Higher TSs result in greater exhaust flow from the engine nozzle vs the LiftFan nozzle. This reduces the strength of the ground sheet from the LiftFan exhaust, compared to the ground sheet from the engine nozzle exhaust that forms the fountain. With an aft FSCG, the point at which the fountain forms underneath the aircraft moves farther forward. Consequently, the fountain flow will angle forward toward the inlets, increasing the likelihood of encountering higher HGI levels for a given condition. With the ground sloping away from the nose of the aircraft (downhill), the fountain will also be angled farther forward toward the inlets. This increases the likelihood of encountering high HGI levels for a given condition.

Subscale wind tunnel testing [10] was used to create an HGI database to use in six-degrees-of-freedom flight simulation tools. With these, we performed sensitivity analyses under piloted conditions and developed flight test HGI limitations for VL envelope expansion, allowing VL flight testing to commence.

A. SHORE-BASED TESTING

Accurately determining the ambient wind conditions in which an aircraft operates is critical to comparing actual and predicted flight test data. In the JB regime, determining the wind conditions in which the aircraft operates is challenging. At JB speeds, the aircraft air data system is ineffective for airspeed determination. Other sources of environmental winds (e.g., control towers, portable wind anemometers) are usually located significantly far from the testing location; however, if these other sources were near the aircraft, they could be influenced by the environment the aircraft propulsion system creates. That would lead to erroneous indications of winds acting upon the aircraft. Therefore, an alternative method for determining wind magnitude and direction was developed.

For the aircraft to remain stationary in JB flight, relative to the ground in a headwind condition, the propulsive jets must be vectored aft to counter the prevailing wind. Similarly, for the aircraft to remain stationary in a crosswind condition, the aircraft must be banked into the wind. This enables the propulsive jets to counter the prevailing wind. Using pretest models, correlations were derived between the headwind component (relative to the aircraft) and the average longitudinal angle of the exhaust plumes. A similar correlation was developed for the crosswind component and the acceleration in the aircraft lateral axis. The correlations were empirically adjusted following the hover pit test and as flight testing progressed. During flight testing in the JB regime, the winds derived from these correlations were regarded as the truth source of winds acting on the aircraft. These allowed for more accurate comparisons to be made between flight test data and prediction.

1. HOVER CAPABILITY

Flight test maneuvers used for validating hover performance typically have tighter tolerances and longer dwell times than those for other disciplines. Each hover required the pilot to establish hover conditions as close to target fuel states as possible, then remain effectively hands-free for at least 30 seconds. The dwell time provided for any transients or feedback loops in either the aircraft or propulsion control systems to dampen out. This provided data with lower levels of scatter (uncertainty). In some cases, the ambient wind was sufficient for a target condition; for others, the pilot had to create a wind relative to the aircraft. The pilot did this by flying the aircraft along a particular ground track into or away from the ambient wind to achieve the desired test conditions. Any gustiness in ambient winds increased the scatter in hover performance results due to aircraft and control system responses to those gusts.

Aircraft gross-weight fidelity in hover was improved by weighing the aircraft postflight. Following a performance hover test point, the engine was shut down as soon as practical after completing the test point. The aircraft was then weighed, and those data were combined with the integration of measured fuel flow. The data set determined the amount of fuel consumed for engine run-time, which we then used to back-calculate the weight of the aircraft at the beginning of the test point.

Dedicated performance flight test hover data were standardized to a nominal set of reference conditions to provide a meaningful comparison with pretest predictions. Because hover maneuvers are conducted at various aircraft mass properties and atmospheric conditions, the data must be adjusted to a set of common reference conditions. Only in this way can we understand the results in a more holistic manner than single point by single point. Analytical adjustments are applied to each measured test point to account for the predicted differences between the test operating conditions and the desired reference conditions. Figure 19 illustrates the results of this effort for hover performance testing. Test aircraft weight is plotted vs propulsion system thrust, with the predicted thrust-vs-weight characteristic shown as the green line. Combining a robust pretest SJE database, accurate inflight thrust, and rigorous determinations of test aircraft weight, we resolved aircraft hover capability to within approximately

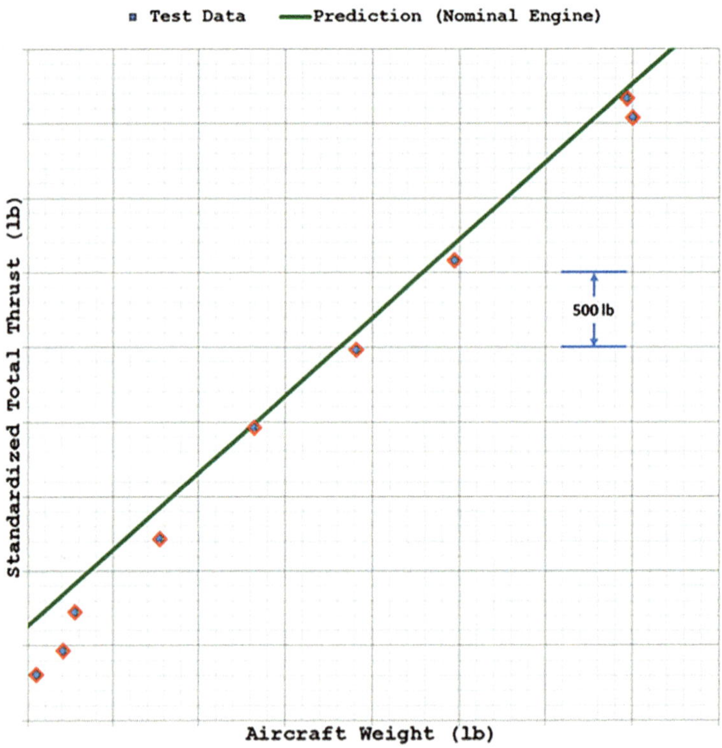

Fig. 19 Thrust-vs-weight prediction accuracy.

200 lb. This provided credence to the validity of the JB SJE database. With the hover performance database validated, maximum hover capability for the VLBB calculation was possible.

2. VERTICAL LANDING

VL flight test began by performing landings in the portion of the flight envelope where HGI was not predicted to be a factor. Once a basic HGI-free understanding of the performance of the aircraft during VL had been established, landings were performed in the region of the flight envelope where HGI was predicted.

During the flight test of VLs, engine and LiftFan inlet temperature rake instrumentation was used to measure temperature rises and temperature distortion levels due to HGI. Performance margin limitations were modified as flight testing progressed based on results observed. Temperature rake instrumentation data from VL testing were used to validate the HGI database. These provided insight into the aircraft control system's response to HGI.

Flight test – measured levels of HGI generally matched predictions from subscale tests. Low levels of HGI were measured in headwind conditions, relative to the aircraft, that were relevant to the VLBB KPP criteria. Therefore, mitigating HGI with performance margin was not pertinent to the VLBB requirement. However, as predicted by subscale test, crosswind and tailwind conditions relative to the aircraft produced moderate to high levels of HGI. As such, they required limitations on aircraft weight for the VL to protect against high-sink-rate landings and/or undesirable handling qualities. These margins are discussed in Section V.D.

A VL can have a preceding In-Ground Effect (IGE) hover, which is a stabilized hover at a low height above ground that is influenced by ground proximity. The VL with this is not, however, a standard maneuver for shore-based operations. Shore-based VLs are typically initiated from heights above ground that are significantly above the IGE/OGE boundary. An IGE hover occurring prior to a VL can happen during shipboard operations and, potentially, during shore-based workups ahead of aircraft carrier operations (field carrier landing practice). The typical hover height over an aircraft carrier deck is low enough such that the aircraft is considered IGE. Therefore, several shore-based IGE hovers were performed prior to ship testing at various wind conditions to ascertain any HGI impacts from the dwell IGE. Results from this testing, as well as nominal VL testing, were used to define the initial limitations for ship testing.

3. PATTERN FUEL

Validating the fuel required to complete the VLBB pattern (and set the hover weight requirement for VLBB) involved performing trim maneuvers. These were executed in the SJ regime at speeds consistent with pattern speeds. The team approached this regime with a plan similar to that for the JB regime. We executed constant-altitude, constant-speed trim maneuvers and exercised inflight thrust methods to validate the contributions of conventional aerodynamics, SJE, and propulsive forces.

Unfortunately, the SJ regime proved to be far more difficult in terms of isolating the individual forces. The higher forward speeds of SJ flight, relative to JB flight, moved the operating conditions of the propulsion system farther from the cases where the inflight thrust tool was calibrated. As a result, the team was unable to utilize inflight thrust in SJ for the primary validation approach.

Even though contributions from individual forces could not be isolated, the team successfully validated the end performance of the model. Engineering judgment was used to adjust various elements of the SJ databases to drive predictions from the model to be consistent with measured flight test results. Figure 20 shows fuel flow vs calibrated airspeed for SJ trim maneuvers. The flight test data have been adjusted to a nominal weight and tropical day conditions. In this way, we evaluated multiple maneuvers at once to identify any potential trends for deviation from predictions. The flight-test-adjusted predictions are shown as the blue line, and the flight test data are the orange circles. The model matches the test data well across the SJ speed range. Most of the SJ pattern for the requirement was conducted in the 150 to 180 kt True Airspeed (KTAS) range, where fuel flow is minimized.

All validation work for the pattern fuel was accomplished with shore-based testing. There is nothing unique about operations at the ship that required further validation during ship-based testing. The fuel required to fly the VLBB specification pattern was calculated using the SJ flight-test-validated databases.

B. SHIP-BASED TESTING

As with the STO, confidence gained from shore-based testing led to successful initial recoveries at the ship. Figure 21 shows an early VL on the USS *Wasp* during DT-1. Three ship trial evolutions were completed to confirm the delivered capability in the more challenging shipborne environment.

1. HOVER CAPABILITY

For hover alongside the ship prior to crossing the deck for the VL, as seen in Fig. 22, the data from shore-based testing in

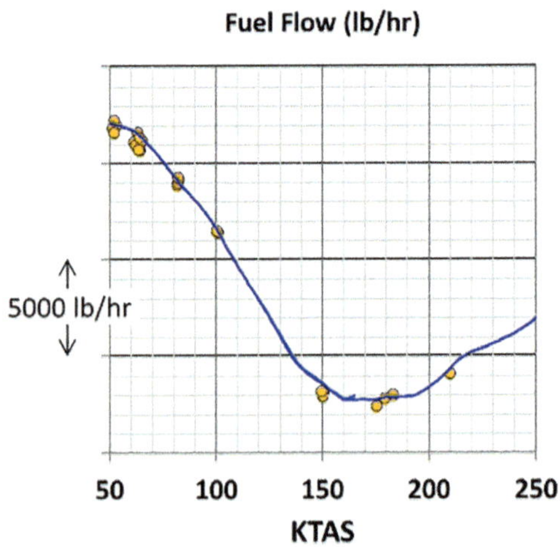

Fig. 20 SJ fuel flow predictions.

Fig. 21 F-35B landing aboard the USS *Wasp* (LHD 1).

OGE conditions apply directly. However, once the aircraft crosses over the deck of the ship, the aircraft immediately enters a near-IGE condition. There, the hover can be more challenging than it is at OGE heights. SJE forces (suck-down) increase as the height above ground decreases. Also, the winds the aircraft experiences above the deck of the ship can be vastly different from the winds experienced in hover over the water next to the ship. This is because the ambient winds are affected by the presence of the ship and the ship superstructure. The winds experienced by the aircraft above the deck are offset from the ambient wind magnitude and direction, and can become turbulent and inconsistent. This phenomenon may also alter the HGI characteristics from those seen during shore-based testing. In such an instance, it would affect the maximum gross weight from which a VL can be initiated. Figure 23 provides two cases of

Fig. 22 OGE hover abeam the USS *Wasp* (LHD 1).

Fig. 23 CFD visualization of the wind flow field around the ship.

Computational Fluid Dynamics (CFD) analyses completed to visualize the flow field around the carrier. The figure also illustrates the potentially turbulent nature of winds around the ship.

The variations in hover capability with wind magnitude, direction, and deck landing spot were evaluated during the three F-35B ship trials. Although not directly applicable to the VLBB specification calculation, these data are necessary to determine the capability provided to the operator via shipboard operating bulletins and mission planning products.

2. VERTICAL LANDING

Testing for VL at the ship began with nominal winds down the deck of the ship, where the expectation for variances in wind from ambient were minimized. The wind envelope was then progressively expanded in both magnitude and direction. VL testing at the ship included landing at different designated landing spot locations. It also had high hovers over the landing spots with winds from the direction of the island superstructure to assess Ship-exhaust Gas Ingestion (SGI) levels. In addition, VL testing involved landing with another aircraft operating on the deck in front of the landing spot to assess the HGI levels created by the on-deck aircraft. Supplementing this, VL testing examined landing with the aircraft facing aft down the deck, and landing with the aircraft facing cross deck, perpendicular to the ship.

From the results of the ship trials testing, no significant effects of the island superstructure or other ship effects exist for wind notionally down the deck. Performance results from shore-based testing are directly applicable to operations at the ship within this envelope. Outside this envelope, the performance margin is required for mitigating HGI, SGI, and other effects resulting from the ship environment. Section V.D describes this in greater detail.

C. VERTICAL LANDING BRING-BACK PERFORMANCE VERIFICATION RESULTS

The validation of the models used to calculate VLBB analytically was completed successfully. The use of inflight thrust methodology was critical for validating aircraft SJE for JB flight. The effort culminated in the resolution of aircraft hover capability to within approximately 200 lb. VL flight testing verified that HGI is not relevant to the VLBB KPP wind envelope and is generally considered only for operational limitations outside the KPP envelope. Test results showed that the pretest models for hover and VL were accurate and did not need adjustment for the VLBB KPP calculation. The fuel required to execute the specification pattern around the ship in SJ conditions was also validated successfully. This was done in a top-level manner rather than isolating the contributions of individual forces. Adjustments to the model for SJ conditions had a very minor effect on the fuel required to perform the pattern. Ultimately, validated VLBB performance exceeded the requirement by more than 7%, as illustrated in Fig. 24.

D. OPERATIONAL CONSIDERATIONS

Outside the VLBB KPP envelope, the operator still has significant performance capability, even though it can be less than the VLBB due to multiple factors. Testing during the System Development and Demonstration program validated the performance capability and highlighted some problems that can affect performance levels provided to the operator.

1. HOVER CAPABILITY

The positioning task in the hover prior to a VL requires the aircraft to be capable of translating forward, sideways, and rearward as the pilot maneuvers the aircraft. The maximum groundspeed for rearward translations is protected by the CLAW, preventing rearward acceleration at the groundspeed limit. The maximum tailwind relative to the aircraft is a pilot-observed limitation. During flying qualities testing to the maximum tailwind limit, ETR consistently increased with increasing tailwind, often to 100%. It was frequently accompanied by uncommanded sink on the order of 10 to 20 ft. The observed behavior was driven by TS moving forward significantly while accelerating into a rearward translation. The predominant cause is intake momentum drag creating a nose-down pitching moment with tailwind conditions. This prompts the flight control system to respond to the nose-down moment by adjusting TS forward. Conversely, intake momentum drag causes a nose-up pitching moment with a headwind. The tailwind condition is depicted in Fig. 25. The individual intake forces (the LiftFan and the auxiliary and main engines) from the intake momentum drag act forward of the aircraft

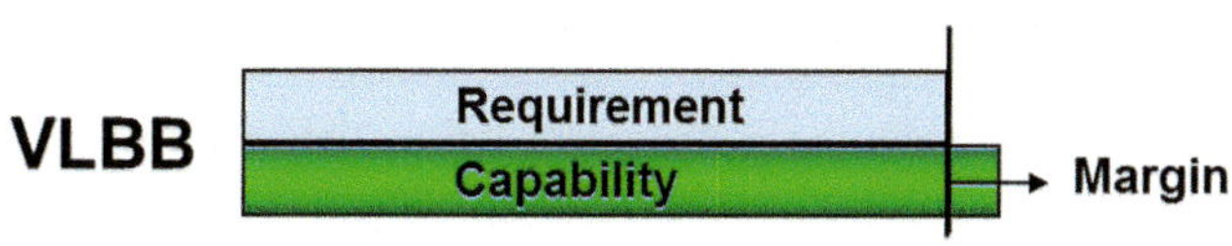

Fig. 24 VLBB verification results.

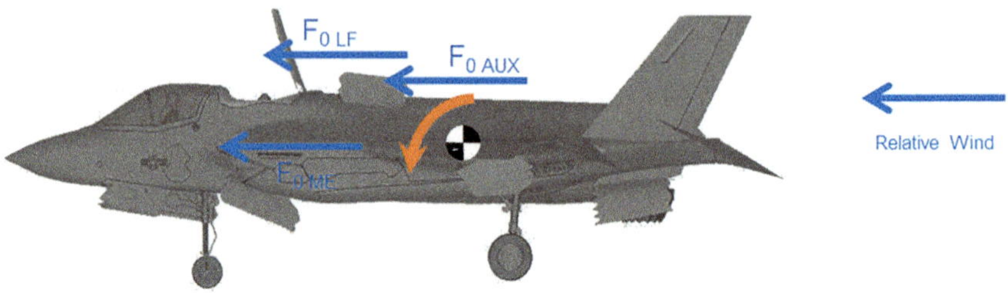

Fig. 25 Intake momentum drag-induced pitching moment.

FSCG. There, they are close to the direction of the freestream creating the nose-down pitching moment.

The forward TS reduces total available system thrust, and the LiftFan nozzle angle will point well forward of the vertical if necessary to achieve rearward acceleration. When combined, these effects reduce the available vertical thrust such that ETR saturates and the aircraft starts to sink.

An aircraft CLAW change mitigated the problem by constraining the LiftFan to less forward nozzle angles when uncommanded sink was detected. This change reduced the rate at which the pilot could accelerate rearward but did not affect the maximum rearward speed that could be achieved. The net result was an improvement in the handling qualities for this area of the flight envelope.

2. VERTICAL LANDING DESCENT RATE

Areas of the operationally cleared wind envelope can experience high levels of HGI, predominantly in crosswind and tailwind conditions relative to the aircraft. Figure 26 shows an example grouping of flight test landings in tailwind conditions relative to the aircraft. Each flight test VL is indicated by a different colored symbol on the wind rosette on the left in Fig. 26. The same symbols are used in the plot on the right in the figure to show temperature rise as a function of height above ground level. The flight test average is indicated by the solid black line, aligning well with the average from subscale testing (shown by the blue line).

Flight test experience confirmed, as expected, that the aircraft descent rate increases over the commanded descent rate during VLs in the presence of significant HGI on warmer days. In the JB regime, the propulsion control system is designed to produce a maximum specified level of thrust. This level is independent of ambient temperature up to a high-temperature theta-break threshold. Once the sensed inlet temperature rises above the theta-break threshold, maximum thrust is reduced. Operations with a sensed inlet temperature above the theta break require an ETR rise to maintain the same level of thrust, as shown in the upper-left plot of Fig. 27.

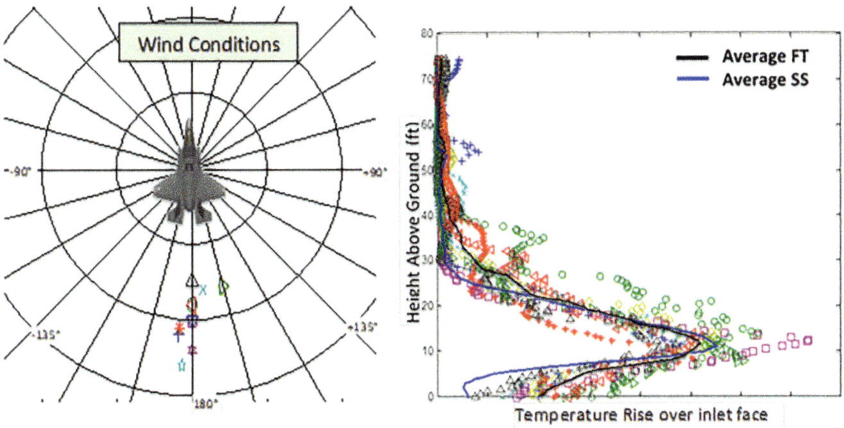

Fig. 26 Example flight test results for winds and associated temperature rises.

The aircraft also experiences a pitch derotation during high-HGI events, and TS is commanded forward to counter the pitch-down moment. Total available thrust is reduced as TS moves forward due to propulsion system tent curve

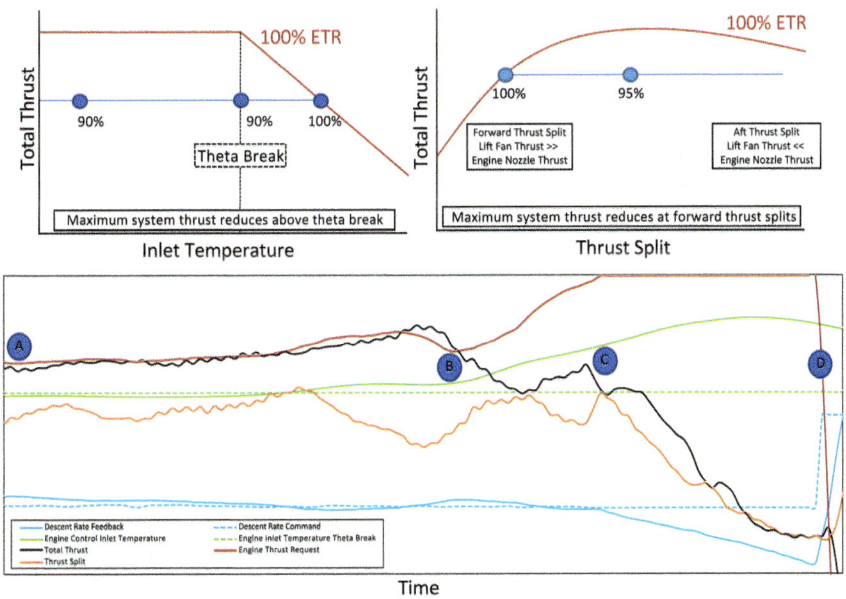

Fig. 27 Theta break and tent curve effects on aircraft in the presence of HGI.

effects. This also leads to a rise in ETR for the same total thrust, as depicted in the upper-right plot of Fig. 27. Once ETR saturates at 100%, any further rise in inlet temperature or decrease in TS will result in a reduction in thrust and an uncommanded increase in descent rate.

Depictions of the theta break and TS-vs-thrust schedules are provided in Fig. 27. There is also an example of their effect on overall system thrust and the aircraft descent rate during a VL in a tailwind condition on a warmer ambient temperature day.

The lower plot in Fig. 27 shows an example VL time history illustrating the effect of HGI on the theta break and TS-vs-thrust schedules. It also depicts the ultimate impact on the aircraft descent rate. The time history begins just before encountering HGI (point A in the in plot) and ends at VL touchdown (point D). Between points A and B in the plot, the sensed inlet temperature (solid green line) is below or near the theta-break threshold (dashed green line). TS (orange line) is relatively constant at a nominal value, and the resultant system total thrust (black line) and ETR (red line) remain relatively constant. The aircraft descent rate (solid blue line) meets the descent rate command (dashed blue line) as well. Between points B and C in the plot, the sensed inlet temperature starts to rise significantly above the theta-break threshold, resulting in an increased ETR for the same total thrust. Between points C and D in the plot, ETR has saturated at 100% and the inlet temperature continues to rise. TS is also commanded forward, resulting in a reduction in total system thrust and an increase in aircraft descent rate above command.

There was an uncommanded decrease in the descent rate during VLs experiencing HGI at colder ambient temperatures. Although this is an unpredicted phenomenon for VL, it is commonly seen during flight test. The mechanism for this is similar to that for the derotation characteristic. On a colder day, a VL with HGI results in sensed inlet temperatures remaining below the theta break. In such instances, the propulsion system attempts to compensate for the HGI to produce the same level of thrust as was present before the HGI event. However, the engine control generally over-reads the inlet temperature during an HGI event. Consequently, the propulsion system momentarily produces more thrust than necessary for the condition, resulting in a decrease in descent rate. Inevitably, the HGI environment is difficult to characterize for *all* possible scenarios with any sensor mechanism short of a full 40-probe inlet rake. As such, some conservatism in control methodology is prudent. The CLAW responds to the reduction in descent rate by decreasing commanded ETR. However, the aircraft experiences HGI at a low enough height that it generally lands before the descent rate increases back to command.

The aircraft descent rate sets the operational performance margin required to mitigate HGI during VL. If the desire were to ensure that HGI would have no noticeable impact on the aircraft descent rate, the resultant limitations would be unnecessarily restrictive. Rather than trying to eliminate the effect of HGI, the effect can be controlled via performance limitations to ensure that the

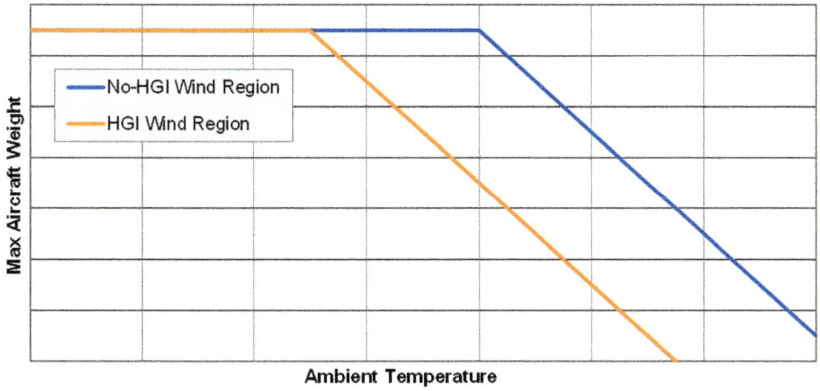

Fig. 28 Example HGI-region limitation.

uncommanded increase in descent rate by touchdown is not excessive. Uncommanded descent rate thresholds are set by handling qualities assessments and landing-gear or aircraft load limits.

Within the wind envelope where HGI is expected, the performance margin for the mitigation of HGI is a function of both aircraft weight and ambient temperature. Additional margin is required, however, for warmer ambient temperatures, as depicted in Fig. 28. This plot also shows the effect of the theta-break thrust schedule within the no-HGI region, as previously discussed.

3. VERTICAL LANDING DEROTATION

VLs with high HGI levels exhibit a pitch-down characteristic just prior to touchdown. HGI levels measured in flight testing for these conditions match those in the subscale prediction; however, the magnitude of the aircraft pitch response was not predicted by models. Posttest analysis indicated that engine inlet temperature sensor lag and HGI compensation logic in the control system were the primary factors contributing to this previously unmodelled characteristic.

During VL in the presence of HGI, the exhaust fountain is circulated into the main engine inlets. It is then directed primarily to the lower portion of the engine inlet face. The engine control receives inlet temperature feedback from two sensors located in the lower portion of the engine inlet face. Therefore, it senses a higher temperature than the average over the engine face during a VL in the presence of HGI. An example is shown in Fig. 29 for a flight test VL with a tailwind relative to the aircraft. In the figure, the T2 sensors measure inlet temperature.

In Fig. 29, the plot on the right shows engine inlet temperature rise vs height above ground for the average face value measured by the inlet rake (blue line). It also indicates the average T2 location value measured by the inlet temperature rake (red line) and the value used by the control system (green line) based on the inlet temperature sensors. The contour plot on the left shows the inlet rake measured temperature over the engine inlet face at the height of the peak average engine face temperature rise. The engine inlet temperature sensor locations are represented by the black circles in the lower half of the engine face. HGI compensation logic within the control system is designed to account for sensor-placement-induced errors resulting from the temperature distortion. It also factors in the attempts to correct the inlet temperature values used in the control for this discrepancy. However, the corrected temperatures are still warmer than the average over the engine face, as measured by the flight test inlet rake instrumentation.

The temperature difference between the average engine inlet face and that sensed by the control system means that core engine thrust is momentarily higher than the control system calculates. This results in a brief nose-down pitching moment. The aircraft CLAW responds to the pitch attitude change by commanding a nose-up TS to counter the pitching moment. However, the derotation pitch rate is low enough that the control loop does not act immediately. An example derotation time history during a flight test VL with a tailwind relative to the aircraft is shown in Fig. 30, which is the same VL shown in Fig. 29.

In Fig. 30, the time history starts just prior to HGI being encountered and ends at landing touchdown. The aircraft pitch attitude (purple line) is at a nominal value until the control system inlet temperature (green line) starts to diverge from the rake average inlet face temperature (blue line). At this point, the aircraft starts to pitch down and TS (orange line) is commanded forward.

This derotation characteristic was consistently observed during VLs and appears to be proportionate to the level of HGI encountered. The pitch attitude can decrease by as much as 2 deg before aircraft touchdown. These changes are

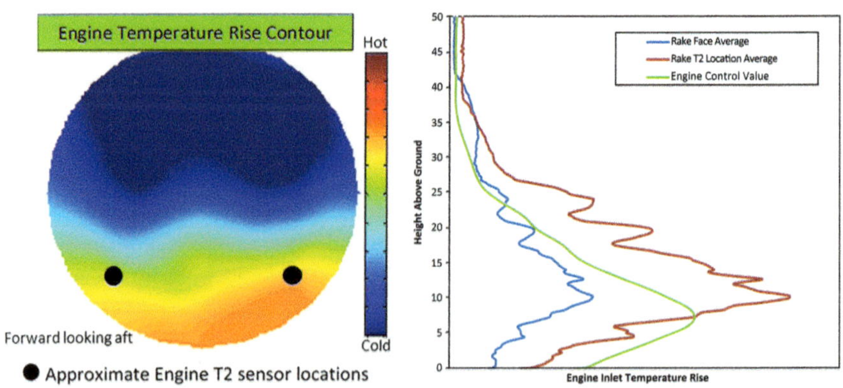

Fig. 29 Engine face distortion impact on sensed temperature rise.

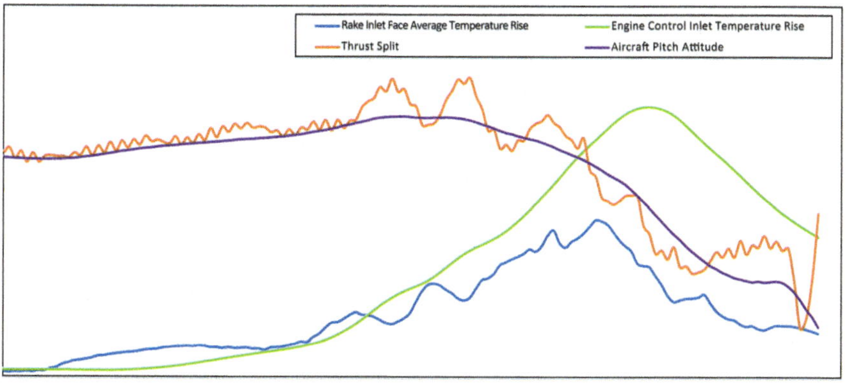

Fig. 30 Derotation illustration.

perceptible to the pilot, but no pilot response is possible because the pilot does not have direct control over pitch attitude during a VL.

A CLAW change was implemented to increase the aircraft pitch attitude during VL descent by 2 deg. The updated VL pitch attitude was intended to prevent nose landing gear touchdown prior to the main landing gear. It also aimed to avoid any load concern accompanying such a maneuver.

4. SHIP OPERATIONS

Ship VL operations with wind down the deck (similar to the VLBB relevant wind envelope) were generally unremarkable. They showed no requirement for additional performance margin over shore-based VLs performed with similar wind conditions. Outside this benign wind envelope, multiple features required additional performance margin for mitigation.

Figure 31 shows the USS *America* and its superstructure. With starboard winds from the direction of the ship superstructure, increased thrust is required to mitigate the effects of SGI. It also counters amplified HGI due to the reflection of the aircraft exhaust ground sheet from the ship's island. Further, it mitigates the downwash effect of the relative winds from the presence of the island superstructure.

The ship's engines release exhaust from stacks on the island superstructure. With an ambient wind relative to the ship that directs the exhaust over the intended landing spot, this exhaust can be ingested by the aircraft. To quantify a worst-case scenario for SGI, we performed high hovers over the intended landing spot. We used winds that placed the aircraft within the ship's exhaust plume and measured the temperature rise at the LiftFan and engine inlet faces using inlet rake instrumentation. Figure 32 shows an F-35B hovering at a nominal height above deck prior to VL, with an illustration (in red) of the

Fig. 31 USS *America* (LHA 6).

ship's exhaust. High hovers were performed at approximately 20 ft higher than the nominal height to gather data in a worst-case scenario for SGI.

HGI and SGI only require mitigation at warmer ambient temperatures where their effect reduces available system thrust above the temperature of the theta break. This can result in a reduced hover-over-deck capability and/or increased descent rate during VL. However, the effect of downwash on the aircraft results

Fig. 32 F-35B and stack gas ingestion for the USS *America* (LHA 6).

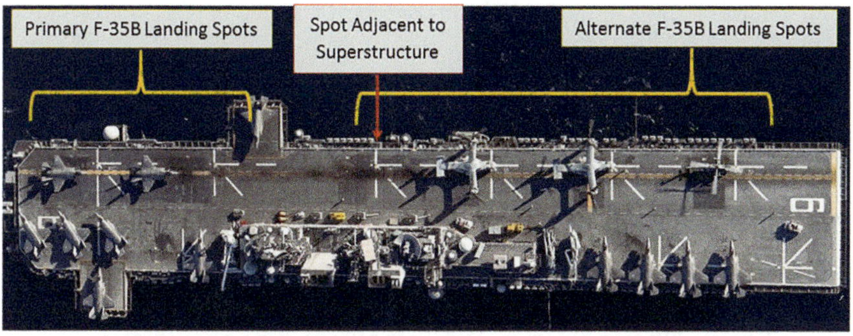

Fig. 33 Overhead view of the layout of the USS *America* (LHA 6).

in a greater required thrust to hover over the deck and, hence, additional performance margin. This effect is independent of ambient temperature. The downwash effects from the island superstructure were clearly seen when operating at the landing spot adjacent to the island superstructure, labelled in Fig. 33.

An F-35B at this landing spot adjacent to the island superstructure is shown in Fig. 34. During one recovery to this landing spot and immediately after the deck cross, the aircraft experienced a significant increase in ETR, as depicted in Fig. 35. The saturation of ETR resulted in an uncommanded descent prior to the VL initiation. The increase in ETR was attributed to an aerodynamic effect from landing close to the island superstructure, and no significant HGI/SGI was

Fig. 34 F-35B at landing spot adjacent to the ship superstructure on the USS *America* (LHA 6).

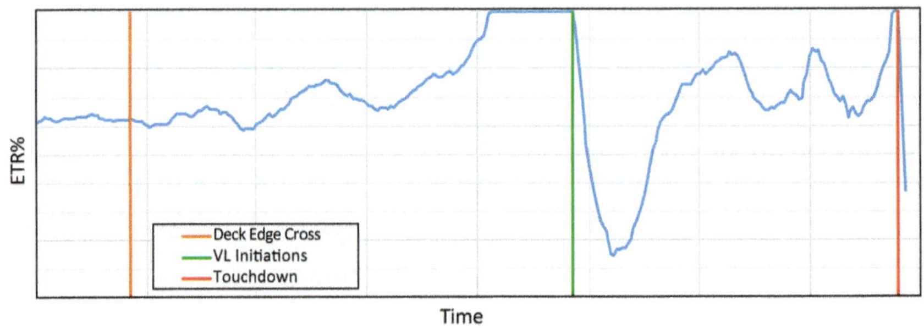

Fig. 35 ETR time history for VL to spot adjacent ship's superstructure.

experienced at the time. The VL was subsequently completed without issue, but the landing highlighted the need to account for downwash on the aircraft.

With port winds, SGI is no longer a concern, and at most landing spots the island superstructure no longer influences the winds that the aircraft experiences. However, influences of the ship operating environment on both the level of HGI experienced and downwash on the aircraft are still seen and require performance margin to mitigate. With winds from port, the proximity of the aircraft to deck edge and the near vertical side of the ship are expected to contribute to a more adverse flow field. This applies to both HGI and aerodynamic effects, and is evaluated through a comparison with similar wind conditions during shore-based operations.

E. PILOT PERSPECTIVE

STOVL JB and VL testing included performance trims and VLs across the functional range of altitudes and wind conditions at various FSCGs and hover-weight ratios. (The ratios are calculated by dividing aircraft gross weight by the maximum hover weight capability.) The trims and VLs provided unexpected challenges from the pilot's perspective. These were a function of nuances in the integration of flight and propulsion controls and engine cutbacks due to the wear of engine components. The effects were also the product of the integration of our testing into a busy airfield at extremely low fuel states.

In the first case, the system is designed to enter JB mode with a 3% thrust margin. In this way, the CLAW will suspend the deceleration from SJ to JB if that margin cannot be maintained. The CLAW feature, a performance deficit protection, is a welcome capability that will prevent entry into JB without sufficient performance. However, it added challenges to performance testing that required JB entry at the highest hover-weight ratio possible. Tolerances were tight and translated to approximately one minute to get from SJ flight to a stabilized hover, providing little room for error or unexpected events.

The importance of working closely with airfield management, Air Traffic Control (ATC), and tower controllers cannot be overstated. JB performance testing frequently put the aircraft in a very low fuel state (approximately two minutes of flight time) with one landing option. Orchestrating the arrival of the aircraft into the complex traffic pattern at Naval Air Station Patuxent River took detailed coordination and communication. Face-to-face discussions, joint simulator sessions with the ATC team, and pre- and postmission briefs were essential elements to the success of this process.

Although JB performance testing presented significant operational challenges, performance VLs were generally unremarkable. The performance VL technique did not differ from that of normal VLs. HGI was encountered on numerous occasions, usually noticeable in the cockpit as a sudden increase in height rate in the last 10 to 20 ft of the descent. Notwithstanding, no pilot response was required or desired.

VI. CONCLUSION

The F-35B STOVL flight test program built upon a solid foundation of preflight modeling and simulation evidence that enabled efficient envelope expansion and posttest results verification. Testing was accomplished over more than seven years, demonstrating that the aircraft provides repeatable performance with an intuitive pilot interface. The interface allowed testing at both minimum fuel states and maximum gross weight capability to feel routine.

The verification of the KPPs of flat-deck STO and VLBB was relatively straightforward. Analyses of data acquired from both shore- and ship-based maneuvers proved that the F-35B exceeded the flat-deck STO requirement. Moreover, it had considerable margin that may be used to accommodate future weapons system growth potential. JB maneuvers (hovers and VL) and SJ maneuvers were utilized to validate the databases used to calculate VLBB performance. Ultimately, these verified that the F-35B design exceeded the requirement.

Verification testing for the flat-deck STO and VLBB requirements was successful. Accordingly, the F-35B flight test program focused on validating the STOVL capability provided to the operator throughout the operating envelope. The unexpected problems the team overcame increased the value of the flight test effort and contributed to making the F-35B more robust. Changes in aircraft CLAW to accommodate unpredictable wind conditions during a STO will help the fleet to avoid nozzle strikes. It will also enhance the fleet's understanding of the effects of HGI on VL performance, which will protect the fleet against the possibility of potentially dangerous touchdown sink rates.

The team planned extensively to conduct safe, efficient, and effective testing to achieve program objectives. Expected problems regarding sink off bow characterization in support of ship-based STOs and HGI concerns with respect to VL performance validation were quantified. They were then effectively mitigated through detailed test hazard analyses and the implementation of a cautious buildup approach. Particularly important was the development and implementation of a recovery strategy in the event of excessive sink off bow.

The end product of the flight test process discussed herein was the capability provided to the operator of the F-35B weapons system. Innovations in flight test and analysis methods contributed to successful STOVL requirements verification and increased pilot confidence in the aircraft's robust performance in this flight regime. The positive performance test results demonstrate that the F-35B provides a considerable upgrade in range, payload, and bring-back capability over the legacy Harrier. Along with this, it retains supersonic speed and 5th-Generation stealth characteristics.

REFERENCES

[1] Wurth, S. P., Smith, M. S., and Celeberti, L., "F-35 Propulsion System Integration, Development & Verification," AIAA Aviation Forum, June 2018 (to be published).

[2] Parsons, D. G., Eckstein, A. G., and Azevedo, J. J., "F-35 Aerodynamic Performance Verification," AIAA Aviation Forum, June 2018 (to be published).

[3] Mason, J. R., "F-35 STOVL Aircraft Performance Model Validation of Short Take-off Flight Test Data," AIAA Aviation Forum, Aug. 2013.

[4] Wurth, S. P., Walker, G. W., and Fuller, J., "F-35B IFPC Development," AIAA Paper 2013-4243, Aug. 2013.

[5] Mange, R. L., and Hoggarth, R., "Highlights of the Lockheed Martin F-35 STOVL Jet Effects Program," Royal Aeronautical Society, International Powered Lift Conference, 2008.

[6] Hoggarth, R., and Stovall, M. B., "F-35 STOVL Mode Flight Test Analysis Techniques and Initial Results," International Powered Lift Conference, Oct. 2010.

[7] Vorwerk, A. V., and Ciszek, R. S., "Use of In-Flight Thrust on JSF Program," American Helicopter Society International, International Powered Lift Conference, Oct. 2010.

[8] Hoggarth, R., and Stovall, M. B., "The Lockheed Martin F-35 STOVL Force and Moment Ground Test Analysis Techniques and Results," International Powered Lift Conference, Oct. 2010.

[9] Pinero, E., "Utilization of the STOVL Operations Test Facility During the F-35 Program," IPLC Paper IPLC2010_30, Sept. 2010.

[10] Cook, R., Curtis, P., and Fenton, P., "State of the Art in Sub-scale STOVL Hot Gas Ingestion Wind Tunnel Test Techniques," SAE International 2005-01-3158.

CHAPTER 18

F-35 Climatic Chamber Testing and System Verification

Victorio J. Rodriguez* **and Billie Flynn**[†]
Lockheed Martin Aeronautics Company, Fort Worth, TX

Marc G. Thompson[‡]
BAE Systems, Warton, United Kingdom

Steven Brelage[§]
Naval Air Warfare Center Aircraft Division, Naval Air Station Patuxent River, MD

The F-35 is expected to withstand the most formidable of weather extremes over its service lifetime. Validating this capability with three unique F-35 variants, most notably the Short Takeoff and Vertical Landing (STOVL) version, was expected to be extremely challenging. The F-35 program was required to create a test environment where a pilot could essentially hover an F-35 indoors, operating at maximum power through extreme weather conditions. To do this, the test team had to stretch ingenuity and imagination to develop never before seen structures and test procedures. The result was a one-of-a-kind test program, housed inside the McKinley Climatic Laboratory (MCL) at Eglin Air Force Base, Niceville, Florida, to prove the F-35 could operate in those climates.

I. INTRODUCTION

The F-35 System Development and Demonstration (SDD) phase included the development and maturation of the aircraft design and configuration as well as the execution of a large and dedicated ground and flight test evaluation program. A major subportion of the SDD test program—and the focus of this paper—was the F-35 Climatic Chamber Testing (CCT), executed in a six-month period from October 2014 through March 2015. The CCT was an on-aircraft test executed at a wide range of environmental conditions to gather data on aircraft

*Manager, F-35 Propulsion and Thermal Analysis.
[†]Senior Experimental Test Pilot, F-35 Flight Test Program.
[‡]Principal Engineering Specialist, F-35 Propulsion Flight Test.
[§]Branch Head, Propulsion and Subsystems Flight Test.

operation. Historically, this phase of testing is conducted early enough in the life of an airplane to ensure that any issues that are discovered can be resolved before the airplane is fielded. This paper will discuss the unique attributes of the F-35 program that challenged traditional testing methods and the difficult decisions required to define, develop, and schedule the testing. The paper will also discuss unique and key capabilities and enablers that allowed successful execution. Finally, the paper will conclude with a high-level review of lessons learned and the relevance of performing complex and expensive full-scale, air-system-level, environmental testing in a world of increased simulation and modeling capabilities and pressure to keep cost and schedule threats to a minimum.

II. F-35 VS PREVIOUS AIRCRAFT CLIMATIC TESTING

The program-level, functional requirements for the F-35 air system are defined in the Joint Strike Fighter (JSF) Contract Specification (JCS). The JCS defines all aspects of the aircraft-level requirements, including the capability to be parked outside of a hangar and operation in a multitude of extreme environments that naturally occur throughout the world. The environmental requirements are based on the guidelines for design established in the MIL-HNBK-310, "Global Climatic Data for Developing Military Products." The requirements were tailored early in the program development for the anticipated F-35 aircraft customers and expected operational usage. These requirements served as the basis for the type and extent of testing required for the F-35 SDD program and included a range of ambient temperatures, humidity levels, rainfall, icing, and other environmental conditions. Nearly all new major military aircraft programs include some sort of climatic environment test and evaluation requirement. The testing to satisfy this requirement is traditionally performed at a dedicated facility, performed in naturally occurring environmental conditions, or a combination of both approaches. In legacy aircraft programs, test methods (including natural weather exposure) and facilities often yielded unpredictable, unrepeatable, and unreliable test conditions and inconclusive results that would not have satisfied the rigorous performance specifications demanded for the F-35. Figure 1 shows examples of the lines of testing performed during the F-35 CCT.

For most aircraft programs there is one design configuration to address and evaluate. The F-35 program, however, consists of an aircraft built in three variants: Conventional Takeoff and Landing (CTOL, F-35A), Short Takeoff and Vertical Landing (STOVL, F-35B), and a Carrier Variant (CV, F-35C). The three variants have significant commonality in mission systems and vehicle systems, and similarities in airframe design despite unique service mission requirements and concept of operations. Development of a "one size fits all" environmental test approach was required to plan a test and evaluation effort that maximized effectivity across all three variants and eliminated the need for multiple variant testing.

Fig. 1 F-35 Climatic Chamber Testing examples.

This led to a challenge to achieve a focused balance of requirements, priorities, objectives, and tough programmatic decisions prior to proceeding to climatic test execution. The unique aspects of the F-35B variant were the major driving factor in the F-35-specific testing that included essentially hovering the aircraft inside the facility.

Advances in computational methods available to the F-35 program, combined with capability improvements to the climatic chamber facility, enabled elimination of the necessity to test in natural environments, which can be both costly and inefficient, often waiting for appropriate weather conditions to be available or located around the world. By slightly expanding the scope of targeted regions of the climatic chamber testing, it was possible to eliminate some originally planned in-air flight testing, resulting in risk reductions to the flight test program and significant program cost and schedule savings. However, the same advances that enabled the usage of the chamber for testing also led to many discussions regarding the possible elimination of the entire line of testing based on perceived fidelity of modeling and computational methods. Ultimately, the critical importance of system-level demonstration of capabilities and data required for the formal validation of models allowed the testing to proceed.

III. MCKINLEY CLIMATIC LABORATORY TEST FACILITY

The F-35 program elected to use a dedicated and controlled environmental facility to execute the primary system-level environmental testing. The McKinley Climatic Laboratory was contracted via the JPO organization to perform the testing. Owned by and operated for the U.S. Air Force (USAF), the MCL is arguably the foremost climatic environmental testing venue in the world and is able to simulate virtually any required environmental condition. Built in 1947 to provide a refrigerated hangar to produce environmental extremes under controlled conditions, the facility was renamed the McKinley Climatic Laboratory in 1972 in honor of Lt. Col. Ashley C. McKinley, one of the initiators of the facility and a major leader behind the requirement to have such a facility available to the U.S. armed forces. Renovated and remodeled between 1993 and 1997, the facility now supports both military and commercial product testing, ranging from low to high ambient temperatures, icing cloud and ice buildup, rainfall, fog, humidity, wind, sand, solar radiation and diurnal cycles, and other environments. It is a full-service facility able to support almost any line of testing or product. The MCL facility has an Air Make-Up Unit (AMU) system that conditions the air supply to specified requirements and allows for extended engine operation within a closed environment. The AMU is capable of supporting up to 1000 lb/second airflow makeup at -65°F in the main test chamber. A total of six chambers are available for specialized evaluation of various systems. The facility, whose motto is "Creating the weather you can't afford to wait for," was designated as a National Historic Mechanical Engineering Landmark in 1987.

The bulk of the F-35 aircraft CCT program was executed in the main chamber, which covers an area measuring $250 \times 262 \times 70$ ft. Additional off-aircraft Logistics Test and Evaluation (LT&E) and Support Equipment (SE) evaluation testing was executed in the smaller equipment test chamber (ECT) ($130 \times 30 \times 25$ ft) in parallel with the aircraft–level testing. Figure 2 shows an aerial photo of the MCL facility taken upon the arrival of the F-35 test aircraft to Eglin AFB. Figure 3 shows the test aircraft in front of the main test chamber of the MCL facility.

IV. CHALLENGES AND DECISIONS IN DEFINING AND SCHEDULING CLIMATIC TEST ACTIVITIES

The decision to execute an air vehicle system–level CCT was established early in the SDD test planning effort with the testing scheduled to occur several years after the formal start of the flight test program once system maturity and capability were established. A program-level decision was made to use only one variant as the means to capture and evaluate all three variant requirements (based on intended commonality), with the F-35B variant selected because the addition of the lift fan, roll post offtakes and nozzles, and three-bearing swivel module

Fig. 2 Arrival of F-35 test aircraft BF-5 at Eglin AFB over MCL.

Fig. 3 F-35 BF-5 positioned in front of the MCL main chamber.

along with the main engine and the STOVL capabilities provided the most complexity and incorporated the greatest number of operating doors and surfaces. As a result, the F-35A and C variants would be cleared via similarity, and any unique systems cleared by a combination of dedicated component-level testing and system-level analyses. In 2013, the decision was made to execute the testing using an interim Mission Systems (MS) and Vehicle Systems (VS) hardware and software configuration that supported both the U.S. Marine Corps (USMC) F-35B Initial Operating Capability (IOC) milestone for 2015 and remained applicable to the planned final fleet configuration. The high-level overall test objective shown as part of the pretest planning briefings and the CCT Test Readiness Review (TRR) meeting held in August 2014 were:

- Evaluate system-level operational capabilities at a range of climatic environments.
- Provide sufficient evaluation of air vehicle effectiveness ahead of major fleet operations.
- Gather required supporting data for
    - Specification criteria compliance
    - Certification criteria
    - Removal or deduction of initial operating limitations
- Update Joint Technical Data (JTD) and Flight Series Database (FSD) information.

Full operation of the aircraft systems (to the extent possible) under each environmental condition was a stated requirement from the outset. This included the operation of the installed propulsion system from low through high power settings. The program selection of the F-35B as the test variant was in no small part to evaluate operation in powered lift mode (wherein the propulsion system effectors are the prime lift and control mechanism) as well as conventional (traditional) mode of operations. As such, a key factor in the test planning was to show the capability for the system operation in powered lift mode at all environmental conditions, including hover and slow speed flight regime. Operations at all power settings within an enclosed setting required sufficient air makeup required to feed the propulsion system and to be able to expel the aircraft exhaust out of the main chamber. This was essential for safety of personnel in the chamber as well as to maintain the specific environmental conditions for extended periods of time. As a direct result of this stated objective, the operation and restraint of the aircraft in both operating modes at all power settings (through full afterburner in conventional mode and hover in powered lift mode operations) was a major consideration in the development of the test objectives, test sequences, and facility requirements and modifications. In the powered lift mode of operations, the lift fan and engine nozzles are vectored downward and forward to provide increasing lift forces required to support the aircraft at slow airspeeds. This characteristic and objective mandated a unique requirement to elevate the aircraft above the main chamber floor to be able to accommodate the necessary exhaust ducting and aircraft support structure to support the aircraft in all modes of operation.

During planning efforts, cost and schedule constraints forced the program to review the lines of testing planned and traditionally performed in climatic evaluation testing vs program objectives and priorities. The formal lines of testing that were accepted and planned for the CCT included:

1. A standard day baseline assessment
2. Incremental testing to elevated ambient temperatures
3. A repeat of the standard day test sequence
4. Incremental testing to colder ambient temperatures
5. Ground and flight icing cloud conditions (including ground vortex icing)
6. Static and blowing rainfall conditions
7. Freezing rain conditions (nonoperating)
8. High relative and absolute humidity
9. A final repeat of standard day test sequence

Several originally planned lines of testing were ultimately eliminated because either sufficient information was achievable via analysis or other means of data gathering and evaluation, or they were constrained by other design and test requirements. One example of this approach was the deletion of the snow loading testing, where analysis could be used to determine structural loads under a given snow accumulation. Additionally, by choosing to use and modify an existing force-and-moment restraint system to secure the aircraft via the landing gear (described in a later section), a planned sequence to swing the landing gear at each environmental condition was also removed from the plans. Though it would have been possible to design a system to restrain the aircraft and allow full movement of the landing gear, this would have involved significant engineering design and analysis efforts and extensive structural modifications to the aircraft, and may have compromised other testing requirements. In the end, this was a good example of the trade-offs that must be made when balancing cost and schedule against testing requirements.

V. AIRCRAFT MODIFICATION, FACILITY, AND TEST PLAN PREPARATION

Initial and subsequent periodic planning was maintained early in the SDD program to develop the high-level structure of the testing objectives and requirements; however, as the SDD phase progressed, design maturation along with normal flight test discoveries combined to delay the scheduled entry of the aircraft into the CCT. For example, improvements to the aircraft thermal management system were identified early in the F-35 design program, and the changes were incorporated into the baseline aircraft design and were ultimately part of the aircraft configuration that was tested during the CCT. As the program schedule and design matured, a more solid test entry date was established, and more frequent and focused test planning efforts began. Between 12 and 18 months ahead of planned test entry, a joint team of contractors—Lockheed Martin, Northrop Grumman, BAE Systems, Pratt & Whitney, and Rolls-Royce, and including JPO, F-35 Integrated Test Force, and MCL personnel—met regularly to discuss and mature the test aircraft configuration, planned objectives, detailed test sequences, detailed schedules, and required documentation and reviews.

Each line of flight testing executed under the F-35 program is governed by a Test Execution Package (TEP). The development of the formal TEP for the CCT was initiated and managed by the F-35 ITF at Naval Air Station (NAS) Patuxent River, Maryland, with support from the ITF at Edwards Air Force Base (AFB), California. The ITF is a combined government and contractor team responsible for the planning, execution, data analysis, and reporting of ground and flight testing for the F-35 program.

The TEP consisted of the Joint Test Plan (JTP), which details the system under test and the required sequences to be executed, and the Test Safety Supplement (TSS), which details system maturity, risk areas, hazards, mitigating procedures, and the overall risk assessment for the planned testing. It should be noted that prior to and during this process, there were numerous concerns about the severity of the proposed testing environments, and the potential for long-term damage to the test aircraft asset as a result of the planned testing.

Attention to detail during this phase of planning, combined with the F-35's unparalleled insight into aircraft subsystem health during testing, allowed the successful reintroduction of aircraft BF-5 into the test fleet post-CCT, including its role as one of the prime assets during F-35B Development Test (DT)-III testing aboard USS *America* (LHA-6) in 2016.

Once the TEP was matured, the package proceeded through two main board review and approval cycles—the Test Readiness Review Board (TRRB) and Executive Review Board (ERB)—prior to final authorization to execute testing. In total, the TEP was developed gradually over approximately nine months from a draft document outline through final ERB approval. The Patuxent River F-35 ITF Test Operations Team also led the development and planning of the joint detachment team that would travel from NAS Patuxent River, supported by other sites and organizations, to Eglin AFB to support the test activities. Initial test schedules planned for a four-to-five-month test period between the aircraft ferry to the MCL and the scheduled aircraft return to NAS Patuxent River.

Testing at each major environmental condition was structured to provide consistent test flow and set expectations for the test team and MCL personnel. Each test event consisted of specific periods defined as: aircraft soak and stabilization period, prestart (maintenance activities), aircraft startup, simulated CTOL operations, simulated STOVL operations, and an aircraft shutdown period. Figure 4

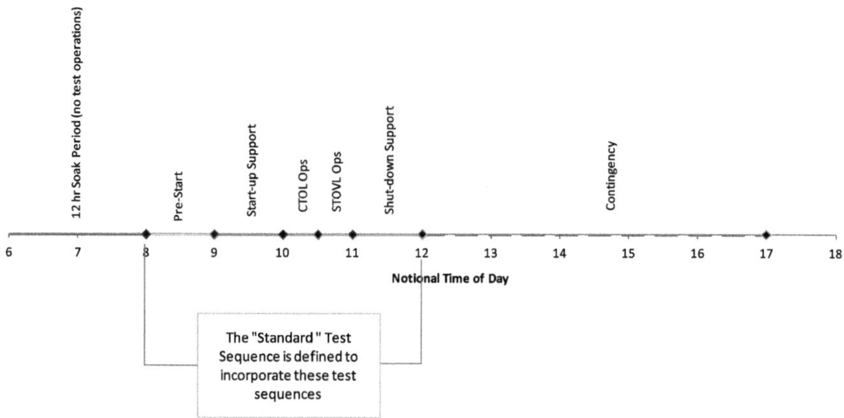

Fig. 4 F-35 CCT test event sequence timeline (notional).

illustrates a generalized planned execution timeline for the testing at each environmental condition.

A Standard Test Sequence (STS) was developed by the joint team for the test execution. This sequence allowed for a routine flow of the testing at each condition, helped better simulate standard ground operations procedures for the fleet aircraft, reduced the overall number of test points, and provided an easier means to compare and evaluate aircraft system performance across the environmental spectrum.

The final STS profile included a normal aircraft startup procedure, a Vehicle System Built-In Test (VSBIT), the execution of two primary Maintenance Built-In Tests (MBITs)—designed to be used in the fleet in place of high power restrained runs after propulsion system maintenance—and simulated conventional mode takeoff maneuvers (high power). Simulated powered lift mode takeoff and landings (that is, short takeoff [STO] and hover/vertical landing [H/VL]) were also scheduled to be performed at each major condition to ensure proper aircraft system operations. This STS was used for most of the testing sequences. For some unique test objectives (e.g., icing evaluation, alert start maneuvers, rain intrusion and corrosion inspections, and dedicated maintenance evaluations), separate specific test sequences were added to the execution plan. Figure 5 illustrates the STS profile that was executed vs time.

Aircraft BF-5, one of the SDD test aircraft, was selected for the CCT. Although BF-5 was a dedicated flight test aircraft, it incorporated many of the production standard system configurations combined with extensive Flight Test Instrumentation (FTI) capabilities. The aircraft underwent a significant pretest modification period at the F-35 Patuxent River ITF facility from June through August 2014 to ensure that the test article had the latest hardware and software packages available. Care was taken to incorporate a sufficient pedigree of components to

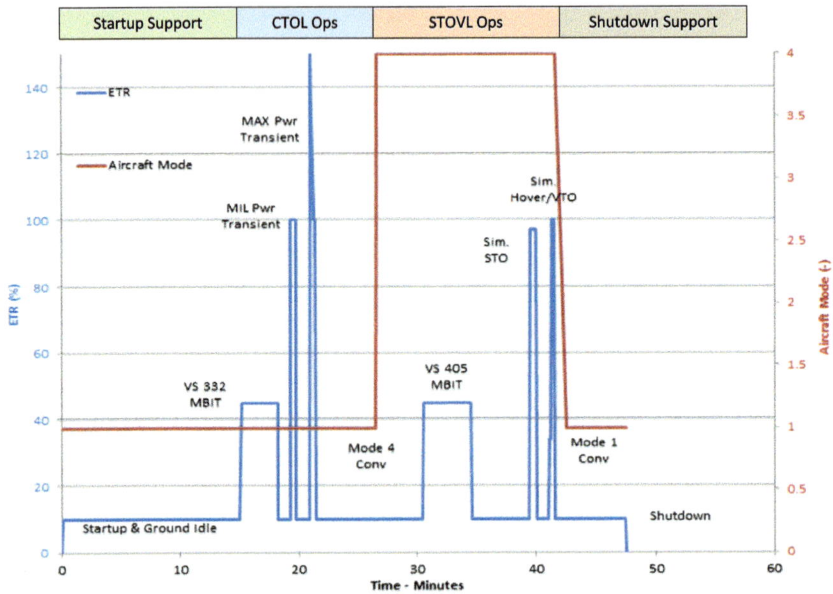

Fig. 5 F-35 CCT STS profile.

adequately evaluate both the Block 2B configuration for the USMC IOC milestone and the Block 3 configuration planned as the final standard for aircraft deliveries. Additionally, dedicated and specific FTI measurement devices were added to the aircraft to serve as primary data collection for validation, reference of environmental conditions, and Safety of Test (SOT) parameters. Most of the modifications were accomplished prior to the aircraft ferry; however, several modifications, including preproduction hardware, nonflight-worthy FTI, and unique test asset installations were completed after aircraft arrival at the MCL.

In preparation for the planned testing at the MCL, the F-35 team coordinated extensively with the MCL facility team to understand and define what setup, modifications, and special equipment would be required to accommodate the test sequences and objectives. A fixture was specifically designed by the MCL team to accommodate the unique requirements for operation of the F-35B aircraft in conventional mode and powered lift mode operations. To allow for full power and articulation planned in both primary modes of operation, the aircraft had to be elevated in the chamber (approximately 13 ft above the floor). The unique main chamber configuration for the F-35 testing consisted of the aircraft restraint system, the exhaust ducting system, the elevated maintenance platform, and aircraft support equipment garages.

The restraint system was a modified version of a system previously designed, built, and used for the aircraft force-and-moment system hover pit testing in Fort

Worth, Texas, in 2007. Reuse of this existing system eliminated the need to physically modify the aircraft for restraint during testing, and the load paths and forces for operation imparted on the aircraft had already been analyzed and established from the previous testing, which significantly reduced the associated cost of design and acquisition. The backbone of this system was made up of six stanchions designed to restrain the aircraft in all axes and for all power settings in both conventional and powered lift mode operations. The six stanchions and attachment hardware were transferred to the MCL team, and the equipment was modified in length to incorporate the MCL-specific load cells and to attach to an I-beam frame anchored to the main chamber floor. The asymmetric nature of the restraint design was such that the system was statically determinant. Real-time monitoring of the load cell values was performed for each run to assess measured load against the designated continuation criteria for testing. Figure 6 illustrates a Computer-Aided Drawing (CAD) model rendering of the aircraft on the modified restraint system for use in the main chamber.

A requirement was also levied to allow sufficient work area and access to the aircraft for pre- and posttest maintenance activities because the aircraft would be significantly above the main chamber floor. This manifested itself in the form of an elevated platform area around the aircraft that was made up of a grated surface, supported from below via columns attached to the main chamber floor. Several iterations were reviewed by all parties before a final agreement was reached regarding the minimum required area. The platform was designed to be freestanding and did not attach to the aircraft or the restraint system in any way and was built from medium grade grating, sufficient for most aircraft maintenance activity. The elevated platform was positioned at a height to closely resemble the normal

Fig. 6 F-35/MCL aircraft restraint system CAD model. (*Illustration courtesy of Dwayne Bell, McKinley Climatic Laboratory.*)

ground plane to the aircraft to accommodate the usage of standard F-35 aircraft support equipment and to perform normal pre- and postrun activities. The area under the platform was then used to route the facility-provided exhaust ducting (see the following paragraph) as well as to provide two environmentally controlled garage areas where common and large support equipment could be housed and stationed during testing.

The operation of the aircraft and propulsion system inside the facility required the aircraft exhaust to be routed out of the facility. This was critical to avoid any issues with exhaust and combustion products buildup and to ensure that the operation of the aircraft propulsion system did not negatively impact the targeted environmental test conditions. To that end, the MCL team designed and manufactured exhaust ducting configuration unique and specific for this F-35 test installation. The design of the exhaust collection system was done in coordination with the Lockheed Martin and Pratt & Whitney Integrated Product Teams (IPTs) and was based in part upon the current systems in use at Pratt & Whitney test facilities and from previous test experience by the MCL team. Several iterations were studied before a final agreement and design was reached. Figure 7 illustrates several pictures of the as-built exhaust ducting in place after aircraft installation in the test fixture.

The ducting was designed to channel the exhaust away from the aircraft and out of the chamber. It was a passive system where the flow moved along via momentum of the exhaust itself and the pressure differential between the front and rear openings in the ducts. All the major exhaust products from the aircraft system were ducted out of the chamber, specifically the main engine, lift fan, roll post, and Power and Thermal Management System (PTMS). There are no combustion products and the temperature of the air is relatively cool compared to all the other exhaust airflows, so the exhaust flow from the lift fan was handled slightly differently than the other exhaust sources. The system was designed to allow partial recirculation of the lift fan exhaust flow directly back into the main chamber during operation. This feature relieved pressure on the main chamber air makeup system, which would have been particularly stressed to support and maintain the required conditioned airflow when simulating the F-35 hover where both the main engine and lift fan are operating at maximum airflows. The implementation scheme allowed improvements in testing efficiency because the team was able to operate for longer durations before the AMU system needed recharging. All exhaust ducting was designed to fit as close to the nozzles and exit planes of the aircraft as possible to minimize induction of the conditioned chamber air into the exhaust ducts, yet still allow for full aircraft operations in both operating modes up to full power setting while remaining clear of all structural movement, operating doors, and major aircraft panel removal paths required for routine maintenance. Figure 8 illustrates a CAD model rendering of the platform surrounding the aircraft while installed on the restraint system and with the exhaust ducting in place. Figure 9 illustrates the position of the aircraft and platform with respect to the entire main chamber facility.

Fig. 7 F-35/MCL exhaust ducting (lift fan, roll nozzle, PTMS, and main engine) overview.

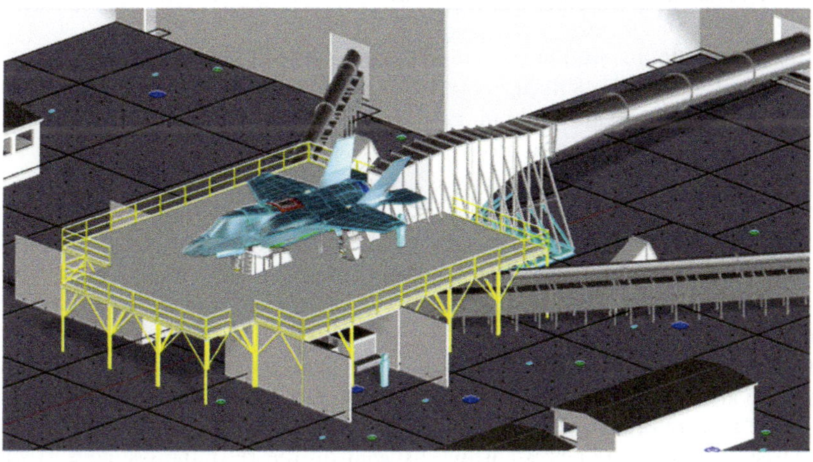

Fig. 8 Aircraft platform and aircraft installation CAD model illustration. (*Illustration courtesy of Dwayne Bell, McKinley Climatic Laboratory.*)

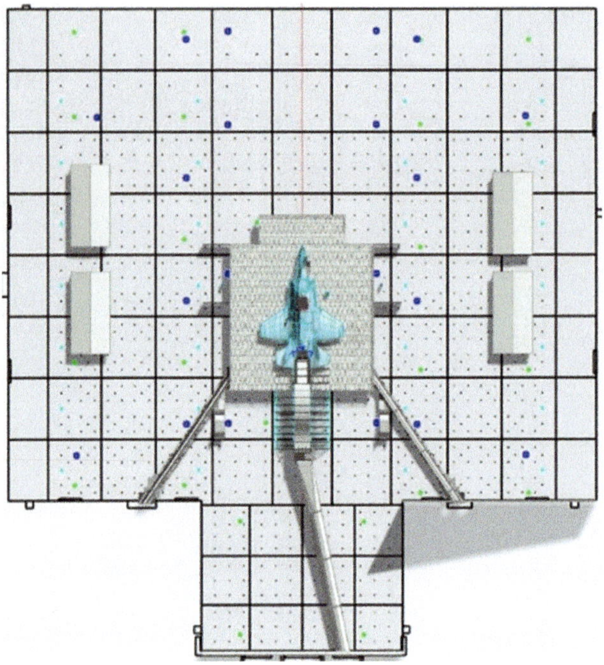

Fig. 9 CAD Model of F-35 aircraft within the MCL main chamber. (*Illustration courtesy of Dwayne Bell, McKinley Climatic Laboratory.*)

VI. ENTRY INTO TEST FACILITY AND TEST EXECUTION

After the ferry and arrival of the aircraft to Eglin AFB, a brief aircraft preparation period was completed to finish required aircraft modifications. On 29 September 2014, the aircraft was moved to the MCL test facility. The aircraft was hoisted and installed into the test fixture (Figs. 10 and 11) and a subsequent checkout and several buildup runs were completed at ambient environmental conditions to gather and assess acoustic data, required to ensure chamber and aircraft acceptability to execute the planned testing. This characterization of the acoustic environment was important because aircraft design requirements do not typically include full power operations inside an enclosed building, and there was a potential risk for structural damage to the aircraft as a result. Based on data gathered in the acoustic surveys, several constraints were established regarding aircraft power settings and STOVL propulsion effector angle combinations that were subsequently monitored during test execution. Additionally, the early testing for the acoustic data collection uncovered some unexpected issues with facility exhaust ducting system performance for the main engine and lift fan that required modifications prior to the start of formal climatic testing.

Fig. 10 F-35 Aircraft installation into MCL main chamber test fixture: part 1.

Fig. 11 F-35 Aircraft installation into MCL main chamber test fixture: part 2.

The formal test events that were executed and completed were the standard day 59°F baseline condition (executed a total of three times in the series of testing), elevated ambient temperatures (including diurnal cycle at 120°F, low ambient temperatures (down to −40°F), icing cloud conditions (ground and simulated flight), ground vortex icing, static and blowing rainfall sequences, freezing rain (nonoperating) and de-icing sequences, and high relative and absolute humidity diurnal (24-hour period) cycle testing. Upon completion of the required testing, the aircraft was removed from the test fixture and exited the main chamber on 1 April 2015. This was followed by a period of aircraft restoration and preparation for ferry back to NAS Patuxent River. The completed time between entry to and exit from the MCL spanned six months. Some initially planned lines of dedicated maintenance and LT&E testing were not able to be performed due to priorities and time constraints.

VII. F-35 ICING CLOUD TESTING

A primary objective of and major driver for F-35 testing at the MCL was the icing environment evaluation for both conventional and powered lift operations. Previous icing conditions testing in the MCL for a full-scale aircraft had been generally limited to conventional ground fog and static operations. The challenge for the F-35 program was to enable sufficient validation that the aircraft and its installed propulsion systems could operate in icing conditions in both conventional and powered lift modes. These two modes of operation required vastly different airflows, icing cloud size and conditions, and location relative to the aircraft. The addition of the objective to evaluate F-35 operating conditions with forward airspeed operations (vs simply ground static conditions) and to gather good quantitative data to compare to icing models further challenged both MCL and the F-35 program engineering teams. Testing required a myriad of new ideas and capabilities that pushed the testing sequences and MCL facilities teams beyond previous experiences. Recent upgrades to the icing spray bar system at the McKinley facility allow simulation of icing conditions far in excess of the relatively benign ground fog conditions previously tested. This new capability, coupled with advances in icing simulation codes, allowed the F-35 program to favor the decision to eliminate planned in-flight icing tanker trials from the follow-on flight test program. The high-level icing test objectives for the icing portion of the CCT were:

- Establish and validate total system-level capabilities at the aircraft level.
- Select conditions to represent JCS/Federal Air Regulation (FAR) Part 25 specifications.
- Perform combination of ground-based and simulated low speed flight conditions in both conventional and powered lift modes of operations.
- Use flight test aids to position aircraft doors, propulsion effectors, and aircraft surfaces to powered lift representative positions.

The MCL was the only facility capable of testing the F-35 aircraft at full scale under closely controlled icing cloud conditions. The facility had the capability to vary ambient temperature, Liquid Water Content (LWC), and water droplet sizes while generating a limited wind speed, all critical to providing a large range of required icing cloud conditions. Close coordination was maintained among Lockheed Martin, Pratt & Whitney, Rolls–Royce, the Propulsion JPO, and the MCL personnel well ahead of the entry into the chamber to define and establish the facility capability vs the specific aircraft requirements. The goal was to develop a test arrangement that provided the greatest combination of desired conditions to assess and evaluate the F-35 capabilities. The icing portion of the chamber testing focused primarily on the functionality and performance of the aircraft inlet-mounted ice detector and the Engine Ice Protection System (EIPS) and Lift Fan Ice Protection System (LFIPS). The secondary objective was the assessment and evaluation of ice accretion on the airframe structure areas that influence the airflow to the propulsion system (e.g., inlet lips, inlet duct, and lift fan inlet bell mouth). A qualitative assessment was also planned of the impact of icing cloud exposure to the canopy visibility. The data gathered from this phase of testing were used for icing model validation and assessment of the installed system capability. The results from the testing were also used by the Lockheed Martin Propulsion, Pratt & Whitney, and Rolls-Royce teams for certification documentation after posttest analysis.

The F-35-specific icing cloud testing required a significant change to the MCL main chamber configuration with the installation of specific test equipment required to generate the icing conditions. The aircraft position, restraint system, and exhaust ducting configurations remained as used for the previous test sequences. The planned testing consisted of both ground (static/low flow speed) and simulated flight (low/medium flow speed) operations in both the conventional and powered lift modes. MCL facility limitations and constraints on the achievable conditions defined the allowable configuration. An open jet icing tunnel fixture was placed centered in front of the aircraft. The tunnel flow was driven by a total of nine fans (three parallel "legs" each with three fans in series) pushing conditioned air from within the main chamber through a duct arrangement fitted with a spray bar system to produce the icing cloud. A CAD illustration of the overall tunnel design configuration and placement relative to the test aircraft is shown in Fig. 12. A photo of the as-built configuration is shown in Fig. 13.

When the icing test was initially conceived, it was intended that there would be only a single position for the open jet icing tunnel that would cover all test conditions for both conventional and powered lift modes of operation. Once the design of the icing tunnel was matured and fixed, Computational Fluid Dynamics (CFD) analysis was used extensively to determine if sufficient coverage of the icing cloud was available to support all the planned test conditions. It quickly became apparent that during the powered lift mode of operation, especially at low forward airspeeds to simulate hover-type conditions, the engine and lift fan combined

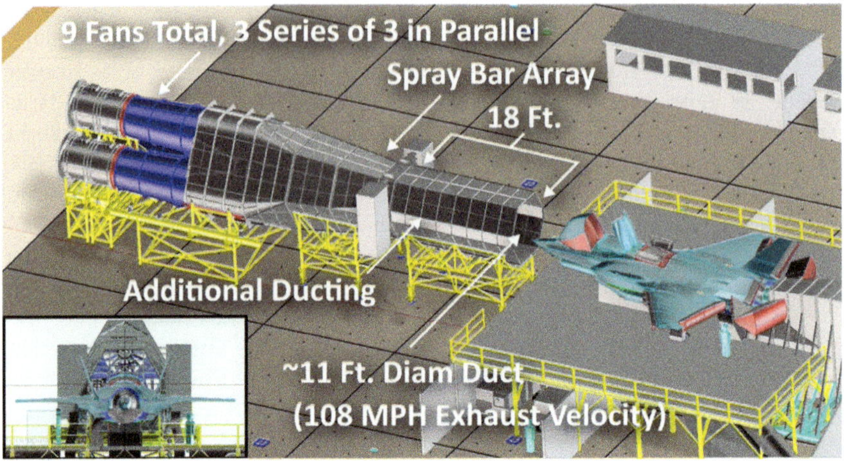

Fig. 12 CAD model of F-35 McKinley chamber icing test setup (overview). (*Illustration courtesy of Dwayne Bell, McKinley Climatic Laboratory.*)

Fig. 13 Photo of F-35 icing test chamber configuration (postsetup).

would pull maximum airflow and the size of the available icing cloud was insufficient to fully ice both the lift fan and engine at the same time. The speed range for the powered lift, low speed testing was increased from 30 kt to 45 kt in an attempt to reduce the required cross-sectional area of the inlet stream tubes but was still not sufficient to provide the necessary icing cloud coverage. The only solution easily achievable was to change the approach to employ two distinct positions for the icing tunnel: one for conventional mode operations and one for powered lift operations. The new positions that were established relative to the original location are shown in Fig. 14. For conventional mode operations, the lower position was utilized, which presented the icing cloud focused on the main engine inlets. For powered lift mode testing, the higher location was used, which delivered the icing cloud to the lift fan inlet and the auxiliary air inlet but did not fully envelope the main engine inlets. This was considered acceptable, because the limiting icing case for the main engine and inlet was determined to be during conventional mode flight operations. The two configurations could not encompass all of the inlet stream tubes for all planned operating cases, but the joint team agreed they were of adequate size and capability to achieve the majority of test requirements.

The installation and checkout of the icing tunnel equipment and the calibration of the tunnel configuration for the targeted icing test conditions was extensive and time consuming, but critical to the establishment of a valid test sequence. The icing cloud calibration equipment provided by the MCL (shown in Fig. 15) consisted of a pitot-static system used for flow speed calculation, a total temperature probe used to establish flow temperature, a multielement probe used to determine LWC, and a Malvern and/or phase Doppler interferometer device used to determine the Mean Volumetric Diameter (MVD) of the water droplets. The initial MVD measuring device was unable to operate consistently at the lowest operating temperatures required for the planned testing, hence the decision was made to switch probe types.

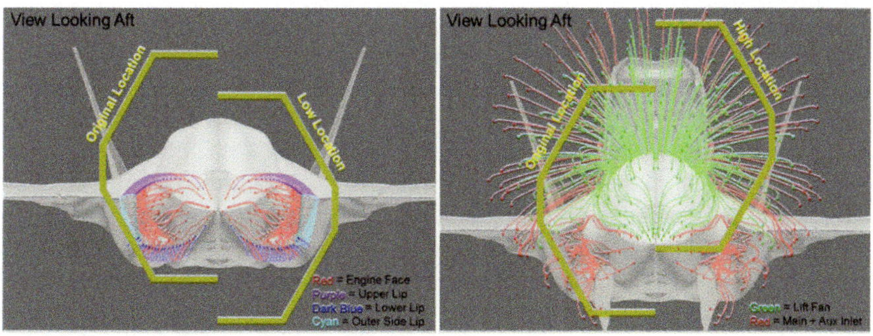

Fig. 14 F-35 McKinley chamber icing test setup—low and high icing tunnel locations.

Fig. 15 F-35 McKinley chamber icing test setup (spray bar and calibration equipment shown).

The conventional mode cloud conditions calibration effort consisted of an iterative process of tweaking facility air and water supply pressure and temperatures to establish the required settings to generate each targeted cloud combination in the test plan. The flow was allowed to stabilize at each condition and flow properties measured and assessed. Adjustments were made to the air and water supply settings if conditions were not within acceptable tolerances based on the MCL icing expert experience and knowledge. The corresponding facility air and water supply conditions required to generate each cloud condition were documented to allow the re-creation of the conditions during the actual test period. Particular care was taken when calibrating for the test conditions targeting ice detector performance and included the additional step of having an engineer standing next to the aircraft with the icing tunnel running, looking into the tunnel to identify cloud nonuniformity and spray nozzle instability from the spray bar close to the location expected to influence the aircraft ice detection system. This inconsistency and instability was more apparent at low LWC conditions, where water pressure in the spray bars tended to be low and could be overpowered by the air pressure required for the target droplet size.

One important test at this stage was the icing calibration grid, where it was possible to measure the thickness of ice that accretes on a predetermined grid throughout the extent of the icing cloud. Ideally this test would take measurements either upstream of or without the test asset in position. Given the sheer size and complexity of the test setup for the F-35, this was not possible, so an icing calibration grid was required that would conform around the aircraft just ahead of the main inlet plane (shown in in Fig. 16). Traditionally an icing calibration grid is constructed by welding together rectangular sections of metal bars; however, these are costly and time consuming to build. Significant savings were achieved by foregoing the traditional construction method and using a "chain link fence" approach. Though not as tightly controlled as the rectangular section bar style of grid, this new approach was deemed adequate. With this type of setup for an icing calibration grid, the measured thickness in the center of each link will not be the same for every link, even for a uniform cloud leaving the icing tunnel exit. This is because there are local accelerations of the air and droplets around the aircraft, and areas where the cloud droplets will not necessarily follow the aerodynamic streamlines. (This is equally true for an aircraft flying through a cloud in the air as it is for this type of testing on the ground.) Therefore, to make an assessment to determine the acceptability of the simulated icing cloud, the team was required to compare the icing grid accretions to the expected local variations in velocity and LWC. CFD and icing analysis cases were used to aid in this comparison.

The calibration process continued until all the achievable cloud conditions had been established. The engine and aircraft systems remained off during the calibration period. The extensive time required to determine the proper flow settings caused a significant (although not realistic normal operating conditions) ice accretion on the aircraft structure. Example photos of the calibration process and

Fig. 16 F-35 conformal icing calibration grid (surrounding aircraft).

Fig. 17 F-35 MCL icing test calibration process—conventional mode.

the associated ice accretion are shown in Fig. 17. The total calibration period was divided into several periods due to a combination of aircraft, facility, and test schedule factors.

The conventional mode icing cloud evaluation test series focused on the aircraft-mounted ice detection system, the installed EIPS integration and performance, and airframe ice accretion characteristics. Two distinct test objectives drove the test sequences: the validation of the ice detection system and the assessment of the integrated aircraft and propulsion systems after exposure to representative ground and flight icing conditions.

The location of the ice detector within the main inlet duct varies slightly among the three aircraft variants. As a result, a dedicated ice detector was added to the BF-5 aircraft to provide data at locations that were representative of all aircraft variants for the verification of ice detector performance across all three variants. The ice detector validation effort was designed to provide the final data required to determine ice detector performance and establish a final ground and flight capability and clearance for the fleet. For performance validation and verification, an icing cloud temperature of 14°F was selected as the test condition, serving as a midpoint within the icing specification envelope. The ice detector validation sequence was performed for both small (15-μ MVD) and large (40-μ MVD) droplet size. Target Engine Thrust Requests (ETRs) for each cloud condition were maintained until sufficient data were captured at the ice detector of focus. This resulted in multiple runs being required to ensure an accurate response and performance characteristics was established.

Evaluation of the integrated operation of ice detection and the installed engine and inlet focused on three representative ground fog icing conditions and targeted specific duration at both ground idle and elevated power settings. Test objectives included operation of the aircraft at each targeted condition for the full engine specification time duration for ground idle while gathering associated airframe accretion information to validate models. The goal was the confirmation and validation of the integrated propulsion system operation for fleet operations. Focus then turned to the evaluation of the integrated operation at three representative flight icing conditions. A tunnel exhaust flow of 100 kt was selected based on the combination of facility capability and cloud size. The flight regime portion of testing concentrated on three engine power settings—flight idle, part power, and military power (full, nonafterburning power)—and targeted maximum continuous and maximum intermittent icing envelope specifications. All testing was executed in a buildup approach to manage risk, starting with the lowest power setting and with the icing conditions considered to be least challenging to the aircraft and propulsion system.

Similar operations in powered lift mode required the reconfiguration of the main chamber to the higher tunnel position, approximately 5 ft vertically. A functional checkout was performed followed by calibration runs to establish the achievable conditions and create a map of the associated facility air and water pressures. The powered lift mode calibration effort of the calibration runs was performed without the aircraft and propulsion system operating. First phase of testing focused on powered lift mode and the evaluation of the integrated lift fan and aircraft inlet at three representative ground fog icing conditions at ground idle and during a powered lift mode MBIT, a preprogrammed sequence used during postmaintenance check of the F-35B's propulsion system. The prescribed icing cloud for each targeted test conditions required a flow of 45 kt from the icing tunnel (based on facility capability and required cloud size). The second phase of powered lift mode testing focused on flight operations and consisted of two primary, simulated flight maneuvers: a part-power flight pattern maneuver and a full-power hover maneuver. Test sequences were performed at maximum continuous and maximum intermittent icing envelope specifications. These testing sequences made use of several predefined Flight Test Aids (FTA), which execute specially programmed aircraft and propulsion system settings to simulate various phases of powered lift mode flight operations and scheduling. The final powered lift mode icing test featured a "graduation exercise" in which a simulated Vertical Landing Bring Back (VLBB) profile was executed in various icing conditions. The VLBB is a defined aircraft flight profile that represents the profile and scenario associated with the return of the aircraft to the ship for analysis and assessment; it represented a key standard for capability assessment for the F-35B.

The icing test was successfully completed in incremental but effective blocks of testing. The ice detection system efforts and results were critical to the validation and calibration of icing code models and analysis for the installed system. On-aircraft test results were key to providing the evidence that analytical predictions of system performance were accurate and therefore able to be used for analysis for all areas of the operating envelope. The testing performed for the assessment of the installed ice protection system was grueling and time consuming, requiring short periods of operation at key conditions, followed by painstakingly detailed documentation of accretion characteristics and measurements at key aircraft and propulsion system areas to validate modeling efforts, analysis tools, and suitability of installed system performance to maximize allowable fleet operations, establish required limitations and guidance, and certify system-level capabilities.

VIII. THERMAL MANAGEMENT SYSTEM TESTING

The Thermal Management System (TMS) is composed of the aircraft subsystems that generate, transport, or dissipate heat from the air vehicle. These systems include the fuel system, hydraulic system, electrical system, propulsion system—including engine and lift fan—and the PTMS. The aircraft fuel system architecture delivers fuel from the feed tank(s) to the main engine. In that process, heat is rejected from the various aircraft system heat exchangers into the fuel supply, which then is also used to cool the engine oil before being supplied to the engine gas generator to be combusted. In exceptionally hot environments, the summation of the heat sources has the potential to exceed the available fuel sink supply, resulting in an aircraft fluid (e.g., fuel or oil) temperature reaching a prescribed limit value. With the addition of the lift fan and subsequent reduction in fuel volume on board, the F-35B is the most challenging configuration. The F-35 contract specification requirements for the aircraft TMS performance were difficult to verify in a normal test environment because of the complex and multiple systems involved coupled with the rare nature of this exceptionally hot environment. As such, the requirements were verified using modeling and simulation methods. Calibration and validation of those models were key to providing final proof of system design and performance.

The hot day testing that was conducted at McKinley provided significant evidence towards the verification of the TMS requirement to operate to the required time and environmental conditions. During the CCT elevated ambient temperature testing sequence, the main chamber was set up to simulate the maximum diurnal cycle (24-hour period, with ambient temperature and solar radiation effects included) as specified in the requirements. By the time the engine was started, fuel feed tank temperatures were 10°F above the TMS specification ground rules and assumptions. Despite this elevated system temperature, the aircraft was still able to perform and complete the required time at engine ground idle, then successfully performed a conversion to powered lift mode and executed a simulated short takeoff maneuver before reaching the maximum allowable preflight fuel system temperature. Figure 18 illustrates the feed tank temperature traces during the executed preflight test run profile.

The main chamber was maintained at the environmental condition present at the end of the preflight segment previously described, and the aircraft was prepared to execute the postflight segment of the requirement. Defueling was performed to remove a majority of fuel load from the takeoff portion of testing to reconfigure the aircraft fuel state at conditions associated with return from a mission. The main chamber was programmed as close to the corresponding

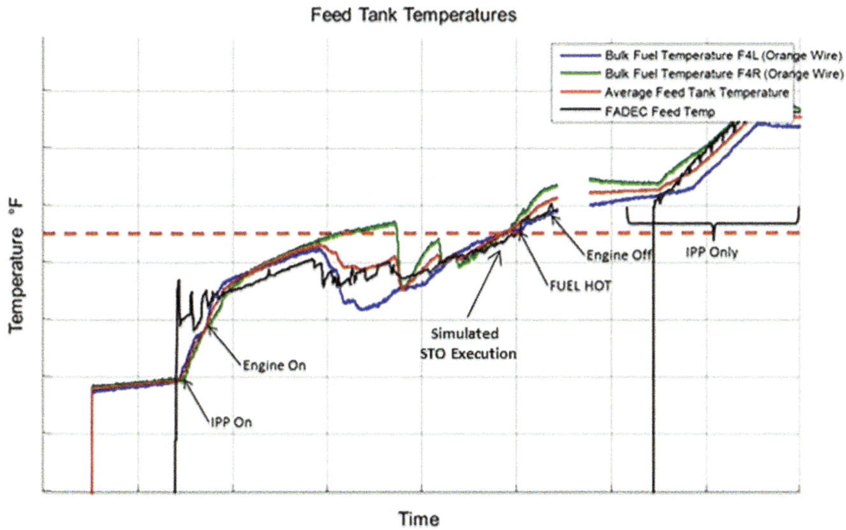

Fig. 18 F-35B TMS evaluation test—preflight profile.

diurnal cycle setting as possible prior to the start of formal testing. The final test sequence involving a simulated hover maneuver, vertical landing, and subsequent postflight ground operation (including shutdown) was performed to simulate the return from the mission to evaluate postflight performance of the TMS. All the required sequences and operating time intervals were successfully completed prior to reaching thermal limit values. Upon engine shutdown, there was less than 700 lb of fuel remaining in the feed tanks, and the fuel feed tank bulk temperature reached the allowable postflight limit. Figure 19 illustrates the feed tank temperature traces during the executed postflight test run profile.

The elevated temperature testing that was conducted at McKinley provided both a demonstration of aircraft system capabilities and significant evidence needed for the verification of the TMS specification requirement to operate on a 1% hot day. For both the preflight and postflight test sequences, system compliance to the timeline requirements was demonstrated, removing any concerns regarding analytical methods, extrapolation of results, and doubt to actual system capabilities. The data served as key input required for the calibration and validation of the model inputs and outputs, enabling the models to be used for validation of the overall system at all required conditions.

IX. F-35 CLIMATIC TESTING RESULTS SUMMARY AND LESSONS LEARNED

All major test objectives were fulfilled, and the aircraft was demonstrated as able to operate at all environmental conditions tested. Simulations and modeling were

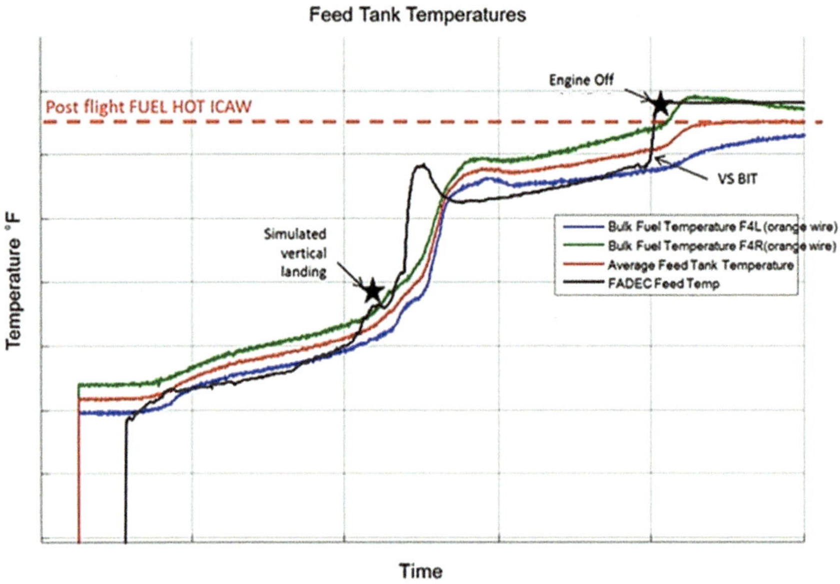

Fig. 19 F-35B TMS evaluation test—postflight profile.

shown to be accurate and predictions very close to actual results for the majority of the systems. The aircraft Thermal Management System (TMS) performance was a major success story coming out of the testing. Testing at elevated ambient temperature was a key factor in the ability to provide the data to correlate and validate the predictions of the TMS to handle heat loads generated by the aircraft and propulsion systems and remain within required guidelines for allowable fuel system temperatures. Sufficient data were gathered in icing conditions to validate modeling and verify the detection of icing conditions and the protection of the propulsion system, and establish aircraft accretion characteristics to help guide aircraft limitations and aircrew guidance. The complex chorography of aircraft doors and control surfaces to move and articulate as scheduled in cold conditions was verified. Ground operating capabilities were demonstrated in all operating conditions, and Mission System (MS) and Vehicle Systems (VS) functionality were characterized and verified. Airframe and structural characteristics for water tightness, intrusion, and drainage were assessed to predictions and documented. Aircrew evaluation of environmental characteristics in the cockpit was noted for all the environmental conditions tested. Joint Technical Data (JTD) and defined maintenance procedures were evaluated, verified, and, where required, updated.

Accurately estimating the time required to conduct both ground and flight testing has always been notoriously difficult to accomplish with any degree of accuracy. This particular CCT required extended time to complete the primary objectives despite a rigorous planning effort. The initial order of testing and notional schedule was established based on legacy fighter aircraft programs and similar climatic test objectives. The schedule was further matured as the F-35-specific operating characteristics, requirements, and unique risks were identified and studied within the planning phases. Considerations included the order of test sequences and environmental conditions, the F-35B's unique configuration and operation, and the complexity and scope of the planned test objectives vs previous experiences (e.g., icing evaluation). Care was taken to account for "unknown unknowns" and discoveries that naturally occur when operating and exploring the extremes of the environmental conditions.

A total of 151 days was allotted to cover the expected test time in the MCL facility, which included the initial entry into the facility, installation of the aircraft into the test fixture, execution of the targeted test sequences, removal from the test fixture, exit from the facility, and aircraft restoration to full flight status. A 20% factor for contingencies and an additional 15% allowance to follow the F-35 SDD flight test program planning factor for retest and discoveries were built into the schedule as well. The actual test effort spanned a total of 184 days from initial entry to the facility to the final removal of the aircraft from the test fixture and departure from the main chamber. The final test execution required a one-month extension of time in the chamber facility despite best efforts to adequately account for nominal contingencies and establish appropriate margins. This one-month schedule extension required a program management assessment of priorities of the planned test objectives that resulted in several of the originally planned test objectives to be removed from the plan (primarily LT&E sequences). The completion of subsequent aircraft restoration and reconfiguration of the aircraft resulted in the ferry flight back to NAS Patuxent River coming two months later than was originally planned.

A second entry was scheduled and performed one year later to complete some specific LT&E objectives that had not been able to be performed in the original test period. Testing was performed over a two-week period in February 2016 by the joint F-35 ITF teams from Edwards AFB and NAS Patuxent River and allowed a more focused evaluation, documentation, analysis, and reporting on the F-35 air system's maintainability and supportability in an extreme climatic environment. The major objective of all-weather testing was to determine how well the weapon system, including its essential support equipment and attendant maintenance crews, could accomplish the aircraft support in the required climatic extremes, using Joint Service Technical Data procedures. A test fleet aircraft asset (BF-07) was used, and testing was performed with the aircraft nonoperating and on the floor of the main chamber (not elevated or restrained).

The CCT is a punishing final exam for complex aircraft systems. As more and more complex and increasingly integrated systems are developed, it becomes crucial to put the system to the test in the extremes of operational environments at an integrated system level. High-level program requirements are often decomposed and flowed from an air system level to lower system and component levels (often featuring many suppliers and subcontractors), whereby the validation of capabilities is left to lower subsystem levels, sometimes performed in isolation of the component itself or assessed under the assumption of perfect operation by other adjoining or integrated systems. Along with the increasing fidelity of computers and modeling codes to perform complex algorithms and to simulate functionalities and change multiple variables, it becomes increasingly easy and common for program management teams to consider actual “total system level” testing to no longer be necessary or value added, or to be expendable in the name of cost and time savings.

The F-35 CCT testing experience and results adds to the existing record of aircraft programs that have found that the actual, full-scale, system-level testing at extreme environmental conditions remains an invaluable means for the understanding and identification of where unperceived or unexpected issues can arise as a result of the complexity of evolving aircraft designs. The MCL facility at Eglin AFB and the associated testing it allows continues to show its capability and value as a proving ground for evaluating design, assembly, and operability/suitability of increasingly complex and costly military systems in real world scenarios ahead of deployment to the end users.

X. SUMMARY

The end result of all the testing performed was a very successful demonstration and confirmation of the F-35 capability to operate at all the required environmental conditions. All the required test objectives were completed, and the aircraft was safely returned to the SDD flight test program at NAS Patuxent River with no major mishaps or damage. A total of 72 test events were conducted corresponding to 101 hours of aircraft operation to complete the scheduled test program. Though no dedicated installed performance data were captured as a primary objective, the thrust values reported by the propulsion system along with the load cell data from the facility restraint system indicated operation and scheduling to be in line with predictions and expectations at each condition. The test period contained many successes in the highly integrated and complex aircraft interactions. This overall success, however, does not mean that there were no unplanned “discoveries.” Results identified several areas requiring additional action via a combination of design changes, scheduling, timing, and logic changes, and some procedural changes that will make the F-35 a better product for the users. The testing was performed early enough in the design phase to allow for the learning and changes to be incorporated into the final aircraft configuration release.

“The proof is in the pudding” is an expression often used to state that the final results will provide the support and evidence that proper technique and due diligence were taken on the path to aircraft fielding. The increasing number of F-35 aircraft being fielded worldwide to the various services and customers will soon enough provide that proof and, in the end, the authors of this paper are confident that the time spent in the planning and the execution as well as the results and the actions taken from the F-35 CCT will in no small part be a key factor in the F-35 program's success for decades to come.

ACKNOWLEDGMENTS

The authors would like to acknowledge and thank all the people who helped in the planning, development, execution, analysis, and completion of the F-35 Climatic Chamber Testing. The number of people involved in the process was extensive and precludes individual naming; however, of specific note, sincere thanks and acknowledgement are required to the many organizations and companies that made this testing and this paper a successful reality. Of note, the Joint Strike Fighter Program Office, the McKinley Climatic Laboratory at Eglin AFB, Lockheed Martin Aeronautics Company, Northrop Grumman, the BAE Systems Plc, Pratt & Whitney, Rolls-Royce, and the F-35 Integrated Test Force organizations at NAS Patuxent River and Edwards AFB.

About the Authors

CHAPTER 1 – F-35 PROGRAM HISTORY – FROM JAST TO IOC, ARTHUR E. SHERIDAN

Art Sheridan has served in numerous roles on the F-35 Joint Strike Fighter program, beginning as the STOVL Chief Engineer and, at various times, leading Aerodynamic Development, Systems Engineering and Integration, Weight Reduction, Air-Vehicle Improvements, and Affordability. Art was Lockheed Martin's Program Manager for teammate, BAE Systems, responsible for the F-35 aft-fuselage and empennage airframe components, as well as the fuel, crew-escape and life-support systems. Recently, Art also managed the F-35 Drag-Chute System program. In his earlier career, Art served as an aerodynamicist, Flight-Sciences lead, and Chief Engineer for several advanced programs. Art is an AIAA Associate Fellow and holds Bachelor and Master of Science degrees in Aeronautical and Astronautical Engineering from the University of Illinois and Stanford.

CHAPTER 2 – F-35 AIR VEHICLE CONFIGURATION DEVELOPMENT, MARK A. COUNTS

Mark Counts led F-35 Configuration Design and Air Vehicle Integration activities from conceptual design, preliminary and detailed design through flight test. Mark currently serves as a Senior Manager in the F-35 Engineering Project Office where he leads a team of project engineers managing proposal and execution activities across the F-35 Line of Business. He has been with Lockheed Martin for 32 years and has conceptual design, preliminary design, and detailed design experience on numerous programs including F-16, YF-22, and F-22. Mark holds Bachelor and Master of Science degrees in Mechanical Engineering from the University of Arkansas.

CHAPTER 3 – F-35 AIR VEHICLE TECHNOLOGY OVERVIEW, CHRIS WIEGAND

Chris Wiegand has over 33 years of experience in aircraft subsystem design development, integration and test, include development and integration of environmental control, thermal and fluids systems, auxiliary power and emergency power systems for the F-16, F-22, F-35 and other aircraft. Chris also has served as Project Engineer responsible for integration of reconnaissance systems in special mission aircraft, and as a Design Engineer responsible for hydrazine and cryogenic fluids design for Atlas/Centaur and other propulsion systems. Chris holds a Bachelor of Science degree in Mechanical Engineering from University of Texas at Austin.

CHAPTER 4 – F-35 DIGITAL THREAD AND ADVANCED MANUFACTURING, DR. DON A. KINARD

Dr. Don Kinard is Lockheed Martin's Senior Technical Fellow for Production Operations, and was responsible for establishing the F-35 Fighter Production System to manage production growth from 1 aircraft per month to an eventual rate of 20 aircraft per month. Don previously served as Director of F-35 Production Engineering and was responsible for JSF Tooling, Planning, Manufacturing Engineering, and Aircraft Systems Testing. Prior to F-35, Don served as Deputy Director of F-22 Production and also held various other positions on F-22 in both Engineering and Manufacturing. Don also leads the Lockheed Martin Corporate Fellow's Manufacturing Team, which is chartered with sharing lessons learned and manufacturing technologies throughout the corporation. Don's academic background is in Chemistry, with a PhD in Polymer Chemistry from Texas A&M University.

CHAPTER 5 – F-35 WEAPONS DESIGN INTEGRATION, DOUGLAS M. HAYWARD

Doug Hayward led the F-35 Weapons Integration Integrated Product Team from the conceptual definition through the detailed design phases. He currently serves as Director of F-35 Systems Engineering and is responsible for leading air system design, integration and verification across the air vehicle and ground system products. Doug previously served as F-35 Air Vehicle Deputy Director, Vehicle Systems Deputy Director, and Specialty Engineering lead. Other prior experiences include weapons and subsystems design and integration on numerous programs including F-16 derivatives, A-12, F-22, and multiple advanced aircraft studies. He is a graduate of Texas Tech University where he earned a Bachelor of Science degree in Mechanical Engineering.

CHAPTER 6 – F-35 SYSTEM DEVELOPMENT AND DEMONSTRATION FLIGHT TESTING AT EDWARDS AIR FORCE BASE AND NAVAL AIR STATION PATUXENT RIVER, DR. MARY L. HUDSON

Dr. Mary Hudson served as the Lockheed Martin Senior Manager of F-35 Flight Test Operations at Edwards Air Force Base, and is currently a Chief Engineer at the Lockheed Martin Skunk Works. Mary has been with Lockheed Martin for over 30 years with the last 16 years on F-35, including roles in F-35 aerodynamics, external environment, store separation, store certification, and as the F-35 Chief Engineer at Edwards. Mary has a Bachelor of Science degree in Mechanical Engineering from New Mexico State, a Master of Science in Mechanical Engineering from Stanford and PhD in Aerospace Engineering from North Carolina State University.

CHAPTER 7 – F-35 STRUCTURAL DESIGN, DEVELOPMENT, AND VERIFICATION, ROBERT M. ELLIS

Bob Ellis led the F-35 Systems Stress Analysis team through the preliminary and detailed design, first flight, and flight test phases of the program. Bob currently serves as the Senior Manager of the F-35 Structures Development Integration team. Bob has been with Lockheed Martin for over 33 years and previously worked in preliminary and detailed structural design on multiple programs including F-16, A-12, and T-50. Bob holds a Bachelor of Science in Mechanical Engineering from Texas A&M University and an MBA in Engineering Management from the University of Dallas.

CHAPTER 8 – F-35 FLIGHT CONTROL LAW DESIGN, DEVELOPMENT AND VERIFICATION, JEFFREY J. HARRIS

Jeff Harris is Sr. Manager of the F-35 Control Law and Air Data Design team. In this role, he led development of the flight control laws for all three versions of the F-35. Jeff has been with Lockheed Martin for thirty years and previously served as lead of the F-22 Control Law Design and Analysis Group. Prior to joining Lockheed Martin, Jeff was a flight test engineer assigned to the Carrier Suitability Branch at the Naval Air Test Center in Patuxent River, Maryland. Mr. Harris holds a Bachelor of Science degree in Mechanical Engineering from Drexel University and a Master of Science in Mechanical Engineering from Stanford University.

CHAPTER 9 – F-35 PROPULSION SYSTEM INTEGRATION, DEVELOPMENT, AND VERIFICATION, STEVEN P. WURTH

Steve Wurth is the Lockheed Martin Fellow for Integrated Flight Propulsion Control, or IFPC. He currently serves as the Propulsion Integration lead on several programs for the Skunk Works. Steve has also served as the F-35 IFPC Deputy and Software lead, the Vehicle System's IFPC Integrator, and IFPC Deputy Chief Engineer, helping to develop the STOVL IFPC systems for the F-35B propulsion system. During the X-35 competition, Steve led the integrated airframe/propulsion contractor team of Lockheed Martin, Pratt & Whitney and Rolls Royce in development and flight demonstration of the X-35B propulsion system. Steve began his career at the Skunk Works in Burbank, California where he performed propulsion installations, designed specialized subsystems, and conducted engine cycle analysis on various programs including the U-2, F-117, SR-71, X-33 and several other programs. Steve holds a Bachelor of Science in Mechanical Engineering from The University of Texas at Austin.

CHAPTER 10 – F-35 SUBSYSTEMS DESIGN, DEVELOPMENT & VERIFICATION, DREW ROBBINS

Drew Robbins is the Deputy Director of F-35 Engineering & Technology. Drew previously served as Senior Manager of F-35 Utilities and Subsystems, in which he led the F-35 Power and Control, Fuel, Fire Protection, Power and Thermal Management, Life Support, and Crew Escape Systems teams. He has also served as the F-35 Power and Control Systems Manager, F-35 Fuel System Integrator, and worked 15 years in the F-35 Control Law Team developing the control strategy and algorithms for all aspects of the flight envelope. Drew received his Bachelor and Master of Science degrees in Aerospace Engineering from the University of Texas and Virginia Polytechnic Institute.

CHAPTER 11 – F-35 MISSION SYSTEMS DESIGN, DEVELOPMENT, AND VERIFICATION, GREG LEMONS

Greg Lemons is Senior Manager of F-35 Mission Systems Requirements and Design, where he leads system level design, architecture and software requirements specification for the F-35 avionics system. Greg has held numerous positions covering a broad area of disciplines on multiple platforms and advanced programs spanning conceptual design to manufacturing. Within the Mission Systems area, he has integrated a wide range of capabilities to include new sensors, automated flight modes and weapons onto fighter platforms. Greg graduated with a Bachelor of Science degree in Aerospace Engineering from Texas A&M University and received a Master's in Engineering Management from Southern Methodist University.

CHAPTER 12 – F-35 INFORMATION FUSION, THOMAS L. FREY JR.

Dr. Tom Frey is the Lockheed Martin Senior Fellow for Information Fusion. He has provided key contributions to the design, development, and testing of the F-35, F-22, and F-16. He holds patents in the areas of adaptive thresholds, digital beamforming, and advanced antenna design. He has been an adjunct professor at Texas Christian University and Columbia College, and currently serves as a member of the Texas A&M External Advisory & Development Council. Tom's has a Bachelor degree from Texas A&M University, and a Master's and PhD from the University of Texas at Arlington.

CHAPTER 13 – F-35 CARRIER SUITABILITY TESTING, TONY WILSON

An experimental test pilot for Lockheed Martin, Tony Wilson – call sign "Brick" - holds the distinct honor of being the first aviator to execute an F-35C shipboard arrestment when he successfully landed aboard the USS Nimitz. A combat veteran, Brick has flown operational missions in multiple variants of the F-18 Hornet. Tony's test pilot tours included serving at VX-23 in Carrier Suitability and F-18 test. Brick has participated in multiple flight test disciplines for a variety of tactical aircraft programs, and has flown over 20 different types of aircraft including the Harrier, the U-2, and the MiG-15. Brick enlisted in the U.S. Navy in 1991 and was later selected to attend Old Dominion University, where he earned a Bachelor of Science in Mechanical Engineering. He also attended the Air Force Institute of Technology and the United States Naval Test Pilot School, earning a Master of Science in Aeronautical Engineering.

CHAPTER 14 – F-35 AERODYNAMIC PERFORMANCE VERIFICATION, DAVID G. PARSONS

David Parsons is the Lockheed Martin Aircraft Performance Flight Test Lead for the F-35 Program, responsible for defining, executing, and analyzing the flight testing required to validate the Aerodynamics and Performance databases across all three F-35 variants. He has more than 30 years of experience in Aerodynamics and Performance flight testing of high performance fighter aircraft, most notably the F-16, the X-35 Concept Demonstrator, and the F-35. David holds a Bachelor of Science degree in Aerospace and Ocean Engineering from Virginia Polytechnic Institute and is a member of the SAE In-Flight Propulsion Measurement committee.

CHAPTER 15 – F-35 HIGH ANGLE OF ATTACK FLIGHT CONTROL DEVELOPMENT AND FLIGHT TEST RESULTS, DANIEL G. CANIN

An experimental test pilot for Lockheed Martin for the past 21 years, Dan Canin - call sign "Dog" - was the lead high angle of attack test pilot for the F-35B and C variants. Dan has over 6,000 flight hours in more than 80 types of aircraft, including over 650 hours in all three variants of the F-35. As a Navy pilot, he completed 369 carrier landings, was a distinguished graduate of the US Naval Test Pilot School, and held flight test leadership positions on the A-12 and T-45A jet trainer programs. In 1997, Dan joined Lockheed Martin Skunk Works, where he was project pilot on the Argentine A-4AR, and performed extensive simulation in support of control law development on the X-35. He's also logged over 700 hours of testing in the F-16 E/F, including loads, flutter, weapons integration, store separation, avionics systems and advanced flight controls development. Dan is a graduate of the California Institute of Technology.

CHAPTER 16 – F-35 WEAPONS SEPARATION TEST AND VERIFICATION, CHRISTOPHER F. HETREED

Chris Hetreed is the Lockheed Martin technical lead for F-35 Store Separation. Chris has been involved in all aspects of store separation on the F-35 program, including software development, modeling and simulation, wind tunnel testing, ground ejection testing, flight testing, and store certification. He has also performed dynamics modeling and simulation of landing gear, arresting gear, pilot seat ejection, and canopy jettison on the F-35 and F-22 programs. Previously, while with MSC Software as an engineering consultant, Mr. Hetreed co-developed the ADAMS/Aircraft modeling and simulation software for rapid and detailed landing gear loads and dynamics analysis. Chris received his Bachelor's and Master of Science degrees in Aerospace Engineering from Virginia Tech.

CHAPTER 17 – F-35 STOVL PERFORMANCE REQUIREMENTS VERIFICATION, DANIEL E. LEVIN

An experimental test pilot for Lockheed Martin, Dan Levin is the lead F-35 pilot at Patuxent River. He has flown development tests in all models and for all disciplines, in particular, STOVL envelope expansion. Dan was previously the lead pilot during development of the F-16 Block 60 aircraft, where he focused on integration of the F110-GE-132 engine and high angle of attack testing. In prior assignments, Dan served as the project pilot for the LANTIRN navigation and targeting system program, served as Flight Dynamics Branch Commander and Operations Officer with the 416th Flight Test Squadron, and performed as the Deputy Operations Group Commander conducting classified flight test. Dan has presented papers on STOVL development for AIAA, the American Helicopter Society, and Society of Experimental Test Pilots. He is a graduate of the US Naval Test Pilot School and a veteran of Desert Storm, with operational tours in the F-16, A-10 and OV-10.

CHAPTER 18 – F-35 CLIMATIC CHAMBER TESTING & SYSTEM VERIFICATION, VICTORIO J. RODRIGUEZ

Vic Rodriguez led the coordination, planning, and execution of the F-35 Climatic Chamber Test Program. He is currently manager of the F-35 Propulsion and Thermodynamics team, with responsibility for propulsion system physical and functional integration, inlet aerodynamics, and the Thermal Management System. In a prior assignment, Vic served as the F-35 Propulsion Flight Test Planning Lead, in which he led the definition and coordination of flight test requirements between Lockheed Martin, the propulsion system contractors, government propulsion organization, and the F-35 Integrated Test Force. Vic previously worked for the Boeing Company as a Test Requirements and Analysis Engineer on the F/A-18 E/F Super Hornet and T-45A programs, both in St. Louis and at Naval Air Station Patuxent River. Vic graduated from the University of Michigan with a Bachelor of Science in Aerospace Engineering.

INDEX

Note: Page numbers followed by "**t**" indicate table, and "*f*" indicate figure.

SUPPORTING MATERIALS

A complete listing of titles in the Progress in Astronautics and Aeronautics series is available from AIAA's electronic library, Aerospace Research Central (ARC) at arc.aiaa.org. Visit ARC frequently to stay abreast of product changes, corrections, special offers, and new publications.

AIAA is committed to devoting resources to the education of both practicing and future aerospace professionals. In 1996, the AIAA Foundation was founded. Its programs enhance scientific literacy and advance the arts and sciences of aerospace. For more information, please visit www.aiaafoundation.org.